Network Mergers and Migrations

WILEY SERIES IN COMMUNICATIONS NETWORKING & DISTRIBUTED SYSTEMS

Series Editor: David Hutchison, *Lancaster University, Lancaster, UK*
Serge Fdida, *Universitè Pierre et Marie Curie, Paris, France*
Joe Sventek, *University of Glasgow, Glasgow, UK*

The 'Wiley Series in Communications Networking & Distributed Systems' is a series of expert-level, technically detailed books covering cutting-edge research, and brand new developments as well as tutorial-style treatments in networking, middleware and software technologies for communications and distributed systems. The books will provide timely and reliable information about the state-of-the-art to researchers, advanced students and development engineers in the Telecommunications and the Computing sectors.

Other titles in the series:

Wright: *Voice over Packet Networks* 0-471-49516-6 (February 2001)
Jepsen: *Java for Telecommunications* 0-471-49826-2 (July 2001)
Sutton: *Secure Communications* 0-471-49904-8 (December 2001)
Stajano: *Security for Ubiquitous Computing* 0-470-84493-0 (February 2002)
Martin-Flatin: *Web-Based Management of IP Networks and Systems* 0-471-48702-3 (September 2002)
Berman, Fox, Hey: *Grid Computing. Making the Global Infrastructure a Reality* 0-470-85319-0 (March 2003)
Turner, Magill, Marples: *Service Provision. Technologies for Next Generation Communications* 0-470-85066-3 (April 2004)
Welzl: *Network Congestion Control: Managing Internet Traffic* 0-470-02528-X (July 2005)
Raz, Juhola, Serrat-Fernandez, Galis: *Fast and Efficient Context-Aware Services* 0-470-01668-X (April 2006)
Heckmann: *The Competitive Internet Service Provider* 0-470-01293-5 (April 2006)
Dressler: *Self-Organization in Sensor and Actor Networks* 0-470-02820-3 (November 2007)
Berndt: *Towards 4G Technologies: Services with Initiative* 0-470-01031-2 (March 2008)
Jacquenet, Bourdon, Boucadair: *Service Automation and Dynamic Provisioning Techniques in IP/MPLS Environments* 0-470-01829-1 (March 2008)
Minei/Lucek: *MPLS-Enabled Applications: Emerging Developments and New Technologies, Second Edition* 0-470-98644-1 (April 2008)
Gurtov: *Host Identity Protocol (HIP): Towards the Secure Mobile Internet* 0-470-99790-7 (June 2008)
Boucadair: *Inter-Asterisk Exchange (IAX): Deployment Scenarios in SIP-enabled Networks* 0-470-77072-4 (January 2009)
Fitzek: *Mobile Peer to Peer (P2P): A Tutorial Guide* 0-470-69992-2 (June 2009)
Shelby: *6LoWPAN: The Wireless Embedded Internet* 0-470-74799-4 (November 2009)
Stavdas: *Core and Metro Networks* 0-470-51274-1 (February 2010)

Network Mergers and Migrations:

Junos® Design and Implementation

Gonzalo Gómez Herrero
Professional Services, Juniper Networks, Inc

Jan Antón Bernal van der Ven
Professional Services, Juniper Networks, Inc

A John Wiley and Sons, Ltd, Publication

This edition first published 2010

Registered office
John Wiley & Sons Ltd, The Atrium, Southern Gate, Chichester, West Sussex, PO19 8SQ, United Kingdom

For details of our global editorial offices, for customer services and for information about how to apply for permission to reuse the copyright material in this book please see our website at www.wiley.com.

Library of Congress Cataloging-in-Publication Data

CIP data to follow

A catalogue record for this book is available from the British Library.

ISBN: 978-0-470-74237-2 (P/B)

Set in 10/12pt Times by Sunrise Setting Ltd, Torquay, UK.
Printed and Bound in Great Britain by CPI Antony Rowe, Chippenham, Wiltshire.

Advanced Praise for Network Mergers and Migrations

"Having been through the network migrations of both SBC-Ameritech, and SBC-AT&T, this book provides valuable direction and advice to an operator that would be useful when dealing with mergers and migrations. It is about time that someone wrote a book covering not just green field network design, but how do you merge networks together while maintaining the needs of the business."

Tom Scholl, Principal IP Network Engineer, IP/MPLS Backbone Design and Development, AT&T Labs

"I have never read such an excellent book detailing so much about Juniper router operation and design knowledge in depth, not only it has helped me with operating all of our Juniper routers in our national backbone network more efficiently, but also on improving my design knowledge as well."

Ji Hui, Senior Manager, China Telecom Group Corporation

"Network Mergers and Migrations gives a clear idea of the difficulty of merging or migrating IP networks, and it is a great help for engineers that have to face this type of operation for the first time or even repeatedly."

Chiara Moriondo, Senior Engineer, Telecom Italia Labs

"Network Mergers and Migrations provides an original approach to IP and MPLS instruction that is right on target. When combined with the depth of technical information and case studies contained within, the result is an irreplaceable resource for any network engineer."

Chris Grundemann, Senior Engineer, tw telecom, inc.

"Network Mergers and Migrations is a thorough guide for anyone involved in a wide range of network consolidation and integration exercises. The authors use Junos as a technical focus for their examples keeping the concepts related to some of the more relevant protocols and topics in current IP and MPLS networks. It's a must-have book for anyone wishing to further enhance their protocol knowledge base."

Shafik Hirjee, Bell Canada, Director, National MPLS/IP Core Network Engineering

"We operate the largest R&E network in the world using Juniper T640s, and my job often involves network planning and design. I have found this book to be just the one I need. I thought I knew Junos, but these authors are at an unprecedented level."

Zheng Zhiyan, Senior Engineer, Network Operation Center, CNGI-6IX/CERNET2

"Recommended reading for networkers who like to look into complex and detailed network migration scenarios, as well as those interested in a better understanding of Junos implementation of common ISP protocols."

Andreas Weiner, Network Engineer/IP Backbone, Telekom Austria TA AG

"Network Mergers and Migrations provides network operators a comprehensive reference for network migration. The level of detail included in this book is exceptional. It's a thorough guide for any network professional who are involved with network migration."

Mazen A. Baragaba & Eid Al Harbi Communications Engineers, Communications Operations Department, Saudi Aramco

"*Efficient IP network planning and management involves more than just the in-depth knowledge of theory. This book helps the reader gain a thorough understanding of the Junos routing architecture via practical and useful case studies, resulting in smooth and clean transitions between different network implementations. The clear and concise exposition make this book a pleasure to read and a useful companion during work.*"

Giorgio Lembo, R&D Director, Tinet S.p.A.

"*Network Mergers and Migrations describes in detail the basics of L3VPN architectures while providing a clear and detailed roadmap for various migration scenarios. Even for telcos (operators) that are not implied in migration phases, a lot of great advice is provided for designing network architectures based on Junos software.*"

David Roy, Orange France, RBCI IP Technical Assistance Center

"*Network Mergers and Migrations is much more than a reference for all relevant IP protocols today, it is a book of practical examples for your daily work and should be on the desk of everyone who has to deal with network migrations and mergers.*"

Christian Kaufmann, Senior Manager Network Architecture, Akamai Technology

"*An invaluable guide for anyone about to undertake a network merger, full of appropriate case studies and Junos tips to back up the concepts discussed.*"

Keith Slater, NOC Manager, Cable&Wireless Worldwide

"*The best book on operating Junos router networks!*"

Jia Yan, Director, Network Maintenance Center, China Telecom Group Beijing Corporation

"*Sooner or later in their careers, network professionals will be faced with a network merger or migration. This book covers all technical aspects of the problems encountered during a migration or merger and provides the best practices and techniques on how to perform the task with the minimal impact on customer services. A must have in any network professional's library.*"

Otto Kreiter, Service Introduction Manager, DANTE Ltd.

Contents

List of Junos Tips

List of Application Notes

List of “What ifs”

About the Authors

Gonzalo Gómez Herrero holds a MS in Telecommunications Engineering from the University of Zaragoza (Spain) and a certificate in Postgraduate Studies in Computer Science from the Technical University of Munich (Germany). He is JNCIE-M #155, JNCIP-E #108, CCIE Routing and Switching #14068 and Juniper Certified Instructor. Prior to joining a Juniper in June 2005, he worked in various technical positions for system integrators and service providers. He is currently a member of the EMEA Professional Services team, having worked in the networking industry for over 10 years.

Juan Antón Bernal van der Ven (JNCI/JNCIE-M #27) holds a BS in Telecommunications and a MS in Electronics Engineering from the Universitat Ramon Llull (Spain). After completing a MSc in Space Telecommunications, he spent two years in the European Space Operations Centre supporting research for ground segment operations. He joined the professional services arm of an ATM switching vendor in 1998, and moved to the Juniper Networks Professional Services team in 2001.

Series Foreword

This book literally gets to the heart of the operation of modern computer networks – on which much of society strongly depends for almost all aspects of life and work. It is concerned with so-called network mergers and migrations which – in the case of migrations more than mergers – are an increasingly frequent occurrence because networks are inevitably under constant change.

Although the imperatives for change come mainly from business requirements, the implications are of course translated to the technical level of the network system – to the routers and their operating environment – and it is solely at the technical level that this book considers the subject.

In particular, the book treats the subject from the viewpoint of Junos® operating system, which is the network operating system developed by Juniper Networks Inc. The two authors are expert practitioners, with a wealth of experience in the field, and from them the reader will gain an in-depth understanding of the issues that face network operators and also network designers when faced with having to carry out changes. Both planning and implementation of solutions are covered in depth for the important network protocols, IGPs, BGP, and MPLS as the main focus.

Although the book is specifically oriented around Junos OS, the treatment is such that readers will learn the principles behind migration as well as a great deal of insight into the practical issues that lie behind the deployment of IGPs, BGP, and MPLS, protocols that are described in many textbooks but rarely from an operations perspective.

A key feature of this book is that each of the five chapters features a case study that takes the reader through the steps involved in the planning and realization of a typical migration involving the protocol(s) covered in that particular chapter. The net effect is a fascinating journey and learning experience through the normally hidden world and work of modern network engineers.

This latest addition to the Wiley CNDS Series is written by two Juniper insiders. It gives a comprehensive and distinctive coverage of this important field and should appeal broadly to researchers and practitioners in the field of communications and computer networks, not just to those interested purely in mergers and migrations.

David Hutchison
Lancaster University

Foreword

Πάντα ῥεῖ καὶ οὐδὲν μένει

-- Heraclitus of Ephesus (c. 535 BC – 475 BC)

Loosely translated, Heraclitus says, "Change is the only constant." Change is one of the hallmarks of life – growth, repair, adaptation. Networks, whether those of service providers, enterprises, or data centers, are not different – they grow, they change: a new port here, a software upgrade there, a customer tweak elsewhere, accompanied by the perennial quest for higher bandwidth, broader services, increased scale, and ultimately, greater utility. These changes, small or large, accrete almost unnoticed, until today's network is unrecognizable from the network a few years ago. These changes are not always planned; they tend to be driven by short-term tactical considerations rather than longer-term strategy. Thus, it is no surprise that the cumulative effect is not always desired – it is what it is.

Network Mergers and Migrations: Junos® Design and Implementation is not about managing the daily changes that occur in networks. Rather, it is about tectonic shifts in networks: how to plan them, how to lay the groundwork for them, and most importantly, how to bring them to fruition effectively and safely. Such tectonic shifts may be needed for a number of reasons:

- as a "course correction" of the accumulation of the smaller changes mentioned above;
- to adapt to significant technological evolution, such as cloud computing;
- to accommodate a major realignment of corporate strategy; and/or
- to optimize new business arrangements, such as mergers or partnerships.

The main characteristics of such tectonic changes are the risk involved and the concomitant apprehension in those who have to orchestrate the change, ranging from concern to terror. Having been in the networking business for a while, I have seen everything from absolute paralysis when faced with scaling the network, to an orderly transition over a couple of weeks from one infrastructure to another, to a (successful!) overnight cut-over. Thus, the human element is a key factor in undertaking such migrations. This book addresses them with a calm appraisal of the tools available and a diverse set of successfully engineered case studies. But the real value of this book lies in its deep technical understanding of the issues associated with migrations. It shows how to minimize the risk, underscores the value of preparation, and describes clearly a systematic process in a variety of scenarios. The authors speak from personal experience, and their advice is the aggregate wisdom of their own efforts as well as those of other network architects.

In the final analysis, this book is more than just a handbook for a one-time "big bang" change that you may be contemplating in your network. This book teaches a rare skill: expertise in planning and executing big bang changes, even if you don't anticipate such a need. The value of this skill is to lighten today's decisions. The fast pace of technological innovation and the unpredictable nature of user demand cast a deep shadow on decisions that must nevertheless be made. There is a fear of getting it wrong, and of having to pay a severe penalty a few years down the road, resulting in these decisions being deferred, less than optimal performance, and dissatisfaction among your customers. *Network Mergers and Migrations: Junos® Design and Implementation* takes away the fear of the unknown and gives you a business advantage that not only makes you more nimble and able to adapt, but also frees you to make those necessary decisions today, without recourse to a crystal ball, resting in the sure knowledge that you can successfully and safely realign your network should the need arise.

Kireeti Kompella
Juniper Fellow

Acknowledgments

The rationale for this book appeared after long discussions with *Patrick Ames* about proposing innovative contents to the internetworking audience from the perspective of engineers working for Juniper Networks Professional Services.

Patrick has been supporting us and proposing ideas focusing on network migrations since the very first conception of this book. He has been fundamental through the complete editorial process until reaching publication and this book would not have come to fruition without him. His insistence, persistence, and encouragement have definitely made this book possible.

Aviva Garrett has been instrumental through the complete review process. Her writing experience has been a cornerstone in this book and her editorial support has been the guiding light through to the final edition. Apart from being our writing style reference guide, she has worked hard to improve the quality of the book and she has constantly encouraged us with her support.

Domiciano Alonso Fernández and *Miguel Cros Cecilia* from Telefónica España have provided tremendously valuable feedback to this book. They have clearly approached the work with the eyes of a service provider and emphasized and corrected numerous topics using their vast knowledge, experience, and lessons learned.

Peter Lundqvist has provided us with extremely helpful reviews throughout all chapters. His comprehensive review on every topic and exercise, and more important, his fantastic sense of humor have been greatly appreciated! ("yes, there are still people out there managing NSSAs!")

We also wish to express our sincere gratitude to *Ina Minei* for her detailed feedback on many technical aspects of the book, *Yakov Rekhter* for his vision on the history of BGP, L3VPNs, and the related technical discussions, *Bruno de Troch* for his review on BGP migrations and the overall case study structure, *Anantharamu Suryanarayana* for his comments on the internal implementation of local-as, and *Hannes Gredler* for being the ultimate technical reference guide for so many topics and questions.

Valuable discussions with *Luca Tosolini* provided a real-life scenario that required migration from BGP confederations to route reflection.

Ignacio Vazquez Tirado, *Rodny Molina Maldonado*, and *Pablo Mosteiro Catoria* helped to describe and understand alternative perspectives for several scenarios and situations in the book.

Becca Nitzan and *Miguel Barreiros* have also contributed detailed and valuable comments on selected chapters.

Senad Palislamovic, *Richard Whitney*, *Majid Ansari*, and *Doughan Turk* have kindly reviewed sections throughout the book and have provided clarifying and helpful feedback on most topics.

A big thank you goes to *Stefano Anchini* and the Juniper Networks EMEA CFTS team for their support in the use of their lab facilities out-of-hours.

Alistair Smith from Sunrise Setting Ltd has been our default gateway for all LaTeX-related questions and has definitely helped us in improving the quality of this book. *Sarah Tilley* from John Wiley & Sons Ltd has been our Project Editor and has kindly supported us with our frequent requests.

Both authors also wish to warmly thank their colleagues from Juniper Networks Professional Services and the Spanish account teams, and their customers, mainly Telefónica España and Telefónica International Wholesale Services, for the lessons learned, for the hard and not-so-hard times, and for the shared experiences. Many thanks for the continuous challenges and for offering us the possibility to learn something new every day.

Finally, we wish personally to express our thanks to some of our well wishers:

Gonzalo – First and most important, I would like to thank my wife, Marta, for her endless patience and support while writing this book. Without her understanding and encouragement, this project would simply not have been possible. Second, many thanks to my son, family, and friends for their loving support, especially during this time. I would also want to extend my heartfelt thanks to my previous employers, particularly my former colleagues at Cable & Wireless, because some of the tidbits included in this book are also due to them. Last but not least, Anton has been the perfect complement in this project. His critical and knowledgeable view, his capacity of abstraction, and his vast experience have been key factors in the development of contents and scenarios.

Anton - I am grateful for the unconditional support of my wife, Rosa Maria, throughout the writing process, and for her understanding during this challenging and sometimes difficult time. She keeps me smiling. My family provided the necessary distractions that such a technical project requires to help me maintain a perspective on the real world. I would like especially to thank Gonzalo for encouraging me to pursue his idea of routing case studies and for giving me the opportunity to participate, and for keeping the project together all along.

Acronyms

ABR	area border router
AFI	address family identifier
AS	autonomous system
ASN	autonomous system number
ASBR	autonomous system border router
BCD	Binary Coded Decimal
BDR	Backup Designated Router
BFD	Bidirectional Forwarding Detection
BGP	Border Gateway Protocol
BoS	Bottom of Stack
CBGP	confederated Border Gateway Protocol
CE	Customer Edge (router)
CLNP	Connectionless Network Protocol
CsC	Carrier Supporting Carrier
DCU	Destination Class Usage
DIS	designated intermediate system
DR	Designated Router
DUAL	Diffuse Update Algorithm
EBGP	exterior Border Gateway Protocol
ECMP	equal-cost multipath
EGP	Exterior Gateway Protocol
EIGRP	Enhanced Interior Gateway Routing Protocol
EL-BGP	external Labeled BGP, see L-EBGP
FA-LSP	Forwarding Adjacency LSP
FEC	Forwarding Equivalence Class
FIB	Forwarding Information Base (forwarding table)
GMPLS	Generalized Multiprotocol Label Switching
GRE	Generic Routing Encapsulation
IANA	Internet Assigned Numbers Authority
IBGP	interior Border Gateway Protocol
IETF	Internet Engineering Task Force
IGP	Interior Gateway Protocol
IGRP	Interior Gateway Routing Protocol
IL-BGP	internal Labeled BGP, see L-IBGP
IS–IS	Intermediate System to Intermediate System routing protocol

ISO	International Standards Organization
L2VPN	Layer 2 Virtual Private Network
L3VPN	Layer 3 Virtual Private Network
L-BGP	Labeled BGP
L-EBGP	Labeled exterior Border Gateway Protocol
L-IBGP	Labeled interior Border Gateway Protocol
LDP	Label Distribution Protocol
LSDB	Link-State Database
LSA	Link-State advertisement
LSI	Label-Switched Interface
LSP	label-switched path
LSP	Link-State packet or Link-State PDU
LSR	label-switched router
MP-BGP	MultiProtocol Border Gateway Protocol
MPLS	Multiprotocol Label Switching
MRAI	Minimum Route Advertisement Interval
MTR	Multi-Topology Routing
MTU	Maximum Transmission Unit
NLRI	Network layer Reachability Information
NET	Network Entity Title
NOC	Network Operating Center
NTP	Network Time Protocol
ORF	outbound Route Filtering
OSPF	Open Shortest Path First routing protocol
PDU	Protocol Data Unit
PE	Provider Edge (router)
PHP	Penultimate Hop Popping
PIM	Protocol-Independent Multicast
RD	Route Distinguisher
RFC	Request for Comments
RIB	Routing Information Base (routing table)
RIP	Routing Information Protocol
RP	Rendezvous Point
RPF	Reverse Path Forwarding
RPM	Remote Performance Measurement
RR	Route Reflector
RR	Routing Registry
RRO	Route Record Object
RSVP	Resource Reservation Protocol
SAFI	subsequent address family identifier
SLA	service-level agreement
SNMP	Simple Network Management Protocol
SPF	Shortest Path First
SRLG	Shared-Risk Link Group
TE	Traffic Engineering
TED	Traffic-Engineering Database

TLV	Type-Length-Value
TTL	Time-To-Live
VPN	Virtual Private Network
VRF	Virtual Routing and Forwarding

Introduction

Business requirements for modern enterprise and service provider networks are in a constant state of flux. From software upgrades to network mergers and acquisitions, the complexity of adapting business requirements to the network spans from trivial cable moves between adjacent devices to dismantling or repurposing a complete location.

Networking has evolved exponentially over the past 30 years. From a business point of view, networking did not exist 30 years ago, except for some private networks or IBM mainframes. Newer technologies have led to the evolution of features so that they are better fits for business applications. Some of our daily technologies such as mobile communications or the worldwide span of the Internet were inconceivable even 15 years ago, and they are now the cornerstone of business-critical services.

The drivers for network migrations are not simply technological evolution, politics, and organizational changes. A simple review of networking wireline and mobile service providers shows how many company acquisitions, joint ventures, integrations, and mergers have taken place across the globe over the last ten years. An indirect result of such business transactions is the need for integration, ranging from complete network mergers to a simple peering or interconnection.

Integration challenges are not limited to service providers. Enterprise networks are also eligible for integration scenarios as a result of internal turbulence and acquisitions, and the requirements for these types of network are often more stringent than for those of service providers. Maintaining *service-level agreements* (SLAs) while performing any kind of integration task can become cumbersome and requires careful planning and execution.

Together with new technologies, more horsepower is needed. Newer features for newer or evolved applications may require a newer software release and hardware to cope with requirements. Even simply by the need of solving software or hardware detects, a minimal migration for maintenance purposes can be vital at some point. Even though routers may not get directly involved, the indirect and network or worldwide effect of certain *Interior Gateway Protocols* (IGPs) or *Border Gateway Protocol* (BGP) require a careful transition scenario and corresponding upgrades to understand newer protocol constructs or to expand scalability burdens.

A recent example is the introduction of 4-byte Autonomous Systems (ASs) into the *Internet*. Such a structural change because of the imminent AS number consumption with 2 bytes has required multiple coordinated activities: Regional Internet Registries (RIRs) defining newer allocation policies to alleviate an orderly transition and coordinating assignment with Local Internet Registries (LIRs); standardization work in the IETF to define transition mechanisms and protocol resources to ensure end-to-end communication with 4-byte AS aware and unaware implementations; routing system vendors implementing approaches for such standardization work, subject to different interpretations and software defects; and service providers controlling deployment of these newer features and potentially expecting 4-byte ASs as customers or peers. In a nutshell, without implementing a new service or feature besides, all major players in the *networking* industry must adapt and migrate mechanisms just for the sake of future-proven sustainable business in a connected world.

Motivation

The internetworking environment grows and evolves rapidly, with new features constantly rendering obsolete old implementations, and new protocols and resources being developed on a regular basis. The network engineering and operations teams must continuously learn and update their network knowledge to achieve successful results when performing their jobs.

This rapid evolution translates into technologies, products, and resources, that, when taken together, set the stage for newer, safer, and more competitive services, which are the core business of Service Providers and the key distinguisher for Enterprise networks. In fact, the traditional barrier between Service Providers and Enterprise networks is becoming more and more diffuse, and both types of entities now share these objectives because continuously growing and improving the network is key to the success of both.

Network migrations are the cornerstone of network evolution. From planning the necessary changes to upgrading network elements with new features, a myriad of activities is needed both to plan properly for natural growth path and to introduce substantial modifications in existing configurations and topologies.

Even when not planning newer service deployments, operations such as dealing with software and hardware defects, consolidating network management across systems, and upgrading operating systems to protect against vulnerabilities, while seemingly simple, can potentially lead to problems because new elements are being introduced into the network topology.

While it is difficult to cover all types of migration, the intention of this book is to provide guidelines for planning, designing, and rolling out a set of common network migration activities. In addition to explaining the concepts related to the migrations, this book includes numerous application notes and case studies based on Junos OS, the core operating system on Juniper Networks routers, to illustrate each concept or idea.

To be sure, this book discusses only some migration topics, and certainly no migration is perfect, each having its own requirements and caveats. In many of the situations presented in this book, different approaches are valid and the proposed process has its pros and cons. Some readers may feel that another option would better fit their needs, and certainly other solutions have been implemented in real networks that have achieved the same goal in a different way.

The underlying purpose of each example is to share lessons that the two authors have learned and to describe clearly the Junos OS resources available so that network engineers can use them as tools to accomplish their goals. Rather than focusing on *what* to achieve, as would be the case in a real network, the focus of the book is *how* to achieve it, by first discussing the general possibilities available for each protocol and then detailing Junos OS specific configurations and knobs.

This book is not purely a network design guide, nor is it purely an operations guide. Rather, it intermingles design approaches and operations tools that can be used at each step to reach the same final milestone: a successful, or at least a minimally intrusive, network migration.

Book audience

Because design guidelines and operations tips are mixed and continuously cross-referenced throughout each chapter, the book is intended for those network designers and operators who need to plan and roll out a migration activity based on some of the protocols and features included in the contents.

Other network engineers who need to validate configurations, perform proof-of-concept testing, or are simply curious and eager to learn about some of the latest networking protocol concepts or ideas, particularly how they are interpreted with Junos OS, may also find it interesting.

The authors assume that readers have basic exposure and knowledge of Junos OS in order to understand the concepts and configurations in this book. This book is also a valuable reference in the Junos OS learning path. Network engineers with no previous experience are encouraged first to read some other available Junos OS books to gain an adequate understanding of the operating system.

The intention here is to maintain a technical focus, mainly on Junos OS, which reflects the expertise of the two authors in their current job roles. This book contains no high-level descriptions or general discussions about the theory of network migrations.

Book structure

Contents

Covering all possible modern network migration types in a single book is unachievable. Furthermore, including all modern operating systems from major router vendors in case studies and application notes would have extended the contents considerably and limited the number of concepts that could be included in the book.

The book includes a wide range of network consolidation and integration concepts related to some of the most relevant protocols and topics in current IP and MPLS networks, keeping Junos OS as a technical focus for examples.

Chapter 1 examines routing infrastructure ideas in Junos OS that are considered to be the fundamental cornerstones for more convoluted protocol migrations. Topics in this chapter are not protocol dependent, but rather, Junos OS specific, and the authors encourage the reader to go through this chapter first to help with proper understanding of the remaining chapters. While this first chapter explains Junos OS resources, the fundamental ideas themselves are independent of any particular routing operating system: how to perform route sharing and leaking between different routing instances. A final Case Study describes this toolbox of Junos OS resources to be utilized in a particular migration scenario.

Chapter 2 targets link-state IGP migrations, principally those related to *Open Shortest Path First* (OSPF) version 2 and *Intermediate System to Intermediate System* (IS–IS). IGP migrations are probably the most relevant and significant network migration type, with more documented cases than any other type of migration. Any migration activities performed on a link-state IGP are particularly sensitive because of the domain-wide scope of link-state IGPs. This chapter is divided into two major sections: hierarchical migrations within a common link-state IGP and migrations from one link-state IGP to another. The first section describes alternatives and processes to split or merge the layering scheme of a given link-state IGP. The second section includes options for performing complete link-state IGP substitution, discussing the interactions and dependencies in each case. Multiple Application Notes throughout the chapter explain in detail specific related topics, and a global Case Study combines link-state IGP hierarchical and protocol migrations in a comprehensive exercise in a common scenario, in which several Junos OS features provide guidance and resolution to integrated networks.

Chapter 3 deals with BGP migrations. A traditional real-life migration Case Study involves AS mergers and breakups, and many public references from providers at forums are available. This chapter also explains the challenges and consequences of any modifications when dealing with changes through BGP, together with Junos OS configuration options that can leverage or alleviate particular migration scenarios.

Chapter 4 focuses on activities related to the *Multiprotocol Label Switching* (MPLS) label distribution protocols, namely the *Label Distribution Protocol* (LDP), the *Resource Reservation Protocol with Traffic Engineering* extensions (RSVP-TE), and the labeled unicast family for BGP, traditionally called *Labeled BGP* (L-BGP). It describes the effects of changes performed with any of these protocols, and discusses their differences, common features, and interactions when distributing label bindings. Because these label distribution protocols are closely related to the link-state IGPs and, in the case of the labeled unicast family, to BGP, this chapter appears in sequence after the IGP and BGP chapters and concepts in it are based on those explained in the previous chapters. A final Case Study illustrates an approach to integrate MPLS label distribution protocols from different domains for end-to-end MPLS services, based on interconnecting these protocols first, and migrating to a unified MPLS label distribution protocol at a later stage.

Finally, Chapter 5 lays out the merger and integration of MPLS *Layer 3 Virtual Private Networks* (L3VPNs), providing two major variants: integrating an existing network as part of a complete L3VPN and integrating L3VPNs using different models. The first variant requires a review of existing communication protocols between *Provider Edge* (PE) devices and *Customer Edge* (CE) devices, along with related concepts and resources to alleviate integration. Multiple Application Notes are included for migration exercises using Junos OS features with each one of the PE–CE communication protocols. The second variant reviews existing L3VPN interconnect models as ways to extend an existing virtual topology through more than one MPLS transport provider. A Case Study reviews all these interconnect mechanisms, illustrating a transition through several models and interacting with MPLS label distribution protocols. For this reason, this chapter is closely related to Chapter 4, and is indirectly related to all preceding chapters.

This incremental approach covers most common network migration types in a modern IP and MPLS environment. Unfortunately, however, many other interesting concepts are beyond the scope of this book.

Chapter structure

Each chapter shares a common and clear structure to ease understanding. This structure is based on a scheme of items providing answers to the following questions:

Motivations: *What is driving this type of migration?*

This section offers an initial analysis of possible reasons that make such a migration necessary. The motivations section helps the reader to understand the rationale for performing such an exercise and to understand other benefits that can be achieved with that kind of migration.

Protocol considerations: *What do standard references say about such migrations?*

A review of the current *state of the art* provides a starting point for determining the resources and possibilities for each type of migration. This initial review is vendor-agnostic and focuses on how each protocol works, what might need to be changed during a migration, and how such changes might impact the network status from a migration perspective. This analysis relies mainly on *Internet Engineering Task Force* (IETF) drafts and *Request for Comments* (RFCs) as general and common references.

Migration strategies: *How can a network engineer drive this type of migration?*

Once general resources and motivations are analyzed, different approaches and alternatives are available to undertake a migration exercise. This section summarizes the general guidelines for using and combining protocol resources and presents options to the reader.

Junos implementation: *How does Junos OS implement the functionality?*

As additional explanation, an overview of how Junos OS implements certain data structures and concepts related to the topic discussed in each chapter is included. This material helps the reader further to understand how the protocols interact with internal constructs and how Junos OS resources can be utilized to achieve changes during the migration.

Junos resources: *What Junos OS specific tools are available for this type of migration?*

This section leverages standard protocol resources from a Junos OS point of view. In addition to including descriptions for protocol features, it also describes protocol-independent and Junos OS specific configuration knobs and snippets that can provide significant assistance during a migration exercise.

Case study: *How can a network engineer implement migration concepts in a network topology?*

Each case study covers planning, design, implementation, verification, monitoring, and possible rollback stages in a migration, making use of protocol concepts and Junos OS resources that ensure successful completion at each phase.

While it is difficult to replicate real-life scenarios in a small laboratory for illustration purposes, both authors have attempted to illustrate concepts in a small-scale network topology. For pedagogical reasons, the same underlying physical setup is constantly

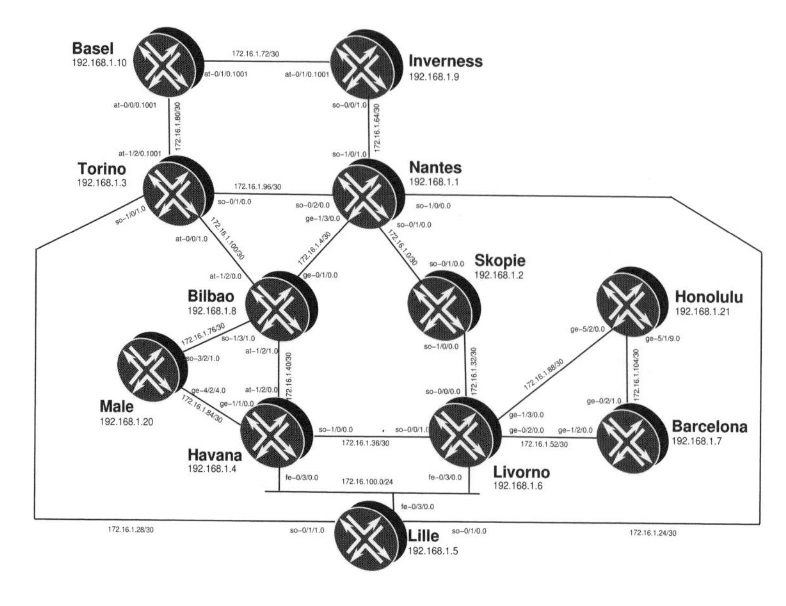

Figure 1: Basic network topology.

reused through all exercises, as shown in Figure 1. This figure is also included on page 528 as a quick reference. The roles of the different routers vary from chapter to chapter and different arbitrary virtual topologies are used in each exercise, but the underlying physical setup remains the same in all cases to ease understanding when describing topics and ideas.

This common physical topology is also used across all chapters to illustrate other various concepts. These concepts, which could not be included in a common case study, are analyzed throughout the book in the form of stand-alone application notes and JUNOS tips.

Application Note: *How can a particular feature be of use?*

Small example cases spread throughout the book represent asides from a current discussion, describing particular changes or functionalities other than those in the main flow. The intention is to provide hints related to migration design options, generally related to a given Junos OS feature, that could be implemented at that time.

Junos tips: *Small operational and design hints*

Tips and tricks are also scattered through the text to provide additional considerations related to a concept, configuration, or output under discussion.

What If . . .: *Brief analysis of design or migration alternatives*

At some decision stages in different migration case studies or application notes, brief evaluation notes are included to describe *what* would have happened *if* another alternative were taken to achieve the integration goal.

Resources

This book has been written in LaTeX and maintained by means of a collaboration system including *GNU gnats* and *CVS*. All diagrams, topologies, and drawings have been accomplished using *Inkscape*.

Junos OS used in different scenarios throughout the book include different 9.5, 9.6, and 10.0 Maintenance releases.

1

Dealing with Routes within a Junos OS Based Router

When migrating services from one network design to the next, it is highly likely that a non-negligible transition period is required. During this time, services may live in two worlds, in which some applications and network sites for those services are partially duplicated to allow the old and new designs to coexist.

From the perspective of an IP network, a service is identified by different abstractions, and a valid concept for a service is a set of IP addresses that are reachable from different network sites. This reachability information, which is exchanged via routing protocols, is collected in routing tables, and the best routing information from routing tables is propagated to the forwarding engine.

This chapter focuses primarily on routing and addressing for the Internet Protocol version 4 (IPv4) and illustrates Junos® operating system (Junos OS) configuration options that can leverage and optimize route manipulation among different routing tables. The underlying concepts presented only for IPv4 here can be generalized to other address families.

1.1 Route Handling Features inside a Junos OS Based Router

Transition scenarios should avoid non-standard routing-protocol behavior, unless a network designer is deliberately planning a migration considering this. The impact of a migration can be reduced by employing specific and often creative behaviors on the routers themselves, behaviors that do not change the expected routing-protocol behavior, or that change that behavior in an identified fashion. These behaviors may have only a local effect on the router, and do not require a redefinition of open standards or modification to routing protocols.

In Junos OS routing architecture, the routing table is the central repository for routing information. Routes are *imported into* the routing table and *exported from* the routing table. This centralized approach avoids direct interaction among routing protocols; rather, protocols interact with the relevant routing table to exchange information outside their domain. Junos OS allows the use of more than one routing table to limit visibility of routing information

Network Mergers and Migrations Gonzalo Gómez Herrero and Jan Antón Bernal van der Ven

to specific protocols only. Junos OS also allows the option for protocols to inject routing information into more than a single routing table.

The example in Listing 1.1 illustrates the process of exchanging a Routing Information Protocol (RIP) route with an Open Shortest Path First (OSPF) protocol, which involves three steps:

Listing 1.1: Redistribution of RIP route into OSPF

```
 1 user@Bilbao-re0> show route receive-protocol rip 10.255.100.77 table inet.0
 2
 3 inet.0: 26 destinations, 28 routes (26 active, 0 holddown, 0 hidden)
 4 + = Active Route, - = Last Active, * = Both
 5
 6 0.0.0.0/0          *[RIP/100] 1d 21:56:47, metric 2, tag 0
 7                     > to 10.255.100.77 via so-1/3/1.0
 8
 9 user@Bilbao-re0>  show route table inet.0 0/0 exact detail
10
11 inet.0: 26 destinations, 28 routes (26 active, 0 holddown, 0 hidden)
12 0.0.0.0/0 (1 entry, 1 announced)
13         *RIP    Preference: 100
14                 Next hop: 10.255.100.77 via so-1/3/1.0, selected
15                 Task: RIPv2
16                 Announcement bits (3): 0-KRT 2-OSPF 4-LDP
17                 Route learned from 10.255.100.77 expires in 174 seconds
18 <...>
19
20 user@Bilbao-re0> show ospf database external lsa-id 0.0.0.0
21     OSPF AS SCOPE link state database
22  Type       ID               Adv Rtr           Seq      Age  Opt  Cksum  Len
23 Extern  *0.0.0.0          192.168.1.8      0x8000003f   488  0x22 0x55aa  36
```

1. The RIP task receives the route from a neighbor (10.255.100.77 in the example). After processing it and deciding that it passes its constraints check as specified by the RIP *import* policy, the route is installed in the routing table (Line 6).

2. The routing table informs interested processes that a new route is available through a notification mechanism (Line 16), whereby protocols register their interest in receiving updates. In this case, OSPF is registered on the announcement list as a recipient of changes in this RIP route.

3. Once notified, and after the RIP route has undergone some changes, the OSPF task evaluates this RIP route from the routing table, analyzes its validity for *exporting* through its own policy, and finally incorporates it into its link-state database in the form of a Type 5 AS External LSA (Line 23).

The Junos OS routing architecture supports multiple routing *instances*, each with its collection of Routing Information Base (RIB) tables, interfaces, and routing protocols. This *RIB group* construct allows the exchange of routes among tables. The remainder of this section covers this concept in more detail.

1.1.1 Instances and RIB tables

Routing information is contained in a collection of routing tables belonging to a routing instance. The term "RIB", for Routing Information Base (and also "FIB," for Forwarding

Information Base) is a tribute to the good old GateD origins of Junos OS and is the internal name for such a routing table.

The content of a RIB does not necessarily need to be a collection of IPv4 routes; in fact, RIBs can store routing information related to a variety of protocols. As an example, a basic configuration on a Service Provider core router with a single routing instance enabling IS–IS, MPLS, and IPv4 features at least three RIBs.

The default, or master, routing instance contains multiple RIBs, each with its specific purposes. Table 1.1 lists some of these RIBs in the master instance and their intended uses. Besides traditional unicast RIBs for IPv4 and IPv6, Junos OS provides next-hop resolution for BGP using an auxiliary RIB, *inet.3* (or *inet6.3* for IPv6) that is generally populated with destinations reachable over MPLS. In an MPLS transit node, the *mpls.0* RIB holds the state for exchanged MPLS labels, that are locally installed in the forwarding plane and used to switch MPLS packets. For IS–IS enabled routes, the *iso.0* table contains the NET Identifier (Net-ID) of the router. MPLS/VPN routes received from remote PEs for Layer 3 or Layer 2 services are stored in the relevant VPN table, from which the real prefix (for IP or Layer 2 connections) is derived. Similarly, multicast routing has its own set of dedicated RIBs, to store (Source, Group) pairs (*inet.1*) or Reverse Path Forwarding (RPF) check information (*inet.2*).

Table 1.1 Default RIBs

RIB name	Use	Family
inet.0	Unicast routing	IPv4
inet6.0	Unicast routing	IPv6
inet.3	Next-hop resolution	IPv4
inet6.3	Next-hop resolution	IPv6
inet.1	Multicast routing	IPv4
inet.2	Multicast RPF	IPv4
mpls.0	MPLS labels	MPLS
iso.0	CLNS routing	ISO
bgp.l3vpn.0	inet-vpn	INET-VPN
bgp.l2vpn.0	l2vpn	L2VPN

Some RIB tables within the master instance.

Note that for optimization purposes, a RIB becomes visible only once it is populated. If no routes from a particular address family exist, the corresponding RIB is not instantiated in Junos OS. This behavior means though that additional supporting RIBs are created for services as needed.

By default, a collection of RIBs is present in the master instance. To provide unique *forwarding contexts*, additional instances can be created, each one with its own set of RIBs. A *forwarding context* is understood as the combination of a *forwarding table* with multiple RIBs feeding it with routing information.

Hence, an instance can contain multiple RIBs, but a RIB can have only one *parent* instance. That is to say, the main routing instance may contain *inet.0*, *inet6.0*, *iso.0*, *mpls.0*, and others, but those RIBs are univocally related to the main routing instance only.

RIBs within a routing instance on the control plane can create companion *forwarding instances*, or can be used purely at the control plane for supporting activities, generally related to other RIBs.

Figure 1.1 depicts a router configured with three instances. For the associated collection of RIBs, only some populate the forwarding table, and the remaining RIBs perform supporting functions. Instance A contains three RIBs, storing IPv4 unicast prefixes (A.inet.0), IPv4 multicast (Source, Group) pairs (A.inet.1), and the multicast RPF check support RIB (A.inet.2). Instance B is likely a Carrier-supporting-Carrier (CsC) VPN Routing and Forwarding (VRF) table implementing both IPv4 unicast prefixes (B.inet.0) and an instance-based MPLS table (B.mpls.0). Similarly, the global instance includes a set of *helper* RIBs for the next-hop resolution (inet.3) and the helper RIBs to support received VPN prefixes (bgp.l3vpn.0 and bgp.l2vpn.0).

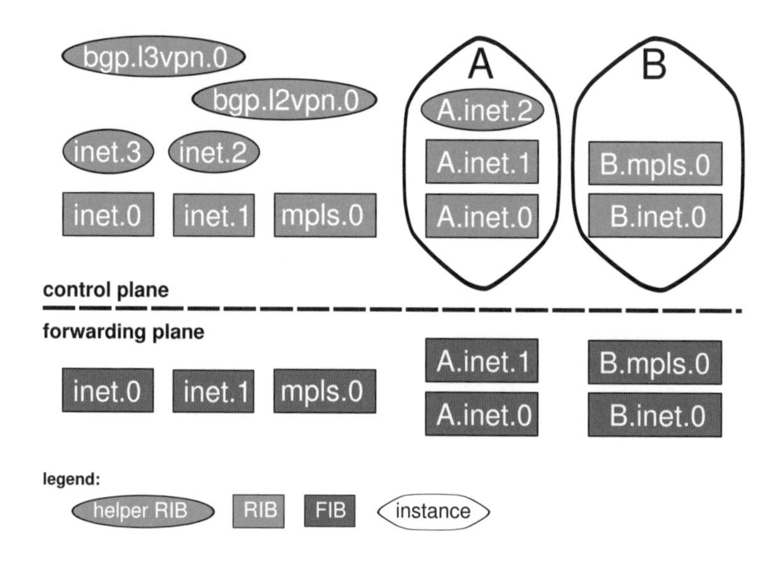

Figure 1.1: Sample distribution of RIBs and instances on a router.

The actual packet-by-packet forwarding of transit data occurs in the forwarding plane. In Junos OS-based architectures, the prefix lookup hardware makes decisions about how to switch the packets based on tables that contain next-hop information. These FIB tables are the forwarding instances and are populated with information derived from best-path selection in the RIBs. The creation of a forwarding instance triggers instantiation in the hardware that populates the *forwarding plane* with FIBs that are consulted by the lookup engine when switching packets.

The *master* routing instance is the only one available for administrators in the default configuration. Additional instances can be defined for interconnect or virtualization purposes, collapsing the functionality of multiple service delivery points into a single platform. Each *type* of instance has a specific behavior, as shown in Table 1.2.

Table 1.2 Instance types

Instance name	Use	Interfaces?
`forwarding`	Filter-based forwarding	no
`non-forwarding`	Constrained route manipulation	no
`vrf`	L3VPN support	yes
`virtual-router`	Simplified VPN, VRF-lite	yes
`l2vpn`	L2 VPN	yes
`vpls`	VPLS	yes
`virtual-switch`	L2 switching	yes
`layer2-control`	L2 loop control support	yes

Types of routing instance.

In Junos OS the default routing instance type, for instances different from the master, is `non-forwarding` for historical reasons. The first customer requirement for virtualization was to support constrained route exchange between two separate OSPF routing tasks, but within the same forwarding instance. This default setting can be overridden with the `routing-instances instance-type` knob.

Listing 1.2 shows sample CLI output for a router with two L3VPN in addition to the default configuration.

Listing 1.2: Looking at available instances

```
 1 user@Livorno> show route instance
 2 Instance             Type
 3          Primary RIB                                  Active/holddown/hidden
 4 master               forwarding
 5          inet.0                                       23/0/1
 6          mpls.0                                       3/0/0
 7          inet6.0                                      3/0/0
 8          l2circuit.0                                  0/0/0
 9 CORP                 vrf
10          CORP.inet.0                                  2/0/0
11          CORP.iso.0                                   0/0/0
12          CORP.inet6.0                                 0/0/0
13 helper               vrf
14          helper.inet.0                                1/0/0
15          helper.iso.0                                 0/0/0
16          helper.inet6.0                               0/0/0
```

Armed with the knowledge that multiple RIBs and various instances hold routing information, we move to the next section to understand how to group RIBs together to allow routing information insertion procedures to act simultaneously on all RIB members in the group.

1.1.2 Grouping RIBs together

Based on customer use cases, Junos OS was enhanced early on to add *RIB groups*, which are a mechanism to provide route information to multiple RIBs. The primary purpose of RIB groups is to allow independent protocol processes to exchange routing information in a constrained fashion, using routing tables to scope route visibility.

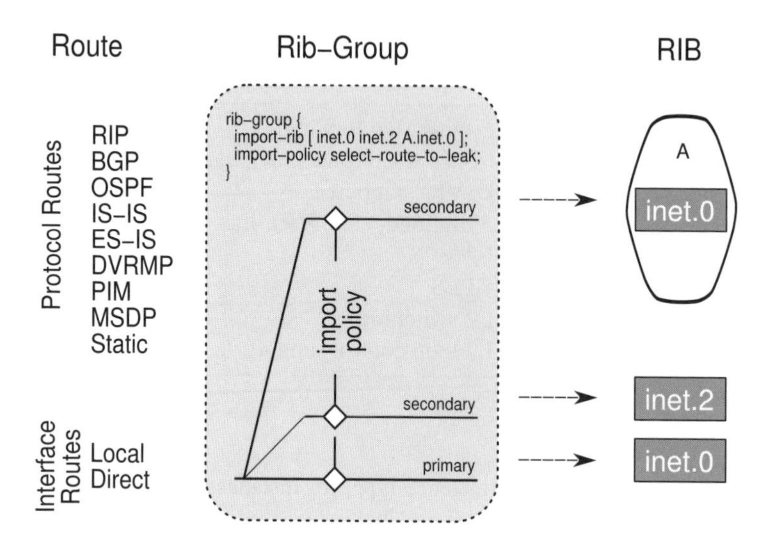

Figure 1.2: RIB group mechanics.

The RIB group construct is a little known cornerstone in Junos OS routing architecture. This *configuration construct* provides a generalized mechanism to distribute routing information locally, without creating new RIBs, as shown in Figure 1.2. The source of the routing information to be distributed can be routes owned by any supported protocols within a RIB, including static routes and routes representing the local interfaces. In essence, RIB groups allow for distribution among RIB tables of routes received dynamically via various protocols or manually through configuration (static routes and direct routes from interfaces).

By default, all routing protocol information in Junos OS is associated with a single RIB; the routes that the protocol provides to the RIB are *primary* routes. A RIB group has a single primary RIB (the one where protocols would store the information if the RIB group were not configured), and optionally a set of constituent *secondary* RIBs. A particular route can be either primary or secondary, not both; however, a RIB can take the primary RIB role for a RIB group, and at the same time can be one of the secondary RIBs in a different RIB group.

Junos Tip: Impact in link-state IGPs of activation of a RIB group configuration

Activation of a RIB group through the `rib-group` feature is considered a reconfiguration event for IS–IS and OSPF in Junos OS. All adjacencies are rebuilt when the configuration is changed to add a RIB group.

When the `rib-group` mechanism is activated, routes that traverse the RIB group and land in the primary RIB are flagged as *primary* routes. Routes that are replicated and installed in other RIBs are flagged as *secondary* routes. These secondary routes are still connected to the primary route and follow the fate of the primary route. Only one RIB, the primary RIB in the RIB group, stores the primary route, and the state of that route drives the state of the set of replicated routes. If a primary route were to disappear or change, all secondary routes dependent on it would follow suit immediately. This dependency facilitates the existence of a single *resolution context* for the entire set of prefixes. Route attributes for the secondary route, including the protocol from which the route was learned, are inherited from the primary route.

Junos Tip: Finding sibling tables for a prefix

The extensive output of `show route` operational command provides hints about the properties for a particular prefix. Listing 1.3 helps to identify primary and secondary routes. For a primary route that is leaked using RIB groups, the list of secondary routes is shown in the *secondary* field on Line 16. For secondary routes, the *state* field includes a flag (Line 25), and an additional legend at the bottom of the output on Line 31 provides a reference to the primary RIB for the prefix.

Listing 1.3: Verifying primary owner instance of routes

```
1  user@male-re0> show route 10.1.1.0/24 exact protocol rip detail
2
3  inet.0: 21 destinations, 24 routes (21 active, 0 holddown, 0 hidden)
4  10.1.1.0/24 (2 entries, 1 announced)
5           RIP     Preference: 180
6                   Next hop type: Router, Next hop index: 1538
7                   Next-hop reference count: 10
8                   Next hop: 10.255.100.78 via so-3/2/1.0, selected
9                   State: <Int>
10                  Inactive reason: Route Preference
11                  Local AS: 65000
12                  Age: 3:54:26    Metric: 2       Tag: 1000
13                  Task: RIPv2
14                  AS path: I
15                  Route learned from 10.255.100.78 expires in 160 seconds
16                  Secondary Tables: inet.2
17
18 inet.2: 4 destinations, 4 routes (4 active, 0 holddown, 0 hidden)
19
20 10.1.1.0/24 (1 entry, 0 announced)
21         *RIP     Preference: 180
22                  Next hop type: Router, Next hop index: 1538
23                  Next-hop reference count: 10
24                  Next hop: 10.255.100.78 via so-3/2/1.0, selected
25                  State: <Secondary Active Int>
26                  Local AS: 65000
27                  Age: 1:51:30    Metric: 2       Tag: 1000
28                  Task: RIPv2
29                  AS path: I
30                  Route learned from 10.255.100.78 expires in 160 seconds
31                  Primary Routing Table inet.0
```

Although the RIB group is applied to all routes received by the protocol or interface, the power of Junos OS policy language can be leveraged to restrict the distribution of specific prefixes to specific RIBs, based on a route filter, a route attribute, or some protocol attribute. Note that to maintain a single resolution context, any changes to the next hop that can be defined as an action of a Junos OS policy are silently ignored.

Junos Tip: Multi-Topology Routing

The Multi-Topology Routing (MTR) feature deviates from the philosophy of the `rib-group` construct providing an *independent* routing resolution context, in which the dependent routes of a primary route may have a different next hop.

This feature does not leverage the RIB group mechanism and is therefore not discussed further. The Junos OS feature guide provides additional information on multiple topologies with independent routing resolution contexts in its Multi Topology Routing Section.

1.1.3 Instructing protocols to use different RIBs

Leaking routes between different RIBs provides a mechanism for populating the RIBs with the desired routing information. This functionality is tremendously powerful for migration scenarios.

In some cases, when the destination RIBs already exist and have a specific purpose, route leaking among RIBs already suffices to achieve the final goal. For some protocols, however, an additional step is required to direct the routes to the proper RIB, the one that is used during the route-lookup process, and to make those routes eligible in that RIB for further population to other systems.

Labeled BGP

[RFC3107] defines an implementation to have BGP carrying label information, therefore binding routing information to a MPLS label. This implementation is widely used as a comprehensive MPLS label distribution protocol and is covered in Chapter 4.

By default, the received routing information with Labeled BGP, or L-BGP, is installed in the main IPv4 table, inet.0. Through configuration under `protocols bgp family inet labeled-unicast rib inet.3`, it is possible to specify that these routes be installed only in the helper RIB table inet.3, used for route resolution. A direct advantage of this configuration is to allow MPLS VPNs, which perform route lookup using inet.3 by default, to access the labeled information straightaway. Hence, prefixes exchanged over a labeled BGP session can provide the transport label for VPN services. Section 4.4.4 provides additional information on the various Junos OS resources available to handle these prefixes.

Protocol Independent Multicast

Protocol Independent Multicast (PIM) traditionally uses the prefixes in the inet.0 table to perform the RPF check to ensure that multicast traffic arrives on the intended upstream interface back towards the source as a standard multicast lookup mechanism. It is possible to create a specific RPF lookup RIB (table inet.2) that contains constrained RPF information; in addition, PIM has to be instructed to use the inet.2 RIB instead of the default (inet.0). Instructing the protocol PIM to use this table is done by pointing to a RIB group. Hence, the `rib-group` construct is leveraged again, but just as a facilitator to indicate which RIB to use.

Application Note: Separate unicast and multicast RPF

Some scenarios may require a different check for multicast traffic relative to the existing unicast topology. These scenarios require two RIBs, each with a different view of the routes. Populating two RIBs with selected routes from a protocol or an interface is the perfect use case for using the `rib-group` construct. That is why the population of RPF information in RIB inet.2 for multicast protocols uses this construct.

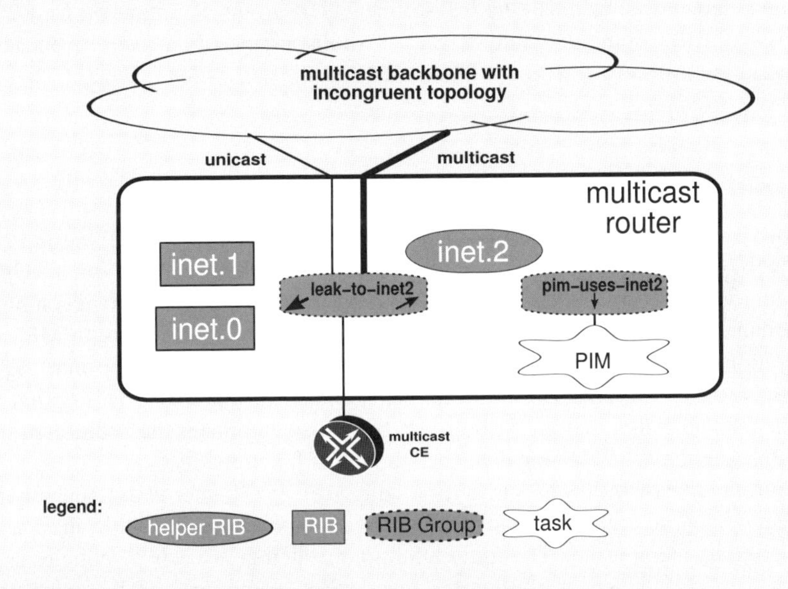

Figure 1.3: Use of inet.2 to constrain the multicast topology.

Figure 1.3 and the configuration displayed in Listing 1.4 provide a sample setup in which interface routes are installed into the default RIB, inet.0. In addition, using the `rib-group` construct (Line 2), the same routes are also installed in the inet.2 RIB in the main routing instance. Through policy language (Line 4), though, only copies of large-capacity core interface routes leak to inet.2, thus providing inet.2 with a partial topology view of the connectivity. Prefixes learned over the core routing protocol (not shown) also leak selectively to both RIBs.

The missing piece in the application is to tell the PIM multicast protocol to look *just* in the inet.2 table, as is done in Line 11, which directs the PIM protocol to use the rib-group for that purpose. Note that the first table in the `rib-group` list is considered as the *Multicast RPF* table. The output starting on Line 14 confirms that the configuration is effective.

Listing 1.4: Use rib-groups to provide an RPF table to PIM

```
1 user@male-re0> show configuration routing-options rib-groups
2 leak-to-inet2 {
3     import-rib [ inet.0 inet.2 ];
4     import-policy big-pipe-only;
5 }
```

```
6  pim-uses-inet2 {
7      import-rib [ inet.2 ];
8  }
9
10 user@male-re0> show configuration protocols pim rib-group
11 inet pim-uses-inet2;  # Only the first RIB taken into account
12
13 user@male-re0> show multicast rpf summary inet
14 Multicast RPF table: inet.2 , 3 entries
```

While it is possible to reuse the leak-to-inet2 RIB group definition on Line 2, which includes both RIBs, doing so would defeat the whole purpose of the exercise because such a configuration instructs PIM to use the first RIB in the list, which is inet.0, when what is wanted is to use inet.2.

1.1.4 Automatic RIB groups and VPN routes

Although visible only indirectly, Junos OS VPN technology uses the RIB group construct internally to build RIB groups dynamically to leak VPN routes received from a BGP peer.

The Junos OS import policy is generic enough in that it allows for matching on any BGP attribute. Therefore, the RIB group concept is of direct applicability to MPLS VPNs. Configuration of the Junos OS `vrf-import` option is a one-stop shop that triggers the automatic creation of internal rib-group elements to distribute prefixes. This option not only avoids the manual configuration of rib-groups but it also automates the addition of members to the VPN. Section 1.1.5 elaborates further about this functionality.

The introduction of MPLS L3VPNs (first with [RFC2547], later with [RFC4364]) added a new requirement to create prefix distribution trees based on extended Route Target (RT) BGP community membership. The Junos OS implementation inspects all policies associated with `instance-type vrf` routing instances when searching for RT communities, and internally builds the associated RIB groups in advance. When routing information for MPLS VPNs arrives, it is funneled through these RIB groups to leak the prefix to all interested RIBs from the corresponding instances.

For MPLS VPNs, Junos OS installs all BGP prefixes received over the MPLS VPN backbone into helper or supporting RIBs. Junos OS maintains one RIB for inet-vpn (*bgp.l3vpn.0*) and another RIB for l2vpn (*bgp.l2vpn.0*). These tables include the full NLRI information, including the Route Distinguisher (RD).

The BGP updates that contain VPN NLRIs have to get rid of the RD part of the NLRI information and become usable forwarding prefixes related to each internal instance. Junos OS performs the appropriate translation for the NLRIs based on the target RIB. No translation is needed to populate the BGP support tables. For customer RIBs, the translation is automatic. Junos OS currently provides no support for *manual* configuration of `rib-groups` in which the source NLRI and destination RIBs are of different families, except for leaking IPv4 prefixes in IPv6 tables for next-hop resolution.

For every VPN set containing multiple VRFs, a distinct RIB group is created. Figure 1.4 depicts how three sample prefixes received over BGP go through the RIB group processing and land in the bgp.l3vpn.0 table. The RT identifies member VRFs, so a distinct RIB group is created for each RT. In the most general case, the prefix has only a single RT community attached; hence, the RIB group contains only two members: the bgp.l3vpn.0 table and the RIB belonging to the relevant VRF.

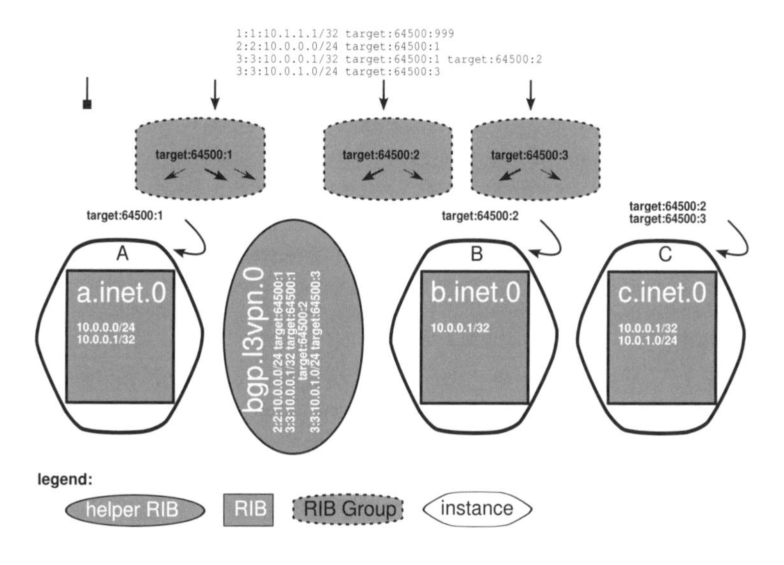

Figure 1.4: Rib-groups to distribute VPN prefixes.

From a scalability perspective, a Provider Edge (PE) device in the MPLS VPN architecture should not need to store route information for VPNs to which it does not belong. By default, Junos OS discards received BGP prefixes with a RT that does not match any of the locally configured targets, when acting as a plain PE. With no matching target in that situation, these prefixes are not installed in any VRF instance and are also not installed in the BGP supporting table in Junos OS unless the `keep-all` feature is configured.

Router reconfiguration that might result in the creation of a new VRF or in the addition of new members to a VRF triggers a BGP route *refresh* message, which requests the peer to resend all VPN information. The router then re-evaluates this information to retrieve any routes that might have been filtered out earlier.

The primary RIB for these automatic RIB groups is the BGP VPN support table, namely bgp.l3vpn.0 in the case of MPLS L3VPNs. One exception to this rule, related to *path selection* behaviors, occurs when the router acts as a Route Reflector or an AS Boundary Router (ASBR). This behavior is discussed in Section 1.2.2.

Listing 1.5 shows the output of the hidden Junos OS command `show bgp targets`, which contains information about the automatic RIB groups that are configured. Starting at Line 3 are details specific to a particular RT community. The example shows that prefixes received with route target target:64500:1 are imported into CORP VRF, and it shows that distribution of prefixes with target:64500:11 are distributed to two other VRFs besides the bgp.l3vpn.0 table.

Listing 1.5: Sample show BGP targets hidden command

```
1 user@Livorno>show bgp targets
2 Target communities:
3 64500:1 :
4   Ribgroups:
```

```
5         vpn-unicast target:64500:1       Family: l3vpn   Refcount: 1
6                 Export RIB: l3vpn.0
7                 Import RIB: bgp.l3vpn.0 Secondary: CORP.inet.0
8   64500:11 :
9     Ribgroups:
10        vpn-unicast target:64500:11      Family: l3vpn   Refcount: 1
11                Export RIB: l3vpn.0
12                Import RIB: bgp.l3vpn.0 Secondary: CORP.inet.0 helper.inet.0
```

1.1.5 Local redistribution using the vrf-import option

Unlike other network operating systems, Junos OS performs route redistribution to and from the relevant routing table (RIB), not between routing protocols. Route exchange between two protocols is feasible only through an intermediate routing table that propagates only the best route information by default (as usual, some concrete exceptions exist to this general rule).

The discussion in Section 1.1.2 relied on manual configuration for each redistribution combination. The configuration complexity increases as more RIBs are added to a RIB group, as illustrated in the sample management L3VPN scenario in Figure 1.5 because more VRFs need to adjust their policies for route import and export actions.

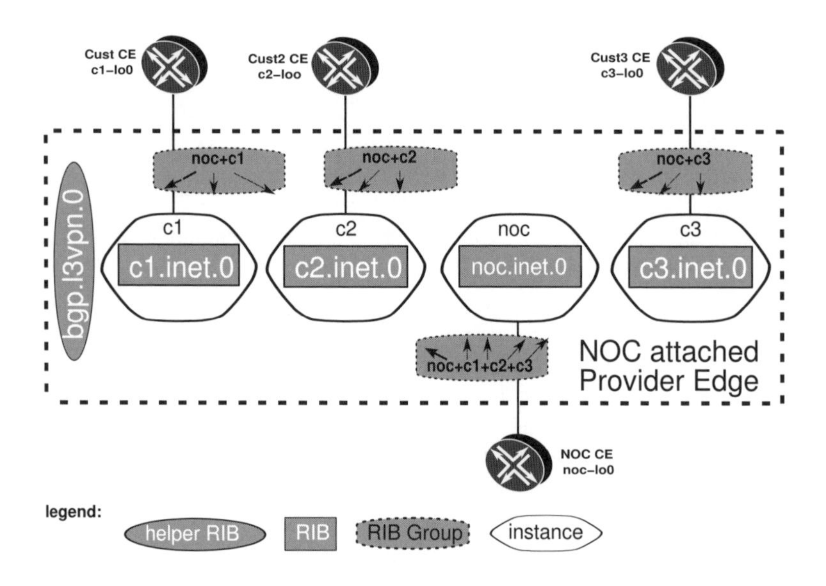

Figure 1.5: Reaching managed CPEs from the NOC.

In this example, a L3VPN *extranet* scenario is created, with multiple RIBs exchanging a subset of the same routes among themselves. Managed CPEs need to be reachable from the NOC VPN. Conversely, the NOC has to be reachable from each VRF. An additional customer VRF (c3) requires the creation of a new RIB group (c3+noc) to install the CPE loopback address in the NOC VRF and to leak the NOC addressing space into the new VRF appropriately by adding this VRF to the NOC RIB group. This topology is representative for a very common setup in MPLS L3VPN service providers that provide managed services for CPEs in customer VPNs, in which management flows need to be intermingled among these VRFs.

Note that these RIB groups apply only to locally attached VRFs. For other VPNs, proper RT tagging suffices, because an automatic RIB group is already in place for prefixes received into the bgp.l3vpn.0 table. One obvious alternative to this design is to dedicate a small PE to attach to the Network Operations Center (NOC) and additional VRFs that home non-managed CPEs.

In light of this inherent complexity, Junos OS has been enhanced to introduce a more intuitive construct that triggers the creation of RIB groups dynamically using the policy language, as is done for VRFs that are not on the same router. A *pseudo-protocol* named *rt-export* allows RIBs on the same router to exchange information. This protocol handles advertisements and processes received routes, delegating all control to the policy language.

Existing customers are protected from this new functionality because it is disabled by default and must be activated through router configuration. Configuring this feature does not conflict with RIB groups and both options can be configured simultaneously, even in the same VRF.

Configuring `routing-options auto-export` on a routing instance binds the *rt-export* protocol module to it. This module tracks changes in RIBs for routes that have to be advertised and ensures that the advertised route is also imported in sibling RIBs on the router if policy so dictates. Traditional `vrf-import` and `vrf-export` policies check for specific matches on other VRFs on the router. These policies trigger local intra-router prefix redistribution based on the RT communities without the need for BGP. The resulting behavior is similar to what would be expected from a prefix received from a remote PE over BGP.

In some cases, routing information between VRFs and regular RIBs must be distributed, as happens when a VPN customer also has Internet access that is available from the main routing instance on the PE.

The same module can be leveraged for *non-forwarding* instances. The `instance-import` and `instance-export` configuration options mimic the intra-router behavior of `vrf-import` and `vrf-export` for VRF for non-forwarding instances. Given that non-forwarding instances lack RT communities, the policy called as argument has to include a `from instance` statement, to control the instances scope of the policy.

This section has focused on populating RIBs with route information. This information is processed locally, and some of the routes become the best alternative within their respective RIB domains. The following section discusses how these behaviors interact with the advertisement of the best selected routes.

1.2 RIB Route Advertisement at MPLS Layer 3 VPNs

For a PE router in the MPLS L3VPN architecture, the routing information that populates its local VRFs and which is learned from remote PE routers has to be advertised to the CEs. The advertisement can be achieved in Junos OS using a configuration similar to that for the main routing instance, but constrained to the specific VRF configuration block. Configuring protocols within a routing-instance constrains the route advertisements to the VRF.

Conversely, the routing information advertised by the CE must be propagated from the PE–CE connection to the MPLS L3VPN cloud by the PE router, which does so performing proper NLRI conversion to include the VPN Label and RD information, eventually tagging the NLRI with RTs.

Figure 1.6 shows how a VRF named *VRFA* is populated with two prefixes coming from two different sources, namely the loopback address from a remotely attached CE, CE2-lo0, and a prefix from the locally attached CE, CE1-lo0. CE2 of customer A advertises its loopback address to PE2, which in turn propagates it to the Route Reflector. The Route Reflector performs regular route-processing with assistance from the helper table (bgp.l3vpn.0 for IPv4 prefixes), using the RD to disambiguate overlapping space for different customers. The prefix is sent to PE1, which inspects its local list of RT matches in the import policy and redistributes the prefix to both the helper RIB and the relevant member VRFs. The advertisement by the local CE1 is received on the CE $\rightarrow$ PE connection at PE1, and the prefix is advertised directly from the VRF table, bypassing the helper RIB completely. For cases where interprovider VPNs are used using VPN NLRI exchange at the border (option B), the behavior of the ASBR is identical to the one performed by a Route Reflector, taking prefixes in and out of the BGP VPN table (bpg.l3vpn.0). More details on the interconnect options can be found in Section 5.2.6 on Page 391.

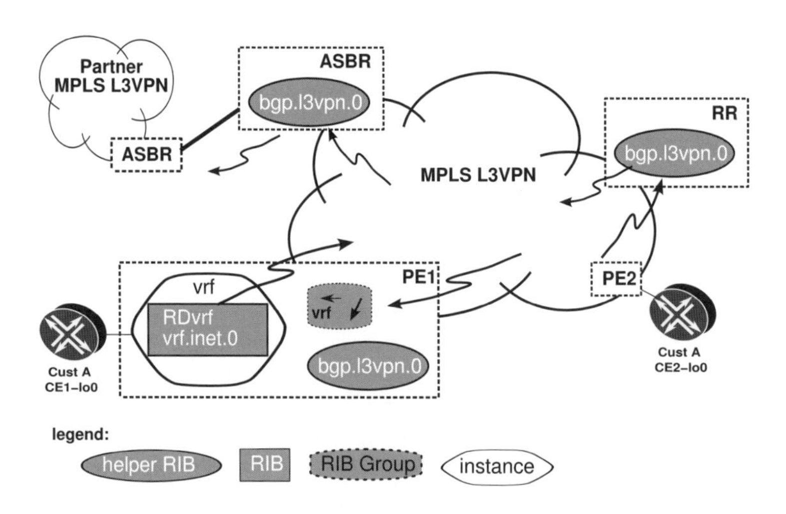

Figure 1.6: Receiving and advertising VPN prefixes.

1.2.1 Levels of advertisement policy – `vpn-apply-export`

By default, exporting routes from a VRF in Junos OS does not take into account the global BGP *export* policy. In interprovider scenarios with combined PE and Inter-AS option B ASBR functionality, blocking VPN advertisements only to certain peer groups may be desired. Also, exposing all PE local VRFs across the partner MPLS L3VPN interconnect, may not be desired, but Junos OS does not allow the destination BGP groups from within the `vrf-export` policy to be specified directly. Therefore, advertisement policy may depend only on the VRF or can combine it with a global exporting policy for the core BGP protocol. Figure 1.7 shows how the instance-level `vrf-export` and global bgp group `export` can be combined to modify attributes. This behavior, which is to be enabled with the `vpn-apply-export` configuration statement, splits the distribution of route information between a VRF part and a global part.

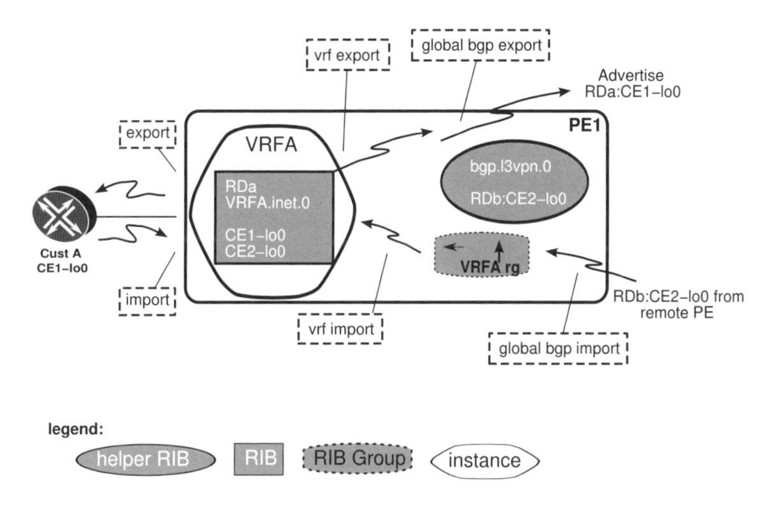

Figure 1.7: Route policy points.

1.2.2 Path selection mode in Junos OS

When the router acts as a combined PE/Inter-AS option B ASBR or PE/Route Reflector device in MPLS VPN architectures, the routing information it propagates includes native VPN routes. As part of the propagation of the multiple routes, a *path selection* process needs to take place to select the best route among the set of comparable routes.

Generally, the RD acts as a differentiator for prefixes belonging to different VPNs that have the same internal address allocation. The RD provides context separation when the same prefix belongs to different VPNs, so the information about the prefix itself is not enough to allow a proper comparison.

Cases exist when a single prefix is available in the same VPN over different PE devices, such as with multihomed sites (see Figure 1.8). The RD may as well be unique for each VRF within a VPN, to provide path diversity to the PEs in a Route Reflector environment. Conversely, when all copies of the same prefix within a VPN share the same RD, overall router memory consumption is reduced because identical VPN prefixes in the form *RD:Prefix* become comparable, and because only the best path is propagated.

Under this *path selection mode*, it is understood that a router can reflect VPN routes if it is acting as an Inter-AS option B boundary router (ASBR) or a Route Reflector, and if the router is also a PE, the best-path comparison for reflected VPN routes also has to look at local

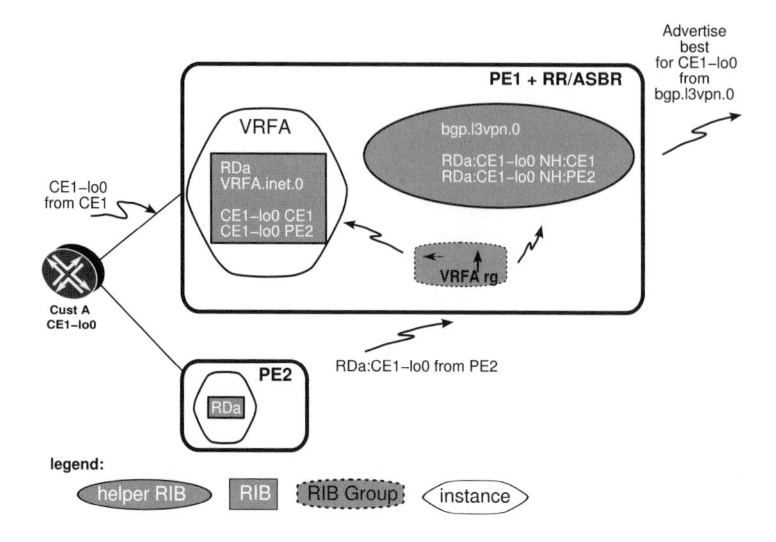

Figure 1.8: Advertisement of comparable NLRIs in PE acting as RR/ASBR.

prefixes to provide consistent routing information. In Junos OS, enabling this *path selection mode* is automatic when the configuration parser detects a MP-BGP session configured that requires route reflection (either internally as a Route Reflector Server, or externally as an ASBR), and triggers a process whereby all local prefixes and BGP paths learned from remote routes are copied to the bgp.l3vpn.0 table.

The example in Figure 1.8 depicts how the local routes are advertised from the helper table bgp.l3vpn.0 instead of using the VRF table. Best-path selection for a reflected route versus a locally learned route is performed correctly, at the expense of duplicating local routes to the bgp.l3vpn.0 RIB. Without this duplication, every VRF would advertise its own version of the best path, which could create non-deterministic inconsistency in the advertised state if, for instance, the VRF advertised its prefix followed by a withdrawal from the bgp.l3vpn.0 table.

Listing 1.6 shows how to display the advertising state of a particular table. Output starting at Line 1 shows the regular PE case, in which prefixes are advertised directly from the VRF. By comparison, Line 7 shows a scenario in which path selection mode is enabled for a PE acting as Route Reflector, and prefixes are advertised from the bgp.l3vpn.0 table instead.

Listing 1.6: Advertising state for bgp.l3vpn.0

```
1  user@Havana> show bgp neighbor 192.168.1.1 | match "table|send"
2    Table bgp.l3vpn.0
3      Send state: not advertising
4    Table CORP.inet.0 Bit: 20000
5      Send state: in sync
6
7  user@Nantes> show bgp neighbor | match "table|send"
8    Table bgp.l3vpn.0 Bit: 10000
9      Send state: in sync
10   Table CORP.inet.0
11     Send state: not advertising
```

Junos Tip: Use of a table filter in path selection mode

With *path selection mode*, both local prefixes and BGP routes with paths from remote PEs are copied to the bgp.l3vpn.0 table and considered for export.

The *table* of the `show route advertising-protocol bgp` command filters for *primary* routes only. If a table qualifier is used and *path selection mode* is in force, the qualifier should refer to the helper table bgp.l3vpn.0 instead of the local VRF that is configured.

Listing 1.7 shows the sample output for router *Nantes*, a PE advertising 10.200.1.0/24 on the CORP VRF, but also behaving as a Route Reflector (in path selection mode).

Listing 1.7: Correct table argument to *show route* in path-selection mode

```
1  user@Nantes> show route advertising-protocol bgp 192.168.1.6 table CORP
2
3  user@Nantes> show route advertising-protocol bgp 192.168.1.6 table bgp.l3vpn.0 terse
4
5  bgp.l3vpn.0: 7 destinations, 7 routes (7 active, 0 holddown, 0 hidden)
6    Prefix                  Nexthop              MED     Lclpref    AS path
7    64500:1:10.200.1.0/24
8  *                         Self                 2       100        I
9    64500:1:10.200.1.9/32
10 *                         Self                 2       100        I
11   64500:2:0.0.0.0/0
12 *                         192.168.1.4                  100        65000 I
```

1.2.3 RIB group versus auto-exported routes

No specific constraints exist regarding advertisement of routes created by the RIB group mechanism. The behavior is similar to any regular prefix. Therefore, if a prefix is received over a CE-PE connection and is leaked to various VRFs, each VRF advertises its own version of the prefix, adding the instance's RD and relevant topology extended communities as necessary. One interesting corollary of this behavior is that, odd as it seems, a secondary OSPF route can be further advertised into OSPF, because the information source is the RIB, not the task. Besides, two separate tasks are involved. An example of this behavior applied to two OSPF tasks is discussed just before Section 1.4.8.

The export behavior differs slightly for *auto-exported* routes. These routes are handled by the *rt-export* module, which behaves like a regular protocol, requiring inclusion of local export and import RTs. Given the flexibility of this approach, to avoid the chance of inconsistent advertisement, constraints are put in place to allow advertisement of only the *primary* route. If advertisements for multiple VPNs are required, care has to be taken to tag the primary route with the relevant RT communities for all participating VPNs, regardless of whether the actual homing of the member VRFs is on the same PE. This is consistent with the typical route-tagging behavior when creating extranets in MPLS VPNs, in which routes are advertised with multiple RTs, one for each remote VRF that requires route import.

1.2.4 RIB selection – `no-vrf-advertise`

For a BGP route that is present in multiple tables, the *primary* route is the one that is selected for advertisement. However, for one special hub-and-spoke scenario, the desire is to advertise the *secondary* route instead. This behavior can be enabled with the `no-vrf-advertise` knob. If multiple secondaries exist, all of them are selected for advertisement.

An equivalent behavior can be obtained by using `rib-groups` to leak the prefix in the desired RIBs, in combination with specific export policy on each RIB to control which of the VRFs actually advertises that prefix. Because the RIB group has to be applied to all prefixes (and protocols) requiring exporting, the `no-vrf-advertise` configuration command is more generic in nature.

Note that the `no-vrf-advertise` behavior is constrained to prevent the BGP advertisements from being sent. When routes are exported by other protocols, including rt-export, each instance of the route can be advertised independently. Exporting routes from the rt-export module, allows for local redistribution of routes into other VRFs.

It is feasible to create a system with multiple RIB tables. These tables can be populated from various sources, and route selection can be constrained within a specific RIB for a particular purpose. Most of the time, however, the desire is for the replicated copy of the route to be installed in a different instance. Particularly during transition stages, traffic belonging to a VRF may need to jump to a different VRF. For the multi-instance case, it is necessary to instruct the *service* to use a particular routing-instance RIB when this service is not already directly connected.

The next section discusses routing-table next hops and filter-based forwarding constructs in Junos OS, which are used to force a lookup in a specific instance.

1.3 Directing Traffic to Forwarding Tables

In the context of [RFC4364], a *site* is homed to a particular routing instance by linking the interface to the VRF. During regular packet flow, a packet coming in over an interface is subject to Layer 2 decapsulation to uncover the packet protocol. As Figure 1.9 shows, packets arriving at the router are first evaluated for the Layer 2 encapsulation, and later a per-protocol instance selection table is consulted. The interface lookup identifies the packet as belonging to a particular site and allows for selection of the proper forwarding instance for prefix lookup. The prefix lookup yields a next hop, which contains precise instructions about how to propagate the packet to the next router.

Junos OS architecture allows packet filters to be applied at different stages during the packet's march through the system. One of the interesting actions of the filters is to *override* the selected forwarding instance.

During migrations, the regular packet flow may require interception so that packets can cross over to a neighboring VRF as part of a transition scenario. Two strategies exist for intercepting this forwarding process: *routing table next hops* and *packet filtering*. Both direct traffic away from the normal forwarding path.

1.3.1 Adding a routing table next hop to a static route

Static routes admit a `next-hop next-table` statement that can be leveraged when the route lookup phase finds that the static is the best possible match. To avoid circular dependencies, Junos OS imposes a generic constraint that prevents a table referred with a *table next hop* to contain itself another route with any *table next hop*. Routing dynamics *might* be such that a packet matches both routes, thus creating an internal forwarding loop.

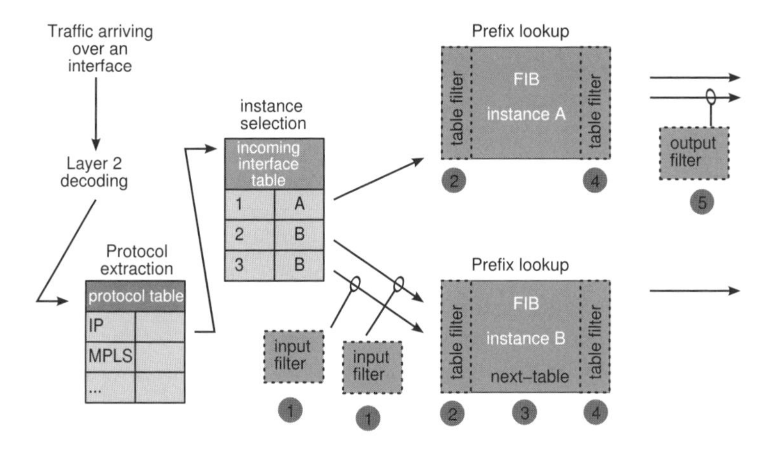

Figure 1.9: Mapping interfaces to instances.

This enforcement is achieved through the Junos OS configuration check, as can be observed in Listing 1.8. Note that this configuration check validates the existence of a table next hop. The target instance should not have any next-table next hop, regardless of whether the actual prefixes have anything in common.

Listing 1.8: Avoiding circular dependencies with next table

```
1  [edit routing-instances]
2  user@Havana# show | display set
3  set routing-instances v4 instance-type virtual-router
4  set routing-instances v4 routing-options static route 2.2.2.2/32 next-table v5.inet.0
5  set routing-instances v5 instance-type virtual-router
6  set routing-instances v5 routing-options static route 1.1.1.1/32 next-table v4.inet.0
7
8  user@Livorno# commit check
9  error: [rib v4.inet.0 routing-options static]
10     next-table may loop
11 error: configuration check-out failed
```

1.3.2 Using packet filtering to control the forwarding process

Junos OS supports a rich set of packet-filtering capabilities, as illustrated in [2000207-001-EN]. The typical filter actions in Junos OS (counting, logging, accepting/discarding the packets) include the possibility of specifying an additional table lookup through routing-instance selection. This table lookup is activated with a `then routing-instance` action statement in the firewall filter.

1.3.3 Usage guidelines

Because filtering can be performed at various points, traffic can also be diverted to a different table at different stages of packet handling. Placement of forwarding filters is very flexible, as depicted in Figure 1.9. Filters can be attached to the interface and to the instance, and in both directions, on incoming and outgoing traffic.

Note that a prefix match in a table at one step is independent of another prefix match in a separate table in a later step. For instance, it is possible to match first on a host (IPv4 /32) route and then jump to another table that matches on a default route.

Using Figure 1.9 as guideline, both table next-hop and firewall strategies are described in the packet-processing sequence as follows:

1. *Input interface filters* can be used when the interface remains homed to the existing instance, but selected traffic needs to be processed first elsewhere. An example is diverting selected source-destination pairs arriving on a specific input interface.

2. *Input forwarding table filters* generalize the previous case for any input interface. Note that these filters take effect before route lookup, on *both traffic directions*. Use of the `interface`, `interface-group`, and `group` qualifiers within the filter may help with scoping of the filter.

3. *Static routes* admit a `next-table next-hop` statement that can be leveraged when the route lookup phase identifies that the static route is the best possible match. Any valid route is allowed, be it a specific destination, a prefix, or a default route. Note that the next-table next hop cannot be applied to dynamic routes.

4. *Output forwarding table filters* work after a lookup is done, when the output interface is already known, and is typically used in combination with the Destination Class Usage (DCU) feature in Junos OS. DCU marking allows the combination of route information (such as BGP communities) with forwarding.

5. *Output interface filters* constrain the above to a single output interface.

1.3.4 Risks of decoupling routing and forwarding

Using some of these resources allows a granular definition of multiple actions on the packet flow in terms of routing lookup and forwarding decisions. In fact, a subset of these features can be combined so that the final forwarding decision for a flow is completely different from the plain routing lookup.

The control plane can be considered to be the *brain* of the system, providing strategy and direction. Standard routing decisions are made as a consequence of information distributed by protocols on the control plane. Tinkering with the forwarding plane could unwittingly have the following consequences on the control plane:

- Static configuration prevents any reaction to protocol dynamics, unless this configuration is carefully planned by having a less preferred static route in combination with dynamic routes. A forwarding loop may occur if adjacent routers have conflicting routing information.

- No redundancy is available for failure conditions, introducing the possibility of traffic blackholes as routes jump to another VRF that has no good destination for the traffic.

1.4 Case Study

This section presents a migration case study that illustrates the different route-sharing and modification technologies outlined in this chapter. Figure 1.10 shows the IP network of a big enterprise that has grown over time connecting all routers in a full mesh. The data center provides access to critical applications, and the hub router connects the office headquarters.

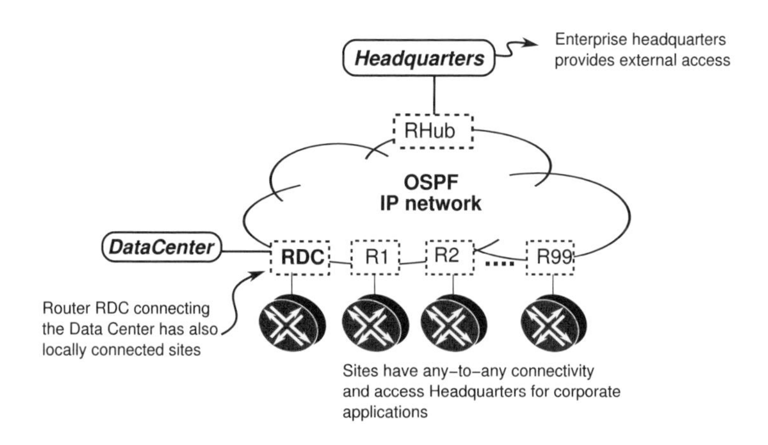

Figure 1.10: Overview of existing enterprise network.

The newly appointed network manager has a good understanding of the business and knows that the network infrastructure is critical. He is also very concerned about the current traffic distribution and wants to maintain control of interoffice user traffic from the headquarters. Therefore, he has set as main priority to rebuild the network to a headquarters-centric, compartmentalized collection of MPLS L3VPNs. The clearest advantage is the analysis of traffic flows and the ability to isolate errors by identifying the VPN reporting applications affected by issues. The migration project consists in moving traffic flows for services for corporate users off the global routing table to a hub-and-spoke VPN, as shown in Figure 1.11.

Throughout this case study only a single VPN, the *CORP* VPN, is analyzed. It is understood that this VPN represents one of the many VPNs of this enterprise.

1.4.1 Original network

Currently, all corporate users, regardless of their department, leverage any-to-any connectivity provided by a flat IP OSPF network. Users have access to headquarters via a default route, and they connect to data center servers located in a special hosting facility that includes backup services for critical applications. The router connecting the data center to the network has additional interconnections to regular users over other router interfaces.

Figure 1.12 illustrates the detailed topology of the network. The most relevant elements to consider in this migration are summarized below:

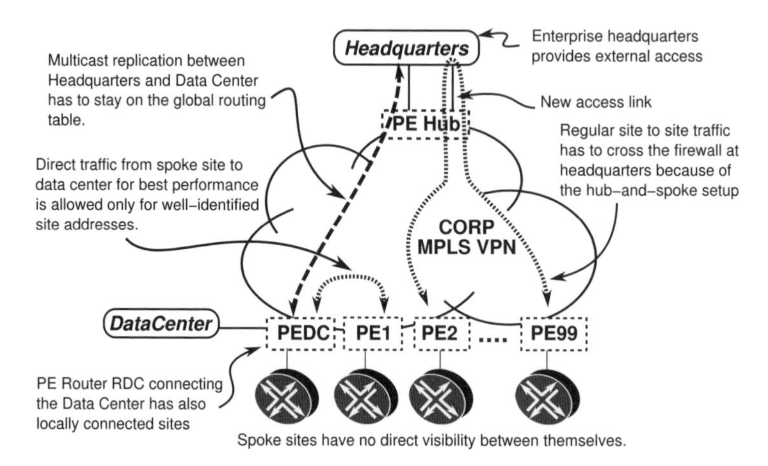

Figure 1.11: Target network migrated to MPLS/VPN with direct flows to data center.

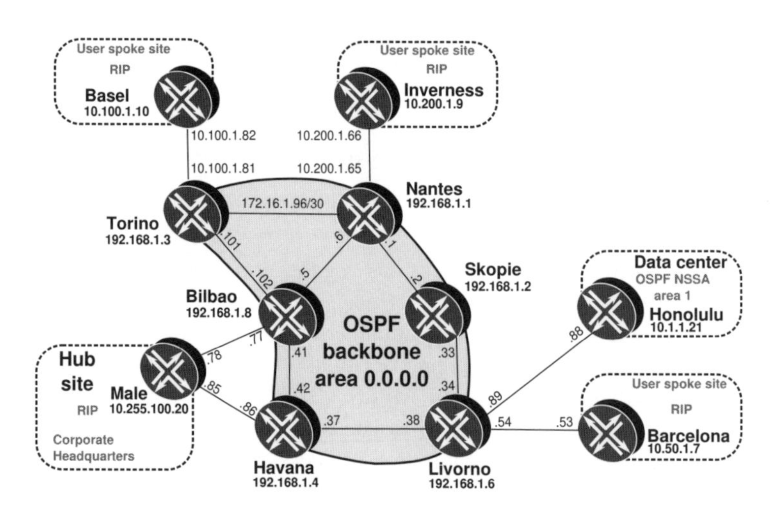

Figure 1.12: Network topology used through the case study.

- Backbone infrastructure addressing is contained within prefix 172.16.1.0/24. PE–CE addressing follows the addressing of the connected site. For instance, the data center WAN address is 10.1.1.88/30.

- User locations connect to the backbone routers using RIP as the routing protocol to advertise the location address space.

- At headquarters, represented by router *Male*, RIP is used to exchange a default route towards remote locations over the link between router *Male* and router *Bilbao* to attract external traffic. In the reverse direction, the headquarters site receives all location prefixes. A corporate firewall controls external access. Note that the link between router *Male* and router *Havana* is added as part of the migration and is originally not carrying any traffic.

- Because of latency constraints, selected corporate applications hosted on the data center (represented by router *Honolulu*) are accessed directly without intervention from headquarters. This access is restricted to a few critical clients whose site-addressing spaces are known. Keeping critical traffic flows to the data center with minimum latency is one of the key requirements for a successful migration.

- As part of the global OSPF domain, the data center is a member of OSPF NSSA Area 1. The data center address ranges are advertised as Type 7 LSAs, which are automatically translated into Type 5 LSAs in the core by the Area Border Router (ABR) *Livorno*. From the core, routers in OSPF Area 1 receive a default route, which suffices for traffic to reach the data center core border router.

- The border router *Havana* towards Headquarters is configured with a set of filters that count traffic going to and from the Data Center for statistics purposes. This counter is currently on the global table and is implemented as an inbound and outbound firewall filter.

- A multicast replication service between headquarters and the data center is independent of customer access. Headquarters uses PIM Sparse Mode dynamically to signal multicast group availability. A PIM rendezvous point is sited at router *Bilbao*. No plans exist to migrate this service to an MPLS L3VPN at this stage. Any migration activities should ensure that this application is not affected.

1.4.2 Target network

The OSPF backbone is extended with LDP as MPLS label distribution protocol, and border routers are promoted to PE devices. The final design calls for a hub-and-spoke VPN configuration, in which corporate users interconnect using the hub located at headquarters. The need for low-latency access to the applications hosted on the data center prompts the use of a fully meshed topology for data center access. The existing any-to-any connectivity is maintained only for data center access; remaining traffic is to reach the hub for accountability and security control. An increased latency in user-to-user traffic is deemed to be acceptable.

As part of the migration, services homed to the VPN at the hub location connect over a new access link, providing separation at the hub for traffic that has already been migrated.

The data center connectivity between router *Livorno* and router *Honolulu* involves a complex Layer 2 infrastructure (not shown in Figure 1.12) that cannot support a new access port at the border router *Livorno*. This restriction poses a challenge supporting a transient state in the migration, in which some sites are connected to the VPN while others are still connected to the global routing table. Access to the data center from the global routing table and to the VPN has to be provided over a single interface.

1.4.3 Migration strategy

Two important requirements for the migration are that it be smooth and that it have only local impact. During the migration, sites can be affected one at a time offering a way to back out of changes if any problems occur. The desire is to avoid changing many sites and components at the same time, and especially to avoid an *all or nothing* approach in which the entire network changes in a single go. Some of the migration elements can be treated separately, while others require a careful sequence of actions to minimize the impact.

During preparatory stages, proper migration planning is envisaged to maintain existing traffic flows.

Maintain multicast on global routing table

Before it receives any routing information from the Hub PE (router *Havana*), router *Male* has to ensure that multicast traffic uses only the existing access on the global routing table for the RPF check using the separation of unicast and multicast topology concept described in Section 1.1.3.

Data center transition challenge

Because of the nature of the networking protocols being used on the end-user hosts, any additional latency in accessing the data center, even a small one, imposes huge penalties in service responsiveness. Relaying critical traffic over headquarters does introduce this unwanted extra latency, which should be avoided in both directions of traffic flow during the migration, both for users moved to the *CORP* VPN and for users still attached to the global routing table. A means to keep direct data center access is needed to maintain service levels during the transition.

Note Some networking protocols defined for corporate environments correctly assume that latency is a negligible consideration for enterprise LANs and implement a retransmission window of one. For every message sent, a confirmation message is expected to acknowledge correct reception. A complete traffic round trip is required for every block of data that needs confirmation. When using the same protocols in the wide area, the user latency increases proportionally to the additional latency experienced. Larger files imply longer latency, because the number of data blocks that need to be exchanged increases, thus introducing a negative impact in end-user experience.

Thus, the desire is to offer data center access both to the global table and to the *CORP* VPN directly, without visiting the hub site for critical applications. One possible approach is to use an additional data center interface during the migration, in the same way as is performed at the hub. The result of this approach is that the data center devices continue to decide which traffic is sent over which interface.

In keeping with the spirit of this case study and showing route-leaking constructs, it is decided for administrative reasons to leave the data center site untouched rather than to add another interface and instead, to explore the possibility of solving the issue on the data center PE router *Livorno*. Looking at the concepts discussed in this chapter, it is apparent that a set of tools can be leveraged to control the migration process from the PE instead of having to configure an additional port.

Using this second approach, the data center site CE is unaware that the site to which it belongs is becoming a member of a VPN. Decisions are made by router *Livorno*, the data center PE. All traffic uses the global routing table to reach critical destinations for non-migrated sites and the *CORP* VPN for already migrated sites.

Migration planning

The proposed approach for the migration timeline is shown in Figure 1.13. Some of the stages require a strict sequencing, while others can be performed in parallel.

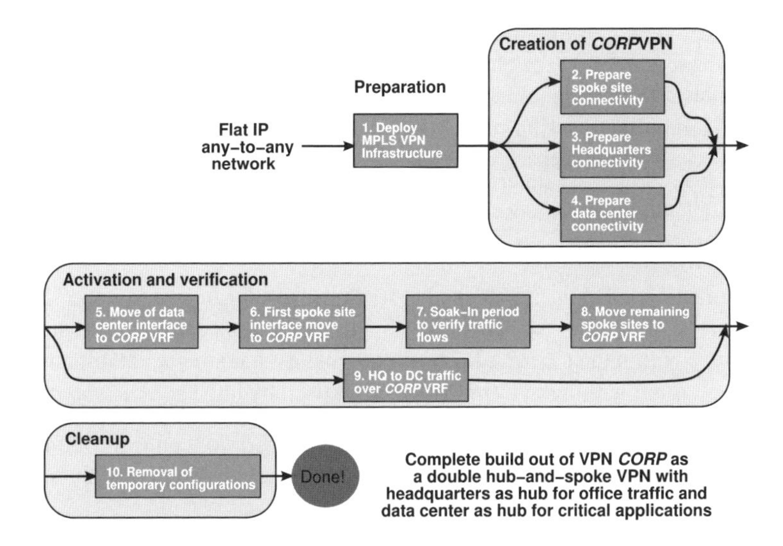

Figure 1.13: Approach to migrate users to *CORP* VPN.

1. As a first stage, the OSPF backbone has to be extended to support MPLS VPN services. Routers that connect spoke sites to the L3VPN are promoted to PE devices. Router *Nantes* is appointed as BGP Route Reflector for the L3VPN service.

2. In stage two, the *CORP* VRF is instantiated in all PE routers with connections to user sites. The PE–CE interface is *not* moved into this VRF because of a dependency with the data center reachability.

3. The third stage involves all preparatory work related to the headquarters location. A new link is provided between headquarters router *Male* and router *Havana*, which becomes headquarters' access to the Hub PE in the hub-and-spoke configuration. As part of headquarters, router *Male* enables routing to router *Havana* to advertise a default route inside the *CORP* VPN, thus readying connectivity to the hub for all members of the VPN. The configuration of the Hub VRF on router *Havana* is modified using the functionality described in Section 1.2.4, `no-vrf-advertise`, to support

the egress IP functionality that already exists at router *Bilbao* for the global routing table. Finally, preparatory work at headquarters finishes modifying the configuration on router *Male* to ensure that multicast traffic uses only the existing access on the global routing table for the RPF check using the incongruent unicast/multicast concepts described in Section 1.1.3.

4. The fourth stage covers preliminary work on the data center. Only one interface connects router *Livorno* and the data center represented by router *Honolulu*. Access over the single access at router *Livorno* presents a challenge in the transient state of the migration because traffic from both migrated and non-migrated sites must bypass the hub router *Havana*. An interim configuration construct using a helper VRF is presented to allow data center traffic to reach the intended destination either over *CORP* VPN if it is available, directly over global table in a meshed fashion, or using the hub as a last resort. To meet this requirement, leaking routes and diverting traffic to the helper VRF as described in Section 1.3 is required.

5. In stage five, the PE–CE interface at the data center PE router *Livorno* is moved to the *CORP* VRF. Preparatory work in previous stages ensures that this step does not change traffic flows. Traffic patterns still follow the original route over the global routing table.

6. A fundamental milestone is achieved in stage six. Router *Inverness* is chosen as the first spoke site to migrate to the *CORP* VPN. Notice that this starts the transition stage, in which sites within the *CORP* VRF reach other sites that are not yet migrated following the default route within the *CORP* VRF advertised by hub router *Havana*.

7. In stage seven, a monitoring period is established to confirm traffic flows and applications behave as expected. Two anomalies are detected and addressed.

8. Stage eight streamlines the migration of spoke sites to the *CORP* VRF. With both the headquarters and the data center locations simultaneously present in both the global table and the *CORP* VPN, user sites can join the VPN one at a time by moving the spoke site access interface from the global routing table to the *CORP* VRF while maintaining the meshed connectivity requirement for critical services. Traffic impact is restricted to the site under migration, which allows for an incremental rollout.

9. Stage nine can take place anywhere after the preparatory stages up to stage four. As soon as the new link to the *CORP* VRF is active at the headquarters and the data center prefix is received over the VPN, router *Male* can choose both the global routing table and the *CORP* VRF for traffic to the data center. This stage moves traffic from headquarters to the data center over the VPN.

10. Stage ten involves cleanup of the old configuration and helper constructs. With the approach presented in this case study, all configuration changes performed in the preparatory stages become final as specified in the target design, except for the helper construct at the data center from stage four and the core protocol configuration that can be removed. Thus, the only cleanup required at the data center PE is to remove the *helper* VRF configuration, which, because it does not impact any service, can be performed once all user sites have been migrated successfully and a prudential monitoring period has elapsed in which post-migration issues have been sorted out.

Notice that removing the helper VRF forces all data center traffic destined to the hub to flow over the *CORP* VRF.

Table 1.3 describes the connectivity matrix between the different locations in the current network. Notice that all traffic is traversing the global routing table.

Table 1.3 Traffic flow between site pairs in original network

	HQ	DC	NMS
Headquarters (HQ)		Global	Global
Data Center (DC)	Global		Global
Non-Migrated Spoke (NMS)	Global	Global	Global

Rows indicate traffic sources. Columns indicate traffic destinations.

1.4.4 Stage one: Building an MPLS L3VPN

The original backbone protocol is OSPFv2, as defined in [RFC2328]. This IGP is reused in the new MPLS VPN design. LDP, as per [RFC5036], is used for MPLS transport label binding between PE devices. Multi-Protocol BGP provides L3VPN prefix connectivity as per [RFC4364]. Router *Skopie* is the BGP Route Reflector and all PE routers establish a MP-BGP session with it including the inet-vpn address family.

For simplicity, connectivity between PE and CE routers leverages RIP version 2, as defined in [RFC2453], except for the data center, which is in OSPF NSSA area 1, and the Hub location, which uses EBGP.

MPLS L3VPN infrastructure

The *CORP* VPN is distributed across all access routers to which a site is connected. The hub site router *Male* connects to the *CORP* VPN in PE router *Havana* using a newly created connection. The old connection remains only to simulate existing services that are already using the global table and are not to be migrated, such as multicast replication between headquarters and the data center.

The target network scenario requires the creation of the *CORP* VRF at the PEs and moving the PE–CE interface of spoke sites that is on the routing table to the *CORP* VRF. This move has to be deferred to stage six, until both the headquarters and the data center have passed a set of preparatory steps.

A sample configuration for a PE router is shown in Listing 1.9. The MPLS configuration defines the interfaces and enables special processing for ICMP messages coming in for VPNs through the `icmp-tunneling` configuration statement shown on Line 3 as per [RFC3032] Section 2.3.2. The BGP `cluster` configuration option shown on Line 18 enables router *Nantes* as a route reflector for the VPNs. The loopback range defined with the BGP `allow` knob shown on Line 19 facilitates the configuration. As a best practice, tracking of IGP metrics for LDP is enabled on Line 36. RIP is used to exchange information with router

Inverness by advertising a default route as shown on Line 46 and prefixes from Inverness are redistributed into the OSPF core as shown on Line 23.

Listing 1.9: Protocols configuration for router *Nantes*

```
1  user@Nantes> show configuration protocols
2  mpls {
3      icmp-tunneling; # Handle ICMP expiration inside VPNs
4      interface so-0/1/0.0;
5      interface so-0/2/0.0;
6      interface ge-1/3/0.0;
7      interface fxp0.0 {
8          disable;
9      }
10 }
11 bgp {
12     group IBGP-RR { # Route reflection cluster
13         type internal;
14         local-address 192.168.1.1;
15         family inet-vpn { # L3VPN application
16             unicast;
17         }
18         cluster 192.168.1.1;
19         allow 192.168.1.0/24;
20     }
21 }
22 ospf {
23     export site-pool;
24     area 0.0.0.0 {
25         interface fxp0.0 {
26             disable;
27         }
28         interface so-0/1/0.0;
29         interface so-0/2/0.0;
30         interface ge-1/3/0.0;
31         interface lo0.0;
32
33     }
34 }
35 ldp {
36     track-igp-metric;
37     interface so-0/1/0.0;
38     interface so-0/2/0.0;
39     interface ge-1/3/0.0;
40     interface fxp0.0 {
41         disable;
42     }
43 }
44 rip {
45     group Inverness {
46         export default;
47         neighbor so-1/0/1.0;
48     }
49 }
```

Instantiation of *CORP* hub-and-spoke VPN

By leveraging the RT community for policy indication, the same VPN can combine multiple virtual topologies. The *CORP* VPN has two overlapping *hub-and-spoke* topologies: one centered at the headquarters and one centered at the data center. Regular user sites are VPN members of the headquarters hub-and-spoke VPN. Selected prefixes within those regular sites carry critical applications and are also members of the data center hub-and-spoke topology.

For regular office work, the *CORP* VPN provides a default route from Headquarters through the hub site PE router *Havana*. Headquarters router *Male* receives all site prefixes to allow connectivity to the spoke sites.

For data center access, the data center router *Honolulu* provides the data center prefix, and it imports selected site prefixes.

Regular interoffice traffic at remote sites traverses the hub to reach headquarters. Critical prefixes at the remote sites have direct access to the data center.

All VRFs in the *CORP* VPN share the same RD (64500:1). An exception is made for the second VRF on the Hub PE on router *Havana*, which is discussed later. To construct the overlapping policy, four RT extended communities are reserved, as illustrated in Table 1.4: a pair to construct a hub-and-spoke centered at router *Havana*, with an RT to advertise prefixes from the headquarters hub site (RT-hub) and one to indicate prefixes from the spoke sites (RT-spoke); and another pair (RT-datacenter, RT-critical) to build a hub-and-spoke topology centered at router *Livorno* connecting the data center site to comply with the meshing requirement between the data center and critical services, represented in the lab setup by the site loopback addresses (/32 prefixes).

Table 1.4 *CORP* VPN settings

Name	Value	Exported by	Imported by
RT-hub	target:64500:1	hub router *Havana*	all
RT-spoke	target:64500:2	all	hub router *Havana*
RT-datacenter	target:64500:10	data center router *Livorno*	all
RT-critical	target:64500:11	spokes critical prefixes	data center router *Livorno*

To allow headquarters connectivity to the data center, the data center PE, router *Livorno*, tags its site prefixes with RT-datacenter when advertising them, and prefixes with this RT are imported by the hub PE. To allow for bidirectional connectivity, the data center PE imports all VPN prefixes tagged with RT-hub. Notice that in this setup the hub advertises only a default route. Router *Livorno*, the data center PE, must take care to apply the tag prefix RT-datacenter only to routes from router *Honolulu*. The router *Barcelona*, which is also attached to router *Livorno*, is not part of the data center, but rather, simulates a spoke customer site and should behave like any regular site.

Data center access from additional VPNs through headquarters

Besides the *CORP* VPN, additional user traffic on the global routing table has to access the data center. In the target network, this traffic is confined to other VPNs (not shown in this case study), and uses the connectivity to headquarters to reach the data center. Users of those VPNs send their traffic to headquarters to get to the applications, and the traffic is relayed over the *CORP* VPN to the data center containing the hosted applications.

1.4.5 Stage two: Preparatory work at spoke sites

Spoke sites such as router *Nantes* and router *Torino* have a standard hub-and-spoke configuration with the particularity that two differentiated hub sites have to be considered: router *Havana* as the hub of headquarters and router *Livorno* as the hub of the data center. In addition, spoke site router *Barcelona* shares PE router *Livorno* and its *CORP* VRF with the data center.

Listing 1.10 shows the global routing configuration. Router *Torino* advertises a default route (Line 29) while routes from router *Basel* learned via RIP are injected in the OSPF backbone in configuration Line 14.

Listing 1.10: Router *Torino* initial protocol configuration

```
1  user@Torino> show configuration protocols
2  mpls { <...> }
3  bgp {
4      group IBGP-VPN {
5          type internal;
6          local-address 192.168.1.3;
7          family inet-vpn {
8              unicast;
9          }
10         neighbor 192.168.1.1;
11     }
12 }
13 ospf {
14     export site-pool;
15     area 0.0.0.0 {
16         interface fxp0.0 {
17             disable;
18         }
19         interface so-0/1/0.0;
20         interface at-0/0/1.0 {
21             metric 2;
22         }
23         interface lo0.0;
24     }
25 }
26 ldp { <...> }
27 rip {
28     group Basel {
29         export default;
30         neighbor at-1/2/0.1001;
31     }
32 }
33 user@Torino> show configuration routing-instances
34 inactive: CORP {
35     instance-type vrf;
36     interface at-1/2/0.1001;
37     route-distinguisher 64500:1;
38     vrf-import [ from-datacenter from-hub ];
39     vrf-export [ tag-critical tag-spoke ];
40     protocols {
41         rip {
42             group Basel {
43                 export default;
44                 neighbor at-1/2/0.1001;
45             }
46         }
47     }
48 }
49
50 user@Torino> show configuration policy-options policy-statement tag-critical
```

```
51 from {
52     protocol rip;
53     route-filter 0.0.0.0/0 prefix-length-range /32-/32;
54 }
55 then {
56     community add RT-critical;
57 }
58
59 user@Torino> show configuration policy-options policy-statement tag-spoke
60 from protocol rip;
61 then {
62     community add RT-spoke;
63     accept;
64 }
```

The *CORP* VRF configuration starting in Line 34 advertises a default route to router *Basel*. VRF policies tag the loopback address of router *Basel* as critical (Line 50) for the data center hub-and-spoke VPN topology and everything (Line 59) as a spoke prefix for the headquarters hub-and-spoke VPN topology.

Notice that the *CORP* VRF is prepared but left inactive at this stage, awaiting its turn for proper activation sequence.

1.4.6 Stage three: Preparatory work at headquarters

The headquarters use a new connection to reach the *CORP* VPN, which allows for work to be performed incrementally. Care has to be taken by router *Male* not to prefer any routing information received from the *CORP* VPN at this preparatory stage. The configuration elements for this work involve mainly the hub PE router *Havana* and the headquarters CE router *Male* as follows:

- Existing accounting functions performed at router *Bilbao* need to be mirrored to the PE router *Havana*. Accounting happens in both directions of traffic. This requirement forces a double VRF configuration to be created at PE router *Havana* to enable VPN egress IP functionality.

- A default route is advertised by router *Male* into the *CORP* VPN. This route attracts traffic to headquarters unless a more specific route exists in the VPN.

- Multicast traffic from router *Male* must be kept on the global table. The migration should maintain multicast on its existing connection to router *Bilbao*.

Details for each of the required configurations follow.

IP accounting on *Havana*

One requirement for the network migration is to retain the ability to count traffic that is exchanged between the data center and the hub site. In the original network, two counters applied to the global routing table on router *Bilbao* collect packet and byte counts of traffic headed towards and from the data center (see Listing 1.11). The filter shown in this listing is applied both inbound and outbound on the interface that faces headquarters on router *Bilbao* (so-1/3/1.0).

Listing 1.11: Data center traffic accounting at hub border

```
user@Bilbao-re0> show configuration firewall family inet
filter count-DC-traffic {
    interface-specific;
    term from-DC {
        from {
            source-address {
                10.1.1.0/24;
            }
        }
        then count from-DC;
    }
    term to-DC {
        from {
            destination-address {
                10.1.1.0/24;
            }
        }
        then count to-DC;
    }
    term default {
        then accept;
    }
}
```

Junos Tip: Sharing IP filters across interfaces

In Junos OS, a filter applied to different units on the same physical interface is shared, with one instance of each of the memory elements (counter, policer). This feature is leveraged in this case study to apply the same filter to two directions of traffic, for each direction incrementing the counter in the respective term. By enabling the `interface-specific` configuration statement, the filter is automatically split and the interface name and traffic direction (i for input and o for output) are appended to the filter name in operational command output, as shown in Listing 1.12.

Listing 1.12: Sharing firewall filter across interfaces

```
user@Bilbao-re0> show firewall

Filter: count-DC-traffic-so-1/3/1.0-i  # Input direction
Counters:
Name                                    Bytes              Packets
from-DC-so-1/3/1.0-i                        0                    0
to-DC-so-1/3/1.0-i                          0                    0
Filter: count-DC-traffic-so-1/3/1.0-o  # Output direction
Counters:
Name                                    Bytes              Packets
from-DC-so-1/3/1.0-o                        0                    0
to-DC-so-1/3/1.0-o                          0                    0
```

Applying the filter on the hub PE works only in the inbound direction by default; traffic in the reverse direction, from the core to the access, is switched by MPLS at the PE and is not processed by IP filters.

Enabling egress IP functionality through the `vrf-table-label` knob, associates a single MPLS *table* label with the whole VRF and prepares the hardware to perform an IP lookup on MPLS-tagged traffic arriving with that label. This topic is discussed at length in Section 5.4.1.

A regular hub-and-spoke PE configuration in which the hub site advertises a default route, as in this case, can generally be implemented with a single VRF and proper tagging of communities, such as the setup carried out in the data center PE router *Livorno*. However, if IP features are enabled at the hub, an IP lookup inside the VRF finds the spoke destinations and spoke-to-spoke traffic bypasses the hub site completely.

To avoid a traffic loop at the hub PE router *Havana*, a second VRF is required to split the two directions of traffic flow. Effectively, each VRF is used in one direction of traffic, although both VRFs share a single PE–CE interface.

Listing 1.13 shows the required configuration. Notice in Line 25 that an additional RD is allocated, because having the same RD on two VRFs of the same PE is not allowed in Junos OS. The hub-and-spoke configuration is split between two VRFs, *CORP-Hub* and *CORP-Hub-Downstream* routing instances, and IP features are enabled at Line 28. Notice that the interface is present only in *CORP-Hub*. The second VRF receives all routes from the first through the *auto-export* mechanism. Line 10 leaks the hub routes from CORP-Hub into CORP-Hub-Downstream, which accepts them on Line 30. The `no-vrf-advertise` configuration on Line 8 blocks advertisement of routes learned from the hub site and delegates this task to *CORP-Hub-Downstream*, which attracts traffic and perform IP functions because of the `vrf-table-label` knob shown on Line 28.

Listing 1.13: IP features at the hub PE router *Havana* requires two VRFs

```
1  user@Havana> show configuration routing-instances
2  CORP-Hub {
3      instance-type vrf;
4      interface ge-1/1/0.0;
5      route-distinguisher 64500:1;
6      vrf-import [ from-spoke from-datacenter ];
7      vrf-export tag-hub;
8      no-vrf-advertise;  # Refrain from advertising local routes in BGP
9      routing-options {
10         auto-export;  # Export local routes to Hub-Downstream VRF
11     }
12     protocols {
13         bgp {
14             group hub {  # EBGP to Hub router Male
15                 type external;
16                 export all;
17                 peer-as 65000;
18                 neighbor 10.255.100.85;
19             }
20         }
21     }
22 }
23 CORP-Hub-Downstream {
24     instance-type vrf;
25     route-distinguisher 64500:2;  # Additional RD
26     vrf-import from-hub;
27     vrf-export tag-hub;
28     vrf-table-label;  # Enable table-label for IP processing
29     routing-options {
30         auto-export;  # Import local routes from Hub VRF
31     }
32 }
```

Establishing the default route on the hub PE *Havana*

Once the *CORP* VRF is configured in router *Havana*, activation of the VRF can take place by establishing a connection of headquarters to PE router *Havana*. The configuration of router

Male to support this new connection including the advertisement of a default route is shown in Listing 1.14. EBGP is used as the PE–CE protocol.

Listing 1.14: Male advertises a default to router *Havana*

```
1  user@male-re0> show configuration protocols bgp
2  group ebgp {
3      type external;
4      export default;
5      neighbor 10.255.100.86 {
6          peer-as 64500;
7      }
8  }
9
10 user@male-re0> show configuration policy-options policy-statement default
11 from {
12     route-filter 0.0.0.0/0 exact;
13 }
14 then accept;
```

Listing 1.15 shows the result of configuration changes in router *Male*, in which router *Havana* receives from the headquarters router *Male* only a default route, which is sent over EBGP. The other prefixes in the *CORP* VRF represent connections via local interfaces.

Listing 1.15: Hub PE *CORP* VRF including the downstream VRF with hub default

```
1  user@Havana> show route table CORP terse 0/0 exact
2
3  CORP-Hub.inet.0: 9 destinations, 9 routes (9 active, 0 holddown, 0 hidden)
4  + = Active Route, - = Last Active, * = Both
5
6  A Destination        P Prf   Metric 1   Metric 2  Next hop        AS path
7  * 0.0.0.0/0          B 170        100             >10.255.100.85  65000 I
8
9  CORP-Hub-Downstream.inet.0: 1 destinations, 1 routes (1 active, 0 holddown, 0 hidden)
10 + = Active Route, - = Last Active, * = Both
11
12 A Destination        P Prf   Metric 1   Metric 2  Next hop        AS path
13 * 0.0.0.0/0          B 170        100             >10.255.100.85  65000 I
```

The `no-vrf-advertise` statement in Line 8 of the CORP-hub configuration shown in Listing 1.13 suppresses advertisement of the primary route, instead advertising the hub tagged, by default, with the CORP-hub-downstream table label, as shown in Listing 1.16 Line 7. The RD in Line 6 also hints at the correct VRF. Note that the `no-vrf-advertise` knob does not prevent the auto-export module from redistributing the routes locally.

Listing 1.16: Advertising hub default with a table label

```
1  user@Havana> show route advertising-protocol bgp 192.168.1.1 detail
2
3  CORP-Hub-Downstream.inet.0: 1 destinations, 1 routes (1 active, 0 holddown, 0 hidden)
4  * 0.0.0.0/0 (1 entry, 1 announced)
5   BGP group IBGP-VPN type Internal
6       Route Distinguisher: 64500:2
7       VPN Label: 16
8       Nexthop: Self
9       Flags: Nexthop Change
10      Localpref: 100
11      AS path: [64500] 65000 I
12      Communities: target:64500:1
13
14 user@Havana> show route table mpls label 16
15
```

```
16 mpls.0: 11 destinations, 11 routes (11 active, 0 holddown, 0 hidden)
17 + = Active Route, - = Last Active, * = Both
18
19 16                 *[VPN/0] 1w0d 15:45:04
20                       to table CORP-Hub-Downstream.inet.0, Pop
```

This default route directs traffic from *CORP* VPN users, via the hub, to destinations in the global routing table or to destinations in other VPNs. Note that this default route is received by all members of the VPN, including the data center.

Maintain multicast away from *CORP* VPN

One of the migration challenges is to keep multicast traffic between the data center and the headquarters in the global routing table, while directing all unicast flows between the headquarters and the data center over the *CORP* VPN by default. Fortunately, PIM in Junos OS can be configured to use a restricted routing view that includes only certain routes. These routes can be leaked through the RIB group mechanism in a RIB that PIM consults for the multicast RPF check.

The configuration for PIM at the onset of the migration is shown in Listing 1.17. Dense groups are defined for the `auto-rp` functionality on Line 7, which allows to discover the *Rendezvous point* (RP) for the sparse mode multicast groups.

Listing 1.17: Original PIM configuration on router *Male*

```
1  pim {
2      dense-groups {
3          224.0.1.39/32;
4          224.0.1.40/32;
5      }
6      rp {
7          auto-rp discovery;
8      }
9      interface so-3/2/1.0 {
10         mode sparse-dense;
11     }
12 }
```

Using the RIB group construct, one additional table is built on router *Male* by redistributing only selected information across the two RIBs. There is no need to create additional instances or RIBs. Two of the existing default RIBs in the master instance (Table 1.2) are leveraged, namely the unicast (inet.0) and multicast (inet.2) RIBs.

The Junos OS configuration for this setup can be seen in Listing 1.18. Two RIB group definitions are required: *leak-dc-to-inet2* (Line 2) distributes routes that pass the policy `dc-n-rp` (Line 4); and *pim-uses-inet2* (Line 6) is used to constrain the RPF view for the PIM protocol, containing just one RIB to override the default lookup in inet.0.

Listing 1.18: RIB groups configuration to provide incongruent unicast/multicast topologies

```
1  user@male-re1> show configuration routing-options rib-groups
2  leak-dc-to-inet2 {
3      import-rib [ inet.0 inet.2 ];
4      import-policy dc-n-rp; # Data Center and Rendezvous Point prefixes
5  }
6  pim-uses-inet2 {
7      import-rib inet.2;
8  }
```

The policy implemented in Listing 1.19 limits population of the inet.2 table to those relevant data center prefixes arriving from the global routing table. It is worth noting that the specific term filter-inet2 (Line 18) blocks additional routes from being leaked to inet.2. The default action in a rib-group policy is to accept; hence, there is no need for an additional term with an *accept* action for routes that have to populate inet.0 (Line 22).

Listing 1.19: Restricting route leak into inet.2 using policy language

```
1  user@male-re1> show configuration policy-options policy-statement dc-n-rp
2  term DC-to-inet0 { # Accept datacenter prefixes
3      from {
4          interface so-3/2/1.0;
5          route-filter 10.1.0.0/16 orlonger;
6      }
7      to rib inet.2; # Into inet.2
8      then accept;
9  }
10 term RP {             # Accept the Rendezvous Point (RP) prefix
11     from {
12         interface so-3/2/1.0;
13         route-filter 192.168.1.8/32 exact;
14     }
15     to rib inet.2; # Into inet.2
16     then accept;
17 }
18 term filter-inet2 { # Block other to inet.2
19     to rib inet.2;
20     then reject;
21 }
22 inactive: term default { # Default policy
23     to rib inet.0;
24     then accept;
25 }
```

Additional configuration shown on Listing 1.20 binds the *leak-dc-to-inet2* RIB group to RIP to leak received prefixes into the inet.2 table on Line 2. In addition, the *pim-uses-inet2* RIB group is bound to PIM on Line 9, to choose only routes in inet.2 for RPF check.

Listing 1.20: Binding RIB groups to protocols

```
1 user@male-re0> show configuration protocols rip
2 rib-group leak-dc-to-inet2;
3 group hub {
4     export default;
5     neighbor so-3/2/1.0;
6 }
7
8 user@male-re1> show configuration protocols pim rib-groups
9 inet pim-uses-inet2;
```

The result of this configuration is shown in Listing 1.21. Notice that there is no difference in the routing information for both inet.0 and inet.2 tables because the data center prefix is on the global routing table and no information is received from *CORP* VPN. Line 38 confirms that PIM is using inet.2 for its RPF check calculation.

Listing 1.21: Leaked routes in inet.2

```
1 user@male-re0> show route 10.1.1/24
2
3 inet.0: 19 destinations, 20 routes (19 active, 0 holddown, 0 hidden)
4 + = Active Route, - = Last Active, * = Both
```

```
5
6   10.1.1.0/24        *[RIP/100] 01:05:32, metric 2, tag 1
7                       > to 10.255.100.78 via so-3/2/1.0
8   10.1.1.21/32       *[RIP/100] 01:05:32, metric 2, tag 0
9                       > to 10.255.100.78 via so-3/2/1.0
10  10.1.1.88/30       *[RIP/100] 01:05:32, metric 2, tag 0
11                      > to 10.255.100.78 via so-3/2/1.0
12
13  inet.2: 4 destinations, 4 routes (4 active, 0 holddown, 0 hidden)
14  + = Active Route, - = Last Active, * = Both
15
16  10.1.1.0/24        *[RIP/100] 00:00:14, metric 2, tag 1
17                      > to 10.255.100.78 via so-3/2/1.0
18  10.1.1.21/32       *[RIP/100] 00:00:14, metric 2, tag 0
19                      > to 10.255.100.78 via so-3/2/1.0
20  10.1.1.88/30       *[RIP/100] 00:00:14, metric 2, tag 0
21                      > to 10.255.100.78 via so-3/2/1.0
22
23  user@male-re0> show route 192.168.1.8/32
24
25  inet.0: 19 destinations, 20 routes (19 active, 0 holddown, 0 hidden)
26  + = Active Route, - = Last Active, * = Both
27
28  192.168.1.8/32     *[RIP/100] 02:44:25, metric 2, tag 0
29                      > to 10.255.100.78 via so-3/2/1.0
30
31  inet.2: 4 destinations, 4 routes (4 active, 0 holddown, 0 hidden)
32  + = Active Route, - = Last Active, * = Both
33
34  192.168.1.8/32     *[RIP/100] 00:00:22, metric 2, tag 0
35                      > to 10.255.100.78 via so-3/2/1.0
36
37  user@male-re0> show multicast rpf
38  Multicast RPF table: inet.2 , 4 entries
39
40  10.1.1.0/24
41      Protocol: RIP
42      Interface: so-3/2/1.0
43      Neighbor: 10.255.100.78
44
45  10.1.1.21/32
46      Protocol: RIP
47      Interface: so-3/2/1.0
48      Neighbor: 10.255.100.78
49
50  10.1.1.88/30
51      Protocol: RIP
52      Interface: so-3/2/1.0
53      Neighbor: 10.255.100.78
54
55  192.168.1.8/32
56      Protocol: RIP
57      Interface: so-3/2/1.0
58      Neighbor: 10.255.100.78
```

Verification of the implemented changes to separate multicast traffic from the *CORP* VRF is discussed in Section 1.4.8.

1.4.7 Stage four: Preparatory work at the data center

Lack of a separate interconnect from the data center to the *CORP* VPN introduces additional complexity because a single access interface shares traffic for both the global routing table and the VPN. During the migration process, the data center access interface has to be homed to both the global routing table and the *CORP* VRF to serve all spoke sites with no intervention from headquarters.

No changes are required in the configuration of the data denter router *Honolulu* because the data center is unaware of the changes to become part of a VPN. Preparatory steps at the data center PE router *Livorno* follow:

- configuration for router *Livorno* as data center PE with single attachment;
- access from all user sites to the data center;
- access from the data center to reach both migrated and non-migrated user sites.

Data center PE configuration

The initial configuration on the PE router *Livorno* related to the protocols on the global routing table is shown in Listing 1.22. Besides the backbone OSPF area (Line 15), router *Livorno* acts as ABR for the data center, sited in NSSA area 1 (Line 24). The access to router *Barcelona* using RIP routing is configured in Line 34.

Listing 1.22: Router *Livorno* initial protocol configuration

```
1  user@Livorno> show configuration protocols
2  mpls { # MPLS core interfaces
3  <...>
4  }
5  bgp { # L3VPN BGP session to the reflector Nantes
6      group IBGP-VPN {
7          type internal;
8          local-address 192.168.1.6;
9          family inet-vpn {
10             unicast;
11         }
12         neighbor 192.168.1.1;
13     }
14 }
15 ospf {
16     area 0.0.0.0 {                # Backbone OSPF
17         interface fxp0.0 {
18             disable;
19         }
20         interface so-0/0/0.0;
21         interface so-0/0/1.0;
22         interface lo0.0;
23     }
24     area 0.0.0.1 {                # Data Center NSSA Area 1
25         nssa {
26             default-lsa default-metric 1;
27             no-summaries;
28         }
29         interface ge-1/3/0.0;
30     }
31 }
32 ldp {                             # LDP for MPLS transport tunnels
33 <...>
34 rip {                             # RIP to site Barcelona
35     group BCN {
36         export default;
37         neighbor ge-0/2/0.0;
38     }
39 }
```

Listing 1.23 provides the preliminary configuration of router *Livorno* with the global routing table protocols and a newly instantiated *CORP* VRF. The data center PE router *Livorno* also homes regular spoke sites like router *Barcelona*, which is reflected in the configuration by having a mixed RT `vrf-import` (Line 9) and `vrf-export` (Line 10) policy and additional RIP protocol configuration for the spoke interface (Line 33).

Listing 1.23: Interim configuration for *CORP* VRF in data center PE router *Livorno*

```
1  user@Livorno# show routing-instances CORP
2   ##
3   ## inactive: routing-instances
4   ##
5  instance-type vrf;
6  interface ge-1/3/0.0;
7  interface ge-0/2/0.0;
8  route-distinguisher 64500:1;
9  vrf-import [ from-critical from-hub ];
10 vrf-export [ tag-datacenter tag-spoke ];
11 routing-options {
12     inactive: interface-routes {
13         rib-group inet leak-to-inet0;
14     }
15     inactive: auto-export {
16         family inet {
17             unicast {
18                 inactive: rib-group leak-to-inet0;
19             }
20         }
21     }
22 }
23 protocols {
24     inactive: ospf {
25         area 0.0.0.1 {
26             nssa {
27                 default-lsa default-metric 1;
28                 no-summaries;
29             }
30             interface ge-1/3/0.0;
31         }
32     }
33     inactive: rip {
34         group BCN {
35             export default;
36             neighbor ge-0/2/0.0;
37         }
38     }
39 }
```

Notice the *inactive:* statement for the VRF as a whole, as well as in Line 12, Line 15, and Line 18. These portions of the configuration will be activated in following stages to illustrate the impact in populating the routing table.

The RIB group reference in Line 18 of Listing 1.23 corresponds to the configuration in Listing 1.24 that includes a policy to leak only the data center interface, but not other interfaces belonging to spoke sites connected on the same PE. Note that the default action for the RIB group policy is to accept all routes. Hence, the *CORP* VRF is populated correctly without requiring a specific term in the policy that accepts routes in order to leak them into CORP.inet.0.

Listing 1.24: Interface leaking rib-group configuration

```
1  [edit]
2  user@Livorno# show routing-options rib-groups
```

```
 3 leak-to-inet0 {
 4     import-rib [ CORP.inet.0 inet.0 ];
 5     import-policy only-datacenter-interface;
 6 }
 7 [edit policy-options]
 8 user@Livorno# show policy-statement only-datacenter-interface
 9 term datacenter {
10     from interface ge-1/3/0.0;
11     to rib inet.0;
12     then accept;
13 }
14 term other {
15     to rib inet.0;
16     then reject;
17 }
```

Access from data center to reach both migrated and non-migrated user sites

Once the migration is finished, the data center router *Honolulu* is unaware of the migration that took place at the PE, except for the reduced network visibility of prefixes from the global routing table that have been replaced by a default route pointing towards the headquarters site.

To reach both migrated and non-migrated sites during the migration, router *Livorno* should first verify whether the destination of traffic coming from the data center is one of the already migrated sites. If this is not the case, router *Livorno* should jump to the global routing table instead of following the headquarter's default inside the *CORP* VRF.

Using a static default route on the *CORP* VRF that points to inet.0 would hide the existence of the hub default route learned from router *Havana* inside the *CORP* VRF. Traffic from the data center to non-migrated sites would effectively be directed always to the global routing table. Notice that ignoring the hub default route would also affect traffic coming from other already migrated spokes on this PE, such as router *Barcelona*.

Therefore, a helper VRF construct is created that is used only for traffic coming from the data center. Its main goal is to provide access to migrated sites and to switch to the global routing table for non-migrated sites. Listing 1.25 shows the required configuration. Figure 1.14 is a graphical view of the proposed solution. Effectively, the data center VRF is used for the reverse direction (traffic from the remote sites towards the data center), while the helper VRF has a constrained routing view that substitutes the headquarters default route for a local bypass. The helper VRF is used only for traffic from the data center to remote sites.

Listing 1.25: The helper VRF

```
 1 routing-instances helper {
 2     instance-type vrf;
 3     route-distinguisher 64500:11;
 4     vrf-import from-critical;
 5     vrf-export null;
 6     routing-options {
 7         static {
 8             route 0.0.0.0/0 next-table inet.0;
 9         }
10     }
11 }
```

Incoming traffic from the datacenter is redirected to use the helper VRF for the duration of the migration of the remote sites. The input firewall filter is attached to the data center

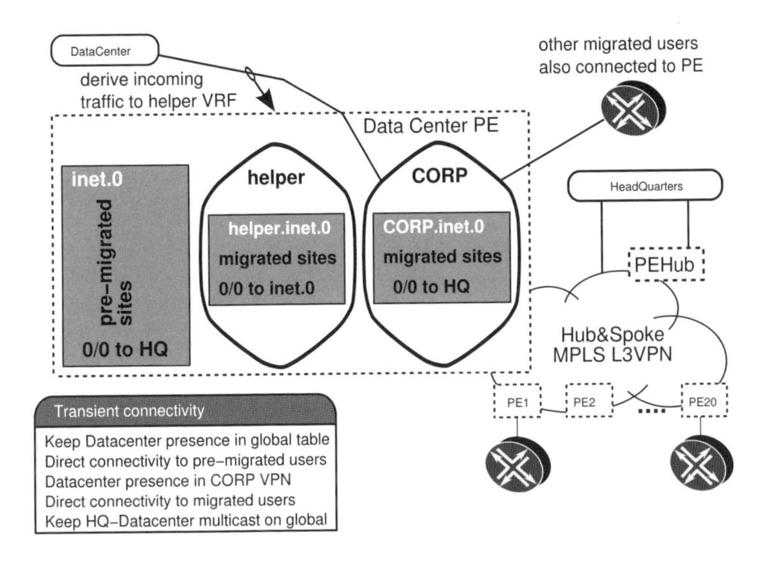

Figure 1.14: Data center PE transition configuration.

interface as shown in Listing 1.26. Care has to be taken to keep the routing protocol traffic homed to the *CORP* VRF, in which the OSPF task is processing updates.

Listing 1.26: Diverting datacenter traffic to the helper VRF

```
1  user@Livorno> show configuration firewall family inet filter redirect-to-helper
2  term ospf {     # Leave control traffic on CORP
3      from {
4          protocol ospf;
5      }
6      then accept;
7  }
8  term traffic { # Other traffic to jump over to helper
9      then {
10         routing-instance helper;
11     }
12 }
13
14 user@Livorno> show configuration interfaces ge-1/3/0
15 description To_Honolulu_ge-5/2/0;
16 unit 0 {
17     family inet {
18         inactive: filter {
19             input redirect-to-helper;
20         }
21         address 10.1.1.89/30;
22     }
23 }
```

Notice that by configuring the filter only on the PE–CE interface to router *Honolulu*, this helper VRF does not participate in traffic from other *CORP* VRF sites attached to router *Livorno* such as router *Barcelona*. Hence, traffic from spoke sites attached to the *CORP* VRF for other sites does not switch back to the global routing table and follows the default route in *CORP* VRF directed to headquarters.

Maintaining access from all user sites to the data center

After the data center interface moves, data center visibility must be maintained simultaneously on both the CORP VPN and the global table for the duration of the migration. Routing information that must be duplicated is both the local interface and the data center addressing learned from OSPF.

Using RIB groups on the PE–CE side as configured in Listing 1.23 on Page 39 offers a perfect solution for this need. As seen in Line 12, it is ensured that the interface is visible in both the *CORP* VRF and in the global routing table. The activation of the auto-export feature with rib-groups (Line 15 and Line 18) triggers leaking of the data center prefixes towards the main table inet.0. More details of the activation are discussed in Section 1.4.8.

Redistribute OSPF area 1 into OSPF backbone

Although the leaking configuration effectively installs the data center OSPF routes into the main table, the OSPF task running on the main instance considers these routes as not belonging to its own task and does not incorporate them into its link-state database by default; that is, these routes are of protocol OSPF, but router *Livorno* is not an ABR after the interface move. The behavior differs from that of the old network, in which the Area 1 NSSA Type 7 LSAs were automatically translated into Type 5 LSAs as part of the ABR function from [RFC2328].

Listing 1.27 shows the additional configuration required to export the data center routes present in the main instance so as to propagate them into the OSPF core.

Listing 1.27: Exporting data center OSPF routes to backbone OSPF

```
1  user@Livorno> show configuration protocols ospf | compare rollback 1
2  + export datacenter;  # Required to inject datacenter OSPF routes
3  user@Livorno> show configuration policy-options policy-statement datacenter
4  from {
5      inactive: protocol ospf;
6      interface ge-1/3/0.0;
7  }
8  then {
9      tag 1000;
10     accept;
11 }
```

1.4.8 Stage five: Move of data center site to *CORP* VRF

Moving the data center interface into the *CORP* VFR provides direct access from the data center to already migrated sites. The data center receives a default route from the hub, which is sufficient to reach the *CORP* VRF at router *Livorno*. The *CORP* VRF at router *Livorno* contains more specific information that is learned from other PEs advertising routing information that belong to critical prefixes at spoke sites.

Note Before performing this step, to avoid breaking the multicast replication flows between the hub and the data center, preparatory work at the hub CE router *Male* has separated the unicast and multicast RIBs, as discussed in Section 1.4.6.

The most critical stage in this migration is the move of the PE–CE interface of the data center. All destinations must keep connectivity to the data center in both directions (from and to the data center). In addition, for critical destinations this connectivity has to be on the shortest path. Heavy preparatory work in previous stages ensures that everything occurs smoothly.

Keeping traffic from the data center on global routing table

Despite the PE–CE interface change, traffic coming from the data center router *Honolulu* should stay on the global routing table at this stage. It is required to activate the firewall filter configured in Section 1.4.7 Listing 1.26 on Page 41 as shown in Listing 1.28 on Line 6 to ensure that traffic does not travel inside the *CORP* VPN for non-migrated sites.

Listing 1.28: Deriving incoming traffic from the data center to the helper VRF

```
1  [edit]
2  user@Livorno# show interfaces ge-1/3/0
3  description To_Honolulu_ge-5/2/0;
4  unit 0 {
5      family inet {
6          inactive: filter { # Derive incoming traffic to helper VRF
7              input redirect-to-helper;
8          }
9          address 10.1.1.89/30;
10     }
11 }
12 [edit]
13 user@Livorno# activate interfaces ge-1/3/0 unit 0 family inet filter
```

Moving the interface to *CORP* VRF

Junos OS protocols perform validation check for interfaces assigned to them. The interface has to be present in the instance where the protocol is running. Because an interface cannot be included in more than one instance, the PE–CE interface for the data center router *Honolulu* must be deactivated in the main instance's protocols, as shown in Listing 1.29, before enabling the *CORP* instance.

Listing 1.29: Moving PE–CE interfaces on router *Livorno*

```
1 user@Livorno# deactivate protocols ospf area 1    # datacenter OSPF
2 user@Livorno# activate routing-instances
3 user@Livorno# activate routing-instances CORP protocols ospf
4 user@Livorno# commit
5 commit complete
```

The ability of Junos OS to group changes into a single *commit* operation optimizes downtime during change scenarios. Not all required configuration has been activated at this stage. The configuration *commit* action shown on Line 4 can be deferred until other inactive groups from the configuration are activated, as discussed henceforth. The intermediate configuration change has been activated for illustration purposes.

Verification of traffic from the data center

A trace route from data center router *Honolulu* to router *Basel* is shown in Listing 1.30. As expected, traffic follows the global routing table bypassing the hub site.

Listing 1.30: Traffic from router *Honolulu* to router *Basel*

```
1 user@honolulu-re0> traceroute 10.100.1.10 source 10.1.1.21
2 traceroute to 10.100.1.10 (10.100.1.10) from 10.1.1.21, 30 hops max, 40 byte packets
3  1 livorno-ge1300 (10.1.1.89)  0.499 ms  0.417 ms  0.402 ms
4  2 skopie-so1000 (172.16.1.33)  0.511 ms  0.462 ms  0.472 ms
5  3 nantes-so0100 (172.16.1.1)  0.521 ms  0.461 ms  0.469 ms
6  4 torino-so0100 (172.16.1.97)  0.571 ms  0.488 ms  0.491 ms
7  5 basel (10.100.1.10)  1.500 ms  1.383 ms  1.579 ms
```

What If... Data center traffic bypassing the helper table

Listing 1.31 illustrates how traffic entering the PE–CE interface from the data center that is *not* derived to the *helper* VRF with a firewall filter (Line 1) follows the default route inside the *CORP* VRF (Line 8). The trace route on Line 12 confirms that traffic is taking a detour over router *Male*, which is undesired.

Listing 1.31: Traffic from router *Honolulu* to router *Basel*

```
1  user@Livorno# deactivate interfaces ge-1/3/0 unit 0 family inet filter input
2
3  user@Livorno> show route table CORP 10.100.1.10
4
5  CORP.inet.0: 11 destinations, 11 routes (11 active, 0 holddown, 0 hidden)
6  + = Active Route, - = Last Active, * = Both
7
8  0.0.0.0/0          *[BGP/170] 00:00:37, localpref 100, from 192.168.1.1
9                        AS path: 65000 I
10                     > via so-0/0/1.0, Push 301504
11
12 user@honolulu-re0> traceroute 10.100.1.10 source 10.1.1.21
13 traceroute to 10.100.1.10 (10.100.1.10) from 10.1.1.21, 30 hops max, 40 byte packets
14  1 livorno-ge1300 (10.1.1.89)  0.496 ms  0.419 ms  0.399 ms
15  2 havana-so1000 (172.16.1.37)  0.721 ms  0.594 ms  0.589 ms
16      MPLS Label=301504 CoS=0 TTL=1 S=1
17  3 male-ge4240 (10.255.100.85)  0.369 ms  0.311 ms  0.311 ms
18  4 bilbao-so1310 (10.255.100.78)  0.503 ms  0.519 ms  0.483 ms
19  5 torino-at0010 (172.16.1.101)  0.649 ms  0.881 ms  1.004 ms
20  6 basel (10.100.1.10)  1.770 ms  1.467 ms  1.493 ms
```

The current CORP VRF configuration (still containing inactive statements for the leaking configuration) yields a routing table with data center prefixes as shown in Listing 1.32. Notice that by moving the PE–CE interface, the data center prefixes are not available on the global table anymore.

Listing 1.32: Data center address space after moving the PE–CE interface

```
1  user@Livorno> show route 10.1/16 terse
2
3  CORP.inet.0: 12 destinations, 12 routes (12 active, 0 holddown, 0 hidden)
4  + = Active Route, - = Last Active, * = Both
5
6  A Destination        P Prf   Metric 1   Metric 2  Next hop        AS path
7  * 10.1.1.0/24        O 150          0            >10.1.1.90
8  * 10.1.1.21/32       O 150          0            >10.1.1.90
9  * 10.1.1.88/30       D   0                       >ge-1/3/0.0
10 * 10.1.1.89/32       L   0                        Local
```

Listing 1.33 shows the address space of the data center at the data center PE router *Livorno* after enabling the RIB group configuration for the data center interface.

Listing 1.33: Activating interface leaking

```
1  user@Livorno# activate routing-instances CORP routing-options interface-routes
2  [edit]
3  user@Livorno# commit and-quit
4  commit complete
5  Exiting configuration mode
6
7  user@Livorno> show route 10.1/16 terse
8
9  inet.0: 40 destinations, 42 routes (39 active, 0 holddown, 1 hidden)
10 + = Active Route, - = Last Active, * = Both
11
12 A Destination        P Prf   Metric 1   Metric 2  Next hop        AS path
13 * 10.1.1.88/30       D   0                        >ge-1/3/0.0
14
15 CORP.inet.0: 11 destinations, 11 routes (11 active, 0 holddown, 0 hidden)
16 + = Active Route, - = Last Active, * = Both
17
18 A Destination        P Prf   Metric 1   Metric 2  Next hop        AS path
19 * 10.1.1.0/24        O 150          0             >10.1.1.90
20 * 10.1.1.21/32       O 150          0             >10.1.1.90
21 * 10.1.1.88/30       D   0                        >ge-1/3/0.0
22 * 10.1.1.89/32       L   0                         Local
```

Activation of the auto-export feature with RIB groups (Listing 1.23, Line 15 and Line 18) triggers leaking of the data center prefixes towards the main table inet.0, as shown in Listing 1.34.

Listing 1.34: Leaking data-center OSPF routes to inet.0

```
1  [edit routing-instances CORP routing-options]
2  user@Livorno# activate auto-export
3  user@Livorno# activate auto-export family inet unicast rib-group
4  user@Livorno> show route 10.1/16 terse
5
6  inet.0: 43 destinations, 45 routes (42 active, 0 holddown, 1 hidden)
7  + = Active Route, - = Last Active, * = Both
8
9  A Destination        P Prf   Metric 1   Metric 2  Next hop        AS path
10 * 10.1.1.0/24        O 150          0             >10.1.1.90
11 * 10.1.1.21/32       O 150          0             >10.1.1.90
12 * 10.1.1.88/30       D   0                        >ge-1/3/0.0
13
14 CORP.inet.0: 12 destinations, 12 routes (12 active, 0 holddown, 0 hidden)
15 + = Active Route, - = Last Active, * = Both
16
17 A Destination        P Prf   Metric 1   Metric 2  Next hop        AS path
18 * 10.1.1.0/24        O 150          0             >10.1.1.90
19 * 10.1.1.21/32       O 150          0             >10.1.1.90
20 * 10.1.1.88/30       D   0                        >ge-1/3/0.0
21 * 10.1.1.89/32       L   0                         Local
```

Listing 1.35 shows the resulting data center OSPF route leaked to the inet.0 in the main instance after the previous configuration has been applied. The announcement bit on Line 12 indicates that the OSPF task in the main instance is interested in changes for this prefix. The OSPF database in the main instance shows the prefix as a Type 5 AS External LSA starting at Line 31.

Listing 1.35: Notification between OSPF tasks

```
1  user@Livorno> show route 10.1.1/24 exact detail table inet.0
2
3  inet.0: 43 destinations, 45 routes (42 active, 0 holddown, 1 hidden)
```

```
4  10.1.1.0/24 (1 entry, 1 announced)
5          *OSPF   Preference: 150
6                  Next hop type: Router, Next hop index: 631
7                  Next-hop reference count: 9
8                  Next hop: 10.1.1.90 via ge-1/3/0.0, selected
9                  State: <Secondary Active Int Ext>
10                 Age: 9:38       Metric: 0       Tag: 1
11                 Task: CORP-OSPF
12                 Announcement bits (3): 0-KRT 2-OSPF 4-LDP
13                 AS path: I
14                 Communities: target:64500:2 target:64500:10 rte-type:0.0.0.1:5:1
15                 Primary Routing Table CORP.inet.0
16
17 user@Livorno> show ospf database
18
19     OSPF database, Area 0.0.0.0
20  Type       ID              Adv Rtr           Seq         Age  Opt  Cksum  Len
21 Router   192.168.1.1      192.168.1.1      0x80000010  2392  0x22 0xb9c0   96
22 Router   192.168.1.2      192.168.1.2      0x80000205  1502  0x22 0x2345   96
23 Router   192.168.1.3      192.168.1.3      0x8000000b  2316  0x22 0x72cf   84
24 Router   192.168.1.4      192.168.1.4      0x8000000a   646  0x22 0x5bd2   84
25 Router  *192.168.1.6      192.168.1.6      0x80000283     4  0x22 0x586a   84
26 Router   192.168.1.8      192.168.1.8      0x8000000b   744  0x22 0xe629   96
27 Network  172.16.1.5       192.168.1.8      0x80000008  1537  0x22 0xa87a   32
28     OSPF AS SCOPE link state database
29  Type       ID              Adv Rtr           Seq         Age  Opt  Cksum  Len
30 Extern   0.0.0.0          192.168.1.8      0x80000007  2373  0x22 0xc572   36
31 Extern  *10.1.1.0         192.168.1.6      0x80000001     4  0x22 0xc386   36
32 Extern  *10.1.1.21        192.168.1.6      0x80000001     4  0x22 0xf044   36
33 Extern  *10.1.1.88        192.168.1.6      0x80000001     4  0x22 0x3eb6   36
34 Extern   10.100.1.0       192.168.1.3      0x80000009  1550  0x22 0x526f   36
35 Extern   10.100.1.10      192.168.1.3      0x80000008   712  0x22 0xefc8   36
36 Extern   10.200.1.0       192.168.1.1      0x80000005  1598  0x22 0xb1b1   36
37 Extern   10.200.1.9       192.168.1.1      0x80000005   824  0x22 0x5703   36
```

Connectivity verification towards the data center

Taking a not-yet migrated user site (router *Basel*), verification of direct communication to the data center is confirmed on Listing 1.36. Traffic bypasses the hub completely.

Listing 1.36: Basel correctly reaches data center in CORP VPN over global table

```
1 user@Basel> traceroute honolulu source basel
2 traceroute to honolulu (10.1.1.21) from basel, 30 hops max, 40 byte packets
3  1  torino-at1201001 (10.100.1.81)  1.424 ms  0.892 ms  0.941 ms
4  2  nantes-so0200 (172.16.1.98)  0.922 ms  1.143 ms  0.934 ms
5  3  skopie-so0100 (172.16.1.2)  0.953 ms  1.147 ms  0.943 ms
6  4  livorno-so0000 (172.16.1.34)  1.447 ms  1.708 ms  1.434 ms
7  5  honolulu (10.1.1.21)  1.469 ms  1.686 ms  1.428 ms
```

What If... Data center not visible in global routing table

If the OSPF redistribution were not taking place at router *Livorno*, router *Basel* would follow the default route coming from router *Male* over the global routing table, as shown in Listing 1.37. This additional latency is undesired for site-critical prefixes. Note that the trace route hop 4 on Line 13 is the hub CE router *Male*. The MPLS label on Line 16 indicates that the data center prefix is being advertised correctly to the hub, which is forwarding traffic over the *CORP* VRF.

Listing 1.37: With no OSPF redistribution, Basel follows the corporate default route

```
user@Livorno# deactivate protocols ospf export

[edit]
user@Livorno# commit and-quit
commit complete
Exiting configuration mode

user@Basel> traceroute honolulu source basel
traceroute to honolulu (10.1.1.21) from basel, 30 hops max, 40 byte packets
 1  torino-at1201001 (10.100.1.81)  1.305 ms  0.899 ms  0.926 ms
 2  nantes-so0200 (172.16.1.98)  0.933 ms  1.137 ms  0.929 ms
 3  bilbao-ge0100 (172.16.1.5)  1.439 ms  1.741 ms  1.436 ms
 4  male-so3210 (10.255.100.77)  0.944 ms  1.151 ms  0.936 ms
 5  havana-ge1100 (10.255.100.86)  1.449 ms  1.714 ms  1.439 ms
 6  livorno-so0010 (172.16.1.38)  1.288 ms  1.028 ms  1.428 ms
     MPLS Label=300640 CoS=0 TTL=1 S=1
 7  honolulu (10.1.1.21)  1.702 ms  1.709 ms  1.436 ms
```

Confirming multicast stays on global routing table

A key requirement of the migration is to maintain multicast on the global routing table. The data center prefix is propagated within the *CORP* VPN and eventually reaches the hub CE router *Male* in which the special preparatory work carried out for multicast in Section 1.4.5 prevents the use of the *CORP* VRF.

The data center prefix on router *Male* as seen in Listing 1.38 shows that the multicast RPF RIB contains only one RIP route coming from neighbor router *Bilbao*, while the unicast RIB inet.0 holds both the BGP route from router *Havana* (over *CORP* VPN) and the RIP route from router *Bilbao*. The RIP route has been configured with a worse preference (180) in order that unicast traffic flows over the *CORP* VPN, if the destination is available. Because the inet.2 table has no BGP route, the RIP route is used, regardless of its preference. This scheme allows migrated user sites to use the *CORP* VPN in both directions, while keeping multicast traffic on the global table.

Listing 1.38: Incongruent routing tables for unicast/multicast

```
user@male-re0> show route 10.1.1.0/24 terse

inet.0: 20 destinations, 23 routes (20 active, 0 holddown, 0 hidden)
+ = Active Route, - = Last Active, * = Both

A Destination        P Prf   Metric 1   Metric 2  Next hop        AS path
* 10.1.1.0/24        B 170        100             >10.255.100.86  64500 I # BGP route
                     R 180          2             >10.255.100.78          # RIP route
* 10.1.1.21/32       B 170        100             >10.255.100.86  64500 I # BGP route
                     R 180          2             >10.255.100.78          # RIP route

inet.2: 3 destinations, 3 routes (3 active, 0 holddown, 0 hidden)
+ = Active Route, - = Last Active, * = Both

A Destination        P Prf   Metric 1   Metric 2  Next hop        AS path
* 10.1.1.0/24        R 180          2             >10.255.100.78
* 10.1.1.21/32       R 180          2             >10.255.100.78

user@male-re0> show multicast rpf | match inet.2
Multicast RPF table: inet.2 , 4 entries
```

With all preparatory work correctly designed, move of the data center PE–CE interface keeps the traffic patterns on the global routing table as originally defined in Table 1.3 on Page 27. This traffic pattern changes as soon as the first user site is migrated.

1.4.9 Stage six: Moving first spoke site router *Inverness* to *CORP* VRF

Replicating the work done for the *CORP* VRF for the PE–CE interface to the data center router *Livorno*, Listing 1.39 shows the deactivation of the global instance RIP protocol and subsequent activation of the *CORP* routing instance for the PE router *Nantes* which connects to CE router *Inverness*. Line 7 shows the population of the *CORP* VRF with the default route to hub router *Havana*, the data center prefix range, and the local prefixes from *Inverness*.

Listing 1.39: Activation of router *Nantes* PE–CE to router *Inverness* into *CORP* VRF

```
1  user@Nantes> show configuration | compare rollback 1
2  [edit protocols]
3  !    inactive: rip { ... }
4  [edit routing-instances]
5  !    active: CORP { ... }
6
7  user@Nantes> show route table CORP
8
9  CORP.inet.0: 8 destinations, 8 routes (8 active, 0 holddown, 0 hidden)
10 + = Active Route, - = Last Active, * = Both
11
12 0.0.0.0/0          *[BGP/170] 01:20:09, localpref 100, from 192.168.1.4
13                      AS path: 65000 I
14                    > to 172.16.1.5 via ge-1/3/0.0, Push 301504, Push 303264(top)
15 10.1.1.0/24        *[BGP/170] 00:11:44, MED 0, localpref 100, from 192.168.1.6
16                      AS path: I
17                    > via so-0/1/0.0, Push 300736, Push 300400(top)
18 10.1.1.21/32       *[BGP/170] 00:11:44, MED 0, localpref 100, from 192.168.1.6
19                      AS path: I
20                    > via so-0/1/0.0, Push 300736, Push 300400(top)
21 10.200.1.0/24      *[RIP/100] 01:20:09, metric 2, tag 1
22                    > to 10.200.1.66 via so-1/0/1.0
23 10.200.1.9/32      *[RIP/100] 01:20:09, metric 2, tag 0
24                    > to 10.200.1.66 via so-1/0/1.0
25 10.200.1.64/30     *[Direct/0] 01:20:09
26                    > via so-1/0/1.0
27 10.200.1.65/32     *[Local/0] 01:20:09
28                      Local via so-1/0/1.0
29 224.0.0.9/32       *[RIP/100] 01:20:09, metric 1
30                      MultiRecv
```

1.4.10 Stage seven: Monitoring and anomaly detection

With the move of the PE–CE interface between router *Nantes* and router *Inverness*, the *CORP* VRF starts carrying traffic. The expected traffic matrix is summarized in Table 1.5. Router *Inverness* is a migrated spoke while all other spoke sites are not yet migrated, as shown in Figure 1.15. As a modification to the original scenario, traffic between spoke sites for non-critical traffic starts using headquarters.

To verify that everything is in place, a matrix of bidirectional traces are issued from every location to confirm that connectivity is as desired.

Table 1.5 Traffic flow matrix halfway through the migration

	HQ	DC	NMS	MS
Headquarters (HQ)		Global	Global	CORP
Data Center (DC)	Global		Global	CORP
Non-Migrated Spoke (NMS)	Global	Global	Global	Hub
Migrated Spoke (MS)	CORP	CORP	Hub	CORP/hub

Connectivity between migrated spokes uses the CORP VRF via the hub.

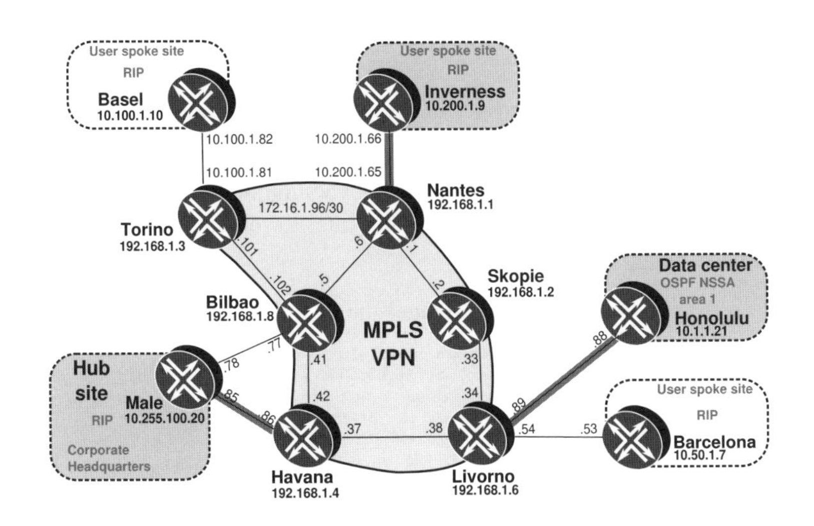

Figure 1.15: Network topology after moving router *Inverness*.

Verification of hub-and-spoke behavior among spoke sites

Correct traffic distributions are verified between router *Inverness* and sample spoke site router *Basel* in Listing 1.40. Because router *Basel* is not yet migrated, traffic from router *Basel* uses the global routing table to reach the hub site router *Male* as shown starting at Line 1. From router *Male*, traffic is forwarded over the *CORP* VPN to the migrated router *Inverness*. The reverse direction is shown starting at Line 14, in which traffic flows within the *CORP* VPN to reach the hub site router *Male* and continues its way over the global routing table to router *Basel*.

Listing 1.40: Trace between router *Inverness* and router *Basel* relays over headquarters

```
1 user@Inverness> traceroute basel source inverness
2 traceroute to basel (10.100.1.10) from inverness, 30 hops max, 40 byte packets
3  1  nantes-so1010 (10.200.1.65)  0.819 ms  0.658 ms  0.590 ms
4  2  bilbao-ge0100 (172.16.1.5)  1.077 ms  1.005 ms  0.961 ms
5      MPLS Label=303264 CoS=0 TTL=1 S=0
6      MPLS Label=301504 CoS=0 TTL=1 S=1
7  3  havana-at1200 (172.16.1.41)  1.062 ms  0.935 ms  0.877 ms
8      MPLS Label=301504 CoS=0 TTL=1 S=1
9  4  male-ge4240 (10.255.100.85)  0.614 ms  0.633 ms  0.590 ms
```

```
10   5  bilbao-so1310 (10.255.100.78)  0.761 ms  0.786 ms  0.743 ms
11   6  torino-at0010 (172.16.1.101)  1.173 ms  1.082 ms  0.935 ms
12   7  basel (10.100.1.10)  2.092 ms  2.256 ms  2.446 ms
13
14 user@Basel> traceroute inverness source basel
15 traceroute to inverness (10.200.1.9) from basel, 30 hops max, 40 byte packets
16   1  torino-at1201001 (10.100.1.81)  1.516 ms  1.204 ms  0.921 ms
17   2  nantes-so0200 (172.16.1.98)  0.931 ms  0.942 ms  11.456 ms
18   3  bilbao-ge0100 (172.16.1.5)  1.429 ms  1.715 ms  1.447 ms
19   4  male-so3210 (10.255.100.77)  1.433 ms  1.470 ms  1.442 ms
20   5  havana-ge1100 (10.255.100.86)  1.444 ms  1.674 ms  1.453 ms
21   6  bilbao-at1210 (172.16.1.42)  1.297 ms  1.573 ms  1.426 ms
22      MPLS Label=303312 CoS=0 TTL=1 S=0
23      MPLS Label=316224 CoS=0 TTL=1 S=1
24   7  nantes-ge1300 (172.16.1.6)  1.508 ms  1.302 ms  1.431 ms
25      MPLS Label=316224 CoS=0 TTL=1 S=1
26   8  inverness (10.200.1.9)  2.253 ms  2.115 ms  1.450 ms
```

Verification of data center connectivity to router *Basel*

Tracing traffic between data center and non-migrated router *Basel* as shown in Listing 1.41 yields the desired result: despite the fact that is not yet migrated, traffic is bypassing the hub site router *Male*. Likewise, traffic from the data center to router *Basel* follows the shortest path, as originally intended.

Listing 1.41: Non-migrated site on the shortest path

```
1  user@honolulu-re0> traceroute basel source honolulu
2  traceroute to basel (10.100.1.10) from honolulu, 30 hops max, 40 byte packets
3    1  livorno-ge1300 (10.1.1.89)  0.458 ms  0.418 ms  0.403 ms
4    2  havana-so1000 (172.16.1.37)  0.508 ms  0.438 ms  0.438 ms
5    3  bilbao-at1210 (172.16.1.42)  0.547 ms  0.506 ms  0.546 ms
6    4  torino-at0010 (172.16.1.101)  25.191 ms  1.001 ms  0.970 ms
7    5  basel (10.100.1.10)  2.026 ms  1.464 ms  1.498 ms
8
9  user@Basel> traceroute honolulu source basel
10 traceroute to honolulu (10.1.1.21) from basel, 30 hops max, 40 byte packets
11   1  torino-at1201001 (10.100.1.81)  1.563 ms  1.224 ms  1.423 ms
12   2  nantes-so0200 (172.16.1.98)  1.438 ms  1.441 ms  1.435 ms
13   3  skopie-so0100 (172.16.1.2)  1.240 ms  1.446 ms  1.427 ms
14   4  livorno-so0000 (172.16.1.34)  1.447 ms  1.116 ms  1.434 ms
15   5  honolulu (10.1.1.21)  1.457 ms  1.131 ms  2.847 ms
```

Verification of data center connectivity to hub

Notice that the traffic between data center and hub is still flowing over the global routing table. As Listing 1.42 shows, this is a direct consequence of the default static route present in the helper VRF pointing to inet.0 as shown on Line 7. This traffic shall move to the *CORP* VRF automatically as part of the cleanup stage.

Listing 1.42: Data center follows the default route to reach the hub

```
1 user@Livorno> show route table helper
2
3 helper.inet.0: 2 destinations, 2 routes (2 active, 0 holddown, 0 hidden)
4 + = Active Route, - = Last Active, * = Both
5
6 0.0.0.0/0          *[Static/5] 07:36:26
7                       to table inet.0
8 10.200.1.9/32      *[BGP/170] 00:26:35, MED 2, localpref 100, from 192.168.1.1
```

```
 9                      AS path: I
10                    > via so-0/0/0.0, Push 316224, Push 300432(top)
11
12
13 user@male-re0> show route 10.1.1.0/24 table inet.0 terse
14
15 inet.0: 21 destinations, 24 routes (21 active, 0 holddown, 0 hidden)
16 + = Active Route, - = Last Active, * = Both
17
18 A Destination        P Prf   Metric 1   Metric 2  Next hop        AS path
19 * 10.1.1.0/24        R 100          2             >10.255.100.78             # RIP to Global
20                      B 170        100             >10.255.100.86   64500 I  # BGP to CORP
21 * 10.1.1.21/32       R 100          2             >10.255.100.78
22                      B 170        100             >10.255.100.86   64500 I
23 * 10.1.1.88/30       R 100          2             >10.255.100.78
```

In the reverse direction, default protocol preferences at hub CE router *Male* select the RIP route, which follows the global routing table, over the BGP route towards the *CORP* VRF, as shown on Line 13.

Router *Male* can change preferences for the RIP route to switch connectivity to the data center over the *CORP* VRF, if desired.

1.4.11 Stage eight: Move of remaining spoke sites to *CORP* VRF

After a conservative monitoring period has elapsed, remaining sites can be incorporated to the *CORP* VPN progressively. A couple of anomalies are observed after integrating router *Barcelona* and are analyzed in the following subsections.

Anomaly with traffic from data center to migrated router *Barcelona*

Although the expectation is for the traffic between router *Honolulu* and the already migrated router *Barcelona* to be one directly connected, traffic is going through the hub, as shown in Listing 1.43.

Listing 1.43: Data center going to the hub to reach the local spoke site Barcelona

```
 1 user@honolulu-re0> traceroute barcelona source honolulu
 2 traceroute to barcelona (10.50.1.7) from honolulu, 30 hops max, 40 byte packets
 3  1  livorno-ge1300 (10.1.1.89)  0.518 ms  0.422 ms  0.409 ms
 4  2  havana-so1000 (172.16.1.37)  0.544 ms  0.474 ms  0.437 ms
 5  3  bilbao-at1210 (172.16.1.42)  0.545 ms  0.521 ms  0.500 ms
 6  4  male-so3210 (10.255.100.77)  0.423 ms  0.399 ms  0.402 ms
 7  5  havana-ge1100 (10.255.100.86)  0.527 ms  0.502 ms  2.307 ms
 8  6  livorno-so0010 (172.16.1.38)  1.949 ms  1.952 ms  1.949 ms
 9     MPLS Label=300784 CoS=0 TTL=1 S=1
10  7  barcelona (10.50.1.7)  1.949 ms  1.951 ms  1.031 ms
```

In the current rollout stage, all incoming traffic from router *Honolulu* is diverted to the helper VRF. The helper VRF contains all critical prefixes, along with a default route to the global table. Listing 1.44 Line 12 shows that the prefix for router *Barcelona* is not in the helper table. This implies that on the forwarding path, the traffic leaks into the global table, in which the data center prefixes reside. The issue is with the return traffic, because the data center traffic follows the default route towards the main instance, then on towards the hub headquarters site, and finally returns over the CORP VPN to the router *Barcelona* site as shown in Listing 1.43. Something must be missing in the configuration for the helper VRF in Listing 1.44 that creates a table as shown starting at Line 12.

Listing 1.44: Helper table missing local spoke critical prefix

```
user@Livorno> show configuration routing-instances helper
instance-type vrf;
route-distinguisher 64500:11;
vrf-import from-critical;
vrf-export null;
routing-options {
    static {
        route 0.0.0.0/0 next-table inet.0;
    }
}

user@Livorno> show route terse table helper

helper.inet.0: 2 destinations, 2 routes (2 active, 0 holddown, 0 hidden)
+ = Active Route, - = Last Active, * = Both

A Destination        P Prf   Metric 1   Metric 2  Next hop        AS path
* 0.0.0.0/0          S   5                         Table
* 10.200.1.9/32      B 170        100           2 >so-0/0/0.0     I
```

What is missing for the helper VRF to import the critical prefix into the helper table? Listing 1.45 shows that the prefix is present in *CORP* VRF. The policies are tagging critical prefixes as shown by the BGP advertisement, and the helper VRF is correctly importing the critical and properly tagged prefix from router *Inverness*.

Listing 1.45: Troubleshooting router *Barcelona* critical prefix

```
user@Livorno> show route 10.50.1.7 exact extensive

CORP.inet.0: 11 destinations, 11 routes (11 active, 0 holddown, 0 hidden)
10.50.1.7/32 (1 entry, 1 announced)
TSI:
KRT in-kernel 10.50.1.7/32 -> {10.50.1.53}
Page 0 idx 0 Type 1 val 87a07e0
        *RIP    Preference: 100
                Next-hop reference count: 5
                Next hop: 10.50.1.53 via ge-0/2/0.0, selected
                State: <Active Int>
                Age: 2:54:25    Metric: 2       Tag: 0
                Task: CORP-RIPv2
                Announcement bits (2): 0-KRT 2-BGP RT Background
                AS path: I
                Route learned from 10.50.1.53 expires in 168 seconds

user@Livorno> show route advertising-protocol bgp 192.168.1.1 10.50.1.7/32 detail

CORP.inet.0: 11 destinations, 11 routes (11 active, 0 holddown, 0 hidden)
* 10.50.1.7/32 (1 entry, 1 announced)
 BGP group IBGP-VPN type Internal
     Route Distinguisher: 64500:1
     VPN Label: 100160
     Nexthop: Self
     Flags: Nexthop Change
     MED: 2
     Localpref: 100
     AS path: I
     Communities: target:64500:2 target:64500:11
```

The issue is apparent only for the locally connected router *Barcelona* site, which is not behaving like a remote site with the same configuration. This information hints at a problem in local redistribution. Another look at Listing 1.44 reveals a missing `auto-export` statement.

Once this element is added, the prefix is redistributed and direct connectivity succeeds. Note that the rt-export *announcement bit* in Listing 1.46 Line 15 was added to the prefix to inform the relevant module of prefix changes.

Listing 1.46: Correct installation of router *Barcelona* prefix

```
1  user@Livorno> show configuration | compare rollback 1
2  [edit routing-instances helper routing-options]
3  +     auto-export;
4
5  user@Livorno> show route 10.50.1.7 exact detail
6
7  CORP.inet.0: 11 destinations, 11 routes (11 active, 0 holddown, 0 hidden)
8  10.50.1.7/32 (1 entry, 1 announced)
9          *RIP    Preference: 100
10                 Next-hop reference count: 7
11                 Next hop: 10.50.1.53 via ge-0/2/0.0, selected
12                 State: <Active Int>
13                 Age: 3:02:37    Metric: 2       Tag: 0
14                 Task: CORP-RIPv2
15                 Announcement bits (3): 0-KRT 2-BGP RT Background 3-rt-export
16                 AS path: I
17                 Route learned from 10.50.1.53 expires in 161 seconds
18
19 helper.inet.0: 4 destinations, 4 routes (4 active, 0 holddown, 0 hidden)
20
21 10.50.1.7/32 (1 entry, 1 announced)
22         *RIP    Preference: 100
23                 Next-hop reference count: 7
24                 Next hop: 10.50.1.53 via ge-0/2/0.0, selected
25                 State: <Secondary Active Int>
26                 Age: 14         Metric: 2       Tag: 0
27                 Task: CORP-RIPv2
28                 Announcement bits (1): 0-KRT
29                 AS path: I
30                 Route learned from 10.50.1.53 expires in 161 seconds
31                 Communities: target:64500:2 target:64500:11
32                 Primary Routing Table CORP.inet.0
```

Anomaly with direct traffic between router *Barcelona* and router *Inverness*

Traffic between router *Inverness* and router *Barcelona* is not going through the hub, as Listing 1.47 shows. This is occurring because the *CORP* VRF at router *Livorno* is configured as a hub for the data center traffic and the spoke site router *Barcelona* is sharing the same VRF. Critical prefixes from router *Inverness* populate the *CORP* VRF, to which router *Barcelona* has direct access.

Listing 1.47: Barcelona reaching router *Inverness* directly over the *CORP* VPN

```
1  user@Barcelona-re0> traceroute source barcelona inverness
2  traceroute to inverness (10.200.1.9) from barcelona, 30 hops max, 40 byte packets
3   1  livorno-ge020 (10.50.1.54)  0.718 ms  0.594 ms  0.568 ms
4   2  skopie-so1000 (172.16.1.33)  1.027 ms  0.927 ms  0.901 ms
5       MPLS Label=299808 CoS=0 TTL=1 S=0
6       MPLS Label=300112 CoS=0 TTL=1 S=1
7   3  inverness (10.200.1.9)  0.928 ms  0.842 ms  0.811 ms
```

One alternative is to create independent routing contexts on the data center PE, by building a regular *spoke CORP* VRF and a dedicated *data center CORP* VRF. Another alternative is to accept this situation as the valid setup chosen for this scenario and which remains in place after the migration completes.

Migrating remaining spoke sites

Additional spoke sites can be migrated, one at a time, in an incremental fashion. Listing 1.48 shows the restricted view of the *CORP* VRF at each of the spoke sites PE. Notice that each VRF does not have visibility of other spoke prefixes.

Listing 1.48: VRF contents for spoke-site PE at end of migration

```
user@Nantes> show route table CORP terse

CORP.inet.0: 8 destinations, 8 routes (8 active, 0 holddown, 0 hidden)
+ = Active Route, - = Last Active, * = Both

A Destination        P Prf   Metric 1   Metric 2  Next hop        AS path
* 0.0.0.0/0          B 170        100             >172.16.1.5     65000 I
* 10.1.1.0/24        B 170        100          0 >so-0/1/0.0      I
* 10.1.1.21/32       B 170        100          0 >so-0/1/0.0      I
* 10.200.1.0/24      R 100          2             >10.200.1.66
* 10.200.1.9/32      R 100          2             >10.200.1.66
* 10.200.1.64/30     D   0                        >so-1/0/1.0
* 10.200.1.65/32     L   0                         Local
* 224.0.0.9/32       R 100          1              MultiRecv

user@Torino> show route table CORP terse

CORP.inet.0: 8 destinations, 8 routes (8 active, 0 holddown, 0 hidden)
+ = Active Route, - = Last Active, * = Both

A Destination        P Prf   Metric 1   Metric 2  Next hop        AS path
* 0.0.0.0/0          B 170        100             >at-0/0/1.0     65000 I
                                                   so-0/1/0.0
* 10.1.1.0/24        B 170        100          0 >so-0/1/0.0      I
* 10.1.1.21/32       B 170        100          0 >so-0/1/0.0      I
* 10.100.1.0/24      R 100          2             >10.100.1.82
* 10.100.1.10/32     R 100          2             >10.100.1.82
* 10.100.1.80/30     D   0                        >at-1/2/0.1001
* 10.100.1.81/32     L   0                         Local
* 224.0.0.9/32       R 100          1              MultiRecv
```

Listing 1.49 shows the data-center view at the end of the migration, with all critical prefixes imported from the spoke sites.

Listing 1.49: VRF contents for data-center PE at end of migration

```
user@Livorno> show route table CORP terse

CORP.inet.0: 13 destinations, 13 routes (13 active, 0 holddown, 0 hidden)
+ = Active Route, - = Last Active, * = Both

A Destination        P Prf   Metric 1   Metric 2  Next hop        AS path
* 0.0.0.0/0          B 170        100             >so-0/0/1.0     65000 I
* 10.1.1.0/24        O 150          0             >10.1.1.90
* 10.1.1.21/32       O 150          0             >10.1.1.90
* 10.1.1.88/30       D   0                        >ge-1/3/0.0
* 10.1.1.89/32       L   0                         Local
* 10.50.1.0/24       R 100          2             >10.50.1.53
* 10.50.1.7/32       R 100          2             >10.50.1.53
* 10.50.1.52/30      D   0                        >ge-0/2/0.0
* 10.50.1.54/32      L   0                         Local
* 10.100.1.10/32     B 170        100          2 >so-0/0/0.0      I
                                                   so-0/0/1.0
* 10.200.1.9/32      B 170        100          2 >so-0/0/0.0      I
* 224.0.0.5/32       O  10          1              MultiRecv
* 224.0.0.9/32       R 100          1              MultiRecv
```

The hub PE router *Havana* has full visibility of all prefixes in *CORP* VRF, as shown in Listing 1.50. Unlike the data center, non-critical prefixes from spoke sites are present in the VRF. The *CORP-Hub-Downstream* table is used to advertise a default route only into the VPN.

Listing 1.50: VRF contents for hub site PE at end of migration

```
1  user@Havana> show route table CORP terse
2
3  CORP-Hub.inet.0: 11 destinations, 11 routes (11 active, 0 holddown, 0 hidden)
4  + = Active Route, - = Last Active, * = Both
5
6  A Destination        P Prf   Metric 1   Metric 2  Next hop         AS path
7  * 0.0.0.0/0          B 170        100              >10.255.100.85   65000 I
8  * 10.1.1.0/24        B 170        100          0  >so-1/0/0.0      I
9  * 10.1.1.21/32       B 170        100          0  >so-1/0/0.0      I
10 * 10.50.1.0/24       B 170        100          2  >so-1/0/0.0      I
11 * 10.50.1.7/32       B 170        100          2  >so-1/0/0.0      I
12 * 10.100.1.0/24      B 170        100          2  >at-1/2/0.0      I
13 * 10.100.1.10/32     B 170        100          2  >at-1/2/0.0      I
14 * 10.200.1.0/24      B 170        100          2  >at-1/2/0.0      I
15 * 10.200.1.9/32      B 170        100          2  >at-1/2/0.0      I
16 * 10.255.100.84/30   D   0                         >ge-1/1/0.0
17 * 10.255.100.86/32   L   0                          Local
18
19 CORP-Hub-Downstream.inet.0: 1 destinations, 1 routes (1 active, 0 holddown, 0 hidden)
20 + = Active Route, - = Last Active, * = Both
21
22 A Destination        P Prf   Metric 1   Metric 2  Next hop         AS path
23 * 0.0.0.0/0          B 170        100              >10.255.100.85   65000 I
```

Once remaining sites are moved, traffic connectivity is over the *CORP* VRF as shown in Table 1.6.

Table 1.6 Traffic flow matrix with all sites migrated

	HQ	DC	MS
Headquarters (HQ)		Global	CORP
Data Center (DC)	Global		CORP
Migrated Spoke (MS)	CORP	CORP	CORP/hub

Data center to hub traffic follows the global routing table.

1.4.12 Stage nine: Hub traffic to data center to follow *CORP* VPN

Raising the RIP protocol precedence in router *Male* as shown in Listing 1.51 derives hub traffic over the *CORP* VRF which yields a modified traffic matrix as shown in Table 1.7.

Listing 1.51: Sending hub traffic over the *CORP* VPN at router *Male*

```
1  user@male-re0# run show route terse 10.1/16
2
3  inet.0: 21 destinations, 24 routes (21 active, 0 holddown, 0 hidden)
4  + = Active Route, - = Last Active, * = Both
5
```

```
 6 A Destination        P Prf   Metric 1   Metric 2  Next hop        AS path
 7 * 10.1.1.0/24        R 100          2            >10.255.100.78
 8                      B 170        100            >10.255.100.86   64500 I
 9 * 10.1.1.21/32       R 100          2            >10.255.100.78
10                      B 170        100            >10.255.100.86   64500 I
11 * 10.1.1.88/30       R 100          2            >10.255.100.78
12
13 inet.2: 4 destinations, 4 routes (4 active, 0 holddown, 0 hidden)
14 + = Active Route, - = Last Active, * = Both
15
16 A Destination        P Prf   Metric 1   Metric 2  Next hop        AS path
17 * 10.1.1.0/24        R 100          2            >10.255.100.78
18 * 10.1.1.21/32       R 100          2            >10.255.100.78
19 * 10.1.1.88/30       R 100          2            >10.255.100.78
20
21 [edit]
22 user@male-re0# set protocols rip group hub preference 180
23
24 user@male-re0> show route terse 10.1/16
25
26 inet.0: 21 destinations, 24 routes (21 active, 0 holddown, 0 hidden)
27 + = Active Route, - = Last Active, * = Both
28
29 A Destination        P Prf   Metric 1   Metric 2  Next hop        AS path
30 * 10.1.1.0/24        B 170        100            >10.255.100.86   64500 I
31                      R 180          2            >10.255.100.78
32 * 10.1.1.21/32       B 170        100            >10.255.100.86   64500 I
33                      R 180          2            >10.255.100.78
34 * 10.1.1.88/30       R 180          2            >10.255.100.78
35
36 inet.2: 4 destinations, 4 routes (4 active, 0 holddown, 0 hidden)
37 + = Active Route, - = Last Active, * = Both
38
39 A Destination        P Prf   Metric 1   Metric 2  Next hop        AS path
40 * 10.1.1.0/24        R 180          2            >10.255.100.78
41 * 10.1.1.21/32       R 180          2            >10.255.100.78
42 * 10.1.1.88/30       R 180          2            >10.255.100.78
43
44 user@male-re0> traceroute honolulu source male
45 traceroute to honolulu (10.1.1.21) from male, 30 hops max, 40 byte packets
46  1  havana-ge1100 (10.255.100.86)  0.588 ms  0.429 ms  0.428 ms
47  2  livorno-so0010 (172.16.1.38)  0.739 ms  0.613 ms  0.619 ms
48      MPLS Label=300816 CoS=0 TTL=1 S=1
49  3  honolulu (10.1.1.21)  0.536 ms  0.501 ms  0.497 ms
```

Table 1.7 Traffic flow matrix in final configuration

	HQ	DC	MS
Headquarters (HQ)		CORP	CORP
Data Center (DC)	Global		CORP
Migrated Spoke (MS)	CORP	CORP	CORP/hub

The data center uses the global routing table to reach headquarters.

1.4.13 Stage ten: Migration cleanup

Once all sites are migrated, the cleanup stage involves removing the interim configuration that was used at the data center PE as well as removing the protocols configuration in the main instance for the spoke sites. CE routers are not aware of the change to a VPN. No cleanup is necessary because nothing was changed on them.

Cleanup of PE router connected to spoke site

Removal of the main instance configuration can be done at any suitable time after the PE–CE interface is activated in the *CORP* VRF. Listing 1.52 shows the cleanup required for router *Nantes*. The main instance RIP configuration block is removed along with the OSPF redistribution policy.

Listing 1.52: Cleanup of router *Nantes*

```
1  user@Nantes# show | compare
2  [edit protocols ospf]
3  -   export site-pool;
4  [edit protocols]
5  -   inactive: rip {
6  -       group Inverness {
7  -           export default;
8  -           neighbor so-1/0/1.0;
9  -       }
10 -   }
```

Cleanup of data center router *Livorno*

The steps that have to be taken at the data center router *Livorno* are shown in Listing 1.53. The summarized list of actions follow.

- Remove the firewall filter attached to the data center interface (ge-1/3/0). This configuration change normalizes the flow of all data center traffic onto the *CORP* VRF.
- Delete the helper VRF, which should contain the same information as the CORP VRF, along with the static default route pointing to inet.0.
- Clean up RIB group references. In the *CORP* VRF, delete interface-routes and auto-export that leak routes to inet.0
- Delete the definitions for firewall filter and rib-group, which are not needed anymore.

Listing 1.53: Data center PE cleanup after all sites migrated

```
1 user@Livorno# delete interfaces ge-1/3/0.0 family inet filter
2 user@Livorno# delete routing-instances helper
3 user@Livorno# delete routing-instances CORP routing-options interface-routes
4 user@Livorno# delete routing-instances CORP routing-options auto-export
5 user@Livorno# delete firewall family inet filter redirect-to-helper
6 user@Livorno# delete routing-options rib-groups leak-to-inet0
```

Removing the helper table activates the default route to the hub site on the *CORP* VRF for traffic from the data center, modifying the traffic matrix to the one shown in Table 1.8.

Table 1.8 Traffic flow matrix in final configuration

	HQ	DC	MS
Headquarters (HQ)		CORP	CORP
Data Center (DC)	CORP		CORP
Migrated Spoke (MS)	CORP	CORP	CORP/hub

All traffic flows use the *CORP* VRF.

1.4.14 Migration summary

The case study presented in this chapter started with a network in a full-mesh traffic matrix that required conversion to hub-and-spoke centered at the headquarters. Additional restrictions with the traffic flows towards the data center forced a second hub-and-spoke centered at the data center.

Different features and behaviors are analyzed in the case study:

- strategy to deploy a VPN progressively maintaining connectivity throughout the migration process;
- hub-and-spoke VPN with IP functionality in PE egress direction. An auxiliary *CORP-Hub-Downstream* VRF with `no-vrf-advertise` in the main VRF separates traffic directions and allows hub-and-spoke functionality together with egress IP functions;
- use of RIB groups to populate multiple tables with prefixes;
- control of multicast RPF check. Leveraging different RIBs for unicast and multicast, it is possible to create a separate topology for multicast traffic;
- binding traffic to tables through firewall filters and table next hops. Incoming traffic can be redirected to a *helper* table for selected traffic flows. In addition, static routes can have another table as next hop.

Bibliography

[2000207-001-EN] Juniper Networks, Inc. Efficient scaling for multiservice networks, February 2009.

[RFC2328] J. Moy. OSPF Version 2. RFC 2328 (Standard), April 1998.

[RFC2453] G. Malkin. RIP Version 2. RFC 2453 (Standard), November 1998. Updated by RFC 4822.

[RFC2547] E. Rosen and Y. Rekhter. BGP/MPLS VPNs. RFC 2547 (Informational), March 1999. Obsoleted by RFC 4364.

[RFC3032] E. Rosen, D. Tappan, G. Fedorkow, Y. Rekhter, D. Farinacci, T. Li, and A. Conta. MPLS Label Stack Encoding. RFC 3032 (Proposed Standard), January 2001. Updated by RFCs 3443, 4182, 5332, 3270, 5129.

[RFC3107] Y. Rekhter and E. Rosen. Carrying Label Information in BGP-4. RFC 3107 (Proposed Standard), May 2001.

[RFC4364] E. Rosen and Y. Rekhter. BGP/MPLS IP Virtual Private Networks (VPNs). RFC 4364 (Proposed Standard), February 2006. Updated by RFCs 4577, 4684.

[RFC5036] L. Andersson, I. Minei, and B. Thomas. LDP Specification. RFC 5036 (Draft Standard), October 2007.

Further Reading

[1] Aviva Garrett. JUNOS Cookbook, April 2006. ISBN 0-596-10014-0.

[2] Matthew C. Kolon and Jeff Doyle. Juniper Networks Routers : The Complete Reference, February 2002. ISBN 0-07-219481-2.

2

Link-State IGP Migrations

The foundation of a mid- or big-size topology in current networks is the *IGP (Interior Gateway Protocol)* that performs core routing and constitutes the basic pillar for further protocols and application.

As per their original definition back in the late 1980s, IGPs interconnect devices *inside* a common network domain. The basic idea at that time was that IGPs would transport all routes within those boundaries and an *EGP (Exterior Gateway Protocol)*, the precursor for the well-known *BGP (Border Gateway Protocol)*, would populate those advertisements between *Autonomous Systems* (ASs).

We have seen how BGP has been developed as a truly scalable multiprotocol concept and was the best fit to package route information, to control the flow of route advertisements and withdrawals, and to transport those information constructs, not only among *ASs* but also within the boundaries of a domain. BGP is a *hard-state* protocol, in which received routes remain present unless a withdrawal is sent or the session with the peer is dropped, without requiring regular routing updates, as opposed to the IGPs. It would be inconceivable nowadays to transport Internet or other external (or internal) routes over an IGP; it would simply not scale.

Nevertheless, an IGP plays a key role in the day-to-day life of an advanced network. It is the glue that binds all these pieces together and allows other protocols, such as IBGP (Internal Border Gateway Protocol), or applications, such as those derived from MPLS, to have a solid and truthful substrate underneath.

Traditionally, IGPs have been classified in to two types: distance-vector and link-state protocols (with Enhanced Interior Gateway Routing Protocol (EIGRP) being the exception to this simplistic differentiation).

Distance-vector protocols or *routing by rumors* rely on information being propagated hop by hop from peer to peer where routing tables are progressively updated following the Bellman–Ford algorithm. Protocols like RIP version 1 and 2 and Interior Gateway Routing Protocol (IGRP) are included in this category.

Link-state protocols or *routing by propaganda* rely on a common and consistent topological database shared among elements inside the network (ultimately inside each network

Network Mergers and Migrations Gonzalo Gómez Herrero and Jan Antón Bernal van der Ven

hierarchy) and the Dijkstra algorithm being run on this information for the sake of path decision. They include OSPF (both versions 2 and 3), IS–IS, IPX NLSP, and ATM PNNI, and even MPLS CSPF setups can be considered as a derivation of a constrained link-state topology.

A noticeable exception to both these classes is EIGRP, baptized as an *advanced distance vector* or *hybrid* protocol. This protocol uses the Diffuse Update Algorithm (DUAL) to achieve loop-free convergence. Every router maintains a neighbor table, as well as the relevant information received from the neighbor. It shares features both from distance-vector and link-state protocols.

Distance-vector protocols suffer from slower convergence (inherent to distributed routing calculation), higher chances of forming single- and multi-hop routing loops, and mechanisms such as counting to infinity or synchronized periodic updates that currently cannot keep up with some of the required service level agreements in modern networks.

As opposed to distance-vector protocols, link-state protocols can also include *traffic engineering (TE) extensions* as part of the information distribution mechanisms that provide additional details for path computation with MPLS applications.

Currently, most advanced and complex networks use a link-state protocol as the IGP for their core network. In the 1990s, it was uncertain whether resource consumption required by a link-state protocol would definitely scale and be controlled by network elements, but different protocol features conceived for this purpose together with a remarkable improvement in the horsepower capabilities of routing platforms made this constraint negligible.

Link-state protocols have proven to be reliable, fast, and more extensible than distance-vector ones and migrating a network towards a scalable link-state IGP has been a best practice in many service providers and enterprises in recent years.

Link-state protocols require higher resource consumption (both inside a router in terms of CPU and memory and in terms of bandwidth consumption due to information flooding) and some of the advanced features require higher complexity such as information sequencing and ageing, periodic database refresh to overwrite stale entries, and more complex adjacency formation in multi-access networks when compared to distance-vector protocols. They are far more simple in terms of configuration and functionalities and precisely for these reasons, they are a better fit for common current network topologies. Link-state protocols have a strong footprint in many networks around the world and will certainly continue to survive.

This chapter focuses on migrations involving the current state-of-the-art link-state core routing IGPs, specifically OSPF and IS–IS.

The reasons for migrating or adapting the core IGP are diverse, ranging from consequences after a simple router upgrade to fix a related software defect to a complete network integration as the result of a company acquisition. Regardless of the rationale, each migration requires careful planning and execution, and each operational change in the core IGP must be carefully considered both because of its domain-wide scope and because of future implications on additional protocols and applications that rely on the IGP. As a simple example, having duplicate loopback addresses or ISO NET IDs in a network can be catastrophic and can lead to network-wide routing churn.

This chapter is divided into two major sections that discuss two major types of migration of link-state IGPs. The first section deals with hierarchical migrations within a link-state IGP, migrations that follow the native layered architecture of link-state IGPs. The second section describes the processes and resources to migrate completely from one link-state

IGP to another. An IGP migration case study at the end of the chapter illustrates that the components of these two types of IGP migration can be easily intertwined.

2.1 Link-state IGP Hierarchical Migrations

One of the first discussions when designing an IGP is to place and adapt a topological hierarchy. Several criteria are used to accommodate layers to a certain environment and this tends naturally to have an influence in IGPs as well.

The IGP layering needs to be modified over time, especially as requirements and routing platforms evolve and as new features and applications need to be offered to customers.

This section first identifies general criteria for defining and designing hierarchical divisions in a link-state IGP. It then describes how to set up these divisions in OSPF and IS–IS, discussing various techniques to implement this topological scheme.

2.1.1 Motivations for link-state IGP hierarchical migrations

Link-state IGPs are divided into areas, forming a hierarchy that allows the network to be scalable. Without division into areas, the growth of a domain could lead to increasing resource consumption for handling traffic and route calculations. Dividing a domain into layers creates an artificial separation inside, allowing disjoint SPF calculation to improve route convergence times while at the same time stitching together route calculations within the domain.

When dividing a backbone into segments or taking a divided domain and flattening it, several factors must be considered. Table 2.1 summarizes some of the most important ones, and these are discussed in more detail later in this section. However, the debate about the rationale for IGP migrations is vigorous, and this table omits many other migration considerations. The final decision about why to migrate layers inside link-state IGPs must always consider the questions of the particular setup and its requirements, the services and applications that must be provided, and other external needs.

Organization structure

Politics can be a major contributor to the need for changing an IGP. Company acquisitions or breakups, changes in management strategies, and dismantling or rolling out a network can lead to migration plans that are cumbersome and often impossible for network engineers to achieve. When integrating or dividing an autonomous system, one of the biggest headaches is how to merge or split up IGPs with minimum service interruption.

Organizational structures very often tend to define network topologies as well. It is fairly common to find different groups in many service providers' engineering and operations departments designing and managing different network hierarchies. This division, sometimes more artificial or political than technical, tends to define network topologies as well (e.g., moving a router from an *access* layer to a *core* layer, or vice versa, depending on which group is responsible for it).

Route manipulation

Traditionally, a *divide-and-conquer* approach is used when defining different hierarchies in a link-state IGP. The scope of routing information propagation for both OSPF and IS–IS is

Table 2.1 Motivations for link-state IGP hierarchical migrations

Evaluation factor	Areas	Flat layer
Organization structure	Alignment with groups	Common scope
Route manipulation	At origin and area boundaries	At origin
Resource limitation	Isolation of low-end devices	Function of the topology itself
Security	Containment of failures	Global impact
Routing optimization	Tuning required, less than optimal routing inherent	Function of the SPF algorithm
Convergence	Reduced routing information, route regeneration, stability due to summarization	Uniform route information flooding
Inter-area TE	Support for additional features (e.g., LSP hierarchies)	Uniform TED and fast convergence capabilities

clearly delimited by establishing areas (in OSPF) or levels (in IS–IS), and both protocols provide enough resources to manipulate route injection to and from specific areas or levels.

In common network designs, this approach concentrates route manipulation at points in the topology where network summarization can be performed, offering a way to scale both the number of routes and stability of the IGP. Because link-state IGPs, by design, require that all systems in each hierarchy have identical information databases, boundaries between the hierarchy layers exist at which information can be aggregated or blocked. This summarization can be performed towards the backbone and towards the internal area, even advertising a simple default route internally. Within each layer, each system builds a routing information database by assembling puzzle pieces from each participating router in the same hierarchy. If such common visibility on route information distribution is artificially broken, the inherent distributed effort paradigm in link-state IGPs also gets broken with it.

This division between hierarchies can be used in network design to manipulate routing information as desired. Link-state IGPs can *hide or transform routes* either at the original injection point or at area boundaries. If this feature is required for a network design, hierarchical division is the perfect way to accomplish it. Routes can be leaked and information flooding constrained when the intention is just to present the necessary prefixes to an endpoint in the topology.

Resource limitation

Some participating network devices come up against *resource limitations*. Link-state IGPs can consume a considerable amount of memory and CPU cycles that could become a burden for some low-end routers, particularly in unstable topologies that have flapping links or frequent route updates.

In this case, a best practice is to place these devices into a type of OSPF or IS–IS area into which restricted route information – possibly simply the default route – is populated. This practice of isolating devices, often in preparation for a platform upgrade, is commonly performed in many environments as a step previous to decommissioning.

Dealing with device resource limitations continues to pose a challenge in the evolution of networks as they grow over time with newer features and requirements.

Security

Establishing hierarchies in link-state IGPs can be considered a *security* premise as well. Routing information is distributed in link-state IGPs in unique data structures in each area or in the backbone, and this information is translated in other data structures at area border systems for further flooding.

This degree of inter-layer opacity indirectly provides a *security* mechanism, because the original system injecting the routing information inside a layer may not be directly visible for routers in other layers.

Therefore, certain effects from malfunctioning systems or software bugs can be constrained within a hierarchy, without affecting the complete domain. *Route manipulation* techniques, such as prefix summarization at area boundaries, can even provide additional *security* by hiding instabilities from one area to the rest of the domain.

Routing optimization

When modifying route information propagation within the domain, there is always a tradeoff with *routing optimization*. Whenever more than a single exit point from an area exists, or when performing route manipulation, a network engineer must be concerned with restricting or changing the reachability of a prefix. These alterations in the routing population can easily lead to less than optimal routing (for example, when metrics are not uniformly translated), a problem that does not exist with a flat hierarchy, which always provides a uniform view of the network. This less than optimal behavior arises from the inherent behavior of hierarchies in the link-state IGPs.

Looking at the OSPF specifications provides several representative examples of this less than optimal routing performance, even without considering the different varieties of stub areas. As OSPF was originally conceived, if the same LSA Type 5 (same destination, cost, and non-zero forwarding address) is injected by more than one Autonomous System Border Router (ASBR), only the one injected by the ASBR with the highest Router Identifier should be used. A similar situation also applies to LSAs Type 7 translated to LSAs Type 5 out of an NSSA by the ABR with a *NSSATranslatorRole* election mechanism and the highest Router Identifier being an ultimate tie-breaker, also creating a potential situation for suboptimal routing to the end prefix.

Concerning IS–IS, [ISO10589] and [RFC1195] originally defined IPv4 related TLVs to encapsulate IPv4 routing information. This included a distinction between internal and external prefixes with TLV 128 Internal IP Reachability and TLV 130 External IP Reachability, and a default route leaking rule that implicitly leads to route hiding: external routes out of a Level 1 Area are not injected to Level 2 and neither internal nor external prefixes from Level 2 are injected into a Level 1 Area by a L1L2 system. External reachability out of a Level 1 Area is granted by the so-called ATTached bit set by L1L2 routers in their

Level 1 LSPs and cause a default route installation at other L1 routers. Junos OS provides enough flexibility to override these default behaviors and implement specific leaking rules for optimal forwarding.

All these default mechanisms are associated with standard default protocol behavior and lead to less than optimal routing.

Convergence

Convergence is often another factor that is considered when deciding whether to implement an IGP in a flat hierarchy or to divide it into layers. Hierarchical layering in a link-state IGP leads indirectly to distance-vector routing between levels, with the known associated drawbacks, but also offers some advantages in terms of stability in some situations.

One might think that reducing the amount of routing information in an area would lead to faster convergence, but it could actually lead to the opposite phenomenon. When comparing convergence times for routes beyond the local area, the route reprocessing, together with the eventual route summarization, at the area boundaries could add a few extra milliseconds to the time it takes information to flood to the rest of the domain.

Apart from eventual delays associated with regenerating information at the area edge, route aggregation can indirectly stabilize a particular setup or also degrade convergence times for applications. Because only summary prefixes are flooded over the rest of the domain, more specific route flaps remain hidden from the rest of the network. On the other hand, hiding more specific routes from an area to the rest of the domain impedes expedited convergence following a routing update from the link-state IGP, requiring the expiration of other applications or protocols, such as BGP, to detect connectivity failures beyond area boundaries.

Traffic Engineering extensions

Traffic Engineering (TE) extensions for OSPF and IS–IS were originally supported just within a single area. This specification was a limitation in scenarios where end-to-end traffic-engineering policies were needed, and the result was a mandatory restriction on the division of an IGP domain into layers.

Inter-area TE standards such as [RFC4105] set a framework of requirements for resource optimization, including fast recovery mechanisms across areas and scalable approaches to preserving the IGP hierarchy. An example of the scalable approach is the extension of existing intra-area TE resources such as OSPF Opaque LSAs and the corresponding sub-TLVs in the new-style IS–IS TLVs, such as TLV 22 Extended IS Reachability, and TLV 135 Extended IP Reachability.

Depending on how vendors implement such interarea requirements, further dependencies arise.

One approach to handling TE extensions with IGP migrations uses MPLS Label-Switched Path (LSP) hierarchies, in which the backbone transparently transports end-to-end paths among routers in different areas. This approach allows implementation of TE extensions domain-wide and grants fast-reroute capabilities within each area. However, it requires support for additional features for TE forwarding adjacency LSPs to settle an overlay model with MPLS LSPs.

To summarize this analysis, in the technical discussions analyzing the traditional *core-distribution-access* models, network engineers commonly deal with the question of whether aggregation should take place within different IGP areas or whether a flat hierarchy would suffice. As usual, *it depends*. Some existing networks run more than 1000 devices in a flat hierarchy, while other legacy devices in the backbones of major service providers are in OSPF totally stubby areas or IS–IS Level 1 areas with just a Level 1 route and a default route installed based on detection of the ATTached bit. As a general conclusion, a flat topology is based on native features of link-state IGPs, while a hierarchical topology introduces additional elements that artificially change these inherent premises of link-state routing protocols, and can present implementation advantages or design flaws in some network topologies.

2.1.2 Generic strategies for link-state IGP hierarchical migrations

When preparing to migrate areas or levels inside link-state IGPs, it is important first to identify the existing hierarchies and at that point, to plan the transition of network adjacencies.

This section discusses these two topics. First, the identification of different hierarchies in OSPF and IS–IS is analyzed and a later subsection analyzes adjacency transition tactics for each protocol.

Hierarchy identification in backbone

In order to avoid potential collisions or overlaps in the rest of the network, an analysis of area references in route information in the backbone is mandatory.

Hierarchy identification in OSPF networks: OSPF can perform *complete* route transformation at the boundaries of an Autonomous System or of an Area within the AS.

For OSPFv2, [RFC2328] Section 3 defines the role of an Area Border Router (ABR) and a certain opacity degree of routes exchanged between areas and the backbone.

OSPF route information is flooded using Link State Advertisements (LSAs). The collection of internal LSAs from each area and all kinds of external LSA is what actually populates the Link-State DataBase (LSDB).

An LSA header is determined by the LS Type, Link State ID, Advertising Router, LS Sequence Number, LS Checksum, and LS Age fields. All these values identify uniquely each type of LSA. Note that there is no reference to a particular area ID contained in the LSA header itself.

[RFC2328] Section 12.4 mandates that ABRs originate a single Type 3 Summary LSA for each known inter-area destination, summarizing Router and Network LSAs within a particular area. Likewise, AS External LSAs are transparently flooded across area boundaries because they have domain scope, but ABRs generate a specific Type 4 Summary LSA to ensure reachability to the router that is advertising External LSAs.

Summary LSAs are flooded throughout a single area only, meaning that the internal prefixes can be different in the Type 3 Summary LSAs in the backbone area (Area 0) and in other participating areas. This *repackaging* process for Type 3 Summary LSAs at the ABR allows network summarization in different directions to take place. Most modern routing operating systems, Junos OS among them, include knobs to allow aggregation when generating new Type 3 Summary LSAs.

Precisely because new information constructs are built at ABRs, the notion of a particular area ID is lost when generating Summary LSAs. LSA headers do not consider any reference to a particular area and the information readvertisement for Type 3 Summary LSAs is based purely on IPv4 prefixes and corresponding masks, not on any area IDs, as shown in Figure 2.1. Type 4 Summary LSAs also include no area ID references.

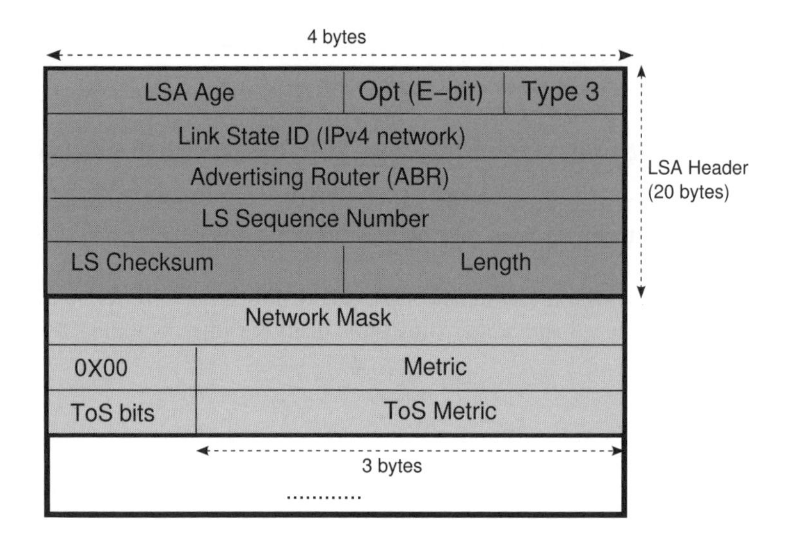

Figure 2.1: OSPF version 2 Type 3 Summary LSA.

The same concepts have been extrapolated to OSPF version 3. Area information aggregation can be performed at area boundaries with similar mechanisms and roles as OSPF version 2. One substantial difference is that OSPFv3 Router and Network LSAs do not contain network addresses but are simply used to convey topology information. They actually contain no address information and are network protocol independent. Intra-Area Prefix LSAs are precisely created as route information containers inside each area; in fact, as a separate protocol data structure.

OSPFv3 Type 0x2003 Inter-Area Prefix LSAs are equivalent to OSPFv2 Type 3 Summary LSAs and follow the same procedures for forwarding information towards different areas. Note that the Link State ID field has lost in OSPFv3 its original semantics from OSPFv2 and in OSPFv3 just serves merely as a distinguisher, as illustrated in Figure 2.2.

OSPFv3 Type 0x2004 Inter-Area Router LSAs are conceptually equivalent to OSPFv2 Type 4 Summary LSAs and retain the same functionality. Note also that neither this LSA nor the Inter-Area Prefix LSA contain any explicit area reference.

Type 6 LSAs were originally conceived in [RFC1584] as multicast extensions for link-state information but their usage is nowadays deprecated. Utilization of Type 8 External Attribute LSAs in current networks has also been made obsolete.

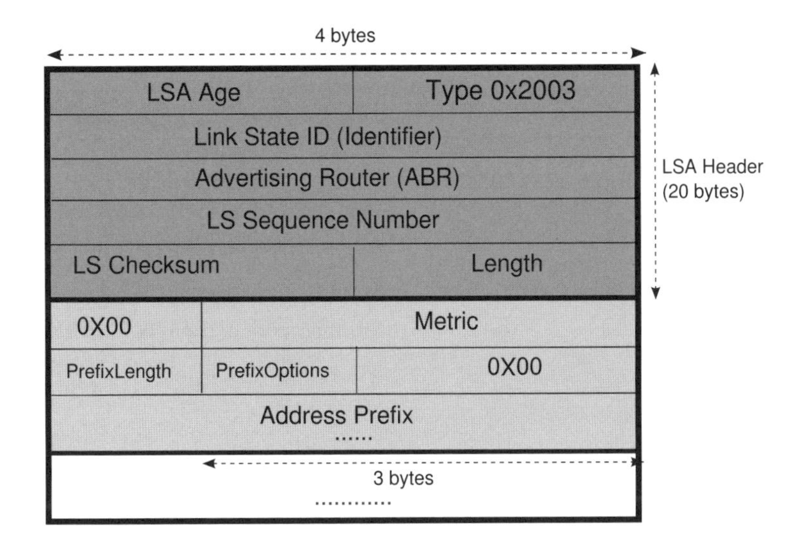

Figure 2.2: OSPF version 3 Type 0x2003 Inter-Area Prefix LSA.

[RFC3101] defined a new type of area, the so-called *Not-So-Stubby* Area (NSSA), to allow certain types of AS external routes to be imported in a limited fashion. This type of area was signaled by a new option bit (the N bit) and a new type of AS External LSA (Type 7 NSSA AS External LSA). Although syntactically similar to Type 5 LSAs, a new LSA type was needed to indicate the need for specific transformation at area boundaries.

In the NSSA, a certain border router translates the Type 7 NSSA AS External LSAs into standard Type 5 AS External LSAs, which are then flooded through the domain. To avoid potentially non-uniform flooding, each area elects a designated router as *NSSATranslatorRole* and this router is the only one that injects the Type 5 LSAs into the domain. When performing such LSA translation, it is enabled potentially to consolidate route advertisement.

[RFC5250] sets boundaries and flooding scopes for Types 9, 10, and 11 LSAs. Type 9 Opaque LSAs are not flooded beyond a common link and Type 10 Opaque LSAs must be populated within the borders of an associated area. Type 10 LSAs received on ABRs are explicitly discarded for translation into other connected areas.

Type 11 Opaque LSAs can be flooded throughout all areas, except for stub or NSSAs similarly to Type 5 AS External LSAs. This means that they rely on information being carried via Type 4 Summary LSAs to advertise ASBR reachability and that the information contained in the LSAs remains transparent and is unchanged from what the injecting router placed in it.

OSPF uses the Options field in LSAs and Hello packets to advertise router capabilities and parameters, and some of the options must match among peers to be able to form an adjacency. In OSPFv2, all bits in the options field have been fully allocated and [RFC4970] advocates the use of opaque LSAs in OSPFv2 and new LSAs in OSPFv3 to advertise router optional capabilities. While existing Types 9, 10, and 11 LSAs can be reused in OSPFv2 to flood the additional information with different scopes, a new Type 12 Router Information (RI) Opaque LSA has been defined for OSPFv3. In any case, no information transformation scheme is yet proposed at area boundaries for this new OSPFv3 LSA.

Table 2.2 summarizes the OSPF transformation rules at area boundaries.

Table 2.2 Analysis of LSA transformation at area boundaries

LSA Type at originating Area	Transformed LSA Type at external Area
Type 1 v2 Router	Summarized in Type 3 LSA
Type 0x2001 v3 Router	Summarized in Type 0x2003 LSA
Type 2 v2 Network	Summarized in Type 3 LSA
Type 0x2002 v3 Network	Summarized in Type 0x2003 LSA
Type 3 v2 Summary	Not applicable
Type 0x2003 v3 Inter-Area Prefix	Not applicable
Type 4 v2 Summary	Not applicable
Type 0x2004 v3 Inter-Area Router	Not applicable
Type 5 v2 AS External	Flooded without manipulation, Type 4 LSA for advertising router
Type 0x4005 v3 AS External	Flooded without manipulation, Type 0x2004 LSA for advertising router
Type 6 v2 Multicast Extensions	Not applicable
Type 0x2006 v3 Multicast Extensions	Deprecated
Type 7 v2 NSSA AS External	Translated to Type 5 LSA, unconditionally or by NSSATranslatorRole
Type 0x2007 v3 NSSA AS External	Translated to Type 0x4005 LSA, unconditionally or by NSSATranslatorRole
Type 8 v2 External Attributes	Not applicable
Type 0x0008 v3 Link	Not applicable
Type 9 v2 Opaque	Link-local scope, not applicable
Type 0x2009 v3 Intra-Area Prefix	Summarized in Type 0x2003 LSA
Type 10 v2 Opaque	Area scope, not applicable
Type 11 v2 Opaque	Domain scope, not flooded to stub or NSSAs
Type 12 v3 Router Information Opaque	Varying scope, no transformation

Hierarchy identification in IS–IS networks: IS–IS has a completely different approach to area correspondence and boundaries when compared to OSPF.

The first fundamental divergence is the placement of area boundaries. In OSPF, the demarcation line crosses the router, in the sense that a logical interface can belong only to one area (with the known exception of multi-area adjacencies) and the router in this case acts as ABR with interfaces *inside* each area. In IS–IS, the Intermediate System is fully inside one (or more than one) area. A router can set up more than one adjacency over the same link, either a Level 1 or a Level 2 adjacency, or an adjacency in both levels.

The IS–IS area identifier is not a compulsorily matching parameter for neighbor relationships as it is in OSPF. Only Level 1 adjacencies need to be circumvented inside a common area, because there is no such concept of a *Level 2 area*. An IS–IS Level 2 is nothing more than a *contiguous* collection of backbone routers that do not necessarily belong to a particular area. This fact notably relaxes connectivity requirements in the backbone when considering a hierarchical alignment, because Level 2 adjacencies do not impose matching areas on participating routers.

Level 1 routers do not actually know the identity of routers or destinations outside of their area per original concept. They install a default route pointing to L1L2 routers that set the ATTached bit in their Level 1 LSPs but ignore possible area duplications outside their island. Level 2 routers also ignore the particular Level 1 area topology and routing information, except for the directly attached Level 1 areas when acting as L1L2 routers.

[RFC1195] already opens the door for the possibility of an area having multiple addresses, something wholly inconceivable in OSPF. The reasoning behind this scheme is to enable graceful transitions for migrating, merging, or partitioning areas. Areas are configured on each intermediate system and in the case of multiple configured areas, the same criteria apply for Level 1 adjacencies to be established: an intermediate system will refuse to become a neighbor with a node if none of its area addresses overlaps its own configured areas. That is to say, a minimum of a single matching area identifier is enough for a *unique* Level 1 adjacency to be set up. From that perspective, IS–IS areas are another Level 1 adjacency setup control mechanism.

In this situation, a single system that maintains more than one IS–IS Level 1 adjacency in different areas in effect performs *area merging*. Flooding scopes from diverse areas are stitched together at the point where the router contains at least one IS–IS Level 1 adjacency in each of them. Because route information is transparently flooded over these interconnect points, such bypasses also remain eligible for path computation.

Area recognition is first performed when inspecting the initial data available with each IS–IS PDU. This information is first referred in the Maximum Area Addresses field, which is included in the IS–IS header of all PDUs, as displayed in Figure 2.3. This provides an initial reference for the maximum number of areas.

The definitive information pointer regarding the IS–IS areas configured in a system is the so-called TLV 1 Area, shown in Figure 2.4. It consists of a sequence of area IDs and their lengths coded in a TLV. As per OSI addressing, an IS–IS area identifier can be 13 bytes long at most, and the theoretical constraint for the number of areas is imposed by a maximum of 255 bytes of TLV data, although routing operating systems tend to impose a lower boundary than this.

Such TLVs are mandatory to appear in all types of IS–IS Hello PDU and LSP (both for Level 1 and Level 2) because they contain substantial information used for establishing adjacencies and performing the SPF computation.

The original TLVs advocated for IPv4 information, defined in [RFC1195], were TLV 128 Internal IP Reachability and TLV 130 External IP Reachability to include internal (local and direct) and external (redistributed) routing information advertised by an intermediate system,

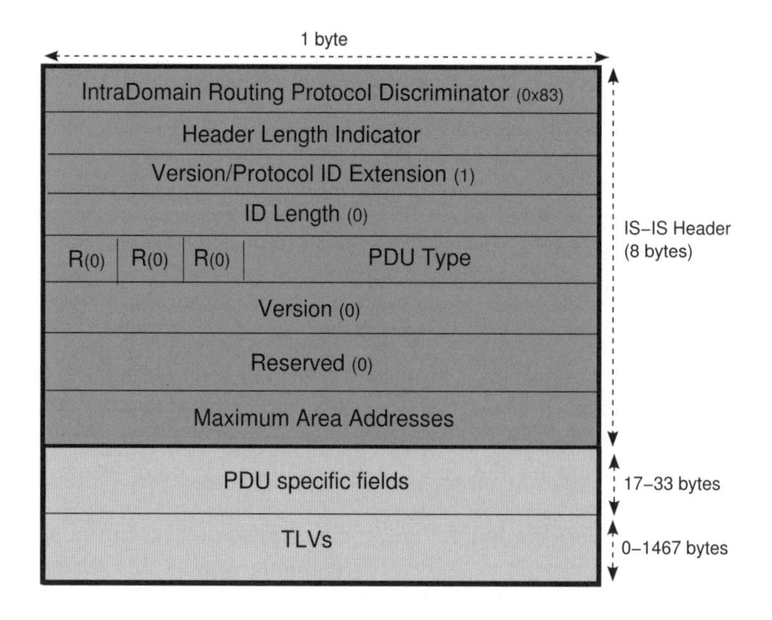

Figure 2.3: Common IS–IS header including Maximum Area Addresses field.

respectively. Figures 2.5 and 2.6 both depict IPv4-related TLVs. The intent is for L1L2 routers to extract and insert IPv4 routes in the corresponding LSPs at area boundaries according to the default protocol behavior, so there is no one-to-one correspondence among TLVs across levels, but these data structures are regenerated following policies and rules at each router.

Actually, by default, route population across IS–IS levels is constrained. IS–IS L1L2 routers do not propagate any routing information from Level 2 to Level 1 areas by default. It is expected that L1 systems will install a default route based on ATTached bit detection and L1L2 routers set it based on their backbone connectivity (meaning in the case of Junos OS reachability of external areas through the backbone) and active IS–IS adjacencies. This mechanism is somewhat equivalent to a Totally Stubby Area in OSPF because in both, the only available external data is a default route pointing to the core exit points.

A further default restriction in the original IS–IS prohibited utilization of TLV 130 External IP Reachability in Level 1 areas. As a result, no external L2 routes could be leaked to Level 1 systems. This opacity between layers was considered a safeguard mechanism when the original IS–IS specification was written, but soon became unpractical in the IP environment.

[RFC2966] relaxed this original constraint by allowing backbone routes to be leaked to L1 areas, reusing the high-order bit in the default metric field in TLVs 128 and 130 to be the so-called Up/Down bit. Routers at level boundaries must set this bit for those TLVs injected in Level 1 areas containing information propagated from the core. In this way, L1L2 routers can determine which route information has been passed to an L1 area and can avoid creating a routing loop by re-injecting it in the backbone.

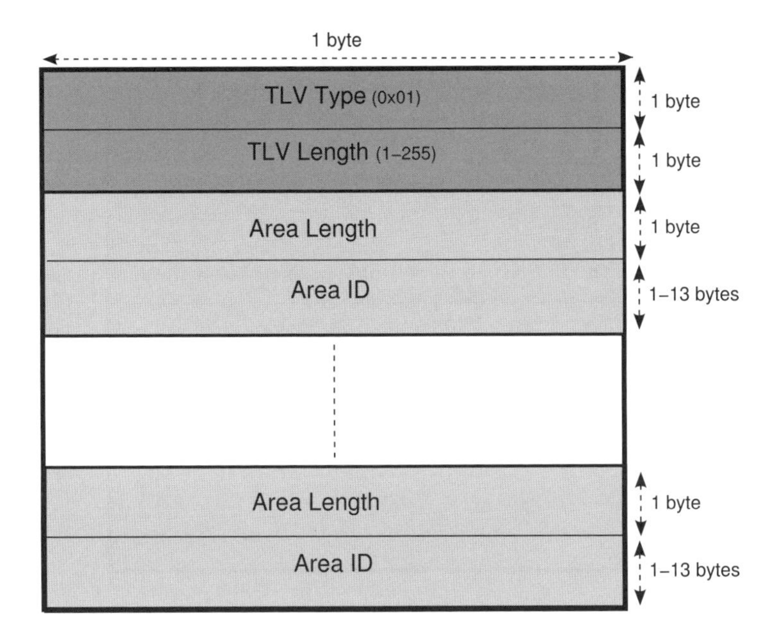

Figure 2.4: IS–IS TLV 1 Area.

This RFC also loosened the initial restriction on using the TLV 130 External IP Reachability in Level 1 areas so that a simple L1 router could now inject route information *external* into the domain, as opposed into a purely Totally Stubby Area in OSPF.

Both these changes considerably influenced IS–IS route population mechanisms and standardized domain-wide prefix distribution with different preferences and redistribution rules.

Nevertheless, there were still two other major restrictions with both TLVs 128 and 130: their definition restricted the metric value to a maximum length of 6 bits and they did not allow further expansibility to attach additional information to each route. These same restrictions applied to TLV 2 IS Reachability which defined intermediate system connectivity, the foundations for the SPF calculation.

[RFC3784] further leveraged both restrictions by means of the definition of additional TLVs, namely TLV 22 Extended IS Reachability and TLV 135 Extended IP Reachability. These new expanded TLVs granted additional metric space, with values up to 24-bit *wide* and added TE Sub-TLVs. However, they did not consider any internal or external prefix distinction. The leaking concept based on setting an Up/Down bit is still valid and mandatory though, in order to avoid potential loops (especially when removing the protocol preference distinction between external and internal prefixes).

In the previous and current implementations of IS–IS, L1L2 routers no longer inject area-specific information from one level into another. The Maximum Area Addresses field in the IS–IS header is set for supportability purposes and although TLV 1 Area is included in LSPs and all Hello PDUs, it is mostly used only for adjacency establishment and ATTached bit setting purposes, not for SPF calculations.

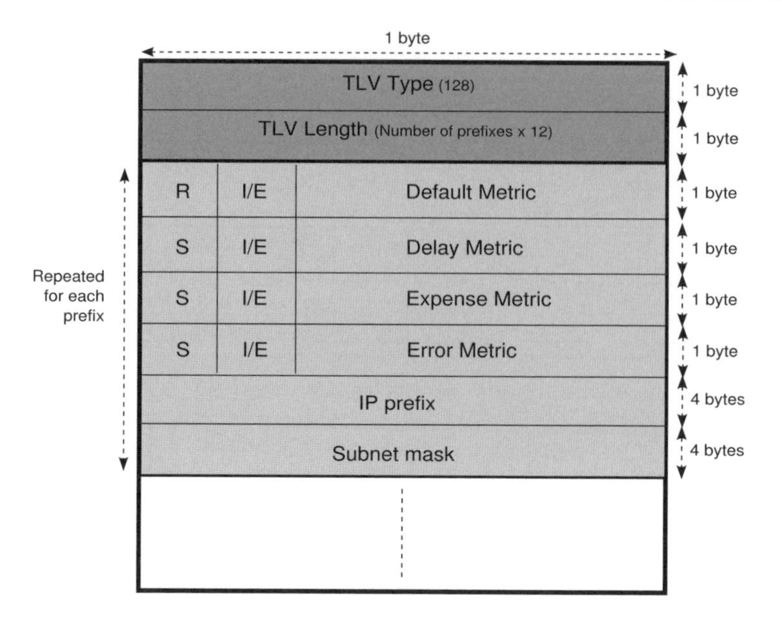

Figure 2.5: IS–IS TLV 128 Internal IP Reachability.

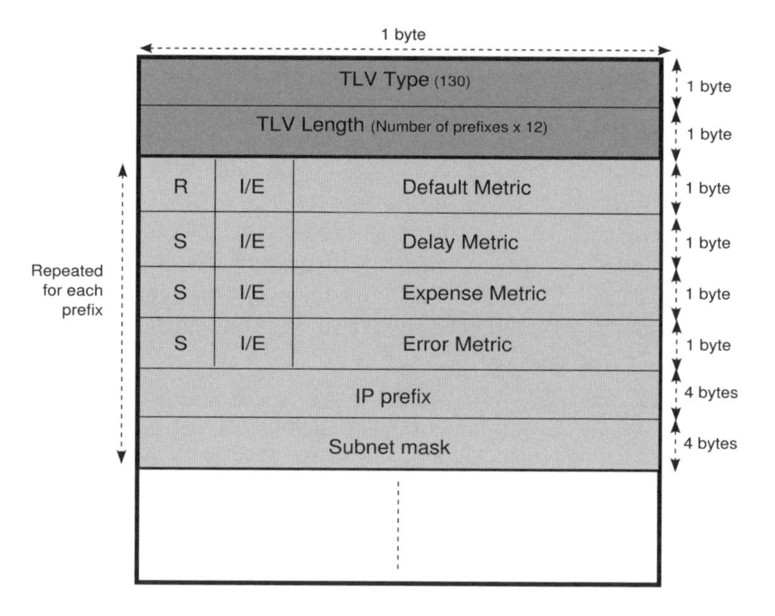

Figure 2.6: IS–IS TLV 130 External IP Reachability.

When manipulating topology information at L1L2 routers, neither the old-style nor new-style IPv4 reachability TLVs contain any area ID reference. It is the responsibility of the L1L2 router to propagate route information between TLVs in different LSPs in both levels.

When evaluating route-leaking across levels, a noticeable exception to routing information population is area merging that occurs by stitching together different IS–IS Level 1 adjacencies in diverse areas. In this case, a formal combination of interconnected areas occurs, as if it were a single area in terms of flooding and route computation.

Because area identification remains relevant for local adjacency establishment and ATTached bit setting, and because Junos OS does not implement the partition repair function, duplication or coincidence in area codes in different parts of the backbone is irrelevant to ensuring end-to-end connectivity.

Adjacency transition

Because both OSPF and IS–IS prefer intra-area paths to inter-area paths, a smooth adjacency migration is vital to the success of any effort to merge, segregate, or renumber a particular area.

Adjacency transition in OSPF: The traditional interface-area association advocated in OSPF poses a problem when extending or shrinking areas in an OSPF backbone, since moving an interface from one area to another (in all media types) becomes an intrusive operation.

[RFC5185] introduces the concept of *multi-area* adjacencies with ABRs being able to establish multiple point-to-point adjacencies belonging to different areas. Each point-to-point link provides a topological connection eligible for flooding and path computation within a particular area. A new Router LSA for each *secondary* area is created once the secondary adjacencies are established, and it contains only *point-to-point* network information for that area to represent the link to another router. However, the Router LSA contains no extra *stub* link information for the common network; nor does it contain any extra Type 3 LSA generation for the associated subnet in order to avoid address overlapping with the information present in the *primary* area.

Because of the nature of this behavior, these multi-area adjacencies are implicitly point-to-point relationships, so broadcast and NBMA OSPF link types cannot join them. Only point-to-point links and LAN media operating in a point-to-point fashion allow multi-area neighbor relationships.

Application Note: OSPF area migrations through secondary adjacencies

Figure 2.7 presents a subset of our well-known topology with the same OSPF area configured in different parts of the network.

In OSPF, duplicate non-backbone areas can coexist in a common backbone with end-to-end connectivity. Listing 2.1 shows the output of the OSPF database from router *Male*'s point of view and Listing 2.2 from router *Barcelona*. Note that although they are internal routers inside an area with the same identifier, their corresponding OSPF databases are completely different.

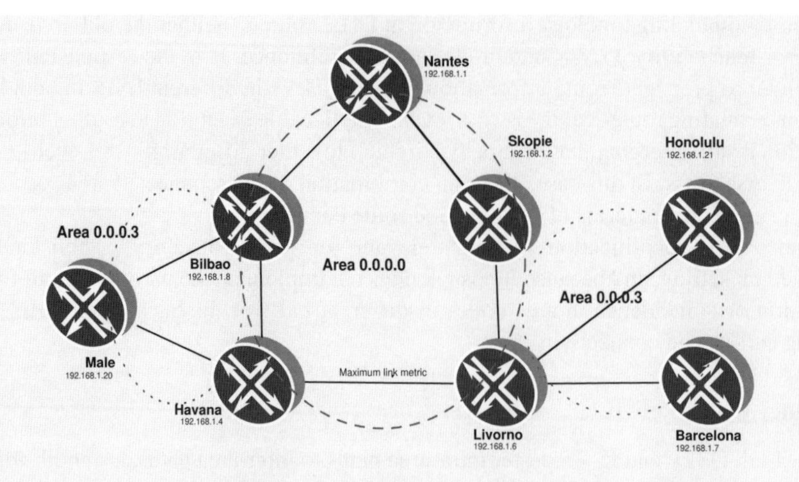

Figure 2.7: Network topology with duplicated Area 0.0.0.3.

Listing 2.1: OSPF database at Male

```
user@male-re0> show ospf database | no-more

    OSPF database, Area 0.0.0.3
 Type       ID               Adv Rtr          Seq         Age  Opt  Cksum  Len
Router   192.168.1.4      192.168.1.4      0x8000000c   428  0x22 0x52d0  36
Router   192.168.1.8      192.168.1.8      0x8000000b   364  0x22 0x4e53  48
Router  *192.168.1.20     192.168.1.20     0x8000000f   102  0x22 0x7c52  72
Network *172.16.1.85      192.168.1.20     0x80000005  1378  0x22 0xe5d4  32
Summary  172.16.1.4       192.168.1.4      0x8000000d   428  0x22 0x8f72  28
Summary  172.16.1.4       192.168.1.8      0x8000000b   364  0x22 0x718f  28
<...>
Summary  172.16.1.52      192.168.1.4      0x80000006   428  0x22 0xbb1d  28
Summary  172.16.1.52      192.168.1.8      0x80000005   364  0x22 0xb91a  28
<...> # Summary LSAs for loopbacks beyond local Area
Summary  192.168.1.1      192.168.1.4      0x80000009   428  0x22 0x91c8  28
Summary  192.168.1.1      192.168.1.8      0x80000008   364  0x22 0x7bdb  28
Summary  192.168.1.2      192.168.1.4      0x80000009   428  0x22 0x87d1  28
Summary  192.168.1.2      192.168.1.8      0x80000008   364  0x22 0x7bd9  28
Summary  192.168.1.3      192.168.1.4      0x80000009   428  0x22 0x7dda  28
Summary  192.168.1.3      192.168.1.8      0x80000008   364  0x22 0x71e2  28
<...>
Summary  192.168.1.21     192.168.1.4      0x80000005   428  0x22 0xda6e  28
Summary  192.168.1.21     192.168.1.8      0x80000004   364  0x22 0xd86b  28
<...>
OpaqArea 1.0.0.1          192.168.1.4      0x80000001   428  0x22 0xbc95  28
OpaqArea 1.0.0.1          192.168.1.8      0x80000001   364  0x22 0xcc7d  28
OpaqArea*1.0.0.1          192.168.1.20     0x80000001   102  0x22 0xfc35  28
OpaqArea 1.0.0.3          192.168.1.4      0x80000002   293  0x22 0x1a39 124
OpaqArea 1.0.0.3          192.168.1.8      0x80000002    41  0x22 0x9dee 124
OpaqArea*1.0.0.3          192.168.1.20     0x80000001   102  0x22 0x9da7 124
OpaqArea*1.0.0.4          192.168.1.20     0x80000001   102  0x22 0x25f2 136
```

Listing 2.2: OSPF database at Barcelona

```
user@Barcelona-re0> show ospf database | no-more

    OSPF link state database, Area 0.0.0.3
```

```
4  Type       ID              Adv Rtr          Seq        Age  Opt  Cksum  Len
5  Router     192.168.1.6     192.168.1.6      0x8000000b  527  0x22 0xa978  48
6  Router    *192.168.1.7     192.168.1.7      0x80000008  139  0x22 0x6335 120
7  Router     192.168.1.21    192.168.1.21     0x80000006  227  0x22 0x7af5  48
8  Network   *172.16.1.53     192.168.1.7      0x80000003 1728  0x22 0x13e1  32
9  Network    172.16.1.89     192.168.1.6      0x80000005  314  0x22 0x744d  32
10 Summary    172.16.1.4      192.168.1.6      0x80000008  526  0x22 0x976c  28
11 <...>
12 Summary    172.16.1.84     192.168.1.6      0x80000006  526  0x22 0x6e48  28
13 <...> # Summary LSAs for loopbacks beyond local Area
14 Summary    192.168.1.1     192.168.1.6      0x80000006  526  0x22 0x95c4  28
15 Summary    192.168.1.2     192.168.1.6      0x80000006  526  0x22 0x81d8  28
16 Summary    192.168.1.3     192.168.1.6      0x80000006  526  0x22 0x81d6  28
17 <...>
18 Summary    192.168.1.20    192.168.1.6      0x80000006  526  0x22 0xd670  28
19 <...>
20 OpaqArea   1.0.0.1         192.168.1.6      0x80000001  527  0x22 0xc489  28
21 OpaqArea*1.0.0.1           192.168.1.7      0x80000001  139  0x22 0xc883  28
22 OpaqArea   1.0.0.1         192.168.1.21     0x80000001  227  0x22 0x12f   28
23 OpaqArea   1.0.0.3         192.168.1.6      0x80000002  258  0x22 0x870a 124
24 OpaqArea*1.0.0.3           192.168.1.7      0x80000001  139  0x22 0x652d 124
25 OpaqArea   1.0.0.3         192.168.1.21     0x80000001  227  0x22 0x86b4 124
26 OpaqArea   1.0.0.4         192.168.1.6      0x80000002  203  0x22 0xb692 124
```

It is interesting to note that both routers see each other via Type 3 Summary LSAs injected by the corresponding ABRs as per Listings 2.3 and 2.4, similar to what would also be expected with different area identifiers.

Listing 2.3: Barcelona loopback address topological information at Male

```
1  user@male-re0> show ospf database extensive lsa-id 192.168.1.7
2
3      OSPF database, Area 0.0.0.3
4   Type       ID              Adv Rtr          Seq        Age  Opt  Cksum  Len
5  Summary    192.168.1.7     192.168.1.4      0x8000000d  168  0x22 0x6dec  28 # Injected by Havana
6    mask 255.255.255.255
7    Topology default (ID 0) -> Metric: 1646
8    Aging timer 00:58:49
9    Installed 00:01:10 ago, expires in 00:58:49, sent 00:01:10 ago
10   Last changed 00:01:10 ago, Change count: 7
11 Summary    192.168.1.7     192.168.1.8      0x8000000c  495  0x22 0x5ff5  28 # Injected by Bilbao
12   mask 255.255.255.255
13   Topology default (ID 0) -> Metric: 1003
14   Aging timer 00:58:49
15   Installed 00:01:10 ago, expires in 00:58:49, sent 00:01:10 ago
16   Last changed 00:01:10 ago, Change count: 6
```

Listing 2.4: Male loopback address topological information at Barcelona

```
1  user@Barcelona-re0> show ospf database extensive lsa-id 192.168.1.20
2
3      OSPF link state database, Area 0.0.0.3
4   Type       ID              Adv Rtr          Seq        Age  Opt  Cksum  Len
5  Summary    192.168.1.20    192.168.1.6      0x8000000d  611  0x22 0xd670  28 # Injected by Livorno
6    mask 255.255.255.255
7    Topology default (ID 0) -> Metric:1546
8    Aging timer 00:57:17
9    Installed 00:02:41 ago, expires in 00:57:18, sent 00:02:41 ago
10   Last changed 00:02:41 ago, Change count: 6
```

These standard translation mechanisms at area boundaries ensure end-to-end connectivity, independently of the area identifier. Assume in this case that the link between router *Havana* and router *Livorno* has a maximum metric, much higher than the path over router *Nantes* and router *Skopie*, so the traffic between both ends will take this alternate path in the backbone, as per Listing 2.5.

Listing 2.5: Traceroute from Male to Barcelona

```
1 user@male-re0> traceroute 192.168.1.7 source 192.168.1.20
2 traceroute to 192.168.1.7 (192.168.1.7) from 192.168.1.20, 30 hops max, 40 byte packets
3  1  bilbao-so1310 (172.16.1.78)  0.503 ms  0.424 ms  0.433 ms
4  2  nantes-ge1300 (172.16.1.6)  0.518 ms  0.441 ms  0.433 ms
5  3  skopie-so0100 (172.16.1.2)  0.540 ms  0.443 ms  0.464 ms
6  4  livorno-so0000 (172.16.1.34)  0.490 ms  0.492 ms  0.457 ms
7  5  barcelona (192.168.1.7)  0.726 ms  0.618 ms  0.617 ms
```

The intention of the migration in this case is to merge both non-backbone areas with the help of an OSPF multi-area adjacency (both in Areas 0.0.0.0 and 0.0.0.3) between router *Havana* and router *Livorno* as in Figure 2.8. In this way, standard routing policies from OSPF, where intra-area paths are preferred against inter-area paths, are enforced to take that particular link despite its maximum metric, preferring that path to the lower cost via router *Skopie* and router *Nantes*.

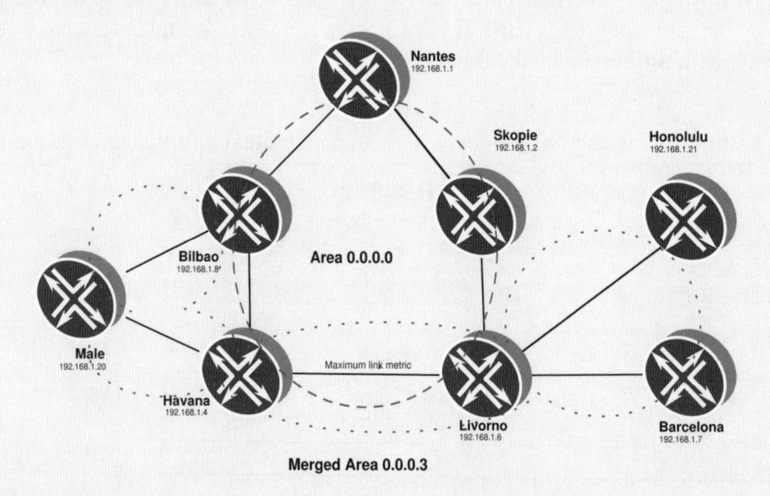

Figure 2.8: Network topology merging Area 0.0.0.3 with multi-area adjacency between Havana and Livorno.

Multi-area adjacencies are configured in Junos OS by adding the `secondary` statement to the link in the desired OSPF area. With this tag, the router attempts to establish a parallel OSPF adjacency over the same point-to-point link with a different area ID. However, other link parameters and attributes, such as timers and authentication, can be configured differently. Listing 2.6 shows the configuration stanzas on router *Havana* and router *Livorno* for this purpose, with the activation knobs on Lines 19 and 41.

Listing 2.6: OSPF multi-area configuration for link between Havana and Livorno

```
1 user@Livorno> show configuration protocols ospf
2 traffic-engineering;
```

```
3  area 0.0.0.0 {
4      interface so-0/0/1.0 {
5          metric 65535;
6      }
7      interface so-0/0/2.0;
8      interface fe-0/3/0.0;
9      interface lo0.0 {
10         passive;
11     }
12     interface so-0/0/0.0;
13 }
14 area 0.0.0.3 {
15     interface ge-0/2/0.0;
16     interface ge-1/3/0.0;
17     interface so-0/0/1.0 {
18         metric 65535;
19         secondary;         # Knob for multi-area adjacency
20     }
21 }
22 user@Havana> show configuration protocols ospf
23 traffic-engineering;
24 area 0.0.0.0 {
25     interface so-0/1/0.0;
26     interface at-0/2/0.1001;
27     interface so-1/0/0.0 {
28         metric 65535;
29     }
30     interface so-1/0/1.0;
31     interface so-1/0/2.0;
32     interface so-1/0/3.0;
33     interface lo0.0 {
34         passive;
35     }
36 }
37 area 0.0.0.3 {
38     interface ge-1/1/0.0;
39     interface so-1/0/0.0 {
40         metric 65535;
41         secondary;         # Knob for multi-area adjacency
42     }
43 }
```

Once this new adjacency is transparently established (the remaining one in Area 0.0.0.0 is not affected), both router *Male* and router *Barcelona* share the same OSPF database, as displayed in Listing 2.7. It is worth mentioning that three ABRs between Area 0.0.0.0 and Area 0.0.0.3 exist at this point and therefore Type 3 Summary LSAs are injected to Area 0.0.0.3 at three different points. For instance, Lines 18, 19, and 20, respectively, show how router *Nantes* is visible via Type 3 Summary LSAs injected at these three ABRs: router *Havana*, router *Livorno*, and router *Bilbao*.

Listing 2.7: OSPF databases in Male and Barcelona after secondary link activation

```
1  user@male-re0> show ospf database | no-more
2
3      OSPF database, Area 0.0.0.3
4   Type       ID               Adv Rtr          Seq         Age  Opt  Cksum  Len
5  Router   192.168.1.4     192.168.1.4      0x80000014    73  0x22 0xbf9a  48
6  Router   192.168.1.6     192.168.1.6      0x80000012    70  0x22 0x4913  60
7  Router   192.168.1.7     192.168.1.7      0x8000000e   879  0x22 0x6335 120
8  Router   192.168.1.8     192.168.1.8      0x8000000f   681  0x22 0x4c54  48
9  Router  *192.168.1.20    192.168.1.20     0x80000017   917  0x22 0x7c52  72
10 Router   192.168.1.21    192.168.1.21     0x8000000e   964  0x22 0x7af5  48
11 Network  172.16.1.53     192.168.1.7      0x80000009   492  0x22 0x11e2  32
12 Network *172.16.1.85     192.168.1.20     0x80000002   106  0x22 0xe3d5  32
```

```
13 Network  172.16.1.89     192.168.1.6      0x80000009   148  0x22 0x724e  32
14 Summary  172.16.1.4      192.168.1.4      0x8000000f    76  0x22 0x8975  28 # Havana ABR
15 Summary  172.16.1.4      192.168.1.6      0x80000013    74  0x22 0x916f  28 # Livorno ABR
16 Summary  172.16.1.4      192.168.1.8      0x8000000e   311  0x22 0x6f90  28 # Bilbao ABR
17 <...>
18 Summary  192.168.1.1     192.168.1.4      0x8000000d    76  0x22 0x8dca  28 # Havana ABR
19 Summary  192.168.1.1     192.168.1.6      0x8000000a    74  0x22 0x91c6  28 # Livorno ABR
20 Summary  192.168.1.1     192.168.1.8      0x80000001  1179  0x22 0x7bdb  28 # Bilbao ABR
21 Summary  192.168.1.2     192.168.1.4      0x8000000d    76  0x22 0x83d3  28
22 Summary  192.168.1.2     192.168.1.6      0x80000009    74  0x22 0x7dda  28
23 Summary  192.168.1.2     192.168.1.8      0x80000001  1179  0x22 0x7bd9  28
24 <...>
25 OpaqArea 1.0.0.1         192.168.1.4      0x800000010   77  0x22 0xb698  28
26 OpaqArea 1.0.0.1         192.168.1.6      0x8000000d    74  0x22 0xbe8c  28
27 OpaqArea 1.0.0.1         192.168.1.7      0x8000000a   879  0x22 0xc883  28
28 OpaqArea 1.0.0.1         192.168.1.8      0x8000000b   672  0x22 0xca7e  28
29 OpaqArea*1.0.0.1         192.168.1.20     0x8000000b   917  0x22 0xfc35  28
30 OpaqArea 1.0.0.1         192.168.1.21     0x8000000a   964  0x22 0x12f   28
31 <...>
```

At this time, both routers see their respective loopback addresses via Router LSAs as they are now part of the same merged area, as shown in Listing 2.8.

Listing 2.8: Barcelona loopback address present at Male via Router LSA

```
1  user@male-re0> show ospf database extensive lsa-id 192.168.1.7
2
3      OSPF database, Area 0.0.0.3
4   Type       ID               Adv Rtr          Seq          Age  Opt  Cksum  Len
5  Router   192.168.1.7      192.168.1.7      0x8000000e   362  0x22 0xae56  72
6    bits 0x2, link count 4
7    id 172.16.1.105, data 172.16.1.106, Type Transit (2)
8      Topology count: 0, Default metric: 100
9    id 172.16.1.54, data 172.16.1.53, Type Transit (2)
10     Topology count: 0, Default metric: 100
11   id 192.168.1.7, data 255.255.255.255, Type Stub (3) # Barcelona's loopback
12     Topology count: 0, Default metric: 0
13   Aging timer 00:53:57
14   Installed 00:05:55 ago, expires in 00:53:58, sent 00:05:53 ago
15   Last changed 00:09:33 ago, Change count: 1
```

For end-to-end connectivity, this intra-area path is selected, independent of any metric values of alternate paths in the backbone, as shown in Listing 2.9.

Listing 2.9: Traceroute from Male to Barcelona with merged Area

```
1 user@male-re0> traceroute 192.168.1.7 source 192.168.1.20
2 traceroute to 192.168.1.7 (192.168.1.7) from 192.168.1.20, 30 hops max, 40 byte packets
3  1  havana-ge1100 (172.16.1.86)  0.540 ms  0.422 ms  0.432 ms
4  2  livorno-so0010 (172.16.1.38)  0.475 ms  0.440 ms  0.427 ms
5  3  barcelona (192.168.1.7)  0.708 ms  0.597 ms  0.595 ms
```

Once a `secondary` adjacency is established, the Router LSA for the particular area does not include any stub network information to represent the link addressing, so as to avoid possible overlaps. That Router LSA in the new area includes only a *point-to-point* network type referring to the *ifIndex* (as if it were an unnumbered link) to represent this additional binding inside the network.

Junos OS enforces the requirement that, at most, only one link definition is defined as the *primary* (there could be no primaries at all), but the configuration can have as many `secondary` adjacencies as desired.

Adding a `secondary` adjacency in Junos OS has no direct impact or side effects on the existing *primary* relationship. The *primary* adjacency remains in FULL state while building up new `secondary` adjacencies.

Junos Tip: Unnumbered point-to-point interfaces in OSPFv2

Unnumbered point-to-point interfaces appear in each Type 1 Router LSA with its corresponding *point-to-point* network type for established adjacencies and in the case of Junos OS, referring to *ifIndex* for the respective interface. Link addressing is actually signified by *stub* network types in Router LSAs.

This behavior has been extended to different LSA types in OSPFv3, where link-addressing information is kept in different data structures from system reachability, similar to IS–IS.

Network engineers tend to define addresses on backbone point-to-point interfaces for various reasons, including the following:

- operational purposes, such as an easy link identification in a traceroute or ping;
- TE purposes with RSVP-TE, such as explicit link addition as part of an RSVP;
- management purposes, such as restricting connections from an address range.

Notice that protocol extensions to traceroute and RSVP are being discussed to facilitate deployment of unnumbered links in the network.

Apart from these applications, a point-to-point-based network could run perfectly on unnumbered interfaces with unique references in each *point-to-point* network type entry in Type 1 Router LSAs.

When transitioning from *primary* to `secondary` or vice versa, OSPF neighbors need to be fully renegotiated and adjacencies need to be established. Listings 2.10 and 2.11 show the Router LSAs from router *Livorno* in each area before the *primary* adjacency is moved from Area 0.0.0.0 to Area 0.0.0.3 (achieved with a single commit and simultaneous OSPF adjacency formation with new data). Note the specific reference to the *point-to-point* network as either the *ifIndex* or stub data in Lines 14 and 36, and the injection of the link addressing via stub network in Line 16 in Listing 2.10 before the migration, and how this is represented on Lines 14, 34, and 36 at Listing 2.11 after the migration.

Listing 2.10: Router LSAs from Livorno with primary adjacency in Area 0.0.0.0

```
1  user@Livorno> show ospf database router lsa-id 192.168.1.6 extensive
2
3     OSPF database, Area 0.0.0.0
4   Type       ID               Adv Rtr           Seq      Age  Opt  Cksum  Len
5  Router   *192.168.1.6     192.168.1.6      0x800006a9   512  0x22 0x7fe9 108
6    bits 0x1, link count 7
7  <...>
8    id 192.168.1.6, data 255.255.255.255, Type Stub (3)
9      Topology count: 0, Default metric: 0
10   id 192.168.1.2, data 172.16.1.34, Type PointToPoint (1)
11     Topology count: 0, Default metric: 643
12   id 172.16.1.32, data 255.255.255.252, Type Stub (3)
```

```
13      Topology count: 0, Default metric: 643
14    id 192.168.1.4, data 172.16.1.38, Type PointToPoint (1) # P2P to Havana
15      Topology count: 0, Default metric: 65535 # Maximum metric
16    id 172.16.1.36, data 255.255.255.252, Type Stub (3) # Link addresses to Havana
17      Topology count: 0, Default metric: 65535 # Maximum metric
18    Topology default (ID 0)
19      Type: PointToPoint, Node ID: 192.168.1.4
20        Metric: 65535, Bidirectional # Maximum metric
21      Type: PointToPoint, Node ID: 192.168.1.2
22        Metric: 643, Bidirectional
23    Gen timer 00:40:14
24    Aging timer 00:51:28
25    Installed 00:08:32 ago, expires in 00:51:28, sent 00:08:30 ago
26    Last changed 00:08:37 ago, Change count: 31, Ours
27
28      OSPF database, Area 0.0.0.3
29   Type       ID               Adv Rtr           Seq      Age  Opt  Cksum  Len
30  Router  *192.168.1.6     192.168.1.6      0x80000014   518  0x22 0xe80    60
31    bits 0x1, link count 3
32    id 172.16.1.54, data 172.16.1.54, Type Transit (2)
33      Topology count: 0, Default metric: 100
34    id 172.16.1.89, data 172.16.1.89, Type Transit (2)
35      Topology count: 0, Default metric: 100
36    id 192.168.1.4, data 0.0.0.70, Type PointToPoint (1) # P2P to Havana
37      Topology count: 0, Default metric: 65535 # Maximum metric
38    Topology default (ID 0)
39      Type: PointToPoint, Node ID: 192.168.1.4
40        Metric: 65535, Bidirectional # Maximum metric
41      Type: Transit, Node ID: 172.16.1.89
42        Metric: 100, Bidirectional
43      Type: Transit, Node ID: 172.16.1.54
44        Metric: 100, Bidirectional
45    Gen timer 00:32:30
46    Aging timer 00:51:22
47    Installed 00:08:38 ago, expires in 00:51:22, sent 00:08:36 ago
48    Last changed 00:26:17 ago, Change count: 5, Ours
```

Listing 2.11: Router LSAs from Livorno with primary adjacency in Area 0.0.0.3

```
1   user@Livorno> show ospf database router lsa-id 192.168.1.6 extensive
2
3       OSPF database, Area 0.0.0.0
4    Type       ID               Adv Rtr           Seq      Age  Opt  Cksum  Len
5   Router  *192.168.1.6     192.168.1.6      0x800006ad    18  0x22 0xba37   96
6     bits 0x1, link count 6
7   <...>
8     id 192.168.1.6, data 255.255.255.255, Type Stub (3)
9       Topology count: 0, Default metric: 0
10    id 192.168.1.2, data 172.16.1.34, Type PointToPoint (1)
11      Topology count: 0, Default metric: 643
12    id 172.16.1.32, data 255.255.255.252, Type Stub (3)
13      Topology count: 0, Default metric: 643
14    id 192.168.1.4, data 0.0.0.70, Type PointToPoint (1) # P2P to Havana
15      Topology count: 0, Default metric: 65535  # Maximum metric
16    Topology default (ID 0)
17      Type: PointToPoint, Node ID: 192.168.1.4
18        Metric: 65535, Bidirectional # Maximum metric
19      Type: PointToPoint, Node ID: 192.168.1.2
20        Metric: 643, Bidirectional
21    Gen timer 00:43:28
22    Aging timer 00:59:41
23    Installed 00:00:18 ago, expires in 00:59:42, sent 00:00:18 ago
24    Last changed 00:00:18 ago, Change count: 35, Ours
25
26      OSPF database, Area 0.0.0.3
27   Type       ID               Adv Rtr           Seq      Age  Opt  Cksum  Len
28  Router  *192.168.1.6     192.168.1.6      0x80000018    18  0x22 0xd429   72
```

```
29    bits 0x1, link count 4
30    id 172.16.1.54, data 172.16.1.54, Type Transit (2)
31      Topology count: 0, Default metric: 100
32    id 172.16.1.89, data 172.16.1.89, Type Transit (2)
33      Topology count: 0, Default metric: 100
34    id 192.168.1.4, data 172.16.1.38, Type PointToPoint (1) # P2P to Havana
35      Topology count: 0, Default metric: 65535 # Maximum metric
36    id 172.16.1.36, data 255.255.255.252, Type Stub (3) # Link addresses to Havana
37      Topology count: 0, Default metric: 65535 # Maximum metric
38    Topology default (ID 0)
39      Type: PointToPoint, Node ID: 192.168.1.4
40        Metric: 65535, Bidirectional # Maximum metric
41      Type: Transit, Node ID: 172.16.1.89
42        Metric: 100, Bidirectional
43      Type: Transit, Node ID: 172.16.1.54
44        Metric: 100, Bidirectional
45    Gen timer 00:44:21
46    Aging timer 00:59:41
47    Installed 00:00:18 ago, expires in 00:59:42, sent 00:00:18 ago
48    Last changed 00:00:18 ago, Change count: 9, Ours
```

Another interesting factor to consider in multi-area adjacencies is that Opaque LSAs are generated with the same TE attributes for the binding in all active areas, thus enabling a similar view of the link in each TE database (TED). Listing 2.12 displays the TE Opaque LSAs for the link at each area after a multi-area adjacency has been established. For instance, the same *admin-group* is included in them as pointed out in Lines 27 and 61 in each area.

Listing 2.12: Opaque LSAs from Livorno for multi-area adjacency link

```
1  user@Livorno> show ospf database opaque-area area 0.0.0.0 extensive
2      advertising-router 192.168.1.6 lsa-id 1.0.0.5
3
4      OSPF database, Area 0.0.0.0
5   Type       ID               Adv Rtr          Seq      Age  Opt  Cksum  Len
6  OpaqArea*1.0.0.5          192.168.1.6      0x80000001   425  0x22 0xe38c 128
7    Area-opaque TE LSA
8    Link (2), length 104:
9      Linktype (1), length 1:
10       1
11     LinkID (2), length 4:
12       192.168.1.4
13     RemIfAdr (4), length 4:
14       172.16.1.37
15     TEMetric (5), length 4:
16       65535  # Maximum metric
17     MaxBW (6), length 4:
18       155.52Mbps
19     MaxRsvBW (7), length 4:
20       155.52Mbps
21     UnRsvBW (8), length 32:
22         Priority 0, 155.52Mbps
23 <...>
24     LinkLocalRemoteIdentifier (11), length 8:
25       Local 70, Remote 0
26     Color (9), length 4:
27       8 # Admin-group
28   Gen timer 00:29:38
29   Aging timer 00:52:54
30   Installed 00:07:05 ago, expires in 00:52:55, sent 00:07:05 ago
31   Last changed 00:07:05 ago, Change count: 1, Ours, TE Link ID: 2147483657
32
33 user@Livorno> show ospf database opaque-area area 0.0.0.3 extensive
34     advertising-router 192.168.1.6 lsa-id 1.0.0.6
35
36     OSPF database, Area 0.0.0.3
```

```
37 Type          ID                 Adv Rtr          Seq        Age  Opt  Cksum  Len
38 OpaqArea*1.0.0.6           192.168.1.6      0x80000001  498  0x22 0xa0d3 136
39   Area-opaque TE LSA
40   Link (2), length 112:
41     Linktype (1), length 1:
42       1
43     LinkID (2), length 4:
44       192.168.1.4
45     LocIfAdr (3), length 4:
46       172.16.1.38
47     RemIfAdr (4), length 4:
48       172.16.1.37
49     TEMetric (5), length 4:
50       65535 # Maximum metric
51     MaxBW (6), length 4:
52       155.52Mbps
53     MaxRsvBW (7), length 4:
54       155.52Mbps
55     UnRsvBW (8), length 32:
56         Priority 0, 155.52Mbps
57 <...>
58     LinkLocalRemoteIdentifier (11), length 8:
59       Local 70, Remote 0
60     Color (9), length 4:
61       8      # Admin-group
62   Gen timer 00:29:18
63   Aging timer 00:51:41
64   Installed 00:08:18 ago, expires in 00:51:42, sent 00:08:18 ago
65   Last changed 00:08:18 ago, Change count: 1, Ours, TE Link ID: 2147483657
```

Despite its fundamental restriction to be applicable only for *point-to-point* link types and the implicit adjacency reset when switching the *primary* and *secondary* roles, using *secondary* adjacencies still provides a powerful migration tool when compared with additional logical interfaces over the existing physical connection or other mechanisms like virtual links:

- The establishment of *secondary* adjacencies is entirely graceful (this may not always be the case with the definition of new logical interfaces, if the existing connection needs to be redefined) and allows OSPF area merging in an entirely non-intrusive fashion.

- TE admin-groups smoothly propagate across areas without further configuration.

Adjacency transition in IS–IS: IS–IS defines area membership not on a per-interface basis, but rather on a per-system basis.

Depending on whether Level 1, Level 2, or both Levels 1 and 2 are active on a logical interface and depending on whether area membership is determined by the configured Net-IDs in the router, IS–IS adjacencies are established either over a specific area or over the Level 2 backbone.

Level 1 Area migration As previously mentioned, a given system can be configured to belong to more than a single Level 1 area for migration purposes and this behavior was already so conceived in the original protocol standards. In that case, the router performs transparent flooding between interconnected IS–IS Level 1 areas; that is to say, the system performs stitching among different area scopes and areas are merged together at that point even though their configuration is completely different.

Application Note: IS–IS additional areas for seamless migrations

Figure 2.9 depicts an IS–IS implementation on top of our standard topology (Figure 1) with the same IS–IS area configured in different parts of the network and in which some systems already belong to more than one area.

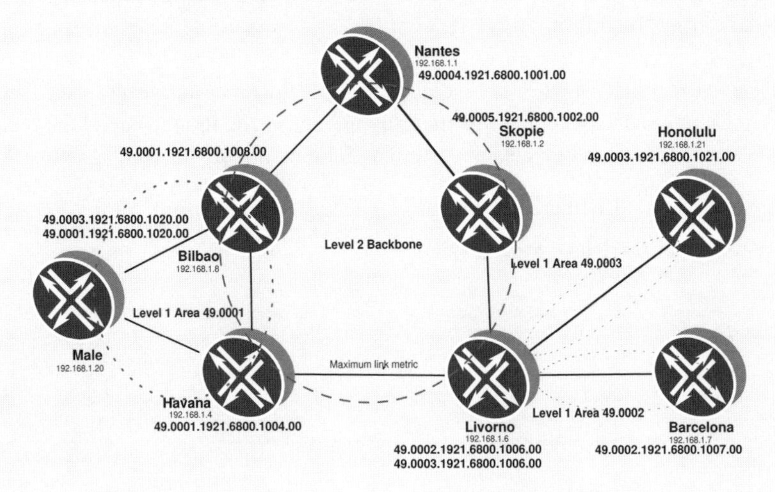

Figure 2.9: Network topology with duplicated Area 49.0003 and multi-area assignment on Livorno and Male.

In this topology, router *Livorno* acts as an L1L2 router and at the same time has two Level 1 adjacencies in different areas. The configuration to achieve this is as simple as including different Net-IDs in both areas, as illustrated in Listing 2.13. When using separate areas, it is important to maintain consistent System ID assignment across the network in order to avoid Net-ID duplication issues that could lead to network failures. Note that there is a certain limit on the maximum number of areas that can be configured, as explained later in this section.

Listing 2.13: Different Net-IDs in Livorno with same System ID and different Area

```
1 user@Livorno> show configuration interfaces lo0.0 family iso
2 address 49.0002.1921.6800.1006.00;
3 address 49.0003.1921.6800.1006.00;
```

Junos Tip: Consequences of duplicate Net-IDs

The consequences of duplicate Net-IDs could be disastrous in a modern network. Just imagine an accidental purging of LSPs for the duplicate system. Removing information from the link-state topology database involves regenerating the LSP with a checksum and a sequence number of zero to indicate that the LSP is a purge. The Net-ID composes the identifier of the LSP, so a *purge storm* ensues when one intermediate system attempts to purge while the other regenerates its information.

However, because of the default lack of transparency between layers, it is possible simultaneously to have duplicate Net-IDs for internal L1 systems with the same configured area but in different physical areas of the network. This can occur because network information is populated by L1L2 systems and Net-IDs are lost at that point.

Merging areas together can uncover a long-standing duplicate ID situation that is only discovered during the migration!

Router *Livorno* has both Level 1 adjacencies and in practice merges both scopes, as depicted in Listing 2.14. Note that this merging happens even though each LSP includes its respective area TLVs to indicate each system's membership, as shown in Listing 2.15.

Listing 2.14: Merged L1 Database at Livorno

```
1 user@Livorno> show isis database level 1
2 IS-IS level 1 link-state database:
3 LSP ID                      Sequence Checksum Lifetime Attributes
4 Livorno.00-00                  0x13d    0xcad6      488 L1 L2 Attached
5 Livorno.02-00                  0x137    0x7a51      502 L1 L2
6 honolulu-re0.00-00      0x148    0x50b       843 L1
7 honolulu-re0.02-00      0x135    0x563b      843 L1
8 Barcelona-re0.00-00     0x14a    0x4ae4      649 L1
9   5 LSPs
```

Listing 2.15: Different Area TLVs at merged L1 Database

```
1 user@Livorno> show isis database extensive level 1 | match "Area|-00 "
2 Livorno.00-00 Sequence: 0x13f, Checksum: 0xc6d8, Lifetime: 1102 secs
3     Packet type: 18, Packet version: 1, Max area: 0
4     Area address: 49.0002 (3)
5     Area address: 49.0003 (3)
6 Livorno.02-00 Sequence: 0x139, Checksum: 0x7653, Lifetime: 1102 secs
7     Packet type: 18, Packet version: 1, Max area: 0
8 honolulu-re0.00-00 Sequence: 0x149, Checksum: 0x30c, Lifetime: 563 secs
9     Packet type: 18, Packet version: 1, Max area: 0
10    Area address: 49.0003 (3)
11 honolulu-re0.02-00 Sequence: 0x136, Checksum: 0x543c, Lifetime: 563 secs
12    Packet type: 18, Packet version: 1, Max area: 0
13 Barcelona-re0.00-00 Sequence: 0x14c, Checksum: 0x46e6, Lifetime: 1165 secs
14    Packet type: 18, Packet version: 1, Max area: 0
15    Area address: 49.0001 (3)
16    Area address: 49.0002 (3)
```

Junos Tip: Suppressing Pseudonode on point-to-point Ethernet segments

A revision of the IS–IS Database shown in Listing 2.14 reveals in Lines 5 and 7 LSPs injected by *Pseudonodes* for each LAN segment.

[RFC5309] proposes a mechanism to avoid generating Pseudonode LSPs when the LAN media between Intermediate Systems is actually a point-to-point circuit. The approach described in [RFC5309] is based on considering such media as an actual point-to-point link for computation and IS–IS PDU processing purposes.

In fact, the point-to-point IIH PDU is encapsulated in an Ethernet frame and no Designated Intermediate System (DIS) is elected for this segment. This implementation has a clear scaling benefit as no superfluous protocol resources are consumed for a simple back-to-back link, regardless of the physical media.

The Junos OS `point-to-point` for each IS–IS interface activates the [RFC5309] implementation for the referred Ethernet media, as shown in Listing 2.16.

Listing 2.16: Configuration of an Ethernet segment as point-to-point in router *Livorno*

```
1 [edit protocols isis]
2 user@Livorno# set interface ge-1/3/0.0 point-to-point
```

Note that configuring widespread Ethernet interfaces as point-to-point is a necessary step to enable them as unnumbered interfaces.

As expected, this de facto merge remains completely transparent for the TED. TE Sub-TLVs are flooded and processed correctly, and the topology is constructed for TE applications as if it were a single area, as shown in Listing 2.17.

Listing 2.17: TED at Livorno for systems in merged L1 Area

```
1  user@Livorno> show ted database detail
2  <...>
3  NodeID: honolulu-re0.00(192.168.1.21)
4    Type: Rtr, Age: 162 secs, LinkIn: 1, LinkOut: 1
5    Protocol: IS-IS(1)
6      To: honolulu-re0.02, Local: 172.16.1.90, Remote: 0.0.0.0
7        Local interface index: 92, Remote interface index: 0
8  NodeID: honolulu-re0.02
9    Type: Net, Age: 161 secs, LinkIn: 2, LinkOut: 2
10   Protocol: IS-IS(1)
11     To: Livorno.00(192.168.1.6), Local: 0.0.0.0, Remote: 0.0.0.0
12       Local interface index: 0, Remote interface index: 0
13     To: honolulu-re0.00(192.168.1.21), Local: 0.0.0.0, Remote: 0.0.0.0
14       Local interface index: 0, Remote interface index: 0
15 NodeID: Barcelona-re0.00(192.168.1.7)
16   Type: Rtr, Age: 441 secs, LinkIn: 1, LinkOut: 1
17   Protocol: IS-IS(1)
18     To: Livorno.02, Local: 172.16.1.53, Remote: 0.0.0.0
19       Local interface index: 82, Remote interface index: 0
```

In such an IS–IS setup, there is no direct visibility among isolated L1 Areas by default. Internal L1 routers achieve connectivity to the rest of the network by installing default routes based on the presence of the ATTached bit from L1L2 routers' LSPs, as indicated in Listing 2.18.

Listing 2.18: Reachability details between L1 Internal routers in different areas

```
1  user@male-re0> show isis database | no-more
2  IS-IS level 1 link-state database:
3  LSP ID                          Sequence Checksum Lifetime Attributes
4  Havana.00-00                       0x2fd   0x2e6c      581 L1 L2 Attached # L1L2 system
5  Havana.02-00                         0x2   0x588e      581 L1 L2
6  Bilbao-re0.00-00                   0x233   0xeeb8      547 L1 L2
7  male-re0.00-00          0x6    0x7386      587 L1
8    4 LSPs
9
10 IS-IS level 2 link-state database:
11   0 LSPs
12
13 user@honolulu-re0> show isis database | no-more
14 IS-IS level 1 link-state database:
15 LSP ID                          Sequence Checksum Lifetime Attributes
16 Livorno.00-00                       0x7f   0xae25      699 L1 L2 Attached # L1L2 system
```

```
17 Livorno.02-00                    0x7c   0xf294        699 L1 L2
18 honolulu-re0.00-00      0x3e4   0x4131       1093 L1
19 honolulu-re0.02-00       0x7c   0xca80        759 L1
20 Barcelona-re0.00-00      0x6a4   0x14bb        710 L1
21   5 LSPs
22
23 IS-IS level 2 link-state database:
24   0 LSPs
25
26 user@male-re0> show isis database extensive | match 192.168.1.21
27
28 user@male-re0> show route 192.168.1.21
29
30 inet.0: 8 destinations, 8 routes (8 active, 0 holddown, 0 hidden)
31 + = Active Route, - = Last Active, * = Both
32
33 0.0.0.0/0          *[IS-IS/15] 00:31:08, metric 10
34                     > to 172.16.1.86 via ge-4/2/4.0
35
36 user@honolulu-re0> show isis database extensive | match 192.168.1.20
37
38 user@honolulu-re0> show route 192.168.1.20
39
40 inet.0: 15 destinations, 15 routes (15 active, 0 holddown, 0 hidden)
41 + = Active Route, - = Last Active, * = Both
42
43 0.0.0.0/0          *[IS-IS/15] 03:31:47, metric 10
44                     > to 172.16.1.89 via ge-5/2/0.0
45
46 user@male-re0> traceroute 192.168.1.21 source 192.168.1.20
47 traceroute to 192.168.1.21 (192.168.1.21) from 192.168.1.20, 30 hops max, 40 byte packets
48  1  havana-ge1100 (172.16.1.86)  1.547 ms  1.090 ms  1.056 ms
49  2  livorno-so0010 (172.16.1.38)  0.470 ms  0.483 ms  0.442 ms
50  3  honolulu (192.168.1.21)  0.495 ms  0.451 ms  0.436 ms
51
52 user@honolulu-re0> traceroute 192.168.1.20 source 192.168.1.21
53 traceroute to 192.168.1.20 (192.168.1.20) from 192.168.1.21, 30 hops max, 40 byte packets
54  1  livorno-ge1300 (172.16.1.89)  0.498 ms  0.431 ms  0.409 ms
55  2  havana-so1000 (172.16.1.37)  0.480 ms  0.453 ms  0.444 ms
56  3  male (192.168.1.20)  0.513 ms  0.497 ms  0.465 ms
```

It must be noted that same area IDs are used in both router *Male* and router *Honolulu*, despite the fact that these systems are located at different sites in the topology. In practice, these two systems behave as if they were different areas, and using the same area IDs causes no connectivity problems.

At this point, the intention is to merge both Level 1 islands in the network by activating an additional Level 1 adjacency between router *Havana* and router *Livorno*. Because a common area ID is compulsory for this purpose, this goal can be easily achieved by activating Level 1 adjacencies over the common link and configuring a joint area ID at one of the two ends (one should be enough), in this case, *49.0002* on router *Havana*. Figure 2.10 illustrates these changes in this IS–IS setup with the Level 1 area migration.

There is a particular side effect of activating additional Level 1 or Level 2 adjacencies over existing ones. Whereas in broadcast link types, IS–IS considers different PDU types for Level 1 and Level 2 Hellos (namely, IIH PDU Types 15 and 16), only a single shared Hello PDU type is conceived for point-to-point circuits (namely, IIH PDU Type 17). This situation is the result of a bandwidth consumption concern in the original [ISO10589] specification, which was based on the traditional lowest-speed links. As a consequence, while Level 1 and Level 2 topologies can be constructed differently over LAN media, hello parameters are shared in non-broadcast links.

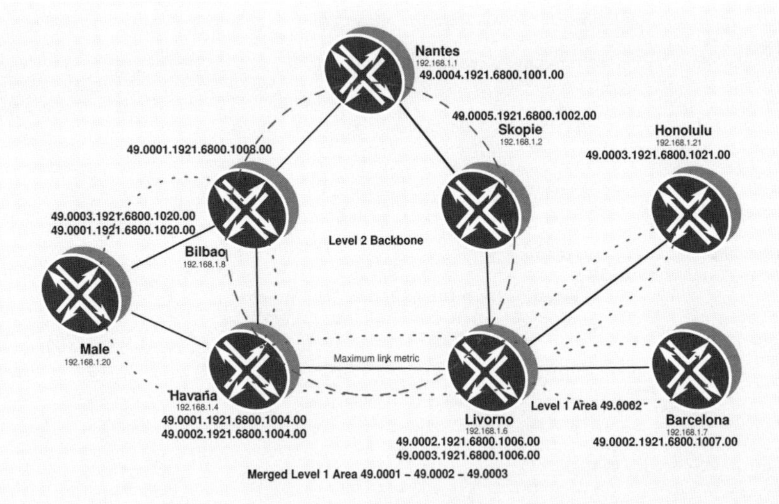

Figure 2.10: Network topology with merged L1 Areas by activating 49.0002 in Havana.

TLV 10 Authentication is attached to a single point-to-point IIH to be shared across Level 1 and Level 2 adjacencies, meaning that authentication details such as clear-text passwords, keys, and activation times need to be common for both levels. This behavior is inherent to the protocol behavior and Junos OS and other routing operating systems provide enough options to circumvent this situation with per-adjacency or hello-only authentication configuration knobs. Later sections explain this behavior in detail.

In any case, activating the additional Level 1 adjacency between router *Havana* and router *Livorno* is entirely graceful, as has already been discussed, and can be as simple as adding Net-IDs with matching areas and enabling Level 1 over the joint link, as shown in Listings 2.19 and 2.20.

Listing 2.19: Configuration changes to activate L1 adjacencies between router *Havana* and router *Livorno*

```
1 [edit interfaces lo0 unit 0 family iso]
2 user@Havana# set address 49.0002.1921.6800.1004.00 # NET-ID in common Area
3 [edit protocols isis]
4 user@Havana# delete interface so-1/0/0.0 level 1 disable
5
6 [edit interfaces lo0 unit 0 family iso]
7 user@Livorno# set address 49.0001.1921.6800.1006.00 # NET-ID in common Area
8 [edit protocols isis]
9 user@Livorno# delete interface so-0/0/1.0 level 1 disable
```

Listing 2.20: Activation of L1 adjacencies between router *Havana* and router *Livorno*

```
1 user@Livorno> show isis adjacency detail Havana
2 Havana
3   Interface: so-0/0/1.0, Level: 3, State: Up, Expires in 20 secs # Level:3 as L1 + L2 adjacencies
4   Priority: 0, Up/Down transitions: 3, Last transition: 00:02:12 ago
5   Circuit type: 3, Speaks: IP, IPv6
6   Topologies: Unicast
```

```
 7    Restart capable: Yes, Adjacency advertisement: Advertise
 8    IP addresses: 172.16.1.37
 9  user@Havana> show isis adjacency detail Livorno
10  Livorno
11    Interface: so-1/0/0.0, Level: 3, State: Up, Expires in 23 secs # Level:3 as L1 + L2 adjacencies
12    Priority: 0, Up/Down transitions: 3, Last transition: 00:02:45 ago
13    Circuit type: 3, Speaks: IP, IPv6
14    Topologies: Unicast
15    Restart capable: Yes, Adjacency advertisement: Advertise
16    IP addresses: 172.16.1.38
17
18  user@Havana> show isis database
19  IS-IS level 1 link-state database:    # Joint IS-IS Level 1 Database
20  LSP ID                      Sequence Checksum Lifetime Attributes
21  Havana.00-00                   0x306   0xa47b     1020 L1 L2 Attached
22  Havana.02-00                     0x8   0x4c94      958 L1 L2
23  Livorno.00-00                   0x87   0xab4b     1025 L1 L2 Attached
24  Livorno.02-00                   0x82   0xe69a     1025 L1 L2
25  Bilbao-re0.00-00               0x238   0xe4bd      826 L1 L2
26  male-re0-00-00         0xb    0x698b       799 L1
27  honolulu-re0-00-00     0x3e8   0x3935       627 L1
28  honolulu-re0-02-00      0x81   0xc085      1133 L1
29  Barcelona-re0-00-00     0x6a8    0xcbf       472 L1
30    9 LSPs
31
32  IS-IS level 2 link-state database:
33  LSP ID                      Sequence Checksum Lifetime Attributes
34  Torino.00-00                  0x16b2   0x386b      381 L1 L2
35  Lille.00-00                    0x38c   0x357a      994 L1 L2
36  Inverness.00-00               0x252e   0x3b50      774 L1 L2
37  Basel.00-00                   0x1555   0x2414      666 L1 L2
38  Nantes.00-00                   0x719   0xe445      742 L1 L2
39  Skopie.00-00                   0x73a   0x72ad     1018 L1 L2
40  Havana.00-00                   0x72d   0xe517     1036 L1 L2
41  Livorno.00-00                   0x8d   0x4d4c     1029 L1 L2
42    8 LSPs
```

Taking the same previous link MPLS configuration from the Application Note 2.1.2 into account, in which a particular `admin-group` was marked on this circuit, it is also worth highlighting that TE information is uniform across levels, as seen in Lines 11 and 20 from Listing 2.21.

Listing 2.21: Propagation of TE details across levels

```
 1  user@Livorno> show isis database extensive Livorno
 2      | match "interface|groups|level|-00 "
 3  IS-IS level 1 link-state database:
 4  Livorno.00-00 Sequence: 0x87, Checksum: 0xab4b, Lifetime: 815 secs
 5      Remaining lifetime: 815 secs, Level: 1, Interface: 0
 6        Local interface index: 74, Remote interface index: 0
 7        Administrative groups:  0 <none>
 8        Local interface index: 81, Remote interface index: 0
 9        Administrative groups:  0 <none>
10        Local interface index: 70, Remote interface index: 72
11        Administrative groups:  0x8 yellow  # Admin group for ifindex 70
12  Livorno.02-00 Sequence: 0x82, Checksum: 0xe69a, Lifetime: 815 secs
13      Remaining lifetime: 815 secs, Level: 1, Interface: 0
14  IS-IS level 2 link-state database:
15  Livorno.00-00 Sequence: 0x8d, Checksum: 0x4d4c, Lifetime: 819 secs
16      Remaining lifetime: 819 secs, Level: 2, Interface: 0
17        Local interface index: 69, Remote interface index: 69
18        Administrative groups:  0 <none>
19        Local interface index: 70, Remote interface index: 72
20        Administrative groups:  0x8 yellow  # Admin group for ifindex 70
```

After this area merge, remote routers in each of the topology sites can now see each other as part of the Level 1 infrastructure. They share common TED information and there is no need for specific route-leaking at L1L2 routers, as shown by a representative example in Listing 2.22 for router *Barcelona*.

Listing 2.22: Level 1 area merge affect on router *Barcelona*

```
1  user@Barcelona-re0> show route 192.168.1.20
2
3  inet.0: 21 destinations, 21 routes (21 active, 0 holddown, 0 hidden)
4  + = Active Route, - = Last Active, * = Both
5
6  192.168.1.20/32     *[IS-IS/15] 00:08:06, metric 30
7                       > to 172.16.1.54 via ge-1/2/0.0
8
9  user@Barcelona-re0> traceroute 192.168.1.20 source 192.168.1.7
10 traceroute to 192.168.1.20 (192.168.1.20) from 192.168.1.7, 30 hops max, 40 byte packets
11  1  livorno-ge020 (172.16.1.54)  0.760 ms  0.711 ms  0.616 ms
12  2  havana-so100100 (172.16.1.37)  1.685 ms  1.352 ms  1.353 ms
13  3  male (192.168.1.20)  0.681 ms  0.721 ms  0.634 ms
14
15
16 user@Barcelona-re0> show isis database | no-more
17 IS-IS level 1 link-state database:
18 LSP ID                      Sequence Checksum Lifetime Attributes
19 Havana.00-00                   0x306   0xa47b      613 L1 L2 Attached
20 Havana.02-00                     0x8   0x4c94      550 L1 L2
21 Livorno.00-00                   0x87   0xab4b      621 L1 L2 Attached
22 Livorno.02-00                   0x82   0xe69a      621 L1 L2
23 Bilbao-re0.00-00               0x239   0xe2be     1157 L1 L2
24 male-re0-00-00          0xb    0x698b       391 L1
25 honolulu-re0-00-00      0x3e9    0x3736      1007 L1
26 honolulu-re0-02-00       0x81    0xc085       729 L1
27 Barcelona-re0-00-00      0x6a9    0xac0        845 L1
28   9 LSPs
29
30 IS-IS level 2 link-state database:
31   0 LSPs
```

By default, Junos OS allows a maximum of *three* different Net-IDs to be configured for each intermediate system, therefore allowing up to *three* separate area assignments. However, the Junos OS statement `max-areas` directly under the `[edit protocols isis]` stanza allows this default number to be expanded to up to 36 different areas that can be configured through different Net-IDs. The Maximum Area Addresses field included in each IS–IS PDU type accurately reflects the configured value, and the presence of Area TLVs in each LSP is limited by this configuration parameter. Note that the Maximum Area Addresses field defaults to 0, thereby indicating support for up to three areas simultaneously, so there is no need explicitly to set this value lower than 3. If more areas are explicitly configured in diverse Net-IDs without the `max-areas` configuration option, only the lowest numerical IDs will be present in the Area TLV.

For instance, in the previous topology from Figure 2.10, router *Livorno* could also act as L1L2 router for more areas than those present in the setup. Listings 2.23 and 2.24 show a situation with five simultaneous Areas. This expansion is seen both in the Maximum Area Addresses field as shown in Lines 5 and 17, and in the number of Area IDs present in the Area TLV, as shown in Lines 6 and 18, respectively, for each IS–IS level.

Listing 2.23: Configuration to expand up to five Level 1 Areas in Livorno

```
1  [edit interfaces lo0 unit 0 family iso]
2  user@Livorno# set address 49.0004.1921.6800.1006.00
3  [edit interfaces lo0 unit 0 family iso]
4  user@Livorno# set address 49.0005.1921.6800.1006.00
5  [edit interfaces lo0 unit 0 family iso]
6  user@Livorno# show
7  address 49.0002.1921.6800.1006.00;
8  address 49.0003.1921.6800.1006.00;
9  address 49.0001.1921.6800.1006.00;
10 address 49.0004.1921.6800.1006.00;
11 address 49.0005.1921.6800.1006.00;
12 [edit protocols isis]
13 user@Livorno# set max-areas 5
```

Listing 2.24: Expansion to five Level 1 Areas in Livorno

```
1  user@Livorno> show isis database extensive Livorno | match "Area|level|-00 "
2  IS-IS level 1 link-state database:
3  Livorno.00-00 Sequence: 0x8a, Checksum: 0xf851, Lifetime: 1196 secs
4      Remaining lifetime: 1196 secs, Level: 1, Interface: 0
5      Packet type: 18, Packet version: 1, Max area: 5 # Maximum Areas field
6      Area address: 49.0001 (3) # 5 Area IDs
7      Area address: 49.0002 (3)
8      Area address: 49.0003 (3)
9      Area address: 49.0004 (3)
10     Area address: 49.0005 (3)
11 Livorno.02-00 Sequence: 0x84, Checksum: 0xe29c, Lifetime: 1196 secs
12     Remaining lifetime: 1196 secs, Level: 1, Interface: 0
13     Packet type: 18, Packet version: 1, Max area: 5
14 IS-IS level 2 link-state database:
15 Livorno.00-00 Sequence: 0x91, Checksum: 0x9358, Lifetime: 1196 secs
16     Remaining lifetime: 1196 secs, Level: 2, Interface: 0
17     Packet type: 20, Packet version: 1, Max area: 5 # Maximum Areas field
18     Area address: 49.0001 (3) # 5 Area IDs
19     Area address: 49.0002 (3)
20     Area address: 49.0003 (3)
21     Area address: 49.0004 (3)
22     Area address: 49.0005 (3)
```

2.1.3 Resources for link-state IGP hierarchical migrations

The previous standard mechanisms are complemented in Junos OS with specific features that can leverage migration activities across IGP hierarchies.

Default route management

A vital factor to consider in migrations is topology reachability. Successful planning should ensure that participating systems have the same or equivalent network visibility after the migration as they did before, unless explicitly intended otherwise. In this sense, *visibility* is not constrained to route propagation via an IGP or BGP, but in modern networks it also includes the population of MPLS label bindings and participation in multicast groups, among others.

In any case, having a default route is always an outstanding option. It can be tremendously useful to ensure network connectivity when restricting a system's vision of the network, but at the same time it can become enormously dangerous by attracting unexpected traffic, especially if it is used as a vehicle to amplify a *denial-of-service* attack or a potential *routing or forwarding loop*. Network designers and operators have defined diverse best practices with regards to default route utilization in their networks, depending on the particular topology, customer exposure, and especially, lessons learned in the past.

When propagating default routes in a network, all these factors need to be considered. While default routes can be tremendously helpful to alleviate routing requirements and can also be constrained to areas or system subsets, they can also open the door to potential problems. When designing a network and considering default route utilization, it is recommended to think twice before making use of these routes, especially in regard to their originating sites and their propagation across hierarchies.

An important factor to consider is the number of destinations that need to be matched by a certain default route. If such destinations need to be served by MPLS applications, the additional challenge of creating a supporting label binding comes in to play. Injecting more label bindings than are strictly needed is not really considered good practice and associating a FEC to a default route is probably the worst-case scenario for this. Some of the latest proposals, such as relaxing exact IGP route matching for label binding acceptance and installation as per [RFC5283], deploying resources for expanded RSVP TE across hierarchies and domains as per [RFC5151], and flexibly controlling label distribution via BGP as per [RFC3107], expose resources to select representative MPLS labels for those destinations, without necessarily needing to bind an MPLS label to a default route.

Several transition stages or temporary scenarios exist where a default route is required and must be carefully managed; these are discussed next.

Default route management in OSPFv2 areas: Apart from allowing redistribution of a default route in an OSPF network as a Type 5 AS External or Type 7 NSSA AS External LSA type, [RFC2328] already defines the so-called *stub* areas.

While information in AS External LSAs has a network-wide scope (even when originated in a specific area as NSSA LSAs), *stub* areas reduce the number of entries in the link-state database in participating systems by removing all AS external information and injecting a default route at the ABRs towards the rest of the area members in the form of a Type 3 Summary LSA. The Type 3 Summary LSA is flooded only inside this area and not beyond. This mechanism is useful in certain environments where routers cannot keep up with the requirements of a complete link-state database visibility or simply do not need it.

Traditional *stub* areas are constrained by configuration in all participating routers that is signaled by means of the E bit in the OSPF Options field. All routers facing a *stub* area must have the E bit cleared in their Hello packets towards it.

Most routing operating systems have gone beyond this original idea allowing the definition of so-called *totally stubby* areas. Although not strictly standardized, the concept still remains valid: why should an internal router in such a network need to know route information outside the area if default route Type 3 Summary LSAs are flooded from the exit points? That is to say, in many setups it is debatable whether specific routes are really needed not only outside the AS but also outside the area. For those networks, this additional constraint on *stub* areas can be a better fit. Most operating systems still offer the possibility to inject default route information in the same way via Type 3 Summary LSAs in such totally stubby areas.

Junos OS has specific configuration knobs to inject such LSAs in those *stub* and *totally stubby* areas, such as `default-metric` to define specific metric values for those LSAs as indicated in Listing 2.25.

Listing 2.25: Configuration to define a metric for a default-route Type 3 Summary LSA for a stub area in Livorno

```
1  [edit protocols ospf]
2  user@Livorno# set area 0.0.0.3 stub default-metric ?
3  Possible completions:
4    <default-metric>     Metric for the default route in this stub area (1..16777215)
```

In this construct, Junos OS does not strictly require a default route to exist in the corresponding routing table for the Type 3 Summary LSA to be generated, which allows greater flexibility to advertise exit points to a stub network. Also, Junos OS performs *active backbone detection* checks for the advertisement. Even though interfaces can be defined inside the backbone area, this configuration option is effective only when an adjacency in Area 0 is properly established, as per [RFC3509]. Therefore, Junos OS offers additional intelligence to this protocol function so that no impact should be expected if an ABR becomes isolated from the backbone or if this knob is accidentally misconfigured in an internal router in the stub area.

NSSAs also impose such a limitation in a specific kind of hierarchy but with the additional degree of freedom that they are able to inject external information inside such an area. Similarly, routing information can be constrained because only dedicated exit points from the NSSA are really needed. This limitation can be further applied to Type 3 Summary LSAs because of the same reason. These areas tend to be called *totally stubby NSSAs* because these are areas in which route-hiding conditions are enforced according to the basic NSSA profile without injecting specific route information from other areas. In fact, [RFC3101] defines default route injection in NSSAs as a Type 3 or a Type 7 LSA; Type 3 is the default in Junos OS, and Type 7 is configurable with the command shown in Listing 2.26. Listing 2.27 illustrates the output for a Type 7 LSA.

Listing 2.26: Configuration to inject a default route for a NSSA in router *Livorno* as a Type 3 or a Type 7 LSA

```
1  [edit protocols ospf]
2  user@Livorno# set area 0.0.0.3 nssa default-lsa default-metric 10 metric-type 2 type-7
```

Listing 2.27: Type 7 LSA for a default route in a NSSA

```
1  user@Livorno> show ospf database area 0.0.0.3 nssa lsa-id 0.0.0.0 extensive
2
3      OSPF link state database, Area 0.0.0.3
4   Type       ID               Adv Rtr           Seq      Age  Opt  Cksum  Len
5  NSSA    *0.0.0.0          192.168.1.6      0x8000006f   943  0x20 0x5376  36
6    mask 0.0.0.0
7    Type 2, TOS 0x0, metric 10, fwd addr 0.0.0.0, tag 0.0.0.0 # Type 2 external metric, not LSA Type
8    Gen timer 00:32:37
9    Aging timer 00:44:17
10   Installed 00:15:43 ago, expires in 00:44:17, sent 1d 13:54:22 ago
11   Last changed 2d 22:26:20 ago, Change count: 1, Ours
```

Because the NSSA function is similar to stub areas, Junos OS does not require a default route to exist explicitly in the ABR's routing table. Junos OS also performs *active backbone detection* with a live adjacency in the backbone for the LSA to be generated.

Default route management in IS–IS areas: By default, IS–IS L1L2 routers set the ATTached bit in their LSPs inside their connected Level 1 areas.

This method makes internal L1 routers without any backbone visibility aware about which other intermediate systems in their area are *attached* to the core and thus have visibility on the rest of the network. No specific default route IPv4 prefix is included in a TLV in an LSP, because this is the standard IS–IS mechanism to indicate *backbone attachment*. Thus, default route installation is decided locally in every internal L1 system and no explicit default route is redistributed in the hierarchy.

The *backbone attachment* concept in IS–IS is interpreted differently by different vendors. Junos OS understands that it not only needs to have an active L2 adjacency in the backbone, but also that it needs to be able to reach external areas via the IS–IS Level 2 CSPF computation, that is, via an area not present in the local Level 1 database.

Because of the requirement for greater flexibility, this original concept soon became obsolete. Most routing systems started to offer adequate machinery to relax this simplistic connectivity option and allow prefix redistribution and leaking between layers, including the internal and external consideration. Junos OS offers enough policy resources arbitrarily to leak internal and external prefixes from Level 2 to Level 1, and vice versa.

This flexibility, provided in Junos OS policy language, can easily empower intended redistribution guidelines without modifying the value of the ATTached bit and can thus install a default route in all internal L1 intermediate systems. The Junos OS option `ignore-attached-bit` prevents installation of the default route in such systems by ignoring its presence in LSPs in the database. Listing 2.28 reviews the IS–IS Level 1 database from router *Barcelona* in our previous setup, and Listing 2.29 shows its installation in the main routing instance table pointing to the system whose LSP is tagged with the ATTached bit.

Listing 2.28: Level 1 IS–IS database in Barcelona

```
1 user@Barcelona-re0> show isis database level 1
2 IS-IS level 1 link-state database:
3 LSP ID                    Sequence Checksum Lifetime Attributes
4 Livorno.00-00                0x5de   0x6bc9      870 L1 L2 Attached
5 Livorno.02-00                0x58b   0xc5ad      865 L1 L2
6 honolulu-re0.00-00     0x8d0   0xa3d3       892 L1 L2
7 honolulu-re0.02-00     0x563   0xef6d       877 L1 L2
8 Barcelona-re0.00-00     0xba4   0x7753      1189 L1
9   5 LSPs
```

Listing 2.29: Default route in Barcelona following L1L2 ATTached bit

```
1 user@Barcelona-re0> show route 0/0 exact
2 inet.0: 19 destinations, 19 routes (15 active, 0 holddown, 4 hidden)
3 + = Active Route, - = Last Active, * = Both
4
5 0.0.0.0/0          *[IS-IS/15] 00:00:02, metric 10
6                     > to 172.16.1.54 via ge-1/2/0.0
```

When the `ignore-attached-bit` command is configured, as shown in Listing 2.30, the IS–IS database remains the same but the system does not install the default route even though there may be LSPs present with the ATTached bit tagged. Note that this local decision made by IS–IS increases the risk of blackholing, so this exception must be treated carefully. Imagine an L1 system with this knob performing an IP lookup in the middle of the path between another internal L1 system and the L1L2 natural exit point. All flows selecting the default route derived from IS–IS as a destination from the internal system will be dropped in the intermediate router if an IP lookup is performed. In the end, this behavior is a clear contradiction to the standard definition of the protocol.

Listing 2.30: Configuration to ignore ATTached bit in router *Barcelona*

```
1 [edit protocols isis]
2 user@Barcelona-re0# set ignore-attached-bit
```

Manipulation of the ATTached bit can also be controlled globally on L1L2 routers with the Junos OS `suppress-attached-bit` feature, as shown in Listing 2.31. Listing 2.32 shows a snapshot from the database in the situation in which router *Livorno*, being an L1L2 intermediate system, is not setting the ATTached bit as per the standard protocol rules mandate, as seen in Line 4.

Listing 2.31: Configuration to suppress ATTached bit marking to L1 DBs in router *Livorno*

```
1 [edit protocols isis]
2 user@Livorno# set suppress-attached-bit
```

Listing 2.32: Level 1 IS–IS database without ATTached bit tagging in Livorno's LSP

```
1 user@Livorno> show isis database level 1
2 IS-IS level 1 link-state database:
3 LSP ID                    Sequence Checksum Lifetime Attributes
4 Livorno.00-00                0x5e1   0x5ddc     1196 L1 L2   # ATTached bit not set
5 Livorno.02-00                0x58e   0xbfb0     1196 L1 L2
6 honolulu-re0.00-00      0x8d4   0x91eb     1108 L1
7 honolulu-re0.02-00      0x567   0xe575     1108 L1
8 Barcelona-re0.00-00      0xba8   0x6f57     1150 L1
9   5 LSPs
```

Although this setting has a global effect and route manipulation is not granular then for each L1 system, its main advantage is a consistent effect across the area, which shares a common database. Also, risks derived from local lookup decisions are minimized, as compared to using the `ignore-attached-bit` knob. The question becomes one of global versus local behavior enforcement.

Figure 2.11 shows the utilization of different configuration options in Junos OS depending on design decisions.

Route manipulation across hierarchies

Dividing a common domain into different layers usually provides the benefit of summarization at boundaries. A remote router sometimes does not strictly need to acquire the complete routing domain topology information, but instead it is sufficient simply to have aggregates heading to its natural exit points. On the other hand, some specific routing information may need to be leaked to this router, for example, to optimize routing paths or consistently to understand label bindings distributed by LDP.

Junos OS offers a wide variety of configuration resources, both for direct interpretation or inside policy-specific language, to manage route-leaking across hierarchies in all link-state routing protocols.

Route manipulation across OSPF hierarchies: Basic network-summary or inter-area-prefix LSA summarization in stub areas or NSSAs can be performed in Junos OS with the `area-range` knob.

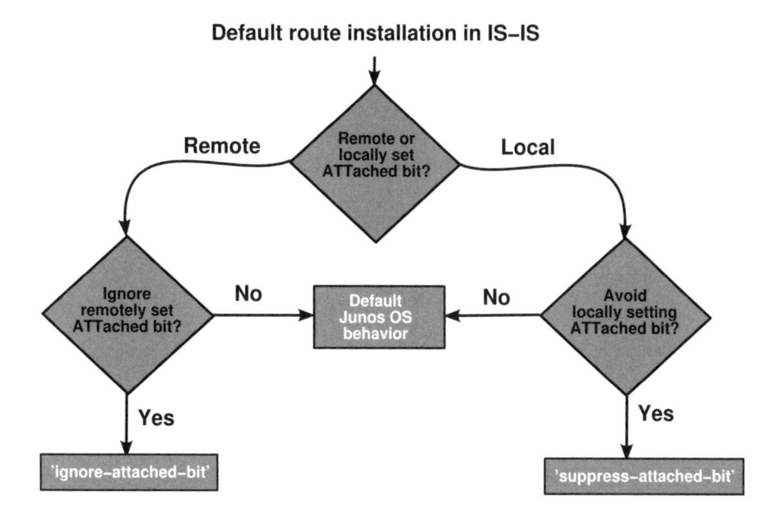

Figure 2.11: IS–IS default route management depending on ATTached bit.

This functionality, illustrated in Listing 2.33, allows aggregation of internal area route information from Type 1, Type 2, or Intra-Area-Prefix LSAs when generating summaries for the backbone and other areas.

Listing 2.33: OSPFv2 and OSPFv3 area-range configuration for Summary and Inter-Area-Prefix LSAs in the backbone

```
1 [edit protocols ospf]
2 user@Livorno# set area 0.0.0.3 area-range ?
3 Possible completions:
4   <area_range>          Range to summarize routes in this area
5 [edit protocols ospf3]
6 user@Livorno# set area 0.0.0.3 area-range 192.168.200.0/24
```

Note that there is also an interesting derivative of this command in NSSAs. As ABRs in NSSAs are in charge of transforming Type 7 LSAs into Type 5, or Type 0x2007 into Type 0x4005 LSAs for the rest of the network, these boundaries also represent a realistic summarization option and Junos OS offers the adequate resource for that by specifying the `area-range` command inside the NSSA, as per Listing 2.34.

Listing 2.34: OSPFv2 and OSPFv3 NSSA area-range configuration for AS External LSAs in the backbone

```
1 [edit protocols ospf]
2 user@Livorno# set area 0.0.0.3 nssa area-range ?
3 Possible completions:
4   <area_range>          Range to summarize NSSA routes in this area
5 [edit protocols ospf3]
6 user@Livorno# set area 0.0.0.3 nssa area-range 192.168.200.0/24
```

In both cases, an administrator can enforce exact matching for this summary by adding the `exact` knob after the area-range aggregate definition. Another available option is to define a certain metric for the summary instead of deriving it from more specific routes by means of the `override-metric` command under `area-range`.

With summarization, it is possible to block at these boundaries all more specific routes to be propagated outside an area. The `restrict` statement can be added to each `area-range` command specifically to signify the opposite intention: block more specific prefixes from being inspected and transformed into Summary, Inter-Area-Prefix, or AS External LSAs for the rest of the network as shown in Listing 2.35.

Listing 2.35: OSPFv2 and OSPFv3 area-range restrict configuration both for standard Summary, Inter-Area-Prefix, and NSSA AS External LSAs generation at ABRs in the backbone

```
1  [edit protocols ospf]
2  user@Livorno# set area 0.0.0.3 area-range 192.168.200.0/24 restrict
3  [edit protocols ospf]
4  user@Livorno# set area 0.0.0.3 nssa area-range 192.168.200.0/24 restrict
5  [edit protocols ospf3]
6  user@Livorno# set area 0.0.0.3 area-range 192.168.200.0/24 restrict
7  [edit protocols ospf3]
8  user@Livorno# set area 0.0.0.3 nssa area-range 192.168.200.0/24 restrict
```

When dealing with LSA propagation across several NSSAs, one particular scenario needs some clarification. When a router running Junos OS is simultaneously acting as ABR for several NSSAs and ASBR to inject external information, the corresponding Type 7 NSSA AS External LSAs or Type 0x2007 NSSA AS External LSAs are injected into each NSSA (on top of existing default summary LSA). The Junos OS `no-nssa-abr` knob allows further reduction of information flooded to each of these NSSAs by limiting the propagation of Type 7 or Type 0x2007 LSAs that correspond to a specific external route redistributed at the ABR for the NSSAs, instead of simply allowing the standard default route population. The `no-nssa-abr` configuration option is global per protocol instance and has been implemented for both for OSPFv2 and OSPFv3, as shown in Listing 2.36.

Listing 2.36: OSPFv2 and OSPFv3 NSSA AS External LSA blocking for self-redistributed routes at ABR to several NSSAs

```
1  [edit protocols ospf]
2  user@Livorno# set no-nssa-abr
3  [edit protocols ospf3]
4  user@Livorno# set no-nssa-abr
```

More granularity can be achieved by making use of specific import and export policies at ABRs to manage the origination and distribution of Network Summary and Inter-Area-Prefix LSAs.

With `network-summary` or `inter-area-prefix` *export* policies, a user can control which LSAs are flooded into an area, including considering those that result from an `area-range`, with a default *reject* action. Listing 2.37 includes configuration snippets for this concept.

Listing 2.37: OSPFv2 and OSPFv3 network-summary or inter-area-prefix export policies for IPv4 prefixes at ABRs in the backbone

```
1  [edit protocols ospf]
2  user@Livorno# set area 0.0.0.3 network-summary-export block-192-168-200-0
3  [edit protocols ospf3]
4  user@Livorno# set realm ipv4-unicast area 0.0.0.3 inter-area-prefix-export
```

```
 5      block-192-168-200-0
 6  [edit policy-options policy-statement block-192-168-200-0]
 7  user@Livorno# show
 8  term block-LSA {
 9      from {
10          route-filter 192.168.200.0/24 exact;
11      }
12      then reject;
13  }
14  term default {
15      then accept;
16  }
```

With `network-summary` or `inter-area-prefix` *import* policies, a user can manage which LSAs learned from an area are used to contribute to the generation of Network Summary and Inter-Area-Prefix LSAs to other areas. If a prefix is rejected by such a policy, Junos OS still adds a route in the respective table, but it neither generates Network Summary or Inter-Area-Prefix LSAs though, nor does it use the route to activate a covering `area-range`. At a minimum, a more specific non-rejected prefix to issue an LSA must exist for an `area-range` command to become active. Similar to other import policies in other stanzas, these policies default to an *accept* action, as indicated in Listing 2.38.

Listing 2.38: OSPFv2 and OSPFv3 network-summary or inter-area-prefix import policies for IPv4 prefixes at ABRs in the backbone

```
 1  [edit protocols ospf]
 2  user@Livorno# set area 0.0.0.3 network-summary-import block-192-168-250-0
 3
 4  [edit protocols ospf3]
 5  user@Livorno# set realm ipv4-unicast area 0.0.0.3 inter-area-prefix-import
 6      block-192-168-250-0
 7
 8  [edit policy-options policy-statement block-192-168-250-0]
 9  user@Livorno# show
10  term block-LSA {
11      from {
12          route-filter 192.168.250.0/24 exact;
13      }
14      then reject;
15  }
```

It is worth remarking here that the Junos OS policy language for such inter-area control policies is not affected by the use of the existing `from area` or `to area` knobs. Policy conditions should match the route information existing on LSAs, but area-related policy conditions are designed to match route information for redistribution to other protocols once the route becomes active in the routing table. These area-related policies do not affect how LSA information is interpreted.

In addition to manipulating LSA transformation across hierarchies, Junos OS also allows route installation for all types of AS external LSAs to be blocked. AS external LSAs are bound to the reachability of the advertising router; they are not a cornerstone of the underlying link-state protocol topology, but rather, provide external information anchored at redistribution points. The Junos OS generic `import` acts on this type of information to reject or allow route installation in the routing table. Note that such `import` policies allow filtering based on exact prefixes, external route types, interfaces, metrics, and tags, but they have a global protocol-wide scope, as depicted in Listing 2.39.

Listing 2.39: OSPFv2 and OSPFv3 import policies for IPv4 prefixes from AS external LSAs

```
1  [edit protocols ospf]
2  user@Livorno# set import block-192-168-150-0
3
4  [edit protocols ospf3]
5  user@Livorno# set realm ipv4-unicast import block-192-168-150-0
6
7  [edit policy-options policy-statement block-192-168-150-0]
8  user@Livorno# show
9  term block-LSA {
10     from {
11         route-filter 192.168.150.0/24 exact;
12     }
13     then reject;
14 }
```

These policies do not lead to any kind of LSA manipulation, but just prevent specific routes, present in the link-state database, from being installed in the routing table. Such policies do not allow any specific route-handling, but instead re-evaluate the installation of the route in the routing table by considering the policy guidelines and then convey a simple *accept* or *reject* action.

Furthermore, these `import` policies apply only to AS external LSAs, which differ from what a user would expect based on `import` policies from other protocols. The rationale for this behavior is that information from internal LSA types is a substantial part of the underlying IGP topology. Filtering internal LSAs fundamentally violates the principles of a link-state routing protocol and would establish the ideal conditions for creating routing loops.

Similarly, because AS external LSAs have domain-wide scope, it is futile to use any `from area` or `to area` conditions inside any `import` policy for route installation, even if the router is just attached to a single area, since LSA information inspected for route installation is not characterized by any OSPF area at all.

Table 2.3 summarizes the available resources for LSA manipulation across OSPF hierarchies.

Figure 2.12 includes a flowchart with design decisions based on available resources in Junos OS for route manipulation.

Route manipulation across IS–IS hierarchies By default, IS–IS has some specific rules with regard to route manipulation across hierarchies.

[RFC1195] defined a primary difference between *internal* routes (those belonging to the same routing domain) and *external* routes (those injected from outside the domain), setting different TLVs and route priorities for them, such that internal route information takes precedence over external route information. [RFC1195] also defined the mechanism for default route installation to be based on detection of the ATTached bit set by L1L2 routers at level boundaries and following the shortest path to such area exit points.

[RFC2966] leveraged this constraint to allow external routes in a given L1 area and to allow arbitrary route-leaking from L1 to L2 areas. This original concept has been traditionally compared to *totally-stubby* OSPF areas, or even *totally-stubby* NSSAs, when allowing external L1 routes inside the area. This distinction between internal and external routes led to the following default route-leaking behavior:

- Internal L1 routes are propagated to L2 backbone.
- External L1 routes are not propagated to L2 backbone.

Table 2.3 Resources available in Junos for route manipulation across OSPF hierarchies

Junos standalone resources	Effect
area-range	Network summary and Inter-area-prefix LSA summarization in stub or NSSA areas
area-range exact	Exact matching enforced for route aggregation
area-range restrict	Restrict prefix from summarization action and population in backbone and other areas
import	Route installation control in OSPFv2 and OSPFv3 AS external LSAs
inter-area-prefix-export	Granular control in OSPFv3 of LSA generation into areas by means of policies
inter-area-prefix-import	Granular control in OSPFv3 of LSAs eligible for translation into inter-area-prefix LSAs by means of policies
network-summary-export	Granular control in OSPFv2 of LSA generation into areas by means of policies
network-summary-import	Granular control in OSPFv2 of LSAs eligible for translation into Network Summary LSAs by means of policies
no-nssa-abr	Further LSA reduction when being simultaneously ABR and ASBR for NSSAs
nssa area-range	Aggregation when transforming Type 7 LSAs into Type 5, or Type 0x2007 into Type 0x4005 LSAs

- Internal L2 routes are not propagated to L1 Area.
- External L2 routes are not propagated to L1 Area.

However, this traditional classification was soon made obsolete. Modern networks needed even more route manipulation flexibility and metric space was scarce with the original old-style TLVs, which had a maximum value of 63. [RFC3784] defined extensions for Traffic-Engineering (TE) and in the process added two modern TLVs to cope with the TLV length requirements, namely TLV 22 Extended IS Reachability and TLV 135 Extended IP Reachability.

In addition to removing the restriction on the length of the metric field, [RFC3784] deprecated the differentiation between internal and external metrics. For practical reasons, configuring *wide metrics* also leverages the internal–external distinction for prefixes injected into a given system. With no external routes, the default propagation rules are simplified:

- Wide-metrics L1 routes are propagated to L2 backbone.
- Wide-metrics L2 routes are not propagated to L1 Area.

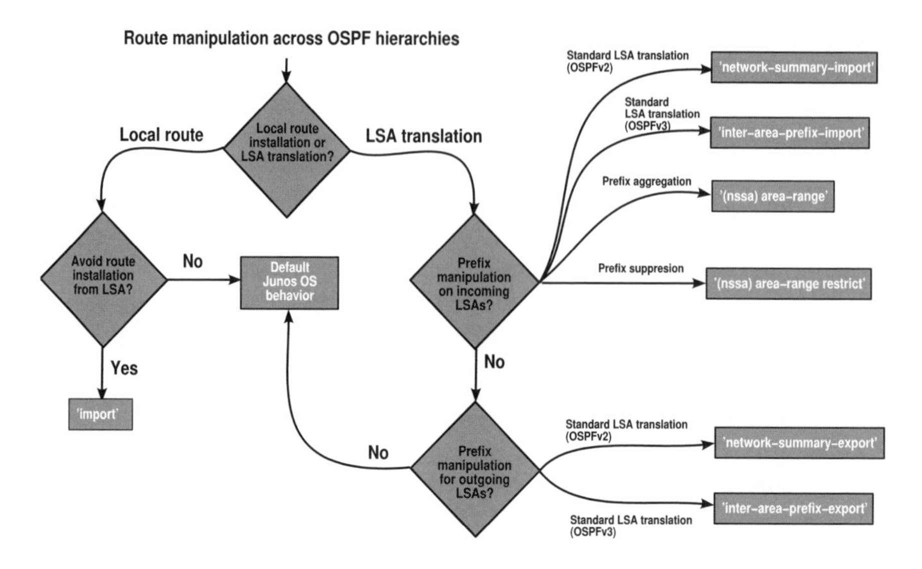

Figure 2.12: OSPF route manipulation across hierarchies.

While keeping all this in mind, an administrator also needs to take into account support for the particular metric style used in a network. A *transitional* stage may be needed, in which systems inject both old-style and new-style TLVs, so that systems can understand either *narrow metrics* or *wide metrics*.

Junos OS accepts both styles, and by default, advertises both styles as well. Sending and accepting both styles is a realistic scenario for a transitional model. If this is not the desired behavior, the `wide-metrics-only` command prevents the generation of old-style TLVs and issues only new-style TLVs on a per-level basis. Note that including this keyword not only has the expected effect of expanding the metric size, but it can also indirectly modify routes leaked to L2 backbone by removing the external distinction.

As compared to OSPF, Junos OS has not implemented specific knobs for route summarization across levels in IS–IS precisely because leaking can be controlled flexibly by policies. From this perspective, the use of Junos OS tailored *aggregate* or *generate* routes can perfectly fit the needs for summarization, being created upon only when more specific prefixes exist, and being perfectly eligible for policy matching conditions.

Junos OS policy language also allows definition of specific policies to be used for route-leaking between levels and in both directions. Specific keywords such as `from level` and `to level` are provided to match criteria precisely in a given policy. Note that according to the definition of *narrow metrics*, routes installed at each level are placed into substantially different databases (except for default internal route-leaking from Level 1 to Level 2, which injects information in both). Additional route population can be defined with fine-grained control to tailor additional inter-level leaking. The example policy in Listing 2.40 illustrates a very simplistic policy to inject all active IS–IS Level 2 prefixes that match host loopback addresses into directly attached Level 1 areas. Such a policy would just need to be added to the standard IS–IS export policy chain. Similar policies could be defined to overcome standard route population limitations across levels.

Listing 2.40: Using level policy keyword to leak IS–IS level 2 routes into level 1 areas

```
 1  [edit policy-options policy-statement Leak-L2-to-L1]
 2 user@Livorno# show
 3 term leak-L2 {
 4     from {
 5         level 2;
 6         route-filter 0.0.0.0/0 prefix-length-range /32-/32;
 7     }
 8     to level 1;
 9     then accept;
10 }
```

Table 2.4 summarizes these available resources for route manipulation across IS–IS hierarchies.

Table 2.4 Resources available in Junos for route manipulation across IS–IS hierarchies

Junos standalone resources	Effect
from level	Policy matching criterion on routes to be exported from a certain level
to level	Policy matching criterion on routes to be exported to a certain level
wide-metrics-only	Enforce metric expansion with newer style TLVs and remove internal/external distinction

Figure 2.13 describes the utilization of Junos OS configuration options related with route manipulation across IS–IS hierarchies.

Point-to-point setups for merging areas

A traditional approach for artificially linking distant network segments has been to establish virtual constructs that are mapped over the existing physical topology, so as to allow indirect stitching of areas or sites. Fairly common benefits of using such resources in link-state routing-protocol-based networks are routing policy enforcement, because intra-area routes are always preferred to inter-area routes.

Such point-to-point virtual setups are usually discouraged by many network designers. Every time a routing topology differs from the physical link topology, operational and troubleshooting challenges and potential divergence issues certainly increase. While these setups can sometimes be the best fit for a transitional period or migration step, they are usually not considered the basis of a good design.

OSPF virtual-links: [RFC2328] initially defined *virtual links* as a method for stitching together a partitioned backbone.

Such unnumbered point-to-point links can be established between backbone routers using a non-backbone area for transport, thus expanding Area 0 beyond its natural boundaries. Routing information for *virtual links* is included in the Type 1 Router LSAs, which carry a

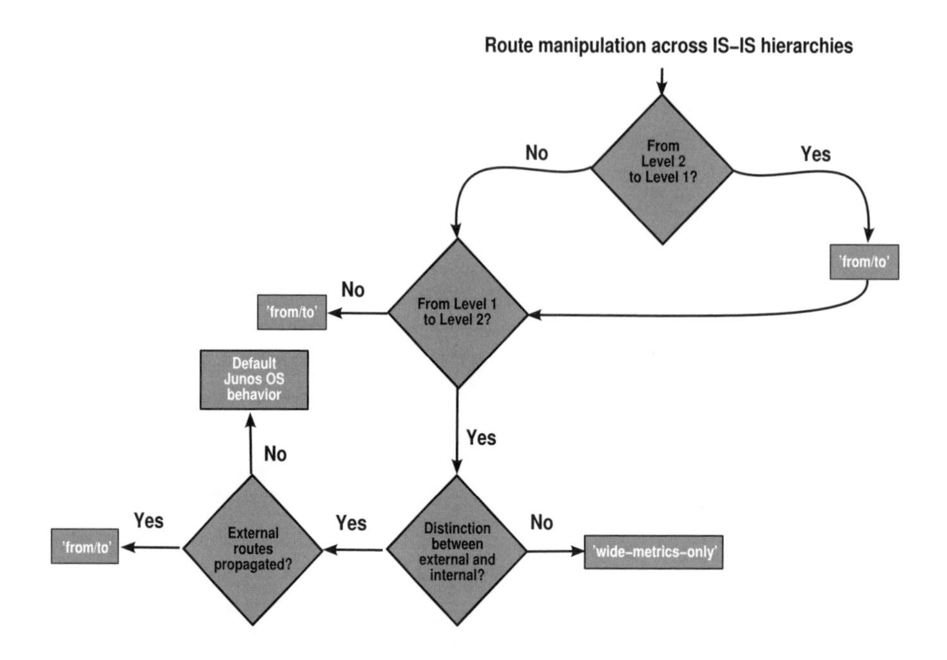

Figure 2.13: IS–IS route manipulation across hierarchies.

metric derived from the intra-area distance between the two routers. LSAs that are normally flooded through the complete AS are never flooded over these virtual adjacencies, being redundant paths in this case because the information also arrives over physical links.

OSPFv3 has followed the same concept since its original definition in [RFC2740]. Neighbor identifiers for the *virtual links* must be site-local or global scope IPv6 addresses to enforce adequate significance and uniqueness in the network.

Junos OS allows configuration of `virtual-links` to repair backbone area partitioning over a certain *transit* area, which, by definition, cannot be stub because it is acting as the transit substrate for those virtual adjacencies. It is worth remarking that these `virtual-links` are therefore configured as part of the Area 0 stanza as shown in Listing 2.41. The configuration includes the transit area in its definition, which is required for the *virtual-link* to work properly. Also, the configuration needs to be set on both ends of the link.

Listing 2.41: OSPFv2 and OSPFv3 definition for virtual-links

```
1  [edit protocols ospf]
2  user@Livorno# set area 0 virtual-link neighbor-id 192.168.1.21 transit-area 0.0.0.3
3  [edit protocols ospf3]
4  user@Livorno# set area 0 virtual-link neighbor-id 192.168.1.21 transit-area 0.0.0.3
```

When establishing *virtual-links*, each router sets the V bit in the Type 1 Router LSA or Type 0x2001 Router LSA for the respective transit areas. After neighbors are detected, routers inject an additional Virtual Link type in the backbone area Router LSA, as shown in Line 15 in Listing 2.42.

Listing 2.42: OSPF Router LSA including Virtual Links

```
user@Livorno> show ospf database extensive router lsa-id 192.168.1.6 area 0

    OSPF link state database, Area 0.0.0.0
 Type       ID               Adv Rtr          Seq      Age  Opt  Cksum  Len
Router   *192.168.1.6      192.168.1.6      0x80000961  219  0x22 0x4595 132
  bits 0x1, link count 9
<...>
  id 192.168.1.6, data 255.255.255.255, Type Stub (3)
  TOS count 0, TOS 0 metric 0
<...>
  id 192.168.1.4, data 172.16.1.38, Type PointToPoint (1)
  TOS count 0, TOS 0 metric 1
  id 172.16.1.36, data 255.255.255.252, Type Stub (3)
  TOS count 0, TOS 0 metric 1
  id 192.168.1.21, data 172.16.1.89, Type Virtual (4)   # Virtual Link Type
  TOS count 0, TOS 0 metric 1
  Gen timer 00:00:13
  Aging timer 00:56:21
  Installed 00:03:39 ago, expires in 00:56:21, sent 00:03:37 ago
  Last changed 00:06:49 ago, Change count: 31, Ours
```

Junos OS identifies such virtual-links as legitimate neighbor relationships and displays them in a similar fashion, as shown in Listing 2.43.

Listing 2.43: OSPF Virtual Link neighbor

```
user@Livorno> show ospf neighbor detail 172.16.1.90
Address          Interface              State     ID               Pri  Dead
172.16.1.90      vl-192.168.1.21        Full      192.168.1.21       0    39
  Area 0.0.0.0, opt 0x42, DR 0.0.0.0, BDR 0.0.0.0
  Up 00:05:08, adjacent 00:05:07
172.16.1.90      ge-1/3/0.0             Full      192.168.1.21     128    34
  Area 0.0.0.3, opt 0x42, DR 172.16.1.89, BDR 172.16.1.90
  Up 00:16:53, adjacent 00:16:53
```

Viewed another way, such virtual relationships are merely *floating connections* that from a TE perspective rely on the definition of parameters in the transit area. They do not actively forward any kind of traffic, but act simply as control-plane connections based on recursive route resolution. When inspecting Opaque LSAs issued for *virtual links* in the backbone area, no bandwidth, metric, or admin-group attributes are associated with them; that is, no *Link TLVs* but just *Router Address TLVs* are present, as defined in [RFC3630]. Because the Opaque LSAs lack these attributes, the virtual links cannot be manipulated for TE purposes, as shown in Listing 2.44.

Listing 2.44: OSPF Opaque LSA referring to a Virtual Link

```
user@Livorno> show ospf database opaque-area area 0 extensive
               advertising-router 192.168.1.21

    OSPF link state database, Area 0.0.0.0
 Type       ID               Adv Rtr          Seq      Age  Opt  Cksum  Len
OpaqArea 1.0.0.1           192.168.1.21     0x80000002  285  0x22 0xfe30  28
  Area-opaque TE LSA
  RtrAddr (1), length 4: 192.168.1.21
  Aging timer 00:55:15
  Installed 00:04:42 ago, expires in 00:55:15, sent 00:04:40 ago
  Last changed 00:09:09 ago, Change count: 1
```

Point-to-point tunnels: Another way to bind separate network regions is based on tunnel constructs between specific end points in the topology.

Such point-to-point tunnels must be able to transport the necessary types of traffic and protocol and must be correctly sized to accomplish this virtual transport, especially in terms of bandwidth and Maximum Transmission Unit (MTU).

Although point-to-point tunnels can be treated from a protocol perspective as another topology link, their establishment is dependent on end-point reachability and routing variations between both sides. This requirement poses the additional challenge of ensuring that routing protocols can run properly over the setup (for instance, support for CLNP or MPLS traffic) and that they cannot easily overwhelm existing bandwidth limitations across the path.

A common concern is the feasible *Maximum Transmission Unit* (MTU) to apply over the tunnel. Even though arbitrary MTU values can be configured at both ends of the tunnel, this construct requires additional encapsulation on top of original data. As a result, MTU limits on the underlying physical topology can be reached when a tunnel is established on top of it.

Another common problem is the potential for a *chicken-and-egg* situation, where point-to-point tunnels are established between real end points but enabling routing protocols over these constructs leads to real end-point reachability over the virtual constructs because of metric election or area segregation. In this case, virtual abstractions alter the perception of the real network topology.

Routing platforms may consider several dependencies and need specific requirements to establish point-to-point tunnels. Junos OS Release Notes include descriptions for different types of point-to-point tunnel and resource needed in each Juniper Networks platform.

For the purpose of enabling a virtual routing topology over a network, Generic Routing Encapsulating (GRE) or *IP-over-IP* point-to-point tunnels are the simplest fits. These types of tunnel also support other protocol families, such as MPLS or IPv6, to be transported over an agnostic underlying IPv4 substrate that establishes an adequate vehicle for migrations. From a TE perspective, such connections are treated as standard logical interfaces with the respective attributes and TLVs.

However, these kinds of tunnel lead to an additional degree of abstraction above the physical link subset. Although they are sometimes the only way to achieve particular design goals, they are certainly not encouraged as a best practice in a permanent network implementation because of all the operational headaches they can introduce and because of the lack of transparency in the design.

MPLS Forwarding Adjacencies: Junos OS supports the use of RSVP-based LSPs to create *virtual* forwarding adjacencies both for OSPFv2 and IS–IS.

The basic idea behind this concept is to advertise such LSPs into link-state protocols as *point-to-point* links so that they can be used in SPF calculations at nodes other than the head end, as opposed to plain IGP shortcuts, in which the only node that knows about the existence of such LSPs is the LSP ingress end.

The configuration steps in Junos OS include a prior definition of the RSVP LSP and an activation of the LSP advertisement under the corresponding IGP stanza, as shown in Listing 2.45. Note that when an LSP stretches across two or more areas, the LSP configuration needs to include the `no-cspf` keyword to trigger proper RSVP signaling and to avoid local path prior calculation considering the TED.

Listing 2.45: Simple definition of LSP advertisement into OSPFv2 or IS–IS

```
[edit protocols mpls]
user@Livorno# set label-switched-path Livorno-Honolulu no-cspf to 192.168.1.21
[edit protocols ospf]
user@Livorno# set area 0 label-switched-path Livorno-Honolulu
[edit protocols isis]
user@Livorno# set label-switched-path Livorno-Honolulu
```

Although Junos OS does not treat these LSPs as virtual interfaces, a valid alternative is to stitch together specific domain end points from an IGP perspective to replicate a particular behavior. Note that Hello messages are not sent over the LSP, and no information flooding takes place over them. These LSPs are virtual structures just for forwarding purposes, but their information is included in the respective protocol constructs. When comparing MPLS *forwarding adjacencies* with the OSPF *virtual link* setup discussed in the previous section on Page 103, if an RSVP-TE LSP is configured between the same end points, a *point-to-point* and a *stub* link type for the LSP remote end are included in the Type 1 Router LSAs, as shown in Listing 2.46. From a TE perspective, this virtual construct has no specific Opaque LSAs. The IS–IS behavior is completely analogous. IS–IS represents this link as another *virtual* IS neighbor lacking specific TE sub-TLVs or other kinds of similar information, as shown in Listing 2.47.

Listing 2.46: OSPFv2 Router LSA including forwarding adjacency link information

```
user@Livorno> show ospf database router extensive lsa-id 192.168.1.6 area 0

    OSPF link state database, Area 0.0.0.0
 Type       ID               Adv Rtr           Seq      Age  Opt  Cksum  Len
Router  *192.168.1.6      192.168.1.6      0x8000096a    61  0x22 0x4e89 144
  bits 0x1, link count 10
  id 192.168.1.21, data 128.0.0.3, Type PointToPoint (1) # LSP advert P2P link
  TOS count 0, TOS 0 metric 1
  id 192.168.1.21, data 255.255.255.255, Type Stub (3) # LSP advert stub link
  TOS count 0, TOS 0 metric 1
  id 172.16.100.0, data 255.255.255.0, Type Stub (3)
  TOS count 0, TOS 0 metric 1
<...>
  id 192.168.1.6, data 255.255.255.255, Type Stub (3)
  TOS count 0, TOS 0 metric 0
<...>
  Gen timer 00:48:58
  Aging timer 00:58:58
  Installed 00:01:01 ago, expires in 00:58:59, sent 00:01:01 ago
  Last changed 00:01:01 ago, Change count: 35, Ours
```

Listing 2.47: IS–IS Level 2 LSP including forwarding adjacency link information

```
user@Livorno> show isis database Livorno.00-00 level 2 extensive
IS-IS level 2 link-state database:

Livorno.00-00 Sequence: 0x4df9, Checksum: 0x936d, Lifetime: 607 secs
   IS neighbor: Skopie.00                     Metric:      643
   IS neighbor: Havana.00                     Metric:      643
   IS neighbor: honolulu-re0.00            Metric:      10
   IP prefix: 172.16.1.32/30                  Metric:      643 Internal Up
   IP prefix: 172.16.1.36/30                  Metric:      643 Internal Up
   IP prefix: 192.168.1.6/32                  Metric:        0 Internal Up
<...>
TLVs:
    Area address: 49.0002 (3)
    Area address: 49.0003 (3)
```

```
15      Speaks: IP
16      Speaks: IPV6
17      IP router id: 192.168.1.6
18      IP address: 192.168.1.6
19      Hostname: Livorno
20      IS extended neighbor: Skopie.00, Metric: default 643
21        IP address: 172.16.1.34
22        Neighbor's IP address: 172.16.1.33
23        Local interface index: 82, Remote interface index: 72
24        Current reservable bandwidth:
25      <...>
26      IS extended neighbor: Havana.00, Metric: default 643
27        IP address: 172.16.1.38
28        Neighbor's IP address: 172.16.1.37
29        Local interface index: 83, Remote interface index: 73
30        Current reservable bandwidth:
31      <...>
32      IS extended neighbor: honolulu-re0.00, Metric: default 10  # IS-neighbor over LSP
33      IP extended prefix: 192.168.1.6/32 metric 0 up
34      IP extended prefix: 172.16.1.36/30 metric 643 up
35      IP extended prefix: 172.16.1.32/30 metric 643 up
36    No queued transmissions
```

Junos OS displays such virtual adjacencies in a similar fashion, including the reference for the advertised LSP, as shown in Listing 2.48 for OSPFv2 and 2.49 for IS–IS. Note that for one of such virtual adjacencies to function, bidirectional LSP reachability is required.

Listing 2.48: OSPFv2 neighbor information including forwarding adjacency

```
1 user@Livorno> show ospf neighbor 192.168.1.21 extensive
2 Address          Interface              State     ID               Pri  Dead
3 192.168.1.21     Livorno-Honolulu       Full      192.168.1.21       0     0 # Interface field
      referring to LSP
4   Area 0.0.0.0, opt 0x0, DR 0.0.0.0, BDR 0.0.0.0
5   Up 00:17:00
6 172.16.1.90      ge-1/3/0.0             Full      192.168.1.21     128    36
7   Area 0.0.0.3, opt 0x42, DR 172.16.1.89, BDR 172.16.1.90
8   Up 02:44:40, adjacent 02:44:40
```

Listing 2.49: IS–IS adjacency information including forwarding adjacency

```
1 user@Livorno>  show isis adjacency extensive honolulu-re0
2 honolulu-re0
3   Interface: Livorno-Honolulu, Level: 2, State: Up, Expires in 0 secs # Interface field referring to
      LSP
4   Priority: 0, Up/Down transitions: 1, Last transition: 00:16:37 ago
5   Circuit type: 2, Speaks: IP
6   Topologies: Unicast
7   Restart capable: No, Adjacency advertisement: Advertise
8   IP addresses: 192.168.1.21
9   Transition log:
10  When                  State        Event           Down reason
11  Sun Feb 15 20:07:22   Up           Seenself
12
13 honolulu-re0
14  Interface: ge-1/3/0.0, Level: 1, State: Up, Expires in 8 secs
15  Priority: 64, Up/Down transitions: 1, Last transition: 08:58:41 ago
16  Circuit type: 1, Speaks: IP, IPv6, MAC address: 0:17:cb:d3:2f:16
17  Topologies: Unicast
18  Restart capable: Yes, Adjacency advertisement: Advertise
19  LAN id: honolulu-re0.02, IP addresses: 172.16.1.90
20  Transition log:
21  When                  State        Event           Down reason
22  Sun Feb 15 11:25:18   Up           Seenself
```

Because these LSPs inject routing information in the standard IGP database, LSPs are available to transport not only MPLS or TE aware applications, but also all kinds of transport traffic, including plain IPv4. This behavior adds opacity to the setup in the sense that the underlying topology is not visible in a straightforward manner for troubleshooting and operational purposes, but it can certainly fulfill requirements for certain migrations or setups.

A final consideration is that for a *forwarding adjacency* to work properly, LSPs must be established in both directions between both ends. Junos OS does not enforce using a single LSP for return traffic in a plain forwarding adjacency; more than one LSP can share the load for revertive reachability, as would happen with standard MPLS traffic. This requirement means that LSPs need to be properly conceived and configured at the design stage to ensure proper LSP pairing for both directions.

This traditional plain *forwarding adjacency* concept has further evolved with [RFC4203], [RFC5305], and [RFC4206] so that this information is simply placed into the TED and is then used to construct LSP hierarchies.

2.2 Link-state IGP Domain Migrations

Another ongoing discussion in the internetworking world has been if OSPF or IS–IS would be the most convenient link-state IGP for core deployments. Many cross-analyses have been done in the networking industry about whether OSPF or IS–IS should be preferred under some circumstances, because both protocols are parallel approaches to running link-state routing from different origins.

The foundations for both protocols are based on common link-state concepts despite the fact that OSPF runs over IP and IS–IS over data link CLNP. In both protocols, an adjacency is created to maintain a neighbor relationship based on agreed parameters and these adjacencies build up a common topology to be considered both for path calculation and information flooding. Routing information is collected into a link-state database, and SPF calculations using the *Dijkstra* algorithm are carried out based on these data.

The terminology in each protocol is indeed different, but the purpose of many of the data structures and router roles is the same. Discussions about formalisms and acronyms apart, certainly some technical features that one or the other protocol presents exist, as well as some behaviors that can be considered as *drawbacks* or as *advantages* in some topologies.

2.2.1 Considerations for a link-state IGP migration

Link-state information chunks are encapsulated in the IS–IS Link-State PDUs (LSPs) which are directly flooded as complete data structures, whereas in OSPF, LSAs are grouped and encapsulated in Link-State Update packets before propagation. IS–IS TLVs inside an LSP would be the equivalent data structures to OSPF LSAs. This means that while information is translated from one LSA format to another in an OSPF ABR or ASBR, the same IS–IS TLV types are used in both L1 and L2 LSPs.

Another well-known difference is area definition and separation: IS–IS considers a Level 2 as a contiguous subdomain of backbone routers, independent of the area identifier included in each Net-ID, whereas OSPF requires a backbone with all adjacencies inside the same Area 0. Area boundaries are therefore different in the sense that an IS–IS L1L2 router is considered as part of the backbone and area edge, and can indistinctly establish both L1 and L2 adjacencies over the same physical link, but OSPF used to define an adjacency for each logical link and hierarchical separation on each ABR. [RFC5185] proposes a new approach to run multi-area adjacencies in OSPF, allowing a single physical point-to-point link to be shared by multiple areas, thus changing this traditional association.

Therefore, Intermediate Systems in Level 2 do not need to have a matching area identifier, and they automatically set the ATTached bit in their L1 LSPs to indicate backbone reachability. [RFC3509] describes different behaviors of an ABR performing LSA generation or transformation, focusing on whether an active adjacency is required in Area 0. Depending on the router operating system, this concept for backbone detection would be another factor to consider in a migration. Junos OS implements *active backbone support* for ABRs, meaning that an active backbone connection is checked in addition to the ABR status to perform ABR functionality, such summarizing LSAs in both directions, or advertising or withdrawing summary LSAs into stub areas and NSSAs.

IS–IS L1 areas traditionally were equivalent to OSPF totally stubby areas, as they have, by default, no visibility on backbone prefixes or routes from other areas, apart from a default route installed based on detection of the ATTached bit, and because external prefixes are not populated to Layer 2 by default. However, route-leaking policies and *wide metrics* can change information distribution in both directions and allow greater flexibility in prefix population. While most routing vendors also include similar route-leaking capability at OSPF ABRs, certain types of area mandate LSA population or blocking. Care must be taken when migrating IS–IS to OSPF areas, not only when adapting redistribution policies but when electing the appropriate type of OSPF area.

When migrating area or system identification from one protocol to the other, numbering schemes must be adapted as well. Area identification is a *32-bit* number in OSPF (exactly the same size as the Router ID), but IS–IS uses an Area ID that can vary from *1 to 13 bytes* and a System ID that is usually *6 bytes* long, both inside the Net-ID.

OSPF supports broadcast, point-to-point, point-to-multipoint, and NBMA media, whereas IS–IS only considers broadcast and point-to-point link types when setting up adjacencies. Migrating from one protocol to the other may also require link transformation or adaption.

Adjacency establishment is also slightly different between the two protocols. Apart from using different timer values, most OSPF hello packet fields need to match for participating neighbors (Area ID, timers, authentication, network mask, options), and there is a Designated Router (DR) and Backup Designated Router (BDR) election in broadcast media. IS–IS just elects a single Designated Intermediate System (DIS) in broadcast links, and some of the hello fields do not agree, such as hello and dead intervals. The DIS role is also preemptive, whereas OSPF DR and BDR are not.

In OSPF broadcast media, an adjacency is formed with a DR and BDR, but IS–IS creates a full adjacency mesh among participating Intermediate Systems on each level.

Likewise, neighbor discovery in IS–IS can happen at different levels, except in point-to-point links, which have a single Point-to-Point Hello PDU for both levels. This means that some of the parameters required to build up an adjacency need to match in both levels, particularly authentication settings.

Because IS–IS is not based on IP, it requires a different inherent mechanism to perform segmentation derived from an initial discovery based on padding hello messages with a so-called TLV 8 Padding. OSPF is based on IP, which directly handles fragmentation and reassembly. [RFC2328] defines an additional check based on the Interface MTU field of

Database Description Packets, so that if this value is larger than the IP datagram size that the router can accept on the receiving interface without fragmentation, the Database Description Packet is rejected and the adjacency does not reach FULL state. Router vendors implement different padding mechanisms with hello messages to confirm an MTU size. An MTU mismatch usually prevents adjacency formation as the Database Description or Link-state Update packets cannot be properly flooded in both directions.

OSPF synchronizes databases with a master and slave election, exchanging those Database Description packets with LSA headers and Link-state Requests and Updates to update each router's topological information. IS–IS considers a much simpler and regular synchronization process, where Complete Sequence Number PDUs (CSNPs) are regularly sent and particular information is retrieved via Partial Sequence Number PDUs (PSNPs). [RFC4811] advocates a new out-of-band Link State Database (LSDB) resynchronization for OSPF to allow the database to be refreshed without topological changes or tearing down adjacencies.

Regular refreshing of information is also handled differently in both protocols. In addition to using timer counters in opposite directions, the maximum link-state information lifetime (MaxAge parameter) is not configurable in OSPF but its equivalent parameter can be changed in IS–IS. Configuring this value can notably help fine-tune background refresh activities.

Metrics are also calculated in different ways. While both protocols allow particular metrics on a per link basis and most operating systems have extended the concept of having a reference bandwidth for link calculation from OSPF to IS–IS automatically to derive metrics depending on link rates, IS–IS has introduced the concept of *wide metrics*, which extends the corresponding field up to 32 bits and automatically allows prefix distribution across the complete domain. *Wide metrics* no longer consider a distinction between *internal* and *external* routes. OSPFv2 Type 1 LSAs include a 16-bit metric field, and Types 3, 4, 5, and 7 LSAs reserve 24 bits.

Because of how LSAs are constructed, OSPF has traditionally required that all neighbors need to understand newer LSA extensions. [RFC3630] and [RFC4970] add a certain flexibility by including TLV coding in Type 10 Opaque LSAs or Type 12 RI Opaque LSAs for OSPFv2 and OSPFv3, potentially allowing future extensibility. However, IS–IS TLVs were originally fashioned in a nested hierarchy and for flexible and transparent extension introduction, in which only unknown TLVs may be discarded by neighbors, not complete LSPs.

The *virtual link* concept was originally described in [RFC1131] as a way to avoid backbone fragmentation. Although [ISO10589] also included a similar concept for IS–IS partition repair, most routing vendors have not implemented it. It is considered a best practice just to define *virtual links* as temporary workarounds to grant Area 0 extensibility. If OSPF *virtual links* need to be migrated to IS–IS, this transition requires full backbone connectivity or use of a similar point-to-point construct.

IS–IS also supports a particular behavior to avoid routing transit traffic over the platform when setting the Overload bit in the LSPs. In this state, intermediate systems signal a lack of transit capability to the rest of the area. [RFC3137] defines a similar behavior for OSPF, in which the cost for all non-stub links is set to infinity. This same concept to advertise maximum metrics has also been extrapolated to Junos OS IS–IS in the sense that transit traffic is not directly prohibited, but rather, remains valid as an alternative path with maximum possible metric. Although the Overload bit was originally conceived to block transit routing as the result of some critical situation, it can also be considered a best practice to set it during intrusive operations on a transit router to minimize service downtime and transit traffic loss. Regardless, correct portability needs to be ensured when moving from one link-state IGP to another.

[RFC2973] standardizes the concept of an IS–IS mesh group by defining restricted flooding topologies as a subset of the full IS–IS domain topology. Highly meshed topologies can lead to excessive information flooding, provided that incoming routing information is reflooded by default on all other outgoing interfaces. While initially this can be seen as a resilient behavior, it generates a considerable amount of additional traffic in sufficiently redundant setups. The *mesh group* concept constrains flooding so that LSPs can be forwarded with enough guarantees, but redundant background noise can be suppressed from the network if the corresponding design premises are taken into account. Note that OSPF has no equivalent concept, which can present a migration drawback.

IPv6 support is also completely different in both protocols. [RFC2740] defines a newer version 3 of OSPF protocol for IPv6 support. The result is that a new protocol flavor needs to be configured with an overlay model when compared with traditional OSPF version 2. OSPFv3 inherits most protocol concepts from version 2, but also defines new data structures and LSAs, requiring a separate database. On the other hand, [RFC5308] defines new IPv6 specific TLVs to be additionally injected in standard IS–IS LSPs. This design ensures soft integration of an IPv6 topology, and no other additional protocol features or databases are required.

TE extensions are supported equally on both link-state IGPs. [RFC3630] redefines the Type 10 Opaque LSA to flood TE information. Interestingly, this Opaque LSA includes a Router TLV that [RFC3630] Section 2.4.1 included so that if IS–IS is also active in the domain, the address referred by this TLV can be used to compute the mapping between the OSPF and IS–IS topologies. There is another Link TLV included in this Opaque LSA, which covers the following sub-TLVs:

- *(Sub-TLV 1)* Link type (1 byte);
- *(Sub-TLV 2)* Link ID (4 bytes);
- *(Sub-TLV 3)* local interface IP address (4 bytes);
- *(Sub-TLV 4)* remote interface IP address (4 bytes);
- *(Sub-TLV 5)* TE metric (4 bytes);
- *(Sub-TLV 6)* maximum bandwidth (4 bytes);
- *(Sub-TLV 7)* maximum reservable bandwidth (4 bytes);
- *(Sub-TLV 8)* unreserved bandwidth (32 bytes);
- *(Sub-TLV 9)* administrative group (4 bytes).

[RFC3784] is the counterpart TE standard for IS–IS. It defines a new TLV 134 TE router ID, that uniquely identifies each router in the TED with a stable identifier and it expands TLV 22 Extended IS Reachability and TLV 135 Extended IP Reachability with newer sub-TLV structures and modified fields that convey TE information.

The TLV 22 Extended IS Reachability is expanded with newer sub-TLV structures that aggregate TE information as follows:

- *(Sub-TLV 3)* administrative group (4 bytes);
- *(Sub-TLV 6)* IPv4 interface address (4 bytes);
- *(Sub-TLV 8)* IPv4 neighbor address (4 bytes);
- *(Sub-TLV 9)* maximum link bandwidth (4 bytes);
- *(Sub-TLV 10)* reservable link bandwidth (4 bytes);
- *(Sub-TLV 11)* unreserved bandwidth (32 bytes);
- *(Sub-TLV 18)* TE default metric (3 bytes);
- *(Sub-TLV 250-254)* reserved for Cisco-specific extensions;
- *(Sub-TLV 255)* reserved for future expansion.

Even though sub-TLV 18 TE Default metric has been referred to above as being 3 octets long, the idea is to preclude overflow within a TE SPF implementation, in which all metrics greater than or equal to a maximum path metric value are considered to have that maximum metric. The rationale is to select a value such that maximum path metric plus a single link metric does not overflow the number of bits for internal metric calculation, which is also 32 bits long. In this way, this IS–IS sub-TLV remains compatible with the OSPF sub-TLV 5 TE metric, which is 4 octets long.

[RFC3784] also expands the TLV 135 Extended IP Reachability with up to 32-bit metric fields and adds a new bit to indicate that a prefix has been redistributed *down* in the hierarchy.

OSPF sub-TLV 9 administrative group allows setting specific groups for links with up to 4 octets, which is the same length from the IS–IS sub-TLV 3 administrative group value field. Migrating *administrative group* values from one protocol to the other is transparent in terms of length.

Therefore, to populate a joint TED in an overlay model with OSPF and IS–IS, the OSPF Router TLV from the Type 10 Opaque LSA and the IS–IS TLV 134 TE router ID must include the same *Router ID* value and link TE metrics must be equivalent.

Another commonality is the ability to support *forwarding adjacencies* by the two protocols. Plain *forwarding adjacencies* have been discussed previously in the case of using RSVP-TE LSP infrastructure to virtualize a point-to-point link available for calculation of the IGP shortest path. The rationale here is to guarantee bidirectional reachability and reuse these adjacencies between Label-Switched Routers (LSRs) to be available for IGPs as virtual connections.

This traditional concept evolved further with GMPLS TE procedures as defined in [RFC4203], [RFC5305], and [RFC4206] to advertise these LSPs (the so-called *Forwarding Adjacency LSPs or FA-LSPs*) as TE links into the same instance of ISIS or OSPF that was used to create the LSP. In this environment, the FA-LSPs represent TE links between two GMPLS nodes whose path is established in the same instance of the GMPLS control plane. Having this virtualization for FA-LSPs available in the TED for path calculations allows an RSVP hierarchical model with different LSP regions in which specific FA-LSPs could be elected as hops in RSVP LSP path computations.

[RFC4203] defines newer sub-TLVs to be added to OSPF Opaque LSA Link TLVs:

- *(Sub-TLV 11)* Link Local/Remote Identifiers (8 bytes);
- *(Sub-TLV 14)* Link Protection Type (4 bytes);
- *(Sub-TLV 15)* Interface Switching Capability Descriptor (variable);
- *(Sub-TLV 16)* Shared Risk Link Group (variable).

Similarly, [RFC5305] conceives parallel mechanisms and information structures for IS–IS. Although *Shared Risk Link Group* information is defined as a new TE TLV 138 Shared Risk Link Group, all other bits are defined as sub-TLVs for the TLV 22 Extended IS Reachability:

- *(Sub-TLV 4)* Link Local/Remote Identifiers (8 bytes);
- *(Sub-TLV 20)* Link Protection Type (2 bytes);
- *(Sub-TLV 21)* Interface Switching Capability Descriptor (variable).

This analysis of concepts and references for TE information, comparing OSPF and IS–IS is summarized in Table 2.5.

Both standards replicate behaviors and information structures for OSPF and IS–IS. However, in the case of an IGP migration, proper system support for one or the other option and adequate information mapping are needed for a smooth transition.

In spite of all the similarities between IS–IS and OSPF, numerous behaviors specific to each exist (specific even to each OSPF version). The discussion in this section has given the most generic technical comparisons between the two protocols when scoping a migration. These are summarized in Table 2.6.

Sometimes, the decision to use one or other IGP is not entirely technical. Political factors (for example, company acquisitions may determine a dominant IGP), hands-on experience with one of them, or considerations about the value of the enormous effort to move to another IGP (together with the potential for area or network-wide service interruption) can easily tilt the balance to one side.

The truth is that both OSPF and IS–IS currently have enough traction in the Internet to ensure their future-proof existence for several years.

2.2.2 Generic strategies for a link-state IGP migration

Protocol migrations require previous in-depth planning and analysis. Careful thinking and proof-of-concept testing are surely part of a successful path when facing such an intrusive operation in nowadays networks.

A preliminary step is the choice of strategy guidelines to migrate protocols. Modern routing operating systems support both coexistence and redistribution among routing protocols, together with specific parameterization or interaction with other routing protocols. But several factors and dependencies determine how and where to define the specific relationship among them.

These strategies, by themselves, do not constitute the unique and sole foundation for a complete link-state IGP migration; neither are they exclusive among themselves. Depending on the setup and circumstances, there may be interaction among the strategies and/or they may apply just to one given scenario.

Table 2.5 TE concept comparison between OSPF and IS–IS

TE resources	OSPF	IS–IS
Administrative group	Type 10 LSA, Link TLV, sub-TLV 9, 4 bytes	TLV 22, sub-TLV 3, 4 bytes
Interface switching capability descriptor	Type 10 LSA, Link TLV, sub-TLV 15, variable	TLV 22, sub-TLV 21, variable
Link local/remote ids	Type 10 LSA, Link TLV, sub-TLV 3, 8 bytes	TLV 22, sub-TLV 6, 8 bytes
Link protection	Type 10 LSA, Link TLV, sub-TLV 14, 4 bytes	TLV 22, sub-TLV 20, 2 bytes
Local IPv4 address	Type 10 LSA, Link TLV, sub-TLV 3, 4 bytes	TLV 22, sub-TLV 6, 4 bytes
Maximum bandwidth	Type 10 LSA, Link TLV, sub-TLV 6, 4 bytes	TLV 22, sub-TLV 9, 4 bytes
Maximum reservable bandwidth	Type 10 LSA, Link TLV, sub-TLV 7, 4 bytes	TLV 22, sub-TLV 10, 4 bytes
Remote IPv4 address	Type 10 LSA, Link TLV, sub-TLV 4, 4 bytes	TLV 22, sub-TLV 8, 4 bytes
SRLG	Type 10 LSA, Link TLV, sub-TLV 16, variable	TLV 138, variable
TE metric	Type 10 LSA, Link TLV, sub-TLV 5, 4 bytes	TLV 22, sub-TLV 18, 3 bytes
Unreserved bandwidth	Type 10 LSA, Link TLV, sub-TLV 8, 32 bytes	TLV 22, sub-TLV 11, 32 bytes

Coexistence of link-state IGPs

Coexistence of link-state IGPs in a network is commonly supported and known as a seamless deployment option.

The basic idea behind the overlay is to populate the link-state database of the final IGP while keeping that protocol non-preferred, with progressive deployment until reaching the milestone of switching over from one protocol to the other. At that point, either a gradual or global transition occurs to move from one active IGP to another.

This approach benefits from the fact that link-state IGPs are based on a common link-state database that is both distributed and domain-wide. This means that although the protocol may not be preferred at a given router, infrastructure information can be flooded once adjacencies or neighbor relationships are established, because computations are local at each system. This situation is usually called *ships in the night*, because protocols work independently one from the other in the same domain. This approach is noticeably different from traditional distance-vector protocols, where routing is decided based on neighbors' routing decisions.

It is fundamental to correctly manipulate the protocol preference at each step and to tweak the values in a timely and coordinated fashion. If protocols running simultaneously contain different information and local route preference changes are not arranged in time, there is

Table 2.6 Rationale for link-state IGP migrations

Factor	OSPF	IS–IS
Information packaging	LSA types	TLVs inside per level LSPs
Area boundaries	Link per area, [RFC5185] for multi-area	Per level per link, authentication shared in P2P
Backbone detection	[RFC3509]	ATTached bit on L1L2
Area definition	Different standard area flavors	Flexible leaking, default totally stubby behavior
Area identification	32-bit integer	Variable 1–13 bytes
Router identification	32-bit integer	6 bytes
Supported media	Broadcast, P2P, point-to-multipoint, and NBMA	Broadcast and P2P
Matching adjacency parameters	Area ID, timers, authentication, network mask, and options	Area ID for L1, authentication, network mask (newer versions)
Fragmentation	IP fragmentation, MTU field in DBD packets	Hello padding variants
Database synchronization	Adjacency establishment, [RFC4811] support	Regular CSNP population
LSA refreshment	Ascending, non-configurable MaxAge	Descending, tunable refresh interval
Metrics	16 bits-24 bits	32 bits, wide-metrics
Extensibility	Network-wide new LSA understanding	Flexible introduction of new TLVs
Virtual links	Implicit	Other P2P constructs
Overload	[RFC3137] support	Implicit
Mesh group	Not present	Implicit
IPv6 support	new protocol OSPFv3, [RFC2740]	Additional TLVs, [RFC5308]
TE	Router sub-TLV, TE-metrics, [RFC3630]	TLV 134 TE router ID, TE-metrics, [RFC3784]
Plain forwarding adjacencies	LSPs as P2P link types in Router LSAs	Additional neighbor in TLV 22 Extended IS Reach
FA-LSPs and RSVP hierarchies	GMPLS TE sub-TLVs, [RFC4203]	GMPLS TE TLV and sub-TLVs, [RFC5305]

certainly a potential for routing loops during the divergence period. This situation may arise as a consequence of changing protocol preferences, as a local decision, despite the link-state IGP domain scope.

This overlay model can also be enforced in some setups in another way, by using different router identifiers for each protocol (for instance, with different loopback addresses advocated for each topology) and with more recent concepts in link-state IGP development, such as *multitopology routing* as defined in [RFC4915] and [RFC5120] to allow different topologies to coexist within the same IGP.

Link-state IGP redistribution

Another interesting concept when approaching a protocol migration is *redistribution*, that is, injecting routing information from one protocol to another.

Redistribution has been supported by most routing vendors since the origins of routing protocols, not only for IGPs but also for any kind of routing or information distribution protocol. The need to pass information from one routing vehicle to another arose as internetworking setups started to become more complex, and nowadays redistribution has become a day-to-day practice in most Internet topologies.

Junos OS makes use of `export` policies to determine which routing information is going to be injected in a given protocol, including information external to that protocol. A central concept in Junos OS is that protocols place routes in the routing table and only the best sibling path becomes instantiated in the forwarding table (note, however, that there could be several simultaneous best paths for load balancing thanks to specific structures). This behavior means that because only the best route is the current valid version for forwarding purposes, for an `export` policy to legitimately pick up a route for redistribution purposes, this route needs to be the best match in the routing table. Existing non-preferred versions of the route in the routing table (not only from other protocols but also due to other attributes or circumstances following Junos OS route selection algorithm rules) are not eligible for route redistribution purposes.

Thus, *redistribution* is not only a question of deciding which routes from which protocol should be elected, but also depends on which routes are selected as the best in the tie-breaking process. This dependency is especially interesting in the case of link-state protocols because information is flooded in specific constructs, and although these details are propagated and present in participating systems, if the route is locally overridden by any other protocol, it is not going to be re-injected although the path is present and it is as valid as at any other system.

Likewise, because the context for TE may be different in the new domain in which the route is redistributed, it is generally understood by most routing vendors that the scope of *redistribution* is constrained to prefix information only and does not include any previously existing *TE attributes*.

To summarize, injecting external routes is purely a local decision as opposed to the domain-wide scope of link-state protocol information flooding (including translation at area boundaries). Therefore, redistribution techniques must be planned with care.

Redistribution is a basic tool for migration purposes, but it can easily lead to unexpected effects. Incorrect redistribution policies are one of the most well-known causes for network outages.

Interaction with other routing protocols

Sometimes, another routing protocol can be used as the vehicle to transport route information between systems.

From pure static routes or aggregates to BGP, specific gateways in each domain can act as representatives to make the routing information available to the external world. Again, this behavior requires careful and planned choice of such systems, because the potential for routing loops increases when another routing protocol is added to the cocktail.

Based on this approach, a well-known technique consists of using BGP among a subset of routers to have each one injecting summary networks representing each domain into the IGP, and vice versa. A possible drawback when advertising network aggregates is less-than-optimal routing as more specific routes are not present in a protocol, but the cost of less-than-optimal routing during a transition period may compensate for the clarity and ease of implementing this strategy.

Link-state IGP parameterization

A link-state IGP domain can be traditionally associated to a subset of systems sharing common IGP parameters and under common management and scope. By using different IGP parameters, an administrator can also differentiate between one domain and another.

While some restrictions exist for this strategy, because some of the IGP attributes have purely *link-local* or *area-local* scope, *authentication* settings are a well-known source for differentiation within the same physical topology. Likewise, stretching or constraining a certain area can be used as a vehicle to expand a whole domain, because area identifiers are an attribute that must match both in OSPF and IS–IS Level 1 adjacencies.

The administrator can control the progressive deployment and constriction of a given link-state IGP by defining particular authentication settings or area identification and can progressively migrate systems by tweaking these attributes. Nevertheless, information may need to be carried from one domain to another, which can be achieved by means of redistribution or another auxiliary protocol.

2.2.3 Resources for a link-state IGP migration

Once technical considerations and strategies have been evaluated, routing operating systems may offer specific features that can leverage link-state IGP migration procedures. Junos OS offers a rich variety of configuration options related with link-state IGP redistribution, transition, and overlay that complement protocol standards.

Injection of IPv4 addressing information by a link-state IGP

By default in Junos OS, each link-state IGP injects route information into its LSAs or LSPs considering all addresses configured over each interface activated for the protocol.

However, certain circumstances exist where it would be desirable just to inject a specific address from an interface in one protocol instead of all available addressing information. One example is a setup in which more than one address is associated with the loopback logical interface in the same instance, because in Junos OS no more than one loopback logical unit can simultaneously be active in each instance. With this requirement, different IGP layers can be overlapped with different loopback addresses or *anchoring points* with the corresponding consequences for LDP, TE information, and other protocols that may depend on that, such as IBGP.

OSPFv2: For OSPFv2, local IPv4 addressing information is injected as Stub link types in Type 1 Router LSAs, and `export` policies control the information that is injected in Type 5 AS External LSAs (or Type 7 NSSA AS External LSAs). This behavior means that `export` policies are not the adequate vehicle to determine which addresses need to be locally injected in Type 1 Router LSAs.

In this case, Junos OS allows the setting of IPv4 addresses directly in the interface configuration within each area's stanza as an option. Thus, the modification would be as simple as just including the respective address instead of the complete interface.

Imagine that the same loopback logical interface has multiple IPv4 addresses, as shown in Listing 2.50. Just including the corresponding address in the OSPFv2 area stanza is enough to inject selectively the address in the system's own Type 1 Router LSA, as shown in Listing 2.51. Lines 24, 26, and 61 show changes in stub networks from the Type 1 Router LSA before and after the change. Note that it is recommended to define the `router-id` explicitly in the `[edit routing-options]` stanza with the most representative address to ensure unique identity across protocols and applications.

Listing 2.50: Multiple addresses in loopback interface at Livorno for reference purposes

```
1 [edit interfaces lo0 unit 0 family inet]
2 user@Livorno# set address 192.168.99.6/32
3 [edit interfaces lo0 unit 0 family inet]
4 user@Livorno# show
5 address 192.168.1.6/32 {
6     primary;
7 }
8 address 192.168.99.6/32;
9 address 127.0.0.1/32;
```

Listing 2.51: Configuration changes in Livorno to restrict 192.168.99.6 injection in Router LSA

```
1  [edit protocols ospf]
2  user@Livorno# show
3  traffic-engineering;
4  reference-bandwidth 100g;
5  area 0.0.0.0 {
6  <...>
7      interface lo0.0;
8  <...>
9  }
10
11 [edit protocols ospf]
12 user@Livorno# run show ospf database lsa-id 192.168.1.6 extensive router
13
14     OSPF database, Area 0.0.0.0
15  Type       ID               Adv Rtr           Seq      Age  Opt  Cksum  Len
16 Router  *192.168.1.6      192.168.1.6       0x800008cf   367  0x22 0xc21c 156
17   bits 0x0, link count 11
18   id 172.16.100.0, data 255.255.255.0, Type Stub (3)
19     Topology count: 0, Default metric: 1000
20   id 172.16.1.53, data 172.16.1.54, Type Transit (2)
21     Topology count: 0, Default metric: 100
22   id 172.16.1.90, data 172.16.1.89, Type Transit (2)
23     Topology count: 0, Default metric: 100
24   id 192.168.99.6, data 255.255.255.255, Type Stub (3)   # Secondary loopback address
25     Topology count: 0, Default metric: 0
26   id 192.168.1.6, data 255.255.255.255, Type Stub (3)    # Primary loopback address
27     Topology count: 0, Default metric: 0
28   id 192.168.1.2, data 172.16.1.34, Type PointToPoint (1)
```

```
29      Topology count: 0, Default metric: 643
30    id 172.16.1.32, data 255.255.255.252, Type Stub (3)
31      Topology count: 0, Default metric: 643
32    id 192.168.1.4, data 172.16.1.38, Type PointToPoint (1)
33      Topology count: 0, Default metric: 643
34    id 172.16.1.36, data 255.255.255.252, Type Stub (3)
35      Topology count: 0, Default metric: 643
36    Gen timer 00:43:52
37    Aging timer 00:53:52
38    Installed 00:06:07 ago, expires in 00:53:53, sent 00:06:05 ago
39    Last changed 4d 21:08:13 ago, Change count: 11, Ours
40
41  [edit protocols ospf]
42 user@Livorno# delete area 0 interface lo0.0
43  [edit protocols ospf]
44 user@Livorno# set area 0 interface 192.168.1.6
45  [edit protocols ospf]
46 user@Livorno# commit
47 commit complete
48  [edit protocols ospf]
49 user@Livorno# run show ospf database lsa-id 192.168.1.6 extensive router
50
51     OSPF database, Area 0.0.0.0
52  Type       ID               Adv Rtr           Seq      Age  Opt  Cksum  Len
53 Router  *192.168.1.6      192.168.1.6      0x800008d0     8  0x22 0x2e17 120
54   bits 0x0, link count 8
55   id 172.16.100.0, data 255.255.255.0, Type Stub (3)
56     Topology count: 0, Default metric: 1000
57   id 172.16.1.53, data 172.16.1.54, Type Transit (2)
58     Topology count: 0, Default metric: 100
59   id 172.16.1.90, data 172.16.1.89, Type Transit (2)
60     Topology count: 0, Default metric: 100
61   id 192.168.1.6, data 255.255.255.255, Type Stub (3)  # Primary loopback address
62     Topology count: 0, Default metric: 0
63   id 192.168.1.2, data 172.16.1.34, Type PointToPoint (1)
64     Topology count: 0, Default metric: 643
65   id 172.16.1.32, data 255.255.255.252, Type Stub (3)
66     Topology count: 0, Default metric: 643
67   id 192.168.1.4, data 172.16.1.38, Type PointToPoint (1)
68     Topology count: 0, Default metric: 643
69   id 172.16.1.36, data 255.255.255.252, Type Stub (3)
70     Topology count: 0, Default metric: 643
71   Gen timer 00:49:51
72   Aging timer 00:59:51
73   Installed 00:00:08 ago, expires in 00:59:52, sent 00:00:08 ago
74   Last changed 00:00:08 ago, Change count: 12, Ours
```

IS–IS: Conceptually speaking, IS–IS is not a protocol directly based on IPv4 but carries IPv4 as topology information in the form of TLV 128 Internal IP Reachability, TLV 130 External IP Reachability, and TLV 135 Extended IP Reachability. This fact means that `export` policies directly control the injection of IPv4 addressing information router's own LSPs in Level 2 or for each area, which is different from what occurs in OSPFv2.

Using the same example from Listing 2.50 with multiple addresses in the loopback unit, Listing 2.52 summarizes policy definition and application needed in IS–IS to block adding secondary addresses to the LSP TLVs. Lines 25, 26, 34, 35, 72, and 82 show TLV information extraction before and after the changes.

Listing 2.52: Configuration changes in Livorno to restrict 192.168.99.6 injection in own LSP

```
1  [edit protocols isis]
2 user@Livorno# show
3 reference-bandwidth 100g;
```

```
 4 level 1 disable;
 5 level 2 wide-metrics-only;
 6 <...>
 7 interface lo0.0;
 8
 9 [edit policy-options policy-statement block-secondary-loopback]
10 user@Livorno# run show isis database Livorno extensive
11 IS-IS level 1 link-state database:
12
13 IS-IS level 2 link-state database:
14
15 Livorno.00-00 Sequence: 0x505e, Checksum: 0xdc73, Lifetime: 1073 secs
16    IS neighbor: Skopie.00                    Metric:      643
17    IS neighbor: Havana.00                    Metric:      643
18    IS neighbor: honolulu-re0.00           Metric:       10
19    IS neighbor: honolulu-re0.02           Metric:      100
20    IS neighbor: Barcelona-re0.02          Metric:      100
21    IP prefix: 172.16.1.32/30                 Metric:      643 Internal Up
22    IP prefix: 172.16.1.36/30                 Metric:      643 Internal Up
23    IP prefix: 172.16.1.52/30                 Metric:      100 Internal Up
24    IP prefix: 172.16.1.88/30                 Metric:      100 Internal Up
25    IP prefix: 192.168.1.6/32                 Metric:        0 Internal Up  # Primary address
26    IP prefix: 192.168.99.6/32                Metric:        0 Internal Up # Secondary address
27 <...>
28   TLVs:
29     Area address: 49.0002 (3)
30     Area address: 49.0003 (3)
31     Speaks: IP
32     Speaks: IPV6
33 <...>
34     IP extended prefix: 192.168.1.6/32 metric 0 up # Primary address
35     IP extended prefix: 192.168.99.6/32 metric 0 up # Secondary address
36     IP extended prefix: 172.16.1.36/30 metric 643 up
37     IP extended prefix: 172.16.1.32/30 metric 643 up
38     IP extended prefix: 172.16.1.88/30 metric 100 up
39     IP extended prefix: 172.16.1.52/30 metric 100 up
40   No queued transmissions
41
42 [edit policy-options policy-statement block-secondary-loopback]
43 user@Livorno# show
44 term sec {
45     from {
46         route-filter 192.168.99.6/32 exact;
47     }
48     then reject;
49 }
50
51 [edit]
52 user@Livorno# set protocols isis export block-secondary-loopback
53 [edit]
54 user@Livorno# commit
55 commit complete
56 [edit]
57 user@Livorno# run show isis database Livorno extensive
58 IS-IS level 1 link-state database:
59
60 IS-IS level 2 link-state database:
61
62 Livorno.00-00 Sequence: 0x5061, Checksum: 0x7eca, Lifetime: 1193 secs
63    IS neighbor: Skopie.00                    Metric:      643
64    IS neighbor: Havana.00                    Metric:      643
65    IS neighbor: honolulu-re0.00           Metric:       10
66    IS neighbor: honolulu-re0.02           Metric:      100
67    IS neighbor: Barcelona-re0.02          Metric:      100
68    IP prefix: 172.16.1.32/30                 Metric:      643 Internal Up
69    IP prefix: 172.16.1.36/30                 Metric:      643 Internal Up
70    IP prefix: 172.16.1.52/30                 Metric:      100 Internal Up
71    IP prefix: 172.16.1.88/30                 Metric:      100 Internal Up
72    IP prefix: 192.168.1.6/32                 Metric:        0 Internal Up # Primary address
```

```
73 <...>
74 TLVs:
75     Area address: 49.0002 (3)
76     Area address: 49.0003 (3)
77     Speaks: IP
78     Speaks: IPV6
79     IP router id: 192.168.1.6
80     IP address: 192.168.1.6
81 <...>
82     IP extended prefix: 192.168.1.6/32 metric 0 up # Primary address
83     IP extended prefix: 172.16.1.36/30 metric 643 up
84     IP extended prefix: 172.16.1.32/30 metric 643 up
85     IP extended prefix: 172.16.1.88/30 metric 100 up
86     IP extended prefix: 172.16.1.52/30 metric 100 up
87   No queued transmissions
```

Control of link-state IGP redistribution

Junos OS includes several policy control and management knobs that leverage the leaking and route redistribution processes between IGPs. Such configuration resources can be helpful in different setups to provide additional mechanisms to safeguard route injection and protect domains against accidental failures or human errors.

General route limitation: `maximum-prefixes` and `maximum-paths`
Junos OS includes specific per-instance knobs to control the maximum number of routes to be installed. The `maximum-prefixes` knob determines a limit on the number of prefixes, whereas `maximum-paths` sets such a limit on the number of paths. Note that in the case of link-state IGPs, routes are expected to have one closest path (with one or more next hops) per prefix decided after SPF calculation.

Such configuration statements apply globally to the instance. Both standalone paths from other protocols and simultaneous advertisement of the same address space by different protocols (which effectively creates different paths for the same prefix) need to be considered as well. Therefore, the *number of paths* requires careful analysis when being influenced by several protocols, because protocols such as RIP only keep the best *path* for a prefix, discarding eventual backup paths from other neighbors, while other protocols such as BGP just select the best *path* as active for the routing table, but keep other backup paths.

In any case, both configuration options can provide another protection barrier in a route redistribution topology. Listing 2.53 summarizes the related configuration options.

Listing 2.53: Instance maximum prefix and path limitation in Junos

```
1  [edit routing-options]
2  user@Livorno# set maximum-prefixes ?
3  Possible completions:
4    <limit>              Maximum number of prefixes (1..4294967295)
5    log-interval         Minimum interval between log messages (5..86400 seconds)
6    log-only             Generate warning messages only
7    threshold            Percentage of limit at which to start generating warnings (1..100)
8  [edit routing-options]
9  user@Livorno# set maximum-paths ?
10 Possible completions:
11   <limit>              Maximum number of paths (1..4294967295)
12   log-interval         Minimum interval between log messages (5..86400 seconds)
13   log-only             Generate warning messages only
14   threshold            Percentage of limit at which to start generating warnings (1..100)
```

Note that if any of these limits is reached, excessive routes are discarded and cannot be retrieved for installation, even if the number of prefixes or paths moves below the limit again. After a network incident related to these events, it is necessary to regenerate completely the routing table for the affected routing instance to restore the service.

External route limitation: `prefix-export-limit` for OSPF and `prefix-export-limit` for IS–IS.

Without any other safety net, route redistribution is controlled by properly applied policies. Although policy definition and application needs careful thought, Junos OS offers an additional resource to limit the maximum number of prefixes that can be exported to a given link-state IGP.

By default, there is no limit on the number of injected routes, and policy misconfiguration or human errors when defining or applying such policies open the door for network-wide meltdown or churn because of the scope-wide nature of external information in link-state IGPs.

A recommended practice is to provide a maximum boundary to the number of prefixes injected by the router in the corresponding IGP. If this limit is reached, the router enters an *overload* state for the IGP until an administrator intervenes. Listings 2.54, 2.55, and 2.56 illustrate its configuration in link-state IGPs. It is worth noting that in IS–IS, this configuration option can be tuned up to the level stanza, which fully covers the needs of a narrow-style topology.

Listing 2.54: OSPFv2 prefix-export-limit configuration in Junos

```
1 [edit protocols ospf]
2 user@Livorno# set prefix-export-limit ?
3 Possible completions:
4   <prefix-export-limit>  Maximum number of prefixes that can be exported (0..4294967295)
```

Listing 2.55: Global and IPv4 address family OSPFv3 prefix-export-limit configuration in Junos

```
1 [edit protocols ospf3]
2 user@Livorno# set prefix-export-limit ?
3 Possible completions:
4   <prefix-export-limit>  Maximum number of prefixes that can be exported (0..4294967295)
5 [edit protocols ospf3]
6 user@Livorno# set realm ipv4-unicast prefix-export-limit ?
7 Possible completions:
8   <prefix-export-limit>  Maximum number of prefixes that can be exported (0..4294967295)
```

Listing 2.56: IS–IS prefix-export-limit configuration in Junos

```
1 [edit protocols isis]
2 user@Livorno# set level 1 prefix-export-limit ?
3 Possible completions:
4   <prefix-export-limit>  Maximum number of external prefixes that can be exported (0..4294967295)
5 [edit protocols isis]
6 user@Livorno# set level 2 prefix-export-limit ?
7 Possible completions:
8   <prefix-export-limit>  Maximum number of external prefixes that can be exported (0..4294967295)
```

OSPF Areas: `from area`

Junos OS considers an OSPF-specific matching condition in route policies to granularly control the area from which routes can be exported to another protocol. Note that this policy does not look into a specific OSPF database, but rather, picks up those *active* routes installed in the routing table as a consequence of an LSA in the corresponding area, which become eligible routes for redistribution into other protocols.

While this is a powerful tool to align route information when migrating from OSPF, it may not cover all existing routing information in the area's LSAs because other route versions may be active in the routing table, such as direct or local routes. Listing 2.57 shows a case in which active OSPF routes from Area 0.0.0.3 are redistributed into IS–IS.

Listing 2.57: OSPF `from area` policy condition

```
1  [edit policy-options policy-statement from-ospf-area-3]
2  user@Livorno# show
3  term 1 {
4      from {
5          protocol ospf;
6          area 0.0.0.3;
7      }
8      then accept;
9  }
10 [edit]
11 user@Livorno# set protocols isis export from-ospf-area-3
```

IS–IS Levels: `from level`, `to level`

With a language similar to policies that control leaking inside a common IS–IS domain, Junos OS includes the same knobs to add further granularity when injecting routes either in IS–IS or another protocol.

Similarly to OSPF, IS–IS does not directly extract all available existing prefix information from TLVs, but rather, applies to existing active routes whose attributes include a specific IS–IS Level. Listing 2.58 illustrates an example that redistributes only IS–IS Level 1 routes (similar to the OSPFv2 export policy) and Listing 2.59 sets up OSPFv2 route injection only in IS–IS' own Level 2 LSP.

Listing 2.58: IS–IS `from level` policy condition

```
1  [edit policy-options policy-statement from-IS-IS-level-1]
2  user@Livorno# show
3  term 1 {
4      from {
5          protocol isis;
6          level 1;
7      }
8      then accept;
9  }
10 [edit]
11 user@Livorno# set protocols ospf export from-IS-IS-level-1
```

Listing 2.59: IS–IS `to level` policy condition

```
1  [edit policy-options policy-statement from-ospf-to-level-2]
2  user@Livorno# show
3  term 1 {
4      from protocol ospf;
5      to level 2;
```

```
6     then accept;
7  }
8  [edit]
9  user@Livorno# set protocols isis export from-ospf-to-level-2
```

External route tagging: `tag`

External route tagging is an easy mechanism for identifying and filtering routes matching a given value. In Junos OS, a `tag` remains both local for the route in the table in Junos OS and is automatically populated to OSPF External LSAs and injected routes in IS–IS LSPs. [RFC2328] Section A.4.5 already defines a 32-bit External Route Tag to be used to propagate information across the domain, and [RFC5130] defined new 32-bit and 64-bit sub-TLVs to be attached to IP prefix TLVs and carry similar information.

Listings 2.60 and 2.61 modify previous policies from Listings 2.58 and 2.59 to add this administrative marking and show corresponding excerpts in the information constructs in Lines 20 of Listing 2.60, and 16, 19, and 22 of Listing 2.61.

Listing 2.60: OSPF tagging AS external routes

```
1   [edit policy-options policy-statement from-IS-IS-level-1]
2  user@Livorno# show
3  term 1 {
4      from {
5          protocol isis;
6          level 2;
7      }
8      then {
9          tag 1000;
10         accept;
11     }
12 }
13  [edit policy-options policy-statement from-IS-IS-level-1]
14 user@Livorno# run show ospf database external extensive advertising-router 192.168.1.6
15     OSPF AS SCOPE link state database
16  Type       ID               Adv Rtr           Seq      Age  Opt  Cksum  Len
17 Extern  *84.16.1.1       192.168.1.6      0x80000004    39  0x22 0x2c5b  36
18   mask 255.255.255.255
19   Topology default (ID 0)
20     Type: 2, Metric: 866, Fwd addr: 0.0.0.0, Tag: 0.0.3.232 # Administrative tag
21   Gen timer 00:49:20
22   Aging timer 00:59:20
23   Installed 00:00:39 ago, expires in 00:59:21, sent 00:00:39 ago
24   Last changed 00:00:39 ago, Change count: 2, Ours
```

Listing 2.61: IS–IS tagging injected route information

```
1   [edit policy-options policy-statement from-ospf-to-level-2]
2  user@Livorno# show
3  term 1 {
4      from protocol ospf;
5      to level 2;
6      then {
7          tag 2000;
8          accept;
9      }
10 }
11  [edit policy-options policy-statement from-ospf-to-level-2]
12 user@Livorno# run show isis database Livorno extensive
13 <...>
14    IP extended prefix: 172.16.1.0/30 metric 803 up
15       6 bytes of subtlvs
```

```
16        Administrative tag 1: 2000 # Administrative tag
17      IP extended prefix: 172.16.1.4/30 metric 903 up
18        6 bytes of subtlvs
19        Administrative tag 1: 2000 # Administrative tag
20      IP extended prefix: 172.16.1.8/30 metric 963 up
21        6 bytes of subtlvs
22        Administrative tag 1: 2000 # Administrative tag
```

These fields can be tremendously helpful in migration scenarios, because related administrative information can be easily attached to routes. Link-state IGP `import` policies can also match on those tags and derive route installation from them. Even though there is no foreseen use for this feature during migrations, having tags for the simple purpose of uniquely tagging and identifying re-injected routes at each point in the topology is well worth the effort of administrating and including an additional tagging action in policies.

Link-state IGP preference

Junos OS has some predefined preferences for link-state IGP routes, even to the point of distinguishing whether they are internal to the infrastructure or external to the domain, and with different values for each IS–IS level. Table 2.7 summarizes the default link-state IGP preferences. Note that a lower value indicates a stronger preference.

Table 2.7 Default link-state IGP route preferences in Junos OS

Route type	Preference value
OSPF internal route	10
IS–IS Level 1 internal route	15
IS–IS Level 2 internal route	18
OSPF AS external route	150
IS–IS Level 1 external route	160
IS–IS Level 2 external route	165

Junos OS includes knobs to manipulate the preference value. In the case of link-state IGPs, Junos OS further distinguishes between internal and external preference, as shown in Listings 2.62 and 2.63. Note that this internal–external distinction can also be further tuned in the case of IS–IS, to isolate Level 1 and Level 2 routes.

Listing 2.62: OSPF preference manipulation in Junos OS

```
1  [edit protocols ospf]
2  user@Livorno# set preference ?
3  Possible completions:
4    <preference>          Preference of internal routes
5  [edit protocols ospf]
6  user@Livorno# set external-preference ?
7  Possible completions:
8    <external-preference>  Preference of external routes
9  [edit protocols ospf3]
```

```
10 user@Livorno# set preference ?
11 Possible completions:
12   <preference>          Preference of internal routes
13 [edit protocols ospf3]
14 user@Livorno# set external-preference ?
15 Possible completions:
16   <external-preference> Preference of external routes
```

Listing 2.63: IS–IS preference manipulation in Junos OS

```
1  [edit protocols isis]
2  user@Livorno# set level 1 preference ?
3  Possible completions:
4    <preference>          Preference of internal routes
5  [edit protocols isis]
6  user@Livorno# set level 1 external-preference ?
7  Possible completions:
8    <external-preference> Preference of external routes
9  [edit protocols isis]
10 user@Livorno# set level 2 preference ?
11 Possible completions:
12   <preference>          Preference of internal routes
13 [edit protocols isis]
14 user@Livorno# set level 2 external-preference ?
15 Possible completions:
16   <external-preference> Preference of external routes
```

Link-state IGP metric correspondence

In the case of deploying coexisting link-static IGP layers, it is fairly obvious that metrics need to be commonly aligned or at least need to be in the same order of magnitude in each protocol but keeping similar ratios, so that the intended behavior in the final scenario will be the same.

From that perspective, OSPFv2 and OSPFv3 have a traditional interface cost model based on a reference value (originally 100000000) divided by the actual link speed rate. This *bandwidth-dependent* calculation made sense for the protocol developers at the time (when compared with more complicated metric references from protocols like IGRP or EIGRP), but this reference value is obsolete in current networks, because link speeds have increased considerably over the last years.

Most routing operating systems are versatile with this parameter and allow different values, so that the *bandwidth reference* is usable in modern network designs. Listings 2.64 and 2.65 show these Junos OS knobs for OSPFv2 and OSPFv3.

Listing 2.64: OSPFv2 reference bandwidth manipulation in Junos OS

```
1 [edit protocols ospf]
2 user@Livorno# set reference-bandwidth ?
3 Possible completions:
4   <reference-bandwidth> Bandwidth for calculating metric defaults (9600..1000000000000)
```

Listing 2.65: OSPFv3 reference bandwidth manipulation in Junos OS

```
1 [edit protocols ospf3]
2 user@Livorno# set reference-bandwidth ?
3 Possible completions:
4   <reference-bandwidth> Bandwidth for calculating metric defaults (9600..1000000000000)
```

In the case of IS–IS, interface-related metrics are not related to the link speed by definition. Most operation systems calculate a default value for all kinds of links and for each level. Another good practice is to acquire the concept from OSPF for default metric calculation based on a reference bandwidth. Junos OS provides a similar knob so that all interface- and neighbor-related metrics are derived from exactly the same calculation, as shown in Listing 2.66.

Listing 2.66: IS–IS reference bandwidth manipulation in Junos OS

```
1 [edit protocols isis]
2 user@Livorno# set reference-bandwidth ?
3 Possible completions:
4   <reference-bandwidth>  Bandwidth for calculating metric defaults (9600..1000000000000)
```

Referencing common metrics in an overlay protocol rollout is an obvious step, but quite often forgotten. While OSPF derives interface costs, by default, from a reference bandwidth, even when it is not configured, IS–IS assigns a default interface cost of 10 for both levels and for all types of media and speed. Even though metrics can be statically assigned at each protocol, it is recommended that engineers think twice when configuring link costs and see whether any default protocol behavior could apply at any link on any system in the network.

Link-state IGP alignment with MPLS label distribution protocols

In such an overlay situation, it is fundamental to plan a smooth transition for all dependent protocols and applications. Migrating an IGP when toggling protocol preferences does not necessarily mean that LDP label bindings or TED information follows the same move.

LDP: *Link-state IGP alignment with LDP*

In Junos OS, it is considered a best practice to include the `track-igp-metric` for the LDP protocol as illustrated in Listing 2.67. In this case, instead of deciding the best route in inet.3 based on a default LDP metric whose unilateral value is 1 for all label bindings, LDP metrics are determined by IGP costs for the best FEC in each case. This step is vital in the case of simultaneous paths or protocol coexistence so as to choose the best path for the optimal label binding.

Listing 2.67: IGP metric tracking activation for LDP in Junos OS

```
1 [edit protocols ldp]
2 user@Livorno# set track-igp-metric
```

In addition to tracking exact IGP metric values, there could be interim periods of time in which an IGP neighbor relationship could be fully established but the parallel LDP session has not reached that status. Because of the exclusive IGP-driven route selection, this situation leads to MPLS traffic blackholing, as MPLS forwarding should follow LDP bindings over a path dictated by the IGP, but there is a given segment on which no label has been allocated.

For such transient or inaccurate states, Junos OS provides a feature called *LDP-IGP synchronization*. The rationale behind this concept is to avoid any such divergence between LDP and the corresponding active IGP by advertising that link with maximum cost in the corresponding IGP until LDP becomes fully operational over it, that is, until the LDP session is established (or until the neighbor for that session is recognized when more than a single link exists for an LDP session peer) and the label database is exchanged. During this period of time, the link has the maximum IGP metric value, which has the clear drawback of affecting traffic which is not MPLS-switched, although this interface with maximum metric is still present and eligible for CSPF computations.

Note that configuration of this feature in Junos OS for both OSPFv2 and IS–IS is not global but per interface and can therefore be further tuned with fine-grained control on a per-link basis. Doing this is as simple as attaching the `ldp-synchronization` knob to the respective interface stanza inside each protocol, as shown in Listing 2.68.

Listing 2.68: LDP–IGP synchronization activation on a given interface in OSPFv2 and IS–IS in Junos OS

```
1 [edit protocols ospf]
2 user@Livorno# set area 0 interface so-0/0/0.0 ldp-synchronization
3 [edit protocols isis]
4 user@Livorno# set interface so-0/0/0.0 ldp-synchronization
```

Considering our lab topology and assuming parallel deployment of flat OSPFv2 and IS–IS Level 2 with default preferences, as shown in Figure 2.14, misconfiguration of *LDP–IGP synchronization* can have an even more dramatic impact that results from misalignment in cost assignment to different links. Listing 2.69 depicts an example situation in this setup in which only OSPFv2 is configured with *LDP–IGP synchronization* and the LDP session over a certain interface is torn down, here with manual intervention (consider flaps or human errors in a live network).

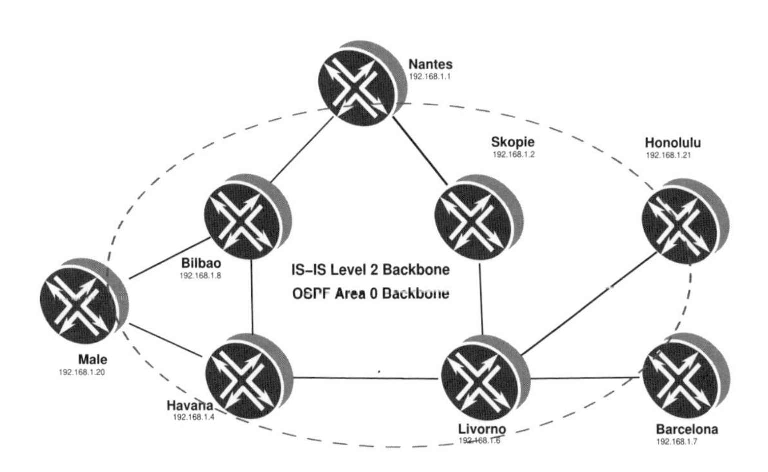

Figure 2.14: Network topology with flat OSPFv2 Area 0 and uniform IS–IS Level 2.

Listing 2.69: LDP–IGP synchronization misalignment effect in Livorno towards Bilbao

```
1 user@Livorno> show route 192.168.1.8
2
3 inet.0: 9830 destinations, 9859 routes (9829 active, 0 holddown, 1 hidden)
4 + = Active Route, - = Last Active, * = Both
```

```
5
6  192.168.1.8/32      *[OSPF/10] 00:07:06, metric 903
7                       > via so-0/0/0.0
8                       [IS-IS/18] 00:07:05, metric 903
9                       > to 172.16.1.33 via so-0/0/0.0
10
11 inet.3: 10 destinations, 11 routes (10 active, 0 holddown, 0 hidden)
12 + = Active Route, - = Last Active, * = Both
13
14 192.168.1.8/32      *[LDP/9] 00:07:06, metric 903
15                      > via so-0/0/0.0, Push 299872
16
17 user@Livorno> configure
18 Entering configuration mode
19
20 [edit]
21 user@Livorno# set protocols ldp interface so-0/0/0.0 disable
22 [edit]
23 user@Livorno# set protocols ospf area 0 interface so-0/0/0.0 ldp-synchronization
24 [edit]
25 user@Livorno# commit and-quit
26 commit complete
27 Exiting configuration mode
28
29 user@Livorno> show route 192.168.1.8
30
31 inet.0: 9830 destinations, 9859 routes (9829 active, 0 holddown, 1 hidden)
32 + = Active Route, - = Last Active, * = Both
33
34 192.168.1.8/32      *[OSPF/10] 00:00:06, metric 1546
35                      > via so-0/0/1.0
36                      [IS-IS/18] 00:08:38, metric 903
37                      > to 172.16.1.33 via so-0/0/0.0
38
39 inet.3: 10 destinations, 11 routes (10 active, 0 holddown, 0 hidden)
40 + = Active Route, - = Last Active, * = Both
41
42 192.168.1.8/32      *[LDP/9] 00:00:06, metric 1546
43                      > via so-0/0/1.0, Push 303280
```

Whenever that interface in router *Livorno* is disabled in LDP and *LDP–IGP synchronization* is active with OSPF (Line 22), router *Livorno* selects another path with a higher metric but keeps the OSPF route active, being a preferred protocol.

In that situation, if router *Havana*, which is on the other side of the LDP-synced link, had already moved to IS–IS Level 2 as the preferred protocol, while router *Livorno* keeps selecting OSPF as the preferred protocol, a routing loop could have been formed because of indirect path selection divergence between both IGPs. In fact, this is the very same effect of not assigning the same or proportional link costs at coexisting IGPs in the same topology.

The bottom line in this situation is that errors in the different cost parameters are not obvious in such transition scenarios and it is therefore recommended that *LDP–IGP synchronization* configuration be aligned on a per-link basis when link-state IGPs are operating as *ships-in-the-night*.

RSVP: *Link-state IGP alignment with RSVP-TE*

TE metrics can be further tuned and even become independent of the simple IGP link cost. Both OSPF and IS–IS TE resources include fields to attach specific *TE metrics* to each link. Listing 2.70 shows the configuration changes needed to specify a different metric on a given link in OSPFv2 and IS–IS (on a per level basis).

Listing 2.70: TE metric configuration in OSPFv2 and IS–IS

```
1 [edit]
2 user@Livorno# set protocols ospf area 0 interface so-0/0/0.0 te-metric 1000
3 [edit]
4 user@Livorno# set protocols isis interface so-0/0/0.0 level 1 te-metric 1000
5 [edit]
6 user@Livorno# set protocols isis interface so-0/0/0.0 level 2 te-metric 1000
```

Also, the transition of IGP route preference does not necessarily mean that information injection in the TED for RSVP calculations needs to be aligned with the IGP preference change. Junos OS can further fine-tune this potential divergence between IGP route preference and TED information with a concept called *TED protocol credibility*.

By default, Junos OS always prefers IS–IS information in the TED in preference to information from other IGPs, even if these IGPs have better route preference. For instance, this situation may occur when IS–IS and OSPF coexist in the same setup, as OSPF is assigned with lower preference values by default, as shown in Table 2.7.

Junos OS has added a specific knob for OSPF version 2 and IS–IS, as seen in Listings 2.71 and 2.72, to tweak this behavior so that TED information is made dependent on the current IGP preference, thus allowing the IGP preference to automatically determine TED credibility.

Listing 2.71: Activation of TED protocol credibility for OSPF version 2 in Junos OS

```
1 [edit protocols ospf]
2 user@Livorno# set traffic-engineering credibility-protocol-preference
```

Listing 2.72: Activation of TED protocol credibility for IS–IS in Junos OS

```
1 [edit protocols isis]
2 user@Livorno# set traffic-engineering credibility-protocol-preference
```

In fact, when the `credibility-protocol-preference` is active, the following calculation is implemented to determine the value:

$$[\textit{TED credibility}] = 512 - [\textit{IGP preference}].$$

In this case, the IGP with the highest *credibility* is chosen. The *credibility* is for the whole TED database and not per link or per node, so this feature has global scope. Note as well that while lower standard IGP preference values indicate stronger predilection, TED credibility follows an ascending scheme and more credible protocols for TED are chosen based on higher values.

In Junos OS, protocols are sorted according to protocol credibility when injecting information in the TED. When a certain IGP adds node and link information to the TED, these details are inserted into a list based on the TED protocol preference, starting with the most preferred protocol. The implementation of this feature does not preclude full repopulation of the TED but simply reordering the protocol list, because the information is already present with the protocol traffic-engineering information extensions.

However, when a given user changes such protocol preference, the old TED information is destroyed and TED is populated with the newly available information for the instances that have traffic-engineering.

Another side effect is that any changes in the standard IGP preference value also influence TED credibility with this knob active and automatically feed the TED with the new information. This means that repopulation not only takes place when activating or deactivating the `credibility-protocol-preference` knob, but also automatically when this knob is on and when any explicit change of IGP preferences occurs.

Looking again at Figure 2.14 with both IGP layers as reference topology, in the case of LDP, paths are aligned as soon as LDP tracks the IGP metric, and the *LDP–IGP synchronization* grants additional resilience for transitory scenarios or human errors. Even though metrics can be lower with another IGP, IGP protocol preference is the final tiebreaker for route selection as soon as it is aligned with LDP. When considering connectivity in this setup from router *Livorno* to router *Bilbao* as shown in Listing 2.73, router *Livorno* selects the OSPFv2 route for label-to-FEC binding on LDP to router *Bilbao*, although a lower cost is visible through IS–IS. This choice has been explicitly forced in this example by not enabling IS–IS *wide-metrics*, thus leading to different cost values in both IGPs.

Listing 2.73: LDP label binding and route selection to Bilbao from Livorno

```
1  user@Livorno> show route 192.168.1.8
2
3  inet.0: 70 destinations, 101 routes (69 active, 0 holddown, 1 hidden)
4  + = Active Route, - = Last Active, * = Both
5
6  192.168.1.8/32      *[OSPF/10] 00:45:03, metric 744
7                       > via so-0/0/0.0
8                       [IS-IS/18] 00:45:03, metric 156
9                       > to 172.16.1.33 via so-0/0/0.0
10
11 inet.3: 10 destinations, 10 routes (10 active, 0 holddown, 0 hidden)
12 + = Active Route, - = Last Active, * = Both
13
14 192.168.1.8/32      *[LDP/9] 00:41:42, metric 744
15                      > via so-0/0/0.0, Push 300224
```

Both OSPFv2 and IS–IS include all previously described traffic-engineering extensions, but only a subset of information prevails in the internal TED. Using the same case, router *Livorno* receives both Opaque LSAs and TE sub-TLVs for data referring to router *Bilbao*, as shown in Listing 2.74.

Listing 2.74: TE IGP constructs for Bilbao

```
1  user@Livorno> show ospf database opaque-area advertising-router 192.168.1.8 extensive
2
3      OSPF database, Area 0.0.0.0
4   Type       ID               Adv Rtr           Seq      Age  Opt  Cksum  Len
5  OpaqArea 1.0.0.1          192.168.1.8      0x80000040  1082  0x22 0x4ebc  28
6    Area-opaque TE LSA
7    RtrAddr (1), length 4: 192.168.1.8
8    Aging timer 00:41:58
9    Installed 00:17:53 ago, expires in 00:41:58, sent 00:17:51 ago
10   Last changed 1d 00:38:36 ago, Change count: 1
11 OpaqArea 1.0.0.3          192.168.1.8      0x80000018   462  0x22 0x4334 124
12   Area-opaque TE LSA
13   Link (2), length 100:
14     Linktype (1), length 1:
15       2
16     LinkID (2), length 4:
17       172.16.1.5
18     LocIfAdr (3), length 4:
19       172.16.1.5
20     RemIfAdr (4), length 4:
21       0.0.0.0
22     TEMetric (5), length 4:
23       100
24     MaxBW (6), length 4:
```

```
25        1000Mbps
26      MaxRsvBW (7), length 4:
27        1000Mbps
28      UnRsvBW (8), length 32:
29  <...>
30      Color (9), length 4:
31        0
32    Aging timer 00:52:18
33    Installed 00:07:33 ago, expires in 00:52:18, sent 00:07:31 ago
34    Last changed 15:26:21 ago, Change count: 1
35
36  user@Livorno> show isis database Bilbao-re0.00-00 extensive
37  IS-IS level 1 link-state database:
38
39  IS-IS level 2 link-state database:
40
41  Bilbao-re0.00-00 Sequence: 0xe95, Checksum: 0x3ce1, Lifetime: 598 secs
42     IS neighbor: Torino.00                    Metric:       63
43     IS neighbor: male.00             Metric:       63
44     IP prefix: 172.16.1.4/30                  Metric:       63 Internal Up
45     IP prefix: 172.16.1.76/30                 Metric:       63 Internal Up
46     IP prefix: 192.168.1.8/32                 Metric:        0 Internal Up
47  <...>
48    TLVs:
49      Area address: 49.0004 (3)
50      Speaks: IP
51      Speaks: IPV6
52      IP router id: 192.168.1.8
53      IP address: 192.168.1.8
54      Hostname: Bilbao-re0
55      IS neighbor: Nantes.02, Internal, Metric: default 63
56      IS neighbor: male.00, Internal, Metric: default 63
57      IS extended neighbor: Nantes.02, Metric: default 63
58        IP address: 172.16.1.5
59        Local interface index: 77, Remote interface index: 0
60        Current reservable bandwidth:
61  <...>
62        Maximum reservable bandwidth: 1000Mbps
63        Maximum bandwidth: 1000Mbps
64        Administrative groups:  0 <none>
65      IS extended neighbor: male.00, Metric: default 63
66        IP address: 172.16.1.78
67        Neighbor's IP address: 172.16.1.77
68        Local interface index: 70, Remote interface index: 95
69      IP prefix: 172.16.1.4/30, Internal, Metric: default 63, Up
70      IP prefix: 192.168.1.8/32, Internal, Metric: default 0, Up
71      IP prefix: 172.16.1.76/30, Internal, Metric: default 63, Up
72      IP extended prefix: 172.16.1.4/30 metric 63 up
73      IP extended prefix: 192.168.1.8/32 metric 0 up
74      IP extended prefix: 172.16.1.76/30 metric 63 up
75    No queued transmissions
```

However, each protocol has a given strict credibility to inject such information in the router TED. Listing 2.75 shows the TED status in our example.

Listing 2.75: TE protocol credibility and database entry for router *Bilbao* in router *Livorno*

```
 1  user@Livorno> show ted protocol
 2  Protocol name        Credibility  Self node
 3  IS-IS(2)             2            Livorno.00(192.168.1.6)
 4  IS-IS(1)             1
 5  OSPF(0)              0            Livorno.00(192.168.1.6)
 6
 7  user@Livorno> show ted database 192.168.1.8 extensive
 8  TED database: 16 ISIS nodes 16 INET nodes
 9  NodeID: Bilbao-re0.00(192.168.1.8)
10    Type: Rtr, Age: 184 secs, LinkIn: 3, LinkOut: 2
11    Protocol: IS-IS(2)                  # Higher credibility for IS-IS injected information
```

```
12      To: Nantes.02, Local: 172.16.1.5, Remote: 0.0.0.0
13        Local interface index: 71, Remote interface index: 0
14        Color: 0 <none>
15        Metric: 63
16        Static BW: 1000Mbps
17        Reservable BW: 1000Mbps
18        Available BW [priority] bps:
19            [0] 1000Mbps      [1] 1000Mbps     [2] 1000Mbps     [3] 1000Mbps
20            [4] 1000Mbps      [5] 1000Mbps     [6] 1000Mbps     [7] 1000Mbps
21        Interface Switching Capability Descriptor(1):
22          Switching type: Packet
23          Encoding type: Packet
24          Maximum LSP BW [priority] bps:
25            [0] 1000Mbps      [1] 1000Mbps     [2] 1000Mbps     [3] 1000Mbps
26            [4] 1000Mbps      [5] 1000Mbps     [6] 1000Mbps     [7] 1000Mbps
27    Protocol: OSPF(0.0.0.0)             # Lower credibility for OSPF injected information
28      To: 172.16.1.5-1, Local: 172.16.1.5, Remote: 0.0.0.0
29        Local interface index: 0, Remote interface index: 0
30        Color: 0 <none>
31        Metric: 100
32        Static BW: 1000Mbps
33        Reservable BW: 1000Mbps
34        Available BW [priority] bps:
35            [0] 1000Mbps      [1] 1000Mbps     [2] 1000Mbps     [3] 1000Mbps
36            [4] 1000Mbps      [5] 1000Mbps     [6] 1000Mbps     [7] 1000Mbps
37        Interface Switching Capability Descriptor(1):
38          Switching type: Packet
39          Encoding type: Packet
40          Maximum LSP BW [priority] bps:
41            [0] 1000Mbps      [1] 1000Mbps     [2] 1000Mbps     [3] 1000Mbps
42            [4] 1000Mbps      [5] 1000Mbps     [6] 1000Mbps     [7] 1000Mbps
```

In this situation, if divergences in the IGP topologies are created by deactivating IS–IS on the best path interface to router *Bilbao* and a simple RSVP-based LSP is defined from router *Livorno*, differences in the path computation become explicit, as shown in Listing 2.76.

Listing 2.76: IGP and TE divergences towards router *Bilbao* in router *Livorno*

```
1  [edit]
2  user@Livorno# set protocols mpls label-switched-path Livorno-to-Bilbao to 192.168.1.8
3  [edit]
4  user@Livorno# set protocols isis interface so-0/0/0.0 disable
5  [edit]
6  user@Livorno# commit and-quit
7  commit complete
8  Exiting configuration mode
9
10 user@Livorno> show route 192.168.1.8
11
12 inet.0: 72 destinations, 105 routes (71 active, 0 holddown, 1 hidden)
13 + = Active Route, - = Last Active, * = Both
14
15 192.168.1.8/32      *[OSPF/10] 00:02:17, metric 744
16                      > via so-0/0/0.0
17                      [IS-IS/18] 00:01:48, metric 189
18                      > to 172.16.1.37 via so-0/0/1.0
19
20 inet.3: 10 destinations, 12 routes (10 active, 0 holddown, 0 hidden)
21 + = Active Route, - = Last Active, * = Both
22
23 192.168.1.8/32      *[RSVP/7] 00:01:57, metric 744
24                      > via so-0/0/1.0, label-switched-path Livorno-to-Bilbao
25                      [LDP/9] 00:02:05, metric 744
26                      > via so-0/0/0.0, Push 300224
27
28 user@Livorno> show ted protocol
29 Protocol name         Credibility  Self node
```

```
30  IS-IS(2)              2            Livorno.00(192.168.1.6)
31  IS-IS(1)              1
32  OSPF(0)               0            Livorno.00(192.168.1.6)
```

The following items are worth noting in Listing 2.76:

- Even though OSPF internal is preferred, RSVP-TE follows the IS–IS path because it is the protocol with the highest TED credibility.
- Despite selecting the IS–IS path, the RSVP-TE route in inet.3 includes the OSPF route metric because such metric is not strictly part of the TE extensions.
- Junos OS default values without considering TED credibility knobs derive from the selection of different IGPs, depending on the application.

The status shown above could easily be a transient or even a permanent situation during a migration period. The `traffic-engineering credibility-protocol-preference` configuration resource allows the IGP preference to align with TED credibility.

If the TED protocol credibility is adapted, OSPFv2 then becomes the preferred protocol with regards to MPLS path selection, as shown in Listing 2.77. By clearing the LSP for optimization, recomputation towards the new route is forced.

Listing 2.77: TED protocol credibility adaption in Bilbao

```
 1  user@Livorno> configure
 2  Entering configuration mode
 3
 4  [edit]
 5  user@Livorno# set protocols ospf traffic-engineering credibility-protocol-preference
 6  [edit]
 7  user@Livorno# set protocols isis traffic-engineering credibility-protocol-preference
 8
 9  user@Livorno> clear mpls lsp optimize
10
11  user@Livorno> show route 192.168.1.8
12
13  inet.0: 72 destinations, 105 routes (71 active, 0 holddown, 1 hidden)
14  + = Active Route, - = Last Active, * = Both
15
16  192.168.1.8/32      *[OSPF/10] 00:16:51, metric 744
17                       > via so-0/0/0.0
18                       [IS-IS/18] 00:16:22, metric 189
19                       > to 172.16.1.37 via so-0/0/1.0
20
21  inet.3: 10 destinations, 12 routes (10 active, 0 holddown, 0 hidden)
22  + = Active Route, - = Last Active, * = Both
23
24  192.168.1.8/32      *[RSVP/7] 00:01:05, metric 744
25                       > via so-0/0/0.0, label-switched-path Livorno-to-Bilbao
26                       [LDP/9] 00:16:39, metric 744
27                       > via so-0/0/0.0, Push 300224
28
29  user@Livorno> show ted protocol
30  Protocol name          Credibility  Self node
31  OSPF(0)                502          Livorno.00(192.168.1.6)
32  IS-IS(1)               497
33  IS-IS(2)               494          Livorno.00(192.168.1.6)
34
35  user@Livorno> show ted database 192.168.1.8 extensive
36  TED database: 16 ISIS nodes 16 INET nodes
37  NodeID: Bilbao-re0.00(192.168.1.8)
```

```
38    Type: Rtr, Age: 53 secs, LinkIn: 3, LinkOut: 2
39    Protocol: OSPF(0.0.0.0)          # Higher credibility for OSPF injected information
40      To: 172.16.1.5-1, Local: 172.16.1.5, Remote: 0.0.0.0
41        Local interface index: 0, Remote interface index: 0
42        Color: 0 <none>
43        Metric: 100
44        Static BW: 1000Mbps
45        Reservable BW: 1000Mbps
46  <...>
47    Protocol: IS-IS(2)               # Lower credibility for IS-IS injected information
48      To: Nantes.02, Local: 172.16.1.5, Remote: 0.0.0.0
49        Local interface index: 71, Remote interface index: 0
50        Color: 0 <none>
51        Metric: 63
52        Static BW: 1000Mbps
53        Reservable BW: 1000Mbps
54  <...>
```

In this setup and with this configuration, both IP and MPLS TE path selection are aligned as dictated by the most preferred and TED-credible IGP. In this context, it is considered a best practice to encompass any IGP preference modification with TED protocol credibility adaptions for route selection to be mirrored in the TED using `traffic-engineering credibility-protocol-preference` knobs in the respective IGPs.

In this state, when tweaking the IS–IS Level 2 internal preference on router *Livorno* to be preferred against OSPF internal routes, this change is immediately followed by preferring that link-state IGP for TED computations as well. Listing 2.78 illustrates the results for this change with an MPLS path reoptimization.

Listing 2.78: IS–IS Level 2 internal preference modified in Livorno

```
1  [edit]
2  user@Livorno# set protocols isis level 2 preference 8
3  [edit]
4  user@Livorno# commit and-quit
5  commit complete
6  Exiting configuration mode
7
8  user@Livorno> clear mpls lsp optimize
9  user@Livorno> show route 192.168.1.8
10
11 inet.0: 72 destinations, 105 routes (71 active, 0 holddown, 1 hidden)
12 + = Active Route, - = Last Active, * = Both
13
14 192.168.1.8/32     *[IS-IS/8] 00:00:31, metric 189
15                     > to 172.16.1.37 via so-0/0/1.0
16                     [OSPF/10] 00:23:11, metric 744
17                     > via so-0/0/0.0
18
19 inet.3: 10 destinations, 12 routes (10 active, 0 holddown, 0 hidden)
20 + = Active Route, - = Last Active, * = Both
21
22 192.168.1.8/32     *[RSVP/7] 00:00:05, metric 189
23                     > via so-0/0/1.0, label-switched-path Livorno-to-Bilbao
24                     [LDP/9] 00:00:31, metric 189
25                     > via so-0/0/1.0, Push 301008
26
27 user@Livorno> show ted protocol
28 Protocol name        Credibility  Self node
29 IS-IS(2)             504          Livorno.00(192.168.1.6)
30 OSPF(0)              502          Livorno.00(192.168.1.6)
31 IS-IS(1)             497
32
33 user@Livorno> show ted database 192.168.1.8 extensive
34 TED database: 16 ISIS nodes 16 INET nodes
```

```
35 NodeID: Bilbao-re0.00(192.168.1.8)
36   Type: Rtr, Age: 403 secs, LinkIn: 3, LinkOut: 2
37   Protocol: IS-IS(2)  # Higher credibility for IS-IS injected information
38     To: Nantes.02, Local: 172.16.1.5, Remote: 0.0.0.0
39       Local interface index: 71, Remote interface index: 0
40       Color: 0 <none>
41       Metric: 63
42       Static BW: 1000Mbps
43       Reservable BW: 1000Mbps
44 <...>
45   Protocol: OSPF(0.0.0.0)  # Lower credibility for OSPF injected information
46     To: 172.16.1.5-1, Local: 172.16.1.5, Remote: 0.0.0.0
47       Local interface index: 0, Remote interface index: 0
48       Color: 0 <none>
49       Metric: 100
50       Static BW: 1000Mbps
51       Reservable BW: 1000Mbps
52 <...>
```

It is worth noting the following from Listing 2.78:

- LDP follows the path dictated by the IS–IS internal route because of the lower preference.
- RSVP-TE follows the path dictated by the IS–IS internal route because it is the most credible protocol for the TED.
- Alignment between both label distribution protocols is automatically achieved when using `traffic-engineering credibility-protocol-preference`.

Link-state IGP adjacency transition for MPLS label distribution protocols

When deploying parallel IGP layers, once topologies have converged to select the same best paths, achieving a smooth transition for applications when moving from one IGP to another so as to pick up the active route is a vital factor for a close-to-hitless migration strategy.

Among those applications, MPLS labels and related protocols rank nowadays as the most expensive customers relying on IGP convergence and stability, especially as more and more MPLS-based services appear in networks. Although IGPs ultimately dictate paths and determine label selection at each ingress and transit node, even when IGPs are selecting the same next hop and determining the same label, some implications exist in MPLS label distribution protocols when preferring a new IGP route against an older one.

RSVP: By default, Junos OS tracks the IS–IS and OSPF adjacency state in RSVP.

Traffic-engineering link and node information are deleted by default when a link-state IGP adjacency is torn down, bringing down MPLS LSPs traversing through it. Also, as Junos OS notifies RSVP modules of the IGP adjacency drop, the RSVP neighbor is brought down as well. Even though IGP preferences and TED credibility can be tweaked without impacting RSVP neighbor relationships, this default behavior poses a challenge to maintaining the MPLS RSVP state and LSP forwarding at a later stage in a migration period, for instance, when removing adjacencies from unpreferred IGPs in a cleanup stage.

The `no-neighbor-down-notification` knob for OSPFv2 interfaces and the `no-adjacency-down-notification` knob for IS–IS interfaces provide a mechanism to ignore the IGP notification to RSVP and therefore ensure a completely hitless migration with regards to RSVP. Under normal circumstances, notifying RSVP upon IGP adjacency loss is considered to be an improvement with regards to convergence.

After attaching the knob to the respective interface, Junos OS requires the interface first to be deactivated at the IGP stanza. Once the adjacency has disappeared, that interface can be safely removed from the corresponding stanza. This transition in two steps is needed so that Junos OS first couples interface deactivation with the lack of an IGP adjacency loss notification to RSVP. This unused configuration snippet can be deleted at a later stage.

This feature also needs symmetric activation on all participating systems in the attached media. This alignment is needed so that no directly connected systems tear down RSVP neighbors and transit LSPs when the IGP adjacency timer expires. If the feature is not configured on all ends of the interface, and not only at ingress or egress LSP endpoints but also on the LSP transit systems, whichever system detects either an OSPF neighbor relationship or an IS–IS adjacency expiration when the interface first gets deactivated, triggers the LSP teardown in its direction.

For all these reasons, this feature is tremendously helpful in the later stages of a migration for a proper and hitless cleanup. However, it needs to be either globally deployed or rolled out carefully in interface groupings of attached routers. Listing 2.79 summarizes the configuration steps needed for both OSPFv2 and IS–IS with the previous scenario and includes steps in router *Skopie*, which is the router on the other end of the interface.

Listing 2.79: Per-interface configuration for OSPFv2 no-neighbor-down-notification and IS–IS no-adjacency-down-notification on router *Livorno* and router *Skopie*

```
1  user@Livorno> show route 192.168.1.8 detail
2
3  inet.0: 78 destinations, 118 routes (77 active, 0 holddown, 1 hidden)
4  192.168.1.8/32 (2 entries, 1 announced)
5          *IS-IS  Preference: 8
6                  Level: 2
7                  Next hop type: Router, Next hop index: 262143
8                  Next-hop reference count: 55
9                  Next hop: 172.16.1.33 via so-0/0/0.0, selected
10                 State: <Active Int>
11                 Local AS: 64530
12                 Age: 1d 1:18:17         Metric: 903
13                 Task: IS-IS
14                 Announcement bits (3): 0-KRT 4-LDP 7-Resolve tree 4
15                 AS path: I
16          OSPF   Preference: 10
17                 Next hop type: Router, Next hop index: 262147
18                 Next-hop reference count: 7
19                 Next hop: via so-0/0/0.0, selected
20                 State: <Int>
21                 Inactive reason: Route Preference
22                 Local AS: 64530
23                 Age: 1d 0:33:04         Metric: 903
24                 Area: 0.0.0.0
25                 Task: OSPF
26                 AS path: I
27
28 inet.3: 9 destinations, 12 routes (9 active, 0 holddown, 0 hidden)
29
30 192.168.1.8/32 (2 entries, 1 announced)
31         State: <FlashAll>
32         *RSVP   Preference: 7
33                 Next hop type: Router
34                 Next-hop reference count: 5
35                 Next hop: via so-0/0/0.0 weight 0x1, selected
36                 Label-switched-path Livorno-to-Bilbao
```

```
37                     Label operation: Push 300544
38                     State: <Active Int>
39                     Local AS: 64530
40                     Age: 1d 0:34:28          Metric: 903
41                     Task: RSVP
42                     Announcement bits (2): 2-Resolve tree 1 4-Resolve tree 4
43                     AS path: I
44             LDP     Preference: 9
45                     Next hop type: Router
46                     Next-hop reference count: 1
47                     Next hop: via so-0/0/0.0, selected
48                     Label operation: Push 299872
49                     State: <Int>
50                     Inactive reason: Route Preference
51                     Local AS: 64530
52                     Age: 1d 1:18:17          Metric: 903
53                     Task: LDP
54                     AS path: I
55  [edit protocols ospf area 0]
56  user@Livorno# set interface so-0/0/0.0 no-neighbor-down-notification
57  [edit]
58  user@Livorno# set protocols isis interface so-0/0/0.0 no-adjacency-down-notification
59  [edit[
60  user@Livorno# commit and-quit
61  commit complete
62  Exiting configuration mode
63
64  [edit]
65  user@Skopie# set protocols isis interface so-1/0/0.0 no-adjacency-down-notification
66  [edit protocols ospf area 0]
67  user@Skopie# set interface so-1/0/0.0 no-neighbor-down-notification
68  [edit[
69  user@Skopie# commit and-quit
70  commit complete
71  Exiting configuration mode
```

Note from Listing 2.79 that it is not strictly necessary to activate the feature at IS–IS, because the final intention is to clean up OSPF configuration and the IS–IS Level 2 internal route is currently preferred. Nevertheless, the activation for IS–IS has been included for illustration purposes. Listing 2.80 summarizes the final steps needed to tear down the OSPF adjacency over the interface in a graceful manner for RSVP.

Listing 2.80: RSVP impact with OSPFv2 no-neighbor-down-notification on Livorno and Skopie

```
1  [edit]
2  user@Livorno# deactivate protocols ospf area 0 interface so-0/0/0.0
3  [edit]
4  user@Livorno# commit and-quit
5  Feb 22 21:36:22.057  Livorno rpd[24428]: RPD_OSPF_NBRDOWN: OSPF neighbor 172.16.1.33 (realm ospf-v2
        so-0/0/0.0 area 0.0.0.0) state changed from Full to Down due to KillNbr (event reason:
        interface went down)
6  commit complete
7  Exiting configuration mode
8
9  user@Livorno> show route 192.168.1.8 detail
10
11 inet.0: 78 destinations, 118 routes (77 active, 0 holddown, 1 hidden)
12 192.168.1.8/32 (2 entries, 1 announced)
13         *IS-IS  Preference: 8
14                 Level: 2
15                 Next hop type: Router, Next hop index: 262143
16                 Next-hop reference count: 55
17                 Next hop: 172.16.1.33 via so-0/0/0.0, selected
18                 State: <Active Int>
19                 Local AS: 64530
```

```
20                 Age: 1d 1:19:36         Metric: 903
21                 Task: IS-IS
22                 Announcement bits (3): 0-KRT 4-LDP 7-Resolve tree 4
23                 AS path: I
24          OSPF   Preference: 10
25                 Next hop type: Router, Next hop index: 701
26                 Next-hop reference count: 36
27                 Next hop: via so-0/0/1.0, selected
28                 State: <Int>
29                 Inactive reason: Route Preference
30                 Local AS: 64530
31                 Age: 46         Metric: 913
32                 Area: 0.0.0.0
33                 Task: OSPF
34                 AS path: I
35
36 inet.3: 9 destinations, 12 routes (9 active, 0 holddown, 0 hidden)
37
38 192.168.1.8/32 (2 entries, 1 announced)
39         State: <FlashAll>
40         *RSVP   Preference: 7
41                 Next hop type: Router
42                 Next-hop reference count: 5
43                 Next hop: via so-0/0/0.0 weight 0x1, selected
44                 Label-switched-path Livorno-to-Bilbao
45                 Label operation: Push 300544
46                 State: <Active Int>
47                 Local AS: 64530
48                 Age: 1d 0:35:47         Metric: 903
49                 Task: RSVP
50                 Announcement bits (2): 2-Resolve tree 1 4-Resolve tree 4
51                 AS path: I
52          LDP    Preference: 9
53                 Next hop type: Router
54                 Next-hop reference count: 1
55                 Next hop: via so-0/0/0.0, selected
56                 Label operation: Push 299872
57                 State: <Int>
58                 Inactive reason: Route Preference
59                 Local AS: 64530
60                 Age: 1d 1:19:36         Metric: 903
61                 Task: LDP
62                 AS path: I
63
64 *** messages ***
65 Feb 22 21:36:56  Skopie rpd[4502]: RPD_OSPF_NBRDOWN: OSPF neighbor 172.16.1.34 (realm ospf-v2 so
       -1/0/0.0 area 0.0.0.0) state changed from Full to Down due to InActiveTimer (event reason:
       neighbor was inactive and declared dead)
```

Listing 2.80 shows that OSPF neighbors have been gracefully decoupled. Note that the OSPF neighbor relationship has expired on router *Skopie* after the standard dead interval (40 seconds) without explicit notification in IS–IS or RSVP.

At that point, interfaces can safely be removed from the OSPF area stanza without further impact. Likewise, the process with IS–IS is completely analogous, requiring first feature activation and later interface deactivation. Once the adjacency expires, the configuration can be safely cleaned up and standardized.

Link-state IGP migration through authentication

All link-state routing protocols include authentication extensions in different flavors to provide additional confidentiality and robustness for adjacency establishment and information distribution.

Although deploying this security paradigm in existing IGP domains is a considerable migration exercise per se, because these parameters can delimit the scope of a given domain, authentication becomes a powerful tool to contain and differentiate link-state IGP domains and transition adjacencies from one domain to another.

This consideration is particularly interesting in cases where networks to be merged or split share the same link-state IGP (even same OSPF area or IS–IS levels), because properly crafted authentication settings can provide adequate containment for each domain while transitioning devices.

OSPFv2: [RFC2328] already included an authentication specification in its Appendix D, discussing the possibility even of using a different authentication procedure for each IP subnet. Each OSPFv2 packet header also includes an authentication type and 64-bit value field, which ensures consistency on all types of OSPFv2 exchange.

Junos OS includes support for all [RFC2328] authentication types inside each area on a per-interface basis with an additional feature that becomes interesting for migration purposes: *soft* authentication-key transition in a synchronized fashion. Note that this does not apply to simple-password authentication, but applies to MD5 authentication. Listing 2.81 illustrates a sample timed activation of MD5 keys and Listing 2.82 shows that this modification is entirely graceful on both sides: there is no adjacency drop or expiration, and routers suddenly start to include newer authentication details in all OSPFv2 packet headers for that adjacency.

Listing 2.81: Configuration of OSPFv2 adjacency MD5 key transition between Livorno and Skopie

```
1 [edit protocols ospf area 0.0.0.0 interface so-0/0/0.0]
2 user@Livorno# set authentication md5 1 key europe start-time 2009-02-27.12:00
3 [edit protocols ospf area 0.0.0.0 interface so-0/0/0.0]
4 user@Skopie# set authentication md5 1 key europe start-time 2009-02-27.12:00
```

Listing 2.82: Soft synchronized OSPFv2 adjacency MD5 key transition between Livorno and Skopie

```
1  # OSPF Database prior to changes
2
3  user@Skopie> show ospf database | match 192.168.1.6
4  Router   192.168.1.6       192.168.1.6       0x80000a93    58  0x22 0xd205 132
5  OpaqArea 1.0.0.1           192.168.1.6       0x80000a90   467  0x22 0xc489  28
6  OpaqArea 1.0.0.3           192.168.1.6       0x80000a90   467  0x22 0x65c9 124
7  OpaqArea 1.0.0.4           192.168.1.6       0x80000a90   467  0x22 0xaa3b 124
8  OpaqArea 1.0.0.5           192.168.1.6       0x80000a90   467  0x22 0x1c57 124
9  OpaqArea 1.0.0.6           192.168.1.6       0x80000a90   467  0x22 0x4f19 124
10
11 # OSPF interface status after activation
12
13 user@Livorno> show ospf interface so-0/0/0.0 extensive
14 Interface            State    Area            DR ID           BDR ID          Nbrs
15 so-0/0/0.0           PtToPt   0.0.0.0         0.0.0.0         0.0.0.0            1
16   Type: P2P, Address: 0.0.0.0, Mask: 0.0.0.0, MTU: 4470, Cost: 643
17   Adj count: 1
18   Hello: 10, Dead: 40, ReXmit: 5, Not Stub
19   Auth type: MD5, Active key ID: 1, Start time: 2009 Feb 27 12:00:00 CET
20 so-0/0/0.0           PtToPt   0.0.0.0         0.0.0.0         0.0.0.0            0
21   Type: P2P, Address: 172.16.1.34, Mask: 255.255.255.252, MTU: 4470, Cost: 643
22   Adj count: 0, , Passive
23   Hello: 10, Dead: 40, ReXmit: 5, Not Stub
24   Auth type: MD5, Active key ID: 1, Start time: 2009 Feb 27 12:00:00 CET
```

```
25
26 # OSPF Database after activation, no new Sequence Numbers
27
28 user@Skopie> show ospf database | match 192.168.1.6
29 Router   192.168.1.6      192.168.1.6      0x80000a93  190  0x22 0xd205 132
30 OpaqArea 1.0.0.1          192.168.1.6      0x80000a90  599  0x22 0xc489  28
31 OpaqArea 1.0.0.3          192.168.1.6      0x80000a90  599  0x22 0x65c9 124
32 OpaqArea 1.0.0.4          192.168.1.6      0x80000a90  599  0x22 0xaa3b 124
33 OpaqArea 1.0.0.5          192.168.1.6      0x80000a90  599  0x22 0x1c57 124
34 OpaqArea 1.0.0.6          192.168.1.6      0x80000a90  599  0x22 0x4f19 124
```

Based on MD5 authentication settings and transitions, this feature allows smooth expansion or constraint of a given domain by progressively activating common or disjoint MD5 keys. This feature is even more interesting in a physical broadcast media shared by different IGP domains, in which participating routers could move from one domain to the other by activating corresponding MD5 keys (and in this case expiring previous and activating new adjacencies) but without the need to physically split or move them from one logical interface to another when transitioning domains.

IS–IS: [ISO10589] advocates the definition of a TLV 10 Authentication to ensure information integrity and confidentiality at each PDU. Similarly to OSPFv2, two authentication types are considered: simple Text and HMAC-MD5, although there is no cryptographic sequence numbering scheme as described in [RFC2328].

The existence of a common Hello PDU type on point-to-point links, as opposed to separate LAN Hello PDUs for Level 1 and Level 2, forces common authentication details to be shared over both L1 and L2 adjacencies in the common link. In a migration scenario, this fact requires either common authentication settings for that link and the two levels, or ignoring authentication information for one of them and complying with it for the other.

With Junos OS, a given administrator can control the creation of adjacencies between neighbors by configuring authentication keys used in Hello PDUs on a per-interface basis. A separate way to control adjacency creation is by database exchange and LSP flooding by configuring same or different authentication keys on a per-system or per-level basis. Commands included in Listing 2.83 summarize IS–IS authentication setting capabilities.

Listing 2.83: IS–IS generic authentication settings for sending PDUs

```
1  [edit protocols isis]
2  user@Livorno# set level 2 authentication-type md5   # or simple for cleartext
3  [edit protocols isis]
4  user@Livorno# set level 2 authentication-key Hurricane
5  [edit protocols isis interface fe-0/3/3.0]
6  user@Livorno# set level 2 hello-authentication-type md5 # or simple for cleartext
7  [edit protocols isis interface fe-0/3/3.0]
8  user@Livorno# set level 2 hello-authentication-key Hurricane
9  [edit protocols isis level 2]
10 user@Livorno# set no-csnp-authentication  # Disable Authentication TLV set in CSNPs
11 [edit protocols isis level 2]
12 user@Livorno# set no-hello-authentication # Disable Authentication TLV set in IIHs
13 [edit protocols isis level 2]
14 user@Livorno# set no-psnp-authentication  # Disable Authentication TLV set in PSNPs
```

This listing shows that all LSP, CSN, and PSN PDUs can include a different TLV 10 Authentication from the Hello PDUs on a certain link and that this can also be disabled in Junos OS on a per-PDU and per-level basis.

Junos OS presents other interesting knobs when processing incoming PDUs with TLV 10 Authentication, that easily allow setup of asymmetric authentication.

The `loose-authentication-check` is a useful resource for progressive authentication deployment and migrations. This knob allows routers to verify MD5 authentication information only if TLV 10 Authentication is present in the PDU; otherwise, processing is done as normal even though the router is locally configured to attach and detect authentication settings. Apart from being able to roll out MD5 authentication incrementally in a network, a system could potentially bridge authenticated and non-authenticated domains without needing to adapt any of either.

Similarly, `no-authentication-check` goes even further by disabling any kind of authentication information parsing in all PDUs. Listing 2.84 depicts those authentication processing exception capabilities.

Listing 2.84: IS–IS global authentication processing exceptions

```
1 [edit protocols isis]
2 user@Havana# set loose-authentication-check
3 [edit protocols isis]
4 user@Havana# set no-authentication-check
```

All these features together enable another dimension to use authentication as a vehicle to merge and separate domains when compared to OSPF, because the distribution of information can be authenticated differently at IIHs, CSNPs, PSNPs, and LSPs, or in a global fashion excluding some of these PDUs. Also, processing such information can be disabled gradually in different stages without impacting topology reachability.

There is a notable remark in the case of IS–IS. Supporting authentication settings, especially MD5, has been a relatively recent development. Because of the incremental nature of IS–IS information constructs, if authentication-unaware systems receive an authenticated PDU, they may still deem the PDU as valid, because TLV 10 Authentication may be ignored or discarded as unknown but its implications are not considered. For that reason, an initial design task when using authentication as a migration vehicle is to potentially identify such corner cases.

The advantage of using full authentication capabilities is that several domains could coexist in the same network without influencing each other's database. Even without using any particular *mesh groups* and because of the flooding nature of IS–IS, a given router in one of the domains still detects LSP headers, but upon processing authentication information, the TLVs are not further processed.

Application Note: Use of IS–IS authentication options to connect multiple domains

Consider our well-known topology as depicted in Figure 2.15 and with respective IS–IS database entries as shown in Listing 2.85.

Listing 2.85: Initial IS–IS database status in each domain

```
1 user@Havana> show isis database
2 IS-IS level 1 link-state database:
3   0 LSPs
4
5 IS-IS level 2 link-state database:
6 LSP ID                      Sequence Checksum Lifetime Attributes
```

```
 7 Nantes.00-00                 0x241f   0x71a6      810 L1 L2
 8 Nantes.02-00                   0x2d   0xa4f8     1060 L1 L2
 9 Havana.00-00                 0x169e   0xb30f      776 L1 L2
10 Havana.02-00                  0x4b5   0x9079      776 L1 L2
11 male-re0.00-00            0x4de   0x1ed5      767 L1 L2
12 Bilbao-re0.00-00             0x16dd   0x5158      614 L1 L2
13   6 LSPs
14 
15 user@Livorno> show isis database
16 IS-IS level 1 link-state database:
17   0 LSPs
18 
19 IS-IS level 2 link-state database:
20 LSP ID                      Sequence Checksum Lifetime Attributes
21 Lille.00-00                   0x290   0x8912     1145 L1 L2
22 Skopie.00-00                 0x1ec1   0x72f5      535 L1 L2
23 Livorno.00-00                0x5691   0x5bb9     1175 L1 L2
24 Livorno.02-00                   0x2   0x65c4     1179 L1 L2
25 honolulu-re0.00-00        0x8d9   0x1134      510 L1 L2
26 honolulu-re0.02-00        0x8ae   0xb667      759 L1 L2
27 Barcelona-re0.00-00       0x8f3   0x2583      729 L1 L2
28 Barcelona-re0.02-00       0x8e5   0x7c7f      767 L1 L2
29   8 LSPs
```

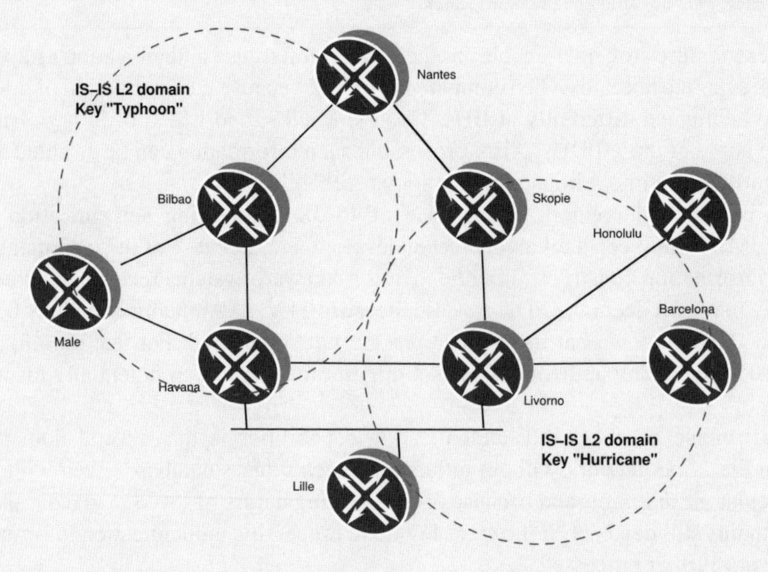

Figure 2.15: Independent IS–IS L2 domains with different authentication MD5 keys.

The `loose-authentication-checks` are activated in all routers in both domains, as indicated in Listing 2.86 for one system. This step needs to be repeated for all participating routers in both domains. Because authentication is still enforced, no routing has changed so far.

Listing 2.86: Initial loose-authentication-check activation in participating systems at both domains

```
1 [edit protocols isis]
2 user@Livorno# set loose-authentication-check
```

Authentication can be gradually deactivated for systems to be integrated in both domains. This process requires both removing any TLV 10 Authentication attachment at PDUs to be sent and ignoring them upon reception.

Considering router *Nantes* as a system to be *shared* between both domains, commands included in Listing 2.87 show configuration steps needed for this particular domain stitching.

Listing 2.87: Configuration steps to integrate Nantes in domain "Typhoon"

```
1  [edit protocols isis]
2  user@Nantes# set no-authentication-check
3  [edit protocols isis]
4  user@Nantes# delete level 2 authentication-key
5  [edit protocols isis]
6  user@Nantes# delete level 2 authentication-type
```

After disabling all authentication settings and checks in router *Nantes*, the system effectively merges locally both IS–IS databases, keeping adjacencies up for both of them, as shown in Listing 2.88.

Listing 2.88: Router *Nantes* integrates locally in both domains

```
1    # Merged IS-IS Databases in Nantes
2  user@Nantes> show isis database
3  IS-IS level 1 link-state database:
4    0 LSPs
5
6  IS-IS level 2 link-state database:
7  LSP ID                      Sequence Checksum Lifetime Attributes
8  Nantes.00-00                  0x2521   0xebb3     1010 L1 L2
9  Nantes.02-00                   0x116   0x828c     1010 L1 L2
10 Skopie.00-00                  0x1fa1   0x43e3      901 L1 L2
11 Havana.00-00                  0x178f   0x3da6     1018 L1 L2
12 Havana.02-00                   0x599   0x5ca6     1018 L1 L2
13 Lille.00-00                      0x4   0xd4d3      951 L1 L2
14 Livorno.00-00                 0x5768   0x10ab     1105 L1 L2
15 Livorno.02-00                   0xdb   0x3d5a     1105 L1 L2
16 male-re0.00-00              0x5bb   0xd71c     1006 L1 L2
17 honolulu-re0.00-00          0x9b1   0x70fb     1172 L1 L2
18 honolulu-re0.02-00          0x985   0x67ce      691 L1 L2
19 Bilbao-re0.00-00              0x17c0   0xb4d9     1008 L1 L2
20 Barcelona-re0.00-00         0x9ca   0x9874      668 L1 L2
21 Barcelona-re0.02-00         0x9bc   0xf10b      917 L1 L2
22   14 LSPs
23
24   # Merged Traffic-Engineering Databases in Nantes
25
26 user@Nantes> show ted database
27 TED database: 17 ISIS nodes 17 INET nodes
28 ID                            Type Age(s) LnkIn LnkOut Protocol
29 Nantes.00(192.168.1.1)        Rtr     297     6      7 IS-IS(2)
30     To: Skopie.00(192.168.1.2), Local: 172.16.1.1, Remote: 172.16.1.2
31       Local interface index: 70, Remote interface index: 67
32     To: Nantes.02, Local: 172.16.1.6, Remote: 0.0.0.0
33       Local interface index: 67, Remote interface index: 0
34 ID                            Type Age(s) LnkIn LnkOut Protocol
35 Nantes.02                     Net     297     2      2 IS-IS(2)
36     To: Nantes.00(192.168.1.1), Local: 0.0.0.0, Remote: 0.0.0.0
37       Local interface index: 0, Remote interface index: 0
38     To: Bilbao-re0.00(192.168.1.8), Local: 0.0.0.0, Remote: 0.0.0.0
39       Local interface index: 0, Remote interface index: 0
40 ID                            Type Age(s) LnkIn LnkOut Protocol
```

```
41 Skopie.00(192.168.1.2)        Rtr     468      4      4 IS-IS(2)
42     To: Nantes.00(192.168.1.1), Local: 172.16.1.2, Remote: 172.16.1.1
43       Local interface index: 67, Remote interface index: 70
44     To: Livorno.00(192.168.1.6), Local: 172.16.1.33, Remote: 172.16.1.34
45       Local interface index: 72, Remote interface index: 0
46 ID                            Type Age(s) LnkIn LnkOut Protocol
47 Havana.00(192.168.1.4)        Rtr     296      7      5 IS-IS(2)
48 Havana.02                     Net     296      1      2 IS-IS(2)
49     To: Havana.00(192.168.1.4), Local: 0.0.0.0, Remote: 0.0.0.0
50       Local interface index: 0, Remote interface index: 0
51     To: male-re0.00(192.168.1.20), Local: 0.0.0.0, Remote: 0.0.0.0
52       Local interface index: 0, Remote interface index: 0
53 <...>
```

Router *Nantes* has established additional adjacencies to border systems, but because the authentication settings from all other systems inside domain "Typhoon" remain different, only this integrated system is visible inside domain "Hurricane". Listing 2.89 illustrates this situation from the representative point of view of router *Livorno*.

Listing 2.89: Router *Livorno* only has additional visibility on router *Nantes*

```
1   # IS-IS Database in domain "Hurricane"
2
3 user@Livorno> show isis database
4 IS-IS level 1 link-state database:
5   0 LSPs
6
7 IS-IS level 2 link-state database:
8 LSP ID                      Sequence Checksum Lifetime Attributes
9 Nantes.00-00                  0x2521   0xebb3     1093 L1 L2
10 Nantes.02-00                  0x116   0x828c     1093 L1 L2
11 Skopie.00-00                 0x1fa1   0x43e3      988 L1 L2
12 Lille.00-00                     0x4   0xd4d3     1034 L1 L2
13 Livorno.00-00                0x5768   0x10ab     1195 L1 L2
14 Livorno.02-00                  0xdb   0x3d5a     1195 L1 L2
15 honolulu-re0.00-00       0x9b0   0x6042      432 L1 L2
16 honolulu-re0.02-00       0x985   0x67ce      777 L1 L2
17 Barcelona-re0.00-00      0x9ca   0x9874      754 L1 L2
18 Barcelona-re0.02-00          0   0xf10b     1004
19   10 LSPs
20
21   # TED Database only reflects domain "Hurricane"
22
23 user@Livorno> show ted database
24 TED database: 9 ISIS nodes 5 INET nodes
25 ID                            Type Age(s) LnkIn LnkOut Protocol
26 Nantes.00(192.168.1.1)        Rtr     739      6      7 IS-IS(2)
27     To: Skopie.00(192.168.1.2), Local: 172.16.1.1, Remote: 172.16.1.2
28     To: Nantes.02, Local: 172.16.1.6, Remote: 0.0.0.0
29 ID                            Type Age(s) LnkIn LnkOut Protocol
30 Nantes.02                     Net     739      1      2 IS-IS(2)
31     To: Nantes.00(192.168.1.1), Local: 0.0.0.0, Remote: 0.0.0.0
32     To: 1921.6816.8008.00(192.168.1.8), Local: 0.0.0.0, Remote: 0.0.0.0
33 ID                            Type Age(s) LnkIn LnkOut Protocol
34 Skopie.00(192.168.1.2)        Rtr     156      4      4 IS-IS(2)
35     To: Livorno.00(192.168.1.6), Local: 172.16.1.33, Remote: 172.16.1.34
36     To: Nantes.00(192.168.1.1), Local: 172.16.1.2, Remote: 172.16.1.1
37 <...>
```

Because router *Nantes* still maintains the adjacencies established to domain "Typhoon" and such systems have been configured with `loose-authentication-check`, systems continue to retain the same visibility as before.

Routers inside domain "Typhoon" now receive flooding through router *Nantes* for all LSPs from domain "Hurricane". However, because `loose-authentication-check` is active in both domains, the authentication mismatch leads to not decoding other TLVs present in each domain "Hurricane" system's LSP and not incorporating routing information.

This result is particularly visible when inspecting the IS–IS database in a given system, because not even the TLV 137 Hostname is decoded and the information present in Junos OS output displays only the LSP headers, as shown in Listing 2.90 from router *Havana*.

Listing 2.90: Havana has no visibility on domain "Hurricane"

```
1   # IS-IS Database in domain "Typhoon"
2
3  user@Havana> show isis database
4  IS-IS level 1 link-state database:
5    0 LSPs
6
7  IS-IS level 2 link-state database:
8  LSP ID                      Sequence Checksum Lifetime Attributes
9  Nantes.00-00                  0x2521   0xebb3     1174 L1 L2
10 Nantes.02-00                   0x116   0x828c     1174 L1 L2
11 1921.6800.1002.00-00               0   0x43e3     1065
12 Havana.00-00                  0x178f   0x3da6     1185 L1 L2
13 Havana.02-00                   0x599   0x5ca6     1185 L1 L2
14 1921.6800.1005.00-00               0   0xd4d3     1115
15 1921.6800.1006.00-00               0   0x4152      603
16 1921.6800.1006.02-00               0   0x8c14     1130
17 male-re0.00-00               0x5bb   0xd71c     1170 L1 L2
18 1921.6800.1021.00-00               0   0x6042      509
19 1921.6800.1021.02-00               0   0x67ce      855
20 Bilbao-re0.00-00              0x17c0   0xb4d9     1172 L1 L2
21 1921.6816.8009.00-00               0   0x9874      831
22 1921.6816.8009.02-00               0   0xf10b     1081
23   14 LSPs
24
25   # Only LSP Headers from domain "Hurricane" decoded, authentication mismatch
26
27 user@Havana> show isis database 1921.6800.1002.00-00 extensive
28 IS-IS level 1 link-state database:
29
30 IS-IS level 2 link-state database:
31
32 1921.6800.1002.00-00 Sequence: 0, Checksum: 0x43e3, Lifetime: 840 secs
33
34   Header: LSP ID: 1921.6800.1002.00-00, Length: 0 bytes
35     Allocated length: 284 bytes, Router ID: 0.0.0.0
36     Remaining lifetime: 840 secs, Level: 2, Interface: 67
37     Estimated free bytes: 238, Actual free bytes: 284
38     Aging timer expires in: 840 secs
39
40   Packet: LSP ID: 0000.0000.0000.00-00, Length: 0 bytes, Lifetime : 0 secs
41     Checksum: 0, Sequence: 0, Attributes: 0 <>
42     NLPID: 0, Fixed length: 0 bytes, Version: 0, Sysid length: 0 bytes
43     Packet type: 0, Packet version: 0, Max area: 0
44   No queued transmissions
45
46   # TED Database only reflects domain "Typhoon"
47
48 user@Havana> show ted database
49 TED database: 6 ISIS nodes 4 INET nodes
50 ID                            Type Age(s) LnkIn LnkOut Protocol
51 Nantes.00(192.168.1.1)        Rtr     321     6      7 IS-IS(2)
52     To: 1921.6800.1002.00(192.168.1.2), Local: 172.16.1.1, Remote: 172.16.1.2
53       Local interface index: 70, Remote interface index: 67
54     To: Nantes.02, Local: 172.16.1.6, Remote: 0.0.0.0
55       Local interface index: 67, Remote interface index: 0
56 ID                            Type Age(s) LnkIn LnkOut Protocol
```

```
57 Nantes.02                     Net      321      2       2 IS-IS(2)
58     To: Nantes.00(192.168.1.1), Local: 0.0.0.0, Remote: 0.0.0.0
59       Local interface index: 0, Remote interface index: 0
60     To: Bilbao-re0.00(192.168.1.8), Local: 0.0.0.0, Remote: 0.0.0.0
61       Local interface index: 0, Remote interface index: 0
62 <...>
```

To sum up, an interesting goal has been achieved just by means of authentication settings and available Junos OS knobs: the ability to circumvent global domain stitching and information flooding between both domains while keeping local control and reachability from migrated systems.

This is powerful for migration purposes as TEDs properly follow IS–IS databases, all transitions are entirely smooth, and not even a single IS–IS adjacency flaps during the process! However, a permanent setup with these divergences would be cumbersome to troubleshoot and operate: having systems with this dual domain visibility is not an orthodox setup that follows standard link-state protocol principles.

Information flooding can be reduced even further with the Junos OS knobs available to activate mesh groups, following the concepts described in [RFC2973] to constrain new LSP information. In particular, utilization of the Junos OS resource `mesh-group blocked` can completely stop flooding for new LSPs over the specific interface. While this path can also be followed for migrations, care must be taken when planning such a resource because legitimate routing details could be blocked unexpectedly and in some scenarios, the migration could fail.

2.3 Case Study

A fairly common real-life situation in which a networking engineer has to face a link-state IGP migration results from a company acquisition. There are many documented cases of the challenges and risks associated with a network merger, many of them even made public.

As a typical case study for link-state IGP migrations, imagine a company acquisition scenario. The multinational Internet service and content provider "Gale Internet" has decided to purchase networks and assets overseas to extend their service portfolio. With these strategic acquisitions, "Gale Internet" acquires stronger regional footprint to offer contents, Internet access and corporate-grade VPNs with each of these networks while interconnecting all of them with newer leased lines and infrastructure, now directly managed from the central location from "Gale Internet".

As a result of these acquisitions, several different link-state IGP domains need to be integrated and connected to their main NOC premises. Top management has decided to centralize network operations and engineering under a common umbrella, managed from the central location, while a parallel mid-term exercise is effectively to integrate those networks.

The real challenge is to retain customers and keep services up during the migration period! The domain-wide scope nature of link-state IGPs can easily exacerbate and propagate any minimal mistake in a remote corner of the network. Those readers who have carried out such migrations in real life will probably acknowledge this and will surely crack a smile when reading this.

2.3.1 Original network

Considering a topology as illustrated in Figures 2.16 and 2.17, which contains several domains, the compelling task is to integrate all these networks into a common IS–IS domain with specific settings. All these domains have been linked by means of leased lines that have been acquired for the purpose of connecting them, and each domain used to have independent management and configuration at each of the small regional service providers.

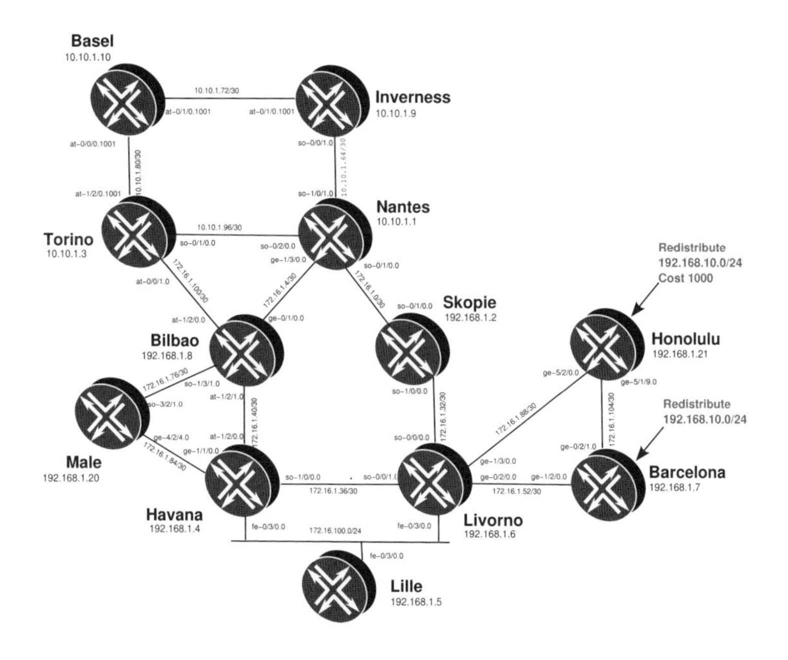

Figure 2.16: Original topology including addressing scheme.

Several aspects in each domain must be remarked in this initial scenario:

- Domain "Cyclone" is an OSPFv2 network based on a single backbone Area 0 and making use of TE extensions to provide information for RSVP-based applications. Both router *Barcelona* and router *Honolulu* are redistributing external address space into OSPFv2.

- Domain "Monsoon" is a small enterprise network based on OSPFv2 and divided in a simple backbone area with both border routers and an NSSA. Domain "Monsoon" uses specific [RFC1918] private address space at loopback and WAN interfaces connecting routers that cannot be directly exposed and collides with other internal services from "Gale Internet" that use the same address ranges.

- Domain "Mistral" represents an IS–IS network based on a plain L1 Area, making use of certain authentication settings. The Area ID for domain "Mistral" differs from the

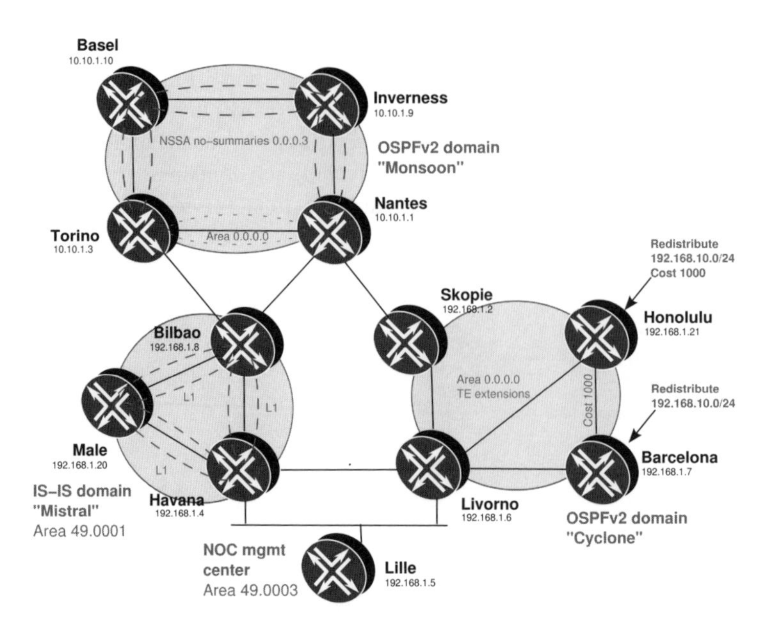

Figure 2.17: Original topology including domain division and basic routing settings.

value to be used in the final topology. Both Area ID and authentication keys need to be adapted as part of this migration exercise.

- The NOC management center is connected to router *Lille*, as core router for the central location from "Gale Internet". Certain security guidelines are implemented on router *Lille*, while redundant connectivity to the final topology is granted by means of an extended broadcast segment connected to domain "Mistral" and domain "Cyclone".

2.3.2 Target network

As usual, the goal is to achieve the migration with the smoothest and cleanest approach and granting as much resilience as possible at each migration step. However, company management wants to standardize operations and monitoring as soon as possible from a centralized NOC, so the migration has two major milestones:

- to ensure reachability and management from the NOC to all devices as a short-term exercise;
- to progressively and smoothly integrate all systems into a common IS–IS domain.

The network topologies after reaching each one of these milestones are illustrated in Figures 2.18 and 2.19.

Some other general requirements have to be taken into account during migration planning and execution:

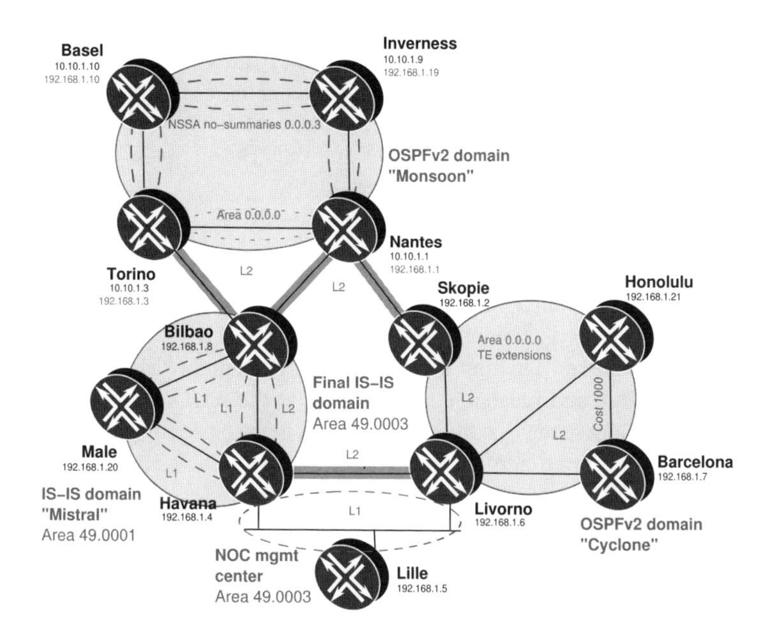

Figure 2.18: First milestone: domain IGP interconnect.

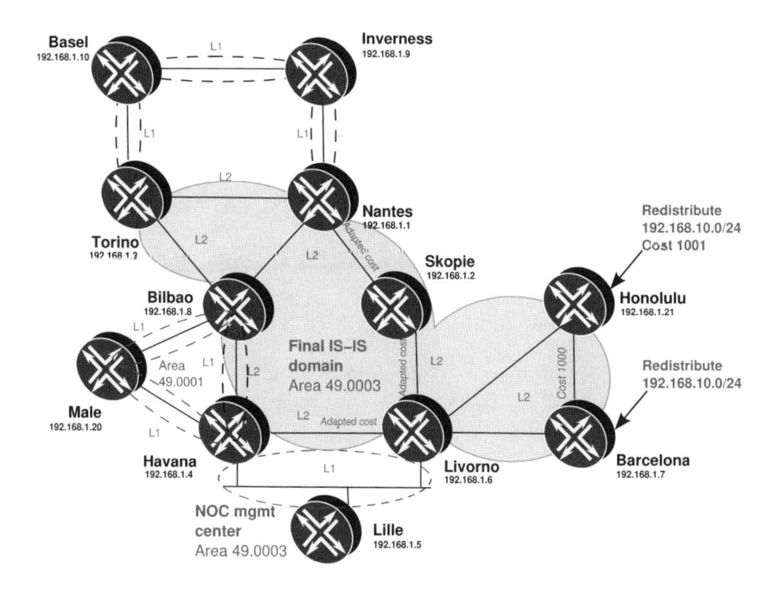

Figure 2.19: Second milestone: final IS–IS topology.

- The NOC management center has been placed in a shared broadcast segment between domain "Mistral" and domain "Cyclone". Router *Lille* is the router providing transit services for the NOC and should remain as a single IS–IS L1 system without any default route being installed. Also, because of its critical role, modifying its configuration should be avoided whenever possible. Both uplinks to domain "Mistral" and domain "Cyclone" should be equally available.

- Domain "Monsoon" is a remote network environment belonging to a service provider whose address space is not directly manageable from the NOC due to address overlapping. In parallel to an integration into a final common IS–IS domain, an address renumbering exercise is compulsory to standardize the global addressing scheme including those routers. The first migration step for domain "Monsoon" will consider adding manageable and reachable addresses from the NOC, to be related with NTP, SNMP, syslog, remote access and other management protocols, and later stages will plan the intrusive migration of each router, one at a time.

- Currently internal OSPF routers inside the totally stubby NSSA at domain "Monsoon" must remain as pure L1 systems in the final scenario.

- Domain "Mistral" is actually a plain IS–IS L1 Area 49.0001 with different authentication parameters, but it should become part of the same IS–IS domain with the same global MD5 authentication-key *Aeris*. Only router *Havana* and router *Bilbao* will become L1L2 systems to be integrated in the backbone.

- Router *Male* will remain as a L1 system in the final scenario. Loopback addresses from the entire domain will need to be leaked from router *Havana* and router *Bilbao* for optimal routing purposes.

- Domain "Mistral" already includes IS–IS area address 49.0001, but the NOC management center at router *Lille* considers it to be 49.0003. This difference needs to be adapted because 49.0003 will be rolled out as standard area in the final scenario, without eliminating 49.0001 from any router on domain "Mistral".

- Domain "Cyclone" requires that IS–IS extensions populate a TED with the same parameters and values as the existing OSPFv2 before removing this protocol. This is to ensure minimal disruption for existing MPLS applications.

- The final IS–IS domain from "Gale Internet" will cover all provider networks, including `wide-metrics-only` both for IS–IS Level 1 and Level 2 with MD5 authentication key *Aeris*.

- External route redistribution from domain "Cyclone" needs to remain untouched in the final scenario. Router *Barcelona* and router *Honolulu* currently inject external route information with different costs as OSPFv2 Type 5 AS External LSA and this redistribution needs to be migrated to IS–IS Level 2 with `wide-metrics-only`. This aggregate represents another network injected to the former service provider that needs to be managed at "Gale Internet" as well and that needs to become reachable for all routers. This prefix summary is primarily connected to router *Barcelona* with a backup connection to router *Honolulu*.

Redundancy and minimal impact are key guidelines during the migration. In the turbulent times of company acquisitions, a global network outage or meltdown that could extend further than a given provider could have fatal consequences and top management are really scrutinizing the effort and the results during this time period. Therefore, all the migration steps need to be planned and carried out with great care.

2.3.3 Migration strategy

An initial phase is needed because company managers have defined as a preliminary objective to have a unified and global device operations and management infrastructure. This requirement means that, as the very first goal, only specific hooks and standardized addresses are added to ensure visibility from the common management center. Other integration actions will be smoothly rolled out in later steps.

From that perspective, the activities that will be carried out until reaching this first milestone will be:

- Domain “Cyclone” does not require complete readdressing, but does need common traffic-engineering information. For that purpose, IS–IS can be deployed as the unpreferred protocol and can start populating the IS–IS database for the global domain while OSPFv2 remains preferred internally. This *ships-in-the-night* temporary model also requires that not only traffic engineering but also link costs and metrics are proportional when comparing both IGPs.

- Domain “Monsoon” includes unusable addresses for the NOC. Rather than completely readdressing all links, the focus in a second stage is on defining new valid loopback addresses for the systems and correctly injecting them in the global domain, and then allowing mutual reachability with systems outside this old domain.

- At domain “Mistral”, IS–IS Level 2 with the final settings is activated to link the area in a redundant fashion to the backbone in a third stage. L1L2 systems will add final area addressing and provide a proper Level 2 to Level 1 leaking policy towards router *Male* and router *Lille* for optimal routing purposes.

- In stage four, all domains are linked by activating IS–IS in those newer leased lines purchased to interconnect the distant system providers. A common IS–IS Level 2 scheme with the final MD5 authentication key *Aeris* is configured at each point. Area addressing is deployed following 49.0003 as standard area ID defined in router *Lille*.

- In order to avoid denial-of-service attacks, router *Lille* must have full reachability to all systems but must not install a default route following IS–IS ATTached bit signaling at any time either now or in the future, as part of a fifth stage to reach the global connectivity milestone.

- Whenever all domains and systems have been adapted, a correct verification in a sixth stage is recommended to confirm the achievement of global connectivity among systems in different locations.

Once these activities have been finished, all systems in participating domains are under common control and management, reaching therefore a first milestone.

Afterwards, other transition stages are compulsory to achieve the final goal of a unified protocol and addressing configuration. This is a gradual process that can be accomplished over time because basic management and connectivity have been ensured in previous stages.

The main activities to be covered for the final topology are:

- As stage seven, domain "Cyclone" is migrated to IS–IS Level 2 as the preferred source for routing and TE information. OSPFv2 can be dismantled at that point at domain "Cyclone" in a smooth fashion, without affecting RSVP LSPs set up based on TE information.
- Stage eight is a progressive address renumbering exercise at domain "Monsoon" to be converted into part of the final IS–IS topology.
- A final stage nine is needed in domain "Mistral" and router *Lille* to adapt Level 1 authentication settings to the final MD5 key *Aeris*. This soft transition is achieved smoothly by means of the `no-authentication-check` Junos OS knob.

After accomplishing all these goals with each participating network, all systems belong to a joint and final topology with consistent addressing and protocol settings.

These stages for the link-state IGP migration strategy and their dependencies are illustrated in Figure 2.20. Note that some stages can be carried out in parallel, but others require strict sequencing.

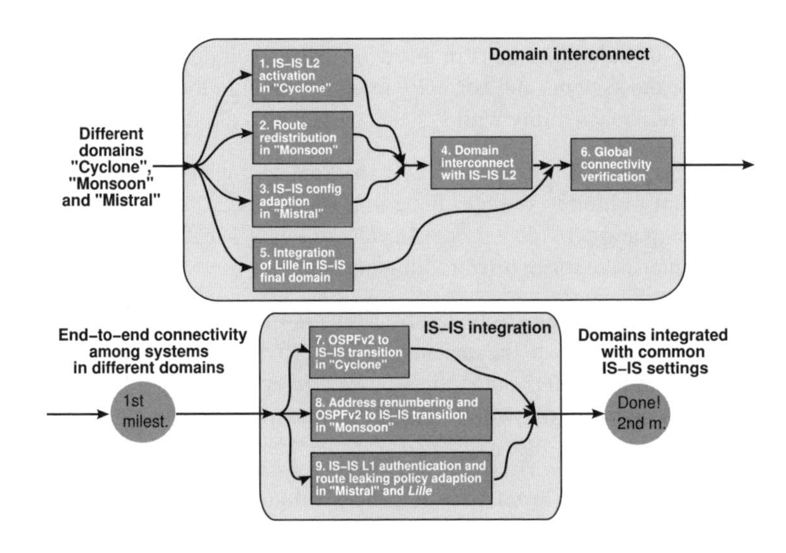

Figure 2.20: Strategy for the link-state IGP migration.

2.3.4 Stage one: IS–IS Level 2 activation in domain "Cyclone"

Before being able to stitch all domains together, IS–IS must be deployed over domain "Cyclone" in a *ships-in-the-night* model. The idea is to start populating the IS–IS and traffic-engineering databases with route information from this domain but still keeping IS–IS unpreferred to avoid modifying current status quo. Listing 2.91 summarizes the initial situation from router *Livorno*'s perspective.

Listing 2.91: Initial OSPFv2 status at Livorno

```
user@Livorno> show configuration protocols ospf
traffic-engineering;
reference-bandwidth 10g;
area 0.0.0.0 {
    interface lo0.0 {
        passive;
    }
    interface so-0/0/0.0 {
        authentication {
            md5 1 key "$9$oKGUiQF/tpBmf0Ihr8Ldbs"; ### SECRET-DATA
        }
    }
    interface ge-0/2/0.0 {
        authentication {
            md5 1 key "$9$kPfQtu1hcl69yKWxwsaZU"; ### SECRET-DATA
        }
    }
    interface ge-1/3/0.0 {
        authentication {
            md5 1 key "$9$S44eKWbs4ZGi7-DkmT/90BI"; ### SECRET-DATA
        }
    }
}

user@Livorno> show ospf database

    OSPF database, Area 0.0.0.0
 Type       ID              Adv Rtr          Seq          Age  Opt  Cksum  Len
Router   192.168.1.2      192.168.1.2      0x80000031  1193  0x22 0x8d67  60
Router  *192.168.1.6      192.168.1.6      0x800003fc   984  0x22 0x3d1b  96
Router   192.168.1.7      192.168.1.7      0x800005e1  1620  0x22 0x9e0f  60
Router   192.168.1.21     192.168.1.21     0x80000062  1279  0x22 0x922c  60
Network  172.16.1.53      192.168.1.7      0x80000036   120  0x22 0xac15  32
Network  172.16.1.90      192.168.1.21     0x8000002d  1579  0x22 0x8306  32
Network  172.16.1.105     192.168.1.21     0x8000002c  1879  0x22 0xfc7d  32
OpaqArea 1.0.0.1          192.168.1.2      0x8000002e   417  0x22 0x5ace  28     # TE extensions
OpaqArea*1.0.0.1          192.168.1.6      0x800003ae   484  0x22 0x603d  28
OpaqArea 1.0.0.1          192.168.1.7      0x800005aa  1320  0x22 0x6637  28
OpaqArea 1.0.0.1          192.168.1.21     0x80000052   979  0x22 0x5e80  28
<...>
    OSPF AS SCOPE link state database
 Type       ID              Adv Rtr          Seq          Age  Opt  Cksum  Len
Extern   192.168.10.0     192.168.1.7      0x8000001a   420  0x22 0x7143  36
Extern   192.168.10.0     192.168.1.21     0x80000006    79  0x22 0x7955  36

user@Livorno> show ospf database extensive | match "0x8| metric|data"
    OSPF database, Area 0.0.0.0
Router   192.168.1.2      192.168.1.2      0x80000031  1214  0x22 0x8d67  60
  id 192.168.1.2, data 255.255.255.255, Type Stub (3)
    Topology count: 0, Default metric: 0
  id 192.168.1.6, data 172.16.1.33, Type PointToPoint (1)
    Topology count: 0, Default metric: 64
  id 172.16.1.32, data 255.255.255.252, Type Stub (3)
    Topology count: 0, Default metric: 64
Router  *192.168.1.6      192.168.1.6      0x800003fc  1005  0x22 0x3d1b  96
  id 172.16.1.53, data 172.16.1.54, Type Transit (2)
    Topology count: 0, Default metric: 10
  id 172.16.1.90, data 172.16.1.89, Type Transit (2)
    Topology count: 0, Default metric: 10
  id 192.168.1.6, data 255.255.255.255, Type Stub (3)
    Topology count: 0, Default metric: 0
  id 192.168.1.2, data 172.16.1.34, Type PointToPoint (1)
```

```
63       Topology count: 0, Default metric: 64
64  Router   192.168.1.7      192.168.1.7      0x800005e1  1641  0x22 0x9e0f  60
65    id 172.16.1.105, data 172.16.1.106, Type Transit (2)
66       Topology count: 0, Default metric: 1000
67    id 172.16.1.53, data 172.16.1.53, Type Transit (2)
68       Topology count: 0, Default metric: 10
69    id 192.168.1.7, data 255.255.255.255, Type Stub (3)
70       Topology count: 0, Default metric: 0
71  Router   192.168.1.21     192.168.1.21     0x80000062  1300  0x22 0x922c  60
72    id 172.16.1.105, data 172.16.1.105, Type Transit (2)
73       Topology count: 0, Default metric: 1000
74    id 172.16.1.90, data 172.16.1.90, Type Transit (2)
75       Topology count: 0, Default metric: 10
76    id 192.168.1.21, data 255.255.255.255, Type Stub (3)
77       Topology count: 0, Default metric: 0
78  Network  172.16.1.53      192.168.1.7      0x80000036   141  0x22 0xac15  32
79  Network  172.16.1.90      192.168.1.21     0x8000002d  1600  0x22 0x8306  32
80  Network  172.16.1.105     192.168.1.21     0x8000002c  1900  0x22 0xfc7d  32
81  OpaqArea 1.0.0.1          192.168.1.2      0x8000002e   438  0x22 0x5ace  28
82  <...>
83      OSPF AS SCOPE link state database
84  Extern   192.168.10.0     192.168.1.7      0x8000001a   441  0x22 0x7143  36
85      Type: 2, Metric: 0, Fwd addr: 0.0.0.0, Tag: 0.0.0.0  # External route injected in Barcelona
86  Extern   192.168.10.0     192.168.1.21     0x80000006   100  0x22 0x7955  36
87      Type: 2, Metric: 1000, Fwd addr: 0.0.0.0, Tag: 0.0.0.0 # External route injected in Honolulu
88
89  user@Livorno> show ted database
90  TED database: 1 ISIS nodes 7 INET nodes
91  ID                           Type Age(s) LnkIn LnkOut Protocol
92  Livorno.00(192.168.1.6)      Rtr      77     3      3 OSPF(0.0.0.0)
93      To: 192.168.1.2, Local: 172.16.1.34, Remote: 172.16.1.33
94        Local interface index: 67, Remote interface index: 0
95      To: 172.16.1.53-1, Local: 172.16.1.54, Remote: 0.0.0.0
96        Local interface index: 0, Remote interface index: 0
97      To: 172.16.1.90-1, Local: 172.16.1.89, Remote: 0.0.0.0
98        Local interface index: 0, Remote interface index: 0
99  ID                           Type Age(s) LnkIn LnkOut Protocol
100                                                       IS-IS(1)
101 192.168.1.2                  Rtr     349     1      1 IS-IS(2)
102                                                       OSPF(0.0.0.0)
103     To: Livorno.00(192.168.1.6), Local: 172.16.1.33, Remote: 172.16.1.34
104       Local interface index: 72, Remote interface index: 0
105 ID                           Type Age(s) LnkIn LnkOut Protocol
106 192.168.1.7                  Rtr      77     2      2 OSPF(0.0.0.0)
107     To: 172.16.1.53-1, Local: 172.16.1.53, Remote: 0.0.0.0
108       Local interface index: 0, Remote interface index: 0
109     To: 172.16.1.105-1, Local: 172.16.1.106, Remote: 0.0.0.0
110       Local interface index: 0, Remote interface index: 0
111 ID                           Type Age(s) LnkIn LnkOut Protocol
112 192.168.1.21                 Rtr      77     2      2 OSPF(0.0.0.0)
113     To: 172.16.1.90-1, Local: 172.16.1.90, Remote: 0.0.0.0
114       Local interface index: 0, Remote interface index: 0
115     To: 172.16.1.105-1, Local: 172.16.1.105, Remote: 0.0.0.0
116       Local interface index: 0, Remote interface index: 0
117 <...>
118
119 user@Livorno> show ted database extensive | match "Node|To:|Metric:"
120 TED database: 1 ISIS nodes 7 INET nodes
121 NodeID: Livorno.00(192.168.1.6)
122     To: 192.168.1.2, Local: 172.16.1.34, Remote: 172.16.1.33
123       Metric: 64
124     To: 172.16.1.53-1, Local: 172.16.1.54, Remote: 0.0.0.0
125       Metric: 10
126     To: 172.16.1.90-1, Local: 172.16.1.89, Remote: 0.0.0.0
127       Metric: 10
128 NodeID: 192.168.1.2
129     To: Livorno.00(192.168.1.6), Local: 172.16.1.33, Remote: 172.16.1.34
130       Metric: 64
131 NodeID: 192.168.1.7
```

```
132     To: 172.16.1.53-1, Local: 172.16.1.53, Remote: 0.0.0.0
133       Metric: 1000
134     To: 172.16.1.105-1, Local: 172.16.1.106, Remote: 0.0.0.0
135       Metric: 10
136 NodeID: 192.168.1.21
137     To: 172.16.1.90-1, Local: 172.16.1.90, Remote: 0.0.0.0
138       Metric: 10
139     To: 172.16.1.105-1, Local: 172.16.1.105, Remote: 0.0.0.0
140       Metric: 1000
141 <...>
142
143 user@Livorno> show route protocol ospf terse table inet.0
144
145 inet.0: 27 destinations, 33 routes (26 active, 0 holddown, 1 hidden)
146 + = Active Route, - = Last Active, * = Both
147
148 A Destination        P Prf   Metric 1   Metric 2  Next hop        AS path
149   172.16.1.32/30     O  10        64               >so-0/0/0.0
150 * 172.16.1.104/30    O  10      1010                172.16.1.53
151                                                    >172.16.1.90
152 * 192.168.1.2/32     O  10        64               >so-0/0/0.0
153 * 192.168.1.7/32     O  10        10               >172.16.1.53
154 * 192.168.1.21/32    O  10        10               >172.16.1.90
155 * 192.168.10.0/24    O 150         0               >172.16.1.53
156 * 224.0.0.5/32       O  10         1                MultiRecv
```

The IS–IS configuration to be deployed in this domain needs to consider definitive settings and to result in adjacency establishment. However, any routing information from IS–IS needs to remain inactive at this point. For that purpose, the following factors need to be evaluated:

- *Route preference*: default preferences in Junos OS result in OSPFv2 internal always being preferred over IS–IS Level 1 and Level 2 internal, and OSPF external being preferred over IS–IS Level 1 and Level 2 external (see Table 2.7).

- *TED protocol credibility*: regardless of protocol preference, IS–IS traffic-engineering information is by default more preferred in the TED than OSPFv2.

- *Metric alignment*: both link standard and TE metrics need to be proportional (if not similar) in IS–IS when compared with OSPFv2 to avoid forwarding divergences.

Although default preference settings match the current purpose, there is still a notable caveat to consider when analyzing the migration requirements: externally redistributed routes into OSPFv2 always results in Type 5 (or Type 7) AS External LSAs, whose preference can be tuned with the `external-preference` knob, but IS–IS `wide-metrics-only` removes the internal or external distinction for redistributed route information and at that point, IS–IS's internal `preference` determines the selection for the routing table.

The prefix 192.168.10.0/24 is injected by router *Barcelona* and router *Honolulu* as Type 5 AS External LSA and is considered as an OSPF *external* route, but when deploying IS–IS in parallel, such redistribution will lead to a Level 2 *internal* route due to `wide-metrics-only`. For that reason, the first migration-specific guideline will be to modify IS–IS internal preferences in this domain (Level 2 in this case study; Level 1 is also modified for illustration purposes) between default OSPF and IS–IS external preferences; that is, compared to Table 2.7, the modifications are:

- IS–IS Level 1 internal preference = 155

- IS–IS Level 2 internal preference = 158

The second configuration guideline derived from the migration strategy is adjustment of *TED protocol credibility*. IS–IS is more credible for the TED than OSPF by default, regardless of protocol preference settings. The step to take at this point is to align *TED protocol credibility* to *route preference*, as illustrated in Listing 2.77. Note that this change needs to be deployed in the existing OSPFv2 configuration *before* issuing any IS–IS changes; otherwise, information injected by IS–IS TE extensions is preferred at the TED. Listing 2.92 shows these previous changes in router *Livorno*, but they need to be done in all domain "Cyclone" routers.

Listing 2.92: Previous OSPFv2 TED protocol credibility changes in Livorno

```
1  user@Livorno> show ted protocol detail
2  Protocol name        Credibility  Self node
3  OSPF(0)              0            192.168.1.6
4
5  [edit]
6  user@Livorno# set protocols ospf traffic-engineering credibility-protocol-preference
7  [edit]
8  user@Livorno# commit and-quit
9  commit complete
10 Exiting configuration mode
11
12 user@Livorno> show ted protocol detail
13 Protocol name        Credibility  Self node
14 OSPF(0)              502          192.168.1.6  # Adapted protocol credibility for TED
```

Another factor is *link cost alignment* between both protocols. In the particular case of domain "Cyclone", the link between *Honolulu* and *Barcelona* diverges from the result of the OSPFv2 reference bandwidth; that is, it is always preferred that both routers communicate to each other through router *Livorno*. Considering that IS–IS predefines a default metric which is bandwidth unaware, two generic options exist to migrate metrics to IS–IS:

- extract *metric* and *TE metric* values from existing OSPF and TEDs;
- use the similar IS–IS `reference-bandwidth` command and manipulate link metrics in the same scenarios as with OSPFv2.

This second option provides a more scalable solution in real life setups, and it is our choice.

Junos Tip: Inspecting IGP and TEDs

The advantage of deploying link-state IGPs using a *ships-in-the-night* model is that information generation is independent at each one and forwarding is a question of route selection derived from protocol preference and TED credibility values.

In Junos OS, `show ted database` command outputs reflect available link TE information from link-state IGPs as collected from their extensions. As opposed to standard IGP databases, `show ted database` commands provide parallel information from same nodes and the equivalent link advertisement values from link-state IGPs, including the credibility for each protocol when ordering outputs.

It is strongly encouraged that database metrics are checked at each system in this scenario. If the previous metric divergence for the link between router *Honolulu* and router *Barcelona* remained unnoticed, a careful inspection of databases in a later step would have revealed the problem. Because IS–IS remains unpreferred, there is no impact at that point, which helps considerably to roll out another link-state protocol in the background without any forwarding impact.

Even when aligning cost assignment, *external* routes need to be considered: the prefix 192.168.10.0/24 is injected by router *Barcelona* and router *Honolulu* as Type 5 AS External LSA with different costs (0 and 1000) and External Type-2 (which is the Junos OS default), as shown in Listing 2.93.

Listing 2.93: Aggregate 192.168.10.0/24 injection in OSPFv2 with different external Type-2 metrics

```
user@Livorno> show ospf database external lsa-id 192.168.10.0 extensive
    OSPF AS SCOPE link state database
 Type       ID              Adv Rtr          Seq        Age  Opt  Cksum  Len
Extern   192.168.10.0    192.168.1.7      0x8000001a  1338  0x22 0x7143  36
  mask 255.255.255.0
  Topology default (ID 0)
    Type: 2, Metric: 0, Fwd addr: 0.0.0.0, Tag: 0.0.0.0 # Type 2 lowest metric
  Aging timer 00:37:41
  Installed 00:22:15 ago, expires in 00:37:42, sent 00:22:13 ago
  Last changed 12:36:18 ago, Change count: 1
Extern   192.168.10.0    192.168.1.21     0x80000006   997  0x22 0x7955  36
  mask 255.255.255.0
  Topology default (ID 0)
    Type: 2, Metric: 1000, Fwd addr: 0.0.0.0, Tag: 0.0.0.0 # Type 2 highest metric
  Aging timer 00:43:23
  Installed 00:16:34 ago, expires in 00:43:23, sent 00:16:32 ago
  Last changed 03:08:04 ago, Change count: 1

user@Livorno> show route 192.168.10.0/24 extensive

inet.0: 27 destinations, 33 routes (26 active, 0 holddown, 1 hidden)
192.168.10.0/24 (2 entries, 1 announced)
TSI:
KRT in-kernel 192.168.10.0/24 -> {172.16.1.53}
        *OSPF   Preference: 150
                Next hop type: Router, Next hop index: 777
                Next-hop reference count: 6
                Next hop: 172.16.1.53 via ge-0/2/0.0, selected
                State: <Active Int Ext>
                Age: 12:35:39   Metric: 0       Tag: 0  # Metric 0 preferred
                Task: OSPF
                Announcement bits (2): 0-KRT 4-LDP
                AS path: I
```

This output highlights another divergence when injecting external information in comparison with final requirements. When activating IS–IS `wide-metrics-only` and removing the external distinction, external information is injected as *IP extended prefix*, and metrics increase as dictated by SPF calculations towards the Intermediate System issuing route injection (from that perspective, this is the same as with other internal prefixes, such as loopback addresses).

On the other hand, [RFC2328] allows similar interpretations in OSPFv2 with *external route types* as dictated by the E bit on AS External LSAs. Type 1 external metrics were designed to be similar to link-state metrics and Type 2 external metrics were designed to express *larger* values as external metrics to other domains. The resulting implementation is that *Type 1* metrics are calculated as the sum of the external route's advertised cost and the distance to the respective ASBR, whereas *Type 2* metrics remain constant and the lowest metric value determined at the redistribution point prevails. Only when several equal-cost Type 2 routes exist is the calculated distance to each ASBR considered as the tiebreaker.

Note that Junos OS allows fine-tuning for both OSPF external metric types through policy language, but considers Type 2 as default redistribution type, whereas OSPF External Type 1 would be closer to the IS–IS `wide-metrics-only` interpretation here, because metric values increase as dictated by SPF results towards the redistribution point. At this stage, the alternatives to align that behavior are:

- Remove `wide-metrics-only` and fall back to an external distinction for prefixes. Metric calculation would be aligned with existing OSPFv2 Type 2.

- Modify existing OSPFv2 configuration to inject information as Type 1 metric. This choice leads to potential forwarding changes when compared with the current forwarding table at each system.

- Align route selection with `wide-metrics-only` as is done with OSPF Type 2. This option does not necessarily mean replicating costs in IS–IS.

While the first option is a safe approach, it flies in the face of the final design requirements. At some point in the future, this behavior will need to be modified to cope with the final integration prerequisites.

The second option is not really supported by company top management, because modifying the existing status quo creates the potential for conflicts and additional unexpected problems, and this is the last thing our managers want to hear when planning a critical change! In the current topology, if the link between router *Barcelona* and router *Livorno* fails, both Type 1 metrics for 192.168.10.0/24 would be the same (because the cost metric between router *Barcelona* and router *Honolulu* is the same as the externally added metric to the route in router *Honolulu*). In that case, router *Honolulu* could be elected as the exit point for that aggregate, although router *Barcelona* remains up and running. Note that this situation would not happen with Type 2 metrics, because values remain uniform across the network and the primary exit would still be selected.

The last option tries to address the real problem: *what is the purpose of having different Type 2 costs for the same prefix?* The answer seems obvious: create a primary/backup scenario. As long as router *Barcelona* is connected to the network, its external connection needs to be preferred and this condition must persist for the entire "Gale Internet" network. The objective is to mimic that routing rationale with IS–IS Level 2 wide metrics and keep that forwarding behavior.

The idea behind this is to make the backup route injection at router *Honolulu* unpreferred compared with router *Barcelona* under all tumultuous topology circumstances. In our scenario and assuming symmetric cost values, this is a very simplistic approach: given the extended metric capabilities with IS–IS wide-metrics and given that both routers are connected to router *Livorno* in a triangular fashion because it is the hub for the rest of the network, this approach assigns a metric to the prefix in router *Honolulu* that should exceed the longest path cost from router *Livorno* to router *Barcelona* minus the lowest calculated metric from router *Livorno* to router *Honolulu*:

$$\text{(Ext metric in Honolulu)} > \text{(longest } \overrightarrow{\text{Livorno-Barcelona}}) - \text{(shortest } \overrightarrow{\text{Livorno-Honolulu}}) \quad (2.1)$$

$$\text{(Ext metric in Honolulu)} > \overline{\text{Livorno-Honolulu}} + \overline{\text{Honolulu-Barcelona}} - \overline{\text{Livorno-Honolulu}} \quad (2.2)$$

$$\text{(Ext metric in Honolulu)} > \overline{\text{Honolulu-Barcelona}} = 1000 \quad (2.3)$$

The situation is summarized in Figure 2.21.

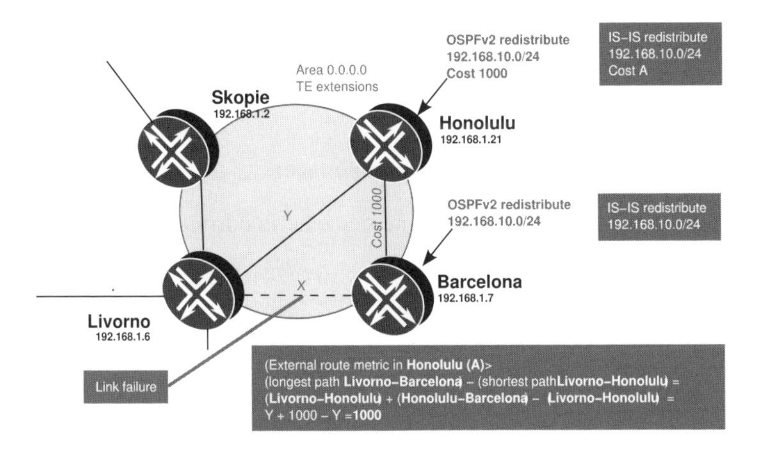

Figure 2.21: Stage one: external route injection metric adjustment at domain "Cyclone".

As long as the external cost for route injection in router *Honolulu* exceeds the link cost between router *Honolulu* and router *Barcelona*, the path pointing to router *Barcelona* is selected while router *Barcelona* remains reachable.

What If... Route selection, MPLS forwarding, and IP forwarding

A parallel discussion here is the way external route information is injected into a link-state database. Unless especially tweaked, Junos OS requires a route to be active in the routing table to be eligible to be exported to other protocols.

Considering the previous scenario, with a failing link between router *Barcelona* and router *Livorno* and route injection at both router *Barcelona* as primary and router *Honolulu* as backup, the forwarding behavior may be completely different depending on how traffic arrives at router *Honolulu*, even though router *Barcelona* is selected as exit point.

If flows arrive at router *Honolulu* as MPLS traffic (previous LSP selection at router *Barcelona* and the network behind it), router *Honolulu* usually performs either Penultimate Hop Popping or MPLS label *swap* with *Explicit Null* as lookup action, and traffic may be forwarded to router *Barcelona*.

If flows arrive at router *Honolulu* as native IPv4, router *Honolulu* performs a standard IP lookup. Under normal conditions, this prefix is externally injected into the IGP. For this redistribution, the route needs to be active in the routing table, meaning it must be instantiated as well being in the forwarding table of router *Honolulu*. In this case, traffic may directly exit the backup link to the external domain as a consequence of a standard IP lookup operation in router *Honolulu*, even though the originally elected exit point was router *Barcelona*.

At this point, the IS–IS configuration is deployed at each node with these settings:

- IS–IS internal preference adjustment;
- IS–IS TED protocol credibility following preference values;
- ISO addressing following a *Binary-Coded Decimal* (BCD) translation scheme and using Area 49.0003;
- IS–IS Level 1 disabled on each internal interface;
- IS–IS reference bandwidth and link metrics inherited from OSPFv2;
- external redistribution policies for 192.168.10.0/24 to add metric 0 in router *Barcelona* and 1001 in router *Honolulu*.

Listing 2.94 includes the IS–IS configuration excerpt in router *Livorno*, which is similar in all domain "Cyclone" routers, with the exception of defining the same metric for the link between router *Honolulu* and router *Barcelona* and tuning external route injection at both of them.

Listing 2.94: IS–IS configuration in Livorno

```
1  [edit protocols isis]
2  user@Livorno# show
3  traffic-engineering credibility-protocol-preference;
4  level 1 {
5      authentication-key "$9$d.bw2ZGiqPQs2T3"; ### SECRET-DATA
6      authentication-type md5;
7      wide-metrics-only;
8      preference 155;
9  }
10 level 2 {
11     authentication-key "$9$.PfQ/9pOIc5Qhr"; ### SECRET-DATA
12     authentication-type md5;
13     wide-metrics-only;
14     preference 158;
15 }
16 interface so-0/0/0.0 {
17     level 1 disable;
18 }
19 interface ge-0/2/0.0 {
20     level 1 disable;
21 }
22 interface ge-1/3/0.0 {
23     level 1 disable;
24 }
25 interface lo0.0 {
26     level 1 disable;
27     level 2 passive;
```

```
28 }
29
30  [edit interfaces]
31 user@Livorno# set lo0.0 family iso address 49.0003.1921.6800.1006.00
32  [edit interfaces]
33 user@Livorno# set so-0/0/0.0 family iso
34  [edit interfaces]
35 user@Livorno# set ge-0/2/0.0 family iso
36  [edit interfaces]
37 user@Livorno# set ge-1/3/0.0 family iso
```

Once the configuration has been fully deployed in the domain, IS–IS adjacencies should get established and databases should start to be populated. Listing 2.95 shows illustrative details for the IS–IS protocol activation from router *Livorno*'s perspective. It is worth noting that because `wide-metrics-only` is applied at IS–IS Level 2, 192.168.10.0/24 is externally injected at both router *Barcelona* and router *Honolulu* but appears only in the form of TLV 135 Extended IP Reachability with the already discussed cost values.

Listing 2.95: IS–IS activation results in Livorno

```
 1 user@Livorno> show isis adjacency
 2 Interface             System         L State        Hold (secs) SNPA
 3 ge-0/2/0.0            Barcelona-re0  2 Up                   19  0:14:f6:84:8c:bc
 4 ge-1/3/0.0            honolulu-re0   2 Up                    6  0:17:cb:d3:2f:16
 5 so-0/0/0.0            Skopie         2  Up                   26
 6
 7 user@Livorno> show isis database
 8 IS-IS level 1 link-state database:
 9 LSP ID                      Sequence Checksum Lifetime Attributes
10 Livorno.00-00                   0x25   0xbaa4      939 L1 L2
11   1 LSPs
12
13 IS-IS level 2 link-state database:
14 LSP ID                      Sequence Checksum Lifetime Attributes
15 Skopie.00-00                    0x2f   0x6372      957 L1 L2
16 Livorno.00-00                   0x29   0xa2d9      939 L1 L2
17 Livorno.02-00                   0x24   0xed0a      939 L1 L2
18 Barcelona-re0.00-00         0x27   0xcac       907 L1 L2
19 honolulu-re0.00-00          0x28   0xc725     1126 L1 L2
20 honolulu-re0.02-00          0x26   0x9893     1126 L1 L2
21 honolulu-re0.03-00          0x25   0x87a2     1126 L1 L2
22   7 LSPs
23
24 user@Livorno> show isis database detail | match "Sequence|prefix"
25 <...>
26 Barcelona-re0.00-00 Sequence: 0x27, Checksum: 0xcac, Lifetime: 893 secs
27    IP prefix: 172.16.1.52/30                  Metric:        10 Internal Up
28    IP prefix: 172.16.1.104/30                 Metric:      1000 Internal Up
29    IP prefix: 192.168.1.7/32                  Metric:         0 Internal Up
30    IP prefix: 192.168.10.0/24                 Metric:         0 Internal Up #  Internal with cost 0
31 honolulu-re0.00-00 Sequence: 0x28, Checksum: 0xc725, Lifetime: 1111 secs
32    IP prefix: 172.16.1.88/30                  Metric:        10 Internal Up
33    IP prefix: 172.16.1.104/30                 Metric:      1000 Internal Up
34    IP prefix: 192.168.1.21/32                 Metric:         0 Internal Up
35    IP prefix: 192.168.10.0/24                 Metric:      1001 Internal Up # Internal with cost 1001
36 honolulu-re0.02-00 Sequence: 0x26, Checksum: 0x9893, Lifetime: 1111 secs
37 honolulu-re0.03-00 Sequence: 0x25, Checksum: 0x87a2, Lifetime: 1111 secs
```

IS–IS information is being flooded with the intended design premises and IS–IS routes still remain as backup paths in the routing tables of all systems. Listing 2.96 shows output comparing various routes on router *Livorno* to confirm that IS–IS activation has had absolutely no impact in the current forwarding state. The situation is similar on all other routers in the domain.

Listing 2.96: IS–IS and OSPFv2 *ships-in-the-night* analysis in router *Livorno*

```
user@Livorno> show route protocol isis active-path table inet.0
  # No active IS-IS routes
inet.0: 27 destinations, 33 routes (26 active, 0 holddown, 1 hidden)

user@Livorno> show route protocol ospf active-path table inet.0
# All IGP selected paths are from OSPFv2

inet.0: 27 destinations, 33 routes (26 active, 0 holddown, 1 hidden)
+ = Active Route, - = Last Active, * = Both

172.16.1.104/30    *[OSPF/10] 11:38:43, metric 1010
                      to 172.16.1.53 via ge-0/2/0.0
                    > to 172.16.1.90 via ge-1/3/0.0
192.168.1.2/32     *[OSPF/10] 1d 00:45:38, metric 64
                    > via so-0/0/0.0
192.168.1.7/32     *[OSPF/10] 1d 00:31:41, metric 10
                    > to 172.16.1.53 via ge-0/2/0.0
192.168.1.21/32    *[OSPF/10] 22:40:04, metric 10
                    > to 172.16.1.90 via ge-1/3/0.0
192.168.10.0/24    *[OSPF/150] 13:31:33, metric 0, tag 0
                    > to 172.16.1.53 via ge-0/2/0.0
224.0.0.5/32       *[OSPF/10] 1d 12:35:27, metric 1
                      MultiRecv

user@Livorno> show route 192.168.1.0/24 terse table inet.0
  # Loopbacks are either from OSPFv2 or direct
inet.0: 27 destinations, 33 routes (26 active, 0 holddown, 1 hidden)
+ = Active Route, - = Last Active, * = Both

A Destination        P Prf   Metric 1   Metric 2  Next hop        AS path
* 192.168.1.2/32     O  10         64             >so-0/0/0.0
                     I 158         64             >172.16.1.33
* 192.168.1.6/32     D   0                        >lo0.0
* 192.168.1.7/32     O  10         10             >172.16.1.53
                     I 158         10             >172.16.1.53
* 192.168.1.21/32    O  10         10             >172.16.1.90
                     I 158         10             >172.16.1.90

user@Livorno> show route 172.16.1.0/24 terse table inet.0
  # WAN addresses are either from OSPFv2 or direct
inet.0: 27 destinations, 33 routes (26 active, 0 holddown, 1 hidden)
+ = Active Route, - = Last Active, * = Both

A Destination        P Prf   Metric 1   Metric 2  Next hop        AS path
* 172.16.1.32/30     D   0                        >so-0/0/0.0
                     O  10         64             >so-0/0/0.0
* 172.16.1.34/32     L   0                         Local
* 172.16.1.36/30     D   0                        >so-0/0/1.0
* 172.16.1.38/32     L   0                         Local
* 172.16.1.52/30     D   0                        >ge-0/2/0.0
* 172.16.1.54/32     L   0                         Local
* 172.16.1.88/30     D   0                        >ge-1/3/0.0
* 172.16.1.89/32     L   0                         Local
* 172.16.1.104/30    O  10       1010              172.16.1.53
                                                  >172.16.1.90
                     I 158       1010             >172.16.1.53
                                                   172.16.1.90

user@Livorno> show ted protocol
  # OSPFv2 is most credible protocol for TED information selection
Protocol name        Credibility  Self node
OSPF(0)              502          Livorno.00(192.168.1.6)
IS-IS(1)             357          Livorno.00(192.168.1.6)
IS-IS(2)             354          Livorno.00(192.168.1.6)

user@Livorno> show ted database extensive | match "OSPF|IS-IS|Node|To:|Metric:"
TED database: 7 ISIS nodes 7 INET nodes
```

```
68 NodeID: Skopie.00(192.168.1.2) # Same OSPFv2 and IS-IS Metric for connection to Livorno
69   Protocol: OSPF(0.0.0.0)
70     To: Livorno.00(192.168.1.6), Local: 172.16.1.33, Remote: 172.16.1.34
71       Metric: 64
72   Protocol: IS-IS(2)
73     To: Livorno.00(192.168.1.6), Local: 172.16.1.33, Remote: 172.16.1.34
74       Metric: 64
75 NodeID: Livorno.00(192.168.1.6) # Same OSPFv2 and IS-IS Metric for connections out of Livorno
76   Protocol: OSPF(0.0.0.0)
77     To: 172.16.1.53-1, Local: 172.16.1.54, Remote: 0.0.0.0
78       Metric: 10
79     To: 172.16.1.90-1, Local: 172.16.1.89, Remote: 0.0.0.0
80       Metric: 10
81     To: Skopie.00(192.168.1.2), Local: 172.16.1.34, Remote: 172.16.1.33
82       Metric: 64
83   Protocol: IS-IS(1)
84   Protocol: IS-IS(2)
85     To: Skopie.00(192.168.1.2), Local: 172.16.1.34, Remote: 172.16.1.33
86       Metric: 64
87     To: Livorno.02, Local: 172.16.1.54, Remote: 0.0.0.0
88       Metric: 10
89     To: honolulu-re0.02, Local: 172.16.1.89, Remote: 0.0.0.0
90       Metric: 10
91 <...>
92 NodeID: Barcelona-re0.00(192.168.1.7) # Same OSPFv2 and IS-IS Metric for connections out of Barcelona
93   Protocol: OSPF(0.0.0.0)
94     To: 172.16.1.53-1, Local: 172.16.1.53, Remote: 0.0.0.0
95       Metric: 10
96     To: 172.16.1.105-1, Local: 172.16.1.106, Remote: 0.0.0.0
97       Metric: 1000
98   Protocol: IS-IS(2)
99     To: Livorno.02, Local: 172.16.1.53, Remote: 0.0.0.0
100      Metric: 10
101    To: honolulu-re0.03, Local: 172.16.1.106, Remote: 0.0.0.0
102      Metric: 1000
103 NodeID: honolulu-re0.00(192.168.1.21)  # Same OSPFv2 and IS-IS Metric for connections out of Honolulu
104   Protocol: OSPF(0.0.0.0)
105     To: 172.16.1.90-1, Local: 172.16.1.90, Remote: 0.0.0.0
106       Metric: 10
107     To: 172.16.1.105-1, Local: 172.16.1.105, Remote: 0.0.0.0
108       Metric: 1000
109   Protocol: IS-IS(2)
110     To: honolulu-re0.02, Local: 172.16.1.90, Remote: 0.0.0.0
111       Metric: 10
112     To: honolulu-re0.03, Local: 172.16.1.105, Remote: 0.0.0.0
113       Metric: 1000
114 <...>
```

Focusing on 192.168.10.0/24 once again, as shown in Listing 2.97, the route is still preferred via OSPFv2 external against IS–IS Level 2 internal and this selection remains independent from the metrics discussion for route injection in IS–IS to mimic the primary/backup setup. Failing to modify standard route preferences would have led to IS–IS Level 2 internal path selection for 192.168.10.0/24.

Listing 2.97: Path availability and selection for 192.168.10.0/24

```
1  user@Livorno> show route 192.168.10.0/24 table inet.0 extensive
2
3  inet.0: 27 destinations, 33 routes (26 active, 0 holddown, 1 hidden)
4  192.168.10.0/24 (2 entries, 1 announced)
5  TSI:
6  KRT in-kernel 192.168.10.0/24 -> {172.16.1.53}
7          *OSPF   Preference: 150 # Standard OSPF external preference
8                  Next hop type: Router, Next hop index: 777
9                  Next-hop reference count: 6
10                 Next hop: 172.16.1.53 via ge-0/2/0.0, selected
```

```
11                  State: <Active Int Ext>
12                  Age: 13:40:51    Metric: 0        Tag: 0
13                  Task: OSPF
14                  Announcement bits (2): 0-KRT 4-LDP
15                  AS path: I
16          IS-IS   Preference: 158 # Modified IS-IS Level 2 internal preference
17                  Level: 2
18                  Next hop type: Router, Next hop index: 777
19                  Next-hop reference count: 6
20                  Next hop: 172.16.1.53 via ge-0/2/0.0, selected
21                  State: <Int>
22                  Inactive reason: Route Preference # Modified value is inactive reason
23                  Age: 7:32:18    Metric: 10
24                  Task: IS-IS
25                  AS path: I
```

2.3.5 Stage two: Route redistribution at domain "Monsoon"

Domain "Monsoon" needs to undergo a full readdressing exercise, but to reach the first integration milestone, only route connectivity will be granted under the unified scheme. The current situation is reflected in Listing 2.98 from router *Torino*, which has visibility to both Area 0.0.0.0 and an internal NSSA 0.0.0.3 with default route injection from both router *Nantes* and router *Torino* with the same default metric 10 (but different calculated costs to each ABR). Note that no authentication type or traffic-engineering extensions previously existed in this domain.

Listing 2.98: Initial OSPFv2 status at Torino

```
1  user@Torino> show configuration protocols ospf
2  reference-bandwidth 100g;
3  area 0.0.0.3 {
4      nssa {
5          default-lsa default-metric 10;
6          no-summaries;
7      }
8      interface at-1/2/0.1001;
9  }
10 area 0.0.0.0 {
11     interface so-0/1/0.0;
12     interface lo0.0 {
13         passive;
14     }
15 }
16 user@Torino> show ospf database
17
18     OSPF database, Area 0.0.0.0
19  Type       ID               Adv Rtr           Seq       Age  Opt  Cksum  Len
20 Router   10.10.1.1        10.10.1.1        0x8000002c   170  0x22 0x3928  60
21 Router  *10.10.1.3        10.10.1.3        0x8000002e   151  0x22 0xe279  60
22 Summary  10.10.1.9        10.10.1.1        0x80000001   120  0x22 0xa6dc  28
23 Summary *10.10.1.9        10.10.1.3        0x80000001   119  0x22 0xdf01  28
24 Summary  10.10.1.10       10.10.1.1        0x80000028    21  0x22 0x8253  28
25 Summary *10.10.1.10       10.10.1.3        0x80000028  1303  0x22 0x53ea  28
26 Summary  10.10.1.64       10.10.1.1        0x80000028   638  0x22 0x1e0a  28
27 Summary *10.10.1.64       10.10.1.3        0x80000028   874  0x22 0x8c73  28
28 Summary  10.10.1.72       10.10.1.1        0x80000027  1924  0x22 0x497   28
29 Summary *10.10.1.72       10.10.1.3        0x80000028   446  0x22 0x776   28
30 Summary  10.10.1.80       10.10.1.1        0x80000027  1495  0x22 0xf9f8  28
31 Summary *10.10.1.80       10.10.1.3        0x80000028    17  0x22 0x8278  28
32
33     OSPF database, Area 0.0.0.3
34  Type       ID               Adv Rtr           Seq       Age  Opt  Cksum  Len
35 Router   10.10.1.1        10.10.1.1        0x8000002c   172  0x20 0x9b5f  48
```

```
36 Router  *10.10.1.3      10.10.1.3       0x8000002b  151  0x20 0xe2bd  48
37 Router   10.10.1.9      10.10.1.9       0x80000029  121  0x20 0x6b5d  84
38 Router   10.10.1.10     10.10.1.10      0x8000002a  823  0x20 0xb2b9  84
39 Summary  0.0.0.0        10.10.1.1       0x80000002  172  0x20 0x52cb  28
40 Summary *0.0.0.0        10.10.1.3       0x80000002  151  0x20 0x46d5  28
41
42 user@Torino> show ospf database extensive | match "0x8| metric|data"
43     OSPF database, Area 0.0.0.0
44 Router   10.10.1.1      10.10.1.1       0x8000002c  184  0x22 0x3928  60
45   id 10.10.1.1, data 255.255.255.255, Type Stub (3)
46     Topology count: 0, Default metric: 0
47   id 10.10.1.3, data 10.10.1.98, Type PointToPoint (1)
48     Topology count: 0, Default metric: 160
49   id 10.10.1.96, data 255.255.255.252, Type Stub (3)
50     Topology count: 0, Default metric: 160
51 Router  *10.10.1.3      10.10.1.3       0x8000002e  165  0x22 0xe279  60
52   id 10.10.1.3, data 255.255.255.255, Type Stub (3)
53     Topology count: 0, Default metric: 0
54   id 10.10.1.1, data 10.10.1.97, Type PointToPoint (1)
55     Topology count: 0, Default metric: 160
56   id 10.10.1.96, data 255.255.255.252, Type Stub (3)
57     Topology count: 0, Default metric: 160
58 Summary  10.10.1.9      10.10.1.1       0x80000001  134  0x22 0xa6dc  28
59   Topology default (ID 0) -> Metric: 643
60 Summary *10.10.1.9      10.10.1.3       0x80000001  133  0x22 0xdf01  28
61   Topology default (ID 0) -> Metric: 803
62 <...>
63     OSPF database, Area 0.0.0.3
64 Router   10.10.1.1      10.10.1.1       0x8000002c  186  0x20 0x9b5f  48
65   id 10.10.1.9, data 10.10.1.65, Type PointToPoint (1)
66     Topology count: 0, Default metric: 643
67   id 10.10.1.64, data 255.255.255.252, Type Stub (3)
68     Topology count: 0, Default metric: 643
69 Router  *10.10.1.3      10.10.1.3       0x8000002b  165  0x20 0xe2bd  48
70   id 10.10.1.10, data 10.10.1.81, Type PointToPoint (1)
71     Topology count: 0, Default metric: 160
72   id 10.10.1.80, data 255.255.255.252, Type Stub (3)
73     Topology count: 0, Default metric: 160
74 Router   10.10.1.9      10.10.1.9       0x80000029  135  0x20 0x6b5d  84
75   id 10.10.1.10, data 10.10.1.73, Type PointToPoint (1)
76     Topology count: 0, Default metric: 643
77   id 10.10.1.72, data 255.255.255.252, Type Stub (3)
78     Topology count: 0, Default metric: 643
79   id 10.10.1.9, data 255.255.255.255, Type Stub (3)
80     Topology count: 0, Default metric: 0
81   id 10.10.1.1, data 10.10.1.66, Type PointToPoint (1)
82     Topology count: 0, Default metric: 643
83   id 10.10.1.64, data 255.255.255.252, Type Stub (3)
84     Topology count: 0, Default metric: 643
85 Router   10.10.1.10     10.10.1.10      0x8000002a  837  0x20 0xb2b9  84
86   id 10.10.1.3, data 10.10.1.82, Type PointToPoint (1)
87     Topology count: 0, Default metric: 160
88   id 10.10.1.80, data 255.255.255.252, Type Stub (3)
89     Topology count: 0, Default metric: 160
90   id 10.10.1.9, data 10.10.1.74, Type PointToPoint (1)
91     Topology count: 0, Default metric: 643
92   id 10.10.1.72, data 255.255.255.252, Type Stub (3)
93     Topology count: 0, Default metric: 643
94   id 10.10.1.10, data 255.255.255.255, Type Stub (3)
95     Topology count: 0, Default metric: 0
96 Summary  0.0.0.0        10.10.1.1       0x80000002  186  0x20 0x52cb  28
97   Topology default (ID 0) -> Metric: 10   # Default route from Nantes
98 Summary *0.0.0.0        10.10.1.3       0x80000002  165  0x20 0x46d5  28
99   Topology default (ID 0) -> Metric: 10  # Default route from Torino
100
101 user@Torino> show route protocol ospf table inet.0
102
103 inet.0: 27 destinations, 29 routes (27 active, 0 holddown, 0 hidden)
104 + = Active Route, - = Last Active, * = Both
```

```
105
106 10.10.1.1/32       *[OSPF/10] 00:03:08, metric 160
107                     > via so-0/1/0.0
108 10.10.1.9/32       *[OSPF/10] 00:02:18, metric 803
109                     > via at-1/2/0.1001
110 10.10.1.10/32      *[OSPF/10] 1d 03:44:01, metric 160
111                     > via at-1/2/0.1001
112 <...>
```

Because of default route injection and because Area 0.0.0.3 is an NSSA with no explicit Summary LSAs, just with those Summary LSAs representing default routes, the resulting routing table on internal systems such as router *Basel* is reduced, as shown in Listing 2.99. Although the amount of backbone routing information in this case study is scarce, this restriction is usually imposed to limit the number of LSAs flooded internally in that area.

Listing 2.99: OSPFv2 routes at Basel

```
1  user@Basel> show route protocol ospf table inet.0
2
3  inet.0: 18 destinations, 20 routes (18 active, 0 holddown, 0 hidden)
4  + = Active Route, - = Last Active, * = Both
5
6  0.0.0.0/0          *[OSPF/10] 00:13:15, metric 170
7                      > via at-0/0/0.1001 # Preferred default exit towards Torino
8  10.10.1.9/32       *[OSPF/10] 00:07:27, metric 643
9                      > via at-0/1/0.1001
10 10.10.1.64/30      *[OSPF/10] 1d 04:21:36, metric 1286
11                     > via at-0/1/0.1001
12 10.10.1.72/30       [OSPF/10] 1d 04:26:50, metric 643
13                     > via at-0/1/0.1001
14 10.10.1.80/30       [OSPF/10] 1d 04:26:50, metric 160
15                     > via at-0/0/0.1001
16 224.0.0.5/32       *[OSPF/10] 2d 01:39:07, metric 1
17                       MultiRecv
```

The migration steps that need to be taken on domain "Monsoon" can be summarized as follows:

- Add legitimate and externally valid loopback addresses to all systems inside the domain "Monsoon".
- Separate OSPFv2 and IS–IS domains in router *Torino* and router *Nantes* by binding different loopback addresses and links to each of them.
- Establish ISO addressing in router *Torino* and router *Nantes* following a BCD translation scheme and using Area 49.0003.
- Disable IS–IS Level 1 on router *Torino* and router *Nantes* on a per-interface basis.
- Leak valid loopback addresses from the backbone into the domain. Flood them as Type 5 LSAs in Area 0.0.0.0 and Type 7 LSAs in Area 0.0.0.3.
- Redistribute externally valid loopback addresses from router *Basel* and router *Inverness* to the final IS–IS domain.

Adding new externally legitimate and valid loopback addresses for "Gale Internet" is a straightforward step, but on router *Torino* and router *Nantes* this implementation is going to be performed slightly differently. While for both router *Basel* and router *Inverness* such loopbacks need to be included as stub networks of their Router LSA, it is desirable to avoid a similar approach at both router *Torino* and router *Nantes* with the intention of separating domains. All externally legitimate loopback addresses are leaked internally into the NSSA, but the goal with domain separation is to make router *Torino* and router *Nantes* prefer each other's legitimate loopback addresses through IS–IS Level 2 internal rather than OSPFv2 internal so as not to have to modify active path selection in the future.

For that purpose, those routes will be externally injected into OSPFv2 (similarly to other loopback addresses) and as a result, router *Torino* and router *Nantes* will select each other's old and non-legitimate loopback addresses via OSPFv2 internal and externally "Gale Internet" valid addresses via IS–IS Level 2 internal with backup paths over OSPFv2 external.

Because the router ID is not changing at this stage, the only necessary modification is to constrain loopback address injection in each link-state IGP at both router *Torino* and router *Nantes*. Selective address injection in OSPFv2 is achieved by removing the complete loopback interface unit as passive interface and just injecting the current internal router ID address in OSPFv2, as shown in Listing 2.100 for router *Torino*. IS–IS requires an `export` policy banning old non-legitimate IP prefixes to be injected in any TLV, as shown in Listing 2.101 for router *Torino*. The same modification needs to take place in router *Nantes* for its loopback addresses.

Listing 2.100: OSPFv2 selective loopback address injection at Torino

```
1  user@Torino> show ospf database router lsa-id 10.10.1.3 extensive
2
3      OSPF database, Area 0.0.0.0
4   Type       ID               Adv Rtr           Seq      Age  Opt  Cksum  Len
5  Router  *10.10.1.3        10.10.1.3        0x8000002f   819  0x22 0xe07a  60
6    bits 0x3, link count 3
7    id 10.10.1.3, data 255.255.255.255, Type Stub (3)
8      Topology count: 0, Default metric: 0
9    id 10.10.1.1, data 10.10.1.97, Type PointToPoint (1)
10     Topology count: 0, Default metric: 160
11   id 10.10.1.96, data 255.255.255.252, Type Stub (3)
12     Topology count: 0, Default metric: 160
13   Gen timer 00:25:29
14   Aging timer 00:46:20
15   Installed 00:13:39 ago, expires in 00:46:21, sent 00:13:37 ago
16   Last changed 1d 04:13:56 ago, Change count: 4, Ours
17
18     OSPF database, Area 0.0.0.3
19  Type       ID               Adv Rtr           Seq      Age  Opt  Cksum  Len
20 Router  *10.10.1.3        10.10.1.3        0x8000002c   487  0x20 0xe0be  48
21   bits 0x3, link count 2
22   id 10.10.1.10, data 10.10.1.81, Type PointToPoint (1)
23     Topology count: 0, Default metric: 160
24   id 10.10.1.80, data 255.255.255.252, Type Stub (3)
25     Topology count: 0, Default metric: 160
26   Gen timer 00:41:52
27   Aging timer 00:51:52
28   Installed 00:08:07 ago, expires in 00:51:53, sent 00:08:05 ago
29   Last changed 1d 04:20:37 ago, Change count: 2, Ours
30
31 [edit protocols ospf]
32 user@Torino# delete area 0 interface lo0.0
33 [edit protocols ospf]
34 user@Torino# set area 0 interface 10.10.1.3
35
36 [edit protocols ospf]
37 user@Torino# show
38 reference-bandwidth 100g;
39 area 0.0.0.3 {
```

```
40      nssa {
41          default-lsa default-metric 10;
42          no-summaries;
43      }
44      interface at-1/2/0.1001;
45  }
46  area 0.0.0.0 {
47      interface so-0/1/0.0;
48      interface 10.10.1.3;
49  }
50
51  [edit protocols ospf]
52  user@Torino# commit
53  commit complete
54  [edit protocols ospf]
55  user@Torino# run show ospf database router lsa-id 10.10.1.3 extensive
56
57   # No stub network created for externally valid loopback address
58
59      OSPF database, Area 0.0.0.0
60   Type       ID               Adv Rtr           Seq      Age  Opt  Cksum  Len
61  Router  *10.10.1.3       10.10.1.3        0x80000030    34  0x22 0xde7b  60
62    bits 0x3, link count 3
63    id 10.10.1.3, data 255.255.255.255, Type Stub (3)
64      Topology count: 0, Default metric: 0
65    id 10.10.1.1, data 10.10.1.97, Type PointToPoint (1)
66      Topology count: 0, Default metric: 160
67    id 10.10.1.96, data 255.255.255.252, Type Stub (3)
68      Topology count: 0, Default metric: 160
69    Gen timer 00:49:26
70    Aging timer 00:59:26
71    Installed 00:00:34 ago, expires in 00:59:26, sent 00:00:32 ago
72    Last changed 1d 04:14:46 ago, Change count: 4, Ours
73
74      OSPF database, Area 0.0.0.3
75   Type       ID               Adv Rtr           Seq      Age  Opt  Cksum  Len
76  Router  *10.10.1.3       10.10.1.3        0x8000002d    34  0x20 0xdebf  48
77    bits 0x3, link count 2
78    id 10.10.1.10, data 10.10.1.81, Type PointToPoint (1)
79      Topology count: 0, Default metric: 160
80    id 10.10.1.80, data 255.255.255.252, Type Stub (3)
81      Topology count: 0, Default metric: 160
82    Gen timer 00:49:26
83    Aging timer 00:59:26
84    Installed 00:00:34 ago, expires in 00:59:26, sent 00:00:32 ago
85    Last changed 1d 04:21:27 ago, Change count: 2, Ours
```

Listing 2.101: IS–IS selective loopback address injection at Torino

```
1   [edit interfaces lo0]
2   user@Torino# show
3   unit 0 {
4       family inet {
5           address 10.10.1.3/32 {
6               primary;
7           }
8       }
9   }
10
11  [edit interfaces lo0]
12  user@Torino# set unit 0 family inet address 192.168.1.3/32
13  [edit interfaces lo0]
14  user@Torino# set unit 0 family iso address 49.0003.1921.6800.1003.00
15  [edit policy-options policy-statement block-non-legitimate-loopback]
16  user@Torino# show
17  from {
18      protocol direct;
19      route-filter 10.10.1.3/32 exact;
```

```
20 }
21 then reject;
22 [edit interfaces so-0/1/0 unit 0]
23 user@Torino# set family iso
24 [edit protocols isis]
25 user@Torino# show
26 export block-non-legitimate-loopback;
27 level 2 {
28     authentication-key "$9$tkmzu01SyKWX-O1Nb"; ### SECRET-DATA
29     authentication-type md5;
30     wide-metrics-only;
31 }
32 level 1 {
33     authentication-key "$9$WtPLXNsYoZDkxNHm"; ### SECRET-DATA
34     authentication-type md5;
35     wide-metrics-only;
36 }
37 interface so-0/1/0.0 {
38     level 1 disable;
39 }
40 interface lo0.0 {
41     level 1 disable;
42     level 2 passive;
43 }
44 [edit protocols isis]
45 user@Torino# commit and-quit
46 commit complete
47 Exiting configuration mode
48
49 user@Torino> show isis adjacency detail
50 Nantes
51   Interface: so-0/1/0.0, Level: 2, State: Up, Expires in 22 secs
52   Priority: 0, Up/Down transitions: 1, Last transition: 00:01:02 ago
53   Circuit type: 2, Speaks: IP, IPv6
54   Topologies: Unicast
55   Restart capable: Yes, Adjacency advertisement: Advertise
56   IP addresses: 10.10.1.98
57
58 user@Torino> show isis database detail
59 IS-IS level 1 link-state database:
60
61 Torino.00-00 Sequence: 0x2, Checksum: 0x11b0, Lifetime: 1092 secs
62
63 IS-IS level 2 link-state database:
64
65 Nantes.00-00 Sequence: 0x3, Checksum: 0x1dbb, Lifetime: 1129 secs
66    IS neighbor: Torino.00                  Metric:      10
67    IP prefix: 10.10.1.96/30                Metric:      10 Internal Up
68    IP prefix: 192.168.1.1/32               Metric:       0 Internal Up
69
70 # No extended IP prefix TLV for internal loopback address
71
72 Torino.00-00 Sequence: 0x3, Checksum: 0x7171, Lifetime: 1131 secs
73    IS neighbor: Nantes.00                  Metric:      10
74    IP prefix: 10.10.1.96/30                Metric:      10 Internal Up
75    IP prefix: 192.168.1.3/32               Metric:       0 Internal Up
76
77 user@Torino> show route protocol isis table inet.0
78
79 inet.0: 31 destinations, 33 routes (31 active, 0 holddown, 0 hidden)
80 + = Active Route, - = Last Active, * = Both
81
82 192.168.1.1/32     *[IS-IS/18] 00:01:17, metric 10
83                     > to 10.10.1.98 via so-0/1/0.0
```

The convenience of activating IS–IS in the directly connecting link between router *Torino* and router *Nantes* is open for discussion. In this analysis, it is considered convenient to grant additional redundancy to router *Torino* in case its newly acquired leased line to router *Bilbao* fails at any time during the migration. This activation overlaps an existing OSPFv2 adjacency in Area 0 but default protocol preferences, selective loopback injection as internal routes in each protocol, and direct route selection ensure that there is no protocol misalignment risk when overlapping both IGPs over this particular link.

Only a single IS–IS adjacency is established between both router *Torino* and router *Nantes* over their direct link now. However, a WAN readdressing exercise is also done at this point to define externally known public addresses for that back-to-back link instead of those internal to the domain. This readdressing exercise can be further tuned with similar IS–IS export policies or OSPF interface commands, as previously described for loopback addresses but it is not shown in this exercise, which assumes domain-wide known addresses in this segment from this point onwards.

The "Gale Internet" legitimate external addresses from router *Basel* and router *Inverness* need to be redistributed from OSPFv2 to IS–IS as part of the route reachability policy. Those public loopback addresses are included as stub networks of their respective Router LSAs and are therefore considered as OSPFv2 internal routes. Because of the IS–IS Level 2 `wide-metrics-only` configuration, such routes are injected in similar TLV 135 Extended IP Reachability structures at both router *Torino* and router *Nantes* and are considered as IS–IS Level 2 internal routes. In this case, default protocol preferences cope with the objective to prefer always an internal route incoming from OSPF NSSA 0.0.0.3 over another path for the same address, redistributed at the other L2 system and considered as an internal route.

What If... OSPF external routes vs IS–IS internal routes via wide-metrics-only

The same earlier discussion from domain "Cyclone" could have appeared in this case if router *Basel* and router *Inverness*, as internal routers, had injected Type 7 LSAs in the network as ASBRs in this NSSA. Whenever this routing information had arrived at router *Torino* and router *Nantes*, it would have been redistributed and considered as IS–IS Level 2 internal at the other border system because of the `wide-metrics-only` policy.

The default protocol preference in that case would have led to one ABR preferring the other to reach such external routes from an NSSA internal system as the first hop would have followed an IS–IS instead of an OSPFv2 path. The foundation for a routing loop would have been formed with that. Protocol preferences would have needed similar adjustment again if current forwarding paths were not required to change.

On the other hand, all legitimate external loopback addresses from the "Gale Internet" IS–IS Level 2 final domain (including router *Lille*'s management loopback) together with 192.168.10.0/24 need to be leaked at both router *Basel* and router *Inverness* to the NSSA for optimal routing purposes.

Mutual protocol redistribution at more than a single point always opens the door for potential routing loops. Loops can be avoided with several techniques such as tweaking protocol preference, tagging and matching prefixes and accurate prefix filtering for redistribution, among others. The advantage in this case is that both OSPFv2 External LSAs and IS–IS TLV 135 Extended IP Reachability propagate tagging with specific fields and sub-TLVs and that, although we are removing the internal/external distinction in the IS–IS domain, these legitimate loopback addresses from router *Basel* and router *Inverness* are received as internal OSPFv2 routes, when available.

Instead of defining a specific route-filter restriction matching legitimate loopbacks from router *Basel* and router *Inverness* in one direction, the policy definition benefits from default protocol preferences, administrative tagging, and the fact that redistributed routes are always *external* in OSPFv2 to make it more generic for a real-life scenario. In the other direction, restricting redistribution to exactly 192.168.10.0/24 and host addresses from the 192.168.1.0/24 range will be enough.

Note that redistributing the local legitimate loopback address at each router is necessary as part of the requirements with the objective of providing optimal routing; otherwise, the IS–IS path from one border router is eligible for redistribution in the other system. Therefore, this policy matches both IS–IS and *direct* routes.

Listing 2.102 shows the definition and application of redistribution policies on router *Torino*, and they are the same on router *Nantes*. Listing 2.103 shows redistribution results on both border routers, highlighting administrative tag propagation, `tag` presence in routing-table information, and routing information present through backup paths.

Listing 2.102: Mutual redistribution policies definition and application at Torino

```
1  [edit policy-options]
2  user@Torino# show
3  policy-statement block-non-legitimate-loopback {
4      from {
5          protocol direct;
6          route-filter 10.10.1.3/32 exact;
7      }
8      then reject;
9  }
10 policy-statement from-domain-to-ospf {
11     from {
12         protocol [ isis direct ];
13         route-filter 192.168.10.0/24 exact;
14         route-filter 192.168.1.0/24 prefix-length-range /32-/32;
15     }
16     then {
17         tag 100; # Tag for redistributed routes
18         accept;
19     }
20 }
21 policy-statement from-ospf-to-domain {
22     term reject-redistributed {
23         from {
24             protocol ospf;
25             external;
26             tag 100; # Tag for redistributed routes
27             route-filter 192.168.1.0/24 prefix-length-range /32-/32;
28         }
29         then reject;
30     }
31     term accept-native {
32         from {
33             protocol ospf;
34             route-filter 192.168.1.0/24 prefix-length-range /32-/32;
35         }
36         then {
37             tag 200; # Tag for native routes
38             accept;
39         }
40     }
41 }
42 [edit]
43 user@Torino# set protocols ospf export from-domain-to-ospf
44 [edit]
45 user@Torino# set protocols isis export from-ospf-to-domain
```

Listing 2.103: Mutual redistribution results in Nantes and Torino

```
1  user@Torino> show ospf database external extensive | match "Extern|Tag"
2  Extern   192.168.1.1      10.10.1.1       0x80000001    22  0x22 0x913b  36
3     Type: 2, Metric: 0, Fwd addr: 0.0.0.0, Tag: 0.0.0.100 # Inject and tag own legitimate loopback
4  Extern  *192.168.1.1      10.10.1.3       0x80000004   882  0x22 0xe3d9  36
5     Type: 2, Metric: 10, Fwd addr: 0.0.0.0, Tag: 0.0.0.100 # Inject and tag peer's legitimate
        loopback learnt by IS-IS
6  Extern   192.168.1.3      10.10.1.1       0x80000002   388  0x22 0xdfdf  36
7     Type: 2, Metric: 10, Fwd addr: 0.0.0.0, Tag: 0.0.0.100 # Inject and tag peer's legitimate
        loopback learnt by IS-IS
8  Extern  *192.168.1.3      10.10.1.3       0x80000001    35  0x22 0x7157  36
9     Type: 2, Metric: 0, Fwd addr: 0.0.0.0, Tag: 0.0.0.100 # Inject and tag own legitimate loopback
10
11 user@Torino> show ospf database nssa extensive | match "NSSA|Tag"
12 NSSA     192.168.1.1      10.10.1.1       0x80000001    36  0x20 0x9339  36
13    Type: 2, Metric: 0, Fwd addr: 0.0.0.0, Tag: 0.0.0.100 # Inject and tag own legitimate loopback
14 NSSA    *192.168.1.1      10.10.1.3       0x80000004   894  0x20 0xe5d7  36
15    Type: 2, Metric: 10, Fwd addr: 0.0.0.0, Tag: 0.0.0.100 # Inject and tag peer's legitimate
        loopback learnt by IS-IS
16 NSSA     192.168.1.3      10.10.1.1       0x80000003   174  0x20 0xdfde  36
17    Type: 2, Metric: 10, Fwd addr: 0.0.0.0, Tag: 0.0.0.100 # Inject and tag peer's legitimate
        loopback learnt by IS-IS
18 NSSA    *192.168.1.3      10.10.1.3       0x80000001    47  0x20 0x7355  36
19    Type: 2, Metric: 0, Fwd addr: 0.0.0.0, Tag: 0.0.0.100 # Inject and tag own legitimate loopback
20
21
22 user@Torino> show isis database extensive level 2
23    | match "00-00 Sequence|tag|IP extended prefix"
24 Nantes.00-00 Sequence: 0x31, Checksum: 0x49d6, Lifetime: 418 secs
25    IP extended prefix: 192.168.1.1/32 metric 0 up
26    IP extended prefix: 10.10.1.96/30 metric 10 up
27    IP extended prefix: 192.168.1.9/32 metric 643 up
28      Administrative tag 1: 200 # Inject and tag Inverness loopback learnt by OSPFv2 internal
29      Administrative tag 2: 3
30    IP extended prefix: 192.168.1.10/32 metric 1286 up
31      Administrative tag 1: 200 # Inject and tag Basel loopback learnt by OSPFv2 internal
32      Administrative tag 2: 3
33 Torino.00-00 Sequence: 0x33, Checksum: 0x2e95, Lifetime: 1178 secs
34    IP extended prefix: 192.168.1.9/32 metric 803 up
35      Administrative tag 1: 200 # Inject and tag Inverness loopback learnt by OSPFv2 internal
36      Administrative tag 2: 3
37    IP extended prefix: 192.168.1.10/32 metric 160 up
38      Administrative tag 1: 200 # Inject and tag Basel loopback learnt by OSPFv2 internal
39      Administrative tag 2: 3
40    IP extended prefix: 192.168.1.3/32 metric 0 up
41    IP extended prefix: 10.10.1.96/30 metric 10 up
42
43 user@Torino> show route 192.168.1.0/24 table inet.0
44
45 inet.0: 31 destinations, 35 routes (31 active, 0 holddown, 0 hidden)
46 + = Active Route, - = Last Active, * = Both
47
48 192.168.1.1/32     *[IS-IS/18] 00:23:42, metric 10 # IS-IS L2 internal preferred
49                     > to 10.10.1.98 via so-0/1/0.0
50                     [OSPF/150] 00:09:21, metric 0, tag 100
51                     > via so-0/1/0.0
52 192.168.1.3/32     *[Direct/0] 08:46:43
53                     > via lo0.0
54                     [OSPF/150] 00:23:58, metric 10, tag 100
55                     > via so-0/1/0.0
56 192.168.1.9/32     *[OSPF/10] 08:55:40, metric 803
57                     > via at-1/2/0.1001
58 192.168.1.10/32    *[OSPF/10] 08:57:13, metric 160
59                     > via at-1/2/0.1001
60
61
62 user@Nantes> show route 192.168.1.0/24 table inet.0
63
```

```
64 inet.0: 33 destinations, 37 routes (33 active, 0 holddown, 0 hidden)
65 + = Active Route, - = Last Active, * = Both
66
67 192.168.1.1/32     *[Direct/0] 00:24:56
68                     > via lo0.0
69                     [OSPF/150] 00:24:23, metric 10, tag 100
70                     > via so-0/2/0.0
71 192.168.1.3/32     *[IS-IS/18] 00:24:47, metric 10 # IS-IS L2 internal preferred
72                     > to 10.10.1.97 via so-0/2/0.0
73                     [OSPF/150] 00:10:15, metric 0, tag 100
74                     > via so-0/2/0.0
75 192.168.1.9/32     *[OSPF/10] 00:24:41, metric 643
76                     > via so-1/0/1.0
77 192.168.1.10/32    *[OSPF/10] 00:24:41, metric 1286
78                     > via so-1/0/1.0
79
80 user@Torino> show isis spf results level 2
81  IS-IS level 2 SPF results:
82 Node              Metric     Interface          Via            SNPA
83 Nantes.00         10         so-0/1/0.0         Nantes
84                   20         10.10.1.96/30
85                   10         192.168.1.1/32
86                   653        192.168.1.9/32
87                   1296       192.168.1.10/32
88 Torino.00         0
89                   10         10.10.1.96/30
90                   0          192.168.1.3/32
91                   803        192.168.1.9/32
92                   160        192.168.1.10/32
93   2 nodes
94
95
96 user@Nantes> show isis spf results level 2
97  IS-IS level 2 SPF results:
98 Node              Metric     Interface          Via            SNPA
99 Torino.00         10         so-0/2/0.0         Torino
100                  20         10.10.1.96/30
101                  10         192.168.1.3/32
102                  813        192.168.1.9/32
103                  170        192.168.1.10/32
104 Nantes.00        0
105                  10         10.10.1.96/30
106                  0          192.168.1.1/32
107                  643        192.168.1.9/32
108                  1286       192.168.1.10/32
109   2 nodes
```

This setup is represented in Figure 2.22.

What If... Mutual protocol redistribution at several points

Mutually redistributing protocols require careful analysis because the potential for loops increases considerably. A recommended design exercise is to determine what could happen in terms of route selection when one or several of these redistribution points fail and recover. The objective should be to identify whether legitimate routes originated in one domain, once available again, could potentially not be selected but their redistributed copies floating in the other domain are selected instead.

In the previous example, we benefit from default Junos OS protocol preferences (OSPFv2 external > IS–IS L2 internal > OSPFv2 internal > direct) and the fact that legitimate "Gale Internet" routes inside the OSPFv2 domain are always internal (and therefore preferred by default preference values) and that we can identify OSPFv2 external routes with tags and kill a double redistribution at that point.

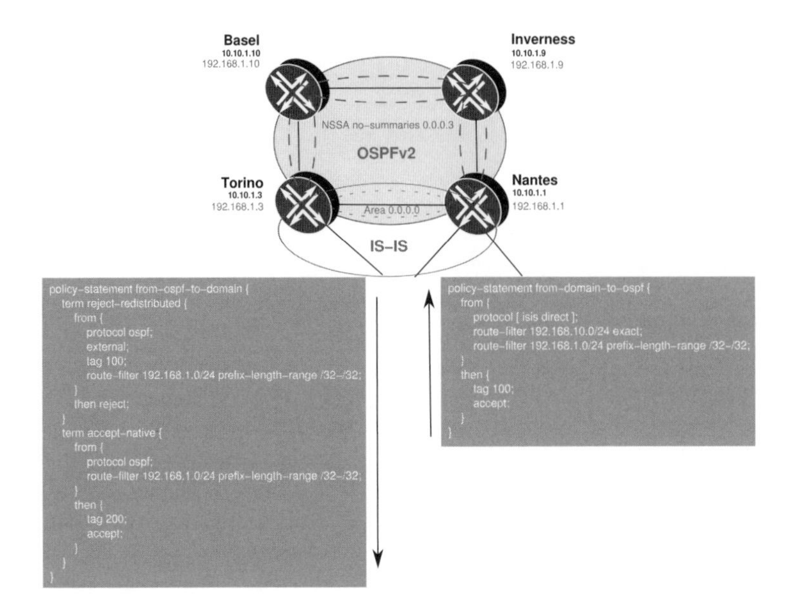

Figure 2.22: Stage two: domain separation and mutual redistribution at domain “Monsoon”.

When issuing routing restarts and other exercises, routing tables change slightly to show *backup* route paths in each case, although the information in each protocol database is the same. This happens whenever one of these paths has been considered for installation. In any case, active paths remain the same, as shown in Listing 2.104.

Listing 2.104: Transient backup route path appearance upon mutual protocol redistribution

```
1  user@Torino> show route 192.168.1.0/24 table inet.0
2
3  inet.0: 31 destinations, 36 routes (31 active, 0 holddown, 0 hidden)
4  + = Active Route, - = Last Active, * = Both
5
6  192.168.1.1/32     *[IS-IS/18] 00:01:47, metric 10
7                      > to 10.10.1.98 via so-0/1/0.0
8                      [OSPF/150] 00:01:58, metric 0, tag 100
9                      > via so-0/1/0.0
10 192.168.1.3/32     *[Direct/0] 00:02:41
11                     > via lo0.0
12                     [OSPF/150] 00:01:58, metric 10, tag 100
13                     > via so-0/1/0.0
14 192.168.1.9/32     *[OSPF/10] 00:02:25, metric 803
15                     > via at-1/2/0.1001
16                     [IS-IS/18] 00:01:47, metric 653, tag 200, tag2 3
17                     > to 10.10.1.98 via so-0/1/0.0
18 192.168.1.10/32    *[OSPF/10] 00:02:25, metric 160
19                     > via at-1/2/0.1001
20
21 user@Nantes> show route 192.168.1.0/24 table inet.0
22
23 inet.0: 33 destinations, 38 routes (33 active, 0 holddown, 0 hidden)
24 + = Active Route, - = Last Active, * = Both
25
26 192.168.1.1/32     *[Direct/0] 00:02:38
```

```
27                            > via lo0.0
28                            [OSPF/150] 00:02:02, metric 10, tag 100
29                            > via so-0/2/0.0
30  192.168.1.3/32           *[IS-IS/18] 00:02:30, metric 10
31                            > to 10.10.1.97 via so-0/2/0.0
32                            [OSPF/150] 00:02:17, metric 0, tag 100
33                            > via so-0/2/0.0
34  192.168.1.9/32           *[OSPF/10] 00:02:22, metric 643
35                            > via so-1/0/1.0
36  192.168.1.10/32          *[OSPF/10] 00:02:22, metric 1286
37                            > via so-1/0/1.0
38                            [IS-IS/18] 00:02:04, metric 170, tag 200, tag2 3
39                            > to 10.10.1.97 via so-0/2/0.0
```

In similar scenarios, just tagging redistribution and matching those tags with the restriction to be a host address inside 192.168.1.0/24 would not be enough, because legitimate routes inside the OSPFv2 domain are also inside the route-filter range. A *chicken-and-egg* situation could be easily created upon no internal/external distinction at the internal OSPFv2 domain or if external OSPFv2 routes were preferred against IS–IS Level 2 internal.

2.3.6 Stage three: IS–IS protocol adaption at domain "Mistral"

Domain "Mistral" includes systems already running IS–IS Level 1 with a different authentication key in a flat area. At the same time, router *Male* must remain in the final topology as internal L1 system acquiring specific routing information from 192.168.10.0/24 and loopback addresses over both its uplinks for the sake of optimal routing, with both router *Bilbao* and router *Havana* as L1L2 systems. Considering router *Bilbao* as an illustrative sample, the situation is shown in Listing 2.105, but because this a flat L1 Area, similar visibility is granted at router *Male* and router *Havana*.

Listing 2.105: Initial IS–IS status at Bilbao

```
1   user@Bilbao-re0> show configuration protocols isis
2   reference-bandwidth 100g;
3   level 1 {
4       authentication-key "$9$ewrvL7g4ZjkPJG6ApuEhX7-"; ### SECRET-DATA
5       authentication-type md5;
6       wide-metrics-only;
7   }
8   interface at-1/2/1.0;
9   interface so-1/3/1.0;
10  interface lo0.0 {
11      level 1 passive;
12  }
13
14  user@Bilbao-re0> show isis adjacency
15  Interface             System         L State        Hold (secs) SNPA
16  at-1/2/1.0            Havana         1  Up                   20
17  so-1/3/1.0            male-re0   1  Up                   23
18
19  user@Bilbao-re0> show isis database detail
20  IS-IS level 1 link-state database:
21
22  Havana.00-00 Sequence: 0xcd, Checksum: 0x6aca, Lifetime: 712 secs
23     IS neighbor: Havana.02                      Metric:       100
24     IS neighbor: Bilbao-re0.00                  Metric:       643
25     IP prefix: 172.16.1.40/30                   Metric:       643 Internal Up
26     IP prefix: 172.16.1.84/30                   Metric:       100 Internal Up
27     IP prefix: 192.168.1.4/32                   Metric:         0 Internal Up
```

```
28
29 Havana.02-00 Sequence: 0xbf, Checksum: 0xf340, Lifetime: 653 secs
30     IS neighbor: Havana.00                   Metric:        0
31     IS neighbor: male-re0.00             Metric:        0
32
33 Bilbao-re0.00-00 Sequence: 0xcc, Checksum: 0xf7fd, Lifetime: 724 secs
34     IS neighbor: Havana.00                   Metric:      643
35     IS neighbor: male-re0.00             Metric:      643
36     IP prefix: 172.16.1.40/30                Metric:      643 Internal Up
37     IP prefix: 172.16.1.76/30                Metric:      643 Internal Up
38     IP prefix: 192.168.1.8/32                Metric:        0 Internal Up
39
40 male-re0.00-00 Sequence: 0xc8, Checksum: 0x67ac, Lifetime: 712 secs
41     IS neighbor: Havana.02                   Metric:      100
42     IS neighbor: Bilbao-re0.00               Metric:      643
43     IP prefix: 172.16.1.76/30                Metric:      643 Internal Up
44     IP prefix: 172.16.1.84/30                Metric:      100 Internal Up
45     IP prefix: 192.168.1.20/32               Metric:        0 Internal Up
46
47 IS-IS level 2 link-state database:
48
49 Havana.00-00 Sequence: 0x9ad, Checksum: 0x1d2e, Lifetime: 228 secs
50     IS neighbor: Bilbao-re0.00               Metric:      643
51     IP prefix: 172.16.1.40/30                Metric:      643 Internal Up
52     IP prefix: 172.16.1.76/30                Metric:      743 Internal Up
53     IP prefix: 172.16.1.84/30                Metric:      100 Internal Up
54     IP prefix: 192.168.1.4/32                Metric:        0 Internal Up
55     IP prefix: 192.168.1.20/32               Metric:      100 Internal Up
56
57 Bilbao-re0.00-00 Sequence: 0x2, Checksum: 0xb2d3, Lifetime: 722 secs
58     IP prefix: 172.16.1.40/30                Metric:       63 Internal Up
59     IP prefix: 172.16.1.76/30                Metric:       63 Internal Up
60     IP prefix: 172.16.1.84/30                Metric:       63 Internal Up
61     IP prefix: 192.168.1.4/32                Metric:       63 Internal Up
62     IP prefix: 192.168.1.8/32                Metric:        0 Internal Up
63     IP prefix: 192.168.1.20/32               Metric:       63 Internal Up
64
65 user@Bilbao-re0> show route protocol isis table inet.0
66
67 inet.0: 16 destinations, 16 routes (16 active, 0 holddown, 0 hidden)
68 + = Active Route, - = Last Active, * = Both
69
70 172.16.1.84/30     *[IS-IS/15] 00:08:19, metric 743
71                       to 172.16.1.41 via at-1/2/1.0
72                     > to 172.16.1.77 via so-1/3/1.0
73 192.168.1.4/32     *[IS-IS/15] 00:08:19, metric 643
74                     > to 172.16.1.41 via at-1/2/1.0
75 192.168.1.20/32    *[IS-IS/15] 00:08:19, metric 643
76                     > to 172.16.1.77 via so-1/3/1.0
```

Considering that the initial short-term objective is to integrate this L1 area in the rest of the domain, the objectives at this stage are summarized as follows:

- Deploy IS–IS Level 2 in router *Bilbao* and router *Havana* with the final domain settings.

- Define route-leaking policies from Level 2 to Level 1 for optimal routing from router *Male*.

Junos OS allows different parameters to be set on a per IS–IS level basis. Thus, the final IS–IS Level 2 authentication key and `wide-metrics-only` could theoretically be directly activated on both router *Bilbao* and router *Havana* as a previous step to integrate this network.

However, after in-depth analysis, there is a notable drawback to this proposal: the adjacency between router *Bilbao* and router *Havana* is based on a *point-to-point* link. For such media, IS–IS issues a joint point-to-point IIH PDU for both Level 1 and Level 2, with a unique and common TLV 10 Authentication. If authentication settings are different between both levels, one of the two adjacencies will not fully form because of the authentication mismatch.

Junos OS includes several related knobs to overcome authentication problems. However, considering that the idea is to grant reachability for all systems and avoid manipulating router *Male*, the following options for the dilemma need to be evaluated:

- Keep the link between router *Bilbao* and router *Havana* as Level 1 only.
- Define the link between router *Bilbao* and router *Havana* as Level 2 only.
- Keep the link available for L1 and L2 adjacencies, but avoid authentication checks in both router *Bilbao* and router *Havana* with the `no-authentication-check` knob or avoid issuing hello authentication checks per level with the `no-hello-authentication` knob.
- Keep the link available for L1 and L2 adjacencies but attach the TLV 10 Authentication with Level 2 settings to point-to-point IIH PDUs over it with the `hello-authentication` knob matching those settings for the Level 1 adjacency.

The first option would mean that the temporary domain would not have additional redundancy (backbone in a ring form) and should not be chosen as an alternative for that reason.

The second option requires more careful thinking: activating IS–IS Level 2 only over that link means that, unless leaked by specific policies or unless manipulating ISO area configuration, the transit network would not be available as a specific route for router *Male* and redundancy between router *Male* and the other systems would be lost in the current setup. The rationale for this behavior resides in how Junos OS interprets *backbone detection* conditions for ATTached bit setting: a system configured for IS–IS L2 goes through all L2 LSPs present in the database for all reachable systems (meaning SPF calculations on IS–IS L2 information) to check whether there is at least one other area entry that does not exist in the conglomerate of all its local L1 areas (this behavior occurs by default, without including the `suppress-attached-bit` in the configuration).

With this implementation for *backbone detection* and ATTached bit setting, the risk is that at this point in the migration only Area 49.0001 is present in the domain, so moving that connection for unique IS–IS L2 would blackhole that backup path and hide routing information to router *Male*.

Junos Tip: IS–IS ATTached bit setting conditions

Backbone detection is basically understood in Junos OS as the presence of a non-L1 area in any of the IS–IS L2 LSPs in the database. Activating IS–IS Level 2 in a domain with the same area identifier may lead to not setting ATTached bits for internal L1 LSPs from L1L2 systems.

Considering the previous example from domain "Mistral", the situation could be even worse if both loopbacks from L1L2 systems are not injected into L1 LSPs (this could be a usual practice when aggregating loopback spaces from areas and backbone in other setups), as shown in Listing 2.106.

Listing 2.106: Lack of loopback address presence and ATTached bit in domain "Mistral"

```
[edit protocols isis]
user@Bilbao-re0# show
reference-bandwidth 100g;
level 1 {
    authentication-key "$9$ewrvL7g4ZjkPJG6ApuEhX7-"; ### SECRET-DATA
    authentication-type md5;
    wide-metrics-only;
}
level 2 {
    authentication-key "$9$RJehrK-ds4JDwYfz3npulKM"; ### SECRET-DATA
    authentication-type md5;
    wide-metrics-only;
}
interface at-1/2/1.0 {
    level 1 disable;
}
interface so-1/3/1.0 {
    level 2 disable;
}
interface lo0.0 {
    level 1 disable;
    level 2 passive;
}

user@Bilbao-re0> show isis database
IS-IS level 1 link-state database:
LSP ID                      Sequence Checksum Lifetime Attributes
Havana.00-00                    0xc9   0x1113      873 L1 L2
Havana.02-00                    0xbd   0xf9e9      768 L1 L2
Bilbao-re0.00-00                0xca   0x94ee      815 L1 L2
male-re0.00-00              0xc5   0x8fee      654 L1
  4 LSPs

IS-IS level 2 link-state database:
LSP ID                      Sequence Checksum Lifetime Attributes
Havana.00-00                   0x9ab   0xdd01      979 L1 L2
Bilbao-re0.00-00                0xce   0xcc36      815 L1 L2
  2 LSPs

user@male-re0> show isis database
IS-IS level 1 link-state database:
LSP ID                      Sequence Checksum Lifetime Attributes
Havana.00-00                    0xad   0xbdaa      997 L1 L2
Havana.02-00                    0xab   0x21f0      998 L1 L2
Bilbao-re0.00-00                0xb0   0x8ae0      555 L1 L2
male-re0.00-00              0xb0   0x3ed6     1193 L1
  4 LSPs

IS-IS level 2 link-state database:
  0 LSPs

user@male-re0> show route 0/0

inet.0: 5 destinations, 5 routes (5 active, 0 holddown, 0 hidden)
+ = Active Route, - = Last Active, * = Both

172.16.1.76/30     *[Direct/0] 1d 13:24:48
                    > via so-3/2/1.0
172.16.1.77/32     *[Local/0] 1d 15:55:56
```

```
60                            Local via so-3/2/1.0
61  172.16.1.84/30       *[Direct/0] 1d 15:55:20
62                          > via ge-4/2/4.0
63  172.16.1.85/32       *[Local/0] 1d 15:55:52
64                            Local via ge-4/2/4.0
65  192.168.1.20/32      *[Direct/0] 1d 15:56:48
66                          > via lo0.0
```

Note that router *Male* has neither any active IS–IS route nor default route tracking the ATTached bit, because this is not set.

The third option listed for overcoming authentication problems would lead to avoidance of any kind of authentication checks on both router *Bilbao* and router *Havana* by means of the `no-authentication-check` knob or skipping any kind of hello authentication checks on a per-level basis via `no-hello-authentication`. Both are valid approaches, but the objective at this stage is to minimize changes and avoid any affect on the L1 systems.

The fourth option considers using TLV 10 Authentication with final settings on point-to-point IIH PDUs over this link between both systems in a common fashion for both levels. This scalpel-like implementation achieves our objective in a more granular fashion: LSPs, CSNPs, and PSNPs for each level will include the respectively defined authentication information for each level, and only IIHs for this link will include settings from Level 2 by modifying hello authentication for Level 1. With that, IS–IS L1 and L2 adjacencies will be set up independently, and routing information will be separately decoded for L1 and L2.

Listing 2.107 summarizes the settings for router *Bilbao* with this latest implementation option, router *Havana* includes a similar excerpt. It is worth noting that there is no ATTached bit set to L1 LSPs despite Level 2 being configured, because no other Areas have been extracted from the IS–IS database. Note that this activation is completely transparent for the current forwarding state as well, because IS–IS Level 1 routes will always be preferred over IS–IS Level 2 routes as per default protocol preference values. No routing changes are expected by this simple activation, because changing the authentication key in Junos OS does not trigger an adjacency re-establishment.

Listing 2.107: IS–IS L2 activation in Bilbao with final authentication settings for Level 1 in link to Havana

```
1  user@Bilbao-re0> show configuration protocols isis
2  reference-bandwidth 100g;
3  level 1 {
4      authentication-key "$9$ewrvL7g4ZjkPJG6ApuEhX7-"; ### SECRET-DATA
5      authentication-type md5;
6      wide-metrics-only;
7  }
8  level 2 {  # Additional IS-IS Level 2 authentication details
9      authentication-key "$9$w2Y2oDjqPT3goF/"; ### SECRET-DATA
10     authentication-type md5;
11     wide-metrics-only;
12 }
13 interface at-1/2/1.0 {
14     level 1 { # Authentication settings from Level 2
15         hello-authentication-key "$9$hl0SyeLX-bYore4Z"; ### SECRET-DATA
16         hello-authentication-type md5;
17     }
18 }
19 interface so-1/3/1.0 {
20     level 2 disable;
21 }
```

```
22 interface lo0.0 {
23     level 1 passive;
24     level 2 passive;
25 }
26
27 user@Bilbao-re0> show route protocol isis table inet.0
28
29 inet.0: 16 destinations, 16 routes (16 active, 0 holddown, 0 hidden)
30 + = Active Route, - = Last Active, * = Both
31
32 172.16.1.84/30     *[IS-IS/15] 01:35:13, metric 743
33                       to 172.16.1.41 via at-1/2/1.0
34                     > to 172.16.1.77 via so-1/3/1.0
35 192.168.1.4/32     *[IS-IS/15] 01:35:13, metric 643
36                     > to 172.16.1.41 via at-1/2/1.0
37 192.168.1.20/32    *[IS-IS/15] 03:56:58, metric 643
38                     > to 172.16.1.77 via so-1/3/1.0
39
40 user@Bilbao-re0> show isis database
41   # Same IS-IS DB as without authentication
42 IS-IS level 1 link-state database:
43 LSP ID                      Sequence Checksum Lifetime Attributes
44 Havana.00-00                    0xe1   0x6db6      328 L1 L2
45 Havana.02-00                    0xd1   0x1770      328 L1 L2
46 Bilbao-re0.00-00                0xdf   0x47c0      423 L1 L2
47 male-re0.00-00              0xda    0x507c     1011 L1
48   4 LSPs
49
50 IS-IS level 2 link-state database:
51 LSP ID                      Sequence Checksum Lifetime Attributes
52 Havana.00-00                     0xa   0x16ec      429 L1 L2
53 Bilbao-re0.00-00                0x16   0xfc2b      694 L1 L2
54   2 LSPs
55
56 user@Bilbao-re0> show isis database detail
57   # Level 2 contains redundant routing information
58 IS-IS level 1 link-state database:
59
60 Havana.00-00 Sequence: 0xe2, Checksum: 0x4405, Lifetime: 1195 secs
61    IS neighbor: Havana.02                      Metric:      100
62    IS neighbor: Bilbao-re0.00                  Metric:      643
63    IP prefix: 172.16.1.40/30                   Metric:      643 Internal Up
64    IP prefix: 172.16.1.84/30                   Metric:      100 Internal Up
65    IP prefix: 192.168.1.4/32                   Metric:        0 Internal Up
66
67 Havana.02-00 Sequence: 0xd2, Checksum: 0xa314, Lifetime: 1195 secs
68    IS neighbor: Havana.00                      Metric:        0
69    IS neighbor: male-re0.00                Metric:        0
70
71 Bilbao-re0.00-00 Sequence: 0xdf, Checksum: 0x47c0, Lifetime: 413 secs
72    IS neighbor: Havana.00                      Metric:      643
73    IS neighbor: male-re0.00                Metric:      643
74    IP prefix: 172.16.1.40/30                   Metric:      643 Internal Up
75    IP prefix: 172.16.1.76/30                   Metric:      643 Internal Up
76    IP prefix: 192.168.1.8/32                   Metric:        0 Internal Up
77
78 male-re0.00-00 Sequence: 0xda, Checksum: 0x507c, Lifetime: 1001 secs
79    IS neighbor: Havana.02                      Metric:      100
80    IS neighbor: Bilbao-re0.00                  Metric:      643
81    IP prefix: 172.16.1.76/30                   Metric:      643 Internal Up
82    IP prefix: 172.16.1.84/30                   Metric:      100 Internal Up
83    IP prefix: 192.168.1.20/32                  Metric:        0 Internal Up
84
85 IS-IS level 2 link-state database:
86
87 Havana.00-00 Sequence: 0xa, Checksum: 0x16ec, Lifetime: 418 secs
88    IS neighbor: Bilbao-re0.00                  Metric:      643
89    IP prefix: 172.16.1.40/30                   Metric:      643 Internal Up
90    IP prefix: 172.16.1.76/30                   Metric:      743 Internal Up
```

```
91      IP prefix: 172.16.1.84/30              Metric:      100 Internal Up
92      IP prefix: 192.168.1.4/32              Metric:        0 Internal Up
93      IP prefix: 192.168.1.8/32              Metric:      643 Internal Up
94      IP prefix: 192.168.1.20/32             Metric:      100 Internal Up
95
96  Bilbao-re0.00-00 Sequence: 0x16, Checksum: 0xfc2b, Lifetime: 684 secs
97      IS neighbor: Havana.00                 Metric:      643
98      IP prefix: 172.16.1.40/30              Metric:      643 Internal Up
99      IP prefix: 172.16.1.76/30              Metric:      643 Internal Up
100     IP prefix: 172.16.1.84/30              Metric:      743 Internal Up
101     IP prefix: 192.168.1.4/32              Metric:      643 Internal Up
102     IP prefix: 192.168.1.8/32              Metric:        0 Internal Up
103     IP prefix: 192.168.1.20/32             Metric:      643 Internal Up
```

Once the Level 2 adjacency has been properly set, route-leaking policies for optimal routing from and to router *Male* can be crafted at this point to prepare backbone connection. In the previous Listing 2.107, loopback interface addresses were injected in both L1 and L2 LSPs for optimal routing purposes. This simplifies route-leaking policies at L1L2 systems by default, because it will just be necessary to match protocol IS–IS and Level 2, together with proper route-filtering, to inject such information to Level 1 systems, and there is no need to match then direct routes.

Listing 2.108 illustrates policy definition, application, and effects from router *Bilbao*'s perspective. Note that router *Bilbao* is not leaking *Havana*'s loopback address, and vice versa, in L1 LSPs at this point, because we enforce that only IS–IS Level 2 routes are matched and such loopback addresses are preferred as IS–IS Level 1 routes. Despite this leaking implementation, the ATTached bit remains unset and router *Male* has no default route pointing to any of both L1L2 systems.

Listing 2.108: IS–IS L2 to L1 leaking policy application and effects in Bilbao

```
1   [edit policy-options policy-statement leak-loopbacks-to-l1]
2   user@Bilbao-re0# show
3   from {
4       protocol isis; # Only match protocol isis
5       level 2; # Only match L2 routes
6       route-filter 192.168.1.0/24 prefix-length-range /32-/32;
7       route-filter 192.168.10.0/24 exact;
8   }
9   to level 1;
10  then {
11      tag 300;
12      accept;
13  }
14
15  [edit]
16  user@Bilbao-re0# set protocols isis export leak-loopbacks-to-l1
17
18  user@Bilbao-re0> show isis database detail
19    # No changes in current IS-IS database yet
20  IS-IS level 1 link-state database:
21
22  Havana.00-00 Sequence: 0xe6, Checksum: 0xaafd, Lifetime: 1069 secs
23     IS neighbor: Havana.02                 Metric:      100
24     IS neighbor: Bilbao-re0.00             Metric:      643
25     IP prefix: 172.16.1.40/30              Metric:      643 Internal Up
26     IP prefix: 172.16.1.84/30              Metric:      100 Internal Up
27     IP prefix: 192.168.1.4/32              Metric:        0 Internal Up
28
29  Havana.02-00 Sequence: 0xd5, Checksum: 0x1470, Lifetime: 1061 secs
30     IS neighbor: Havana.00                 Metric:        0
31     IS neighbor: male-re0.00           Metric:        0
32
```

```
33 Bilbao-re0.00-00 Sequence: 0xe5, Checksum: 0x90f6, Lifetime: 897 secs
34    IS neighbor: Havana.00                   Metric:      643
35    IS neighbor: male-re0.00              Metric:      643
36    IP prefix: 172.16.1.40/30                Metric:      643 Internal Up
37    IP prefix: 172.16.1.76/30                Metric:      643 Internal Up
38    IP prefix: 192.168.1.8/32                Metric:        0 Internal Up
39
40 male-re0.00-00 Sequence: 0xdd, Checksum: 0x2a66, Lifetime: 825 secs
41    IS neighbor: Havana.02                   Metric:      100
42    IS neighbor: Bilbao-re0.00               Metric:      643
43    IP prefix: 172.16.1.76/30                Metric:      643 Internal Up
44    IP prefix: 172.16.1.84/30                Metric:      100 Internal Up
45    IP prefix: 192.168.1.20/32               Metric:        0 Internal Up
46
47 IS-IS level 2 link-state database:
48
49 Havana.00-00 Sequence: 0xf, Checksum: 0x5be3, Lifetime: 1069 secs
50    IS neighbor: Bilbao-re0.00               Metric:      643
51    IP prefix: 172.16.1.40/30                Metric:      643 Internal Up
52    IP prefix: 172.16.1.76/30                Metric:      743 Internal Up
53    IP prefix: 172.16.1.84/30                Metric:      100 Internal Up
54    IP prefix: 192.168.1.4/32                Metric:        0 Internal Up
55    IP prefix: 192.168.1.8/32                Metric:      643 Internal Up
56    IP prefix: 192.168.1.20/32               Metric:      100 Internal Up
57
58 Bilbao-re0.00-00 Sequence: 0x1b, Checksum: 0x47e8, Lifetime: 896 secs
59    IS neighbor: Havana.00                   Metric:      643
60    IP prefix: 172.16.1.40/30                Metric:      643 Internal Up
61    IP prefix: 172.16.1.76/30                Metric:      643 Internal Up
62    IP prefix: 172.16.1.84/30                Metric:      743 Internal Up
63    IP prefix: 192.168.1.4/32                Metric:      643 Internal Up
64    IP prefix: 192.168.1.8/32                Metric:        0 Internal Up
65    IP prefix: 192.168.1.20/32               Metric:      643 Internal Up
```

This situation is illustrated in Figure 2.23.

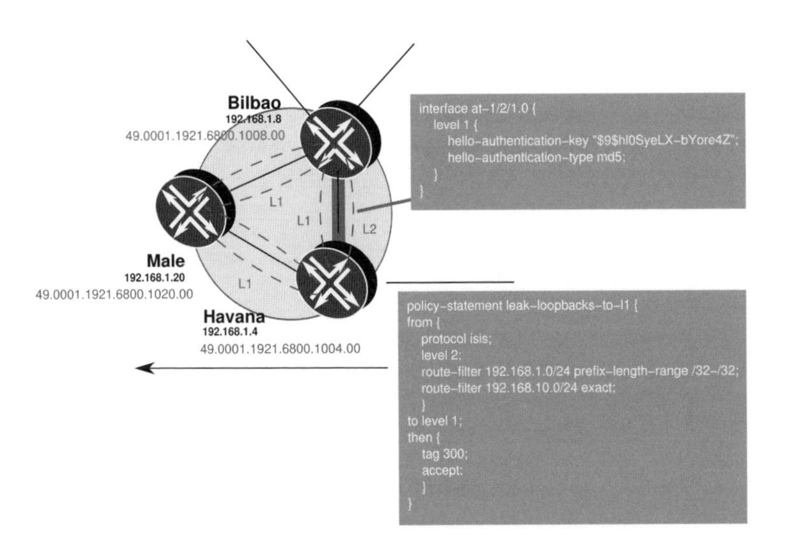

Figure 2.23: Stage three: IS–IS L2 activation and redistribution at domain "Mistral".

2.3.7 Stage four: Domain interconnection via IS–IS Level 2

All three domains have been prepared to advertise and receive public loopback addresses. At this point, border systems from each domain just need to activate IS–IS Level 2 over the newly acquired leased lines with the intention of stitching together domains and allowing route propagation.

Before effectively merging domains, another factor should be evaluated: *link costs*. Domain "Monsoon" inherits `reference-bandwidth` 100 g from OSPFv2 and domain "Mistral" includes the same reference per original definition, but domain "Cyclone" has mirrored in IS–IS `reference-bandwidth` 10 g from OSPFv2.

Under current circumstances, metric values are an order of magnitude lower in domain "Cyclone" than automatically derived values from other domains. Keeping the same scheme could lead to routing asymmetries, although all outgoing and incoming costs to this domain will be of the same magnitude. While it makes sense to keep the same reference bandwidth internally in domain "Cyclone", it is recommended to align costs with other leased lines to and from other domains to ease operation and troubleshooting.

In addition, the domain interconnect topology results in router *Skopie* being single-homed to router *Livorno* as internal connection in domain "Cyclone", but this link becomes part of the internal ring that can act as transit for traffic between other remote end points. Modifying the cost of that link to be aligned with the reference bandwidth for other domains and external leased lines would diverge from the current OSPFv2 link cost, but would not influence routing decisions internally in that domain (as it is single-homed).

The configuration snippet in Listing 2.109 illustrates the changes needed in router *Livorno* to interconnect domain "Cyclone" with domain "Mistral" over a specific leased line, together with the ad-hoc metric setting for that link and the interface to router *Skopie*. The configuration changes are similar in all border systems for all external connections, apart from particular link metrics. It is important to note that IS–IS Level 1 should be explicitly disabled from those links to avoid any IS–IS Level 1 adjacency that could lead to undesired effects.

Listing 2.109: IS–IS L2 deployment over leased-line and metric adaption in router *Livorno* for domain interconnection

```
1  [edit protocols isis]
2  user@Nantes# set interface so-0/0/1.0 level 1 disable
3  [edit protocols isis]
4  user@Livorno# set interface so-0/0/1.0 level 2 metric 643
5  [edit protocols isis]
6  user@Livorno# set interface so-0/0/0.0 level 2 metric 643
7  [edit]
8  user@Livorno# set interfaces so-0/0/1.0 family iso
9
10 user@Livorno> show isis adjacency
11 Interface             System          L State        Hold (secs) SNPA
12 ge-0/2/0.0            Barcelona-re0 2 Up                  23  0:14:f6:84:8c:bc
13 ge-1/3/0.0            honolulu-re0 2 Up                    8  0:17:cb:d3:2f:16
14 so-0/0/0.0            Skopie          2  Up                   23
15 so-0/0/1.0            Havana          2  Up                   22
16
17 user@Livorno> show isis database
18 IS-IS level 1 link-state database:
19 LSP ID                      Sequence Checksum Lifetime Attributes
20 Livorno.00-00                  0x94   0x4bdd      1132 L1 L2 Attached
21   1 LSPs
22
```

```
23 IS-IS level 2 link-state database:
24 LSP ID                     Sequence Checksum Lifetime Attributes
25 Nantes.00-00                   0x79   0x5476     1176 L1 L2
26 Nantes.02-00                    0x3   0x2fbc     1176 L1 L2
27 Skopie.00-00                   0xa2    0xecd     1153 L1 L2
28 Torino.00-00                   0x7c   0xc3b9      678 L1 L2
29 Havana.00-00                   0x1e   0xd0c4      899 L1 L2
30 Livorno.00-00                  0xa0   0xcd10     1139 L1 L2
31 Livorno.02-00                  0x93   0x924e     1132 L1 L2
32 Barcelona-re0.00-00        0x95    0xe531     838 L1 L2
33 Bilbao-re0.00-00               0x31   0x84cb     1190 L1 L2
34 honolulu-re0.00-00         0x94    0xfc5f     550 L1 L2
35 honolulu-re002-00         0x92    0xc68f     695 L1 L2
36 honolulu-re003-00         0x91    0x656f    1036 L1 L2
37   12 LSPs
```

Because domains are interconnected, IS–IS Level 2 routing information is flooded. Before proceeding to a preliminary interconnect analysis, router *Lille* needs to be linked to the currently built topology. At this stage, note that router *Livorno* is setting ATTached bit because it is detecting an external area in its IS–IS Level 2 Database, namely 49.0003.

2.3.8 Stage five: Integration of router *Lille* in the IS–IS domain

As per design prerequisite, router *Lille* needs to be connected to both router *Livorno* and router *Havana* via an IS–IS Level 1 adjacency without modifying its area identifier. Router *Livorno*'s Net-ID includes Area ID 49.0003, but router *Havana* was previously in Area 49.0001.

Also, prefix leaking is needed for optimal routing; the exact same policy defined in Listing 2.108 can be used for this. This policy is already implemented in router *Havana* for the internal Area 49.0001, but needs to be addressed in router *Livorno*. Also, this system is part of a domain with a different `reference-bandwidth`, and it would be preferable to adapt metric costs to the same value as router *Havana* to have optimal routing from remote end points.

Router *Lille* is a very sensitive system with direct connections to the NOC and to all management systems, and as a protection mechanism, no default route should be installed as a consequence of any IS–IS ATTached bits.

On top of all these obvious requirements, another subtle situation needs careful thought: router *Havana* is already L1L2 system for another L1 area with *different* authentication settings than the final key values. Authentication keys for all IS–IS PDUs cannot be configured on a per-area basis, but on a per-level basis, which means that modifying these values would have further implications inside the domain "Mistral". Effectively, because both Area 49.0001 and Area 49.0003 are included in IS–IS Level 1 adjacencies to router *Havana*, both areas are stitched together at that point.

The available options in this stage are the following:

- Migrate the entire Area 49.0001 Level 1 authentication key to the final value, *Aeris*.

- Use the `loose-authentication-check` knob at router *Livorno* and router *Havana* and do not implement any authentication on router *Lille*.

- Implement Area 49.0001 authentication settings at this Area 49.0003 in router *Lille*, router *Livorno*, and router *Havana*. This will need to be migrated again at a later step.

While the first option is the realistic final scenario, the intent at this point is not to modify the authentication settings, because this would require migrating authentication for all systems inside the domain "Mistral" (imagine a much larger domain than this simplistic case study).

The second option allows authentication checks on router *Havana* for all currently existing adjacencies protected with authentication, except for the newest one to router *Lille*, and it also allows the final authentication settings on router *Livorno* to be configured with only minimal configuration of loose authentication checks in each case. However, IS–IS Level 1 is actually not constrained only to these three systems, because router *Havana* effectively connects two different areas together in the same Level 1. This means that the scope of authentication parameterization really goes beyond router *Havana* up to router *Male* and router *Bilbao* (or to a more populated area as a more realistic scenario). While a given combination of loose and no authentication checks can provide the intended result, being able to fully decode LSPs in one L1 system but not on the other is certainly not as optimal for troubleshooting as the next option.

Despite going in an opposite direction to the final scenario, the preference here is the third option provided that router *Livorno* is not connected to any other L1 area and that this modification just affects router *Lille* and router *Livorno* at this time. A step back can sometimes be the best approach to move forward!

All these requirements can be summarized in the following migration guidelines:

- Add an Area ID 49.0003 derived Net-ID on router *Havana* and activate the IS–IS L1 adjacency towards router *Lille* with the same old authentication settings.

- Implement L2-to-L1 leaking policies in router *Livorno* and adapt the link cost calculation. Deploy authentication settings from the domain "Mistral" in router *Livorno* for IS–IS Level 1.

- Consider that some active loopback routes in router *Livorno* are still OSPFv2 routes and would need to be leaked down as well for optimal routing purposes.

- Activate L1 adjacencies over the shared broadcast domain and ignore ATTached bit setting in router *Lille*. Deploy authentication settings from domain "Mistral" in router *Lille* for IS–IS Level 1.

Adding a new area identifier to an existing intermediate system is a completely graceful operation in Junos OS. Listing 2.110 summarizes this simple configuration step together with IS–IS L1 activation for the Ethernet link to router *Lille*.

Listing 2.110: IS–IS L1 adjacency activation steps at Havana

```
1 [edit]
2 user@Havana# set interfaces lo0.0 family iso address 49.0003.1921.6800.1004.00
3 [edit]
4 user@Havana# set interfaces fe-0/3/0.0 family iso
5 [edit]
6 user@Havana# set protocols isis interface fe-0/3/0.0 level 2 disable
```

On router *Livorno* a similar route-leaking policy needs to be defined and implemented, together with parallel link metric setting. Likewise, IS–IS Level 1 includes *temporary* authentication values from domain "Mistral" at this stage. Similarly to our previous policy in router *Havana*, the intent is not to match direct routes for leaking to IS–IS Level 1, but rather, the loopback interface is included as `passive` in the configuration.

However, there is an additional circumstance at this stage in the migration. Router *Livorno* has fully deployed IS–IS, but this protocol remains unpreferred against OSPFv2 internal or external routes. As Junos OS `export` policies look in link-state IGPs at active versions of a route in the routing table, not at link-state IGP databases, it is necessary to add OSPFv2 active routes in this combination, as long as OSPFv2 remains *preferred* in domain "Cyclone". All these changes are illustrated in Listing 2.111.

Listing 2.111: IS–IS L1 adjacency activation and leaking steps at Livorno

```
1  [edit policy-options policy-statement leak-loopbacks-to-l1]
2  user@Livorno# show
3  term isis {
4      from {
5          protocol isis;  # IS-IS active loopback routes
6          level 2;
7          route-filter 192.168.1.0/24 prefix-length-range /32-/32;
8          route-filter 192.168.10.0/24 exact;
9      }
10     to level 1;
11     then {
12         tag 300;
13         accept;
14     }
15 }
16 term ospf {
17     from {
18         protocol ospf;  # Currently OSPFv2 active routes
19         route-filter 192.168.1.0/24 prefix-length-range /32-/32;
20         route-filter 192.168.10.0/24 exact;
21     }
22     to level 1;  # Only to single existing L1 adjacency
23     then {
24         tag 300;
25         accept;
26     }
27 }
28 [edit]
29 user@Livorno# set interfaces fe-0/3/0.0 family iso
30 [edit]
31 user@Livorno# set protocols isis interface fe-0/3/0.0 level 2 disable
32 [edit]
33 user@Livorno# delete protocols isis interface lo0.0 level 1 disable
34 [edit]
35 user@Livorno# set protocols isis interface lo0.0 level 1 passive
36 [edit]
37 user@Livorno# set protocols isis export leak-loopbacks-to-l1
38 [edit]
39 user@Livorno# set protocols isis level 1 authentication-key "Mistral"
40 [edit]
41 user@Livorno# set protocols isis interface fe-0/3/0.0 level 1 metric 1000
```

Finally, IS–IS Level 1 with similar `wide-metrics-only` and domain "Mistral" authentication key is implemented in router *Lille*, together with explicitly ignoring the ATTached bit setting with respect to default route installation, as shown in Listing 2.112.

Listing 2.112: IS–IS L1 adjacency configuration at Lille

```
1  user@Lille> show configuration protocols isis
2  reference-bandwidth 100g;
3  ignore-attached-bit;
4  level 2 disable;
```

```
5  level 1 {
6      authentication-key "$9$M2887daJDkmTUjCuOBSyNdb"; ### SECRET-DATA
7      authentication-type md5;
8      wide-metrics-only;
9  }
10 interface fe-0/3/0.0 {
11     level 2 disable;
12 }
13 interface lo0.0 {
14     level 2 disable;
15 }
```

As expected, both Level 1 Areas are stitched together on router *Havana* and router *Lille* has full visibility now on all loopback addresses and the external network range. This scenario is illustrated in Figure 2.24.

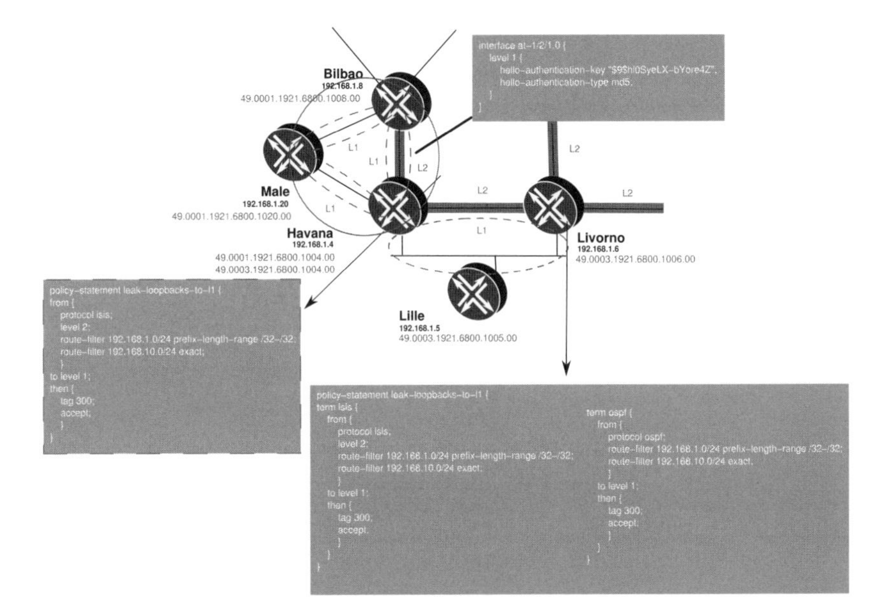

Figure 2.24: Stage five: IS–IS L1 Area merging and activation at router *Lille*.

Nevertheless, the route injection policy from OSPFv2 at router *Livorno* introduces another change when compared with the previous IS–IS database state. OSPFv2 routes are preferred and redistributed into IS–IS Level 1 with `wide-metrics-only`. All other systems attached to this merged Level 1 prefer those injected routes via Level 1 versus the Level 2 version natively present in their intended Level 2 LSPs. This situation also occurs because of the lack of distinction between internal and external prefixes in Level 1 with that knob. Note also that prefixes injected from OSPFv2 into IS–IS Level 1 do not have the Up/Down bit set as per [RFC3784], as opposed to prefixes leaked from Level 2. As a result, those prefixes could be eligible for further redistribution to Level 2 at some other L1L2 systems.

Although this situation could initially appear to be counterintuitive, this setup properly handles forwarding decisions at this transient point in the migration. From the point of view of metrics, this setup is closer to the desired behavior as well, in this case providing a backup path through IS–IS Level 1 during the transition, until OSPFv2 is removed from the domain “Cyclone”. As a result, no further fine-tuning is required at this point.

This scenario is shown in Listing 2.113 from router *Lille*'s perspective. Lines 39, 41, 42, 45, and 46 show how the Up/Down bit is not set in prefixes injected by router *Livorno*, being preferred in domain “Cyclone” as OSPFv2 routes.

Listing 2.113: IS–IS L1 adjacency activation results at Lille

```
1  user@Lille> show isis adjacency
2  Interface             System         L State        Hold (secs) SNPA
3  fe-0/3/0.0            Havana         1  Up                    6  0:90:69:b4:bc:5d
4  fe-0/3/0.0            Livorno        1  Up                   20  0:14:f6:85:40:5d
5
6  user@Lille> show isis database detail
7  IS-IS level 1 link-state database:
8
9  Havana.00-00 Sequence: 0xc, Checksum: 0x81f1, Lifetime: 831 secs
10    IS neighbor: Havana.02                     Metric:     1000
11    IS neighbor: Havana.03                     Metric:      100
12    IS neighbor: Bilbao-re0.00                 Metric:      643
13    IP prefix: 172.16.1.40/30                  Metric:      643 Internal Up    # Up/Down bit not set
14    IP prefix: 172.16.1.84/30                  Metric:      100 Internal Up    # Up/Down bit not set
15    IP prefix: 172.16.100.0/24                 Metric:     1000 Internal Up  # Up/Down bit not set
16    IP prefix: 192.168.1.1/32                  Metric:      743 Internal Down
17    IP prefix: 192.168.1.3/32                  Metric:      903 Internal Down
18    IP prefix: 192.168.1.4/32                  Metric:        0 Internal Up  # Up/Down bit not set
19    IP prefix: 192.168.1.9/32                  Metric:     1386 Internal Down
20    IP prefix: 192.168.1.10/32                 Metric:     2029 Internal Down
21
22 Havana.02-00 Sequence: 0x2, Checksum: 0x3192, Lifetime: 632 secs
23    IS neighbor: Havana.00                     Metric:        0
24    IS neighbor: Livorno.00                    Metric:        0
25
26 Havana.03-00 Sequence: 0x2, Checksum: 0x6922, Lifetime: 706 secs
27    IS neighbor: Havana.00                     Metric:        0
28    IS neighbor: male-re0s.00              Metric:        0
29
30 Lille.00-00 Sequence: 0xeb, Checksum: 0x668b, Lifetime: 674 secs
31    IS neighbor: Havana.03                     Metric:     1000
32    IP prefix: 172.16.100.0/24                 Metric:     1000 Internal Up
33    IP prefix: 192.168.1.5/32                  Metric:        0 Internal Up
34
35 Livorno.00-00 Sequence: 0xd, Checksum: 0xff42, Lifetime: 829 secs
36    IS neighbor: Havana.02                     Metric:     1000
37    IP prefix: 172.16.100.0/24                 Metric:     1000 Internal Up
38    IP prefix: 192.168.1.1/32                  Metric:     1286 Internal Down
39    IP prefix: 192.168.1.2/32                  Metric:       64 Internal Up    # From OSPFv2
40    IP prefix: 192.168.1.3/32                  Metric:     1446 Internal Down
41    IP prefix: 192.168.1.6/32                  Metric:        0 Internal Up    # From OSPFv2
42    IP prefix: 192.168.1.7/32                  Metric:       10 Internal Up    # From OSPFv2
43    IP prefix: 192.168.1.9/32                  Metric:     1929 Internal Down
44    IP prefix: 192.168.1.10/32                 Metric:     2572 Internal Down
45    IP prefix: 192.168.1.21/32                 Metric:       10 Internal Up    # From OSPFv2
46    IP prefix: 192.168.10.0/24                 Metric:        0 Internal Up  # From OSPFv2
47
48 Bilbao-re0.00-00 Sequence: 0x13, Checksum: 0xafa5, Lifetime: 829 secs
49    IS neighbor: Havana.00                     Metric:      643
50    IS neighbor: male-re0.00                Metric:      643
51    IP prefix: 172.16.1.40/30                  Metric:      643 Internal Up
52    IP prefix: 172.16.1.76/30                  Metric:      643 Internal Up
53    IP prefix: 192.168.1.1/32                  Metric:      100 Internal Down
54    IP prefix: 192.168.1.3/32                  Metric:      260 Internal Down
55    IP prefix: 192.168.1.8/32                  Metric:        0 Internal Up
56    IP prefix: 192.168.1.9/32                  Metric:      743 Internal Down
57    IP prefix: 192.168.1.10/32                 Metric:     1386 Internal Down
58
59 male-re0.00-00 Sequence: 0x4, Checksum: 0x7c87, Lifetime: 552 secs
```

```
60      IS neighbor: Havana.03                       Metric:      100
61      IS neighbor: Bilbao-re0.00                   Metric:      643
62      IP prefix: 172.16.1.76/30                    Metric:      643 Internal Up
63      IP prefix: 172.16.1.84/30                    Metric:      100 Internal Up
64      IP prefix: 192.168.1.20/32                   Metric:        0 Internal Up
65
66  IS-IS level 2 link-state database:
67
68  user@Lille> show route protocol isis table inet.0
69
70  inet.0: 36 destinations, 36 routes (36 active, 0 holddown, 0 hidden)
71  + = Active Route, - = Last Active, * = Both
```

Junos Tip: IS–IS Up/Down bit setting conditions

As per [RFC3784], the Up/Down bit is preserved in TLV 135 Extended IP Reachability to avoid routing loops from forming when transitioning IS–IS levels. However, [RFC3784] enforces that this bit SHALL be set to 0 when a prefix is first injected into IS–IS.

When deploying a *ships-in-the-night* model between IGPs and exporting routes from a given protocol into IS–IS Level 1, Junos OS considers them to be brand new for IS–IS and hence the bit is not set. This behavior happens even when IS–IS Level 2 is present on the systems and such prefixes are included in TLVs 135 Extended IP Reachability from Level 2 LSPs, but are not selected as active paths.

Extra care must be taken because when those leaked internal prefixes are eligible for L2 default propagation at other exit points of a Level 1 area and after that, they are eligible again for leaking to other L1 areas. This is not necessarily dangerous and can achieve the desired effect as in this example, but it indirectly transits down hierarchy boundaries twice: first when injected from active OSPFv2 and second when leaked from Level 2 to Level 1 at some other L1L2 system.

2.3.9 Stage six: Global connectivity verification

After all respective actions have been undertaken to cope with the initial requirements to have unified management on all networks and before a full system integration, as many tests as needed must be issued on all systems to ensure full end-to-end visibility before moving on to the next migration stage.

Considering router *Livorno* as a representative system of domain "Cyclone", Listing 2.114 depicts different checks and tests to confirm our objectives. Forwarding testing has not been included here for the sake of brevity. The following items are remarkable from these tests:

- IS–IS L1 database includes all L1 LSPs from merged L1 Areas 49.0001 and 49.0003 (starting from Line 2).
- No ATTached bit is set at any L1 LSPs in the IS–IS L1 database because both areas are merged and reachable via IS–IS Level 1 (starting from Line 2).
- Level 2 LSPs are visible for all domain "Cyclone" systems, along with border routers from other domains (starting from Line 13).
- The OSPFv2 Database remains the same as before the migration in domain "Cyclone" (starting from Line 31).

- OSPFv2 routes are preferred for all internal domain "Cyclone" destinations, including 192.168.10.0/24 (starting from Line 58).
- All destinations external to domain "Cyclone" are reachable via IS–IS L1 or L2 with modified protocol preferences (starting from Line 73).
- Leaking from OSPFv2 is performed for all domain "Cyclone" systems to IS–IS Level 1 Area.
- The difference between TED information injected by OSPFv2 and IS–IS refers to our well-known decision to adapt costs for the link between router *Livorno* and router *Skopie* (Lines 128 and 134 for the same link).
- Forwarding to all destinations from the corresponding loopback addresses works, following optimal paths.

Listing 2.114: Initial domain integration tests at Livorno

```
1  user@Livorno> show isis database
2  IS-IS level 1 link-state database:
3  LSP ID                      Sequence Checksum Lifetime Attributes
4  Havana.00-00                   0x113    0x276      485 L1 L2
5  Havana.02-00                    0xf2   0xf638      632 L1 L2
6  Havana.03-00                     0xc    0x194      632 L1 L2
7  Lille.00-00                     0xef   0x7fd0      458 L1
8  Livorno.00-00                   0xab   0xb3d7      665 L1 L2
9  Bilbao-re0.00-00               0x113   0xe49f      553 L1 L2
10 male-re0.00-00             0xf8    0xf3c3      603 L1
11   7 LSPs
12
13 IS-IS level 2 link-state database:
14 LSP ID                      Sequence Checksum Lifetime Attributes
15 Nantes.00-00                    0x87   0x557a     1109 L1 L2
16 Nantes.02-00                    0x10   0xd225      549 L1 L2
17 Skopie.00-00                    0xb0   0x7cbd     1163 L1 L2
18 Torino.00-00                    0x8a   0x5801      385 L1 L2
19 Havana.00-00                    0x37   0x95df     1147 L1 L2
20 Livorno.00-00                   0xbd   0xaa83      665 L1 L2
21 Livorno.02-00                   0xa2   0x1002      817 L1 L2
22 Barcelona-re0.00-00        0xa3    0x27a5      819 L1 L2
23 Bilbao-re0.00-00                0x42   0x41e3      984 L1 L2
24 honolulu-re0.00-00         0xa2    0x6629      505 L1 L2
25 honolulu-re002-00         0xa0    0x9ce7      933 L1 L2
26 honolulu-re003-00         0x9f    0x8086     1186 L1 L2
27   12 LSPs
28
29 user@Livorno> show ospf database
30
31     OSPF database, Area 0.0.0.0
32  Type       ID               Adv Rtr           Seq      Age  Opt  Cksum  Len
33 Router   192.168.1.2      192.168.1.2      0x8000005f   629  0x22 0x3195  60
34 Router  *192.168.1.6      192.168.1.6      0x8000042d  1283  0x22 0xd94d  96
35 Router   192.168.1.7      192.168.1.7      0x80000619   642  0x22 0x2d48  60
36 Router   192.168.1.21     192.168.1.21     0x80000093   300  0x22 0x305d  60
37 Network  172.16.1.53      192.168.1.7      0x8000006d   942  0x22 0x3e4c  32
38 Network  172.16.1.90      192.168.1.21     0x8000005d   600  0x22 0x2336  32
39 Network  172.16.1.105     192.168.1.21     0x8000005c   900  0x22 0x9cad  32
40 OpaqArea 1.0.0.1          192.168.1.2      0x8000005a  2173  0x22 0x2fa   28
41 OpaqArea*1.0.0.1          192.168.1.6      0x800003da   779  0x22 0x869   28
42 OpaqArea 1.0.0.1          192.168.1.7      0x800005e2   342  0x22 0xf56f  28
43 OpaqArea 1.0.0.1          192.168.1.21     0x80000082  2100  0x22 0xfdb0  28
44 OpaqArea 1.0.0.3          192.168.1.2      0x80000056  1429  0x22 0xfbf5 136
45 OpaqArea*1.0.0.3          192.168.1.6      0x8000004b  2265  0x22 0xe313 124
```

```
46 OpaqArea 1.0.0.3         192.168.1.7     0x8000006b   42  0x22 0x7c57 124
47 OpaqArea 1.0.0.3         192.168.1.21    0x8000007e 1500  0x22 0x4172 124
48 OpaqArea*1.0.0.4         192.168.1.6     0x80000015 1783  0x22 0x8e98 136
49 OpaqArea 1.0.0.4         192.168.1.7     0x8000006e 1542  0x22 0x20fa 124
50 OpaqArea 1.0.0.4         192.168.1.21    0x8000007f 1800  0x22 0xd7d9 124
51 OpaqArea*1.0.0.5         192.168.1.6     0x80000053  280  0x22 0x70c4 124
52     OSPF AS SCOPE link state database
53  Type       ID              Adv Rtr           Seq      Age  Opt  Cksum  Len
54 Extern   192.168.10.0    192.168.1.7     0x80000051 1242  0x22 0x37a   36
55 Extern   192.168.10.0    192.168.1.21    0x80000035 1200  0x22 0x1b84  36
56
57 user@Livorno> show route protocol ospf table inet.0 terse
58
59 inet.0: 42 destinations, 48 routes (41 active, 0 holddown, 1 hidden)
60 + = Active Route, - = Last Active, * = Both
61
62 A Destination        P Prf   Metric 1   Metric 2  Next hop        AS path
63   172.16.1.32/30     O  10        64             >so-0/0/0.0
64 * 172.16.1.104/30    O  10      1010              172.16.1.53
65                                                  >172.16.1.90
66 * 192.168.1.2/32     O  10        64             >so-0/0/0.0
67 * 192.168.1.7/32     O  10        10             >172.16.1.53
68 * 192.168.1.21/32    O  10        10             >172.16.1.90
69 * 192.168.10.0/24    O 150         0             >172.16.1.53
70 * 224.0.0.5/32       O  10         1              MultiRecv
71
72 user@Livorno> show route protocol isis table inet.0 terse
73
74 inet.0: 42 destinations, 48 routes (41 active, 0 holddown, 1 hidden)
75 + = Active Route, - = Last Active, * = Both
76
77 A Destination        P Prf   Metric 1   Metric 2  Next hop        AS path
78 * 10.10.1.96/30      I 158      1446             >172.16.1.33
79 * 172.16.1.0/30      I 158      1286             >172.16.1.33
80 * 172.16.1.4/30      I 158      1386             >172.16.1.33
81                                                   172.16.1.37
82 * 172.16.1.40/30     I 155      1643             >172.16.100.1
83 * 172.16.1.76/30     I 155      1743             >172.16.100.1
84 * 172.16.1.84/30     I 155      1100             >172.16.100.1
85 * 172.16.1.100/30    I 158      1929             >172.16.1.37
86   172.16.1.104/30    I 158      1010             >172.16.1.53
87                                                   172.16.1.90
88 * 192.168.1.1/32     I 158      1286             >172.16.1.33
89   192.168.1.2/32     I 158       643             >172.16.1.33
90 * 192.168.1.3/32     I 158      1446             >172.16.1.33
91 * 192.168.1.4/32     I 155      1000             >172.16.100.1
92 * 192.168.1.5/32     I 155      1000             >172.16.100.5
93   192.168.1.7/32     I 158        10             >172.16.1.53
94 * 192.168.1.8/32     I 155      1643             >172.16.100.1
95 * 192.168.1.9/32     I 158      1929             >172.16.1.33
96 * 192.168.1.10/32    I 158      1606             >172.16.1.33
97 * 192.168.1.20/32    I 155      1100             >172.16.100.1
98   192.168.1.21/32    I 158        10             >172.16.1.90
99   192.168.10.0/24    I 158        10             >172.16.1.53
100
101 user@Livorno> show ted database extensive | match "NodeID|Protocol|Local:|Metric"
102 NodeID: Nantes.00(10.10.1.1)
103   Protocol: IS-IS(2)
104     To: Skopie.00(192.168.1.2), Local: 172.16.1.1, Remote: 172.16.1.2
105       Metric: 160
106     To: Torino.00(10.10.1.3), Local: 10.10.1.98, Remote: 10.10.1.97
107       Metric: 160
108     To: Nantes.02, Local: 172.16.1.6, Remote: 0.0.0.0
109       Metric: 100
110 <...>
111 NodeID: Skopie.00(192.168.1.2)
112   Protocol: OSPF(0.0.0.0)
113     To: Livorno.00(192.168.1.6), Local: 172.16.1.33, Remote: 172.16.1.34
114       Metric: 64
```

```
115   Protocol: IS-IS(2)
116     To: Livorno.00(192.168.1.6), Local: 172.16.1.33, Remote: 172.16.1.34
117       Metric: 643
118     To: Nantes.00(10.10.1.1), Local: 172.16.1.2, Remote: 172.16.1.1
119       Metric: 643
120 <...>
121 NodeID: Livorno.00(192.168.1.6)
122   Protocol: OSPF(0.0.0.0)
123     To: 172.16.1.53-1, Local: 172.16.1.54, Remote: 0.0.0.0
124       Metric: 10
125     To: 172.16.1.90-1, Local: 172.16.1.89, Remote: 0.0.0.0
126       Metric: 10
127     To: Skopie.00(192.168.1.2), Local: 172.16.1.34, Remote: 172.16.1.33
128       Metric: 64
129   Protocol: IS-IS(1)
130     To: Havana.03, Local: 172.16.100.8, Remote: 0.0.0.0
131       Metric: 1000
132   Protocol: IS-IS(2)
133     To: Skopie.00(192.168.1.2), Local: 172.16.1.34, Remote: 172.16.1.33
134       Metric: 643
135     To: Livorno.02, Local: 172.16.1.54, Remote: 0.0.0.0
136       Metric: 10
137     To: honolulu-re002, Local: 172.16.1.89, Remote: 0.0.0.0
138       Metric: 10
139     To: Havana.00(192.168.1.4), Local: 172.16.1.38, Remote: 172.16.1.37
140       Metric: 643
141 NodeID: Livorno.02
142   Protocol: IS-IS(2)
143     To: Livorno.00(192.168.1.6), Local: 0.0.0.0, Remote: 0.0.0.0
144       Metric: 0
145     To: Barcelona-re0.00(192.168.1.7), Local: 0.0.0.0, Remote: 0.0.0.0
146       Metric: 0
147 NodeID: Barcelona-re0.00(192.168.1.7)
148   Protocol: OSPF(0.0.0.0)
149     To: 172.16.1.53-1, Local: 172.16.1.53, Remote: 0.0.0.0
150       Metric: 10
151     To: 172.16.1.105-1, Local: 172.16.1.106, Remote: 0.0.0.0
152       Metric: 1000
153   Protocol: IS-IS(2)
154     To: Livorno.02, Local: 172.16.1.53, Remote: 0.0.0.0
155       Metric: 10
156     To: honolulu-re003, Local: 172.16.1.106, Remote: 0.0.0.0
157       Metric: 1000
158 <...>
159 NodeID: honolulu-re0.00(192.168.1.21)
160   Protocol: OSPF(0.0.0.0)
161     To: 172.16.1.90-1, Local: 172.16.1.90, Remote: 0.0.0.0
162       Metric: 10
163     To: 172.16.1.105-1, Local: 172.16.1.105, Remote: 0.0.0.0
164       Metric: 1000
165   Protocol: IS-IS(2)
166     To: honolulu-re002, Local: 172.16.1.90, Remote: 0.0.0.0
167       Metric: 10
168     To: honolulu-re003, Local: 172.16.1.105, Remote: 0.0.0.0
169       Metric: 1000
170 NodeID: honolulu-re002
171   Protocol: IS-IS(2)
172     To: Livorno.00(192.168.1.6), Local: 0.0.0.0, Remote: 0.0.0.0
173       Metric: 0
174     To: honolulu-re0.00(192.168.1.21), Local: 0.0.0.0, Remote: 0.0.0.0
175       Metric: 0
176 NodeID: honolulu-re003
177   Protocol: IS-IS(2)
178     To: honolulu-re0.00(192.168.1.21), Local: 0.0.0.0, Remote: 0.0.0.0
179       Metric: 0
180     To: Barcelona-re0.00(192.168.1.7), Local: 0.0.0.0, Remote: 0.0.0.0
181       Metric: 0
182 <...>
```

As for domain "Monsoon", router *Basel* and router *Torino* are selected as illustrative examples of purely internal systems, which are still unaware of the upcoming IS–IS deployment, and a border system between both domains.

Listing 2.115 shows test results and inspections at an internal router in the OSPFv2 stubby NSSA, excluding forwarding test results. Note the following about the output:

- Each loopback address from each participating device is represented by a Type 7 NSSA AS External LSA per ASBR, together with the externally injected route in domain "Cyclone" (starting from Line 11).
- "Gale Internet" legitimate public addresses from ABRs are visible via Type 7 NSSA AS External LSAs and not as stub networks from their Type 1 Router LSAs (starting from Line 11).
- Although the topology is symmetric, with two ABRs and two internal routers, router *Nantes* is metrically closer to the backbone than router *Torino*, because router *Bilbao* has a GigabitEthernet link to router *Nantes* and an ATM OC-3 link to router *Torino*. This backbone topology and correct metric propagation across protocol redistribution leads the link between router *Basel* and router *Inverness* to be the preferred egress and ingress traffic point for router *Basel* (Line 27).

Listing 2.115: Initial domain integration tests at Basel

```
 1  user@Basel> show ospf database
 2
 3      OSPF database, Area 0.0.0.3
 4   Type       ID               Adv Rtr           Seq      Age  Opt  Cksum  Len
 5  Router   10.10.1.1        10.10.1.1        0x80000005   394  0x20 0xe938  48
 6  Router   10.10.1.3        10.10.1.3        0x80000004   354  0x20 0x3196  48
 7  Router   10.10.1.9        10.10.1.9        0x80000006   507  0x20 0x4523  96
 8  Router  *10.10.1.10       10.10.1.10       0x80000004   521  0x20 0x42cb  96
 9  Summary  0.0.0.0          10.10.1.1        0x80000003   587  0x20 0x50cc  28
10  Summary  0.0.0.0          10.10.1.3        0x80000003  2065  0x20 0x44d6  28
11  NSSA     192.168.1.1      10.10.1.1        0x80000005   490  0x20 0x8b3d  36
12  NSSA     192.168.1.1      10.10.1.3        0x80000002  1584  0x20 0xcb5d  36
13  NSSA     192.168.1.2      10.10.1.1        0x80000003   781  0x20 0xcb5d  36
14  NSSA     192.168.1.2      10.10.1.3        0x80000003   741  0x20 0xc581  36
15  NSSA     192.168.1.3      10.10.1.1        0x80000002  1602  0x20 0xc365  36
16  NSSA     192.168.1.3      10.10.1.3        0x80000003   547  0x20 0x6f57  36
17  <...>
18  NSSA     192.168.10.0     10.10.1.1        0x80000003  1473  0x20 0x157d  36
19  NSSA     192.168.10.0     10.10.1.3        0x80000003  1472  0x20 0x4fa0  36
20
21  user@Basel> show route protocol ospf table inet.0
22
23  inet.0: 31 destinations, 33 routes (31 active, 0 holddown, 0 hidden)
24  + = Active Route, - = Last Active, * = Both
25
26  0.0.0.0/0          *[OSPF/10] 00:58:43, metric 170
27                      > via at-0/0/0.1001
28  10.10.1.9/32       *[OSPF/10] 00:59:16, metric 643
29                      > via at-0/1/0.1001
30  10.10.1.64/30      *[OSPF/10] 00:59:16, metric 1286
31                      > via at-0/1/0.1001
32  10.10.1.72/30       [OSPF/10] 00:59:35, metric 643
33                      > via at-0/1/0.1001
34  10.10.1.80/30       [OSPF/10] 00:59:35, metric 160
35                      > via at-0/0/0.1001
36  192.168.1.1/32     *[OSPF/150] 00:58:33, metric 0, tag 100
37                      > via at-0/1/0.1001
```

```
38 192.168.1.2/32      *[OSPF/150] 00:54:01, metric 160, tag 100
39                      > via at-0/1/0.1001
40 192.168.1.3/32      *[OSPF/150] 00:58:48, metric 0, tag 100
41                      > via at-0/0/0.1001
42 192.168.1.4/32      *[OSPF/150] 00:55:48, metric 743, tag 100
43                      > via at-0/1/0.1001
44 192.168.1.5/32      *[OSPF/150] 00:01:28, metric 1743, tag 100
45                      > via at-0/1/0.1001
46 192.168.1.6/32      *[OSPF/150] 00:24:38, metric 803, tag 100
47                      > via at-0/1/0.1001
48 192.168.1.7/32      *[OSPF/150] 00:24:37, metric 813, tag 100
49                      > via at-0/1/0.1001
50 192.168.1.8/32      *[OSPF/150] 00:57:05, metric 100, tag 100
51                      > via at-0/1/0.1001
52 192.168.1.9/32      *[OSPF/10] 00:59:16, metric 643
53                      > via at-0/1/0.1001
54 192.168.1.20/32     *[OSPF/150] 00:57:05, metric 743, tag 100
55                      > via at-0/1/0.1001
56 192.168.1.21/32     *[OSPF/150] 00:24:38, metric 813, tag 100
57                      > via at-0/1/0.1001
58 192.168.10.0/24     *[OSPF/150] 00:24:37, metric 813, tag 100
59                      > via at-0/1/0.1001
60 224.0.0.5/32        *[OSPF/10] 00:59:40, metric 1
61                        MultiRecv
```

Listing 2.116 shows results from router *Torino*'s perspective, again excluding forwarding checks. Some aspects to highlight:

- Each loopback address from each participating device is represented by a Type 7 NSSA AS External LSA and a Type 5 AS External LSA per ASBR, together with the externally injected route in domain "Cyclone" (starting from Line 51 for Type 5 and Line 40 for Type 7 LSAs).
- "Gale Internet" legitimate public addresses from internal systems from domain "Monsoon" are visible via stub networks at their Type 1 Router LSAs and via Type 3 Summary LSAs (starting from Line 17).
- Despite closer distance to the parallel ABR router *Nantes*, intra-area routes are always preferred inside NSSA 0.0.0.3.

Listing 2.116: Initial domain integration tests at Torino

```
1  user@Torino> show isis adjacency
2  Interface             System          L State        Hold (secs) SNPA
3  at-0/0/1.0            Bilbao-re0      2  Up                   18
4  so-0/1/0.0            Nantes          2  Up                   22
5
6  user@Torino> show ospf neighbor
7  Address          Interface              State      ID               Pri  Dead
8  10.10.1.98       so-0/1/0.0             Full       10.10.1.1        128    32
9  10.10.1.82       at-1/2/0.1001          Full       10.10.1.10       128    33
10
11 user@Torino> show ospf database
12
13     OSPF database, Area 0.0.0.0
14  Type       ID               Adv Rtr           Seq      Age  Opt  Cksum  Len
15 Router   10.10.1.1        10.10.1.1        0x80000005  2272  0x22 0x8701  60
16 Router  *10.10.1.3        10.10.1.3        0x80000007  2238  0x22 0x3152  60
17 Summary  10.10.1.9        10.10.1.1        0x80000004    82  0x22 0xa0df  28
18 Summary *10.10.1.9        10.10.1.3        0x80000004   138  0x22 0xd904  28
19 Summary  10.10.1.10       10.10.1.1        0x80000003  2748  0x22 0xcc2e  28
20 Summary *10.10.1.10       10.10.1.3        0x80000004    48  0x22 0x9bc6  28
```

```
21 Summary  10.10.1.64      10.10.1.1       0x80000004  173  0x22 0x66e5  28
22 Summary *10.10.1.64      10.10.1.3       0x80000004 2331  0x22 0xd44f  28
23 Summary  10.10.1.72      10.10.1.1       0x80000003 2651  0x22 0x4c73  28
24 Summary *10.10.1.72      10.10.1.3       0x80000003 2708  0x22 0x5151  28
25 Summary  10.10.1.80      10.10.1.1       0x80000003 2554  0x22 0x42d4  28
26 Summary *10.10.1.80      10.10.1.3       0x80000004  320  0x22 0xca54  28
27 Summary  192.168.1.9     10.10.1.1       0x80000003 2457  0x22 0xeb40  28
28 Summary *192.168.1.9     10.10.1.3       0x80000003 2611  0x22 0x2564  28
29 Summary  192.168.1.10    10.10.1.1       0x80000003 2364  0x22 0x168f  28
30 Summary *192.168.1.10    10.10.1.3       0x80000003 2518  0x22 0xe627  28
31
32     OSPF database, Area 0.0.0.3
33  Type       ID              Adv Rtr         Seq         Age  Opt  Cksum  Len
34 Router   10.10.1.1       10.10.1.1       0x80000006  266  0x20 0xe739  48
35 Router  *10.10.1.3       10.10.1.3       0x80000005  229  0x20 0x2f97  48
36 Router   10.10.1.9       10.10.1.9       0x80000007  153  0x20 0x4324  96
37 Router   10.10.1.10      10.10.1.10      0x80000005  167  0x20 0x40cc  96
38 Summary  0.0.0.0         10.10.1.1       0x80000004  448  0x20 0x4ecd  28
39 Summary *0.0.0.0         10.10.1.3       0x80000004 2424  0x20 0x42d7  28
40 NSSA     192.168.1.1     10.10.1.1       0x80000006  357  0x20 0x893e  36
41 NSSA    *192.168.1.1     10.10.1.3       0x80000003 2053  0x20 0xc95e  36
42 NSSA     192.168.1.2     10.10.1.1       0x80000004  629  0x20 0xc95e  36
43 NSSA    *192.168.1.2     10.10.1.3       0x80000004  593  0x20 0x382   36
44 NSSA     192.168.1.3     10.10.1.1       0x80000003 2089  0x20 0xc166  36
45 NSSA    *192.168.1.3     10.10.1.3       0x80000004  411  0x20 0x6d58  36
46 <...>
47 NSSA     192.168.10.0    10.10.1.1       0x80000004 1265  0x20 0x137e  36
48 NSSA    *192.168.10.0    10.10.1.3       0x80000004 1320  0x20 0x4da1  36
49     OSPF AS SCOPE link state database # AS-external LSAs for all external legitimate loopback
       addresses
50  Type       ID              Adv Rtr         Seq         Age  Opt  Cksum  Len
51 Extern   192.168.1.1     10.10.1.1       0x80000004  537  0x22 0x8b3e  36
52 Extern  *192.168.1.1     10.10.1.3       0x80000003 2146  0x22 0xc760  36
53 Extern   192.168.1.2     10.10.1.1       0x80000004  718  0x22 0xc760  36
54 Extern  *192.168.1.2     10.10.1.3       0x80000004  684  0x22 0x184   36
55 Extern   192.168.1.3     10.10.1.1       0x80000003 2179  0x22 0xbf68  36
56 Extern  *192.168.1.3     10.10.1.3       0x80000004  502  0x22 0x6b5a  36
57 <...>
58 Extern   192.168.10.0    10.10.1.1       0x80000004 1354  0x22 0x1180  36
59 Extern  *192.168.10.0    10.10.1.3       0x80000004 1411  0x22 0x4ba3  36
60
61 user@Torino> show isis database
62 IS-IS level 1 link-state database:
63 LSP ID                      Sequence Checksum Lifetime Attributes
64 Torino.00-00                     0xa   0x311b      606 L1 L2 Attached
65   1 LSPs
66
67 IS-IS level 2 link-state database:  # Full visibility on all domain systems running IS-IS
68 LSP ID                      Sequence Checksum Lifetime Attributes
69 Nantes.00-00                     0xf   0x811d      781 L1 L2
70 Nantes.02-00                     0x8   0x9a45      781 L1 L2
71 Skopie.00-00                     0xc   0x3e56      965 L1 L2
72 Torino.00-00                     0xb   0xc1f0      744 L1 L2
73 Havana.00-00                    0x12   0xcc65      996 L1 L2
74 Livorno.00-00                   0x14   0x8d4f     1002 L1 L2
75 Livorno.02-00                    0x6   0xe232     1002 L1 L2
76 Barcelona-re0.00-00              0xc   0xdf75      789 L1 L2
77 Bilbao-re0.00-00                0x10   0x5e6f      774 L1 L2
78 honolulu-re0.00-00               0xb   0xa849      850 L1 L2
79 honolulu-re0.02-00               0x8   0x23af      840 L1 L2
80 honolulu-re0.03-00               0x6   0x4d43     1086 L1 L2
81   12 LSPs
82
83 user@Torino> show isis database detail
84 IS-IS level 1 link-state database:
85
86 Torino.00-00 Sequence: 0xa, Checksum: 0x311b, Lifetime: 604 secs
87    IP prefix: 192.168.1.9/32                 Metric:      803 Internal Up
88    IP prefix: 192.168.1.10/32                Metric:      160 Internal Up
```

```
89
90  IS-IS level 2 link-state database:
91
92  Nantes.00-00 Sequence: 0xf, Checksum: 0x811d, Lifetime: 778 secs
93     IS neighbor: Nantes.02                    Metric:      100
94     IS neighbor: Skopie.00                    Metric:      160
95     IS neighbor: Torino.00                    Metric:      160
96     IP prefix: 10.10.1.96/30                  Metric:      160 Internal Up
97     IP prefix: 172.16.1.0/30                  Metric:      160 Internal Up
98     IP prefix: 172.16.1.4/30                  Metric:      100 Internal Up
99     IP prefix: 192.168.1.1/32                 Metric:        0 Internal Up
100    IP prefix: 192.168.1.9/32                 Metric:      643 Internal Up # Inverness' loopback
101    IP prefix: 192.168.1.10/32                Metric:     1286 Internal Up # Basel's loopback
102
103 Nantes.02-00 Sequence: 0x8, Checksum: 0x9a45, Lifetime: 778 secs
104    IS neighbor: Nantes.00                    Metric:        0
105    IS neighbor: Bilbao-re0.00                Metric:        0
106 <...>
107 Torino.00-00 Sequence: 0xb, Checksum: 0xc1f0, Lifetime: 741 secs
108    IS neighbor: Nantes.00                    Metric:      160
109    IS neighbor: Bilbao-re0.00                Metric:      643
110    IP prefix: 10.10.1.96/30                  Metric:      160 Internal Up
111    IP prefix: 172.16.1.100/30                Metric:      643 Internal Up
112    IP prefix: 192.168.1.3/32                 Metric:        0 Internal Up
113    IP prefix: 192.168.1.9/32                 Metric:      803 Internal Up # Inverness' loopback
114    IP prefix: 192.168.1.10/32                Metric:      160 Internal Up # Basel's loopback
115 <...>
116 user@Torino> show route protocol ospf table inet.0
117
118 inet.0: 50 destinations, 65 routes (50 active, 0 holddown, 0 hidden)
119 + = Active Route, - = Last Active, * = Both
120
121 10.10.1.1/32       *[OSPF/10] 01:42:52, metric 160
122                     > via so-0/1/0.0
123 10.10.1.9/32       *[OSPF/10] 01:43:00, metric 803
124                     > via at-1/2/0.1001
125 <...>
126 192.168.1.1/32      [OSPF/150] 01:42:52, metric 0, tag 100
127                     > via so-0/1/0.0
128 192.168.1.2/32      [OSPF/150] 01:38:15, metric 160, tag 100
129                     > via so-0/1/0.0
130 192.168.1.3/32      [OSPF/150] 01:41:40, metric 160, tag 100
131                     > via so-0/1/0.0
132 192.168.1.4/32      [OSPF/150] 01:40:07, metric 743, tag 100
133                     > via so-0/1/0.0
134 192.168.1.5/32      [OSPF/150] 00:45:42, metric 1743, tag 100
135                     > via so-0/1/0.0
136 192.168.1.6/32      [OSPF/150] 01:08:52, metric 803, tag 100
137                     > via so-0/1/0.0
138 192.168.1.7/32      [OSPF/150] 01:08:52, metric 813, tag 100
139                     > via so-0/1/0.0
140 192.168.1.8/32      [OSPF/150] 01:41:21, metric 100, tag 100
141                     > via so-0/1/0.0
142 192.168.1.9/32     *[OSPF/10] 01:43:00, metric 803
143                     > via at-1/2/0.1001
144 192.168.1.10/32    *[OSPF/10] 01:43:00, metric 160
145                     > via at-1/2/0.1001
146 192.168.1.20/32     [OSPF/150] 01:41:21, metric 743, tag 100
147                     > via so-0/1/0.0
148 192.168.1.21/32     [OSPF/150] 01:08:52, metric 813, tag 100
149                     > via so-0/1/0.0
150 192.168.10.0/24     [OSPF/150] 01:08:52, metric 813, tag 100
151                     > via so-0/1/0.0
152 224.0.0.5/32       *[OSPF/10] 04:15:08, metric 1
153                       MultiRecv
154
155 user@Torino> show route protocol isis table inet.0
156
157 inet.0: 50 destinations, 65 routes (50 active, 0 holddown, 0 hidden)
```

```
158 + = Active Route, - = Last Active, * = Both
159
160 172.16.1.0/30      *[IS-IS/18] 01:41:48, metric 320
161                      > to 10.10.1.98 via so-0/1/0.0
162 172.16.1.4/30      *[IS-IS/18] 01:41:48, metric 260
163                      > to 10.10.1.98 via so-0/1/0.0
164 <...>
165 192.168.1.1/32     *[IS-IS/18] 01:41:48, metric 160
166                      > to 10.10.1.98 via so-0/1/0.0
167 192.168.1.2/32     *[IS-IS/18] 01:38:23, metric 320
168                      > to 10.10.1.98 via so-0/1/0.0
169 192.168.1.4/32     *[IS-IS/18] 01:40:15, metric 903
170                      > to 10.10.1.98 via so-0/1/0.0
171 192.168.1.5/32     *[IS-IS/18] 00:45:50, metric 1903
172                      > to 10.10.1.98 via so-0/1/0.0
173 192.168.1.6/32     *[IS-IS/18] 01:09:00, metric 963
174                      > to 10.10.1.98 via so-0/1/0.0
175 192.168.1.7/32     *[IS-IS/18] 01:09:00, metric 973
176                      > to 10.10.1.98 via so-0/1/0.0
177 192.168.1.8/32     *[IS-IS/18] 01:41:34, metric 260
178                      > to 10.10.1.98 via so-0/1/0.0
179 192.168.1.9/32      [IS-IS/18] 00:02:46, metric 803, tag 200, tag2 3
180                      > to 10.10.1.98 via so-0/1/0.0
181 192.168.1.20/32    *[IS-IS/18] 01:41:34, metric 903
182                      > to 10.10.1.98 via so-0/1/0.0
183 192.168.1.21/32    *[IS-IS/18] 01:09:00, metric 973
184                      > to 10.10.1.98 via so-0/1/0.0
185 192.168.10.0/24    *[IS-IS/18] 01:09:00, metric 973
186                      > to 10.10.1.98 via so-0/1/0.0
```

Listing 2.117 describes router *Havana*'s view, except for forwarding test results. It is worth emphasizing some topics as follows:

- Router *Havana* is not setting the ATTached bit on L1 LSPs, because all existing areas are visible through Level 1 (Line 12);
- The Up/Down bit is properly marked inside respective L1 LSP TLVs for routing information leaked from Level 2 to Level 1 to avoid Level 2 reinjection (starting from Line 47);
- Domain "Monsoon" injected prefixes are easily identified via route Tag 200 (Lines 118 and 120).

Listing 2.117: Initial domain integration tests at Havana

```
1  user@Havana> show isis adjacency
2  Interface          System          L State        Hold (secs) SNPA
3  at-1/2/0.1001      Bilbao-re0      3  Up                 26  # L1 and L2 adjacency over same link
4  fe-0/3/0.0         Lille           1  Up                 24  0:14:f6:81:48:5d
5  fe-0/3/0.0         Livorno         1  Up                 20  0:14:f6:85:40:5d
6  ge-1/1/0.0         male-re0        1  Up                 25  0:19:e2:2b:38:fa
7  so-1/0/0.0         Livorno         2  Up                 24
8
9  user@Havana> show isis database
10 IS-IS level 1 link-state database:
11 LSP ID                      Sequence Checksum Lifetime Attributes
12 Havana.00-00                    0x14   0x9177     1030 L1 L2  # No ATTached bit
13 Havana.02-00                     0x8   0x1fa3      838 L1 L2
14 Havana.03-00                     0x8   0x962e      838 L1 L2
15 Lille.00-00                      0x6   0x4171      633 L1
16 Livorno.00-00                   0x16   0xa8d8      676 L1 L2
17 Bilbao-re0.00-00                0x15   0x5e1b      988 L1 L2
18 male-re0.00-00                   0xb   0x6c38     1072 L1
19   7 LSPs
20
```

```
21 IS-IS level 2 link-state database:
22 LSP ID                      Sequence Checksum Lifetime Attributes
23 Nantes.00-00                     0xf   0x811d      617 L1 L2
24 Nantes.02-00                     0x8   0x9a45      617 L1 L2
25 Skopie.00-00                     0xc   0x3e56      802 L1 L2
26 Torino.00-00                     0xb   0xc1f0      577 L1 L2
27 Havana.00-00                    0x12   0xcc65      838 L1 L2
28 Livorno.00-00                   0x14   0x8d4f      843 L1 L2
29 Livorno.02-00                    0x6   0xe232      843 L1 L2
30 Barcelona-re0.00-00              0xc   0xdf75      631 L1 L2
31 Bilbao-re0.00-00                0x10   0x5e6f      611 L1 L2
32 honolulu-re0.00-00               0xb   0xa849      691 L1 L2
33 honolulu-re002-00                0x8   0x23af      682 L1 L2
34 honolulu-re003-00                0x6   0x4d43      928 L1 L2
35   12 LSPs
36
37 user@Havana> show isis database detail
38 IS-IS level 1 link-state database:
39
40 Havana.00-00 Sequence: 0x12, Checksum: 0x1a06, Lifetime: 388 secs
41    IS neighbor: Havana.02                    Metric:     1000
42    IS neighbor: Havana.03                    Metric:      100
43    IS neighbor: Bilbao-re0.00                Metric:      643
44    IP prefix: 172.16.1.40/30                 Metric:      643 Internal Up
45    IP prefix: 172.16.1.84/30                 Metric:      100 Internal Up
46    IP prefix: 172.16.100.0/24                Metric:     1000 Internal Up
47    IP prefix: 192.168.1.1/32                 Metric:      743 Internal Down   # From IS-IS Level 2
48    IP prefix: 192.168.1.3/32                 Metric:      903 Internal Down   # From IS-IS Level 2
49    IP prefix: 192.168.1.4/32                 Metric:        0 Internal Up
50    IP prefix: 192.168.1.9/32                 Metric:     1386 Internal Down   # From IS-IS Level 2
51    IP prefix: 192.168.1.10/32                Metric:     1063 Internal Down   # From IS-IS Level 2
52
53 Havana.02-00 Sequence: 0x8, Checksum: 0x1fa3, Lifetime: 835 secs
54    IS neighbor: Havana.00                    Metric:        0
55    IS neighbor: male-re0.00                  Metric:        0
56
57 Havana.03-00 Sequence: 0x8, Checksum: 0x962e, Lifetime: 835 secs
58    IS neighbor: Havana.00                    Metric:        0
59    IS neighbor: Lille.00                     Metric:        0
60    IS neighbor: Livorno.00                   Metric:        0
61
62 Lille.00-00 Sequence: 0x6, Checksum: 0x4171, Lifetime: 631 secs
63    IS neighbor: Havana.03                    Metric:     1000
64    IP prefix: 172.16.100.0/24                Metric:     1000 Internal Up
65    IP prefix: 192.168.1.5/32                 Metric:        0 Internal Up
66
67 Livorno.00-00 Sequence: 0x13, Checksum: 0x32fe, Lifetime: 585 secs
68    IS neighbor: Havana.02                    Metric:     1000
69    IP prefix: 172.16.100.0/24                Metric:     1000 Internal Up
70    IP prefix: 192.168.1.1/32                 Metric:     1286 Internal Down   # From IS-IS Level 2
71    IP prefix: 192.168.1.2/32                 Metric:       64 Internal Up     # From OSPFv2
72    IP prefix: 192.168.1.3/32                 Metric:     1446 Internal Down   # From IS-IS Level 2
73    IP prefix: 192.168.1.6/32                 Metric:        0 Internal Up
74    IP prefix: 192.168.1.7/32                 Metric:       10 Internal Up     # From OSPFv2
75    IP prefix: 192.168.1.9/32                 Metric:     1929 Internal Down   # From IS-IS Level 2
76    IP prefix: 192.168.1.10/32                Metric:     1606 Internal Down   # From IS-IS Level 2
77    IP prefix: 192.168.1.21/32                Metric:       10 Internal Up     # From OSPFv2
78    IP prefix: 192.168.10.0/24                Metric:        0 Internal Up
79
80 Bilbao-re0.00-00 Sequence: 0x19, Checksum: 0xf990, Lifetime: 548 secs
81    IS neighbor: Havana.00                    Metric:      643
82    IS neighbor: male-re0.00                  Metric:      643
83    IP prefix: 172.16.1.40/30                 Metric:      643 Internal Up
84    IP prefix: 172.16.1.76/30                 Metric:      643 Internal Up
85    IP prefix: 192.168.1.1/32                 Metric:      100 Internal Down   # From IS-IS Level 2
86    IP prefix: 192.168.1.3/32                 Metric:      260 Internal Down   # From IS-IS Level 2
87    IP prefix: 192.168.1.8/32                 Metric:        0 Internal Up
88    IP prefix: 192.168.1.9/32                 Metric:      743 Internal Down   # From IS-IS Level 2
89    IP prefix: 192.168.1.10/32                Metric:      420 Internal Down   # From IS-IS Level 2
```

```
90
91  <...>
92
93  IS-IS level 2 link-state database:
94  <...>
95
96  user@Havana> show route protocol isis table inet.0
97
98  inet.0: 46 destinations, 46 routes (46 active, 0 holddown, 0 hidden)
99  + = Active Route, - = Last Active, * = Both
100
101 <...>
102 172.16.1.104/30    *[IS-IS/18] 01:11:35, metric 1653
103                     > to 172.16.1.38 via so-1/0/0.0
104 192.168.1.1/32     *[IS-IS/18] 01:42:40, metric 743
105                     > to 172.16.1.42 via at-1/2/0.1001
106 192.168.1.2/32     *[IS-IS/15] 01:14:53, metric 1064, tag 300  # From domain Cyclone
107                     > to 172.16.100.8 via fe-0/3/0.0
108 192.168.1.3/32     *[IS-IS/18] 01:42:40, metric 903
109                     > to 172.16.1.42 via at-1/2/0.1001
110 192.168.1.5/32     *[IS-IS/15] 00:48:24, metric 1000
111                     > to 172.16.100.5 via fe-0/3/0.0
112 192.168.1.6/32     *[IS-IS/15] 01:12:03, metric 1000
113                     > to 172.16.100.8 via fe-0/3/0.0
114 192.168.1.7/32     *[IS-IS/15] 01:14:53, metric 1010, tag 300  # From domain Cyclone
115                     > to 172.16.100.8 via fe-0/3/0.0
116 192.168.1.8/32     *[IS-IS/15] 01:42:50, metric 643
117                     > to 172.16.1.42 via at-1/2/0.1001
118 192.168.1.9/32     *[IS-IS/18] 01:42:40, metric 1386, tag 200, tag2 3  # From domain Monsoon
119                     > to 172.16.1.42 via at-1/2/0.1001
120 192.168.1.10/32    *[IS-IS/18] 01:42:40, metric 1063, tag 200, tag2 3  # From domain Monsoon
121                     > to 172.16.1.42 via at-1/2/0.1001
122 192.168.1.20/32    *[IS-IS/15] 01:43:21, metric 100
123                     > to 172.16.1.85 via ge-1/1/0.0
124 192.168.1.21/32    *[IS-IS/15] 01:14:53, metric 1010, tag 300          # From domain Cyclone
125                     > to 172.16.100.8 via fe-0/3/0.0
126 192.168.10.0/24    *[IS-IS/15] 01:14:53, metric 1000, tag 300          # From domain Cyclone
127                     > to 172.16.100.8 via fe-0/3/0.0
```

Finally, Listing 2.118 shows the domain view at router *Lille* (forwarding testing not included). Some points worth mentioning:

- Neither router *Havana* nor router *Livorno* nor router *Bilbao* are setting the ATTached bit on L1 LSPs, as all existing areas are visible through Level 1 (starting from Line 9).
- The Up/Down bit is properly marked inside respective L1 LSP TLVs for routing information leaked from Level 2 to Level 1 to avoid Level 2 reinjection (starting from Line 31).
- Router *Livorno* leaks also routes actively present in its routing table via OSPFv2 to Level 1 and they do not have the Up/Down bit set (starting from Line 68).

Listing 2.118: Initial domain integration tests at Lille

```
1 user@Lille> show isis adjacency
2 Interface           System         L State        Hold (secs) SNPA
3 fe-0/3/0.0          Havana         1  Up                   8  0:90:69:b4:bc:5d
4 fe-0/3/0.0          Livorno        1  Up                  24  0:14:f6:85:40:5d
5
6 user@Lille> show isis database
7 IS-IS level 1 link-state database:
8 LSP ID                      Sequence Checksum Lifetime Attributes
9 Havana.00-00                    0x14   0x9177      805 L1 L2 # No ATTached bit set
```

```
10 Havana.02-00                  0x8    0x1fa3      612 L1 L2
11 Havana.03-00                  0x8    0x962e      613 L1 L2
12 Lille.00-00                   0x6    0x4171      412 L1
13 Livorno.00-00                 0x16   0xa8d8      452 L1 L2 # No ATTached bit set
14 Bilbao-re0.00-00              0x15   0x5e1b      763 L1 L2 # No ATTached bit set
15 male-re0.00-00                0xb    0x6c38      847 L1
16   7 LSPs
17
18 IS-IS level 2 link-state database:
19   0 LSPs
20
21 user@Lille> show isis database detail
22 IS-IS level 1 link-state database:
23
24 Havana.00-00 Sequence: 0x12, Checksum: 0x1a06, Lifetime: 388 secs
25    IS neighbor: Havana.02                  Metric:     1000
26    IS neighbor: Havana.03                  Metric:      100
27    IS neighbor: Bilbao-re0.00              Metric:      643
28    IP prefix: 172.16.1.40/30               Metric:      643 Internal Up
29    IP prefix: 172.16.1.84/30               Metric:      100 Internal Up
30    IP prefix: 172.16.100.0/24              Metric:     1000 Internal Up
31    IP prefix: 192.168.1.1/32               Metric:      743 Internal Down  # From IS-IS Level 2
32    IP prefix: 192.168.1.3/32               Metric:      903 Internal Down  # From IS-IS Level 2
33    IP prefix: 192.168.1.4/32               Metric:        0 Internal Up
34    IP prefix: 192.168.1.9/32               Metric:     1386 Internal Down  # From IS-IS Level 2
35    IP prefix: 192.168.1.10/32              Metric:     1063 Internal Down  # From IS-IS Level 2
36 <...>
37 Livorno.00-00 Sequence: 0x13, Checksum: 0x32fe, Lifetime: 585 secs
38    IS neighbor: Havana.02                  Metric:     1000
39    IP prefix: 172.16.100.0/24              Metric:     1000 Internal Up
40    IP prefix: 192.168.1.1/32               Metric:     1286 Internal Down  # From IS-IS Level 2
41    IP prefix: 192.168.1.2/32               Metric:       64 Internal Up    # From OSPFv2
42    IP prefix: 192.168.1.3/32               Metric:     1446 Internal Down  # From IS-IS Level 2
43    IP prefix: 192.168.1.6/32               Metric:        0 Internal Up
44    IP prefix: 192.168.1.7/32               Metric:       10 Internal Up    # From OSPFv2
45    IP prefix: 192.168.1.9/32               Metric:     1929 Internal Down  # From IS-IS Level 2
46    IP prefix: 192.168.1.10/32              Metric:     1606 Internal Down  # From IS-IS Level 2
47    IP prefix: 192.168.1.21/32              Metric:       10 Internal Up    # From OSPFv2
48    IP prefix: 192.168.10.0/24              Metric:        0 Internal Up
49
50 Bilbao-re0.00-00 Sequence: 0x19, Checksum: 0xf990, Lifetime: 548 secs
51    IS neighbor: Havana.00                  Metric:      643
52    IS neighbor: male-re0.00                Metric:      643
53    IP prefix: 172.16.1.40/30               Metric:      643 Internal Up
54    IP prefix: 172.16.1.76/30               Metric:      643 Internal Up
55    IP prefix: 192.168.1.1/32               Metric:      100 Internal Down  # From IS-IS Level 2
56    IP prefix: 192.168.1.3/32               Metric:      260 Internal Down  # From IS-IS Level 2
57    IP prefix: 192.168.1.8/32               Metric:        0 Internal Up
58    IP prefix: 192.168.1.9/32               Metric:      743 Internal Down  # From IS-IS Level 2
59    IP prefix: 192.168.1.10/32              Metric:      420 Internal Down  # From IS-IS Level 2
60 <...>
61
62 user@Lille> show route protocol isis table inet.0
63
64 inet.0: 35 destinations, 35 routes (35 active, 0 holddown, 0 hidden)
65 + = Active Route, - = Last Active, * = Both
66
67 <...>
68 192.168.1.1/32     *[IS-IS/18] 00:51:55, metric 1743, tag 300  # Leaked due to policy
69                     > to 172.16.100.1 via fe-0/3/0.0
70 192.168.1.2/32     *[IS-IS/18] 00:51:55, metric 1903, tag 300  # Leaked due to policy
71                     > to 172.16.100.1 via fe-0/3/0.0
72 192.168.1.3/32     *[IS-IS/18] 00:51:55, metric 1903, tag 300  # Leaked due to policy
73                     > to 172.16.100.1 via fe-0/3/0.0
74 192.168.1.4/32     *[IS-IS/15] 00:51:55, metric 1000
75                     > to 172.16.100.1 via fe-0/3/0.0
76 192.168.1.6/32     *[IS-IS/15] 00:51:39, metric 1000
77                     > to 172.16.100.8 via fe-0/3/0.0
78 192.168.1.7/32     *[IS-IS/18] 00:51:55, metric 1653, tag 300  # Leaked due to policy
```

```
79                            > to 172.16.100.1 via fe-0/3/0.0
80 192.168.1.8/32            *[IS-IS/15] 00:51:54, metric 1643
81                            > to 172.16.100.1 via fe-0/3/0.0
82 192.168.1.9/32            *[IS-IS/18] 00:51:55, metric 2386, tag 300, tag2 3  # Leaked due to policy
83                            > to 172.16.100.1 via fe-0/3/0.0
84 192.168.1.10/32           *[IS-IS/18] 00:51:55, metric 2063, tag 300, tag2 3  # Leaked due to policy
85                            > to 172.16.100.1 via fe-0/3/0.0
86 192.168.1.20/32           *[IS-IS/15] 00:51:54, metric 1100
87                            > to 172.16.100.1 via fe-0/3/0.0
88 192.168.1.21/32           *[IS-IS/18] 00:51:55, metric 1653, tag 300  # Leaked due to policy
89                            > to 172.16.100.1 via fe-0/3/0.0
90 192.168.10.0/24           *[IS-IS/18] 00:51:55, metric 1653, tag 300  # Leaked due to policy
91                            > to 172.16.100.1 via fe-0/3/0.0
```

At this stage, basic connectivity to all systems is granted, and the very first milestone in our migration has been successfully achieved without modifying any previously existing forwarding paths in any domain and without any other kind of service disruption. The state of the global setup after reaching this first milestone is summarized in Figure 2.25.

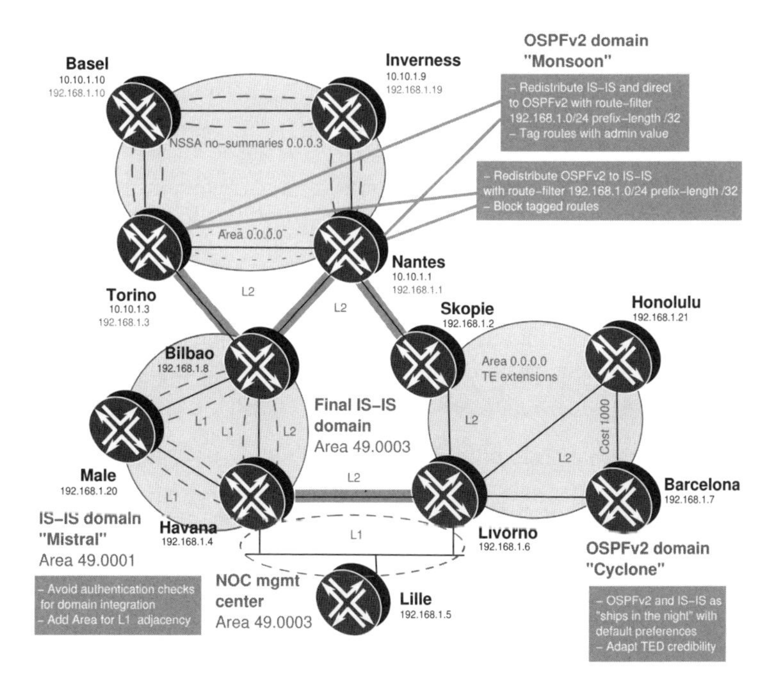

Figure 2.25: Stage six: domain IGP interconnect.

2.3.10 Stage seven: OSPFv2 to IS–IS Level 2 transition in domain "Cyclone"

At this stage, IS–IS should transition to become the most preferred and TED-credible protocol. Because of the previous configuration of `traffic-engineering credibility-protocol-preference`, changing IS–IS Level 2 and 1 preferences in participating systems at this point directly affects the credibility values for the TED.

Because the final objective in the migration is to dismantle OSPFv2 completely, it is necessary at some point to completely remove OSPFv2 adjacencies. Even if OSPFv2 becomes unpreferred, by default, RSVP is notified when OSPFv2 adjacencies drop, so as to trigger the corresponding notification for expedited MPLS convergence.

However, in our case, connectivity is ensured by IS–IS at that point, so dropping RSVP neighbors and generating the corresponding notifications upon removal of OSPFv2 is not desired. As previously described, the Junos OS OSPFv2 knob `no-neighbor-down-notification` skips that notification, with the intention fully to transition in a *ships-in-the-night* model from OSPFv2 to another preferred protocol. Note that this knob needs to be activated on the corresponding links before dismantling their adjacencies so that the router is previously aware of that operation. Activating such a configuration option in advance is entirely graceful for the system and simply acts as previous advice, while also unprotecting the setup from RSVP reaction against legitimate IGP failures. Thus, it is recommended to keep it active only for a transient period.

To summarize, this migration round through the systems in domain "Cyclone" should cover the following activities:

- protocol preference transition to automatically consider IS–IS Levels 1 and 2 as most *credible* protocols towards routing table and TED;
- activation of `no-neighbor-down-notification` for existing OSPFv2 adjacencies.

Because internal and external preferences for OSPFv2 and IS–IS Level 1 and Level 2 need to be tuned, the rationale for the preference adaption this time is to restore default values for IS–IS and to mark OSPFv2 as unpreferred, all with the same Junos OS commit. With the idea of avoiding any other collisions with BGP or other protocols, the following values have been selected:

- IS–IS Level 1 internal preference = 15 (default);
- IS–IS Level 2 internal preference = 18 (default);
- IS–IS Level 1 external preference = 160 (default);
- IS–IS Level 2 external preference = 165 (default);
- OSPF internal route = 167;
- OSPF AS external route = 169.

Once corresponding checks are issued in the complete domain and no OSPFv2 paths remain active in the routing table, OSPFv2 can be safely removed from routers without further impact.

Modifying protocol preferences and adding `no-neighbor-down-notification` to existing OSPFv2 adjacencies are completely graceful actions for existing IGP adjacencies in this setup, as shown in representative Listing 2.119 for router *Livorno*. At the same time, OSPFv2 preferred routes are not leaked in router *Livorno*, by deactivating the referring leaking policy term.

Listing 2.119: Protocol preference adaption and RSVP notification drop avoidance in Livorno

```
[edit]
user@Livorno# set protocols ospf preference 167
[edit]
user@Livorno# set protocols ospf external-preference 169
[edit]
user@Livorno# delete protocols isis level 1 preference
[edit]
user@Livorno# delete protocols isis level 2 preference
[edit protocols ospf area 0]
user@Livorno# set interface so-0/0/0.0 no-neighbor-down-notification
[edit protocols ospf area 0]
user@Livorno# set interface ge-0/2/0.0 no-neighbor-down-notification
[edit protocols ospf area 0]
user@Livorno# set interface ge-1/3/0.0 no-neighbor-down-notification
[edit]
user@Livorno# deactivate policy-options policy-statement leak-loopbacks-to-l1 term ospf

user@Livorno> show route protocol ospf active-path table inet.0

inet.0: 41 destinations, 47 routes (40 active, 0 holddown, 1 hidden)
+ = Active Route, - = Last Active, * = Both

224.0.0.5/32       *[OSPF/10] 02:39:56, metric 1 # No active routes apart from allOSPFrouters multicast
                    MultiRecv

user@Livorno> show route protocol isis terse table inet.0
  # Fully integrated domain in IS-IS

inet.0: 41 destinations, 47 routes (40 active, 0 holddown, 1 hidden)
+ = Active Route, - = Last Active, * = Both

A Destination        P Prf   Metric 1   Metric 2  Next hop        AS path
<...>
* 172.16.1.40/30     I  15       1643             >172.16.100.1
* 172.16.1.76/30     I  15       1743             >172.16.100.1
* 172.16.1.84/30     I  15       1100             >172.16.100.1
* 172.16.1.100/30    I  18       1929             >172.16.1.37
* 172.16.1.104/30    I  18       1010             >172.16.1.53
                                                   172.16.1.90
* 192.168.1.1/32     I  18       1286             >172.16.1.33
* 192.168.1.2/32     I  18        643             >172.16.1.33
* 192.168.1.3/32     I  18       1446             >172.16.1.33
* 192.168.1.4/32     I  15       1000             >172.16.100.1
* 192.168.1.5/32     I  15       1000             >172.16.100.5
* 192.168.1.7/32     I  18         10             >172.16.1.53
* 192.168.1.8/32     I  15       1643             >172.16.100.1
* 192.168.1.9/32     I  18       1929             >172.16.1.33
* 192.168.1.10/32    I  18       1606             >172.16.1.33
* 192.168.1.20/32    I  15       1100             >172.16.100.1
* 192.168.1.21/32    I  18         10             >172.16.1.90
* 192.168.10.0/24    I  18         10             >172.16.1.53

user@Livorno> show isis database detail Livorno
IS-IS level 1 link-state database:

Livorno.00-00 Sequence: 0x19, Checksum: 0x93d7, Lifetime: 890 secs
   IS neighbor: Havana.02                    Metric:     1000
   IP prefix: 172.16.100.0/24                Metric:     1000 Internal Up
   IP prefix: 192.168.1.1/32                 Metric:     1286 Internal Down
   IP prefix: 192.168.1.2/32                 Metric:      643 Internal Down # Leaking from IS-IS L2
   IP prefix: 192.168.1.3/32                 Metric:     1446 Internal Down
   IP prefix: 192.168.1.6/32                 Metric:        0 Internal Up
   IP prefix: 192.168.1.7/32                 Metric:       10 Internal Down # Leaking from IS-IS L2
   IP prefix: 192.168.1.9/32                 Metric:     1929 Internal Down
   IP prefix: 192.168.1.10/32                Metric:     1606 Internal Down
   IP prefix: 192.168.1.21/32                Metric:       10 Internal Down # Leaking from IS-IS L2
   IP prefix: 192.168.10.0/24                Metric:       10 Internal Down
```

```
69 IS-IS level 2 link-state database:
70
71 Livorno.00-00 Sequence: 0x1f, Checksum: 0x23b0, Lifetime: 890 secs
72    IS neighbor: Skopie.00                   Metric:      643
73    IS neighbor: Havana.00                   Metric:      643
74    IS neighbor: Livorno.02                  Metric:       10
75    IS neighbor: honolulu-re0.02             Metric:       10
76    IP prefix: 172.16.1.32/30                Metric:      643 Internal Up
77    IP prefix: 172.16.1.36/30                Metric:      643 Internal Up
78    IP prefix: 172.16.1.40/30                Metric:     1643 Internal Up
79    IP prefix: 172.16.1.52/30                Metric:       10 Internal Up
80    IP prefix: 172.16.1.76/30                Metric:     1743 Internal Up
81    IP prefix: 172.16.1.84/30                Metric:     1100 Internal Up
82    IP prefix: 172.16.1.88/30                Metric:       10 Internal Up
83    IP prefix: 172.16.100.0/24               Metric:     1000 Internal Up
84    IP prefix: 192.168.1.4/32                Metric:     1000 Internal Up
85    IP prefix: 192.168.1.6/32                Metric:        0 Internal Up
86    IP prefix: 192.168.1.8/32                Metric:     1643 Internal Up
87    IP prefix: 192.168.1.20/32               Metric:     1100 Internal Up
88
89 Livorno.02-00 Sequence: 0xc, Checksum: 0x6bc2, Lifetime: 890 secs
90    IS neighbor: Livorno.00                  Metric:        0
91    IS neighbor: Barcelona-re0.00            Metric:        0
92
93 user@Livorno> show ted protocol
94 Protocol name        Credibility  Self node
95 IS-IS(1)             497          Livorno.00(192.168.1.6)
96 IS-IS(2)             494          Livorno.00(192.168.1.6)
97 OSPF(0)              345          Livorno.00(192.168.1.6)
```

Similar graceful protocol transitions take place on all other systems in this domain, although it is recommended that the existing IS–IS Level 2 be expanded simultaneously for operational purposes.

What If... Route preference for locally injected prefixes with IS–IS Up/Down bit set

A particular corner case deserves further comments. Even though IS–IS becomes the more preferred protocol, if the route injection via `export` policy is not deactivated or removed, OSPFv2 routes still remain preferred in Junos OS. Listing 2.120 illustrates the previous case at router *Livorno* without deactivating injection.

Listing 2.120: Routing table and IS–IS Database status at Livorno with OSPF external route injection

```
1  user@Livorno> show route protocol ospf active-path table inet.0
2
3  inet.0: 41 destinations, 43 routes (40 active, 0 holddown, 1 hidden)
4  + = Active Route, - = Last Active, * = Both
5
6  192.168.1.2/32     *[OSPF/167] 00:06:11, metric 64
7                      > via so-0/0/0.0
8  192.168.1.7/32     *[OSPF/167] 00:06:11, metric 10
9                      > to 172.16.1.53 via ge-0/2/0.0
10 192.168.1.21/32    *[OSPF/167] 00:06:11, metric 10
11                     > to 172.16.1.90 via ge-1/3/0.0
12 192.168.10.0/24    *[OSPF/169] 00:06:11, metric 0, tag 0
13                     > to 172.16.1.53 via ge-0/2/0.0
14 224.0.0.5/32       *[OSPF/10] 02:19:09, metric 1
15                       MultiRecv
16
17 user@Livorno> show isis database detail Livorno level 1
```

```
18 Livorno.00-00 Sequence: 0x16, Checksum: 0xfa4c, Lifetime: 806 secs
19    IS neighbor: Havana.02                      Metric:     1000
20    IP prefix: 172.16.100.0/24                  Metric:     1000 Internal Up
21    IP prefix: 192.168.1.1/32                   Metric:     1286 Internal Down # Leaking retained
22    IP prefix: 192.168.1.2/32                   Metric:       64 Internal Up
23    IP prefix: 192.168.1.3/32                   Metric:     1446 Internal Down # Leaking retained
24    IP prefix: 192.168.1.6/32                   Metric:        0 Internal Up
25    IP prefix: 192.168.1.7/32                   Metric:       10 Internal Up
26    IP prefix: 192.168.1.9/32                   Metric:     1929 Internal Down # Leaking retained
27    IP prefix: 192.168.1.10/32                  Metric:     1606 Internal Down # Leaking retained
28    IP prefix: 192.168.1.21/32                  Metric:       10 Internal Up
29    IP prefix: 192.168.10.0/24                  Metric:        0 Internal Up
30
31
32 user@Livorno> show route 192.168.1.2 extensive table inet.0
33
34 inet.0: 41 destinations, 43 routes (40 active, 0 holddown, 1 hidden)
35 192.168.1.2/32 (1 entry, 1 announced)
36 TSI:
37 KRT in-kernel 192.168.1.2/32 -> {so-0/0/0.0}
38 IS-IS level 1, LSP fragment 0
39         *OSPF   Preference: 167 # IS-IS route does not appear, only modified OSPF preference
40                 Next hop type: Router, Next hop index: 568
41                 Next-hop reference count: 3
42                 Next hop: via so-0/0/0.0, selected
43                 State: <Active Int>
44                 Age: 7:11   Metric: 64
45                 Area: 0.0.0.0
46                 Task: OSPF
47                 Announcement bits (3): 0-KRT 4-LDP 5-IS-IS
48                 AS path: I
```

This situation, where the IS–IS route does not appear, is a built-in security mechanism when injecting routes into Level 1 without the Up/Down bit set to prevent an indirect backdoor for a routing loop when leaking if the Up/Down bit were not set.

Once the policy term is deactivated, OSPFv2 active routes disappear and leaked prefixes are tagged with the Up/Down bit, as shown in Listing 2.121.

Listing 2.121: Routing table and IS–IS Database status at Livorno without OSPF external route injection

```
1 [edit policy-options policy-statement leak-loopbacks to l1]
2 user@Livorno# deactivate term ospf
3
4 user@Livorno> show route 192.168.1.2 extensive table inet.0
5
6 inet.0: 41 destinations, 47 routes (40 active, 0 holddown, 1 hidden)
7 192.168.1.2/32 (2 entries, 1 announced)
8 TSI:
9 KRT in-kernel 192.168.1.2/32 -> {172.16.1.33}
10 IS-IS level 1, LSP fragment 0
11         *IS-IS  Preference: 18 # IS-IS route is preferred now
12                 Level: 2
13                 Next hop type: Router, Next hop index: 518
14                 Next-hop reference count: 14
15                 Next hop: 172.16.1.33 via so-0/0/0.0, selected
16                 State: <Active Int>
17                 Age: 4  Metric: 643
18                 Task: IS-IS
19                 Announcement bits (3): 0-KRT 4-LDP 5-IS-IS
20                 AS path: I
21          OSPF   Preference: 167 # OSPF route unpreferred
22                 Next hop type: Router, Next hop index: 568
23                 Next-hop reference count: 2
24                 Next hop: via so-0/0/0.0, selected
```

```
25                  State: <Int>
26                  Inactive reason: Route Preference
27                  Age: 7:59    Metric: 64
28                  Area: 0.0.0.0
29                  Task: OSPF
30                  AS path: I
```

Once these changes are completely deployed on all systems, after checking that no OSPFv2 routes remain as *active* paths in any routing table, OSPFv2 can be deleted and systems can be considered as completely integrated in the final domain. Listing 2.122 illustrates those changes in router *Livorno*, which has an RSVP-TE LSP full mesh towards all systems in the previous domain "Cyclone", that remains stable upon these modifications.

Listing 2.122: OSPFv2 final removal in Livorno

```
1  user@Livorno> monitor start messages
2  user@Livorno> monitor start rsvp-transition
3  user@Livorno> monitor start mpls-transition
4
5   [edit protocols mpls]
6  user@Livorno# show
7  traceoptions {
8      file mpls-transition;
9      flag state;
10     flag error;
11 }
12 label-switched-path Livorno-to-Skopie {
13     to 192.168.1.2;
14 }
15 label-switched-path Livorno-to-Barcelona {
16     to 192.168.1.7;
17 }
18 label-switched-path Livorno-to-Honolulu {
19     to 192.168.1.21;
20 }
21 interface fxp0.0 {
22     disable;
23 }
24 interface so-0/0/0.0;
25 interface ge-0/2/0.0;
26 interface ge-1/3/0.0;
27
28  [edit protocols rsvp]
29 user@Livorno# show
30 traceoptions {
31     file rsvp-transition;
32     flag error;
33     flag event;
34     flag pathtear;
35     flag state;
36 }
37 interface fxp0.0 {
38     disable;
39 }
40 interface so-0/0/0.0 {
41     link-protection;
42 }
43 interface ge-0/2/0.0 {
44     link-protection;
45 }
46 interface ge-1/3/0.0 {
47     link-protection;
48 }
49
```

```
50 [edit]
51 user@Livorno# delete protocols ospf
52 [edit]
53 user@Livorno# commit and-quit
54 *** mpls-transition ***
55 Apr  6 01:09:53.834589 TED free LINK Livorno.00(192.168.1.6)->0000.0000.0000.00(172.16.1.53); Local
       Node: 172.16.1.54(192.168.1.6:0) Remote Node: 0.0.0.0(172.16.1.53:0)
56 Apr  6 01:09:53.835064 TED free LINK Livorno.00(192.168.1.6)->0000.0000.0000.00(172.16.1.90); Local
       Node: 172.16.1.89(192.168.1.6:0) Remote Node: 0.0.0.0(172.16.1.90:0)
57 Apr  6 01:09:53.835119 TED free LINK Barcelona-re0.00(192.168.1.7)->0000.0000.0000.00(172.16.1.105);
       Local Node: 172.16.1.106(192.168.1.7:0) Remote Node: 0.0.0.0(172.16.1.105:0)
58 Apr  6 01:09:53.835168 TED free LINK Barcelona-re0.00(192.168.1.7)->0000.0000.0000.00(172.16.1.53);
       Local Node: 172.16.1.53(192.168.1.7:0) Remote Node: 0.0.0.0(172.16.1.53:0)
59 Apr  6 01:09:53.835222 TED free LINK honolulu-re0.00(192.168.1.21)->0000.0000.0000.00(172.16.1.105);
       Local Node: 172.16.1.105(192.168.1.21:0) Remote Node: 0.0.0.0(172.16.1.105:0)
60 <...>
61 *** messages ***
62 Apr  6 01:09:53.833  Livorno rpd[1102]: RPD_OSPF_NBRDOWN: OSPF neighbor 172.16.1.53 (realm ospf-v2 ge
       -0/2/0.0 area 0.0.0.0) state changed from Full to Down due to KillNbr (event reason: interface
       went down)
63 Apr  6 01:09:53.833  Livorno rpd[1102]: RPD_OSPF_NBRDOWN: OSPF neighbor 172.16.1.90 (realm ospf-v2 ge
       -1/3/0.0 area 0.0.0.0) state changed from Full to Down due to KillNbr (event reason: interface
       went down)
64 Apr  6 01:09:53.834  Livorno rpd[1102]: RPD_OSPF_NBRDOWN: OSPF neighbor 172.16.1.33 (realm ospf-v2 so
       -0/0/0.0 area 0.0.0.0) state changed from Full to Down due to KillNbr (event reason: interface
       went down)
```

Junos Tip: Configuration of `no-neighbor-down-notification`

The `no-neighbor-down-notification` configuration option needs to be issued not only at the RSVP-TE LSP ingress point, but also at all transit and end devices for an RSVP-TE LSP. Failing to do so will trigger the corresponding notification to RSVP as well.

In the previous case, if router *Barcelona* is not configured with this knob, when deleting OSPFv2 in router *Livorno*, no immediate LSP teardown occurs, but when the former OSPFv2 neighbor relationship expires on router *Barcelona*, it will issue the corresponding RSVP PathErr message towards ingress as per Listing 2.123.

Listing 2.123: PathErr generation upon incomplete no-neighbor-down-notification configuration

```
1 Apr  6 01:10:38.842  Livorno rpd[1102]: RPD_RSVP_NBRDOWN: RSVP neighbor 172.16.1.53 down on
      interface ge-0/2/0.0, neighbor seq number change
2 Apr  6 01:10:38.844  Livorno rpd[1102]: RPD_MPLS_PATH_DOWN: MPLS path  down on LSP Livorno-to-
      Barcelona
3 Apr  6 01:10:38.846  Livorno rpd[1102]: RPD_MPLS_LSP_DOWN: MPLS LSP Livorno-to-Barcelona down on
      primary()
4 Apr  6 01:10:38.862  Livorno rpd[1102]: RPD_MPLS_PATH_UP: MPLS path  up on LSP Livorno-to-
      Barcelona path bandwidth 0 bps
5 <...>
6 *** mpls-transition ***
7 Apr  6 01:10:38.844201 mpls lsp Livorno-to-Barcelona primary 172.16.1.54: Down
8 Apr  6 01:10:38.844264 RPD_MPLS_PATH_DOWN: MPLS path  down on LSP Livorno-to-Barcelona
9 Apr  6 01:10:38.844388 tag_path_status called for path Livorno-to-Barcelona(primary ) PSB code 2:0
10 Apr  6 01:10:38.844420 Receive PathErr from 0.0.0.0, can't locate sender node in database
11 Apr  6 01:10:38.844442 mpls lsp Livorno-to-Barcelona primary Session preempted
12 Apr  6 01:10:38.846095 mpls lsp Livorno-to-Barcelona primary Deselected as active
13 Apr  6 01:10:38.846210 RPD_MPLS_LSP_DOWN: MPLS LSP Livorno-to-Barcelona down on primary()
14 Apr  6 01:10:38.862826 tag_path_status called for path Livorno-to-Barcelona(primary ) PSB code 5:0
15 Apr  6 01:10:38.862870 mpls lsp Livorno-to-Barcelona primary Up
16 Apr  6 01:10:38.862969 RPD_MPLS_PATH_UP: MPLS path  up on LSP Livorno-to-Barcelona path bandwidth
      0 bps
17 <...>
```

```
18 *** rsvp-transition ***
19 Apr  6 01:10:38.842533 RSVP Nbr 172.16.1.53 (ge-0/2/0.0): state changed [Up to Down]
20 Apr  6 01:10:38.842602 RPD_RSVP_NBRDOWN: RSVP neighbor 172.16.1.53 down on interface ge-0/2/0.0,
        neighbor seq number change
21 Apr  6 01:10:38.842627 RSVP Nbr 172.16.1.53 (ge-0/2/0.0): state changed [Down to Restarting]
22 Apr  6 01:10:38.842648 RSVP Nbr 172.16.1.53 (ge-0/2/0.0): state changed [Restarting to Dead]
23 Apr  6 01:10:38.844321 RSVP delete resv state, session 192.168.1.7(port/tunnel ID 51927 Ext-ID
        192.168.1.6) Proto 0
24 Apr  6 01:10:38.844368 RSVP PathErr to client
```

This transition is depicted in Figure 2.26, in which router *Skopie* is fully integrated and the rest of the systems are pending migration.

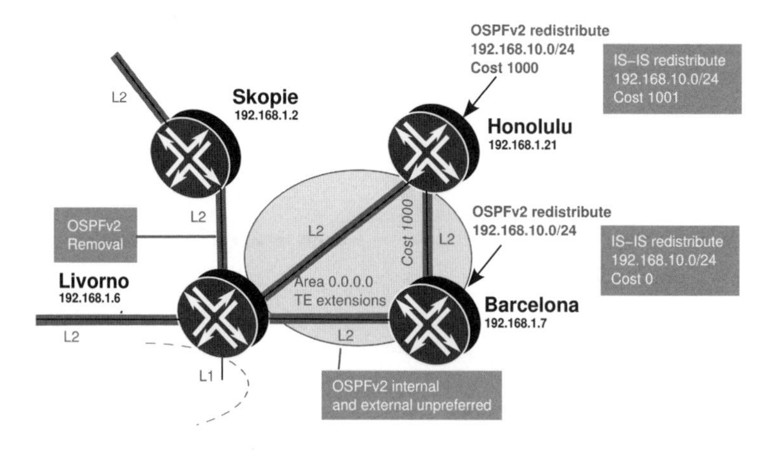

Figure 2.26: Stage seven: OSPFv2 dismantling in router *Skopie* after IS–IS L2 preference activation.

2.3.11 Stage eight: Address renumbering and OSPFv2 replacement with IS–IS in domain "Monsoon"

The renumbering exercise throughout domain "Cyclone" is inherently intrusive in the sense that router identifiers and WAN addressing need to be adapted to the global "Gale Internet" scheme.

Although obviated in this migration case study, renumbering addresses in a router is an entirely invasive operation. Not only are IGPs affected, but router management also needs to be completely adapted for protocols. Applications such as NTP, SNMP, and syslog, and other related protocols, such as BGP and RSVP-TE, may need to change next-hop settings, session peers, or LSP destinations, not only for the local system being migrated, but also for other protocol peers that undergo this transition. Likewise, access lists or firewall filters that may be applied to different interfaces or resources need to be addressed in parallel for this prefix substitution.

The guidelines in this stage are designed to minimize downtime during the gradual transition and to grant redundancy and resilience for internal systems, while expanding IS–IS with legitimate and final "Gale Internet" addresses.

When carrying out this exercise in each system, the following actions are considered:

- Renumber router identifier and WAN addresses towards the final "Gale Internet" domain.
- Retain OSPFv2 neighbor relationships to non-migrated systems, injecting only old non-legitimate loopback addresses in the shrinking old domain.
- Activate new IS–IS adjacencies in the corresponding level, but inject legitimate and valid loopback addresses to both new and old domains.
- Internal routers will be IS–IS L1 systems only and will install a similar default route following ATTached bit setting.
- Redistribute legitimate loopback routes from OSPFv2 internal systems up to IS–IS.
- Leak legitimate loopback routes and 192.168.10.0/24 from IS–IS down to remaining OSPFv2 internal systems. This includes leaking from L1L2 systems to L1 and migrated L1 systems to OSPFv2.

At this stage, each internal system has native connectivity to the final domain and full management integration should occur. Once this has happened, migration tactics are based on moving this borderline one step ahead inside the shrinking OSPFv2 cloud and performing a similar integration exercise in the next router from domain "Monsoon".

Our final objective is to replicate in the end scenario the same routing decisions and forwarding paths that were previously derived from OSPFv2 costs and route injection at ABRs, while at the same time considering that router *Basel* and router *Inverness* must remain as internal L1 systems from the same Area 49.0003. Optimal routing to loopback addresses from the complete domain should be achieved.

This condition requires that one of the border systems first be renumbered and converted into a fully IS–IS integrated system, while the remaining router acts as the gateway for OSPFv2 route injection. Internal systems are then progressively migrated until OSPFv2 completely disappears.

Several approaches for considering the order and sequence of this system migration exist. Because router *Nantes* is closer to the backbone in terms of metrics than router *Torino* and the main global premise is to ensure stability, the considered migration order is: router *Torino* → router *Basel* → router *Inverness* → router *Nantes*, with a conservative approach: keep the currently preferred exit point for most flows for systems still to be migrated.

On the other hand, the premise is to keep OSPFv2 on those systems with an interface to non-migrated routers. This requires another policy to inject those active routes now via IS–IS Level 1 into OSPFv2, instead of the previous policy version from L1L2 systems.

Protocol preference also needs careful thought. Although route leaking between IS–IS and OSPFv2 is crafted with route filters, a given border system should not prefer externally injected loopback addresses in OSPFv2 over IS–IS Level 1 or Level 2 internal, and legitimate loopback addresses still native to the OSPFv2 domain should persist as active paths against redistributed paths in IS–IS Level 1 or Level 2 internal with the `wide-metrics-only`.

The same criteria apply for default route installation, although in this case it is worth mentioning that the default route is injected into NSSA 0.0.0.3 in the form of a Type 3 Summary LSA and hence remains internal for route selection purposes (if it were injected as Type 7 NSSA AS External LSA, some extra adjustment would be needed).

To summarize, default protocol preference values lead to our goal, and no further tweaks are needed here.

However, default route injection needs to be engineered carefully. By default, Junos OS performs *active backbone detection* when generating default route LSAs into *totally stubby areas* or *totally stubby NSSAs*, as per [RFC3509]. If OSPFv2 at router *Torino* is completely dismantled at this stage, router *Nantes* lacks any kind of active OSPFv2 neighbor relationship in the backbone area, which means that it becomes *detached* from the backbone. While one could argue whether the current default route is really necessary, after performing more specific prefix leaking, IS–IS L1 installs a default route based on detection of the ATTached bit. Two design options are possible to maintain native default route propagation inside OSPFv2 during the migration:

- Do not dismantle OSPFv2 between both router *Torino* and router *Livorno*.
- Use the Junos OS *hidden* knob `no-active-backbone` to override *active backbone detection* in OSPF.

While the first alternative could suffice, this OSPFv2 adjacency would be useless because there is no OSPFv2 meaningful routing information beyond the adjacency.

Using `no-active-backbone` in router *Nantes* provides an elegant workaround to circumvent the transient situation, because backbone attachment is purely decided based on Area 0.0.0.0 configuration and not on an active neighbor. Notice that the `no-active-backbone` *hidden* Junos OS configuration command is considered unsupported and should be used with care.

At this point, it is considered a better approach to previously prepare router *Nantes* for unilateral default route injection with `no-active-backbone`, while fully migrating router *Torino* without further headaches, because this is only a temporary stage. This modification is shown in Listing 2.124.

Listing 2.124: Disabling OSPFv2 active backbone detection in Nantes

```
1 [edit protocols ospf]
2 user@Nantes# set no-active-backbone
```

Router *Torino* is the first gateway to be fully integrated into IS–IS in an intrusive fashion, as a result of adopting the router identifier for the public "Gale Internet" legitimate loopback address. To avoid any kind of transient disruption, the IS–IS Overload bit and a similar OSPF *overload* implementation in Junos OS are set in the previous stage and removed after the configuration changes. Although our topology is quite simple, it is always considered a best practice to detour any kind of remaining link-state IGP-driven transit flows during an intrusive operation. To simplify the standard IS–IS Level 2 to Level 1 loopback address policy, the legitimate "Gale Internet" loopback address is set as `passive` in Level 1 for optimal reachability. These configuration changes in router *Torino* are shown in Listing 2.125.

Listing 2.125: Complete integration into IS–IS final domain in Torino

```
[edit]
user@Torino# set protocols isis overload
[edit]
user@Torino# set protocols ospf overload
[edit]
user@Torino# commit
commit complete
[edit]
user@Torino# set routing-options router-id 192.168.1.3
[edit]
user@Torino# set protocols isis interface at-1/2/0.1001 level 2 disable
[edit]
user@Torino# delete interfaces at-1/2/0 unit 1001 family inet address 10.10.1.81/30
[edit]
user@Torino# set interfaces at-1/2/0 unit 1001 family inet address 172.16.1.81/30
[edit]
user@Torino# set interfaces at-1/2/0 unit 1001 family iso
[edit]
user@Torino# set protocols isis interface lo0.0 level 1 passive
[edit]
user@Torino# delete protocols isis export from-ospf-to-domain
[edit]
user@Torino# set protocols isis export leak-loopbacks-to-l1
[edit]
user@Torino# delete protocols ospf
[edit]
user@Torino# edit policy-options policy-statement leak-loopbacks-to-l1
[edit policy-options policy-statement leak-loopbacks-to-l1]
user@Torino# show
from {
    protocol isis;
    level 2;
    route-filter 192.168.1.0/24 prefix-length-range /32-/32;
    route-filter 192.168.10.0/24 exact;
}
to level 1;
then {
    tag 400;
    accept;
}
[edit]
user@Torino# commit
commit complete
[edit]
user@Torino# delete protocols isis overload
[edit]
user@Torino# commit and-quit
commit complete
Exiting configuration mode
```

Junos Tip: Link-state IGP overload bit utilization for intrusive works

The overload bit was originally defined in IS–IS to represent a specific situation in which a router was running out of fragments and could not properly forward any transit traffic. This function has traditionally been adopted by many network operators as a smooth method to avoid transit traffic disruptions in redundant networks by gracefully raising this flag to the rest of the domain to let other systems recompute alternative IGP paths in a very elegant fashion.

This concept was *inherited* by many operating systems for OSPF, Junos OS OSPF among them, by maximizing the attached link metric after configuring the similar knob.

Some points need clarification in this case:

- The Overload bit is a hard-coded setting in IS–IS. Routers that are single-homed in IS–IS to a system with the Overload bit, become isolated at that moment with an effective blackhole. The Junos OS extension `advertise-high-metrics` does not hard-code the Overload bit in the LSP header but maximizes IS neighbor and transit link prefix metrics to achieve a similar effect, while still allowing single-homed setups to remain connected, if any exist.
- The `overload` bit and derivatives in Junos OS act purely from a link-state IGP transit perspective. Other kinds of routes (such as BGP) remain reachable with the same criteria and the knob does not stop forwarding alone by itself for those destinations.

Once router *Torino* has been fully migrated to IS–IS, Listing 2.126 shows the intermediate stage where the frontier between OSPFv2 domain "Monsoon" and the final IS–IS domain has been moved down to router *Basel*. The same `overload` bit settings are in effect during the transition. Also, a particular IS–IS leaking policy will be newly defined to match Level 1 routes tailored for the known ranges to be redistributed into OSPFv2, together with tailored loopback address injection in each domain. This situation is illustrated in Figure 2.27.

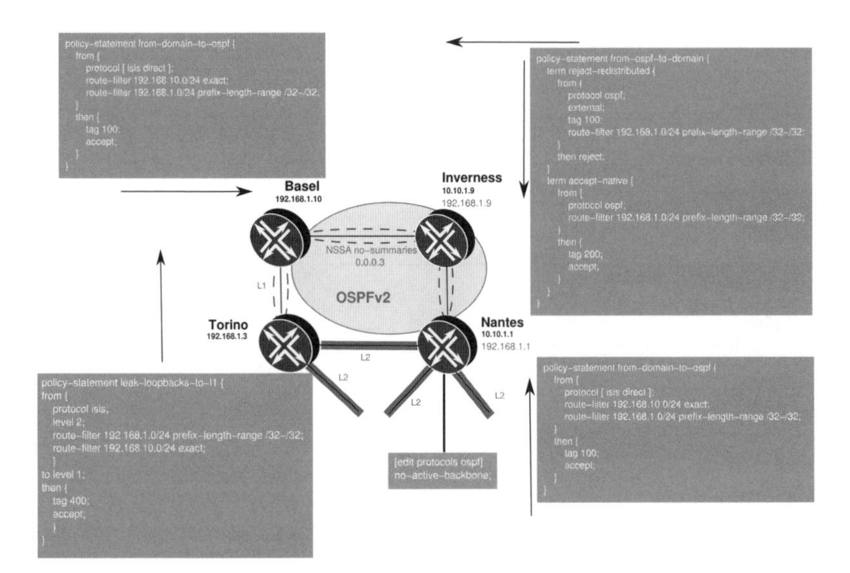

Figure 2.27: Stage eight: IS–IS L1 migration at router *Basel* and mutual redistribution with OSPFv2.

Listing 2.126: IS–IS integration and mutual protocol redistribution in Basel

```
 1 [edit]
 2 user@Basel# set protocols ospf overload
 3 [edit]
 4 user@Basel# commit
 5 [edit policy-options policy-statement block-non-legitimate-loopback]
 6 user@Basel# show
 7 from {
 8     protocol direct;
 9     route-filter 10.10.1.10/32 exact; # Avoid own old loopback address
10 }
11 then reject;
12 [edit policy-options policy-statement from-ospf-to-domain]
13 user@Basel# show
14 term reject-redistributed {
15     from {
16         protocol ospf;
17         external;
18         tag 100;
19         route-filter 192.168.1.0/24 prefix-length-range /32-/32;
20     }
21     then reject; # Avoid already redistributed address to be injected back
22 }
23 term accept-native {
24     from {
25         protocol ospf;
26         route-filter 192.168.1.0/24 prefix-length-range /32-/32;
27     }
28     then {
29         tag 200;
30         accept; # Redistribute native legitimate loopback addresses from domain
31     }
32 }
33 [edit policy-options policy-statement from-domain-to-ospf]
34 user@Basel# show
35 from {
36     protocol [ isis direct ];
37     route-filter 192.168.10.0/24 exact; # Remote domain prefix
38     route-filter 192.168.1.0/24 prefix-length-range /32-/32; # "Gale Internet" loopback address range
39 }
40 then {
41     tag 100;
42     accept;
43 }
44
45 [edit] # Renumber system and migrate to IS-IS
46 user@Basel# set routing-options router-id 192.168.1.10
47 [edit]
48 user@Basel# delete protocols ospf area 0.0.0.3 interface lo0.0
49 [edit]
50 user@Basel# delete protocols ospf area 0.0.0.3 interface at-0/0/0.1001
51 [edit protocols ospf]
52 user@Basel# set area 0.0.0.3 interface 10.10.1.10
53 [edit protocols ospf]
54 user@Basel# set export from-domain-to-ospf
55 [edit]
56 user@Basel# set overload
57 [edit]
58 user@Basel# set protocols isis level 2 disable
59 [edit]
60 user@Basel# set protocols isis level 1 authentication-type md5
61 [edit]
62 user@Basel# set protocols isis level 1 authentication-key "Aeris"
63 [edit]
64 user@Basel# set protocols isis interface lo0.0 level 1 passive
65 [edit]
66 user@Basel# set protocols isis interface at-0/0/0.1001
67 [edit]
```

```
68 user@Basel# set protocols isis level 1 wide-metrics-only
69 [edit protocols isis]
70 user@Basel# set export [ block-non-legitimate-loopback from-ospf-to-domain ]
71 [edit]
72 user@Basel# set protocols isis reference-bandwidth 100g
73 [edit] # Migrate previous link to outer world from OSPF to IS-IS
74 user@Basel# delete interfaces at-0/0/0.1001 family inet address 10.10.1.82/30
75 [edit]
76 user@Basel# set interfaces at-0/0/0.1001 family inet address 172.16.1.82/30
77 [edit]
78 user@Basel# set interfaces at-0/0/0.1001 family iso
79 [edit] # Add BCD-coded NET-Id
80 user@Basel# set interfaces lo0 unit 0 family iso address 49.0003.1921.6800.1010.00
81 [edit]
82 user@Basel# show protocols isis
83 export [ block-non-legitimate-loopback from-ospf-to-domain ];
84 reference-bandwidth 100g;
85 overload;
86 level 2 disable;
87 level 1 {
88     authentication-key "$9$I7/EhyMWx-b2cyYo"; ### SECRET-DATA
89     authentication-type md5;
90     wide-metrics-only;
91 }
92 interface at-0/0/0.1001;
93 interface lo0.0 {
94     level 1 passive;
95 }
96 [edit]
97 user@Basel# show protocols ospf
98 overload;
99 export from-domain-to-ospf;
100 reference-bandwidth 100g;
101 area 0.0.0.3 {
102     nssa no-summaries;
103     interface at-0/1/0.1001;
104     interface 10.10.1.10;
105 }
106 [edit]
107 user@Basel# commit
108 commit complete
109 [edit] # Allow back transit traffic after integration
110 user@Basel# delete protocols ospf overload
111 [edit]
112 user@Basel# delete protocols isis overload
113 [edit]
114 user@Basel# commit and-quit
115 commit complete
116 Exiting configuration mode
117
118 user@Basel> show isis adjacency
119 Interface             System         L State        Hold (secs) SNPA
120 at-0/0/0.1001         Torino         1 Up                    20
121
122 user@Basel> show ospf neighbor
123 Address          Interface              State     ID               Pri  Dead
124 10.10.1.73       at-0/1/0.1001          Full      10.10.1.9        128    34
125
126 user@Basel> show ospf database detail router lsa-id 192.168.1.10
127
128     OSPF database, Area 0.0.0.3
129  Type       ID               Adv Rtr           Seq      Age  Opt  Cksum  Len
130 Router  *192.168.1.10     192.168.1.10     0x80000007    51  0x20 0xeb38  60
131   bits 0x2, link count 3
132   id 10.10.1.9, data 10.10.1.74, Type PointToPoint (1)
133     Topology count: 0, Default metric: 643
134   id 10.10.1.72, data 255.255.255.252, Type Stub (3)
135     Topology count: 0, Default metric: 643
136   id 10.10.1.10, data 255.255.255.255, Type Stub (3)  # Router ID not as Stub route
```

```
137     Topology count: 0, Default metric: 0
138
139 user@Basel> show ospf database
140
141     OSPF database, Area 0.0.0.3
142  Type       ID              Adv Rtr          Seq      Age  Opt  Cksum  Len
143 Router   10.10.1.1        10.10.1.1        0x80000008  2089  0x20 0xe33b  48
144 Router   10.10.1.9        10.10.1.9        0x80000019  2500  0x20 0x13ec  96
145 Router   10.10.1.10       10.10.1.10       0x80000018  3538  0x20 0x7d30  84
146 Router  *192.168.1.10     192.168.1.10     0x80000006  1708  0x20 0xed37  60
147 Summary  0.0.0.0          10.10.1.1        0x80000001  2089  0x20 0x54ca  28  # Default route from
        Nantes
148 NSSA     192.168.1.1      10.10.1.1        0x80000008  1322  0x20 0x8540  36
149 NSSA    *192.168.1.1      192.168.1.10     0x80000002  1445  0x28 0xacee  36
150 NSSA     192.168.1.3      10.10.1.1        0x80000007  1422  0x20 0xb96a  36
151 NSSA    *192.168.1.3      192.168.1.10     0x80000002  1198  0x28 0x52e7  36
152 NSSA     192.168.1.4      10.10.1.1        0x80000007   623  0x20 0x8a4f  36
153 NSSA    *192.168.1.4      192.168.1.10     0x80000002   951  0x28 0xaeff  36
154 <...>
155
156 user@Basel> show isis database
157 IS-IS level 1 link-state database:
158 LSP ID                      Sequence Checksum Lifetime Attributes
159 Torino.00-00                    0x54   0x7aab      454 L1 L2 Attached
160 Basel.00-00                      0x3   0x3393      513 L1
161   2 LSPs
162
163 IS-IS level 2 link-state database:
164   0 LSPs
165
166 user@Basel> show isis database detail
167 IS-IS level 1 link-state database:
168
169 Torino.00-00 Sequence: 0x54, Checksum: 0x7aab, Lifetime: 452 secs
170    IS neighbor: Basel.00                     Metric:      160
171    IP prefix: 172.16.1.80/30                 Metric:      160 Internal Up
172    IP prefix: 192.168.1.1/32                 Metric:      160 Internal Down
173    IP prefix: 192.168.1.3/32                 Metric:        0 Internal Up
174    IP prefix: 192.168.1.4/32                 Metric:      903 Internal Down
175    IP prefix: 192.168.1.6/32                 Metric:     1546 Internal Down
176    IP prefix: 192.168.1.7/32                 Metric:     1556 Internal Down
177    IP prefix: 192.168.1.8/32                 Metric:      260 Internal Down
178    IP prefix: 192.168.1.20/32                Metric:      903 Internal Down
179    IP prefix: 192.168.1.21/32                Metric:     1556 Internal Down
180    IP prefix: 192.168.10.0/24                Metric:     1556 Internal Down
181
182 Basel.00-00 Sequence: 0x3, Checksum: 0x3393, Lifetime: 510 secs
183    IS neighbor: Torino.00                    Metric:       10
184    IP prefix: 172.16.1.80/30                 Metric:       10 Internal Up
185    IP prefix: 192.168.1.9/32                 Metric:      643 Internal Up
186    IP prefix: 192.168.1.10/32                Metric:        0 Internal Up
187
188 IS-IS level 2 link-state database:
189
190 user@Basel> show route protocol ospf active-path table inet.0
191
192 inet.0: 27 destinations, 39 routes (27 active, 0 holddown, 0 hidden)
193 + = Active Route, - = Last Active, * = Both
194
195 0.0.0.0/0          *[OSPF/10] 00:35:05, metric 1296 # Active route preferred from Nantes
196                     > via at-0/1/0.1001
197 10.10.1.9/32       *[OSPF/10] 00:41:45, metric 643
198                     > via at-0/1/0.1001
199 10.10.1.64/30      *[OSPF/10] 00:41:45, metric 1286
200                     > via at-0/1/0.1001
201 192.168.1.9/32     *[OSPF/10] 00:41:45, metric 643
202                     > via at-0/1/0.1001
203 224.0.0.5/32       *[OSPF/10] 15:22:34, metric 1
204                       MultiRecv
```

```
205
206 user@Basel> show route protocol isis active-path table inet.0
207
208 inet.0: 27 destinations, 39 routes (27 active, 0 holddown, 0 hidden)
209 + = Active Route, - = Last Active, * = Both
210
211 # External routes preferred via IS-IS
212
213 192.168.1.1/32      *[IS-IS/18] 00:39:26, metric 170, tag 400
214                      > to 172.16.1.81 via at-0/0/0.1001
215 192.168.1.3/32      *[IS-IS/15] 00:39:26, metric 10
216                      > to 172.16.1.81 via at-0/0/0.1001
217 192.168.1.4/32      *[IS-IS/18] 00:39:26, metric 913, tag 400
218                      > to 172.16.1.81 via at-0/0/0.1001
219 192.168.1.6/32      *[IS-IS/18] 00:39:26, metric 1556, tag 400
220                      > to 172.16.1.81 via at-0/0/0.1001
221 192.168.1.7/32      *[IS-IS/18] 00:39:26, metric 1566, tag 400
222                      > to 172.16.1.81 via at-0/0/0.1001
223 192.168.1.8/32      *[IS-IS/18] 00:39:26, metric 270, tag 400
224                      > to 172.16.1.81 via at-0/0/0.1001
225 192.168.1.20/32     *[IS-IS/18] 00:39:26, metric 913, tag 400
226                      > to 172.16.1.81 via at-0/0/0.1001
227 192.168.1.21/32     *[IS-IS/18] 00:39:26, metric 1566, tag 400
228                      > to 172.16.1.81 via at-0/0/0.1001
229 192.168.10.0/24     *[IS-IS/18] 00:39:26, metric 1566, tag 400
230                      > to 172.16.1.81 via at-0/0/0.1001
```

Note that LSAs injected by router *Basel* include the legitimate "Gale Internet" loopback address as Advertising Router, which is not injected as part of its Type 1 Router LSA but as another separate Type 7 NSSA AS External LSA, by means of the OSPFv2 `export` policy at router *Basel*.

While it could be argued that a valid transient situation could be built up with Type 7 NSSA AS External LSAs relying on other Type 7 NSSA AS External LSAs for reachability, our selective loopback injection has avoided that discussion, as shown in Listing 2.127 from router *Inverness*' perspective, because the Forwarding Address field is properly filled with the internal loopback address as per [RFC2328] (Line 12) and such Type 7 NSSA AS External LSAs remain eligible for forwarding, independently of the situation of the Advertising Router Type 7 NSSA AS External LSAs.

Listing 2.127: Route selection in Inverness during IS–IS transition

```
1  user@Inverness> show ospf database nssa lsa-id 192.168.1.1 detail
2
3      OSPF database, Area 0.0.0.3
4   Type       ID               Adv Rtr           Seq      Age  Opt  Cksum  Len
5  NSSA     192.168.1.1      10.10.1.1        0x80000009   363  0x20 0x8341  36
6    mask 255.255.255.255
7    Topology default (ID 0)
8      Type: 2, Metric: 0, Fwd addr: 0.0.0.0, Tag: 0.0.0.100
9  NSSA     192.168.1.1      192.168.1.10     0x80000003   363  0x28 0xaaef  36
10   mask 255.255.255.255
11   Topology default (ID 0)
12     Type: 2, Metric: 170, Fwd addr: 10.10.1.10, Tag: 0.0.0.100
13                                # Forwarding Address pointing to internal valid loopback
14
15 user@Inverness> show route protocol ospf table inet.0
16
17 inet.0: 32 destinations, 34 routes (32 active, 0 holddown, 0 hidden)
18 + = Active Route, - = Last Active, * = Both
19
20 0.0.0.0/0          *[OSPF/10] 00:42:31, metric 653
21                     > via so-0/0/1.0
22 10.10.1.10/32      *[OSPF/10] 15:26:50, metric 643
```

```
23                          > via at-0/1/0.1001
24  10.10.1.64/30           [OSPF/10] 04:20:21, metric 643
25                          > via so-0/0/1.0
26  10.10.1.72/30           [OSPF/10] 15:26:50, metric 643
27                          > via at-0/1/0.1001
28  192.168.1.1/32         *[OSPF/150] 04:20:20, metric 0, tag 100
29                          > via so-0/0/1.0
30  192.168.1.3/32         *[OSPF/150] 00:46:45, metric 10, tag 100
31                          > via at-0/1/0.1001
32  192.168.1.4/32         *[OSPF/150] 04:20:05, metric 743, tag 100
33                          > via so-0/0/1.0
34  192.168.1.6/32         *[OSPF/150] 04:19:57, metric 1386, tag 100
35                          > via so-0/0/1.0
36  192.168.1.7/32         *[OSPF/150] 04:19:56, metric 1396, tag 100
37                          > via so-0/0/1.0
38  192.168.1.8/32         *[OSPF/150] 04:20:05, metric 100, tag 100
39                          > via so-0/0/1.0
40  192.168.1.10/32        *[OSPF/150] 00:49:11, metric 0, tag 100
41                          > via at-0/1/0.1001
42  192.168.1.20/32        *[OSPF/150] 04:20:05, metric 743, tag 100
43                          > via so-0/0/1.0
44  192.168.1.21/32        *[OSPF/150] 04:19:51, metric 1396, tag 100
45                          > via so-0/0/1.0
46  192.168.10.0/24        *[OSPF/150] 04:19:56, metric 1396, tag 100
47                          > via so-0/0/1.0
48  224.0.0.5/32           *[OSPF/10] 15:26:55, metric 1
49                            MultiRecv
```

Junos Tip: Leaking L1 routes with wide-metrics-only

The indirect effect of `wide-metrics-only` in our topology is that all redistributed routes from domain "Monsoon" OSPFv2 into IS–IS Level 1 are considered to be *internal*. This means that default route population policies across IS–IS hierarchies suffice to be able to flood them to the rest of the network.

In case we did not have `wide-metrics-only` for IS–IS Level 1, an additional route-leaking policy would be needed at L1L2 systems, router *Torino* in this intermediate case, to override default behavior and flood loopbacks across the complete setup.

Similar steps are taken until domain "Monsoon" is fully integrated into the final IS–IS domain. Progress can be easily identified by monitoring the remaining active OSPFv2 routes on router *Nantes*, as per Listing 2.128 when only the remaining link to router *Inverness* was in OSPFv2 NSSA 0.0.0.3 (the OSPFv2 backbone connection in Area 0.0.0.0 to router *Torino* was previously dismantled).

Listing 2.128: Latest OSPFv2 remaining information in Nantes prior to dismantling

```
1  user@Nantes> show ospf database
2
3      OSPF database, Area 0.0.0.0 # Area 0 only includes own LSAs
4   Type       ID               Adv Rtr           Seq      Age  Opt  Cksum  Len
5  Router  *10.10.1.1        10.10.1.1        0x8000000e    47  0x22 0x922b  48
6  Summary *10.10.1.9        10.10.1.1        0x80000004    45  0x22 0xa0df  28
7  Summary *10.10.1.64       10.10.1.1        0x80000002    47  0x22 0x6ae3  28
8  ASBRSum *192.168.1.9      10.10.1.1        0x80000004    45  0x22 0xdb4e  28
9
10     OSPF database, Area 0.0.0.3 # Area 3 just includes own and Inverness' LSAs from legitimate
       loopback
11  Type       ID               Adv Rtr           Seq      Age  Opt  Cksum  Len
12 Router  *10.10.1.1        10.10.1.1        0x80000010    47  0x20 0xc7f9  48
13 Router   192.168.1.9      192.168.1.9      0x8000000a    28  0x20 0x7147  72
```

```
14 Summary *0.0.0.0         10.10.1.1       0x80000005  47  0x20 0x4cce  28
15 NSSA    *192.168.1.1     10.10.1.1       0x8000000e  47  0x20 0x7946  36
16 NSSA     192.168.1.1     192.168.1.9     0x80000002  48  0x28 0x52eb  36
17 NSSA    *192.168.1.3     10.10.1.1       0x8000000d  47  0x20 0xad70  36
18 NSSA     192.168.1.3     192.168.1.9     0x80000002  48  0x28 0xf7e4  36
19 <...>
20 NSSA    *192.168.1.21    10.10.1.1       0x8000000b  47  0x20 0x70c4  36
21 NSSA     192.168.1.21    192.168.1.9     0x80000002  48  0x28 0x426e  36
22 NSSA    *192.168.10.0    10.10.1.1       0x8000000b  47  0x20 0xdf61  36
23 NSSA     192.168.10.0    192.168.1.9     0x80000002  48  0x28 0xb10b  36
24     OSPF AS SCOPE link state database
25  Type       ID              Adv Rtr          Seq      Age  Opt  Cksum  Len
26 Extern  *192.168.1.1     10.10.1.1       0x80000013  45  0x22 0x385a  36
27 Extern  *192.168.1.3     10.10.1.1       0x80000013  45  0x22 0xdd53  36
28 Extern  *192.168.1.4     10.10.1.1       0x80000013  45  0x22 0x3a6b  36
29 <...>
30 Extern  *192.168.1.20    10.10.1.1       0x80000012  45  0x22 0x9bfa  36
31 Extern  *192.168.1.21    10.10.1.1       0x80000011  45  0x22 0x2cda  36
32 Extern  *192.168.10.0    10.10.1.1       0x80000011  45  0x22 0x9b77  36
33
34 user@Nantes> show route protocol ospf active-path table inet.0
35
36 inet.0: 45 destinations, 58 routes (45 active, 0 holddown, 0 hidden)
37 + = Active Route, - = Last Active, * = Both
38
39 10.10.1.9/32       *[OSPF/10] 00:00:49, metric 643
40                     > via so-1/0/1.0
41 224.0.0.5/32       *[OSPF/10] 4d 23:49:14, metric 1
42                      MultiRecv
```

At this point, non-legitimate loopback addresses and IS–IS export policies to inject valid loopback addresses or redistribute OSPFv2 can safely be removed without any impact. Note that specific IS–IS Level 2 to Level 1 route-leaking policies need to remain for optimal routing purposes. This is depicted in Figure 2.28.

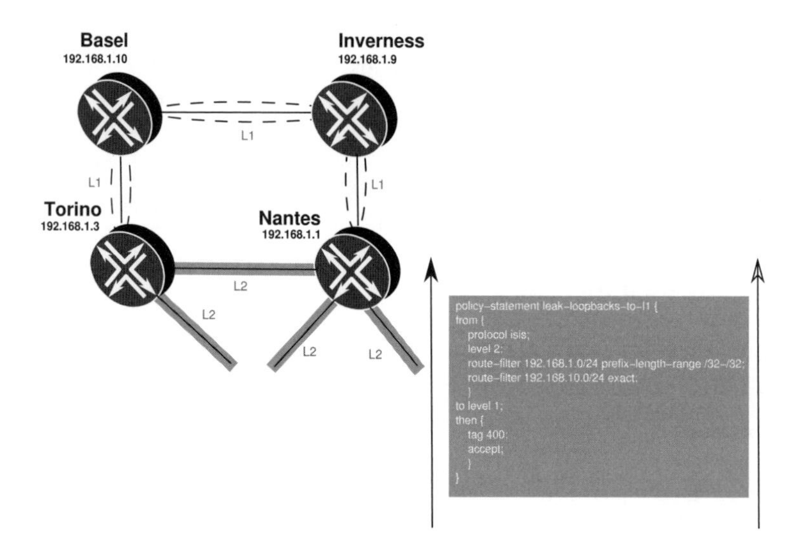

Figure 2.28: Stage eight: final IS–IS status and L2 to L1 leaking at domain “Monsoon”.

2.3.12 Stage nine: IS–IS Level 1 authentication and route-leaking adaption in domain "Mistral" and router *Lille*

This last migration step mainly consists of adapting the IS–IS Level 1 authentication key to the final value *Aeris* and fine-tuning the remaining leaking policy at router *Livorno* to remove OSPFv2 route injection, given that no more OSPFv2 active routes are in place.

Keeping the same rationale in mind, the idea is to perform authentication transition in a smooth manner, which can be achieved by means of the IS–IS `no-authentication-check` knob in Junos OS in the following sequence:

- Activate `no-authentication-check` knob in all participating systems, including router *Lille*.
- Adapt Level 1 authentication key to *Aeris* gradually in each system. Remove specific authentication settings for IIHs on the point-to-point link between router *Havana* and router *Bilbao* for a joint key *Aeris* for both levels.
- Remove `no-authentication-check` in each router, resulting in final key *Aeris* being active the entire domain.

The main drawback to this method is that all systems are exposed to accepting IS–IS unauthenticated information (`no-authentication-check` applies for both levels) during the transition time frame.

Although avoiding authentication checks can be considered as weakness and a potential backdoor for attacks, it is assumed that this will be just a transient period of time. There is no impact when adding or removing `no-authentication-check`, and these three iterations can almost take place consecutively, once sufficient checks are done. Listing 2.129 illustrates how adapting authentication information in the system is harmless as long as `no-authentication-check` remains active (Lines 51, 56, 61, and 66 show a single transition when activating the adjacency, which occurred several hours ago as indicated in the *Last Transition* field).

Listing 2.129: IS–IS Level 1 authentication transition in Havana

```
1  [edit]
2  user@Havana# edit protocols isis
3  [edit protocols isis]
4  user@Havana# show
5  export leak-loopbacks-to-l1;
6  reference-bandwidth 100g;
7  no-authentication-check;
8  level 1 {
9      authentication-key "$9$ZAUH.n6A01hCtMX7NY2k.P"; ### SECRET-DATA
10     authentication-type md5;
11     wide-metrics-only;
12 }
13 level 2 {
14     authentication-key "$9$JjGUiPfzn9pDitO"; ### SECRET-DATA
15     authentication-type md5;
16     wide-metrics-only;
17 }
18 interface fe-0/3/0.0 {
19     level 2 disable;
20 }
21 interface so-1/0/0.0 {
22     level 1 disable;
```

```
23 }
24 interface ge-1/1/0.0 {
25     level 2 disable;
26 }
27 interface at-1/2/0.1001 {
28     level 1 {
29         hello-authentication-key "$9$7W-dwoaUiqfVwPQ"; ### SECRET-DATA
30         hello-authentication-type md5;
31     }
32 }
33 interface lo0.0 {
34     level 1 passive;
35     level 2 passive;
36 }
37
38 [edit protocols isis]
39 user@Havana# set level 1 authentication-key "Aeris"
40 [edit protocols isis]
41 user@Havana# delete interface at-1/2/0.1001 level 1
42 [edit protocols isis]
43 user@Havana# commit and-quit
44
45 # "Last transition field" indicates that adjacencies remained unaffected
46 # Only 1 Up/Down transition for all adjacencies
47
48 user@Havana> show isis adjacency detail
49 Bilbao-re0
50   Interface: at-1/2/0.1001, Level: 3, State: Up, Expires in 21 secs
51   Priority: 0, Up/Down transitions: 1, Last transition: 17:22:06 ago
52   Circuit type: 3, Speaks: IP, IPv6
53 <...>
54 Livorno
55   Interface: fe-0/3/0.0, Level: 1, State: Up, Expires in 19 secs
56   Priority: 64, Up/Down transitions: 1, Last transition: 06:46:16 ago
57   Circuit type: 1, Speaks: IP, IPv6, MAC address: 0:14:f6:85:40:5d
58 <...>
59 male-re0
60   Interface: ge-1/1/0.0, Level: 1, State: Up, Expires in 22 secs
61   Priority: 64, Up/Down transitions: 1, Last transition: 17:17:20 ago
62   Circuit type: 1, Speaks: IP, IPv6, MAC address: 0:19:e2:2b:38:fa
63 <...>
64 Livorno
65   Interface: so-1/0/0.0, Level: 2, State: Up, Expires in 23 secs
66   Priority: 0, Up/Down transitions: 1, Last transition: 17:22:34 ago
67   Circuit type: 2, Speaks: IP, IPv6
68 <...>
```

Once this final authentication key has been deployed in all participating systems, `no-authentication-check` can be safely deleted.

Finally, because no OSPFv2 active routes exist on router *Livorno* any longer, the existing route-leaking policy can be normalized and adapted to inject only Level 2 loopback addresses into IS–IS Level 1 for optimal routing purposes from all internal L1 systems. This is shown in Listing 2.130.

Listing 2.130: Final route-leaking policy adaption at Livorno

```
1 [edit policy-options policy-statement leak-loopbacks-to-l1]
2 lab@Livorno# show
3 term isis {
4     from {
5         protocol isis;
6         level 2;
7         route-filter 192.168.1.0/24 prefix-length-range /32-/32;
8         route-filter 192.168.10.0/24 exact;
9     }
10    to level 1;
```

```
11        then {
12            tag 300;
13            accept;
14        }
15    }
16    term ospf {
17        from {
18            protocol ospf;
19            route-filter 192.168.1.0/24 prefix-length-range /32-/32;
20            route-filter 192.168.10.0/24 exact;
21        }
22        to level 1;
23        then {
24            tag 300;
25            accept;
26        }
27    }
28    [edit policy-options policy-statement leak-loopbacks-to-l1]
29    lab@Livorno# delete term ospf
30    [edit protocols isis]
31    lab@Livorno# delete no-authentication-check
32
33    user@Livorno> show isis database detail Livorno level 1
34    IS-IS level 1 link-state database:
35
36    Livorno.00-00 Sequence: 0x87, Checksum: 0x81b0, Lifetime: 1154 secs
37       IS neighbor: Havana.02                  Metric:     1000
38       IP prefix: 172.16.100.0/24              Metric:     1000 Internal Up
39       IP prefix: 192.168.1.1/32               Metric:     1386 Internal Down
40       IP prefix: 192.168.1.3/32               Metric:     1546 Internal Down
41       IP prefix: 192.168.1.6/32               Metric:        0 Internal Up # Up bit just for own
           loopback
42       IP prefix: 192.168.1.7/32               Metric:       10 Internal Down # Down bit marked now
43       IP prefix: 192.168.1.9/32               Metric:     2029 Internal Down
44       IP prefix: 192.168.1.10/32              Metric:     1706 Internal Down
45       IP prefix: 192.168.1.21/32              Metric:       10 Internal Down # Down bit marked now
46       IP prefix: 192.168.10.0/24              Metric:       10 Internal Down
```

All these changes for the final stage are depicted in Figure 2.29.

2.3.13 Migration summary

The goal of the case study after all migration steps is summarized in Figure 2.30.

While the migration rollout has been lengthy, no customers suffered any impact, except for the compulsory address renumbering exercises in domain "Monsoon", which might have been executed during a maintenance window on a weekend night.

A parallel exercise that has not been addressed directly in this case study is the enforcement of configuration homogeneity across all domains. While some divergences have been analyzed in some stages, such as the adaption of `reference-bandwidth` and link cost manipulation for external links from domain "Cyclone", other important link-state IGP configuration aspects, such as timer values or general protocol statements such as SPF options or multitopology support, are beyond the scope of this case study, because an in-depth evaluation of each configuration knob would be required. However, this standardization effort is particularly relevant in live networks.

Apart from this, all requirements have been satisfied in a thoughtful manner and meeting deadlines. While initial connectivity was rapidly deployed, systems have been progressively integrated in a unique and unified IS–IS domain, including internal L1, L1L2, and pure L2 systems as per the design.

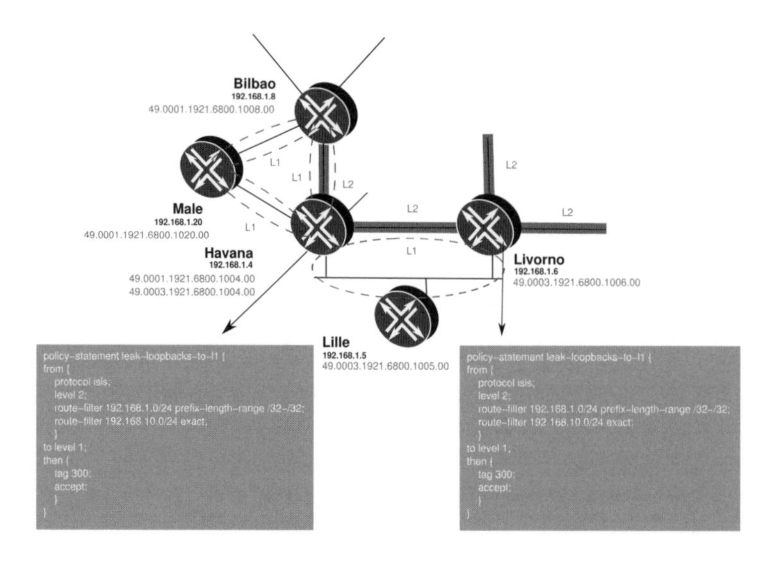

Figure 2.29: Stage nine: final IS–IS status at domain “Mistral” and Lille.

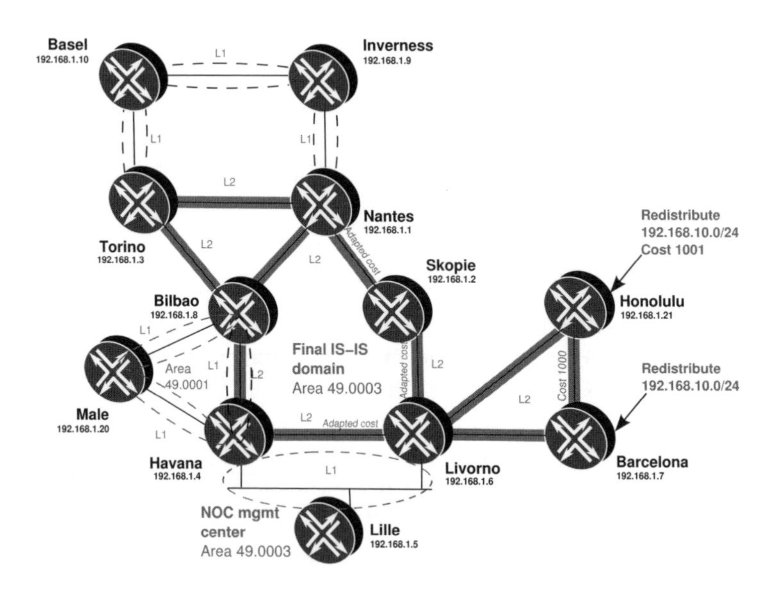

Figure 2.30: Final IS–IS topology including metric adaption and redistribution.

Top management of “Gale Internet” are clearly proud of their network engineers! Challenges like such a sensitive link-state IGP integration are not for the faint hearted. But more important are the instructive lessons learned that one can apply to such real-life migrations. Among other concepts, the following topics have been evaluated, revisited, and put in practice:

- link-state IGP preference adjustment both for internal and external routes;
- external route injection differences when using IS–IS `wide-metrics-only` when compared with standard OSPFv2 rules;
- replication of link metric values in IS–IS and OSPFv2, including bandwidth-derived references;
- IS–IS metric calculation and adjustment to adapt OSPFv2 external Type 2 redistribution design;
- TED protocol credibility deployment for encompassed TED information transition;
- selective loopback address injection in OSPFv2 Type 1 Router LSA and IS–IS LSP;
- definition of multiple leaking policies using administrative tag matching and setting;
- control of mutual protocol redistribution in more than a single leaking point;
- alignment of authentication settings for IS–IS Level 1 and Level 2 in a point-to-point link and tweaking per-level hello-authentication;
- identification of IS–IS ATTached bit setting conditions;
- area stitching in a joint IS–IS Level 1 and respective route selection;
- identification of IS–IS Up/Down bit setting conditions and possible consequences;
- protocol preference and TED credibility transition without affecting RSVP neighbor relationships or LSPs;
- identification of a safeguard mechanism against protocol transition to IS–IS to avoid accidental leaking;
- progressive intrusive migration into an existing domain, including route redistribution and protocol redefinition to mirror previous area definition and forwarding paths;
- replication of standalone default route selection by tweaking OSPF *active backbone detection* during migration;
- utilization of both OSPFv2 and IS–IS *overload* settings for intrusive maintenance works;
- IS–IS per-level authentication changes by means of transient deactivation and progressive rollout of a new MD5 key.

Bibliography

[ISO10589] ISO. ISO/IEC 10589:2002, Intermediate system to intermediate system routing information exchange protocol for use in conjunction with the protocol for providing the connectionless-mode network service (ISO 8473), 2002.

[RFC1131] J. Moy. OSPF specification. RFC 1131 (Proposed Standard), October 1989. Obsoleted by RFC 1247.

[RFC1195] R. W. Callon. Use of OSI IS-IS for routing in TCP/IP and dual environments. RFC 1195 (Proposed Standard), December 1990. Updated by RFCs 1349, 5302, 5304.

[RFC1584] J. Moy. Multicast Extensions to OSPF. RFC 1584 (Historic), March 1994.

[RFC1918] Y. Rekhter, B. Moskowitz, D. Karrenberg, G. J. de Groot, and E. Lear. Address Allocation for Private Internets. RFC 1918 (Best Current Practice), February 1996.

[RFC2328] J. Moy. OSPF Version 2. RFC 2328 (Standard), April 1998.

[RFC2740] R. Coltun, D. Ferguson, and J. Moy. OSPF for IPv6. RFC 2740 (Proposed Standard), December 1999. Obsoleted by RFC 5340.

[RFC2966] T. Li, T. Przygienda, and H. Smit. Domain-wide Prefix Distribution with Two-Level IS-IS. RFC 2966 (Informational), October 2000. Obsoleted by RFC 5302.

[RFC2973] R. Balay, D. Katz, and J. Parker. IS-IS Mesh Groups. RFC 2973 (Informational), October 2000.

[RFC3101] P. Murphy. The OSPF Not-So-Stubby Area (NSSA) Option. RFC 3101 (Proposed Standard), January 2003.

[RFC3107] Y. Rekhter and E. Rosen. Carrying Label Information in BGP-4. RFC 3107 (Proposed Standard), May 2001.

[RFC3137] A. Retana, L. Nguyen, R. White, A. Zinin, and D. McPherson. OSPF Stub Router Advertisement. RFC 3137 (Informational), June 2001.

[RFC3509] A. Zinin, A. Lindem, and D. Yeung. Alternative Implementations of OSPF Area Border Routers. RFC 3509 (Informational), April 2003.

[RFC3630] D. Katz, K. Kompella, and D. Yeung. Traffic Engineering (TE) Extensions to OSPF Version 2. RFC 3630 (Proposed Standard), September 2003. Updated by RFC 4203.

[RFC3784] H. Smit and T. Li. Intermediate System to Intermediate System (IS-IS) Extensions for Traffic Engineering (TE). RFC 3784 (Informational), June 2004. Obsoleted by RFC 5305, updated by RFC 4205.

[RFC4105] J.-L. Le Roux, J.-P. Vasseur, and J. Boyle. Requirements for Inter-Area MPLS Traffic Engineering. RFC 4105 (Informational), June 2005.

[RFC4203] K. Kompella and Y. Rekhter. OSPF Extensions in Support of Generalized Multi-Protocol Label Switching (GMPLS). RFC 4203 (Proposed Standard), October 2005.

[RFC4206] K. Kompella and Y. Rekhter. Label Switched Paths (LSP) Hierarchy with Generalized Multi-Protocol Label Switching (GMPLS) Traffic Engineering (TE). RFC 4206 (Proposed Standard), October 2005.

[RFC4811] L. Nguyen, A. Roy, and A. Zinin. OSPF Out-of-Band Link State Database (LSDB) Resynchronization. RFC 4811 (Informational), March 2007.

[RFC4915] P. Psenak, S. Mirtorabi, A. Roy, L. Nguyen, and P. Pillay-Esnault. Multi-Topology (MT) Routing in OSPF. RFC 4915 (Proposed Standard), June 2007.

[RFC4970] A. Lindem, N. Shen, J.-P. Vasseur, R. Aggarwal, and S. Shaffer. Extensions to OSPF for Advertising Optional Router Capabilities. RFC 4970 (Proposed Standard), July 2007.

[RFC5120] T. Przygienda, N. Shen, and N. Sheth. M-ISIS: Multi Topology (MT) Routing in Intermediate System to Intermediate Systems (IS-ISs). RFC 5120 (Proposed Standard), February 2008.

[RFC5130] S. Previdi, M. Shand, and C. Martin. A Policy Control Mechanism in IS-IS Using Administrative Tags. RFC 5130 (Proposed Standard), February 2008.

[RFC5151] A. Farrel, A. Ayyangar, and J.-P. Vasseur. Inter-Domain MPLS and GMPLS Traffic Engineering – Resource Reservation Protocol-Traffic Engineering (RSVP-TE) Extensions. RFC 5151 (Proposed Standard), February 2008.

[RFC5185] S. Mirtorabi, P. Psenak, A. Lindem, and A. Oswal. OSPF Multi-Area Adjacency. RFC 5185 (Proposed Standard), May 2008.

[RFC5250] L. Berger, I. Bryskin, A. Zinin, and R. Coltun. The OSPF Opaque LSA Option. RFC 5250 (Proposed Standard), July 2008.

[RFC5283] B. Decraene, J.-L. Le Roux, and I. Minei. LDP Extension for Inter-Area Label Switched Paths (LSPs). RFC 5283 (Proposed Standard), July 2008.

[RFC5305] T. Li and H. Smit. IS-IS Extensions for Traffic Engineering. RFC 5305 (Proposed Standard), October 2008. Updated by RFC 5307.

[RFC5308] C. Hopps. Routing IPv6 with IS-IS. RFC 5308 (Proposed Standard), October 2008.

[RFC5309] N. Shen and A. Zinin. Point-to-Point Operation over LAN in Link State Routing Protocols. RFC 5309 (Informational), October 2008.

Further Reading

[1] Aviva Garrett. JUNOS Cookbook, April 2006. ISBN 0-596-10014-0.

[2] Matthew C. Kolon and Jeff Doyle. Juniper Networks Routers : The Complete Reference, February 2002. ISBN 0-07-219481-2.

[3] James Sonderegger, Orin Blomberg, Kieran Milne, and Senad Palislamovic. JUNOS High Availability, August 2009. ISBN 978-0-596-52304-6.

[4] Hannes Gredler and Walter Goralski. The Complete IS-IS Routing Protocol, January 2005. ISBN 1-85233-822-9.

[5] Jeff Doyle and Jennifer Carroll. Routing TCP/IP, Volume 1, October 2005. ISBN 978-1587052026.

[6] Jeff Doyle. OSPF and IS-IS: Choosing an IGP for Large-Scale Networks, November 2005. ISBN 978-0321168795.

3

BGP Migrations

The Border Gateway Protocol (BGP) came to standardized life in 1989 with [RFC1105]. After more than 20 years of its initial specification, BGP, meant to be a short-term solution awaiting some yet-to-be-determined long-term solution for the inter-domain routing, is still very young and healthy. Production implementations of the protocol have been deployed since 1990. The most current specification of the BGP protocol at the time of this writing is still considered to be version 4 as per [RFC4271], thanks to the embedded extensibility. This chapter considers this version to be the current and most widely deployed one.

Figure 3.1 shows an example of a BGP network composed of interconnected Autonomous Systems (ASs), each with its own address space, that exchange reachability information. The AS number is carried with the advertisements in the form of the AS_PATH attribute, providing loop capability detection.

Integration or division of organizations implies changes in the AS number of the BGP-speaking network devices interconnecting with external peers. AS renumbering is an infrequent yet cumbersome task, with most expertise residing in the technical departments of network integrators or Service Providers that merged smaller ISPs during the consolidation phase after the Internet boom.

Various internal peering topologies are possible. Smaller ASs (or networks with highly scalable equipment) follow a full-mesh arrangement. Following the example in Figure 3.1, the *confederation* built within AS64501 is composed of two sub-AS numbers not exposed to other domains and that prevent internal loops. AS65550 implements *route reflection*, allowing routers within the domain to have routing visibility with just peering sessions to the reflectors.

The success of BGP as a vehicle to interconnect *routing administrations* identified by an AS has required the extension of the 16-bit AS number (two bytes, up to AS65535) to 32-bit. Interaction between 2-byte ASs and 4-byte ASs has been embedded in the protocol extension to allow migration and coexistence.

Renumbering ASs and switching internal topologies considering *Confederations*, *Route Reflectors*, or flat full mesh, are the most common migration scenarios.

Network Mergers and Migrations Gonzalo Gómez Herrero and Jan Antón Bernal van der Ven

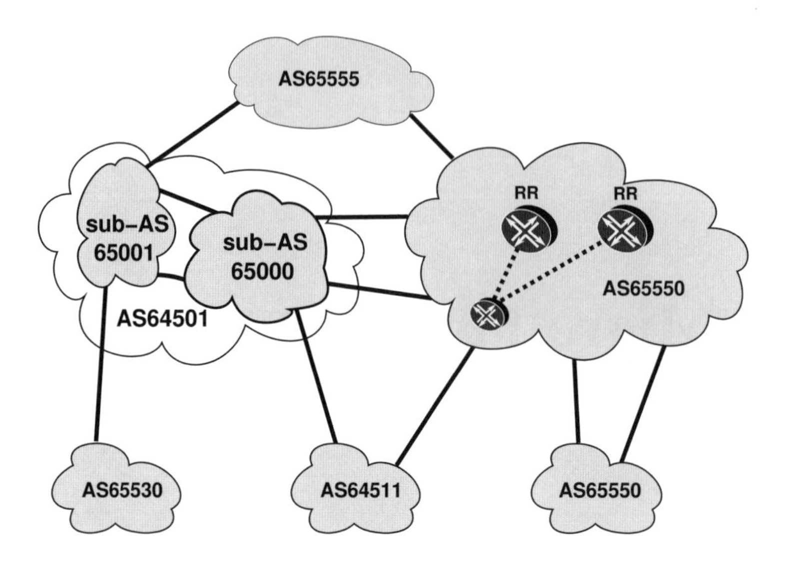

Figure 3.1: Reliable and loop-free information distribution with BGP.

Application Note: Use of AS numbers in documentation

The reader may have noticed that most of the AS numbers used throughout this book span the autonomous system number range used for documentation purposes, [RFC5398], which includes provisions for both 2-byte and 4-byte numbers. The 2-byte range is adjacent to the private AS range defined in [RFC1930] and is typically used as a confederation member sub-AS or for private inter-AS arrangements. Figure 3.2 summarizes the allocation.

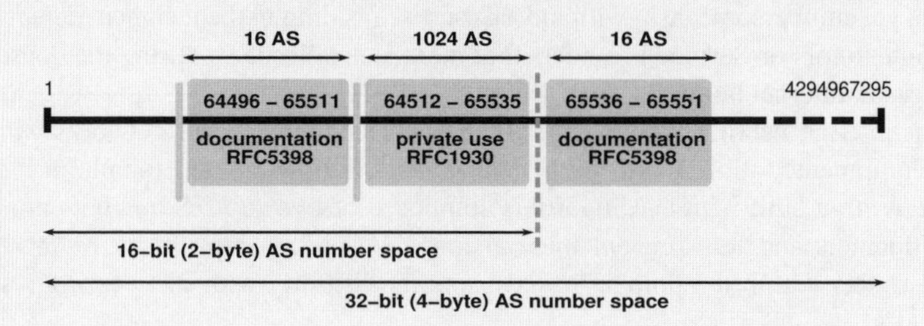

Figure 3.2: Special allocations for autonomous system numbers.

Use of this reserved space of AS numbers prevents accidental configuration of already assigned AS numbers in production networks. Level3 communications (AS1), University of Delaware (AS2), and Massachusetts Institute of Technology (AS3) are some obvious beneficiaries.

Following the general structure of the book, this chapter first describes the motivations for BGP migrations, giving practical cases that might justify the migration effort. A section on considerations analyzes specific protocol behaviors that have to be understood before tackling a migration, followed by strategies to consider when undertaking this effort. Details about Junos OS implementation provide a glimpse of the internals of the system. Thanks to the history of Junos OS in BGP networks, a bag of tools described in the resource section is made available to satisfy corner-case migration requirements.

The case study in this chapter is aligned with case studies in other chapters of the book and simulates the acquisition of a Service Provider, using BGP confederations for a fast integration in an initial stage, followed by a subsequent autonomous system integration using route reflection.

3.1 Motivations for BGP Migrations

From a control plane perspective, BGP changes may involve software and router configuration: new software may be required to support new or enhanced functionality (the latest implementation of an RFC); configuration changes may be a consequence of new or altered internal topological arrangements, or external factors such as business interconnections. No matter why, the underlying reason for a BGP protocol change is usually *because it matters from a business perspective*. Operational efficiency, new services, and scalability are just some examples that justify modifying existing environments.

The following sections, while by no means comprehensive, provide an overall view of the elements that are worth considering.

Change in AS identifier

This item is one of the most tedious changes to face during asset acquisition. The book maintains a strong focus on the BGP behaviors, covering administrative considerations only briefly. This motivation is one of the pillars of the case study to follow.

Modification of internal topologies

If route reflection [RFC4456] or confederations [RFC5065] or both, are not already introduced in the original network design, a successful business may have to migrate to one or both to support growth of the business. All these BGP scalability measures accommodate network growth.

Internal BGP reconfiguration after a merger or acquisition may be required. Merging together two ASs might yield a mixed infrastructure, which may require the aforementioned scalability measures. The same measures may be worth taking if any of the networks are not fully meshed internally, building a confederation for fast integration, dismantling it to provide a homogeneous result, or using route reflection clusters for direct integration of common services that require minimal infrastructure deployment.

Robustness and performance

Keeping all eggs in the same basket for different services might not be a wise decision. The additional cost of introducing multiple route reflectors might be offset against the cost of losing a critical service as a result of the misbehavior of a standard service. As an example, having each route reflector be responsible for a single service may be a good lesson learned from past incidents where heavy routing churn of the Internet service affected business customers using VPN services. Besides robustness, per-service scalability of the control plane provides performance benefits that allow separate growth paths for different services.

Support of new applications

BGP can be used as the delivery channel for new services. Once the foundation layer for the routing *database* is in place, additional *data sets* can peruse the infrastructure in a straightforward way. The BGP capabilities option as per [RFC3392], used during BGP session establishment, allows for a negotiated agreement of features. Support of BGP MPLS VPNs in the enterprise is one such example. Proposals of data center interconnections are slowly moving away from Layer 2 switching with Spanning Tree in the wide area to an MPLS backbone with BGP as the control plane supporting a Layer 2 (L2) VPN or Virtual Private LAN Service (VPLS), incorporating the data center requirements into a stable and reliable MPLS VPN core.

Support of 4-byte AS numbers

Public AS identifiers were extended to 32 bits with [RFC4893]. Although the specification calls for a transition mechanism, adding full support may likely imply changing the software implementation. A one-shot upgrade of the network is unrealistic, so the specification includes embedded support for a long transition period.

Routing policy modifications

Changes in routing policies can be perceived as a modification of the *database contents*. In Junos OS, this routing database is the RIB. For a critical network in continuous operation, any BGP routing change that involves altering traffic is carefully analyzed, and the proper strategy that considers minimal impact to be a critical metric is evaluated. The most common change in routing policy consists of moving traffic to follow either the closest or furthest exit for a specific destination. In the trade, this is called hot potato (hand off as soon as possible) and cold potato (hand off as late as possible) routing. Interestingly enough, routing policies are outside the scope of the protocol specification; lack of standardization provides a lot of flexibility, along with some side effects, as seen in [BGP-WEDGE].

Summarizing from a BGP transport protocol perspective, a change in the internal BGP infrastructure (such as the session topology or the introduction of a new service) and a change in the global identifier (AS Number) are the most common motivations for considering a migration of the BGP protocol. Changes in the contents of the exchanged information are part of using the protocol itself and are considered in the broader context of migration of the services provided by the protocol.

3.2 Considerations for BGP Migrations

BGP is a database exchange protocol, featuring peer flow control and information flooding with loop avoidance. As a lesson learned from the past, the structure of the information that is distributed follows a variable encoding paradigm using TLV elements, which allows for extensibility. The information is qualified with an *Address Family Identifier* (AFI) and a *Subsequent Address Family Identifier* (SAFI) to identify the payload (for example, IPv4, IPv6, and NSAP), which makes BGP suitable as the transport protocol for a wide variety of applications.

For *vanilla* BGP, the data exchanged in this database is *reachability* information, tagged with a flexible list of attributes. New applications, such as VPLS and Route Target (RT) constraints, give different semantics to the exchanged data. Instead of reachability, VPLS exchanges membership information for autodiscovery and signaling. RT constraints exchange RT membership information. Hence, the data exchanged for these new applications is not reachability, but rather *membership*.

The BGP protocol focuses on the exchange of information and relies on TCP to provide reliable transport and flow control. As a hard-state protocol, BGP exchanges information only when needed, with no periodic exchanges, unlike protocols such as PIM, RSVP, or RIP. BGP designers considered that periodically readvertising known information using duplicate updates in a database of this size would not be scalable. A bare-bones keepalive message tickles the TCP connection to ensure that the session is still alive, but it is quickly suppressed if other information is exchanged within the response interval.

Avoiding routing loops is instrumental to ensuring a stable, loop-free packet forwarding environment. BGP avoids loops with the introduction of an attribute that records the AS path that the information has traversed.

Although BGP has grown up over the years, it is still evolving. The base specification has been complemented with extensions for authentication with [RFC2385], capability advertisement mechanisms with [RFC3392], route refresh techniques with [RFC2918], and multiprotocol extensions with [RFC4760], to name a few, with additional proposals still being considered, such as [ID-ADVISORY]. The IETF Interdomain Routing charter [IETF-IDR] contains a complete listing of work, both standardized and in progress, related to BGP.

New feature introduction can be addressed incrementally thanks to BGP capability advertisement, as defined in [RFC3392]. This flexible mechanism negotiates supported applications during connection setup, yielding as a result a common set of capabilities to be used for the life of the long-lived session. The capability advertisement feature is the underlying reason that keeps version 4 current, because it loosely decouples combined feature support from the protocol version number.

BGP follows a destination-based routing paradigm. In migration scenarios, it is very attractive to enforce routing policy beyond destination reachability. This is unfortunately (but consciously) not supported by BGP, because designing the protocol with this added complexity was left to the long-term solution. Hence, any migration that involves circumventing the destination-based paradigm during a transition period entails a fair amount of complexity as additional route views or packets jump between forwarding tables within the system. The case study in Section 1.4 on Page 21 offers an example of this complexity.

With BGP's continuing support for new applications (such as MPLS L3VPNs, VPLS, and VPN autodiscovery), some network engineers still claim that the protocol was devised as a *routing* protocol and was not meant to carry these types of application. Changing the definition of the BGP protocol to a "highly scalable and extensible database distribution protocol" yields "routing information" as a subset of the possible data types to store in the database; that is, BGP can be used to distribute more information than *just* IP prefixes.

3.2.1 Protocol messages

BGP uses five messages, namely OPEN, UPDATE, KEEPALIVE, NOTIFICATION, and the late sibling, REFRESH. Figures 3.3 and 3.4 show the layout of these messages. All the messages start with a common header. Additional extensions to the protocol have managed to squeeze themselves into one of these five messages, reflecting the flexibility of the original design.

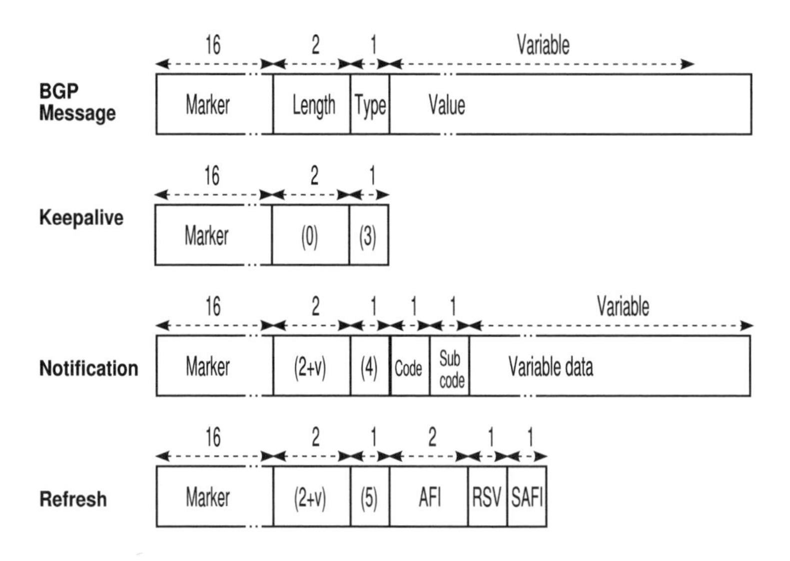

Figure 3.3: KEEPALIVE, NOTIFICATION, and REFRESH BGP protocol messages.

The KEEPALIVE message (Figure 3.3) consists of a BGP header with no further contents, used to indicate liveness of the session. The pace at which these messages are sent out is negotiated in the OPEN message through the *hold time* parameter. The protocol has provision for using UPDATE messages as a liveness indicator, thus suppressing a KEEPALIVE message for busy sessions. A packet protocol capture of a BGP session in the middle of a full routing advertisement to a BGP client would show UPDATE messages in the forward direction, and, in the *reverse* direction, empty TCP acknowledgment segments would confirm the data as being received correctly, interleaved with periodic KEEPALIVE messages from the client side to maintain the established session. As long as communication occurs in both directions before the negotiated hold time, the session is kept alive. From a migration perspective, the *hold time* indirectly defines the pace of keepalives and is negotiated as the lower of both proposals. Changing the hold time affects both sides of the peering session after session establishment and does not require the configutation of a symmetric value on both peers.

As an exception mechanism, the specification of the NOTIFICATION message (Figure 3.3) aims at detailed reporting of the nature of a problem. All notifications are considered fatal and terminate the session. The appearance of this message during migrations, other than as the results of a subcode 4 *Administratively Reset*, probably indicates some undesired protocol violation. An example of such notification is the report of an unexpected AS number by the peer. Further examples of notifications are documented in [RFC4271].

The REFRESH message (Figure 3.3), not defined in the original BGP specification, is a new capability that peers can signal, as per [RFC2918]. One peer can request retransmission of routing information from another peer, and the request can be limited to a subset of the negotiated applications exchanged. An implementation may keep unused inbound information in its RIB-In, which allows a refresh to be performed as a local operation without peer intervention. Compared to the original alternative of restarting a session, the operational advantage of the REFRESH mechanism is twofold:

- A policy change on one end may result in no change in the other direction, avoiding the need for readvertisement in both directions after a session flap. An example of the advantage is a service provider that receives customer routes and advertises full routing. If the customer receives a new address block, it has to ask the provider to change its import policy to allow the additional prefix range. Changing the policy and sending a refresh message to receive the new prefix blocks is all that is required by the provider, instead of restarting the BGP protocol session and readvertising the full route table.

- Policy changes affecting only one of the protocols do not require total re-initialization of the session when multiprotocol sessions are in use. A typical scenario involves a BGP session sharing Internet and MPLS L3VPN routing information, in which a new import policy (a new VRF with a new route target community) requires receiving VPN routes anew.

The OPEN message (Figure 3.4) is the first one sent out after the TCP session is established. Its main purpose is to exchange configuration information and common parameters to be used throughout the peering session. There needs to be agreement on what version the session will use. All modern systems implement version 4, so the version field is not likely to be a factor during migrations. Besides the hold time used by BGP keepalive messages to verify continuous TCP connectivity, the two elements of special interest in a migration context are the *My Autonomous System* field, to decide what peering relationship is set up, and the *optional capabilities* parameter.

Information is maintained by means of UPDATE messages (Figure 3.4). A single message can contain both advertisements (for new or updated information) as well as withdrawals. The protocol assumes that a more recent advertisement is an implicit withdrawal for previously signaled information. The original message considered only IPv4 reachability information. A multiprotocol extension, as described in [RFC4760], added an extra MP_UNREACH attribute to complement the *withdrawn routes* field and an MP_REACH attribute to specify the non-IPv4 protocol NLRI information. Because of the reliable nature of the TCP transport layer, an advertisement is not refreshed periodically. Although the protocol is robust with regards to receiving multiple duplicate update information, this situation should not occur.

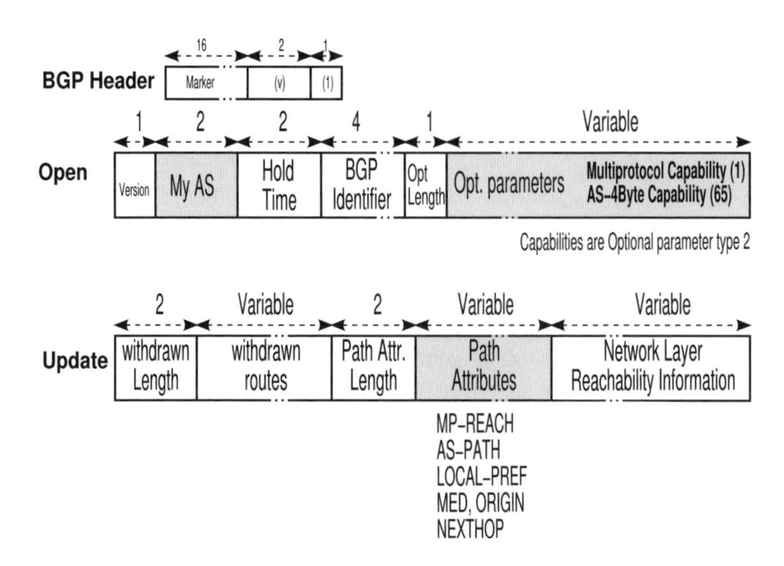

Figure 3.4: OPEN and UPDATE BGP protocol messages.

3.2.2 Capability advertisement

Learning from past experiences, BGP protocol designers wanted to avoid redefining the protocol to accommodate new business requirements. Bumping up the version number of the BGP protocol every time a new feature was defined was the minor problem. The challenge was to allow for independent feature support without requiring a combinatorial mix of protocol versions. Use of a different version for N features would require 2^N version numbers to accommodate all possibilities, yielding a major roadblock for new feature introduction. Adding feature number M would mean that 2^{M-1} additional versions would need to be defined, to cover for all combinations of feature M and all previous $(M - 1)$ features enabled simultaneously.

Through flexible capability advertisement from [RFC3392] procedures, each side of a BGP session advertises its desired interest for the various supported features. Capabilities are encoded in the *Optional Parameters* field of the BGP OPEN message (Figure 3.4), a field that can hold multiple optional parameters. From a migration perspective, knowing the capabilities of the peer introduces the possibility for incremental change, so this feature is instrumental to progressive rollout.

Listing 3.1 shows a sample Junos OS BGP protocol trace in which the initial OPEN message is sent to the peer reporting a list of supported capabilities. Note that this support can be restricted, by configuration, to agree only on selected functionality.

Listing 3.1: Indicating peer capabilities in OPEN message

```
1 Jun 27 19:15:42.295730 BGP SEND 192.168.1.5+49305 -> 192.168.1.20+179
2 Jun 27 19:15:42.295766 BGP SEND message type 1 (Open) length 123
3 Jun 27 19:15:42.295793 BGP SEND version 4 as 65000 holdtime 90 id 192.168.1.5 parmlen 94
4 Jun 27 19:15:42.295817 BGP SEND MP capability AFI=1, SAFI=1                # Inet   unicast
```

```
5 Jun 27 19:15:42.296001 BGP SEND Refresh capability, code=128                    # Refresh (legacy)
6 Jun 27 19:15:42.296022 BGP SEND Refresh capability, code=2                      # Refresh (standard)
7 Jun 27 19:15:42.296045 BGP SEND Restart capability, code=64, time=120, flags=   # Graceful restart
8 Jun 27 19:15:42.296069 BGP SEND 4 Byte AS-Path capability (65), as_num 65000    # AS4Byte
```

A partial summary of the capabilities supported by Junos OS along with the standard that describes the capability is shown in Table 3.1. The capabilities referenced as *legacy* indicate pre-standard implementations using private capability identifiers greater than 127.

Table 3.1 Some Junos OS BGP capabilities

Capability ID	Description	Reference
1	Multiprotocol Extensions	[RFC4760]
2	Route Refresh message	[RFC2918]
3	Outbound Route Filtering	[RFC5291]
64	Graceful Restart	[RFC3478]
65	4 Byte AS PATH	[RFC4893]
128	Route Refresh message	legacy
130	Outbound Route Filtering	legacy

3.2.3 Address families

The Multiprotocol Extension for BGP-4 defined with [RFC4760] was the first capability to be introduced. Its capability identifier type code 1 highlights this fact. This extension allows BGP to carry information not just for IPv4 and IPv6 routing, but also for other applications (L3VPN, VPLS, and many more). The BGP protocol originally assumed that fixed format IPv4 prefixes were part of the *withdrawn routes* and the NLRI fields in the UPDATE message (Figure 3.4). To allow for new applications, the BGP protocol was extended through the introduction of two new attributes, MP_REACH and MP_UNREACH, which allowed for variable-length data types.

Signaling of the different supported applications is performed as different capabilities in the OPEN message, using the capabilities advertisement procedures discussed previously. A set of supported applications can be signaled, allowing a common session to share L3VPN, IPv4, IPv6, and any other application. The applications whose information is to be carried are identified using the AFIs [IANA-AFI] and SAFIs [IANA-SAFI], for which the Internet Assigned Numbers Authority (IANA) holds a regulated registry. Table 3.2 shows a subset of address families supported by Junos OS.

As part of capability advertisement, the common set of supported families is agreed upon between both parties in the BGP session. A sample exchange is shown in Listing 3.2. The received UPDATE shows in Line 14 that the peer desires to establish a labeled–unicast relationship, but this is not configured by the sender, as hinted by the resulting log as shown in Line 19.

After session bringup, the common set of supported capabilities is used for the life of the session, and the extract can be observed by the operational command starting on Line 25.

Listing 3.2: Agreeing on supported applications

```
1  Jun 26 12:43:13.245764 BGP SEND 172.16.100.8+60939 -> 172.16.100.5+179
2  Jun 26 12:43:13.245793 BGP SEND message type 1 (Open) length 59
3  Jun 26 12:43:13.245817 BGP SEND version 4 as 65511 holdtime 90 id 192.168.1.6 parmlen 30
4  Jun 26 12:43:13.245836 BGP SEND MP capability AFI=1, SAFI=1                # Family inet
5  Jun 26 12:43:13.245853 BGP SEND Refresh capability, code=128
6  Jun 26 12:43:13.245870 BGP SEND Refresh capability, code=2
7  Jun 26 12:43:13.245889 BGP SEND Restart capability, code=64, time=120, flags=
8  Jun 26 12:43:13.245909 BGP SEND 4 Byte AS-Path capability (65), as_num 65511}
9  Jun 26 12:43:13.247772
10 Jun 26 12:43:13.247772 BGP RECV 172.16.100.5+179 -> 172.16.100.8+60939
11 Jun 26 12:43:13.247833 BGP RECV message type 1 (Open) length 67
12 Jun 26 12:43:13.247856 BGP RECV version 4 as 65500 holdtime 90 id 192.168.1.5 parmlen 38
13 Jun 26 12:43:13.247875 BGP RECV MP capability AFI=1, SAFI=1
14 Jun 26 12:43:13.247892 BGP RECV MP capability AFI=1, SAFI=4                # Labeled unicast
15 Jun 26 12:43:13.247909 BGP RECV Refresh capability, code=128
16 Jun 26 12:43:13.247925 BGP RECV Refresh capability, code=2
17 Jun 26 12:43:13.247944 BGP RECV Restart capability, code=64, time=120, flags=
18 Jun 26 12:43:13.247963 BGP RECV 4 Byte AS-Path capability (65), as_num 65500
19 Jun 26 12:43:13.248074 bgp_process_caps:
20   mismatch NLRI with 172.16.100.5 (External AS 65500):
21   peer: <inet-unicast inet-labeled-unicast>(257)
22   us: <inet-unicast>(1)
23 /* output formatted for clarity */
24
25 user@Lille>show bgp neighbor 192.168.1.20 | match nlri
26   NLRI for restart configured on peer: inet-unicast inet-labeled-unicast
27   NLRI advertised by peer: inet-unicast
28   NLRI for this session: inet-unicast
```

Table 3.2 Selection of address families supported by Junos OS

AFI	SAFI	BGP family configuration
1	1	inet unicast
1	2	inet multicast
1	3	inet anycast
1	4	inet labeled-unicast
1	5	inet-mvpn signaling
1	66	inet-mdt unicast
1	132	route-target
2	2	inet6 multicast
2	4	inet6 labeled-unicast
3	128	iso-vpn unicast
25	65	l2vpn signaling

Because capabilities are agreed with the details contained in the OPEN message, reconfiguring the set of families supported by a peer triggers a session flap, which has to be considered during the planning phases of a maintenance activity.

3.2.4 Implications of public AS change

An AS is a connected group of one or more IP prefixes run by one or more network operators that has a *single* and *clearly defined* routing policy (see [RFC1930]). Routing policies refer to how the rest of the network makes routing decisions based on NLRI from this autonomous system.

Public ASs have a globally unique AS number (ASN) associated with it; this number is used in both the exchange of exterior routing information (between neighboring ASs) and as an identifier of the AS itself. The existence of the AS number used in the BGP protocol assists in avoiding routing loops as primary purpose. Within a particular AS domain, the ASN has little significance for routing policy. Therefore, although one might be tempted to use public ASN allocations to infer the physical location of prefixes, by building a geographical map of ASs per continent, giving the ASN such semantic was never the intention.

The IP addressing space is tied to the ASN. Changing this association between a prefix and an ASN necessarily affects how partners see that information, now tied to a different ASN. The administrative activities that occur as part of an AS migration may involve requesting a new public AS to the corresponding Internet Registry, transferring the existing address space to the new AS, and providing peers with necessary notice to allow for changes in their import filtering rules.

The public Internet makes no provisions for an AS transfer. Every Routing Registry has an allocation for a range of ASNs, which are not portable. If the new AS falls outside the domain of the current Registry, a new ASN has to be requested from a different Registry. It is understood that an AS represents a collection of networks under a single administration; hence, its address allocation has to move to the new AS, presenting a transition stage in which the addresses may appear as being sourced from two ASs simultaneously. The BGP protocol imposes no restriction for a prefix to be sourced from different ASs, but this circumvents the standard loop prevention mechanism based on AS detection and opens a potential backdoor for problems, in case route advertisement is not properly controlled. Conversely, the introduction of route aggregation allows for a prefix to appear as residing in more than one AS. However, this is the exception rather than the rule.

It is common practice at peering points to include explicit policy checks to receive updates from a peer only if the peer AS is present in the AS_PATH. Changing an AS involves notifying peers to update these security filters accordingly. Similar to AS checks, prefix filters verify that a specific block of addresses is valid by performing a cross-check with the public allocation information.

Modifying the AS_PATH during a migration increases the chance of routing loops at intermediate steps where the update traverses old and new ASs. A protection measure worth considering is to define specific filter rules preventing updates from the AS that is in transition from coming back to their home domain.

For operators providing MPLS L3VPN services, care must be taken to map relevant parameters that use the AS value to the new AS. These parameters include communities (both normal and extended) and Route Distinguishers (RDs).

3.2.5 AS numbers in route advertisements

From an information exchange perspective, AS information is present in two messages of the BGP protocol, OPEN and UPDATE. The former affects session establishment, while the latter may impact propagation of routing information over the network. The OPEN message contains AS information as part of the message header or within an optional capability.

Valid AS numbers

Two BGP peers attempt to set up a BGP session, agreeing on common settings through negotiation. Initial BGP session negotiation occurs through the exchange of OPEN messages. As described in Section 4.2 of [RFC4271], the OPEN message includes a *My Autonomous System* field. A valid peer AS is required to bring up the session. Deployed versions of Junos OS validate that the content of the *My Autonomous System* is a non-zero 16-bit AS number.

The extension to 32-bit AS numbers specified in [RFC4893] could not modify this fixed-size field of the OPEN message, and decided to allocate the reserved AS_TRANS value ("23456") instead, while the real 32-bit AS number is stored as part of a capability. It is considered an error to present AS_TRANS as the AS number without an additional 4-byte AS encoded in a capability, and the capability should not include AS_TRANS as the 4-byte AS. When the local end has a 2-byte AS, the AS4 capability is still signaled, including the 2-byte AS in the 4-byte capability information. In this case, the *My Autonomous System* field does include the true 2-byte AS.

Table 3.3 summarizes the valid AS numbers that can be exchanged during an OPEN session.

Table 3.3 Valid AS Numbers in OPEN message

Configuration	My AS	AS4 capability	Validity
AS-2Byte	AS-2Byte	AS-2Byte	valid
AS-4Byte	AS "23456"	AS-4Byte	valid
0	–	–	error
AS "23456"	–	–	error
Any	–	0	error
Any	–	AS "23456"	error

AS numbers and BGP peering relationship

The peering type (internal, external, or confederation) is relevant when deciding to accept an OPEN message. Processing of UPDATE messages over an established session also takes into account the peering relationship to decide whether the path attributes have to be modified.

For an internal peering, the negotiated local AS and the remote AS have to match; for an external peering, the received value has to match the local configuration for the `peer-as`. A confederation peering has an additional behavior that is seen when the global AS of both peers is the same, which is that it also considers the confederation sub-AS at both sides.

Changes in the AS for a session require resignaling of the OPEN message, thus affecting establishment of a neighbor relationship.

3.2.6 AS-related attributes in advertisements

Reachability information is propagated using the UPDATE message, which contains the AS number parameter in various attributes. The main purpose of AS numbers is to ensure a loop-free packet forwarding environment. This means that changes to the AS require careful analysis of the various attributes to avoid blocking information as a result of loop detection or injection of inconsistent AS parameters.

AS_PATH attribute

The AS_PATH attribute contains a list of ASs that this information has traversed. It is a key element of the loop-prevention mechanism for external BGP, and is also leveraged for loop prevention for internal BGP when using confederations, as per [RFC5065].

Table 3.4 summarizes the elements that constitute AS_PATH. A *sequence* is considered to be a continuous path of traversed ASs with the most recent AS presented first in the list.

The possibility of knowing about aggregated prefixes in BGP was added in version 4 of the protocol. The concept of a continuous path of traversed ASs had to be extended to include a *set* of ASs that contain a subset of the addressing represented by a prefix. A set is to be understood as the mathematical set of ASs that all member prefixes traversed. It may not necessarily mean that a prefix belongs to all the ASs, because prefix aggregation can create a set from prefixes crossing different ASs.

Introduction of confederations further extended the AS_PATH attribute to mimic the behavior of the global AS for the sub-AS within the confederation.

Table 3.4 AS_PATH attributes

Name	Purpose
AS_SEQUENCE	Sequence of global ASs
AS_SET	Unordered set of global ASs
AS_CONFED_SEQUENCE	Sequence of confederation sub-ASs
AS_CONFED_SET	Unordered set of confederation sub-ASs

Based on the BGP peering relationship (confederation, external, or internal), the AS_PATH information is modified when routing information is propagated. This is relevant for mergers with regard to identifying troubling points during an AS migration. Table 3.5 summarizes the options. For cases in which the outbound session is internal BGP (IBGP), no modification is made to the AS_PATH attribute. Notice that adding the AS to the AS_PATH would trip AS loop-detection checks at the peering router in the same domain. If the outbound session is external BGP (EBGP), any CONFED elements are stripped off and the global confederation AS is added. Updates from a confederation peer (CBGP) do not have its AS modified when talking to internal peers. Conversely, routes going out over a confederation peering modify only the sub-AS component in the AS_PATH.

Under the assumption that more AS hops imply higher cost, the AS_PATH provides an inter-AS metric that influences best-path selection. A modification to the AS number should not in itself trigger a metric change, but insertion or removal of an AS (through company mergers or divestitures) may influence overall routing, especially for TE policies at other ends expecting a given number of ASs.

Table 3.5 Outbound modification of AS_PATH

From	To	AS_PATH modification
EBGP	IBGP	No handling
CBGP	IBGP	No handling
Any	EBGP	Prepend local AS in AS_SEQUENCE
Any	CBGP	Prepend confederation sub-AS in AS_CONFED_SEQUENCE

AGGREGATOR attribute

Aggregate routes may have contributed with prefixes sourced from different ASs. The AGGREGATOR attribute hints at the last router (identified by the BGP Identifier in the OPEN message) and the AS that performed the aggregation function. During a transient stage, some devices may indicate disparity in the source AS for the aggregate route. The BGP Identifier may indicate which systems are yet to be migrated.

List of COMMUNITIES attribute

This attribute was not part of the original BGP specification, but rather, was added later with [RFC1997]. The attribute allows marking or coloring a set of prefixes based on policy language. With the introduction of communities, it becomes possible to constrain connectivity or perform a certain degree of traffic engineering. Connectivity can be constrained by limiting the flow of routing information, and the flow of routing information can be constrained by marking a certain set of routes with a particular community and then filtering out routes that have such a community. This attribute improves the description of the intended routing distribution policy.

Except for well-known community allocations, the semantics of the community marker are left for private use. The private community is based on the AS number plus a locally significant identifier. This allows widespread usage because each AS domain can define its own set of private communities, with the possibility of discrimination when overlapping the use of the identifier across domains.

The need for a broader range of communities because of the introduction of L3VPNs with [RFC4360] promoted a new format called the *extended community* that also includes the AS number in one of its variants.

Note that multiple communities can be attached to a route. Junos OS flexible policy language allows the addition, removal, setting, and checking of communities, or a combination of these actions, to be used to translate (map) community conventions between domains.

3.2.7 Internal peering topologies

BGP as a control plane is formed by a set of point-to-point TCP sessions between two peers. Within an AS, three topologies can be combined to provide reachability:

- *Full mesh*, as originally defined in the BGP specification. Any two routers connect directly and exchange both the local information and relay external information.

- *Confederation*, as defined in [RFC5065], in which a set of sub-ASs are connected together to form a confederation. The full-mesh requirement is lifted, and additional attributes are added to identify routing loops. The confederation BGP session acts like a regular external BGP session, with two modifications: no change in some attributes (protocol NEXTHOP, MED, and LOCAL_PREFERENCE), and handling of the AS_CONFED variants within the AS_PATH.
- *Route Reflection*, as described in [RFC4456], in which one *reflector* takes the role of relaying routing information between internal peers. From a migration perspective, it is worth noting that this scaling technique included *easy transition* as one of the design criteria.

Demarcation points

Existing deployments for both confederation and reflection topologies achieve the same goal of scalability by means of hiding information, but like multi-area OSPF as compared to multilevel IS–IS, they differ in their demarcation point, as shown in Figure 3.5.

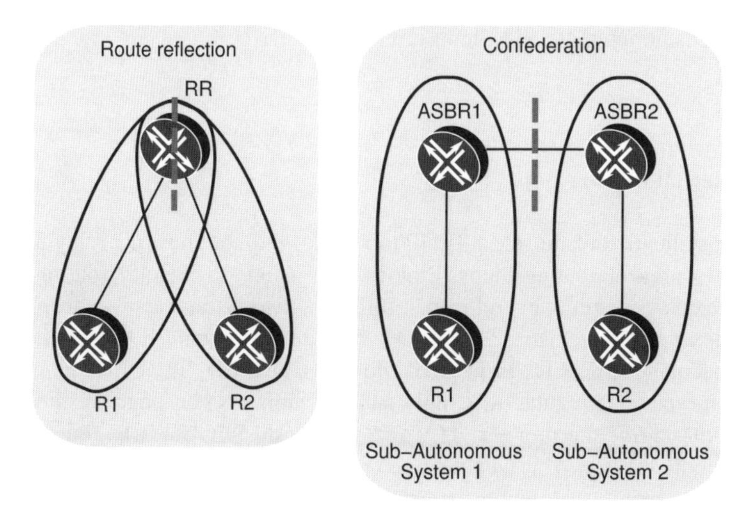

Figure 3.5: Comparing reflection and confederation boundaries.

Notice that a remarkable difference between confederations and reflection is the fact that in a confederation, internal routers can belong to a single confederation, but in reflection, internal routers can be simultaneously homed to different reflection clusters.

The typical use case for confederations is to interconnect ASs that have different administrative boundaries and in which a clear policy has to be enforced between the sub-ASs. It is common for each sub-AS to have a different IGP, and in that case, the standard procedure of the protocol setting the next hop is overridden by using policies.

The link between the sub-ASs provides a clear boundary. This behavior mirrors the multilevel approach in multilevel IS–IS, in which each intermediate system belongs to a different area and a Level 2 adjacency is used to join areas. To join domains using confederations typically involves two confederation border routers with a *confederation BGP* session between them, a slightly modified version of external BGP. From a network planning perspective, additional ports and network capacity must be allocated for the confederation interconnect. Router ports are devoted to an internal function instead of serving customer access demands; this is not required with route reflection.

Route reflection, on the other hand, mimics the logic of an OSPF ABR with multiple areas homed to the same router. A BGP route reflector client can simultaneously join different routers, each one acting as route reflector for a standalone or shared cluster with another reflector server. Joining domains using reflection might involve a single route reflector homing all reflector clients, with a common IGP that does not require changes to the defaults.

Over time, again as with OSPF and IS–IS, customer requirements and practical enhancements have improved simultaneous support of features within a single router, while remaining within the bounds of relevant standards. The end result is that functionality between the two scaling approaches becomes equivalent, and the decision between reflection or confederation is sometimes a matter of historic perspective. Merging of ASs in which one AS is already confederated may justify extension of the confederation; pre-existence of reflection is a good head start when continuing to add clients or clusters.

Avoiding internal routing loops

The initial specification of full-mesh IBGP topologies did not allow the relaying of internal information to prevent routing loops. Protocol extensions to identify routing loops were added to both confederations and reflectors, following two distinct approaches.

In confederations, the existing AS_PATH attribute was extended by defining two additional TLV elements, AS_CONFED_SET and AS_CONFED_SEQUENCE. Albeit within the same namespace as regular ASs, only sub-AS numbers can populate these elements. These elements mimic the previous TLV (AS_SET, AS_SEQUENCE), but their scope is limited to a confederation. The specification requires these new elements to be visible only within the confederation, stripping them when advertising routes to external BGP peers. Being for private use, the sub-AS numbering typically involves ASs from the private range as per [RFC1930]. As part of troubleshooting during a migration, messages that contain these invalid elements may indicate peering between a confederated and a non-confederated router.

The approach of reflection introduced the notion of a *reflection cluster* that takes care of a set of reflector clients. This notion requires no new extensions and so eases migration. The ORIGINATOR_ID (with the router ID of the originator of the prefix) ensures that a client, which is otherwise unaware of reflection extensions, does not receive an update that it originated, as might happen with two route reflectors homing the same client. As an additional precautionary measure to cover for misconfiguration, the CLUSTER_LIST attribute includes the list of cluster identifiers traversed by the update. A reflector must filter an incoming update with its own identifier as a loop prevention mechanism. The specifics of reflection attributes are relevant when merging or splitting reflection clusters.

Integration of IBGP peering roles

By carefully designing the IBGP roles at a granular level, a good opportunity exists for controlling migration between the various topologies using Junos OS features:

- For the same AS, confederation peering sessions with a different local sub-AS are feasible, simulating a hub as in the reflection case.
- It is possible to apply policy language to IBGP sessions. Confederation peerings are considered external sessions, but route reflection sessions may also be a benefit for inbound and outbound policies.
- Route reflectors can apply outbound policy that changes the protocol next hop. Within the same AS, two route reflectors can establish a *trust* relationship similar to what confederations achieve. With this functionality, it is possible to block incoming or outgoing announcements as it is done with confederations.
- An EBGP session can be simulated with IBGP by peering over the interface addresses and enabling a next-hop change on a route reflection cluster.
- The protocol next-hop can be configured to remain unchanged, which is not the default behavior. A use case for this situation is the interconnection of reflectors over external peerings for L3VPNs.
- A single Junos OS router can act as route reflector on one session, simulate membership to a confederation on another, and further establish a regular internal or external peering. This flexibility is very attractive when using a helper device supporting a transition.

3.2.8 Keep last active route with BGP equal cost external paths

For densely meshed ASs, it is common to receive the same reachability information from different peers. Depending on policy, best-path selection may end up tie-breaking at the last possible step, the peer IP address. The inactive path shows *Update source* as the reason.

For EBGP peerings, the local system has to rely on an identifier (the peer's IP address) that is beyond the control of the AS. For equal-cost paths coming from multiple external peerings to the same AS, the local best-path selection may change based on the presence of these additional paths.

A subset of the well-documented persistent route oscillation case is defined under [RFC3345] and can be resolved by slightly modifying the last step in the decision process, avoiding a path change because of the update source and forcing a prefix to stay on the *last active* path, as per [RFC5004].

Migration scenarios involving changes in peering sessions and path advertisements are likely affected by this behavior, keeping prefixes on the old path while a new equal cost path is already available.

This behavior can be tackled by properly changing BGP attributes for the old path (giving it an overall lower cost, for instance) or removing the old path altogether.

This situation is exemplified in a particular scenario at a later section on page 362.

3.2.9 Grouping policy with communities

BGP communities are a powerful tool to group a set of prefixes that share a common property. This type of grouping can be a local decision, or it can be made in agreement with third parties to implement policies at an inter-AS level.

Migrations may usually involve moving routing information from the old to the new stage. The ability to tag the migration stage on a granular level (per prefix) with BGP may benefit transition and rollback procedures.

Following are two examples in which policy grouping may make sense:

- *Redistribution control*: decide which parts of the routing domain have to transition into the new scenario. A use case for *redistribution control* is the migration of critical services requiring a maintenance window, with the bulk of standard services being performed during the day.

- *Filtering control*: during an AS migration, loop avoidance mechanisms inherently provided by the AS_PATH may not be feasible because the AS is not present in prefixes coming from not-yet migrated nodes. Tagging legacy AS information with a community allows for inbound filtering of these prefixes that would pass the AS_PATH check but may trigger a forwarding loop.

While both examples cover significant uses for BGP communities as a control tool, many other migration scenarios can greatly benefit from these BGP attributes as a management and policy implementation mechanism.

3.2.10 Handling of protocol next-hop changes

One of the key additions to general routing in the BGP specification is the introduction of a level of *indirection* in route resolution. The NEXTHOP attribute provides a reference of the BGP speaker that needs to be reached to forward the packet for the advertised NLRI, but this BGP speaker does not need to be directly connected. As a constraint check, a feasible path must exist in the router to reach the NEXTHOP to consider the received NLRI valid. This NEXTHOP *reachability* enforcement is performed by most routing operating systems, Junos OS among them.

A similar indirection strategy is taken in multi-area OSPF for external reachability information contained in Type 5 AS External LSAs, as described in [RFC2328]. In intra-area scenarios, knowing the local Type 1 Router LSA is enough to resolve external information. Because the Type 1 LSA does not leave an area, reachability to the ASBR is made available by the ABR through the utilization of other Type 4 ASBR Summary LSAs. These Type 4 LSAs provide reachability information for the ASBR in foreign areas. A recursive lookup resolves the Type 5 prefix information over the Type 4 router reference, which provides the relevant Type 1 ABR information. Unlike BGP, which uses the complete routing table for resolution, this type of resolution process is self-contained in the OSPF protocol.

The BGP specification provides fixed semantics for the NEXTHOP attribute and the NLRI component of the UPDATE message. Both fields represent IPv4 addresses. To support new applications, the multiprotocol extension defined in [RFC4760] introduces the concept of an *address family* and also includes the next hop in the MP_REACH attribute along with the NLRI, for each of these address families. The next-hop information is meaningless in the route withdrawal and is not included in the MP_UNREACH, because only one path can be advertised on a session for an NLRI.

The peering relationship (internal, external, or confederation) influences the automatic modification of the next hop. The rationale behind this concept is that leaving a domain is likely to hide the internal infrastructure, while interconnecting within a domain might benefit from extended visibility of the final destination for optimal forwarding (as the result of a lower route metric value).

A summary of the default behavior regarding next-hop changes is given in Table 3.6. Notice that the default behavior for the first case is different for IP and MPLS NLRIs. The default behavior in IP is to leave next-hop unchanged for information traveling from EBGP to IBGP. MPLS routes differ in that they set next-hop self by default, with the understanding that the MPLS label domain for each of the ASs differs.

Table 3.6 Default next-hop changes for BGP updates

From	To	Next-hop change for IPv4 NLRI	Next-hop change for MPLS NLRI
external	internal	No	Yes
internal	internal	No	No
internal	external	Yes	Yes
external	external	Yes	Yes

Confederations and route reflectors propagate updates within an internal BGP domain, and as such are considered to be under a common infrastructure. Hence, by default, advertisements to a reflector client or a confederation peer in another sub-AS do not change the next-hop. Advertisements from reflectors and confederations in different sub-ASs are treated as advertisements coming from internal peers for the purposes of next-hop handling.

For updates transitioning from external to external session on the same local interface, it is possible to optimize next-hop handling: receiving an update from one party on a subnet intended for another party in the same subnet can be advertised with a next hop of the originating party. This optimization is called *third party next hop*, and is an implementation that short-cuts forwarding that would otherwise involve traffic in and out of the same interface on the intermediate router.

Junos Tip: Avoiding default BGP next-hop changes

The Junos OS policy language includes the action `then next-hop` that is leveraged both for IP and MPLS routes when them.

The default behavior follows the scheme listed in Table 3.6. Overriding the default behavior to *prevent* the next hop from changing can be achieved by resetting the defaults to never change the next hop and then relying on the policy language to set the next hop when required. Section 5.6.9 includes further details about the Junos OS `no-nexthop-change` knob.

Policy language flexibility in Junos OS assists in interconnection scenarios where not modifying the next hop for IP or MPLS routes makes sense, leveraging the level of indirection provided by BGP. When an MPLS labeled path exists end to end between ASs, it can be interesting to interconnect route reflectors that are not in the forwarding path over a multihop EBGP session that keeps the original next hop in the UPDATE, advertising the final end points as next hops.

A practical example of a L3VPN inter-AS connection with ASBRs relaying updates to the reflectors is provided in Section 5.6.9 of the case study in Chapter 5.

Figure 3.6 summarizes the decision process for next-hop handling.

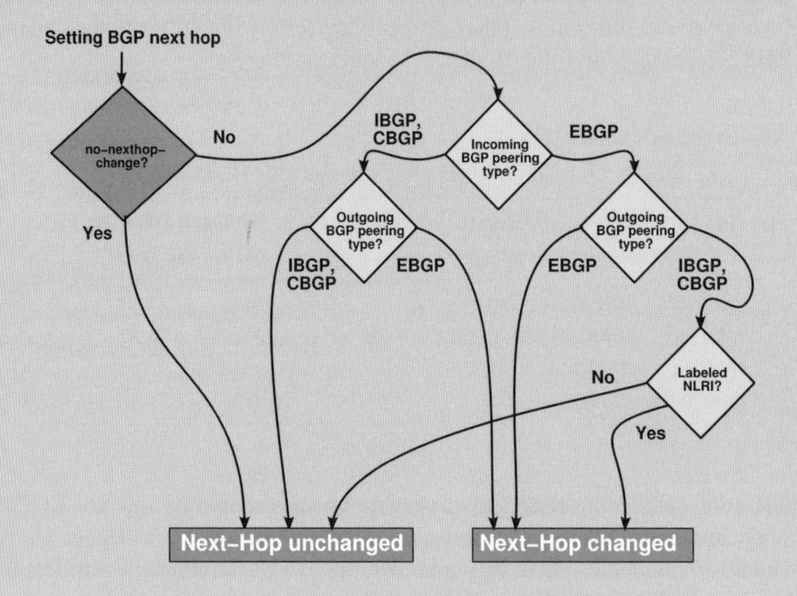

Figure 3.6: Decision process Setting next hop to self.

Local policy can modify the default next-hop change behavior, by declaring one of the local IP addresses or *self* as the next hop for an update that is being propagated. When a next-hop *self* change is in effect, the default is to use the address of the peering session as the next hop.

In migration scenarios, it may be required to set up parallel BGP sessions between two systems on different pairs of IP addresses. The next-hop change for the same prefix yields similar advertisements with differing next hops.

3.3 Generic Strategies for BGP Migrations

As described in the motivations section, Section 3.1, changes to the IBGP topology and AS renumbering are the main reasons to undergo modifications in the BGP infrastructure. This section discusses strategies to address these types of migration.

3.3.1 Leverage protocol layering

It is common practice to leave infrastructure routes to the IGP, leaving to IBGP the task of carrying the bulk of services identified by end-customer prefixes. BGP leverages the IGP protocol to set up its control plane, establishing TCP sessions between router loopback addresses, and relies on this IGP information to perform recursive BGP next-hop resolution. Splitting or merging IGPs modifies the reachability information required for the BGP control plane. A good strategy is to *divide and conquer* to simplify the migration exercise, by maintaining independence of IGP and IBGP migration activities whenever possible.

Multihop external sessions also benefit from protocol layering. In Junos OS, except for regular EBGP over directly connected sessions, a next-hop resolution process takes place to resolve the BGP next hop for prefixes learned over internal and multihop external sessions.

3.3.2 Adding redundancy

BGP rides on top of TCP, a connection-oriented transport protocol. From a graceful migration perspective, little can be done to keep a TCP session open if one of the constituent elements of the TCP socket needs modification, as is the case with a peering address change. For a network in continuous operation, the major operational concern for any network change is to keep downtime and routing churn to a minimum during migration activities. A typical strategy to minimize traffic loss because of prefix unreachability is to create more state by adding duplicate information. Both multiple advertisements of the same reachability information from different points and additional BGP peering sessions provide the necessary redundancy that may cover for routing information that disappears during transient phases in the middle of a migration.

Unfortunately, there is a connectivity impact that affects locally originated prefixes and single-homed customers that have no alternate paths, but this can typically be managed in a controlled environment through a maintenance window agreement with the end customer.

From an overall design perspective, besides improving maintenance operations, it is also beneficial to have multiple paths to a destination to improve restoration times, leaving convergence to a local decision when multiple precomputed backup paths may be available before the newly advertised best path is received. This design may also reduce overall churn in certain topologies, and it is a recommended workaround to prevent a persistent route oscillation condition, such as those described in [RFC3345].

However, having multiple alternate paths for the same prefix might not be desirable in certain situations in which a prefix is disconnected and a path withdrawal triggers *path hunting*, because different peers provide their new best-path information in successive updates until everything converges to "this prefix was disconnected." This tradeoff must be considered seriously during transient migration phases so as not to severely impact service convergence during regular operation.

3.3.3 Scalability

Adding peering sessions necessarily influences control plane scaling in the number of sessions and the number of available paths. The desire for redundancy has to be carefully balanced with the appropriate control plane capacity limits to comply with the business requirement; that is, minimizing downtime while maintaining connectivity during the migration.

3.3.4 Beware of IP routers in transit

The migration of BGP-aware routers might affect forwarding of transit traffic. Transit routers should be kept out of the forwarding path (using IGP metric techniques such as *overload* with high metrics as shown in Junos Tip on Page 213) if BGP changes are planned. An alternative is to use tunneling mechanisms (such as MPLS forwarding or Generic Routing Encapsulation (GRE) tunneling) to keep end-to-end traffic transparent to BGP changes on border routers that are also acting as transit routers. For MPLS services such as L3VPNs, this alternative is embedded in the design when using TE paths, as packets are always encapsulated and are immune to non-infrastructure IP changes in the transit layer.

Application Note: Reflectors on the forwarding path and the cluster ID

For a route to be reflected, the route reflector has to have a route that passes constraint checks and has a valid next hop to the destination. Effectively, if the reflector considers that it can reach the destination, it can advertise the prefix.

Scalability advantages of defining a pair of route reflectors within the same Cluster-ID are well understood. Client routes received from the peer reflector in the same cluster are filtered during inbound processing, and as such are not advertised further.

In designs with reflectors on the IP forwarding path, the tradeoff with such a design is that misconfiguration or a transient loss of connectivity from a client to one of the reflectors may blackhole traffic traversing that reflector whose destination is that route reflection client.

Notice that in this design, despite the partial control plane connectivity loss, the alternate reflector still advertises the prefix with a valid next hop to the rest of the network. Traffic following shortest-path metrics may be directed to the reflector that learned the prefix only from the companion route reflector but discarded it because its Cluster-ID is present in the received CLUSTER-LIST attribute, indicating a possible loop formation.

If this risk of blackholing is likely to happen in a given topology, one of the following alternatives may be used:

- Enable a different Cluster-ID on each reflector, with the scaling implications this has.
- Ensure that a tunneling mechanism is in place to avoid IP lookups on the reflector. This includes MPLS forwarding for standard IPv4 traffic flows.
- Remove the route reflector from the forwarding path.

3.3.5 Coordinating external peering changes

EBGP migrations usually involve a change in the AS number. During a transient phase, reachability information is announced over both the old and new ASs while the migration is in progress. Because EBGP migrations involve external sessions with peers, customers, and transit providers, it may be a lengthy process. Leveraging the Junos OS `alias` knob removes the need for strict timing when coordinating the AS change on the peering between both operation teams, thus allowing the focus to remain on updating security filters and loop avoidance.

3.3.6 Best-path selection

Through its various attributes, BGP provides a plethora of options to uniquely identify one best path from among a set of routes. For cases in which the change has to remain transparent to the rest of the network, it may be desirable to keep these attributes unchanged in the outgoing messages. Best-path selection can be enforced through local policy language using Junos OS protocol preference, which is useful to control single-router peering migrations that should otherwise not be noticeable to the rest of the network. A sample use of such scenario is covered in the inter-AS peering change from Option A to Option B in Section 5.6.5 of the case study in Chapter 5.

3.3.7 Changing IBGP topologies

This section discusses the modifications in the arrangement of internal peering sessions, describing the modifications depending on the desired target setup (confederation, reflection, or full mesh).

Moving to a confederation topology

The existing topology may be either a full-mesh or a route reflector setup. Interestingly enough, for existing networks with a common IGP, there may be little incentive to move from a reflection to a confederation topology, because both scaling mechanisms achieve the same goal of information reduction and the differences between the two can be compensated for by smart implementations. Support for both designs in the future is guaranteed by widespread use and by proven operational deployments.

The goal is to move a router from a regular AS to a confederated AS with minimal impact. The migration can be summarized as follows:

- identifying the confederation border router or routers. The administrative boundary is typically aligned with a change in domain control, such as an IGP or a demarcation point for the forwarding path of most of the traffic. If the source topology includes route reflection, the route reflectors might already be in the forwarding path and might be suitable candidates to become confederation border routers;

- assigning sub-AS numbers. Although these numbers are of local, intra-domain significance, Junos OS takes these AS numbers into account when performing AS loop prevention checks. A common approach is to assign AS numbers in the private range as defined in [RFC1930];

- changing routers into confederated devices one at a time, especially those connected to external peerings. The confederation border represents the AS to the rest of the network. The first elements to introduce in the confederation are the border routers that have external peerings where AS_PATH handling is performed;

- bringing up confederation peerings;
- verifying the availability of routing information;
- decommissioning old sessions. Before removing sessions, care has to be taken to verify that the active information is as expected. Incongruent information on both old and new sessions will likely influence the actual forwarding because of the next-hop reachability changes.

Moving to a route reflector topology

The route reflection technique relies on a new behavior added to a subset of routers that is independent of the pre-existing routing information. For redundancy, at least one pair of routers is selected to be reflectors. These can be existing devices or new elements to be dedicated to this task. Old BGP speakers that do not support reflection ignore the new attributes but still process the routing information.

Changing to a reflection topology requires careful selection in placement of the reflector. Because of the information reduction that reflectors introduce, care has to be taken to choose the proper location in the network to avoid oscillation, with situations described in [RFC3345]. Oscillation can be avoided by ensuring that the decisions of the reflector mimic the ones the reflector client would make. For this consistent path selection to happen, it is desirable to keep the metrics used by the reflector comparable to the ones used by the clients.

Maintaining reflectors on the forwarding path of traffic outside the cluster is a good approach to keeping the decisions within the cluster homogeneous. Dedicated reflectors tend to combine a strong control plane with a limited forwarding plane and as such, network designers tend to avoid placing them in the forwarding path.

Migration to a route reflector topology can be summarized as follows:

- identifying the route reflectors. If the original topology is a confederation, the suggested approach is to start by enabling reflection on the confederation borders, which keeps reflectors on the forwarding path and allows the appropriate path selection to occur;
- defining a cluster identifier on the reflector;
- peering with all reflector clients. Starting from a full-mesh scenario, enabling reflection effectively adds redundancy to the existing information. Notice that BGP path selection considers a path worse, the longer the contents of the CLUSTER-LIST attribute. Not having a CLUSTER-LIST attribute, as is the case in a full mesh, is considered better, and hence provides a natural "depreference" mechanism when moving to Reflection;
- verifying that additional routing information is available. No conflicts should appear related to the reachability of both BGP update versions, unless any changes in the actual forwarding behavior because of next-hop reachability are desired;
- decommissioning the old session. Given the added redundancy, if the original stage was full mesh, pre-existing sessions can be torn down progressively with no loss of connectivity.

Moving to a Full Mesh

This type of migration involves adding the necessary sessions across routers. If the initial state is a confederation, multiple personalities can be configured on the router through AS duality to maintain both types of peering simultaneously and to allow for incremental changes.

Because reflection and confederation provide scalability through reduction of information, adding direct sessions adds new information. No changes in traffic patterns occur in those cases in which the new information is always worse than the already available information. One approach would be to unprefer information over the new sessions to allow for analysis of best-path information changes after carrying out the migration.

3.4 Junos OS Implementation of BGP

Having a database exchange protocol running relentlessly for several years on a router with very long up times requires careful design. As in any protocol, information is received, processed, and propagated, but unlike other protocols, the scale at which this has to happen for BGP is significant. Traditional Internet designs involve multiple peers advertising the same routing information, requiring storage of the *active* routing table multiple times. Implementations of BGP that run on the Internet require managing a datastore of millions of routes, to be distributed to hundreds of interconnecting peers. With the advent of BGP as a supporting vehicle to provide additional services (as in the case of customer VPN routes), the size of the database may be even larger.

A base implementation that band-aids, or temporarily fixes, scalability by adding processing power is short-sighted. Here, Junos OS comes to rescue with a sound, scalable, and proven implementation, albeit one that has continuously to incorporate lessons learned from the Internet classroom to keep fit.

The following sections escort a BGP update within a router, from the packet's arrival from a BGP peer to its final propagation to a neighboring router. This discussion essentially traces the *life of a packet* in the forwarding plane world. This discussion assumes that the BGP session is already established.

3.4.1 Inbound

When the remote peer device sends a packet to the local peer, the final message it sends is an UPDATE message. At the session layer, the TCP socket, which has a large enough sized window allows this message to cross over the transmission media and the first 4 KB of the message are placed into the receive buffers. Syntactic and semantic checks following Postel's Law from [RFC793] are liberal in their controls of the received message. The BGP process extracts Network Layer Reachability Information (NLRI) and initializes new data structures. In congestion situations, BGP leverages TCP transport layer *backpressure* capabilities to ensure that inbound updates can actually be processed. Under high load, the TCP session is throttled to ensure that the sender does not flood the receiver with updates that worsen the processing requirements. In addition, the sender implementation optimizes outbound updates by means of *state compression*, suppressing older information when newer information overrides it.

3.4.2 Adj-RIB-In

The BGP protocol specification defines RIB-in, referring to the information that has not yet been processed by inbound route policy. The BGP protocol implementation can examine the update to determine whether it is appropriate and can make the decision either to keep the route or to filter it out to save memory.

Junos Tip: Retaining invalid routes

Through the configurable `keep` option in Junos OS, it is possible to maintain received inappropriate routes in memory for troubleshooting purposes. These routes remain in a *hidden* state and, as such, cannot propagate further.

By default, RIB-in filtering occurs for paths that fail the AS loop prevention check (that is, when one of the local ASs is present in the AS_PATH of the route more times than the number of configured loops). Paths that are rejected in import policy are still maintained in memory as hidden routes to avoid the need for path advertisement if local inbound policy changes.

To improve scalability in L3VPN applications, path filtering also occurs by default in routers that are acting as regular PE devices for those paths in which no RT extended communities match any of the locally configured VRFs. To allow proper route propagation, an exception is made to this automatic filtering for routers performing either a route reflector or an ASBR function.

Use of the `keep none` alternative modifies the default behavior to remove all traces of paths that fail consistency checks (AS, invalid next hop) as well as inbound policy. When this option is enabled, the local BGP peer sends a REFRESH message to the remote peer after any configuration policy changes to request that the remote peer send a fresh copy of the full set of paths.

From a generic troubleshooting perspective, the `keep all` knob may assist in identifying specific reasons why a path is remaining hidden during an AS migration. Listing 3.3 illustrates this concept by showing two prefixes that are hidden because of an AS loop. Activating `keep all` flaps all BGP sessions.

Listing 3.3: Troubleshooting looped paths with *keep all*

```
1  user@Livorno# run show route receive-protocol bgp 172.16.1.37 table inet.0 all
2
3  inet.0: 36 destinations, 38 routes (35 active, 0 holddown, 1 hidden)
4    Prefix                  Nexthop              MED     Lclpref    AS path
5  * 192.168.52.0/24         172.16.1.37                  100        (65002) I
6
7  user@Livorno# set protocols bgp keep all
8
9  # All sessions flap after this configuration change
10
11 user@Livorno# run show route receive-protocol bgp 172.16.1.37 table inet.0 all
12
13 inet.0: 36 destinations, 42 routes (35 active, 0 holddown, 5 hidden)
14   Prefix                  Nexthop              MED     Lclpref    AS path
15   192.168.51.0/24         172.16.1.37                  100        (65002 65001) I        # Looped
16 * 192.168.52.0/24         172.16.1.37                  100        (65002) I
17   192.168.53.0/24         172.16.1.37                  100        (65002 65001) 64503 I # Looped
```

3.4.3 Loc-RIB

Information selected by means of the BGP decision process is collected into the Local RIB. In Junos OS, the Loc-RIB corresponds to the regular routing table, a destination-based prefix table containing routes from all protocols. Multiple routes for the same prefix are grouped together, including both routes learned from BGP and those learned from other protocols. The decision process in selecting an active route involves a best-path selection algorithm that considers all routes that are present in the routing table. This process is not constrained to a particular protocol. The result of the best-path selection process may yield a BGP route that is *inactive*, and hence the route is not to be advertised further.

The decision process is triggered when a new update arrives or when a state change occurs, triggering the underlying transport to resolve the BGP next hop anew. These changes do not occur at scheduled times. An example of the decision process running is when an IGP change occurs that makes a BGP next hop less preferable.

3.4.4 Adj-RIB-Out

The outbound RIB includes routing information that is pending advertisement by means of UPDATE messages that are sent from the local peer to its remote peers. To optimize memory usage, BGP neighbors are *grouped* in sets that share a common advertising policy (*export*, in Junos OS parlance).

Junos Tip: Grouping together BGP peers

Because of the flexibility of Junos OS configuration, BGP properties can be configured at the neighbor level as well as at the group level, possibly creating different advertisement policies within the same group. This additional Junos OS logic automatically splits the groups into smaller groups if neighbors are configured with properties that differ from the group properties.

Listing 3.4 provides an example in which an IBGP group has three configured neighbors, one of which has a local preference setting that deviates from the default value 100 (Line 8). The resulting groups that are created detect the change in outbound policy, splitting the group accordingly (Line 28).

Listing 3.4: Automatic splitting of BGP groups

```
1  user@Livorno> show configuration protocols bgp group IBGP
2  type internal;
3  local-address 192.168.1.6;
4  family inet-vpn {
5      unicast;
6  }
7  neighbor 192.168.1.2 {
8      local-preference 10; # Change outbound default
9  }
10 neighbor 192.168.1.7;
11 neighbor 192.168.1.21;
12
13 user@Livorno> show bgp group IBGP
14 Group Type: Internal    AS: 64501                    Local AS: 64501
15   Name: IBGP            Index: 0                     Flags: <Export Eval>
16   Holdtime: 0
17   Total peers: 2        Established: 1
```

```
18   192.168.1.7+54434
19   192.168.1.21
20   bgp.l3vpn.0: 0/0/0/0
21   NMM.inet.0: 0/0/0/0
22
23 Group Type: Internal    AS: 64501                    Local AS: 64501
24   Name: IBGP            Index: 1                     Flags: <Export Eval>
25   Options: <Localpref> # Local Preference option differs for this peer
26   Holdtime: 0 Localpref: 10
27   Total peers: 1        Established: 1
28   192.168.1.2+179 # Peer split off from configured group
29   bgp.l3vpn.0: 5/5/5/0
30   NMM.inet.0: 4/5/5/0
```

Given that the nature of a group is to have a consistent export policy, outbound changes are processed on a per-peer-group basis. The multiprotocol nature of BGP is addressed by enqueuing the various protocol families in sequence on a session. During the evaluation of the export policy (Junos OS flags routes in this state with the *Exportpending* status), route sending is deferred to allow for possible changes. Efficient packing of information grouped by a common set of attributes is a further optimization that may re-merge different NLRIs in the same UPDATE message.

A further verification step checks whether the route meets the *Outbound Route Filtering* (ORF) or the route-target filter constraints, preventing it from being advertised if it meets these constraints.

The advertised information is marked to avoid any duplicate advertisements. An example that could trigger such a duplicate advertisement is a change in a route's CLUSTER-LIST attribute that has to be sent to an external peer. External advertisements do not include this attribute, so the update can be suppressed to improve efficiency.

These three types of optimization (deferring, remerging, and removing information) leverage a per-group route queue that is closely coupled to the TCP socket. Routes are added or removed from the advertisement queue, providing *state compression* and ensuring that only the most current information is advertised only when needed.

Sending updates regarding a group is optimized for slow peers. In cases when a large number of routes change, the TCP sessions of high-performance BGP peers capable of handling more routes receive a higher volume of these updates and slower peers eventually block their receiving side of the TCP session. Handling all these updates involves accommodating diverse load and latency capabilities, including moving slow peers to the *out-of-sync* state. Once the TCP session is unblocked, the updating sequence continues, but only with the latest relevant information.

Junos Tip: BGP Minimum Route Advertisement Interval

To somewhat maintain prefix churn under control, BGP defines in [RFC4271] Section 9.2.1.1 the parameter *MinRouteAdvertisementInterval* as the amount of time that must elapse between route advertisements and withdrawals.

Junos OS implementation does implement a Minimum Route Advertisement Interval (MRAI) of 0. The MRAI values specified in [RFC4271] are 5 seconds for IBGP and 30 seconds for EBGP.

The BGP configuration command `out-delay` can be used to group prefix activity and send at most one prefix change for the specified interval. The command is applied at the group and neighbor levels, but tracking is performed per destination prefix. If configured, two updates for a destination prefix are separated at least *out-delay* seconds apart. This configuration command is also applicable to VPN prefixes exchanged using BGP on a PE–CE interface.

3.5 Resources for Junos OS BGP Migrations

This section discusses specific Junos OS details and behavior to assist in migrations. Advertising an inactive route due to protocol preferences, AS identification, and allowing for controlled AS loops are discussed next.

3.5.1 Advertisement of inactive BGP routes

Some migration scenarios can result in a prefix being present in the routing table both as an IGP route and a BGP route. Default or manipulated protocol preferences in Junos OS may activate the IGP route, stopping propagation of the BGP version that contains BGP attributes. The `advertise-inactive` configuration knob assists in this scenario, allowing for a non-active BGP route to be advertised into BGP. This knob applies only for the case in which the reason for not advertising the BGP route is the existence of the same route from another protocol.

3.5.2 Flexible identification of the local AS

Bigger and better routers with enhanced scalability in the control plane hint towards a consolidation based on devices that perform multiple roles simultaneously. As described in Section 3.2.5, BGP identifies a *My Autonomous System* field in exchanges of the BGP OPEN message, but it is likely that the identification of the local AS number may require tweaking during mergers, migrations, or to accommodate new applications in which routing information traverses the same AS on different planes.

Originally intended as a straightforward way to merge ASs, the `local-as` knob in Junos OS has undergone several enhancements based on customer demand. From the onset, its aim was to provide a way to simulate an old AS in the path, allowing the acquisition of an AS to be incrementally accommodated in the network by simulating the existence of the old AS.

To better illustrate the operation of the different variants, the example in Figure 3.7 describes the AS_PATH information for two updates originating in AS64501 and AS64502 as they traverse the core AS64500, which is configured with a local-as of 64510.

The following sections detail the different alternatives to the `local-as` configuration statement, which can be used alone or in combination with additional parameters: *loops* (not discussed further), *private*, *alias*, and *no-prepend-global-as*.

Default `local-as` behavior

The default functionality of the `local-as` knob is to insert the local AS number along with the global AS number in both directions of the updates. From an AS_PATH perspective, this behavior truly simulates an external domain that is connected by EBGP at both ends. In the

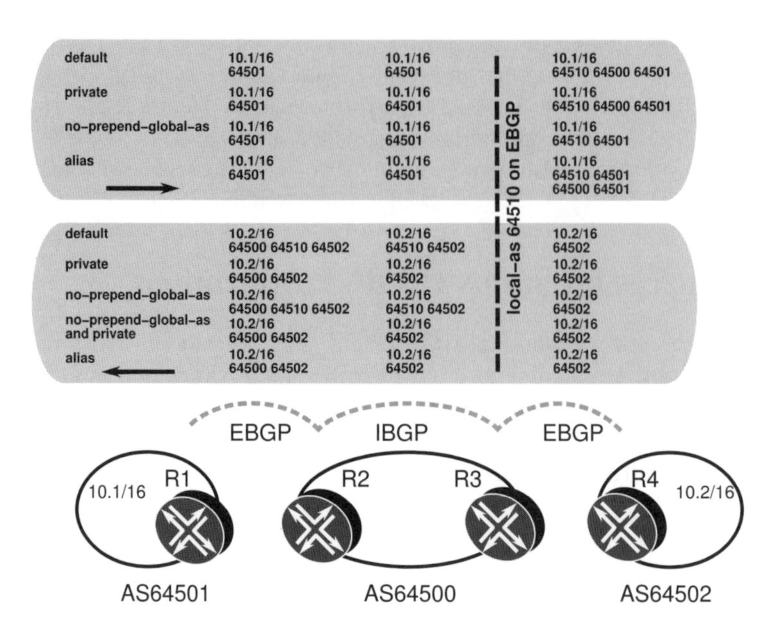

Figure 3.7: Influence of local-as alternatives in AS_PATH.

example in Figure 3.7, the prefix 10.1.0.0/16 in AS64501 on router R1 is advertised to router R2 in AS64500. The routing table contains this prefix with an AS_PATH of AS64501 (R2 does not add its own AS in IBGP peerings). R3 adds the local-as AS64510 information before advertising it to R4. In the reverse direction, the prefix 10.2.0.0/16 in AS64502 is advertised by R4, and is installed in the routing table of R3 with an additional AS, AS64510.

`private` option The local-as knob was quickly enhanced with a `private` option, which keeps the local-as identifier only in updates sent towards the peer, stripping the information internally. The BGP peering session is established on the *private* AS, and this private AS is added to outgoing updates. However, the private AS is not added to incoming updates. The rationale behind this behavior is to acknowledge to the rest of the peering neighbors that the clients behind the (now *private*) AS have been integrated successfully.

`alias` option For large AS domains with multiple external peers, coordination of both ends of the external peering connection can hinder rollout. An extension through the `alias` knob implements an alternative behavior, switching between the old AS and the new AS. Once the session is brought up with one of both ASs selected as peer AS, the identifier of the AS that is not selected for use plays no role in AS_PATH handling procedures for this neighbor. Notice that this knob cannot be used in combination with the `private` option.

The Junos Tip on Page 274 details the case in which the `alias` knob is configured on both ends of the peering arrangement, introducing a lockstep behavior as both peers fight endlessly to find the proper AS to use.

`no-prepend-global-as` option Rolling out new applications can require that the core infrastructure BGP protocol be decoupled from the access BGP protocol. This need fostered the introduction of `no-prepend-global-as` knob, which is an alternative to the use of *independent domains* discussed in Section 5.2.5. The knob effectively produces the same behavior as the `alias` command in choosing the local-as, but the two are incompatible and cannot be used together. The Application Note on Page 451 provides a use case to integrate route distribution into a L3VPN without prepending the transport AS. Notice that `no-prepend-global-as` cannot be used in combination with the `alias` configuration command, but it is possible to define `private` to also filter out the local-as in inbound updates.

Use of `local-as` with internal peerings

One of the most interesting side effects of the `local-as` knob is its use with IBGP peering sessions. These peerings require that both ends of the session agree on the same AS value at the time that the session is established. The advantage of the `local-as` knob is that it allows a per-neighbor selection for what is to be considered to be the local-as. The result is that a single router can be an IBGP speaker in two different ASs simultaneously. If the alternating flexibility of the `alias` knob is added to the mix, a smooth AS migration path can be accomplished by transitioning internal peering ASs for neighbors one by one. Note that there is no specific AS_PATH handling for routes learned through IBGP sessions.

Application Note: Migration of an IBGP AS

Figure 3.8 shows an example of a staged IBGP migration from AS64500 to 64501 using `local-as alias` and a route reflector. The key element is the usage of the knob on the route reflector group, combined with progressive AS changes on a router-by-router basis. Because internal peerings perform no modifications to the AS_PATH, it is possible to build a session between two different *system* ASs. The back-and-forth effect of the `alias` knob means that no further coordination is required between the reflector and the clients.

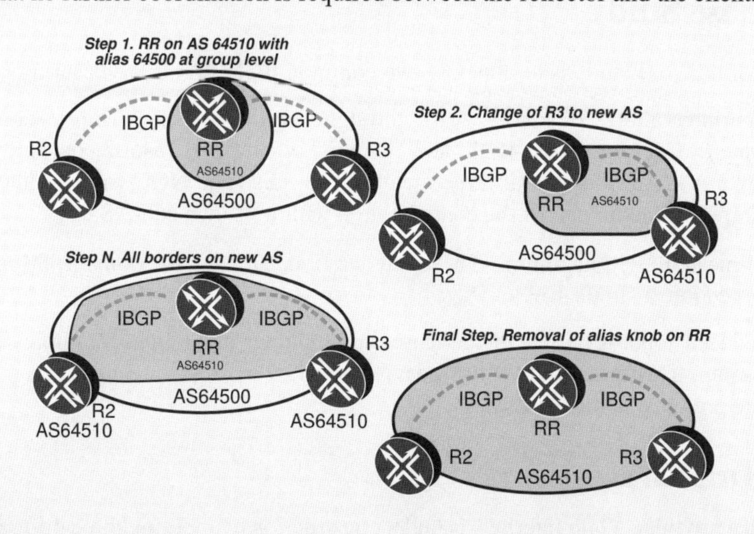

Figure 3.8: Smooth migration of IBGP AS.

The second stage of the case study in Section 3.6.2 provides a practical example that leverages this functionality.

A similar approach can be achieved for a full-mesh scenario, but it involves configuring `alias` on each router in the domain to allow for incremental session changes from the old AS to the new AS.

3.5.3 Allowing for AS loops

To avoid routing loops, Junos OS implements AS filtering on incoming updates. Routes including an AS that is configured locally are considered to be *unusable*. By default, the route is verified in Junos OS against *any* AS that is configured locally as part of a common *AS-path domain*, no matter in what routing instance, and not just for the local neighbor. No distinction is made with regard to a confederation AS or a privately configured local-as. This validation can be scoped to a set of locally configured ASs through the independent-domain functionality. Section 5.2.5 provides additional details.

In some applications, the granularity of the AS identifier verification is too coarse, and a design including a limited number of loops for certain peers is desired. As a protection measure, specifying a limit to the number of loops is required. The default validation is then relaxed to allow for up to the configured number of locally configured ASs. In Junos OS, the granularity of this configuration option ranges from the whole router all the way in the BGP configuration hierarchy down to a specific family under a neighbor `(family inet unicast loops)`.

A sample scenario that justifies the introduction of this functionality is the existence of a L3VPN application in which the same Service Provider (the same AS) provided both secure and unsecure VPN services.

3.6 Case Study

The goal of this case study is to illustrate two common BGP migration scenarios:

- fast integration of two ASs through confederation. The existing BGP domain "Cyclone," AS64501, and domain "Mistral" AS64502, become a single, 4-byte AS, AS65550, with three member sub-ASs. The NOC router, router *Lille*, is incorporated to be part of the confederation within its own sub-AS;
- full-mesh BGP integration. The confederation, along with domain "Monsoon," is merged into a single 4-byte AS.

The reduced topology used in this case study shown in Figure 3.9 mimics a real-life network setup. This case study relies heavily on the `local-as` configuration statement to simulate presence in different ASs.

3.6.1 Original network

The service provider "Gale Internet" is under construction by merging three different routing domains, as shown in Figure 3.10. The following considerations have to be taken into account before addressing this migration:

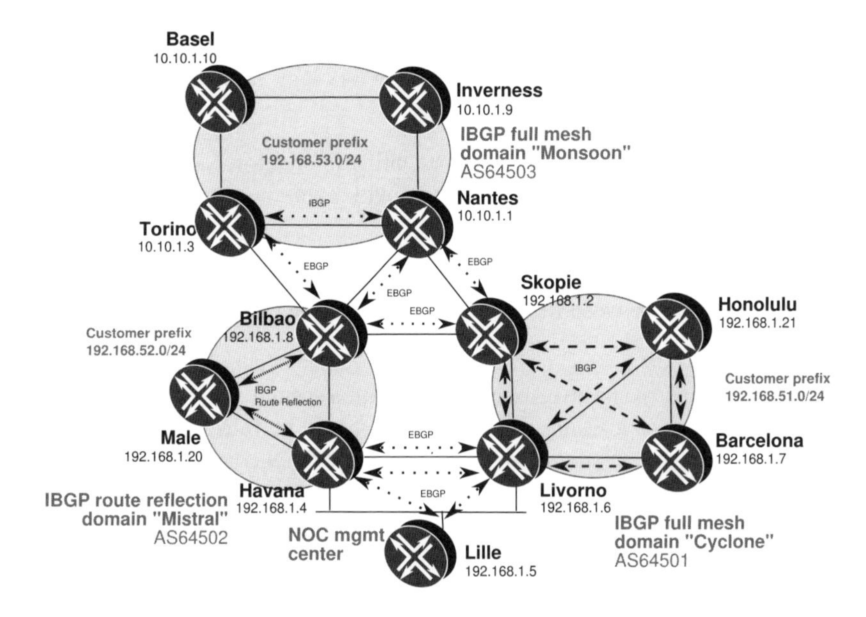

Figure 3.9: Original topology including BGP AS allocations.

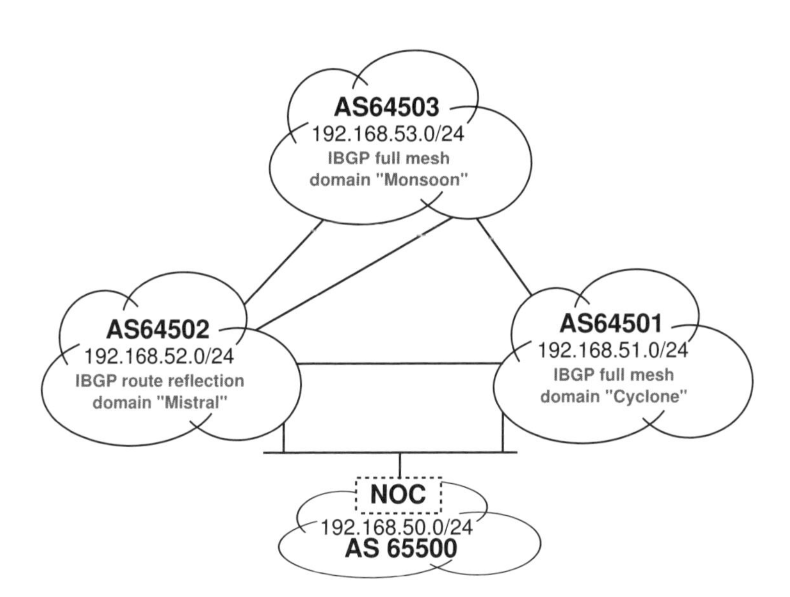

Figure 3.10: Initial BGP topology before starting the migration.

- Domain "Cyclone" is the service provider AS 64501, with a full-mesh IBGP topology.
- Domain "Mistral" has a BGP route reflection topology with router *Male* as the Route Reflector. Domain "Mistral" is AS 64502.
- Four customer addressing pools are identified, one for each AS plus the NOC infrastructure. Each pool has been tagged with community-name *customer* of value 65535:0 for traceability purposes.
- At the onset of the migration, the IGP in each domain is self-contained within the AS. Connectivity between the ASs is performed using EBGP peering over the WAN addresses.
- Each border router tags advertised prefixes with a special community identifying both the router that is advertising and the AS, as shown in Table 3.7. router *Havana* and router *Livorno* have an additional peering over the LAN connecting to the NOC.

Table 3.7 BGP communities used in the case study

Community	Description
65535:0	Customer route
64501:1	Customer of domain "Cyclone"
64502:1	Customer of domain "Mistral"
64503:1	Customer of domain "Monsoon"
64503:101	Advertised by router *Nantes*
64501:102	Advertised by router *Skopie*
64502:104	Advertised by router *Havana*
64502:1104	Advertised by router *Havana* over LAN
64501:106	Advertised by router *Livorno*
64501:1106	Advertised by router *Livorno* over LAN
64502:108	Advertised by router *Bilbao*

An additional session exists between router *Livorno* and router *Havana* over the NOC LAN.

Listing 3.5 shows the customer routes before the start of the migration as seen from router *Nantes*. The output is filtered to show only the route's AS path, community, and the path selection state. For example, prefix 192.168.52.0/24 on Line 25 belonging to domain "Mistral" has an active path from the connection to router *Bilbao*, a path learned from router *Torino*, and a path from router *Skopie* in domain "Cyclone" with a longer AS_PATH as shown on Line 35.

Listing 3.5: Initial view of customer routes at router *Nantes*

```
1 user@Nantes> show route table inet.0 community-name customer detail |
2              match "^[^ ]|Communities|AS path|reason|Proto|via"
3
4 inet.0: 40 destinations, 47 routes (40 active, 0 holddown, 0 hidden)
```

```
192.168.50.0/24 (3 entries, 1 announced)
                Next hop: 172.16.1.2 via so-0/1/0.0, selected
                AS path: 64501 65500 I
                Communities: 64501:1 64501:102 65535:0
                Next hop: 172.16.1.5 via ge-1/3/0.0, selected
                Inactive reason: Active preferred
                AS path: 64502 65500 I
                Communities: 64502:1 64502:108 65535:0
                Next hop: via so-0/2/0.0, selected
                Inactive reason: Not Best in its group - Interior > Exterior > Exterior via Interior
                AS path: 64502 65500 I
                Communities: 64502:1 64502:108 65535:0
192.168.51.0/24 (2 entries, 1 announced)
                Next hop: 172.16.1.2 via so-0/1/0.0, selected
                AS path: 64501 I
                Communities: 64501:1 64501:102 65535:0
                Next hop: 172.16.1.5 via ge-1/3/0.0, selected
                Inactive reason: AS path
                AS path: 64502 64501 I
                Communities: 64501:1 64501:106 64502:1 64502:108 65535:0
192.168.52.0/24 (3 entries, 1 announced)
                Next hop: 172.16.1.5 via ge-1/3/0.0, selected  # Router Bilbao
                AS path: 64502 I
                Communities: 64502:1 64502:108 65535:0
                Next hop: via so-0/2/0.0, selected              # Router Torino
                Inactive reason: Not Best in its group - Interior > Exterior > Exterior via Interior
                AS path: 64502 I
                Communities: 64502:1 64502:108 65535:0
                Next hop: 172.16.1.2 via so-0/1/0.0, selected  # Router Skopie
                Inactive reason: AS path
                AS path: 64501 64502 I
                Communities: 64501:1 64501:102 64502:1 64502:104 65535:0
192.168.53.0/24 (1 entry, 1 announced)
                Next hop: via so-1/0/1.0, selected
                AS path: I
                Communities: 65535:0
```

A similar view is provided for router *Havana* in domain "Mistral" in Listing 3.6. Notice that the same 192.168.52.0/24 prefix is originated within the AS64502, and as such one path is learned from router *Male* only as shown on Line 13. Route communities attached to the different paths are shown starting on Line 18.

Listing 3.6: Initial view of customer routes at router *Havana*

```
user@Havana> show route table inet.0 community-name customer terse

inet.0: 36 destinations, 43 routes (36 active, 0 holddown, 0 hidden)
+ = Active Route, - = Last Active, * = Both

A Destination        P Prf   Metric 1   Metric 2  Next hop        AS path
* 192.168.50.0/24    B 170        100             >172.16.100.5   65500 I
                     B 170        100             >172.16.1.38    64501 65500 I
                     B 170        100             >172.16.100.5   64501 65500 I
* 192.168.51.0/24    B 170        100             >172.16.1.38    64501 I
                     B 170        100             >172.16.100.8   64501 I
                     B 170        100             >172.16.100.8   65500 64501 I
* 192.168.52.0/24    B 170        100             >172.16.1.85    I
* 192.168.53.0/24    B 170        100             >172.16.1.85    64503 I
                     B 170        100             >172.16.1.38    64501 64503 I
                     B 170        100             >172.16.100.8   64501 64503 I

user@Havana> show route table inet.0 community-name customer detail |
              match "^[^ ]|comm"

inet.0: 36 destinations, 43 routes (36 active, 0 holddown, 0 hidden)
192.168.50.0/24 (3 entries, 1 announced)
```

```
23                   Communities: 65535:0
24                   Communities: 64501:1 64501:106 65535:0
25                   Communities: 64501:1 64501:1106 65535:0
26 192.168.51.0/24 (3 entries, 1 announced)
27                   Communities: 64501:1 64501:106 65535:0
28                   Communities: 64501:1 64501:1106 65535:0
29                   Communities: 65535:0
30 192.168.52.0/24 (1 entry, 1 announced)
31                   Communities: 65535:0
32 192.168.53.0/24 (3 entries, 1 announced)
33                   Communities: 64503:1 64503:101 65535:0
34                   Communities: 64501:1 64501:106 64503:1 64503:101 65535:0
35                   Communities: 64501:1 64501:1106 64503:1 64503:101 65535:0
```

The view within domain "Cyclone" is shown for router *Livorno* in Listing 3.7. Line 22 and Line 25 show two paths for the prefix 192.168.52.0/24, one over a direct leased line (WAN) connection, and another over the shared media with the upcoming NOC router. Notice that the prefix for the NOC pool 192.168.50.0/24 is also present.

Listing 3.7: Initial view of customer routes at router *Livorno*

```
1  user@Livorno> show route table inet.0 community-name customer detail |
2                 match "^[^ ]|Communities|AS path|reason|Proto|via"
3
4  inet.0: 38 destinations, 47 routes (37 active, 0 holddown, 1 hidden)
5  192.168.50.0/24 (3 entries, 1 announced)
6                   Next hop: 172.16.100.5 via fe-0/3/0.0, selected
7                   AS path: 65500 I
8                   Communities: 65535:0
9                   Next hop: 172.16.1.37 via so-0/0/1.0, selected
10                  Inactive reason: AS path
11                  AS path: 64502 65500 I
12                  Communities: 64502:1 64502:104 65535:0
13                  Next hop: 172.16.100.5 via fe-0/3/0.0, selected
14                  Inactive reason: Not Best in its group - Update source
15                  AS path: 64502 65500 I
16                  Communities: 64502:1 64502:1104 65535:0
17 192.168.51.0/24 (1 entry, 1 announced)
18                  Next hop: 172.16.1.53 via ge-0/2/0.0 weight 0x1, selected
19                  AS path: I
20                  Communities: 65535:0
21 192.168.52.0/24 (3 entries, 1 announced)
22                  Next hop: 172.16.1.37 via so-0/0/1.0, selected
23                  AS path: 64502 I
24                  Communities: 64502:1 64502:104 65535:0
25                  Next hop: 172.16.100.1 via fe-0/3/0.0, selected
26                  Inactive reason: Not Best in its group - Update source
27                  AS path: 64502 I
28                  Communities: 64502:1 64502:1104 65535:0
29                  Next hop: 172.16.100.1 via fe-0/3/0.0, selected
30                  Inactive reason: AS path
31                  AS path: 65500 64502 I
32                  Communities: 64502:1 64502:1104 65535:0
33 192.168.53.0/24 (4 entries, 1 announced)
34                  Next hop: via so-0/0/0.0 weight 0x1, selected
35                  AS path: 64503 I
36                  Communities: 64503:1 64503:101 65535:0
37                  Next hop: 172.16.1.37 via so-0/0/1.0, selected
38                  Inactive reason: AS path
39                  AS path: 64502 64503 I
40                  Communities: 64502:1 64502:104 64503:1 64503:101 65535:0
41                  Next hop: 172.16.100.1 via fe-0/3/0.0, selected
42                  Inactive reason: Not Best in its group - Update source
43                  AS path: 64502 64503 I
44                  Communities: 64502:1 64502:1104 64503:1 64503:101 65535:0
45                  Next hop: 172.16.100.1 via fe-0/3/0.0, selected
```

```
46              Inactive reason: AS path
47              AS path: 65500 64502 64503 I
48              Communities: 64502:1 64502:1104 64503:1 64503:101 65535:0
```

3.6.2 Target network

The dependence on the parallel IGP migration to a unified and common IS–IS topology blocks a direct full-mesh integration. Therefore, the approach taken is to merge two of the ASs in a confederation first, to later integrate all three domains into a full-mesh IBGP.

First milestone: Build confederation AS65550

The goal of the first phase in the migration is to transition the two independent ASs in domain “Cyclone” and domain “Mistral” into a single confederation. Figure 3.11 shows the target network scenario after reaching the first milestone. The independent AS64501 and AS64502 are merged into a confederation.

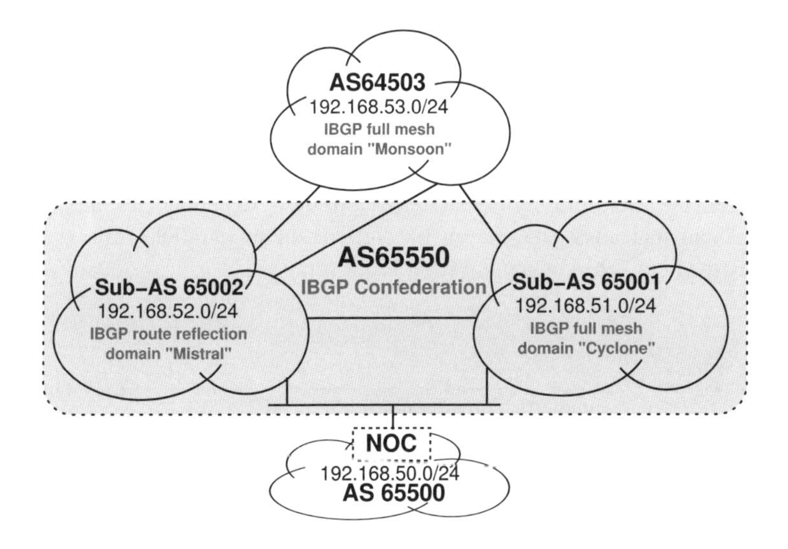

Figure 3.11: Network state after first milestone with BGP confederation.

Second milestone: Full-mesh IBGP

Migration at the IGP layer is progressing smoothly to integrate all domains into a unified IS–IS topology. After the first milestone of the IGP migration is achieved, as described in the IGP case study in Section 2.3.9, all routers have full visibility of the loopback addresses within the domain.

Business requirements impose a global merge into a single 4-byte AS65550 by transitioning the confederation to a full IBGP mesh. The second phase in the BGP migration is to move to a full-mesh IBGP topology, only after IS–IS reachability between all routers in the three domains is accomplished, as shown in Figure 3.12.

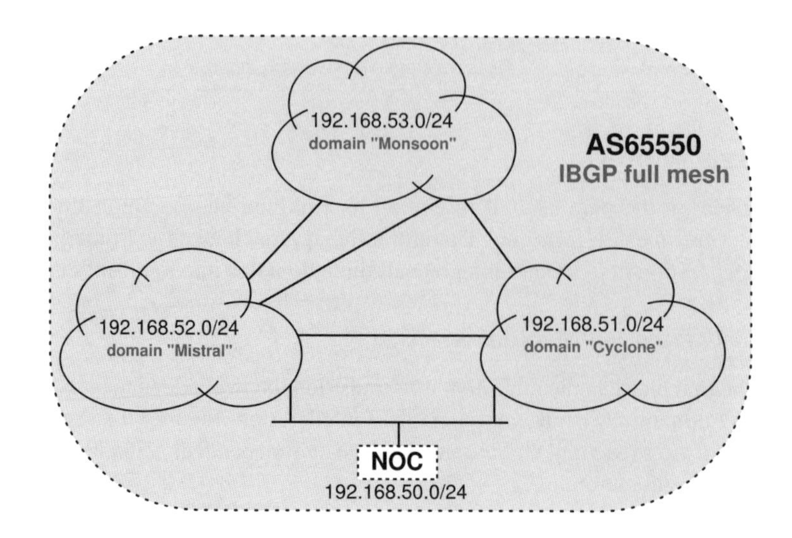

Figure 3.12: Final state after migration to full-mesh IBGP.

3.6.3 Migration strategy

The stages for the BGP migration strategy and their dependencies are illustrated in Figure 3.13. Note that some stages can be carried out in parallel while others such as confederating member sub-ASs require strict sequencing.

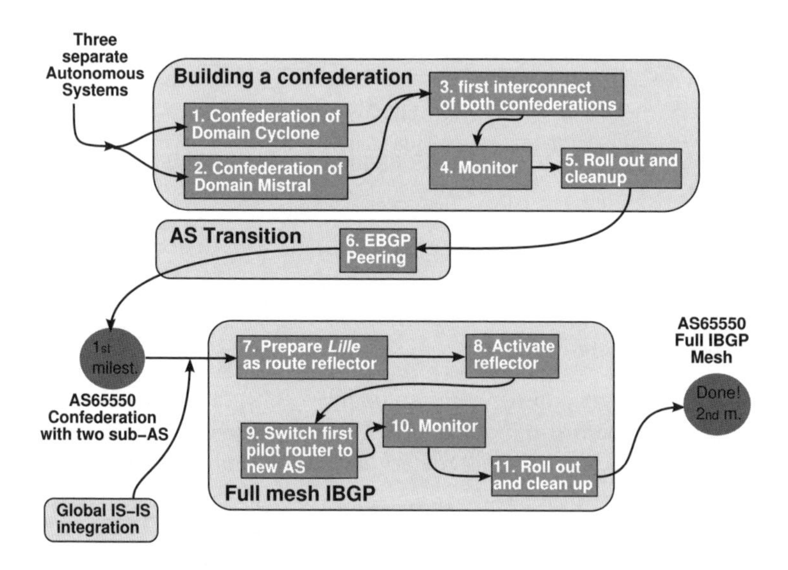

Figure 3.13: Approach taken for the BGP migration.

The activities to carry out to achieve the migration follow:

- Build a confederation out of AS64501 and AS64502;
  - In stage one, domain "Cyclone" is confederated into a single AS65550 with only one sub-AS 65001. Despite being confederated, all external peerings continue to use the original AS 64501 for peering.
  - In stage two, domain "Mistral" is also confederated into the same AS65550 with only one sub-AS 65002. External peerings follow the original AS 64502. Notice that not doing so and having AS65550 in the AS_PATH would trip AS loop detection mechanisms. Prefixes would be filtered between the two domains, which is undesirable.
  - After this preparatory work, stage three establishes the first confederation interconnect between domain "Mistral" and domain "Cyclone." The second peering session over the LAN between router *Havana* and router *Livorno* is left on the original ASs to simulate an incremental transition involving multiple interconnects.
  - Stage four monitors the transient state of both networks, which are confederated into a single AS65550 and at the same time separated into two ASs.
  - Once migration issues are addressed, stage five undertakes the rollout to the remaining routers in both domains, followed by cleanup of interim configuration.
- Present the confederation prefixes as belonging to AS65550 to external peers;
  - Previous work did not modify how external ASs view both domains, other than because of changes in AS_PATH length. Stage six modifies how other ASs view the customers' prefixes within domain "Mistral" and domain "Cyclone" by peering on the 4-byte AS 65550. Customer prefixes in these two domains are advertised as originating from AS65550.
- Full-mesh IBGP of the IBGP confederation AS 65550 with full-mesh AS 65503 and the NOC. The ultimate goal is to become a single AS. After the IS–IS migration described in the case study of Chapter 2, the staged approach taken to incorporate domain "Monsoon" and building a BGP full mesh follows:
  - Preparatory work is performed in stage seven. The NOC router *Lille* is prepared as a route reflector for AS 65550, albeit with an inactive (`bgp cluster`) reflection configuration. One reflector group is configured for each of the ASs or sub-ASs to be merged. All routers are configured with a new IBGP group, which eventually becomes the final configuration. Configured policies ensure that the reflected routing information is not considered "best." The evaluation of the reflected information should confirm that best-path selection is appropriate (based on the IGP cost).
  - At stage eight, reflection is activated for router *Lille*, which becomes a relay for the prefixes received from within the domain.
  - At stage nine, the first router is transitioned to the new AS65550. Router *Nantes* is taken as a representative router for the change.

- Stage nine allows for a monitoring period to ensure that everything is working as expected.
- Finally, stage ten releases full rollout of the changes and establishes clean up procedures. Routers within the confederation maintain the 4-byte AS but are reconfigured so as not to belong to a confederation. Routers in domain "Monsoon" are reconfigured to the new AS.

3.6.4 Stage one: Confederating domain "Cyclone"

Building a confederation involves renumbering the router AS into the confederation AS along with identifying the member sub-AS. As AS numbers are negotiated in OPEN messages, BGP sessions need to re-initialize for changes to take effect. Because re-initialization is a disruptive operation, the goal is to minimize the impact by performing an incremental, router-by-router migration. Hence, a router will be required to represent two different ASs, an old one and a new one, depending on the specific peering session. Confederating an AS introduces a new sub-AS in the AS_PATH, that is also considered in loop validation checks.

The incremental change to confederate domain "Cyclone" follows these steps:

- confederate router *Skopie*, leaving internal and external peerings untouched. The effect on external AS domain "Monsoon" as well as on the existing internal peers is evaluated;
- confederate router *Livorno* into the same sub-AS as router *Skopie*;
- confederate remaining internal routers in domain "Cyclone."

Confederating Skopie

The BGP peerings at Skopie before the start of the migration are shown in Listing 3.8. The sessions carry prefixes for regular Internet destinations, as well as the loopback addresses within the domain using labeled BGP. This is discussed further in Chapter 4. The command starting on Line 20 shows the AS numbers used in the peering sessions.

Listing 3.8: Original BGP peerings at Skopie

```
1  user@Skopie> show bgp summary
2  Groups: 2 Peers: 4 Down peers: 0
3  Table          Tot Paths  Act Paths Suppressed    History Damp State    Pending
4  inet.3                11          4          0          0          0          0
5  inet.0                 5          4          0          0          0          0
6  Peer                     AS      InPkt     OutPkt    OutQ   Flaps Last Up/Dwn State|#Active/Received/
       Accepted/Damped...
7  172.16.1.1            64503        227        229       0       0     1:35:44 Establ
8    inet.3: 0/4/4/0
9    inet.0: 1/2/2/0
10 192.168.1.6           64501        227        218       0       0     1:35:56 Establ
11   inet.3: 4/5/5/0
12   inet.0: 2/2/2/0
13 192.168.1.7           64501        214        217       0       0     1:35:48 Establ
14   inet.3: 0/1/1/0
15   inet.0: 1/1/1/0
16 192.168.1.21          64501        214        217       0       0     1:35:52 Establ
17   inet.3: 0/1/1/0
18   inet.0: 0/0/0/0
19
```

```
20 user@Skopie> show bgp neighbor | match ^Peer
21 Peer: 172.16.1.1+179 AS 64503  Local: 172.16.1.2+61782 AS 64501
22 Peer: 192.168.1.6+179 AS 64501 Local: 192.168.1.2+58608 AS 64501
23 Peer: 192.168.1.7+179 AS 64501 Local: 192.168.1.2+53273 AS 64501
24 Peer: 192.168.1.21+179 AS 64501 Local: 192.168.1.2+53061 AS 64501
```

Listing 3.9 shows the existing configuration at router *Skopie*. Additional configuration to become a member of a confederation starts at Line 4, by setting up the sub-AS number and the list of confederation members along with the confederation AS.

Notice on Line 22 that the local-as command is applied to the global level affecting all BGP groups to ensure at this preparatory stage that no change occurs to the internal or external peering by activating this configuration.

Listing 3.9: Activating the confederation

```
1  user@Skopie> show configuration routing-options autonomous-system
2  64501;
3
4  user@Skopie# set routing-options autonomous-system 65001
5  user@Skopie# set routing-options confederation 65550 members [ 65000 65001 65002 65003 ]
6
7  user@Skopie# show routing-options | find autonomous
8  autonomous-system 65001;
9  confederation 65550 members [ 65001 65002 ];
10
11 user@Skopie# set protocols bgp local-as 64501 alias
12 user@Skopie> show configuration protocols bgp
13 family inet {
14     unicast;
15     labeled-unicast {
16         rib {
17             inet.3;
18         }
19     }
20 }
21 export [ nhs own-loopback ];
22 local-as 64501 alias;          # Maintain peerings on original AS64501
23 group IBGP-Mesh {
24     type internal;
25     local-address 192.168.1.2;
26     export [ nhs own-loopback ];
27     neighbor 192.168.1.6;
28     neighbor 192.168.1.21;
29     neighbor 192.168.1.7;
30 }
31 group EBGP {
32     export [ tag-64501 tag-skopie customer-routes own-loopback ];
33     neighbor 172.16.1.1 {
34         peer-as 64503;
35     }
36     inactive: neighbor 172.16.100.99 {
37         peer-as 64502;
38     }
39 }
```

Committing these changes in the configuration involves a BGP session flap. All peerings reconfigure to the new settings, using the confederation sub-AS (65001) for the internal peerings, and the confederation AS (65550) for the external peering. This is shown in Listing 3.10. Note that both the local and remote ASs show no indication of a TCP transport port as shown starting at Line 12, hinting that the peerings are down.

Listing 3.10: Reconfigured sessions after confederation settings

```
1  user@Skopie> show bgp summary
2  Groups: 2 Peers: 4 Down peers: 4
3  Table          Tot Paths  Act Paths Suppressed    History Damp State    Pending
4  inet.3                 0          0          0          0          0          0
5  inet.0                 0          0          0          0          0          0
6  Peer                     AS      InPkt     OutPkt    OutQ   Flaps Last Up/Dwn State|#Active/Received/
       Accepted/Damped...
7  172.16.1.1           64503          0          0       0       0           2 Active
8  192.168.1.6          65001          0          0       0       0           2 Active
9  192.168.1.7          65001          0          0       0       0           2 Active
10 192.168.1.21         65001          0          0       0       0           2 Active
11
12 user@Skopie> show bgp neighbor | match ^Peer
13 Peer: 172.16.1.1 AS 64503      Local: 172.16.1.2 AS 65550
14 Peer: 192.168.1.6 AS 65001     Local: 192.168.1.2 AS 65001
15 Peer: 192.168.1.7 AS 65001     Local: 192.168.1.2 AS 65001
16 Peer: 192.168.1.21 AS 65001    Local: 192.168.1.2 AS 65001
```

With the above settings, the peers reject session establishment because of incorrect AS numbers. This rejection trips the local-as logic to switch to the original AS number, 64501. A sample exchange is shown in Listing 3.11.

Listing 3.11: Sample AS negotiation in IBGP with alias

```
1  Jun 29 06:03:20.260834 bgp_send: sending 59 bytes to 192.168.1.21 (Internal AS 65001)
2  Jun 29 06:03:20.260870
3  Jun 29 06:03:20.260870 BGP SEND 192.168.1.2+59769 -> 192.168.1.21+179
4  Jun 29 06:03:20.260918 BGP SEND message type 1 (Open) length 59
5  Jun 29 06:03:20.260955 BGP SEND version 4 as 65001 holdtime 90 id 192.168.1.2 parmlen 30
6  Jun 29 06:03:20.260985 BGP SEND MP capability AFI=1, SAFI=1
7  Jun 29 06:03:20.261013 BGP SEND Refresh capability, code=128
8  Jun 29 06:03:20.261040 BGP SEND Refresh capability, code=2
9  Jun 29 06:03:20.261071 BGP SEND Restart capability, code=64, time=120, flags=
10 Jun 29 06:03:20.261102 BGP SEND 4 Byte AS-Path capability (65), as_num 65001
11 Jun 29 06:03:20.262303
12 Jun 29 06:03:20.262303 BGP RECV 192.168.1.21+179 -> 192.168.1.2+59769
13 Jun 29 06:03:20.262379 BGP RECV message type 1 (Open) length 59
14 Jun 29 06:03:20.262416 BGP RECV version 4 as 64501 holdtime 90 id 192.168.1.21 parmlen 30
15 Jun 29 06:03:20.262446 BGP RECV MP capability AFI=1, SAFI=1
16 Jun 29 06:03:20.262473 BGP RECV Refresh capability, code=128
17 Jun 29 06:03:20.262500 BGP RECV Refresh capability, code=2
18 Jun 29 06:03:20.262531 BGP RECV Restart capability, code=64, time=120, flags=
19 Jun 29 06:03:20.262562 BGP RECV 4 Byte AS-Path capability (65), as_num 64501
20 Jun 29 06:03:20.262683
21 Jun 29 06:03:20.262683 BGP RECV 192.168.1.21+179 -> 192.168.1.2+59769
22 Jun 29 06:03:20.262747 BGP RECV message type 3 (Notification) length 21
23 Jun 29 06:03:20.262780 BGP RECV Notification code 2 (Open Message Error)
24     subcode 2 (bad peer AS number)
25 <...>
26 Jun 29 06:03:36.317318
27 Jun 29 06:03:36.317318 BGP RECV 192.168.1.21+51147 -> 192.168.1.2+179
28 Jun 29 06:03:36.317588 BGP RECV message type 1 (Open) length 59
29 Jun 29 06:03:36.317627 BGP RECV version 4 as 64501 holdtime 90 id 192.168.1.21 parmlen 30
30 Jun 29 06:03:36.317658 BGP RECV MP capability AFI=1, SAFI=1
31 Jun 29 06:03:36.317686 BGP RECV Refresh capability, code=128
32 Jun 29 06:03:36.317712 BGP RECV Refresh capability, code=2
33 Jun 29 06:03:36.317744 BGP RECV Restart capability, code=64, time=120, flags=
34 Jun 29 06:03:36.317774 BGP RECV 4 Byte AS-Path capability (65), as_num 64501
35 Jun 29 06:03:36.317995 advertising receiving-speaker only capability to neighbor 192.168.1.21 (
       Internal AS 65001)
36 Jun 29 06:03:36.318040 bgp_send: sending 59 bytes to 192.168.1.21 (Internal AS 65001)
37 Jun 29 06:03:36.318074
38 Jun 29 06:03:36.318074 BGP SEND 192.168.1.2+179 -> 192.168.1.21+51147
39 Jun 29 06:03:36.318192 BGP SEND message type 1 (Open) length 59
40 Jun 29 06:03:36.318233 BGP SEND version 4 as 64501 holdtime 90 id 192.168.1.2 parmlen 30
```

```
41 Jun 29 06:03:36.318262 BGP SEND MP capability AFI=1, SAFI=1
42 Jun 29 06:03:36.318290 BGP SEND Refresh capability, code=128
43 Jun 29 06:03:36.318317 BGP SEND Refresh capability, code=2
44 Jun 29 06:03:36.318347 BGP SEND Restart capability, code=64, time=120, flags=
45 Jun 29 06:03:36.318378 BGP SEND 4 Byte AS-Path capability (65), as_num 64501
46 Jun 29 06:03:36.318932
```

Listing 3.12 shows the result after AS negotiation, with peering ASs listed starting at Line 19. This time the ports bound to the IP addresses indicate a properly established TCP session.

Listing 3.12: Result on router *Skopie* after AS negotiation

```
1  user@Skopie> show bgp summary
2  Groups: 2 Peers: 4 Down peers: 0
3  Table          Tot Paths  Act Paths Suppressed    History Damp State    Pending
4  inet.3                11          4          0          0          0          0
5  inet.0                 6          4          0          0          0          0
6  Peer                     AS      InPkt     OutPkt    OutQ   Flaps Last Up/Dwn State|#Active/Received/
      Accepted/Damped...
7  172.16.1.1           64503         15         18       0       0          33 Establ
8    inet.3: 0/4/4/0
9    inet.0: 1/3/3/0
10 192.168.1.6          64501         14          8       0       0        1:37 Establ
11   inet.3: 4/5/5/0
12   inet.0: 2/2/2/0
13 192.168.1.7          64501          8          7       0       0        1:37 Establ
14   inet.3: 0/1/1/0
15   inet.0: 1/1/1/0
16 192.168.1.21         64501          6          6       0       0        1:37 Establ
17   inet.3: 0/1/1/0
18
19 user@Skopie> show bgp neighbor | match ^Peer
20 Peer: 172.16.1.1+179 AS 64503  Local: 172.16.1.2+51633 AS 64501
21 Peer: 192.168.1.6+62149 AS 64501 Local: 192.168.1.2+179 AS 64501
22 Peer: 192.168.1.7+64096 AS 64501 Local: 192.168.1.2+179 AS 64501
23 Peer: 192.168.1.21+64771 AS 64501 Local: 192.168.1.2+179 AS 64501
```

Because the peering AS negotiated during the OPEN message exchange decides the contents of the AS_PATH in updates, and the peering AS did not change, the routing information still matches the situation at the start of this stage.

A sample output from router *Livorno* and router *Nantes* reflects this, as shown in Listing 3.13.

Listing 3.13: Routing view in router *Nantes* and router *Livorno* after confederating skopie

```
1  user@Nantes> show route table inet.0 community-name customer terse
2
3  inet.0: 40 destinations, 47 routes (40 active, 0 holddown, 0 hidden)
4  + = Active Route, - = Last Active, * = Both
5
6  A Destination        P Prf   Metric 1   Metric 2  Next hop        AS path
7  * 192.168.50.0/24    B 170        100             >172.16.1.5     64502 65500 I
8                       B 170        100             >so-0/2/0.0     64502 65500 I
9                       B 170        100             >172.16.1.2     64501 65500 I
10 * 192.168.51.0/24    B 170        100             >172.16.1.2     64501 I
11                      B 170        100             >172.16.1.5     64502 64501 I
12 * 192.168.52.0/24    B 170        100             >172.16.1.5     64502 I
13                      B 170        100             >so-0/2/0.0     64502 I
14                      B 170        100             >172.16.1.2     64501 64502 I
15 * 192.168.53.0/24    B 170        100             >so-1/0/1.0     I
16
17 user@Livorno> show route table inet.0 community-name customer terse
18
```

```
19 inet.0: 39 destinations, 48 routes (38 active, 0 holddown, 1 hidden)
20 + = Active Route, - = Last Active, * = Both
21
22 A Destination        P Prf   Metric 1   Metric 2  Next hop          AS path
23 * 192.168.50.0/24    B 170        100             >172.16.100.5     65500 I
24                      B 170        100             >172.16.1.37      64502 65500 I
25                      B 170        100             >172.16.100.5     64502 65500 I
26 * 192.168.51.0/24    B 170        100             >172.16.1.53      I
27 * 192.168.52.0/24    B 170        100             >172.16.1.37      64502 I
28                      B 170        100             >172.16.100.1     64502 I
29                      B 170        100             >172.16.100.1     65500 64502 I
30 * 192.168.53.0/24    B 170        100             >so-0/0/0.0       64503 I
31                      B 170        100             >172.16.1.37      64502 64503 I
32                      B 170        100             >172.16.100.1     64502 64503 I
33                      B 170        100             >172.16.100.1     65500 64502 64503 I
```

Confederating Livorno

The same configuration added on router *Skopie* in Listing 3.9 is also implemented on router *Livorno*. The AS negotiation stage agrees to use the global AS, 64501, for the peerings to external neighbors as well as for the peerings to the internal neighbors that are not yet migrated.

After committing the preliminary configuration, the BGP peerings flap, as shown in Listing 3.14. Initially, all peers are assumed to be on the confederation sub-AS for internal peerings and global AS for external peerings, as shown on Line 14.

Listing 3.14: Peerings flap at router *Livorno* after confederating

```
 1 user@Livorno> show bgp summary
 2 Groups: 4 Peers: 6 Down peers: 6
 3 Table          Tot Paths  Act Paths Suppressed    History Damp State    Pending
 4 inet.0                 0          0          0          0          0          0
 5 inet.3                 0          0          0          0          0          0
 6 Peer                     AS      InPkt     OutPkt    OutQ   Flaps Last Up/Dwn State|#Active/Received/
        Accepted/Damped...
 7 172.16.1.37           64502          0          0       0       0           7 Active
 8 172.16.100.1          64502          0          0       0       0           7 Active
 9 172.16.100.5          65500          0          0       0       0           7 Active
10 192.168.1.2           65550          0          0       0       0           7 Active
11 192.168.1.7           65550          0          0       0       0           7 Active
12 192.168.1.21          65550          0          0       0       0           7 Active
13
14 user@Livorno> show bgp neighbor | match ^Peer
15 Peer: 172.16.1.37 AS 64502      Local: 172.16.1.38 AS 65550
16 Peer: 172.16.100.1 AS 64502     Local: 172.16.100.8 AS 65550
17 Peer: 172.16.100.5 AS 65500     Local: 172.16.100.8 AS 65550
18 Peer: 192.168.1.2 AS 65001      Local: 192.168.1.6 AS 65001
19 Peer: 192.168.1.7 AS 65001      Local: 192.168.1.6 AS 65001
20 Peer: 192.168.1.21 AS 65001     Local: 192.168.1.6 AS 65001
```

The result of AS negotiation is shown in Listing 3.15. Two AS numbers are used for IBGP peering: the old global AS, for non-migrated peers, and the new confederation sub-AS (65001) towards router *Skopie* (Line 5).

Listing 3.15: New peerings at router *Livorno* after confederating

```
1 user@Livorno> show bgp neighbor | match ^Peer
2 Peer: 172.16.1.37+62508 AS 64502 Local: 172.16.1.38+179 AS 64501
3 Peer: 172.16.100.1+64757 AS 64502 Local: 172.16.100.8+179 AS 64501
4 Peer: 172.16.100.5+58576 AS 65500 Local: 172.16.100.8+179 AS 64501
5 Peer: 192.168.1.2+179 AS 64501 Local: 192.168.1.6+51045 AS 65001   # IBGP Peering on sub-AS
```

```
6 Peer: 192.168.1.7+50246 AS 64501 Local: 192.168.1.6+179 AS 64501   # Still on old AS
7 Peer: 192.168.1.21+58473 AS 64501 Local: 192.168.1.6+179 AS 64501  # Still on old AS
```

Confederating internal peers

In the same fashion, the remainder of routers in AS64501 in domain "Cyclone" are progressively confederated into sub-AS 65001, one router at a time.

What If... Keeping internal peers on the old AS

If the AS migration is considered to be a transitional effort, it might not require changing all internal routers. Taking router *Honolulu* as an example of an internal router, there is no requirement to confederate it to the new AS. Advertisement over internal peerings do not care about the locally configured AS, so ensuring establishment of the session (by keeping local-as at the peers) suffices. A similar approach could be taken for router *Male* in domain "Mistral".

For ISPs with a few external connections on a couple of border routers, AS renumbering may be as simple as modifying just these border routers!

3.6.5 Stage two: Confederating domain "Mistral"

The strategy for the domain "Mistral" is a replica of the one taken for domain "Cyclone," this time for sub-AS 65002. At this stage, care has to be taken to ensure that router *Havana* and router *Livorno* peer only on the original AS numbers, and to avoid a sub-AS confederation BGP peering until both domains have properly migrated all border routers with external peerings.

Configuration changes to confederate domain "Mistral"

Listing 3.16 summarizes the changes in domain "Mistral". The configuration for router *Bilbao* mimics the one of router *Skopie*, mapped to the new sub-AS. Note that router *Male* has no alias configuration, because it is assumed that all internal peering connections to the already confederated border routers are made with the new subconfederation AS number. As an alternative, router *Male* could keep the original AS number and stick to the `alias` setting to connect to the border routers, or it could modify its own AS number but keep the knob to maintain service to additional internal border routers.

Listing 3.16: Changes to confederate AS 64502

```
1  # Bilbao, similar to Skopie
2  user@Bilbao# set routing-options autonomous-system 65002;
3  user@Bilbao# set routing-options confederation 65550 members [ 65001 65002 ]
4  user@Bilbao# set protocols bgp local-as 64502 alias
5
6  # Male, similar to Honolulu and Barcelona
7  user@Male# set routing-options autonomous-system 65002;
8  user@Male# set routing-options confederation 65550 members [ 65001 65002 ]
9
10 # Havana, similar to Livorno
11 user@Havana# set routing-options autonomous-system 65002;
12 user@Havana# set routing-options confederation 65550 members [ 65001 65002 ]
13 user@Havana# set protocols bgp local-as 64502 alias
```

The configuration for router *Havana* in Listing 3.16 mimics the one from router *Livorno*. However, because both ends of the peering session have `alias` configured (Line 13), additional configuration shown in Listing 3.17 is necessary to cover a corner case in which alias negotiation never succeeds.

Listing 3.17: Changes to confederate AS64502

```
1 [ edit protocols bgp group EBGP ]
2 user@Havana# set neighbor 172.16.1.38 local-as 64502 no-prepend-global-as
3 user@Havana# set neighbor 172.16.100.8 local-as 64502 no-prepend-global-as
```

What If... AS selection when both sides use alias

The `alias` knob alternates the local AS value between the router AS and the configured local-as. When both ends of the session implement this behavior, as is done by the neighbor on Line 19 of Listing 3.21, there may be a condition where the two sides get stuck at the global AS number, as shown in Listing 3.18. Both sides attempt to establish a session, but fail to switch ASs as they both bring down the TCP connection before *receiving* the notification with the bad peer-as number information, as shown on Line 29 and Line 56.

Having a BGP peer behave as a *passive* side (thus never initiating the TCP session) does not help because `alias` requires the OPEN message exchange to start on the local side for the AS to switch.

Listing 3.18: Stuck alias negotiation

```
1  # Havana to Livorno
2  Jun 30 05:21:30.712125 bgp_send: sending 59 bytes to 172.16.100.8 (External AS 64501)
3  Jun 30 05:21:30.712162
4  Jun 30 05:21:30.712162 BGP SEND 172.16.100.1+64956 -> 172.16.100.8+179
5  Jun 30 05:21:30.712209 BGP SEND message type 1 (Open) length 59
6  Jun 30 05:21:30.712246 BGP SEND version 4 as 23456 holdtime 90 id 192.168.1.4 parmlen 30
7  Jun 30 05:21:30.712313 BGP SEND MP capability AFI=1, SAFI=1
8  Jun 30 05:21:30.712343 BGP SEND Refresh capability, code=128
9  Jun 30 05:21:30.712370 BGP SEND Refresh capability, code=2
10 Jun 30 05:21:30.712401 BGP SEND Restart capability, code=64, time=120, flags=
11 Jun 30 05:21:30.712431 BGP SEND 4 Byte AS-Path capability (65), as_num 65550
12 Jun 30 05:21:30.714514
13 Jun 30 05:21:30.714514 BGP RECV 172.16.100.8+179 -> 172.16.100.1+64956
14 Jun 30 05:21:30.714588 BGP RECV message type 1 (Open) length 59
15 Jun 30 05:21:30.714622 BGP RECV version 4 as 23456 holdtime 90 id 192.168.1.6 parmlen 30
16 Jun 30 05:21:30.714689 BGP RECV MP capability AFI=1, SAFI=1
17 Jun 30 05:21:30.714717 BGP RECV Refresh capability, code=128
18 Jun 30 05:21:30.714743 BGP RECV Refresh capability, code=2
19 Jun 30 05:21:30.714774 BGP RECV Restart capability, code=64, time=120, flags=
20 Jun 30 05:21:30.714804 BGP RECV 4 Byte AS-Path capability (65), as_num 65550
21 Jun 30 05:21:30.714960 bgp_process_open:2587: NOTIFICATION sent to 172.16.100.8 (External AS 64501):
22                        code 2 (Open Message Error)
23                        subcode 2 (bad peer AS number),
24                        Reason: peer 172.16.100.8 (External AS 64501) claims 65550, 64501 configured
25 Jun 30 05:21:30.715070 bgp_send: sending 21 bytes to 172.16.100.8 (External AS 64501)
26 Jun 30 05:21:30.715107
27 Jun 30 05:21:30.715107 BGP SEND 172.16.100.1+64956 -> 172.16.100.8+179
28 Jun 30 05:21:30.715155 BGP SEND message type 3 (Notification) length 21
29 Jun 30 05:21:30.715186 BGP SEND Notification code 2 (Open Message Error)
30                        subcode 2 (bad peer AS number)
31
32 # Livorno to Havana
33 Jun 30 05:21:30.713454
34 Jun 30 05:21:30.713454 BGP RECV 172.16.100.1+64956 -> 172.16.100.8+179
35 Jun 30 05:21:30.713581 BGP RECV message type 1 (Open) length 59
```

```
36 Jun 30 05:21:30.713604 BGP RECV version 4 as 23456 holdtime 90 id 192.168.1.4 parmlen 30
37 Jun 30 05:21:30.713623 BGP RECV MP capability AFI=1, SAFI=1
38 Jun 30 05:21:30.713640 BGP RECV Refresh capability, code=128
39 Jun 30 05:21:30.713656 BGP RECV Refresh capability, code=2
40 Jun 30 05:21:30.713675 BGP RECV Restart capability, code=64, time=120, flags=
41 Jun 30 05:21:30.713730 BGP RECV 4 Byte AS-Path capability (65), as_num 65550
42 Jun 30 05:21:30.713795 advertising receiving-speaker only capability
43                        to neighbor 172.16.100.1 (External AS 64502)
44 Jun 30 05:21:30.713824
45 Jun 30 05:21:30.713824 BGP SEND 172.16.100.8+179 -> 172.16.100.1+64956
46 Jun 30 05:21:30.713852 BGP SEND message type 1 (Open) length 59
47 Jun 30 05:21:30.713873 BGP SEND version 4 as 23456 holdtime 90 id 192.168.1.6 parmlen 30
48 Jun 30 05:21:30.713890 BGP SEND MP capability AFI=1, SAFI=1
49 Jun 30 05:21:30.713907 BGP SEND Refresh capability, code=128
50 Jun 30 05:21:30.713922 BGP SEND Refresh capability, code=2
51 Jun 30 05:21:30.713940 BGP SEND Restart capability, code=64, time=120, flags=
52 Jun 30 05:21:30.713958 BGP SEND 4 Byte AS-Path capability (65), as_num 65550
53 Jun 30 05:21:30.713977
54 Jun 30 05:21:30.713977 BGP SEND 172.16.100.8+179 -> 172.16.100.1+64956
55 Jun 30 05:21:30.714003 BGP SEND message type 3 (Notification) length 21
56 Jun 30 05:21:30.714022 BGP SEND Notification code 2 (Open Message Error)
57                        subcode 2 (bad peer AS number)
58 Jun 30 05:21:30.714128 bgp_pp_recv:2860: NOTIFICATION sent to 172.16.100.1+64956 (proto):
59                        code 2 (Open Message Error)
60                        subcode 2 (bad peer AS number),
61                        Reason: no group for 172.16.100.1+64956 (proto)
62                        from AS 65550 found (peer as mismatch), dropping him
```

If the timing is right, such that one of the peers has already switched on the alias AS, the peering is brought up correctly. One possible way to avoid this race condition is to force one of the sides to always use the local-as (with the `no-prepend- global-as` option) instead of bouncing back and forth between the two ASs.

The configuration change is shown in Listing 3.19, where router *Livorno* has the additional flag on Line 4. Traces for router *Havana* starting on Line 10 show the resulting bringup.

Listing 3.19: Use of no-prepend-global-as to counter negotiation deadlock

```
1  user@Livorno# show protocols bgp group EBGP neighbor 172.16.100.1
2  export [ tag-64501 tag-Livorno-Lan customer-routes own-loopback ];
3  peer-as 64502;
4  local-as 64501 no-prepend-global-as; # Stick to this AS
5
6  user@Havana# show protocols bgp group EBGP neighbor 172.16.100.8
7  export [ tag-64502 tag-havana-lan customer-routes ];
8  peer-as 64501;
9
10
11 Jun 30 05:53:01.966119 BGP RECV 172.16.100.8+57819 -> 172.16.100.1+179
12 Jun 30 05:53:01.966241 BGP RECV message type 1 (Open) length 59
13 Jun 30 05:53:01.966286 BGP RECV version 4 as 64501 holdtime 90 id 192.168.1.6 parmlen 30
14 Jun 30 05:53:01.966317 BGP RECV MP capability AFI=1, SAFI=1
15 Jun 30 05:53:01.966344 BGP RECV Refresh capability, code=128
16 Jun 30 05:53:01.966371 BGP RECV Refresh capability, code=2
17 Jun 30 05:53:01.966402 BGP RECV Restart capability, code=64, time=120, flags=
18 Jun 30 05:53:01.966432 BGP RECV 4 Byte AS-Path capability (65), as_num 64501
19 Jun 30 05:53:01.966678 advertising receiving-speaker only capability
20                        to neighbor 172.16.100.8 (External AS 64501)
21 Jun 30 05:53:01.966723 bgp_send: sending 59 bytes to 172.16.100.8 (External AS 64501)
22 Jun 30 05:53:01.966757
23 Jun 30 05:53:01.966757 BGP SEND 172.16.100.1+179 -> 172.16.100.8+57819
24 Jun 30 05:53:01.966803 BGP SEND message type 1 (Open) length 59
25 Jun 30 05:53:01.966838 BGP SEND version 4 as 23456 holdtime 90 id 192.168.1.4 parmlen 30
26 Jun 30 05:53:01.966866 BGP SEND MP capability AFI=1, SAFI=1
27 Jun 30 05:53:01.966893 BGP SEND Refresh capability, code=128
```

```
28 Jun 30 05:53:01.966920 BGP SEND Refresh capability, code=2
29 Jun 30 05:53:01.966948 BGP SEND Restart capability, code=64, time=120, flags=
30 Jun 30 05:53:01.966978 BGP SEND 4 Byte AS-Path capability (65), as_num 65550
31 Jun 30 05:53:01.968733
32 Jun 30 05:53:01.968733 BGP RECV 172.16.100.8+57819 -> 172.16.100.1+179
33 Jun 30 05:53:01.968811 BGP RECV message type 3 (Notification) length 21
34 Jun 30 05:53:01.968843 BGP RECV Notification code 2 (Open Message Error)
35                          subcode 2 (bad peer AS number)
36 Jun 30 05:53:33.971997
37 Jun 30 05:53:33.971997 BGP RECV 172.16.100.8+49936 -> 172.16.100.1+179
38 Jun 30 05:53:33.972162 BGP RECV message type 1 (Open) length 59
39 Jun 30 05:53:33.972200 BGP RECV version 4 as 64501 holdtime 90 id 192.168.1.6 parmlen 30
40 Jun 30 05:53:33.972230 BGP RECV MP capability AFI=1, SAFI=1
41 Jun 30 05:53:33.972258 BGP RECV Refresh capability, code=128
42 Jun 30 05:53:33.972284 BGP RECV Refresh capability, code=2
43 Jun 30 05:53:33.972315 BGP RECV Restart capability, code=64, time=120, flags=
44 Jun 30 05:53:33.972346 BGP RECV 4 Byte AS-Path capability (65), as_num 64501
45 Jun 30 05:53:33.972595 advertising receiving-speaker only capability
46                          to neighbor 172.16.100.8 (External AS 64501)
47 Jun 30 05:53:33.972639 bgp_send: sending 59 bytes to 172.16.100.8 (External AS 64501)
48 Jun 30 05:53:33.972674
49 Jun 30 05:53:33.972674 BGP SEND 172.16.100.1+179 -> 172.16.100.8+49936
50 Jun 30 05:53:33.972721 BGP SEND message type 1 (Open) length 59
51 Jun 30 05:53:33.972755 BGP SEND version 4 as 64502 holdtime 90 id 192.168.1.4 parmlen 30
52 Jun 30 05:53:33.972783 BGP SEND MP capability AFI=1, SAFI=1
53 Jun 30 05:53:33.972810 BGP SEND Refresh capability, code=128
54 Jun 30 05:53:33.972837 BGP SEND Refresh capability, code=2
55 Jun 30 05:53:33.972866 BGP SEND Restart capability, code=64, time=120, flags=
56 Jun 30 05:53:33.972906 BGP SEND 4 Byte AS-Path capability (65), as_num 64502
```

IBGP activation inside domain “Mistral”

Because the session between router *Havana* and router *Male* is an IBGP one, the AS_PATH is not modified in the exchanged advertisements. Hence, there is no net effect to the routing information other than to the loop-detection mechanisms for received updates, which now include the confederation AS and member sub-ASs. Listing 3.20 shows no change in the routing view for router *Male*, despite the anomalous peering arrangement.

Listing 3.20: Routing views after confederating router *Havana*

```
1  user@male-re0> show bgp neighbor | match ^Peer
2  Peer: 192.168.1.4+179 AS 65002 Local: 192.168.1.20+61582 AS 65002
3  Peer: 192.168.1.8+179 AS 65002 Local: 192.168.1.20+51381 AS 65002
4
5  user@male-re0> show route table inet.0 community-name customer detail |
6            match "^[^ ]|comm|AS path|reason|via"
7
8  inet.0: 12 destinations, 12 routes (12 active, 0 holddown, 0 hidden)
9  192.168.50.0/24 (1 entry, 1 announced)
10                 Next hop: 172.16.1.86 via ge-4/2/4.0, selected
11                 AS path: 65500 I
12                 Communities: 65535:0
13 192.168.51.0/24 (1 entry, 1 announced)
14                 Next hop: 172.16.1.86 via ge-4/2/4.0, selected
15                 AS path: 64501 I
16                 Communities: 64501:1 64501:106 65535:0
17 192.168.52.0/24 (1 entry, 1 announced)
18                 AS path: I
19                 Communities: 65535:0
20 192.168.53.0/24 (1 entry, 1 announced)
21                 Next hop: via so-3/2/1.0, selected
22                 AS path: 64503 I
23                 Communities: 64503:1 64503:101 65535:0
```

3.6.6 Stage three: First confederation peering to bind both sub-AS domains together

After all the border routers are confederated in each domain, the two independent ASs with a confederation sub-AS each can be merged into a single confederation AS with two sub-ASs. This part of the transition has to be a coordinated operation on both ends of the confederation border. The existing peering between router *Livorno* and router *Havana* is modified into a confederation peering by configuring both sides of the peering session simultaneously.

Listing 3.21 shows the changes on both ends of the peering between router *Livorno* and router *Havana*. The neighbor on Line 15 is deactivated, and a new group is added for the confederation peering on Line 27. The second session between the two ASs on Line 19 is maintained to illustrate a transitional stage in which sessions use both types of arrangement, EBGP and CBGP.

Listing 3.21: Bringing up the confederation BGP external peering

```
1  user@Livorno# show protocols bgp
2  <...>
3  local-as 64501 alias;
4  group IBGP {
5      type internal;
6      local-address 192.168.1.6;
7      export [ nhs own-loopback ];
8      neighbor 192.168.1.2;
9      neighbor 192.168.1.7;
10     neighbor 192.168.1.21;
11 }
12 group EBGP {
13     type external;
14     export own-loopback;
15     inactive: neighbor 172.16.1.37 {
16         export [ tag-64501 tag-Livorno customer-routes own-loopback ];
17         peer-as 64502;
18     }
19     neighbor 172.16.100.1 {
20         export [ tag-64501 tag-Livorno-Lan customer-routes own-loopback ];
21         peer-as 64502;
22     }
23     neighbor 172.16.100.5 {
24         peer-as 65500;
25     }
26 }
27 group EBGP-Confed {
28     type external;
29     export [ nhs tag-64501 tag-Livorno customer-routes own-loopback ];
30     neighbor 172.16.1.37 {
31         peer-as 65002;
32     }
33 }
```

The confederation peering rationale assumes an intra-organizational setup, in which both organizations have full visibility. By default, the protocol mechanics do not change the next hop in updates egressing a confederation border. Given that, at this stage, visibility is limited because both domains are still independent at the IGP layer, a *next-hop self* policy is added in Line 29 to allow for proper route resolution.

Committing the changes in Listing 3.21 brings up the confederation peering on the sub-AS numbers, as shown on Line 3 of Listing 3.22. Note that the parallel session on Line 4 is still using the old AS numbers.

Listing 3.22: External confederation peering

```
1 user@Livorno> show bgp neighbor | match ^Peer
2
3 Peer: 172.16.1.37+52323 AS 65002 Local: 172.16.1.38+179 AS 65001
4 Peer: 172.16.100.1+179 AS 64502 Local: 172.16.100.8+63329 AS 64501
5 Peer: 172.16.100.5+58798 AS 65500 Local: 172.16.100.8+179 AS 64501
6 Peer: 192.168.1.2+56608 AS 64501 Local: 192.168.1.6+179 AS 65001
7 Peer: 192.168.1.7+52918 AS 64501 Local: 192.168.1.6+179 AS 64501
8 Peer: 192.168.1.21+58262 AS 64501 Local: 192.168.1.6+179 AS 64501
```

3.6.7 Stage four: Monitoring period

A cautionary period is defined to monitor this interim stage where some of the peerings are on the original AS using EBGP and one peering is already CBGP.

Prefixes originated in two ASs

Having two ASs that interconnect sessions with both old and new arrangements simultaneously may be the case when confederation peerings are brought up incrementally in the network and both ASs has multiple peering routers. During the transition, routing information has two different origins. The advertisements in Listing 3.23 show this dichotomy for the now internal customer prefix from sub-AS 65002 on router *Livorno*. For instance, looking at the customer prefix from domain "Mistral," 192.168.52.0/24 has a path with the sub-AS (Line 12), and the paths learned from the original external AS (direct on Line 13 and through router *Lille* on Line 14).

Listing 3.23: Prefixes with duplicate origins

```
1  user@Livorno> show route table inet.0 community-name customer terse
2
3  inet.0: 38 destinations, 48 routes (37 active, 0 holddown, 2 hidden)
4  + = Active Route, - = Last Active, * = Both
5
6  A Destination        P Prf   Metric 1   Metric 2  Next hop        AS path
7  * 192.168.50.0/24    B 170        100             >172.16.100.5   65500 I
8                       B 170        100             >172.16.1.37    (65002) 65500 I
9                       B 170        100             >172.16.100.5   64502 65500 I
10 * 192.168.51.0/24    B 170        100             >172.16.1.53    I
11                      B 170        100             >172.16.100.1   64502 I
12 * 192.168.52.0/24    B 170        100             >172.16.1.37    (65002) I
13                      B 170        100             >172.16.100.1   64502 I
14                      B 170        100             >172.16.100.1   65500 64502 I
15 * 192.168.53.0/24    B 170        100             >172.16.1.37    (65002) 64503 I
16                      B 170        100             >so-0/0/0.0     64503 I
17                      B 170        100             >172.16.100.1   64502 64503 I
18                      B 170        100             >172.16.100.1   65500 64502 64503 I
```

Routing loop analysis

Analysis of Listing 3.23 on Line 11 shows the customer route 192.168.51/24 belonging to domain "Cyclone" as originated by domain "Mistral". This route is originated in domain "Cyclone" and advertised to domain "Mistral" over the new confederated session, as shown in Figure 3.14. The advertisement includes the confederation sub-AS, 65001, and the information about the old AS, AS64501, is lost. Domain "Mistral" considers this prefix as originated within its own global AS65550, and sends this route back over the pre-existing EBGP session. Because it is an external peering, router *Havana* removes the confederation sub-AS information and prepends its own AS, AS64502, in the update to router *Livorno*.

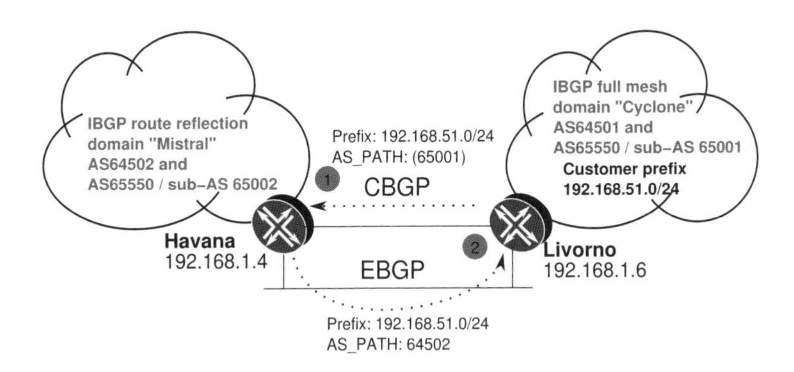

Figure 3.14: Routing loop for prefix 192.168.51.0/24 during transition.

Having both types of peering (confederated and non-confederated) introduces a potential routing loop and the default protection against using this route is the AS_PATH length rule (a prefix is better if it has a shorter AS_PATH). The experienced reader may already raise his or her eyebrows at the risk this *valid* path is introducing.

Care has to be taken to compensate for the lack of natural AS loop protection. Given that all advertisements are tagged with the originating AS, border routes can match on this community as a protection mechanism. Prefix filtering can also be implemented in cases where the community attribute is filtered by a transit BGP peer. The additional configuration policy in Listing 3.24 gives an example of this approach. A community *from64501* is tagged to all prefixes traversing domain "Cyclone" and an import policy blocks updates with this community.

Listing 3.24: Blocking self advertisements to compensate for the lack of AS loop detection

```
1  [edit]
2  user@Livorno# show policy-options policy-statement tag-64501
3  from protocol bgp;
4  then {
5      community add from64501;
6  }
7  [edit]
8  user@Livorno# show configuration policy-options policy-statement block-own
9  term community-based {
10     from community from64501;
11     then reject;
12 }
13 term prefix-based {
14     from {
15         route-filter 192.168.51.0/24 orlonger;
16     }
17     then reject;
18 }
19
20 [edit]
```

```
user@Livorno# show protocols bgp | compare
[edit protocols bgp group EBGP]
+  import block-own;
[edit protocols bgp group EBGP-Confed]
+  import block-own;
```

However, implementing this change might eliminate more than just the customer prefixes if a third-party AS is receiving transit service. Looking at the prefix from domain "Monsoon," 192.168.53.0/24, and comparing Listing 3.23, Line 17 and Listing 3.25, Line 10, the missing path for prefix 192.168.53/24 can be explained as seen in Listing 3.25, Line 13, which shows that router *Livorno* tagged this update when advertising it to router *Havana* (Line 38).

Listing 3.25: Blocking looped third-party updates

```
user@Livorno# run show route table inet.0 terse 192.168.53.0/24

inet.0: 38 destinations, 45 routes (37 active, 0 holddown, 5 hidden)
+ = Active Route, - = Last Active, * = Both

A Destination        P Prf   Metric 1   Metric 2  Next hop        AS path
* 192.168.51.0/24    B 170        100            >172.16.1.53     I
* 192.168.52.0/24    B 170        100            >172.16.1.37     (65002) I
                     B 170        100            >172.16.100.1    64502 I
* 192.168.53.0/24    B 170        100            >so-0/0/0.0      64503 I

user@Livorno# run show route table inet.0 community-name customer all 192.168.53.0/24
              detail | match "^[^ ]|state|BGP|AS path|commu"

inet.0: 37 destinations, 46 routes (36 active, 0 holddown, 5 hidden)
192.168.53.0/24 (3 entries, 1 announced)
        *BGP    Preference: 170/-101
                Label-switched-path Livorno-to-Skopie
                State: <Active Int Ext>
                Local AS: 65001 Peer AS: 64501
                Task: BGP_65001.192.168.1.2+56608
                Announcement bits (3): 0-KRT 4-BGP RT Background 5-Resolve tree 2
                AS path: 64503 I
                Communities: 64503:1 64503:101 65535:0
         BGP                /-101
                State: <Hidden Ext>
                Inactive reason: Unusable path     # Hidden route because of policy
                Local AS: 65001 Peer AS: 64502
                Task: BGP_64502.172.16.100.1+179
                AS path: 64502 64503 I
                Communities: 64501:1 64501:106 64502:1 64502:1104 64503:1 64503:101 65535:0
         BGP
                State: <Hidden Int Ext>
                Inactive reason: Unusable path     # Hidden route because of AS loop
                Local AS: 65001 Peer AS: 65002
                Task: BGP_65002.172.16.1.37+52323
                AS path: (65002 65001) 64503 I (Looped: 65001)
                Communities: 64501:1 64501:106 64502:1 64502:104 64503:1 64503:101 65535:0
```

Hidden routes for labeled paths

Changing from single-hop EBGP to CBGP activates the route resolution process. No next-hop resolution is necessary in single-hop EBGP because a local connection exists. For CBGP and IBGP, the next-hop can be far away.

In regular IPv4 prefixes, BGP next-hop resolution occurs on the inet.3 RIB first, then on inet.0. For labeled paths, like VPN-IPv4 and Labeled-IPv4, resolution occurs using inet.3 only. Even though the CBGP session peers on the interface address, the resolution process looks at inet.3 for the appropriate next hop. This concept is discussed further in Section 4.4.4.

The paths received from the CBGP session in router *Livorno* shown on Listing 3.26 indicate that the IPv4 paths are correctly received and installed while Labeled-IPv4 paths remain hidden (Line 11).

Listing 3.26: Change from EBGP to CBGP hides routes in inet.3

```
1  user@Livorno> show route receive-protocol bgp 172.16.1.37
2
3  inet.0: 36 destinations, 42 routes (35 active, 0 holddown, 3 hidden)
4    Prefix              Nexthop              MED     Lclpref    AS path
5  * 192.168.52.0/24     192.168.1.20                 100        (65002) I
6  * 192.168.53.0/24     192.168.1.8                  100        (65002) 64503 I
7
8  inet.3: 7 destinations, 16 routes (6 active, 0 holddown, 7 hidden)  # Hidden routes in inet.3
9
10
11 user@Livorno> show route receive-protocol bgp 172.16.1.37 hidden
12
13 inet.0: 36 destinations, 42 routes (35 active, 0 holddown, 3 hidden)
14   Prefix              Nexthop              MED     Lclpref    AS path
15   192.168.51.0/24     192.168.1.7                  100        (65002 65001) I
16
17 inet.3: 7 destinations, 16 routes (6 active, 0 holddown, 7 hidden)
18   Prefix              Nexthop              MED     Lclpref    AS path
19   192.168.1.2/32      172.16.1.37                  100        (65002) 64502 64501 I
20   192.168.1.6/32      172.16.1.37                  100        (65002) 64502 64501 I
21   192.168.1.7/32      172.16.1.37                  100        (65002) 64502 64501 I
22   192.168.1.8/32      192.168.1.8                  100        (65002) I     # No next hop
23   192.168.1.20/32     192.168.1.20                 100        (65002) I     # No next hop
24   192.168.1.21/32     172.16.1.37                  100        (65002) 64502 64501 I
```

To activate these labeled paths it is required to perform an additional configuration step as described in the Junos Tip on Section 4.5.4.

Listing 3.27 shows the configuration to leak the interface in inet.3 using RIB groups along with the corrected view.

Listing 3.27: Leak interface in inet.3 to allow resolution

```
1  user@Livorno# show | compare
2  [edit policy-options policy-statement interface-to-inet3]
3       term loopback { ... }
4  +    term cbgp {
5  +        from interface so-0/0/1.0;
6  +        then accept;
7  +    }
8       term no-more-inet3 { ... }
9
10 user@Livorno# show routing-options rib-groups int-to-inet3
11 import-rib [ inet.0 inet.3 ];
12 import-policy interface-to-inet3;
13
14 user@Livorno> show route receive-protocol bgp 172.16.1.37
15
16 inet.0: 36 destinations, 42 routes (35 active, 0 holddown, 3 hidden)
17   Prefix              Nexthop              MED     Lclpref    AS path
18 * 192.168.52.0/24     172.16.1.37                  100        (65002) I
19 * 192.168.53.0/24     172.16.1.37                  100        (65002) 64503 I
20
21 inet.3: 9 destinations, 18 routes (8 active, 0 holddown, 5 hidden)
22   Prefix              Nexthop              MED     Lclpref    AS path
23 * 192.168.1.4/32      172.16.1.37                  100        (65002) I
24 * 192.168.1.8/32      172.16.1.37                  100        (65002) I
25 * 192.168.1.20/32     172.16.1.37                  100        (65002) I
```

Confederation impact on third-party ASs

External ASs such as domain "Monsoon" should not care about the confederation segments, because confederation handling is purely internal. However, as Listing 3.28 shows, the change to confederation is likely to modify the forwarding pattern, because now both ASs advertise all routes within the confederation as originated by themselves.

All information is now *one hop away*. Before bringing up the confederation peering, the command in Line 1 shows the customer-routing information from both confederation sub-ASs as received only from the closest AS.

The same command output in Line 19 shows the same prefixes after setting up the confederation peering. Note that the customer prefixes 192.168.51/24 and 192.168.52/24 belonging to the confederation AS65550 are equally close from any peering connection.

Listing 3.28: External impact of confederation peering activation on third-party AS

```
1  user@Nantes> show route table inet.0 community-name customer terse
2
3  inet.0: 40 destinations, 47 routes (40 active, 0 holddown, 0 hidden)
4  + = Active Route, - = Last Active, * = Both
5
6  A Destination        P Prf   Metric 1   Metric 2  Next hop        AS path
7  * 192.168.50.0/24    B 170        100             >172.16.1.2     64501 65500 I
8                       B 170        100             >172.16.1.5     64502 65500 I
9                       B 170        100             >so-0/2/0.0     64502 65500 I
10 * 192.168.51.0/24    B 170        100             >172.16.1.2     64501 I
11                      B 170        100             >172.16.1.5     64502 64501 I
12 * 192.168.52.0/24    B 170        100             >172.16.1.5     64502 I
13                      B 170        100             >so-0/2/0.0     64502 I
14                      B 170        100             >172.16.1.2     64501 64502 I
15 * 192.168.53.0/24    B 170        100             >so-1/0/1.0     I
16
17 # bringup of confederation session between Livorno and Havana
18
19 user@Nantes> show route table inet.0 community-name customer terse
20
21 inet.0: 40 destinations, 48 routes (40 active, 0 holddown, 0 hidden)
22 + = Active Route, - = Last Active, * = Both
23
24 A Destination        P Prf   Metric 1   Metric 2  Next hop        AS path
25 * 192.168.50.0/24    B 170        100             >172.16.1.2     64501 65500 I
26                      B 170        100             >172.16.1.5     64502 65500 I
27                      B 170        100             >so-0/2/0.0     64502 65500 I
28 * 192.168.51.0/24    B 170        100             >172.16.1.2     64501 I
29                      B 170        100             >172.16.1.5     64502 I
30                      B 170        100             >so-0/2/0.0     64502 I
31 * 192.168.52.0/24    B 170        100             >172.16.1.2     64501 I
32                      B 170        100             >172.16.1.5     64502 I
33                      B 170        100             >so-0/2/0.0     64502 I
34 * 192.168.53.0/24    B 170        100             >so-1/0/1.0     I
```

Taking as an example the customer routes of AS64501 in Listing 3.29, the path received from AS64502 (Line 30) does not include AS64501. Forwarding did not change, as indicated on Line 26, only because of the *active preferred* feature.

Listing 3.29: New routing information after confederation bringup

```
1 user@Nantes> show route table inet.0 192.168.51.0/24 detail
2
3 inet.0: 40 destinations, 48 routes (40 active, 0 holddown, 0 hidden)
4 192.168.51.0/24 (3 entries, 1 announced)
5         *BGP    Preference: 170/-101
6                 Next hop type: Router, Next hop index: 567
```

```
 7                Next-hop reference count: 8
 8                Source: 172.16.1.2
 9                Next hop: 172.16.1.2 via so-0/1/0.0, selected
10                State: <Active Ext>
11                Local AS: 64503 Peer AS: 64501
12                Age: 5:35:50
13                Task: BGP_64501.172.16.1.2+51633
14                Announcement bits (3): 0-KRT 4-BGP RT Background 5-Resolve tree 1
15                AS path: 64501 I
16                Communities: 64501:1 64501:102 65535:0
17                Accepted
18                Localpref: 100
19                Router ID: 192.168.1.2
20         BGP    Preference: 170/-101
21                Next hop type: Router, Next hop index: 574
22                Next-hop reference count: 7
23                Source: 172.16.1.5
24                Next hop: 172.16.1.5 via ge-1/3/0.0, selected
25                State: <Ext>
26                Inactive reason: Active preferred
27                Local AS: 64503 Peer AS: 64502
28                Age: 1:01
29                Task: BGP_64502.172.16.1.5+179
30                AS path: 64502 I  # AS Path does not include 64501
31                Communities: 64501:1 64501:106 64502:1 64502:108 65535:0
32                Accepted
33                Localpref: 100
34                Router ID: 192.168.1.8
35         BGP    Preference: 170/-101
36                Next hop type: Indirect
37                Next-hop reference count: 3
38                Source: 192.168.1.3
39                Next hop type: Router, Next hop index: 564
40                Next hop: via so-0/2/0.0, selected
41                Protocol next hop: 192.168.1.3
42                Indirect next hop: 8c930a8 -
43                State: <NotBest Int Ext>
44                Inactive reason: Not Best in its group - Interior > Exterior > Exterior via Interior
45                Local AS: 64503 Peer AS: 64503
46                Age: 1:01       Metric2: 160
47                Task: BGP_64503.192.168.1.3+64364
48                AS path: 64502 I
49                Communities: 64501:1 64501:106 64502:1 64502:108 65535:0
50                Accepted
51                Localpref: 100
52                Router ID: 10.10.1.3
```

The analysis of external AS shows that traffic may change for prefixes belonging to confederation members. This change is also expected whenever an AS peers in multiple locations with another AS and is not specific to confederations.

Junos Tip: Detecting a leak of confederation segments

In Junos OS, the EBGP session strips confederation segments only if the router is configured as a confederation member.

If the confederation peering between sub-AS 65001 and sub-AS 65002 is established before all border routers are confederated, leaking of the AS_CONFED elements in the AS_PATH occurs. This leak can be detected easily on the receiving peer, because the offending route flags the route as *hidden* and *unusable* and contains the invalid confederation member in the AS_PATH.

Listing 3.30 is the route view on router *Nantes* when router *Bilbao* is not yet configured as a confederation member, but when the peering between router *Livorno* and router

Havana is a confederation BGP session. The AS_PATH information of the update from router *Bilbao* for customers in domain "Cyclone" (192.168.51.0/24) includes the confederation AS 65001, which is unexpected on an external peering session. Line 16 reports this fact.

Listing 3.30: Illegal confederation segment received over EBGP

```
1  user@Nantes> show route hidden table inet.0 detail
2
3  inet.0: 43 destinations, 49 routes (43 active, 0 holddown, 1 hidden)
4  <...>
5  192.168.51.0/24 (2 entries, 1 announced)
6          BGP
7                  Next hop type: Router, Next hop index: 563
8                  Next-hop reference count: 6
9                  Source: 172.16.1.5
10                 Next hop: 172.16.1.5 via ge-1/3/0.0, selected
11                 State: <Hidden Ext>
12                 Inactive reason: Unusable path
13                 Local AS: 64503 Peer AS: 64502
14                 Age: 2:07
15                 Task: BGP_64502.172.16.1.5+179
16                 AS path: 64502 (65001) I (InvalidConfedSegment)
17                 Communities: 64501:1 64501:106 64502:1 64502:108 65535:0
18                 Router ID: 192.168.1.8
```

3.6.8 Stage five: CBGP deployment and EBGP cleanup between domain "Cyclone" and domain "Mistral"

General rollout and subsequent cleanup of legacy configuration is carried out in this stage.

General deployment of CBGP

Interesting effects showed up during the monitoring stage. Once clarified, general rollout of remaining CBGP peerings can take place in a coordinated fashion. Routing changes can be handled using BGP attributes to prioritize some routes over others, which is a common operation in the internetworking industry.

Cleanup of EBGP

After domain "Cyclone" and domain "Mistral" have been confederated and when all the confederation peerings are up, there is no further need to keep the original AS for internal peerings. An example is shown in Listing 3.31. Line 1 shows the confederation peerings at router *Livorno* over the internal AS65001.

Because `alias` is active for IBGP, the peering between router *Skopie* and router *Livorno* shown in Line 5 and Line 15 illustrates diverse AS settings based on how the AS negotiation takes place.

Listing 3.31: IBGP peering on confederation AS

```
1  user@Skopie# run show bgp neighbor | match ^Peer
2  Peer: 172.16.1.37+51276 AS 65002 Local: 172.16.1.38+179 AS 65001
3  Peer: 172.16.100.1+179 AS 65002 Local: 172.16.100.8+65414 AS 65001
4  Peer: 172.16.100.5+58798 AS 65500 Local: 172.16.100.8+179 AS 64501
```

```
5  Peer: 192.168.1.2+179 AS 64501 Local: 192.168.1.6+54176 AS 65001 # Real-AS
6  Peer: 192.168.1.7+60913 AS 65001 Local: 192.168.1.6+179 AS 65001
7  Peer: 192.168.1.21+179 AS 65001 Local: 192.168.1.6+55362 AS 65001
8
9  # peering flaps, router restart
10
11 user@Livorno> show bgp neighbor | match ^Peer
12 Peer: 172.16.1.37+179 AS 65002 Local: 172.16.1.38+60039 AS 65001
13 Peer: 172.16.100.1+179 AS 65002 Local: 172.16.100.8+59415 AS 65001
14 Peer: 172.16.100.5+64620 AS 65500 Local: 172.16.100.8+179 AS 64501
15 Peer: 192.168.1.2+62535 AS 65001 Local: 192.168.1.6+179 AS 65001 # Sub-AS
16 Peer: 192.168.1.7+62835 AS 65001 Local: 192.168.1.6+179 AS 65001
17 Peer: 192.168.1.21+179 AS 65001 Local: 192.168.1.6+55843 AS 65001
```

The `alias` command can be removed for IBGP, but it has to stay active for external peerings, discussed in Section 3.6.9. Configuration changes are shown in Listing 3.32. Notice that the removal of the `alias` knob is not disruptive, as revealed by the *Last Up/Down* column starting at Line 8.

Listing 3.32: Keep alias command for EBGP

```
1  [edit]
2  user@Skopie# show | compare
3  [edit protocols bgp]
4  -    local-as 64501 alias;
5  [edit protocols bgp group EBGP]
6  +     local-as 64501 alias;
7
8  user@Skopie> show bgp summary
9  Groups: 2 Peers: 4 Down peers: 0
10 Table          Tot Paths  Act Paths Suppressed    History Damp State    Pending
11 inet.0                 7          4          0          0          0          0
12 inet.3                11          4          0          0          0          0
13 Peer                     AS      InPkt     OutPkt    OutQ   Flaps Last Up/Dwn State|#Active/Received/
       Accepted/Damped...
14 172.16.1.1            64503         26         29       0       0        7:57 Establ
15   inet.0: 1/3/3/0
16   inet.3: 3/4/4/0
17 192.168.1.6           65001         26         27       0       0        7:52 Establ
18   inet.0: 2/3/3/0
19   inet.3: 1/5/5/0
20 192.168.1.7           65001         22         28       0       0        7:57 Establ
21   inet.0: 1/1/1/0
22   inet.3: 0/1/1/0
23 192.168.1.21          65001         21         29       0       0        7:57 Establ
24   inet.0: 0/0/0/0
25   inet.3: 0/1/1/0
```

3.6.9 Stage six: External peering on new 4-byte AS

At this point, the confederation has been successfully migrated. However, the external view still sees the old AS, awaiting customers, peers, and service providers to change their external peering configuration. The intuitive change at the external AS is the peer-as setting, but the changes usually also involve tweaking policies related to the updates themselves, including routing policy and placing additional prefix rules in forwarding filters.

Use of domain "Monsoon" as transit

The goal when merging an AS is to consolidate different ASs into one. If the AS to merge is receiving transit from a third party, this transit service breaks as soon as the AS_PATH includes the AS number of the consolidated AS. At the onset of the migration, domain "Mistral" and domain "Cyclone" can get transit service through router *Nantes* in domain "Monsoon." As soon as the external peering uses the confederation AS 65550, transit breaks because of AS loop detection at router *Skopie* and router *Bilbao*. Mechanisms exist to allow leaking of such prefixes through the loops configuration construct, but this involves a change in the whole domain. After all, an AS should not require its confederation to be repaired through a transit AS!

Changing EBGP peering at domain "Monsoon"

Listing 3.33 is an example of a change in router *Nantes* as per configuration on Line 18 that takes effect immediately by router *Skopie* and router *Bilbao* after AS negotiation occurs.

Listing 3.33: External AS border Nantes to peer to 4-byte confederation AS

```
1  user@Nantes> show bgp summary
2  Groups: 2 Peers: 5 Down peers: 0
3  Table          Tot Paths  Act Paths Suppressed    History Damp State    Pending
4  inet.3                 9          8          0          0          0          0
5  inet.0                10          4          0          0          0          0
6  Peer                     AS      InPkt     OutPkt    OutQ   Flaps Last Up/Dwn State|#Active/Received/
       Accepted/Damped...
7  172.16.1.2           64501       1019        937       0       4       12:23 Establ
8    inet.3: 4/5/5/0
9    inet.0: 1/3/3/0
10 172.16.1.5           64502       1370       1327       0       4     1:18:57 Establ
11   inet.3: 4/4/4/0
12   inet.0: 2/3/3/0
13 192.168.1.3          64503       1811       1805       0       0    13:14:47 0/3/3/0
       0/0/0/0
14 192.168.1.9          64503       1755       1807       0       0    13:15:26 0/0/0/0
       0/0/0/0
15 192.168.1.10         64503       1759       1808       0       0    13:15:39 1/1/1/0
       0/0/0/0
16
17  [edit]
18 user@Nantes# show | compare
19  [edit protocols bgp group EBGP neighbor 172.16.1.5]
20 -      peer-as 64502;
21 +      peer-as 65550;
22  [edit protocols bgp group EBGP neighbor 172.16.1.2]
23 -      peer-as 64501;
24 +      peer-as 65550;
25
26 user@Nantes> show bgp summary
27 Groups: 2 Peers: 5 Down peers: 0
28 Table          Tot Paths  Act Paths Suppressed    History Damp State    Pending
29 inet.3                16          8          0          0          0          0
30 inet.0                10          4          0          0          0          0
31 Peer                     AS      InPkt     OutPkt    OutQ   Flaps Last Up/Dwn State|#Active/Received/
       Accepted/Damped...
32 172.16.1.2           65550         24         16       0       0        5:59 Establ
33   inet.3: 4/8/8/0
34   inet.0: 0/3/3/0
35 172.16.1.5           65550         22         24       0       0        6:03 Establ
36   inet.3: 4/8/8/0
37   inet.0: 3/3/3/0
38 192.168.1.3          64503       1829       1829       0       0    13:22:56 0/3/3/0
       0/0/0/0
39 192.168.1.9          64503       1773       1830       0       0    13:23:35 0/0/0/0
       0/0/0/0
40 192.168.1.10         64503       1777       1832       0       0    13:23:48 1/1/1/0
       0/0/0/0
```

Junos Tip: Handling of unsupported 4-byte AS

All border Routers in domain "Monsoon" need to change their external peerings to the new 4-byte confederated AS 65550. The caveat is that router *Torino* runs a BGP implementation that does not support 4-byte AS numbers. Listing 3.34 shows the peering and the route view on router *Torino* before and after router *Nantes* migrates its external peering to the confederation. The negotiation for the 4-byte AS capability fails between router *Nantes* and router *Torino* (Line 15). Although router *Nantes* shows that it wants to advertise the 4-byte AS numbers in the AS_PATH for the prefixes learned from the confederation AS (Line 24), it detects that router *Livorno* does not support the AS-4-byte extension. The 4-byte AS numbers are masked in the optional, transitive AS4_PATH attribute by router *Nantes*, which provides the AS_TRANS as an alternative for router *Torino* (Line 28).

Listing 3.34: Torino does not support 4-byte AS numbers

```
1  user@Torino> show route community-name customer terse table inet.0
2
3  inet.0: 40 destinations, 45 routes (40 active, 0 holddown, 0 hidden)
4  + = Active Route, - = Last Active, * = Both
5
6  A Destination        P Prf   Metric 1   Metric 2  Next hop         AS path
7  * 192.168.51.0/24    B 170        100             >172.16.1.102    64502 I
8                       B 170        100             >so-0/1/0.0      64502 I
9  * 192.168.52.0/24    B 170        100             >172.16.1.102    64502 I
10                      B 170        100             >so-0/1/0.0      64502 I
11 * 192.168.53.0/24    B 170        100             >at-1/2/0.1001   I
12                      B 170        100             >at-1/2/0.1001   I
13
14 user@Nantes> show bgp neighbor 192.168.1.3 | match "4 byte"
15   Peer does not support 4 byte AS extension
16
17 user@Nantes> show route advertising-protocol bgp 192.168.1.3
18
19 user@Nantes> show route advertising-protocol bgp 192.168.1.3
20
21 inet.0: 40 destinations, 48 routes (40 active, 0 holddown, 0 hidden)
22   Prefix                 Nexthop              MED     Lclpref    AS path
23 * 192.168.50.0/24        Self                         100        65550 65500 I
24 * 192.168.51.0/24        Self                         100        65550 I
25 * 192.168.52.0/24        Self                         100        65550 I
26
27 user@Torino> show route receive-protocol bgp 192.168.1.1 table inet.0 community-name
28     customer
29
30 inet.0: 38 destinations, 43 routes (38 active, 0 holddown, 0 hidden)
31   Prefix                 Nexthop              MED     Lclpref    AS path
32   192.168.50.0/24        192.168.1.1                  100        23456 65500 I
33   192.168.51.0/24        192.168.1.1                  100        23456 I
34   192.168.52.0/24        192.168.1.1                  100        23456 I
```

Junos OS provides a generalized mechanism to aid in troubleshooting BGP attributes, by inserting the unknown attributes as an additional field in the *extensive* route output. A sample output of this field is provided in Listing 3.35, with the attribute sent by a 4-byte AS-capable device in Line 8 and the corresponding unrecognized attribute being reported on the receiver at Line 32.

Listing 3.35: Handling unsupported AS4_PATH attributes

```
1  Trace information
2  Jul  5 00:26:53.047371 bgp_send: sending 84 bytes to 192.168.1.3 (Internal AS 64503)
```

```
3  Jul  5 00:26:53.047499
4  Jul  5 00:26:53.047499 BGP SEND 192.168.1.1+179 -> 192.168.1.3+63331
5  Jul  5 00:26:53.047562 BGP SEND message type 2 (Update) length 84
6  Jul  5 00:26:53.047599 BGP SEND flags 0x40 code Origin(1): IGP
7  Jul  5 00:26:53.047638 BGP SEND flags 0x40 code ASPath(2) length 4: 23456
8  Jul  5 00:26:53.047673 BGP SEND flags 0xc0 code AS4Path(17) length 6: 65550 # AS4Path attribute
9  Jul  5 00:26:53.047705 BGP SEND flags 0x40 code NextHop(3): 192.168.1.1
10 Jul  5 00:26:53.047736 BGP SEND flags 0x40 code LocalPref(5): 100
11 Jul  5 00:26:53.047776 BGP SEND flags 0xc0 code Communities(8): 64501:1 64501:106 65535:0
12 Jul  5 00:26:53.047818 BGP SEND           192.168.51.0/24
13
14 user@Torino> show route 192.168.51/24 detail inactive-path
15
16 inet.0: 40 destinations, 45 routes (40 active, 0 holddown, 0 hidden)
17 192.168.51.0/24 (2 entries, 1 announced)
18          BGP    Preference: 170/-101
19                 Next hop type: Indirect
20                 Next-hop reference count: 2
21                 Source: 192.168.1.1
22                 Next hop type: Router, Next hop index: 491
23                 Next hop: via so-0/1/0.0, selected
24                 Protocol next hop: 192.168.1.1
25                 Indirect next hop: 8a73138 -
26                 State: <Int Ext>
27                 Inactive reason: Interior > Exterior > Exterior via Interior
28                 Local AS: 64503 Peer AS: 64503
29                 Age: 12:26      Metric2: 160
30                 Task: BGP_64503.192.168.1.1+179
31                 AS path: 23456 I Unrecognized Attributes: 9 bytes
32                 AS path:  Attr flags e0 code 11: 02 01 00 01 00 0e          # AS4Path attribute
33                 Communities: 64501:1 64501:106 64502:1 64502:108 65535:0
34                 Localpref: 100
35                 Router ID: 10.10.1.1
```

In this case, the EBGP peering between router *Torino* and router *Bilbao* needs to stick to the old AS, AS64502, a bit longer, until router *Torino* is upgraded to a release with 4-byte AS support.

With this last step, the first milestone has been accomplished. The network is confederated and external peerings see the confederation as the 4-byte AS65550.

Meanwhile, the parallel link-state IGP migration is progressing successfully. Visibility of all routers in the three domains is available through a unified and common IS–IS topology, which allows the second part of the migration towards an IBGP full mesh to start.

3.6.10 Stage seven: Bringup IBGP peerings to Lille

Listing 3.36 shows the initial BGP configuration required for router *Lille* to become a route reflector for all routers to be migrated. A generic group with the `allow` setting enables any router with AS 65550 to peer with router *Lille* and to obtain all routing information. Groups with specific neighborship information are required to align the proper alias AS with the sub-AS domain. One such group is defined for domain "Cyclone" (Line 29), domain "Mistral" (Line 39), and an additional group for the external domain "Monsoon" (Line 48). The existing group used for the transport of labeled BGP paths in the MPLS transport case study (Chapter 4) is left untouched, maintaining the `alias` knob for the AS 65500 (Line 13).

Notice at this stage that the configuration on router *Lille* does *not* act as a reflector (that is, the `cluster` statement is not active). This is done on purpose as a safety measure to control which routing information is being received.

Listing 3.36: BGP setup for Lille

```
1  user@Lille# show protocols bgp
2  family inet {
3      unicast;
4      labeled-unicast {
5          rib {
6              inet.3;
7          }
8      }
9  }
10 group EBGP {
11     local-address 172.16.100.5;
12     export [ customer-routes from-loopback-to-LBGP ];
13     local-as 65500 alias;
14     neighbor 172.16.100.1 {
15         peer-as 64502;
16     }
17     neighbor 172.16.100.8 {
18         peer-as 64501;
19     }
20 }
21 group IBGP-RR-Cluster {
22     type internal;
23     family inet {
24         unicast;
25     }
26     inactive: cluster 192.168.1.5;
27     allow 192.168.0.0/16;
28 }
29 group IBGPCluster-Cyclone {
30     type internal;
31     local-address 192.168.1.5;
32     local-as 65001 alias;
33     inactive: cluster 192.168.1.5;
34     neighbor 192.168.1.2;
35     neighbor 192.168.1.6;
36     neighbor 192.168.1.7;
37     neighbor 192.168.1.21;
38 }
39 group IBGPCluster-Mistral {
40     type internal;
41     local-address 192.168.1.5;
42     local-as 65002 alias;
43     inactive: cluster 192.168.1.5;
44     neighbor 192.168.1.4;
45     neighbor 192.168.1.20;
46     neighbor 192.168.1.8;
47 }
48 group IBGPCluster-Monsoon {
49     type internal;
50     local-address 192.168.1.5;
51     local-as 64503 alias;
52     inactive: cluster 192.168.1.5;
53     neighbor 192.168.1.1;
54     neighbor 192.168.1.9;
55     neighbor 192.168.1.10;
56     neighbor 192.168.1.3;
57 }
58
59
60 user@Lille# show routing-options
61 <...>
62 router-id 192.168.1.5;
63 autonomous-system 65550;
```

Each router to be migrated is configured with an additional BGP group, as the configuration template in Listing 3.37 shows. Note that for internal peering, each router uses the value configured under the autonomous system, which is the confederation sub-AS for confederation members in domain "Mistral" and domain "Cyclone" and the domain AS for domain "Monsoon."

Listing 3.37: IBGP route reflector client template configuration

```
1 inactive: group IBGP-RRC {
2     type internal;
3     local-address 192.168.1.X; # Varies per router
4     neighbor 192.168.1.5;      # Lille
5 }
```

Listing 3.38 illustrates the received customer prefixes from various border routers from the viewpoint of router *Lille*. The active paths correspond to the redistribution points of the customer infrastructure routes.

Listing 3.38: Lille as a regular IBGP router homed to all routers

```
1  user@Lille> show bgp summary | match "^[^ ]|inet.0"
2
3  Groups: 4 Peers: 13 Down peers: 0
4  Table          Tot Paths  Act Paths Suppressed    History Damp State    Pending
5  inet.0                21          3          0          0          0          0
6  inet.3                20          7          0          0          0          0
7  Peer                     AS      InPkt     OutPkt    OutQ   Flaps Last Up/Dwn State|#Active/Received/
        Accepted/Damped...
8  172.16.100.1          64502        121        120       0       0       47:26 Establ
9    inet.0: 0/2/2/0
10 172.16.100.8          64501        124        127       0       0       47:22 Establ
11   inet.0: 0/2/2/0
12 192.168.1.1           64503         71         66       0       0       28:36 Establ
13   inet.0: 0/0/0/0
14 192.168.1.2           65001         68         67       0       0       27:36 Establ
15   inet.0: 0/1/0/0
16 192.168.1.3           64503         38         30       0       0       13:27 Establ
17   inet.0: 0/3/2/0
18 192.168.1.4           65002         84         78       0       0       31:26 Establ
19   inet.0: 0/3/0/0
20 192.168.1.6           65001         92         78       0       1       31:41 Establ
21   inet.0: 0/3/0/0
22 192.168.1.7           65001         80         83       0       0       33:28 Establ
23   inet.0: 1/1/1/0
24 192.168.1.8           65002         79         75       0       0       32:01 Establ
25   inet.0: 0/1/0/0
26 192.168.1.9           64503         57         57       0       0       24:40 Establ
27   inet.0: 0/0/0/0
28 192.168.1.10          64503         70         74       0       1       18:01 Establ
29   inet.0: 1/1/1/0
30 192.168.1.20          65002        103         84       0       0       33:59 Establ
31   inet.0: 1/4/1/0
32 192.168.1.21          65001         52         50       0       0       22:31 Establ
33   inet.0: 0/0/0/0
```

Most of the hidden routes are updates received over the pre-existing EBGP or CBGP peerings that are propagated back to router *Lille*. Their AS_PATH hints at a loop, and so they are hidden, as shown in Listing 3.39.

Listing 3.39: Hidden routes on Lille because of AS loop

```
1 user@Lille> show route table inet.0 community-name customer all terse
2
3 inet.0: 34 destinations, 53 routes (34 active, 0 holddown, 12 hidden)
```

```
4  + = Active Route, - = Last Active, * = Both
5
6  A Destination        P Prf   Metric 1   Metric 2  Next hop        AS path
7  * 192.168.50.0/24    S   5                         Receive
8                       B                            >172.16.100.1   65500 I          # Loop
9                       B                            >172.16.100.1   65500 I          # Loop
10                      B                            >172.16.100.8   65500 I          # Loop
11                      B                             172.16.100.1   64502 65500 I # Loop
12                                                   >172.16.100.8
13 * 192.168.51.0/24    B 170        100             >172.16.100.8   I
14                      B 170        100             >172.16.100.1   64502 I
15                      B 170        100              172.16.100.1   64502 I
16                                                   >172.16.100.8
17                      B 170        100             >172.16.100.8   64501 I
18                      B                            >172.16.100.1   (65001) I
19                      B                            >172.16.100.1   (65001) I
20 * 192.168.52.0/24    B 170        100             >172.16.100.1   I
21                      B 170        100             >172.16.100.1   64502 I
22                      B 170        100              172.16.100.1   64502 I
23                                                   >172.16.100.8
24                      B 170        100             >172.16.100.8   64501 I
25                      B                            >172.16.100.8   (65002) I
26 * 192.168.53.0/24    B 170        100              172.16.100.1   I
27                                                   >172.16.100.8
28                      B                            >172.16.100.1   65500 I          # Loop
29                      B                            >172.16.100.1   64503 I
30                      B                            >172.16.100.1   65500 I          # Loop
31                      B                            >172.16.100.8   64503 I
32                      B                            >172.16.100.8   65500 I          # Loop
```

Once router *Lille* peers with all routers, it can advertise its own address space as an internal route. The view of this address space as seen by router *Livorno* is shown in Listing 3.40. Notice at Line 7 that the confederation AS is not counting towards the total AS_PATH length.

Listing 3.40: Selection after advertising the NOC space

```
1  user@Livorno> show route 192.168.50/24 detail | match "^[^ ]|AS|comm|reason|BGP"
2
3  inet.0: 37 destinations, 49 routes (36 active, 0 holddown, 4 hidden)
4  192.168.50.0/24 (4 entries, 1 announced)
5          *BGP    Preference: 170/-101
6                  Local AS: 65001 Peer AS: 65002
7                  Task: BGP_65002.172.16.1.37+179
8                  Announcement bits (3): 0-KRT 4-BGP RT Background 5-Resolve tree 1
9                  AS path: (65002) I
10                 Communities: 64502:1 64502:104 65535:0
11          BGP    Preference: 170/-101
12                 Inactive reason: Not Best in its group - Update source
13                 Local AS: 65001 Peer AS: 65002
14                 Task: BGP_65002.172.16.100.1+179
15                 AS path: (65002) I
16                 Communities: 64502:1 64502:104 65535:0
17          BGP    Preference: 170/-101
18                 Inactive reason: Not Best in its group - Router ID
19                 Local AS: 65001 Peer AS: 65001
20                 Task: BGP_65001.192.168.1.5+179
21                 AS path: I
22                 Communities: 65535:0
23          BGP    Preference: 170/-101
24                 Inactive reason: AS path
25                 Local AS: 65001 Peer AS: 65500
26                 Task: BGP_65500.172.16.100.5+65042
27                 AS path: 65500 I
28                 Communities: 65535:0
```

3.6.11 Stage eight: Enable route reflection on router *Lille*

The updates from the NOC towards the routers are now directly reachable over the IGP. Updates from other routers are received at router *Lille*, but reflection needs to be activated to relay those updates, as shown in Listing 3.41.

Listing 3.41: Configuration of router *Lille* as a route reflector

```
1 user@Lille> show configuration | compare rollback 1
2 [edit protocols bgp group IBGP-RR-Cluster]
3 !      active: cluster 192.168.1.5;
4 [edit protocols bgp group IBGPCluster-Cyclone]
5 !      active: cluster 192.168.1.5;
6 [edit protocols bgp group IBGPCluster-Mistral]
7 !      active: cluster 192.168.1.5;
8 [edit protocols bgp group IBGPCluster-Monsoon]
9 !      active: cluster 192.168.1.5;
```

Listing 3.42 shows the customer routes as learned by the NOC router *Lille*.

Listing 3.42: View on router *Lille* and border router *Nantes*

```
1  user@Lille> show route community-name customer terse
2
3  inet.0: 34 destinations, 40 routes (34 active, 0 holddown, 4 hidden)
4  + = Active Route, - = Last Active, * = Both
5
6  A Destination        P Prf   Metric 1   Metric 2  Next hop        AS path
7  * 192.168.0.0/24     S   5                        Receive
8  * 192.168.51.0/24    B 170        100            >172.16.100.8    I
9  * 192.168.52.0/24    B 170        100            >172.16.100.1    I
10 * 192.168.53.0/24    B 170        100             172.16.100.1    I
11                                                  >172.16.100.8
12                      B 170        100             172.16.100.1    I
13                                                  >172.16.100.8
14
15 user@Lille>  show route 192.168.48/21 terse
16
17 inet.0: 34 destinations, 52 routes (34 active, 0 holddown, 11 hidden)
18 + = Active Route, - = Last Active, * = Both
19
20 A Destination        P Prf   Metric 1   Metric 2  Next hop        AS path
21 * 192.168.50.0/24    S   5                        Receive
22 * 192.168.51.0/24    B 170        100            >172.16.100.8    I
23                      B 170        100            >172.16.100.1    64502 I
24                      B 170        100             172.16.100.1    64502 I
25                                                  >172.16.100.8
26                      B 170        100            >172.16.100.8    64501 I
27 * 192.168.52.0/24    B 170        100            >172.16.100.1    I
28                      B 170        100            >172.16.100.1    64502 I
29                      B 170        100             172.16.100.1    64502 I
30                                                  >172.16.100.8
31                      B 170        100            >172.16.100.8    64501 I
32 * 192.168.53.0/24    B 170        100             172.16.100.1    I
33                                                  >172.16.100.8
```

3.6.12 Stage nine: Switch router *Nantes* to the 4-byte AS

By changing the AS to the global AS 65550, any router in the domain that has a direct impact in introduces a flap in all BGP neighbors. The session is only established correctly with router *Lille*, which belongs to the 4-byte AS, and with other already migrated routers, all of which are in the global AS. For border routers with external peerings, the `local-as` configuration statement is suggested for asynchronous migrations, as usual.

Taking router *Nantes* as the first router to move to the AS 65550, Listing 3.43 shows the peerings before the configurations change. Line 25 provides the delta in changes. The result is shown on Line 32. Notice on Line 39 that only the peering with router *Lille* can be established properly.

Listing 3.43: Nantes on new AS

```
1  [edit]
2  user@Nantes# run show route terse table inet.0 community-name customer
3
4  inet.0: 46 destinations, 58 routes (46 active, 0 holddown, 0 hidden)
5  + = Active Route, - = Last Active, * = Both
6
7  A Destination        P Prf   Metric 1   Metric 2  Next hop        AS path
8  * 192.168.0.0/24     B 170        100             >172.16.1.2     I
9                                                     172.16.1.5
10                      B 170        100             >172.16.1.2     65550 I
11                      B 170        100             >172.16.1.5     64502 I
12 * 192.168.51.0/24    B 170        100             >172.16.1.2     I
13                      B 170        100             >172.16.1.2     65550 I
14                      B 170        100             >172.16.1.5     64502 I
15 * 192.168.52.0/24    B 170        100             >172.16.1.5     I
16                      B 170        100             >172.16.1.2     65550 I
17                      B 170        100             >172.16.1.5     64502 I
18 * 192.168.53.0/24    B 170        100             >so-1/0/1.0     I
19                      B 170        100             >so-1/0/1.0     I
20                      B 170        100             >so-1/0/1.0     I
21                      B 170        100             >172.16.1.2     65550 I
22                      B 170        100             >172.16.1.5     64502 I
23
24 [edit]
25 user@Nantes# show | compare
26 [edit routing-options]
27 -   autonomous-system 64503;
28 +   autonomous-system 65550;
29 [edit protocols bgp]
30 !     inactive: group EBGP { ... }
31
32 user@Nantes> show bgp summary
33 Groups: 2 Peers: 4 Down peers: 3
34 Table          Tot Paths  Act Paths Suppressed    History Damp State    Pending
35 inet.3                 0          0          0          0          0          0
36 inet.0                 4          4          0          0          0          0
37 Peer                     AS      InPkt     OutPkt    OutQ   Flaps Last Up/Dwn State
38 192.168.1.3           65550          0         12       0       0        5:28 Active
39 192.168.1.5           65550         15         13       0       0        4:32 4/4/4/0
40 192.168.1.9           65550          0         12       0       0        5:28 Active
41 192.168.1.10          65550          0         12       0       0        5:28 Active
42
43 user@Nantes> show route terse table inet.0 community-name customer
44
45 inet.0: 42 destinations, 44 routes (42 active, 0 holddown, 0 hidden)
46 + = Active Route, - = Last Active, * = Both
47
48 A Destination        P Prf   Metric 1   Metric 2  Next hop        AS path
49 * 192.168.50.0/24    B 170        100              172.16.1.5     I
50                                                   >172.16.1.2
51 * 192.168.51.0/24    B 170        100             >172.16.1.2     I
52 * 192.168.52.0/24    B 170        100             >172.16.1.5     I
53 * 192.168.53.0/24    B 170        100             >so-1/0/1.0     I
```

3.6.13 Stage ten: Monitor

No major issues are observed during the monitoring period.

3.6.14 Stage eleven: Roll out and cleanup

Additional routers can now be changed to become true AS65550 members. Once the routing information is in place, additional IBGP sessions can be established between migrated routers. Eventually, this rollout builds a full-mesh IBGP for AS65550.

As routers complete full peering on the new AS, the `local-as` configuration command can be gradually removed. When all routers are fully meshed, router *Lille* becomes redundant in the distribution of IBGP information and its reflector function can be disabled. As part of the full IBGP mesh, router *Lille* maintains advertisement of the NOC pool to the rest of the network.

3.6.15 Migration summary

Because of the full visibility requirement, the BGP migration for "Gale" Internet took a two-phased approach.

In the first phase, a BGP confederation integrated domain "Mistral" and domain "Cyclone," showing joint customers that integration was a fact, under a single AS.

The successful IS–IS migration allowed a move to a full-mesh IBGP using an incremental approach.

Depending on the sequencing in which routers are migrated, traffic changes are likely. BGP routing policy, which was in place separating the different domains, becomes useless within a full-mesh IBGP. Intra-domain policy control needs to take place.

Bibliography

[BGP-WEDGE] T. G. Griffin. BGP Wedgies: Bad Policy Interactions that Cannot be Debugged, 2004.

[IANA-AFI] Internet Assigned Numbers Authority. Address Family Identifiers, December 2009.

[IANA-SAFI] Internet Assigned Numbers Authority. Subsequent Address Family Identifiers, December 2009.

[ID-ADVISORY] T. Scholl and J. Scudder. BGP advisory message, March 2009.

[IETF-IDR] Internet Engineering Task Force. Inter Domain Routing charter, 2009.

[RFC793] J. Postel. Transmission Control Protocol. RFC 793 (Standard), September 1981. Updated by RFCs 1122, 3168.

[RFC1105] K. Lougheed and Y. Rekhter. Border Gateway Protocol (BGP). RFC 1105 (Experimental), June 1989. Obsoleted by RFC 1163.

[RFC1930] J. Hawkinson and T. Bates. Guidelines for Creation, Selection, and Registration of an Autonomous System (AS). RFC 1930 (Best Current Practice), March 1996.

[RFC1997] R. Chandra, P. Traina, and T. Li. BGP Communities Attribute. RFC 1997 (Proposed Standard), August 1996.

[RFC2328] J. Moy. OSPF Version 2. RFC 2328 (Standard), April 1998.

[RFC2385] A. Heffernan. Protection of BGP Sessions via the TCP MD5 Signature Option. RFC 2385 (Proposed Standard), August 1998.

[RFC2918] E. Chen. Route Refresh Capability for BGP-4. RFC 2918 (Proposed Standard), September 2000.

[RFC3345] D. McPherson, V. Gill, D. Walton, and A. Retana. Border Gateway Protocol (BGP) Persistent Route Oscillation Condition. RFC 3345 (Informational), August 2002.

[RFC3392] R. Chandra and J. Scudder. Capabilities Advertisement with BGP-4. RFC 3392 (Draft Standard), November 2002.

[RFC3478] M. Leelanivas, Y. Rekhter, and R. Aggarwal. Graceful Restart Mechanism for Label Distribution Protocol. RFC 3478 (Proposed Standard), February 2003.

[RFC4271] Y. Rekhter, T. Li, and S. Hares. A Border Gateway Protocol 4 (BGP-4). RFC 4271 (Draft Standard), January 2006.

[RFC4360] S. Sangli, D. Tappan, and Y. Rekhter. BGP Extended Communities Attribute. RFC 4360 (Proposed Standard), February 2006.

[RFC4456] T. Bates, E. Chen, and R. Chandra. BGP Route Reflection: An Alternative to Full Mesh Internal BGP (IBGP). RFC 4456 (Draft Standard), April 2006.

[RFC4760] T. Bates, R. Chandra, D. Katz, and Y. Rekhter. Multiprotocol Extensions for BGP-4. RFC 4760 (Draft Standard), January 2007.

[RFC4893] Q. Vohra and E. Chen. BGP Support for Four-octet AS Number Space. RFC 4893 (Proposed Standard), May 2007.

[RFC5004] E. Chen and S. Sangli. Avoid BGP Best Path Transitions from One External to Another. RFC 5004 (Proposed Standard), September 2007.

[RFC5065] P. Traina, D. McPherson, and J. Scudder. Autonomous System Confederations for BGP. RFC 5065 (Draft Standard), August 2007.

[RFC5291] E. Chen and Y. Rekhter. Outbound Route Filtering Capability for BGP-4. RFC 5291 (Proposed Standard), August 2008.

[RFC5398] G. Huston. Autonomous System (AS) Number Reservation for Documentation Use. RFC 5398 (Informational), December 2008.

Further Reading

[1] Aviva Garrett. JUNOS Cookbook, April 2006. ISBN 0-596-10014-0.

[2] Matthew C. Kolon and Jeff Doyle. Juniper Networks Routers : The Complete Reference, February 2002. ISBN 0-07-219481-2.

[3] James Sonderegger, Orin Blomberg, Kieran Milne, and Senad Palislamovic. JUNOS High Availability, August 2009. ISBN 978-0-596-52304-6.

[4] Sam Halabi. Internet Routing Architectures, September 2000. ISBN 978-1578702336.

[5] John W. Stewart. BGP4: Inter-Domain Routing in the Internet, December 1998. ISBN 978-1578702336.

[6] Randy Zhang and Micah Bartell. BGP Design and Implementation, December 2009. ISBN 978-1587051098.

4

MPLS label distribution Migrations

MPLS has revolutionized the internetworking industry in recent years. Originally envisioned as a flexible tag-switching architecture that could span beyond an ATM or Frame Relay cloud, MPLS currently constitutes a reference framework to deploy multiple services from optical and Ethernet backhaul to Traffic Engineering or private Layer 3 VPNs.

Multiple enterprises and service providers have had to deploy MPLS in their networks as a moving-forward strategy to better traffic-engineer their networks or offer MPLS-based services, and soon thereafter the question about merging MPLS label distribution domains arose. Different label distribution protocols or even diverse adoptions from an early standard have caused integration headaches for many network engineers.

4.1 Motivations for MPLS label distribution Migrations

Foundations for successful deployment of MPLS-based applications have not only been integrating MPLS with other protocols such as an IGP or BGP, but also the election of the correct strategy for MPLS label allocation and distribution. Those topics have been addressed since the very early origins of MPLS.

[RFC3035], [RFC3036] (made obsolete by [RFC5036]), and [RFC3037] soon advocated the concept of a brand new *Label Distribution Protocol* (LDP). This new protocol was exclusively defined to distribute MPLS label bindings with a set of mechanisms by which Label-Switched Paths (LSPs) may get established by mapping Forwarding Equivalence Classes (FECs) to MPLS label bindings. LDP included several discovery procedures as well, so that not only directly connected peers, but also, by means of an extended discovery mechanism based on LDP targeted hellos, distant LSRs could set up an LDP session. This machinery has been the cornerstone for many MPLS-based applications and needs to be considered carefully for migration purposes.

On the other hand, [RFC3107] redefined the utilization of a well-known transport-capable protocol to carry FECs to label mappings: *BGP*. By means of a specific new capability

Network Mergers and Migrations Gonzalo Gómez Herrero and Jan Antón Bernal van der Ven

and AFI/SAFI combination, BGP was expanded to be able to advertise MPLS labels using standard BGP flow-control and establishment mechanisms. The maturity and future-proof transport scalability features of this protocol made this specific flavor, commonly known as Labeled BGP (L-BGP), a successful option to distribute labels with minimal protocol expansion. *L-BGP* has also become a very popular protocol to act as substrate for specific inter-AS VPN interconnections.

In parallel, efforts were made to cope with Traffic-Engineering (TE) requirements for MPLS, as described in [RFC2702]. The biggest headache for engineers at that time was determining the best way to overcome the limitations of standard IGP control mechanisms and habilitate resource-oriented LSPs explicitly. Soon, [RFC3209] and [RFC3936] emerged with new extensions and procedures for an old and known Resource Reservation protocol (RSVP), leading to *RSVP-TE*. With such extensions, RSVP was virtually extended to support those prerequisites for MPLS. Other approaches such as expanding LDP with Constrained-Routing LDP (CR-LDP) defined in [RFC3212] were futile in the end and the IETF and networking industry finally reached consensus with [RFC3468], moving *CR-LDP* to informational status and stopping further work on it.

This evolution has finally left three such protocol flavors, namely LDP, L-BGP, and RSVP-TE as references for MPLS label distribution. Each protocol provides specific features and has caveats. Each protocol may be the best or worst match in a certain environment or circumstances. This chapter evaluates these protocols and describes MPLS label distribution domain integration from the perspective of how these protocols interact and handle MPLS label transport, particularly from the Junos OS perspective.

4.2 Considerations for MPLS label distribution Migrations

Each label distribution protocol provides different features that may suit some design needs at a given time but can also present some drawbacks.

While RSVP-TE is explicitly a *downstream-on-demand* protocol, L-BGP works in a *downstream-unsolicited* fashion, meaning that label bindings are distributed as soon as they become available. LDP presents both options with different label *request* and *mapping* mechanisms and with control modes within the same protocol definition. Standard transport label distribution in LDP occurs in a *downstream-unsolicited* fashion as well, although there are some setups in which it is done on demand, and both can perfectly coexist.

LDP presents information in the form of TLV structures and includes *notification* and *advisory* mechanisms, so that future extensibility is guaranteed, with specific bits in each TLV indicating how error handling should be used for unknown TLVs. RSVP-TE expanded the original RSVP standard with additional objects but [RFC3209] states that unknown Label objects or Label Request objects should be handled with ResvErr messages or PathErr messages, the same ones sent by transit LSRs when they do not recognize other objects such us Explicit Route Objects or Record Route objects. Only Hello extensions are defined as backward-compatible and are able to be silently discarded. On the other hand, L-BGP is enforced by standard BGP extensibility with new Capabilities and Families. Initial BGP session negotiation determines, therefore, which capabilities and address families are supported, and protocol extensibility is conditioned in this sense by handling transitive attributes that may be unknown.

By means of all of its extensions, RSVP-TE LSPs can be instantiated with *explicit* routes (that is to say, not strictly following the best IGP path) with different options, resources such as *bandwidth* can be explicitly allocated to LSPs (routing vendors tend to define applications based on this), both fast and ingress rerouting of established LSPs are available with RSVP-TE in a smooth *make-before-break* fashion, and the protocol allows tracking of the actual route traversed by each LSP, as well as diagnostics and bandwidth-driven updates on the fixed paths. Such path protection and rerouting can also follow a given pre-emption and hold scheme to determine at each point in time which path the traffic is taking, depending on resource and node availability. All these concepts are not available in LDP or L-BGP. LDP enforces label bindings, but it is up to the IGP to determine LSP selection without any end-to-end guarantees (other mechanisms such as *Bidirectional Forwarding Detection* or BFD may be coupled for that), and L-BGP makes use of all standard flow-control BGP mechanisms but does not perform any resource allocation or transit route tracking.

[RFC4090] defines *fast-reroute* mechanisms for RSVP-TE-based LSPs so that local repairs can be pre-established and become active in tens of milliseconds. There are no similar built-in protocol mechanisms at LDP or L-BGP, although the possibility to tunnel those sessions over an RSVP-TE exists.

However, all these extra mechanisms do not come at no cost. RSVP-TE requires more state consumption than LDP or L-BGP. Usually Path state blocks and Reservation state blocks are maintained not only at ingress and egress LSRs, but also at RSVP-TE LSP transit routers, so as to determine when the request needs to be forwarded and where a label needs to be mapped (this state needs to be linked to further resource reservation or fast reroute information each time). LDP and L-BGP are completely agnostic to this. Because of their downstream-unsolicited nature, label bindings are active at the moment the forwarding table is populated with such transport labels.

Also, most routing vendors have developed the dedicated RSVP-TE LSP concept to make it available as a Forwarding Adjacency that can act as a virtual tunnel for the IGP. In that sense, not only MPLS traffic but also plain IPv4 traffic can be transported over such *Forwarding Adjacencies* because they act merely as another virtual IGP link. This concept was further developed with [RFC4203], [RFC4205], and [RFC4206] to allow construction of MPLS LSP hierarchies by creating Forwarding Adjacency LSPs. In this way, the underlying RSVP-TE LSPs present virtualized links in the Traffic-Engineering Database (TED), thus making an underlying LSP substrate available only for other end-to-end LSPs but not visible in the standard IGP database for plain IPv4 or IPv6 traffic. Neither of these features is available in LDP or L-BGP, so both these protocols must collaborate with RSVP-TE at some point to be able to build up such hierarchies.

Considering a *domain* to be a number of network elements within a common path computation responsibility and an *area* as the traditional division of an IGP, interarea and interdomain LSPs are another differentiating topic. [RFC5283] proposes a relaxation of the traditional LDP longest-match prefix approach to activate a label binding and therefore to leverage IGP route propagation across area boundaries to encompass the label binding FEC: aggregate prefixes for more specific FECs become valid arguments to activate label bindings. However, if the traditional IGP-LDP dependence across domains is not maintained, additional machinery is needed to populate label bindings through the interconnect. [RFC5151] specifies signaling mechanisms for contiguous, nested, or stitched LSPs across domains considering a given degree of TE visibility among them. On the other hand, L-BGP can set up interdomain LSPs with all TE resources available in EBGP and if support for [RFC5543] is granted, LSP-specific TE information can be carried as an additional attribute.

All these factors, along with many others, need to be evaluated to make the right decision to integrate MPLS label distribution protocols, as summarized in Table 4.1.

4.3 Generic Strategies for an MPLS label distribution protocol Migration

A failure or an outage during an MPLS label distribution protocol migration, or simply misalignment with another protocol such as a coupled link-state IGP, can have severe consequences for MPLS-based applications.

Apart from defining the migration premises and performing proofs of concept, carrying out MPLS label distribution protocol migration activities requires additional thinking beyond the label distribution protocols themselves, due to other external dependencies that affect selection of the best MPLS label. Label selection is not determined by the label distribution protocols alone, but rather, by whoever picks the best route, leading to MPLS label binding with available distributed labels. In this sense, IGPs, BGP, static and direct routes, and other protocols injecting route information must be analyzed for label-binding purposes and must be considered together in a migration framework.

4.3.1 MPLS label distribution protocol coexistence

Parallel *coexistence* of MPLS label distribution protocols in a network is common practice and is the preferred option at many service provider and enterprise networks.

Because these MPLS label distribution protocols have different and orthogonal binding methods, it is ultimately a question of maintaining each one at the control plane and selectively choosing the adequate label binding at a time by means of preferences or administrative distances.

Each MPLS label distribution protocol populates its label database and distributes such information according to its own methods. At each ingress LSR, one label binding is selected depending on a given preference.

If the proper label distribution infrastructure is in place, route selection is performed only at ingress LSRs, because MPLS label switching takes place along the LSP as lookup operation. This behavior means that ingress LSRs determine which LSP is taken as well as the preference of each label distribution protocol. Once a label distributed from a given protocol is selected at the ingress, it follows the LSP and other label distribution protocols do not interfere (unless tunneling techniques are used).

The bottom line is that MPLS label distribution protocols can coexist as *ships-in-the-night* and that their control plane mechanisms are independent. Each LSR makes its own decision to pick up the proper label binding from the respective protocol, and this selection remains along the LSP.

Table 4.1 Rationale for MPLS label distribution protocol migrations

Factor	LDP	RSVP-TE	L-BGP
Label distribution mode	Downstream Unsolicited and on-demand	Downstream On-demand	Downstream Unsolicited
Information packaging	TLVs inside well-known messages	Objects with different treatment	Controlled by BGP capabilities except for transitive optional attributes
Path establishment	IGP-driven	IGP-driven, controlled by ERO	Depending on BGP neighbor reachability
Diagnostics	IGP-driven or message from LDP peer	End-to-end in both directions	From BGP peer
Resource reservation	None	Controlled by ERO	Depending on BGP neighbor reachability
Rerouting	Depending on IGP or over RSVP-TE	Make-before-break, fast reroute	Depending on underlying substrate or over RSVP-TE
Pre-emption	Depending on best IGP route	Comprehensive scheme, objects	Depending on best BGP route per path selection
State consumption	LDP database population	Path and Resv blocks, linking labels, resources and other LSPs	BGP routes
Hierarchies	Targeted sessions over RSVP-TE	[RFC4206]	Over RSVP-TE or LDP
Inter-domain interconnect	Targeted sessions over RSVP-TE	[RFC5151]	Native in EBGP
Inter-area interconnect	[RFC5283]	[RFC5151] or without prior CSPF	Transparent
Traffic-Engineering	Tunneling over RSVP-TE	Present since [RFC2702]	[RFC5543]

4.3.2 MPLS label distribution protocol redistribution

Redistribution in MPLS label distribution protocols has implications beyond those of other protocols such as IGPs.

Because each protocol has different requirements, when selecting label bindings to be identified from one protocol and created in another for a simple label *swap* (or *pop* in the case of penultimate or ultimate routers), an engineer needs to consider that the matching prefix, which identifies the FEC, also needs to be redistributed or must follow the same redistribution path.

For MPLS label distribution protocols in Junos OS, the IGPs also play a key role. LDP and RSVP-TE require a matching route in inet.0, globally determined by an existing IGP (or a static or a direct route on rare occasions). L-BGP allows flexible mechanisms to inject routes into one or another table, but recursive resolution for the next hop requires a previously present route in inet.3. From that point of view, redistributing label bindings from one protocol into another may also require IGP redistribution or at least the need to enforce the requirement that routes resolving the LSP are present in a certain table for the label binding to become effective.

Similar to other types of redistribution, only the best sibling paths are eligible for redistribution into another protocol. The behavior here is slightly different among the protocols:

- *LDP*: tends to create and distribute label bindings without split-horizon control and with *ordered* control, unsolicited distribution in Junos OS. The task of selecting the best path is left to the IGP, which ultimately defines label selection from the LDP database.
- *RSVP-TE*: can create more than a single MPLS LSP to the same destination. Ultimately, the internal pre-emption and hold RSVP schemes, together with constrained SPF execution, define the active LSP to which to forward traffic.
- *L-BGP*: implements the standard BGP path selection mechanisms, with all tie-breaking comparisons, to define the best path.

For each protocol, when selecting the active label bindings to use to redistribute from one into another, the implications of these path-selection mechanisms need to be considered. Usually, a route that covers the path must be present in the other domains as well for the label binding to remain active.

Interaction with other protocols

MPLS label distribution protocol transport label bindings for FECs and may require other routing protocols to inject parallel plain routing information for the same prefixes.

LDP poses no specific requirement for the parallel route for the FEC. This route does not even need to be specifically present in a link-state IGP. In addition, static, aggregate, or BGP routes in the main routing table can act as substrate for the label binding to become locally active. The advantage for LDP in having a uniform IGP leading to consistent route selection for the FEC is that the inherited paths are best from an SPF calculation point of view and are free of loops, because LDP inherently has neither route calculation nor built-in loop detection mechanisms.

RSVP-TE first looks into the *TED*. As seen in previous chapters, this particular database is populated by specific IGP extensions with additional information that is considered by RSVP-TE for best path selection. The *IGP metric* is used as a tie-breaker as well, but other factors that constrain path determination can also be configured at the ingress LSR.

L-BGP is a specific family in BGP that associates label bindings to routes. This means that other families can be negotiated and other types of route can also be exchanged over the same session. L-BGP routes do not specifically need a parallel route from an IGP, but route activation first requires a proper resolution and next-hop reachability, as with other BGP routes. Also, parallel *unlabeled* routes representing FECs can be perfectly exchanged over the same session as standard AFI 1 SAFI 1 routes if needed.

4.3.3 MPLS label distribution protocol overlay

Another feasible integration scenario is based on an *overlay* model of MPLS label distribution protocols. An *overlay* means that a certain protocol can *transport* and *resolve* its label bindings over another protocol that is underneath.

This model is particularly useful when the intent is to scale with *downstream on-demand* label distributions or to distribute labels in a *downstream-unsolicited* fashion over a primary *downstream on-demand* label distribution topology. While the latter provides basic connectivity for participating systems, with a label binding assigned to each of them as a FEC, the overlying protocol eases transport and distribution for label bindings associated with FECs unknown to this core topology.

The result is that the underlying layer acts as *resolution* for a flexible upper label transport. Three common techniques are representative of this case:

- *RSVP-TE over RSVP-TE*: RSVP-TE distributes *downstream on-demand* labels over and beyond the existing RSVP-TE LSP mesh acting as transport or core. Thanks to GMPLS extensions and mechanisms defined in [RFC4203], [RFC4205], and [RFC4206] to create Forwarding Adjacency LSPs (FA-LSPs), the external LSPs can resolve over internal or transport FA-LSPs. This setup grants scalability in the sense that fewer states and resources are consumed when creating hierarchies with RSVP-TE, while providing features like fast reroute or TE end to end.

- *LDP tunneling over RSVP-TE*: LDP transports label bindings in a *downstream-unsolicited* fashion over a given RSVP-TE LSP mesh. LDP routes resolve over RSVP-TE driven next hops in the corresponding table.

- *Internal L-BGP over RSVP-TE or LDP*: L-BGP transports label bindings in a *downstream-unsolicited* fashion over a given RSVP-TE LSP mesh or an LDP domain. LDP or RSVP-TE routes provide the necessary resolution substrate for independent L-BGP routes to become active. In this case, no parallel route in the main routing table may be needed.

An MPLS label distribution protocol *overlay* model requires additional label *push* actions at ingress LSRs and relevant *pop* and *swap* actions when entering or exiting one layer, with the corresponding requirements on the forwarding plane. However, it creates a recursive and scalable scenario that is particularly useful in network migrations in which some layers may need to remain opaque to others or label bindings may need to be transparently transported over other protocols.

4.3.4 MPLS label distribution protocol parameterization

Similar to IGPs, an MPLS domain can be defined as a myriad of systems under common MPLS scoping and parameterization. This means that MPLS label distribution domains can be split up and contained by using specific parameters and features.

Junos OS supports authentication extensions to RSVP-TE, LDP, and BGP and such settings can easily accommodate domain transition or integration depending on the selected keys. LDP and BGP can be configured with a particular key for HMAC-MD5 on a per-session basis, whereas RSVP-TE enforces configuration on a per-interface basis, requiring that all received and transmitted RSVP messages with all neighbors are authenticated on the interface.

In addition to protecting against replay attacks and message spoofing, these authentication parameters can also help to differentiate among adjacencies and neighbors within the same physical topology. Progressive deployment or shrinking of a given MPLS domain can be steered with authentication settings that allow or restrict neighbor relationships at the desired pace.

As a summary for all possible MPLS label distribution integration strategies, Figure 4.1 depicts all feasible redistribution, overlay, and interconnect options supported in Junos OS for MPLS label distribution protocols in different domains.

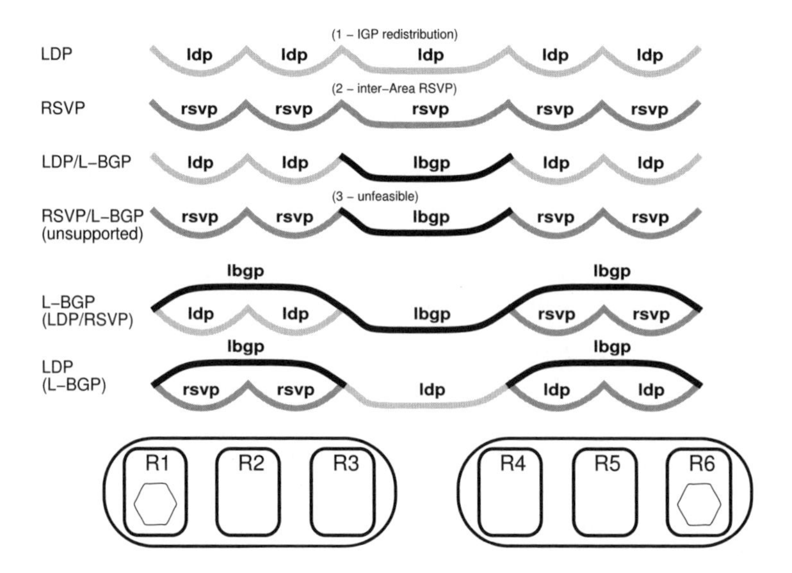

Figure 4.1: MPLS label distribution protocol integration options in Junos OS.

4.4 Resources for an MPLS label distribution protocol Migration

For all these generic MPLS label distribution migration strategies, routing operating systems offer resources and features for implementation. Certainly, these features are sometimes also dependent on resources outside MPLS label distribution protocols, such as IGP redistribution to inject routes representing FECs from one domain to another.

Junos OS extends the generic preference-based protocol scheme to MPLS label distribution protocols. This global model aligns routes from protocols to determine active label bindings in inet.3.

Apart from the global preference scheme, Junos OS offers a wide range of configuration options related to control for LDP label distribution and FEC selection, label binding population filtering, IGP shortcuts for RSVP-TE LSPs, additional FEC association to RSVP-TE LSPs, LDP tunneling, route resolution, and label advertisement for L-BGP, among others.

4.4.1 MPLS label distribution protocol preference

Junos OS has some default preferences for MPLS label distribution protocols that determine active label bindings for inet.3 in each case. Table 4.2 illustrates the default preferences for MPLS label distribution protocols. In the case of L-BGP, the standard default preference for BGP is inherited automatically, keeping the same value for internal and external L-BGP. Note that a lower value indicates a stronger preference.

Table 4.2 MPLS LDP preferences in Junos

Route type	Preference value
RSVP-TE	7
LDP	9
Internal and external L-BGP	170

These default values are automatically assumed, but Junos OS includes configuration options to manipulate the preference value for each of these protocols. Listing 4.1 indicates the global configuration knobs to determine a different preference from the default value. However, Junos OS policy language also offers the possibility to change the preference value as a feasible action.

Listing 4.1: MPLS LDP preference manipulation in Junos OS

```
 1  [edit protocols ldp]
 2  user@Livorno# set preference ?
 3  Possible completions:
 4    <preference>          Route preference
 5  [edit protocols mpls]
 6  user@Livorno# set preference ?
 7  Possible completions:
 8    <preference>          Preference value
 9  [edit protocols bgp]
10  user@Livorno# set preference ?
11  Possible completions:
12    <preference>          Preference value
```

An adequate and ordered protocol preference scheme is a key factor for a redistribution or *ships-in-the-night* model among MPLS label distribution protocols.

4.4.2 Resources to control LDP label distribution

FEC eligibility for LDP

Despite injecting routes into the Junos OS inet.3 table once label bindings are flooded, LDP looks into inet.0 only to *create* new label bindings. The default `egress-policy` in Junos OS is to create label bindings for addresses present on *lo0.0* only, as opposed to other operating systems that issue specific bindings for all LDP-enabled interfaces. This default behavior enforces the idea that loopback addresses are also used by default as next hops for other applications using that label as transport, without a specific need to allocate unnecessary labels for other addresses as FECs.

This behavior can be adjusted, though, for more granular or coarser label allocation in LDP. Junos OS `egress-policies` create MPLS label allocations for the prefixes that match the policy conditions. It is worth remarking that these policies control new label allocation for local prefixes on the router, when the router is acting as egress LSR for all these prefixes as destinations (the router advertises either *explicit* or *implicit* null), but these policies are not used to control label binding flooding between neighbors.

LDP looks in Junos OS into inet.0 to identify prefixes as FECs for new label allocations. This is practical for direct or IGP routes present in inet.0, but poses a challenge when redistributing label bindings from another MPLS label distribution protocol, such as L-BGP or RSVP-TE, that may have placed label bindings in inet.3. In this case, `rib-groups` can provide adequate machinery for programming a label *swap* between protocols. Using `from rib inet.3` in an LDP `egress-policy` is futile, because this match is not effective as per LDP design.

Considering the default LDP label database status depicted in Listing 4.2, Listing 4.3 shows the effects of applying a specific `egress-policy` on router *Male* to allocate labels to all direct routes. Lines 26 and 27, and 51 and 52 show how additional *implicit-null* labels are allocated for the own loopback address and also transit networks, and are both advertised to router *Havana* and router *Bilbao*. Also, router *Havana* and router *Bilbao* are advertising back another label for the non-local transit networks following that advertisement from router *Male*, as shown in Line 19 and Line 44.

It is worth remarking that `egress-policies` completely override the default Junos OS LDP label allocation behavior, and if the loopback address is not covered, an implicit or explicit-null advertisement is no longer made for it. In this case, the loopback address is nothing more than another direct route.

Listing 4.2: LDP database at Male

```
1  user@male-re0> show ldp database
2  Input label database, 192.168.1.20:0--192.168.1.4:0
3    Label     Prefix
4   100064     192.168.1.1/32
5  <...>
6
7  Output label database, 192.168.1.20:0--192.168.1.4:0
8    Label     Prefix
9   308592     192.168.1.1/32
10  308544     192.168.1.2/32
11  308672     192.168.1.3/32
12  308560     192.168.1.4/32
13  308656     192.168.1.5/32
14  308576     192.168.1.6/32
15  308688     192.168.1.7/32
```

```
16  308528      192.168.1.8/32
17  308608      192.168.1.9/32
18  308624      192.168.1.10/32
19       3      192.168.1.20/32
20  308640      192.168.1.21/32
21  308640      192.168.10.1/32
22
23 Input label database, 192.168.1.20:0--192.168.1.8:0
24   Label      Prefix
25  299840      192.168.1.1/32
26 <...>
27
28 Output label database, 192.168.1.20:0--192.168.1.8:0
29   Label      Prefix
30  308592      192.168.1.1/32
31  308544      192.168.1.2/32
32  308672      192.168.1.3/32
33  308560      192.168.1.4/32
34  308656      192.168.1.5/32
35  308576      192.168.1.6/32
36  308688      192.168.1.7/32
37  308528      192.168.1.8/32
38  308608      192.168.1.9/32
39  308624      192.168.1.10/32
40       3      192.168.1.20/32
41  308640      192.168.1.21/32
42  308640      192.168.10.1/32
```

Listing 4.3: LDP egress policies at Male

```
1  [edit policy-options policy-statement allocate-label-direct]
2 user@male-re0# show
3 term direct {
4     from protocol direct;
5     then accept;
6 }
7
8  [edit]
9 user@male-re0# set protocols ldp egress-policy allocate-label-direct
10
11  [edit]
12 user@male-re0# commit and-quit
13 commit complete
14 Exiting configuration mode
15
16 user@male-re0> show ldp database
17 Input label database, 192.168.1.20:0--192.168.1.4:0
18   Label      Prefix
19  100032      172.16.1.76/30 # Transit Male-Bilbao
20  100064      192.168.1.1/32
21 <...>
22
23
24 Output label database, 192.168.1.20:0--192.168.1.4:0
25   Label      Prefix
26       3      172.16.1.76/30 # Transit Male-Bilbao
27       3      172.16.1.84/30 # Transit Male-Havana
28  308592      192.168.1.1/32
29  308544      192.168.1.2/32
30  308672      192.168.1.3/32
31  308560      192.168.1.4/32
32  308656      192.168.1.5/32
33  308576      192.168.1.6/32
34  308688      192.168.1.7/32
35  308528      192.168.1.8/32
36  308608      192.168.1.9/32
37  308624      192.168.1.10/32
38  308640      192.168.1.21/32
```

```
39  308640      192.168.10.1/32
40       3      192.168.20.1/32 # Male's loopback
41
42 Input label database, 192.168.1.20:0--192.168.1.8:0
43   Label      Prefix
44  299776      172.16.1.84/30 # Transit Male-Havana
45  299840      192.168.1.1/32
46 <...>
47
48
49 Output label database, 192.168.1.20:0--192.168.1.8:0
50   Label      Prefix
51       3      172.16.1.76/30 # Transit Male-Bilbao
52       3      172.16.1.84/30 # Transit Male-Havana
53  308592      192.168.1.1/32
54  308544      192.168.1.2/32
55  308672      192.168.1.3/32
56  308560      192.168.1.4/32
57  308656      192.168.1.5/32
58  308576      192.168.1.6/32
59  308688      192.168.1.7/32
60  308528      192.168.1.8/32
61  308608      192.168.1.9/32
62  308624      192.168.1.10/32
63  308640      192.168.1.21/32
64  308640      192.168.10.1/32
65       3      192.168.20.1/32 # Male's loopback
```

Label binding population filtering at LDP

Both *export* and *import* filtering are available for LDP in Junos OS with the intention to filter outgoing label bindings and control route installation for incoming label bindings, respectively.

Export filtering: Junos OS allows filtering of outbound LDP label bindings by applying policies to block them from being advertised to neighboring routers.

These policies block or accept population of label bindings in the LDP database and act at that level only. This means that they can match on `route-filter`, `interface`, `neighbor`, or `next-hop`, but the only valid actions are `accept` or `reject`. They cannot modify any specific LDP route attributes, because they are applied to outgoing advertisements.

Also, the unique `from` condition accepted is a `route-filter`, while it can be applied `to` a specific `interface`, `neighbor` or `next-hop`. This granular scalpel can filter outbound label advertisements matching a specific prefix to LDP peers in different fashions.

Using the same default LDP label database status shown in Listing 4.2, Listing 4.4 shows the policy definition to block a label advertisement for a FEC representing router *Havana* to router *Bilbao* as LDP neighbor. Line 50 indicates how filtered LDP bindings appear, showing that the label binding is filtered towards router *Bilbao*, despite installing the LDP route locally as per Line 66.

Listing 4.4: LDP export policies at Male

```
1 [edit policy-options policy-statement block-Havana-to-Bilbao-ldp-label-bindings]
2 user@male-re0# show
3 term block-Havana-to-Bilbao {
4     from {
5         route-filter 192.168.1.4/32 exact;  # FEC for Havana
```

```
6       }
7       to neighbor 192.168.1.8; # Bilbao LDP peer
8       then reject;
9   }
10  term default {
11      then accept;
12  }
13
14  [edit]
15  user@male-re0# set protocols ldp export block-Havana-to-Bilbao-ldp-label-bindings
16
17
18  user@male-re0> show ldp database
19  Input label database, 192.168.1.20:0--192.168.1.4:0
20    Label     Prefix
21   100064     192.168.1.1/32
22  <...>
23
24  Output label database, 192.168.1.20:0--192.168.1.4:0
25    Label     Prefix
26   308592     192.168.1.1/32
27   308544     192.168.1.2/32
28   308672     192.168.1.3/32
29   308560     192.168.1.4/32
30   308656     192.168.1.5/32
31   308576     192.168.1.6/32
32   308688     192.168.1.7/32
33   308528     192.168.1.8/32
34   308608     192.168.1.9/32
35   308624     192.168.1.10/32
36        3     192.168.1.20/32
37   308640     192.168.1.21/32
38   308640     192.168.10.1/32
39
40  Input label database, 192.168.1.20:0--192.168.1.8:0
41    Label     Prefix
42   299840     192.168.1.1/32
43  <...>
44
45  Output label database, 192.168.1.20:0--192.168.1.8:0
46    Label     Prefix
47   308592     192.168.1.1/32
48   308544     192.168.1.2/32
49   308672     192.168.1.3/32
50   308560     192.168.1.4/32 (Filtered) # Filtered label binding to Bilbao
51   308656     192.168.1.5/32
52   308576     192.168.1.6/32
53   308688     192.168.1.7/32
54   308528     192.168.1.8/32
55   308608     192.168.1.9/32
56   308624     192.168.1.10/32
57        3     192.168.1.20/32
58   308640     192.168.1.21/32
59   308640     192.168.10.1/32
60
61  user@male-re0> show route table inet.3 192.168.1.4/32
62
63  inet.3: 11 destinations, 11 routes (11 active, 0 holddown, 0 hidden)
64  + = Active Route, - = Last Active, * = Both
65
66  192.168.1.4/32     *[LDP/9] 08:19:21, metric 100 # LDP route to Havana
67                      > to 172.16.1.86 via ge-4/2/4.0
68
69  user@Bilbao-re0> show ldp database session 192.168.1.20
70  Input label database, 192.168.1.8:0--192.168.1.20:0
71    Label     Prefix
72   308592     192.168.1.1/32
73   308544     192.168.1.2/32
74   308672     192.168.1.3/32
```

```
75  308656     192.168.1.5/32 # Missing FEC-label binding for Havana
76  308576     192.168.1.6/32
77  308528     192.168.1.8/32
78  308608     192.168.1.9/32
79  308624     192.168.1.10/32
80       3     192.168.1.20/32
81  308640     192.168.1.21/32
82  308640     192.168.10.1/32
83
84 Output label database, 192.168.1.8:0--192.168.1.20:0
85   Label     Prefix
86 <...>
```

With regard to migrations, the consequences of LDP label advertisement filtering are that LSPs can be practically blackholed at the moment an outbound LDP filter is placed over a best IGP path. This LDP export filtering does not detour the establishment of an LSP path, because this is indirectly decided by the IGP, but can in practice activate or suppress a given LDP-based LSP whether or not the label is present.

Nevertheless, LDP export filtering is able to craft LDP LSP path creation whether or not the IGP path follows a given direction. For instance, Listing 4.4 indirectly enforces that an LDP-based LSP path from router *Bilbao* to router *Havana* only exists in our topology when an IS–IS L1 adjacency exists between them both. The reason is that intra-area L1 routes are first preferred and router *Male* blocks that LSP with this LDP `export` policy. This happens even though router *Male* installs the LSP to router *Havana* and router *Bilbao*, so no internal LSPs inside the IS–IS L1 Area are affected.

Import filtering: Junos OS allows received LDP label bindings to be filtered by applying policies to accept or deny them from neighboring routers for local label installation.

Note that the label bindings can only be accepted or rejected as a consequence of `import` policies, but no other route attributes can be manipulated with those policies. Therefore, when compared to `export` policies, import filtering directly affects locally installed LDP routes as well, as routes are either accepted or rejected for installation. This means that `import` policies can match on `route-filter`, `interface`, `neighbor` or `next-hop`, or a combination of these, to first locally influence LDP route installation with `accept` or `reject` actions and second, affect LDP advertisements in possible downstream paths. LDP import filtering is another coarse tool to determine whether or not label advertisements for a specific FEC or from a particular LDP peer are valid. However, because LDP import filtering is applied inbound, it affects local route installation as well.

Taking Listing 4.2 as the default LDP database scenario, Listing 4.5 describes the policy definition and inbound application to block the LDP label binding for the loopback address from router *Havana* as representative FEC, when this label binding is received directly from router *Havana*. Line 23 shows how the LDP label binding is filtered inbound and as a consequence, there is no local route for it.

Listing 4.5: LDP import policies at Male

```
1 [edit policy-options policy-statement block-Havana-from-Havana-ldp-binding]
2 user@male-re0# show
3 term block-Havana {
4     from {
5         neighbor 192.168.1.4; # When advertised from Havana
6         route-filter 192.168.1.4/32 exact; # Block Havana's FEC
7     }
```

```
    then reject;
}
term default {
    then accept;
}

[edit]
user@male-re0# set protocols ldp import block-Havana-from-Havana-ldp-binding

user@male-re0> show ldp database
Input label database, 192.168.1.20:0--192.168.1.4:0
  Label     Prefix
 100064     192.168.1.1/32
 100016     192.168.1.2/32
 100144     192.168.1.3/32
      3     192.168.1.4/32 (Filtered)
 100128     192.168.1.5/32
<...>

Output label database, 192.168.1.20:0--192.168.1.4:0
  Label     Prefix
 308592     192.168.1.1/32
 308544     192.168.1.2/32
 308672     192.168.1.3/32
 308656     192.168.1.5/32 # Missing label binding for Havana
 308576     192.168.1.6/32
 308688     192.168.1.7/32
 308528     192.168.1.8/32
 308608     192.168.1.9/32
 308624     192.168.1.10/32
      3     192.168.1.20/32
 308640     192.168.1.21/32
 308640     192.168.10.1/32

Input label database, 192.168.1.20:0--192.168.1.8:0
  Label     Prefix
 299840     192.168.1.1/32
 299824     192.168.1.2/32
 299920     192.168.1.3/32
 299792     192.168.1.4/32 # Not filtered
 299904     192.168.1.5/32
<...>

Output label database, 192.168.1.20:0--192.168.1.8:0
  Label     Prefix
 308592     192.168.1.1/32
 308544     192.168.1.2/32
 308672     192.168.1.3/32
 308656     192.168.1.5/32 # Missing label binding for Havana
 308576     192.168.1.6/32
 308688     192.168.1.7/32
 308528     192.168.1.8/32
 308608     192.168.1.9/32
 308624     192.168.1.10/32
      3     192.168.1.20/32
 308640     192.168.1.21/32
 308640     192.168.10.1/32

user@male-re0> show route table inet.3 192.168.1.4/32
```

With regard to migrations, other implementation possibilities are available with this feature, but principally LDP import filtering allows the possibility of controlling, independently of the IGP metric, whether an LDP-based LSP is established through a peer. LDP `import` filters do not steer LSP path establishment through one peer or another, something that is controlled by the IGP, but the option to determine whether LSPs towards given FECs are not feasible through an LDP peer (and also for all downstream LSRs) can be implemented with this. For example, Listing 4.5 shows that an LDP path is established only internally to router *Havana* whenever the IGP calculation points to router *Bilbao* as next hop, a behavior that can allow other less-preferred protocol label bindings (for example, L-BGP by default) to be selected without needing to tweak protocol preferences. In the case of *equal-cost multiple paths* to a destination, LDP import filtering can also determine which of the available sibling paths is selected, rather than allowing the selection to be done based on other criteria.

All these resources for manipulating and controlling label distribution and for installing labels with LDP are summarized in the workflow shown in Figure 4.2, which has been simplified to emphasize the most important tie-breaking design drivers.

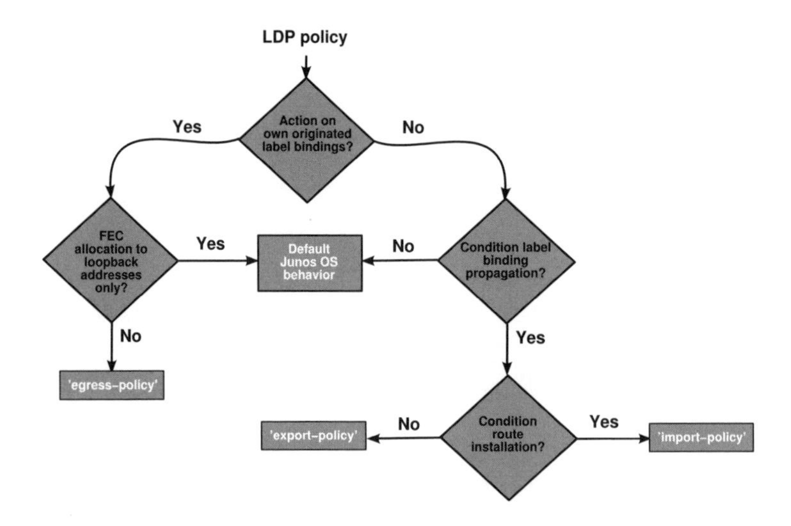

Figure 4.2: LDP policy design feature workflow.

4.4.3 Resources for route installation with RSVP-TE

IGP shortcuts for TE

As previously described, Junos OS has a dedicated routing table, inet.3, that contains MPLS LSP destinations, which are routes representing FECs with a legitimate and valid label binding. Junos OS considers first by default routes in inet.3 to be representative as next hops for all kinds of BGP routes, and only if these next hops are not found in inet.3, Junos OS uses afterwards routes from inet.0 as eligible BGP next hops. Other kinds of traffic, such as routing protocol flows between IGP-routed addresses, do not make use of this resolution infrastructure under those conditions and directly look into inet.0 by default. The rationale for this default behavior is that BGP is the scalable routing information transport protocol par excellence, which requires recursive resolution, whereas IGPs are focused on determining ultimate next hops for system reachability, rather than transporting routes.

This conservative behavior can be relaxed by expanding inet.3 route eligibility to all types of traffic, not only to BGP routes, by activating *IGP shortcuts*. In that situation, all MPLS next hops are moved from inet.3 to inet.0 and therefore become feasible for other types of traffic, not only BGP. With that strategy, inet.3 is completely emptied and inet.0 becomes the repository for IP and MPLS lookups. The consequence is that MPLS LSPs are also eligible next hops in inet.0 (emulating default behavior for other operating systems) for all types of lookup.

In fact, Junos OS performs SPF calculations for route entries and these LSPs are selected as next hops for destinations located *downstream from the egress LSR*. This means that LSP selection is not prone to loops for destinations closer to the egress LSRs, because the ingress point decides whether traffic to such a destination has to be switched over that LSP by looking into the IGP database. This is a *sledge-hammer* approach to using MPLS LSPs for all kinds of traffic whenever they are available, based just on a single local decision at the ingress LSR. In the end, there are no modifications to any of the standard protocol operations, but rather, it is a purely local routing table decision. One advantage to this method is that it provides the ability to activate this behavior without influencing any other systems in the IGP area or domain.

Considering domain "Cyclone" as reference scenario, Listing 4.6 shows the consequences of activating `traffic-engineering bgp-igp`, with RSVP routes becoming active paths in inet.0

Apart from replicating other routing vendor's default behavior to reach IGP routes over MPLS LSPs, *IGP shortcuts* can be helpful when facing migrations to *hide* an IGP core with the MPLS substrate and, especially, to indirectly avoid internal IP lookups inside the core towards destinations beyond the LSP egress point. Those routes selected by the ingress LSRs to be beyond the LSPs are automatically MPLS-switched with no need to inject a specific label binding for them using any MPLS LSP, that is to say, traffic can be MPLS-switched to destinations without any specific label binding in any MPLS LDP.

However, this feature is not granular: it is all or nothing! Despite being a purely local decision without involving any other systems in the domain, these *IGP shortcuts* cannot be fine-tuned by destination, FEC, VPN or even by LSP.

Listing 4.6: Traffic-engineering bgp-igp at Livorno

```
1  [edit protocols mpls]
2  user@Livorno# show
3  traffic-engineering bgp-igp;
4  label-switched-path Livorno-to-Skopie {
5      to 192.168.1.2;
6  }
7  label-switched-path Livorno-to-Barcelona {
8      to 192.168.1.7;
9  }
10 label-switched-path Livorno-to-Honolulu {
11     to 192.168.1.21;
12 }
13 <...>
14
15 user@Livorno> show route table inet.0 192.168.1.0/24
16
17 inet.0: 38 destinations, 48 routes (37 active, 0 holddown, 3 hidden)
18 + = Active Route, - = Last Active, * = Both
19
20 <...>
21 192.168.1.2/32     *[RSVP/7] 00:00:39, metric 64 # RSVP route as active path in inet.0
22                     > via so-0/0/0.0, label-switched-path Livorno-to-Skopie
23                     [OSPF/10] 00:00:39, metric 64
```

```
24                       > via so-0/0/0.0
25 192.168.1.6/32       *[Direct/0] 2d 19:03:56
26                       > via lo0.0
27 192.168.1.7/32       *[RSVP/7] 00:00:39, metric 10 # RSVP route as active path in inet.0
28                       > to 172.16.1.53 via ge-0/2/0.0, label-switched-path Livorno-to-Barcelona
29                       [OSPF/10] 00:00:39, metric 10
30                       > to 172.16.1.53 via ge-0/2/0.0
31 192.168.1.21/32      *[RSVP/7] 00:00:39, metric 10 # RSVP route as active path in inet.0
32                       > to 172.16.1.90 via ge-1/3/0.0, label-switched-path Livorno-to-Honolulu
33                       [OSPF/10] 00:00:39, metric 10
34                       > to 172.16.1.90 via ge-1/3/0.0
35 <...>
```

A collateral effect occurs in Junos OS when electing the routes to be reachable via MPLS LSPs: LSP FECs are injected into inet.0 thanks to the label binding and *inet.3 is emptied by this action*.

Because inet.3 is designed to contain MPLS LSP destinations, Junos OS properly resolves any kind of MPLS VPN route when the next hop for the advertisement is present in inet.3. Indirectly, this behavior makes other destinations beyond MPLS LSP egress points reachable through them, but also causes these label bindings to be no longer eligible as next hops for any kind of VPN route.

In an attempt to have the best of both worlds, another knob was conceived: `traffic-engineering bgp-igp-both-ribs`. With it, both objectives are achieved: MPLS LSP destinations are copied to inet.0 to allow routes beyond to be reachable through them, and these same destinations are kept in inet.3 for proper VPN route resolution. In the end, the same routes derived from active label bindings are present in both tables! This is also another *all-or-nothing* feature: it cannot be granularly tweaked on a per-LSP-destination or VPN basis; it applies to all active paths with label bindings.

Listing 4.7 illustrates the differential effect in domain "Cyclone" when activating that feature in *Livorno*. Note that the same RSVP routes are present in both inet.0 and inet.3 with the same attributes (in fact, similar mechanisms to `rib-groups` are internally used for this).

Listing 4.7: Traffic-engineering bgp-igp-both-ribs at Livorno

```
1  [edit protocols mpls]
2  user@Livorno# show
3  traffic-engineering bgp-igp-both-ribs;
4  label-switched-path Livorno-to-Skopie {
5      to 192.168.1.2;
6  }
7  label-switched-path Livorno-to-Barcelona {
8      to 192.168.1.7;
9  }
10 label-switched-path Livorno-to-Honolulu {
11     to 192.168.1.21;
12 }
13 <...>
14
15 user@Livorno> show route table inet.0 192.168.1.0/24
16
17   # RSVP routes in inet.0
18
19 inet.0: 38 destinations, 48 routes (37 active, 0 holddown, 3 hidden)
20 + = Active Route, - = Last Active, * = Both
21
22 <...>
23 192.168.1.2/32       *[RSVP/7] 00:00:07, metric 64
24                       > via so-0/0/0.0, label-switched-path Livorno-to-Skopie
25                       [OSPF/10] 00:02:52, metric 64
```

```
26                           > via so-0/0/0.0
27  192.168.1.6/32        *[Direct/0] 2d 19:06:09
28                           > via lo0.0
29  192.168.1.7/32        *[RSVP/7] 00:00:07, metric 10
30                           > to 172.16.1.53 via ge-0/2/0.0, label-switched-path Livorno-to-Barcelona
31                           [OSPF/10] 00:02:52, metric 10
32                           > to 172.16.1.53 via ge-0/2/0.0
33  192.168.1.21/32       *[RSVP/7] 00:00:07, metric 10
34                           > to 172.16.1.90 via ge-1/3/0.0, label-switched-path Livorno-to-Honolulu
35                           [OSPF/10] 00:02:52, metric 10
36                           > to 172.16.1.90 via ge-1/3/0.0
37  <...>
38
39  user@Livorno> show route table inet.3 192.168.1.0/24
40    # Same routes present in inet.3
41
42  inet.3: 7 destinations, 8 routes (6 active, 0 holddown, 1 hidden)
43  + = Active Route, - = Last Active, * = Both
44
45  192.168.1.2/32        *[RSVP/7] 00:00:12, metric 64
46                           > via so-0/0/0.0, label-switched-path Livorno-to-Skopie
47  192.168.1.6/32        *[Direct/0] 2d 18:50:51
48                           > via lo0.0
49  192.168.1.7/32        *[RSVP/7] 00:00:12, metric 10
50                           > to 172.16.1.53 via ge-0/2/0.0, label-switched-path Livorno-to-Barcelona
51  192.168.1.21/32       *[RSVP/7] 00:00:12, metric 10
52                           > to 172.16.1.90 via ge-1/3/0.0, label-switched-path Livorno-to-Honolulu
```

Nevertheless, there is still another consideration. Think about our IGP migration scenarios from Section 2.3 and the situations above. Because Junos OS selects only the best sibling path from the routing table for `export` policies in protocols, these features are *overshadowing* the underlying IGP-calculated paths in the routing table.

In both Listings 4.6 and 4.7, the RSVP routes have lower preference than the original OSPF-calculated routes in inet.0, although these latter ones are really the foundations for the MPLS LSPs. Other redistribution policies that may be expecting OSPF routes to be matched (as extensively tested in some Link-state IGP migrations from Section 2.3) are fooled by the local activation of any of the two knobs.

Another concept was created in Junos OS to address this situation: `traffic-engineering mpls-forwarding`. This feature goes even further by keeping MPLS LSP routes active in inet.3 for VPN resolution and inet.0 *for MPLS forwarding purposes only*. The overshadowed IGP routes that are the real foundation for the MPLS LSPs remain eligible for any kind of policy selection but are not downloaded to the Forwarding Engine as active paths but rather, the MPLS LSPs.

Put another way, despite being another *all-or-nothing* knob that cannot be granularly adjusted, `traffic-engineering mpls-forwarding` does not affect any previous IGP interaction setup and keeps MPLS LSP destinations active in inet.3 for VPN resolution. Listing 4.8 describes the incremental difference when compared with the previous outputs. Note that Junos OS utilizes the "@" symbol to represent the overshadowed but control-plane active IGP route and "#" to indicate the active MPLS LSP used in the forwarding plane, as shown in Lines 21, 23, 27, 29, 31, and 33, respectively, for router *Skopie*, router *Barcelona*, and router *Honolulu* in inet.0.

Listing 4.8: Traffic-engineering mpls-forwarding at Livorno

```
1  [edit protocols mpls]
2  user@Livorno# show
3  traffic-engineering mpls-forwarding;
```

```
 4 label-switched-path Livorno-to-Skopie {
 5     to 192.168.1.2;
 6 }
 7 label-switched-path Livorno-to-Barcelona {
 8     to 192.168.1.7;
 9 }
10 label-switched-path Livorno-to-Honolulu {
11     to 192.168.1.21;
12 }
13 <...>
14
15 user@Livorno> show route table inet.0 192.168.1.0/24
16
17 inet.0: 38 destinations, 48 routes (37 active, 0 holddown, 3 hidden)
18 + = Active Route, - = Last Active, * = Both
19
20 <...>
21 192.168.1.2/32     @[OSPF/10] 00:04:47, metric 64  # IGP route for Livorno
22                     > via so-0/0/0.0
23                    #[RSVP/7] 00:00:08, metric 64  # MPLS LSP for Livorno
24                     > via so-0/0/0.0, label-switched-path Livorno-to-Skopie
25 192.168.1.6/32     *[Direct/0] 2d 19:06:09
26                     > via lo0.0
27 192.168.1.7/32     @[OSPF/10] 00:04:47, metric 10  # IGP route for Barcelona
28                     > to 172.16.1.53 via ge-0/2/0.0
29                    #[RSVP/7] 00:00:08, metric 10  # MPLS LSP for Barcelona
30                     > to 172.16.1.53 via ge-0/2/0.0, label-switched-path Livorno-to-Barcelona
31 192.168.1.21/32    @[OSPF/10] 00:04:47, metric 10  # IGP route for Honolulu
32                     > to 172.16.1.90 via ge-1/3/0.0
33                    #[RSVP/7] 00:00:07, metric 10  # MPLS LSP for Honolulu
34                     > to 172.16.1.90 via ge-1/3/0.0, label-switched-path Livorno-to-Honolulu
35 <...>
36
37 user@Livorno> show route table inet.3 192.168.1.0/24
38
39 inet.3: 7 destinations, 8 routes (6 active, 0 holddown, 1 hidden)
40 + = Active Route, - = Last Active, * = Both
41
42 192.168.1.2/32     *[RSVP/7] 00:00:13, metric 64
43                     > via so-0/0/0.0, label-switched-path Livorno-to-Skopie
44 192.168.1.6/32     *[Direct/0] 2d 18:52:46
45                     > via lo0.0
46 192.168.1.7/32     *[RSVP/7] 00:00:13, metric 10
47                     > to 172.16.1.53 via ge-0/2/0.0, label-switched-path Livorno-to-Barcelona
48 192.168.1.21/32    *[RSVP/7] 00:00:12, metric 10
49                     > to 172.16.1.90 via ge-1/3/0.0, label-switched-path Livorno-to-Honolulu
```

Additional prefix association to RSVP-TE MPLS LSPs

As previously seen, Junos OS only associates the configured FEC with a particular RSVP-TE-steered LSP by default. FECs tend to correspond to egress LSR host addresses and these routes may not suffice in certain scenarios, in which it is well known that other destinations are reachable beyond the LSP destination by design.

However, Junos OS has another granular knob to install additional routes over a common RSVP-TE LSP: `install`. This feature allows the configuration of additional destination prefixes over an LSP. These prefixes therefore appear in inet.3, together with the MPLS LSP egress point, and become eligible next hops for MPLS VPNs.

This behavior may not suffice in some scenarios that need actually to *forward* traffic to those destinations, not only to make them feasible next hops. In Junos OS, this means that such destinations should not appear in inet.3, but in inet.0. The `install` knob has been complemented with an `active` suffix to indicate the desire to inject the prefix into inet.0

instead of inet.3. In this way, MPLS switching can take place for that destination aggregate until the LSP egress point, and at that point, an IP lookup should ensure final reachability, instead of making this additional prefix a next hop available beyond the LSP.

This feature provides an additional tool to control traffic association with an LSP: destinations can be associated on a granular, per-LSP basis. This is also a purely local decision at the ingress point and is not explicitly related to any protocol mechanism: prefixes are statically bound at the ingress point for resolution in inet.3 or for forwarding in inet.0 with the `active` knob. It provides, therefore, the benefits of traffic control at the ingress LSR bound to a *downstream on-demand* label solicitation for the LSP FEC, along with the collateral effects of such granularity. For a global domain-wide association, this behavior needs to be replicated at all ingress points, rather than the *downstream unsolicited* behavior of LDP tunneling over RSVP-TE, in which the egress LSR controls label association.

Listing 4.9 describes the effect and configuration changes needed to activate the association of 192.168.10.1/32 at router *Livorno* over its LSP to router *Barcelona* with the `install` knob and the `active` suffix. Note that this prefix is not included as part of any RSVP object, as depicted starting from Line 20 with a trace options excerpt for the RSVP-TE Path message. It is also worth highlighting that the `active` knob moves the prefix destination to inet.0, without keeping a copy into inet.3: it has the equivalent effect of setting a static route in inet.0 to an RSVP-TE LSP, as seen in Line 44.

Listing 4.9: Installing external aggregate over LSP to Barcelona in Livorno

```
1  [edit protocols mpls label-switched-path Livorno-to-Barcelona]
2  user@Livorno# set install 192.168.10.1/32
3
4  user@Livorno> show route 192.168.10.1/32 exact
5
6  inet.0: 28 destinations, 30 routes (27 active, 0 holddown, 1 hidden)
7  + = Active Route, - = Last Active, * = Both
8
9  192.168.10.1/32    *[OSPF/150] 00:10:07, metric 0, tag 0
10                     > to 172.16.1.53 via ge-0/2/0.0
11
12 inet.3: 5 destinations, 5 routes (4 active, 0 holddown, 1 hidden)
13 + = Active Route, - = Last Active, * = Both
14
15 192.168.10.1/32    *[RSVP/7] 00:00:05, metric 10
16                     > to 172.16.1.53 via ge-0/2/0.0, label-switched-path Livorno-to-Barcelona
17
18 *** 'rsvp-traceoptions' has been created ***
19 Jul 12 17:27:25.420391 task_timer_dispatch: calling RSVP_RSVP SESSION, late by 0.001
20 Jul 12 17:27:25.420591 RSVP send Path 192.168.1.6->192.168.1.7 Len=212 ge-0/2/0.0 # Path message
       from Livorno to Barcelona
21 Jul 12 17:27:25.420620   Session7 Len 16 192.168.1.7(port/tunnel ID 52104 Ext-ID 192.168.1.6) Proto 0
22 Jul 12 17:27:25.420644   Hop       Len 12 172.16.1.53/0x08e33248
23 Jul 12 17:27:25.420662   Time      Len  8 30000 ms
24 Jul 12 17:27:25.420688   SrcRoute Len 12  172.16.1.54 S
25 Jul 12 17:27:25.420707   LabelRequest Len  8 EtherType 0x800
26 Jul 12 17:27:25.420728   Properties Len 12 Primary path
27 Jul 12 17:27:25.420749   SessionAttribute Len 28 Prio (7,0) flag 0x0 ''Livorno-to-Barcelona''
28 Jul 12 17:27:25.420770   Sender7  Len 12 192.168.1.6(port/lsp ID  17)
29 Jul 12 17:27:25.420809   Tspec     Len 36 rate 0bps size 0bps peak Infbps m 20 M 1500
30 Jul 12 17:27:25.420829   ADspec    Len 48 MTU 1500
31 Jul 12 17:27:25.420850   RecRoute Len 12  172.16.1.53
32 Jul 12 17:27:25.421019 task_send_msg: task RSVP socket 11 length 236 flags MSG_DONTROUTE(4) to
       172.16.1.90 out interface ge-1/3/0.0
33 Jul 12 17:27:25.421077 task_timer_uset: timer RSVP_RSVP SESSION <Touched Processing> set to offset 45
       at 17:28:10
34 Jul 12 17:27:25.421151 task_timer_dispatch: returned from RSVP_RSVP SESSION, rescheduled in 45
35
```

```
36  [edit protocols mpls label-switched-path Livorno-to-Honolulu]
37  user@Livorno# set install 192.168.10.1/32 active
38
39  user@Livorno> show route 192.168.10.1/32 exact
40
41  inet.0: 28 destinations, 31 routes (27 active, 0 holddown, 1 hidden)
42  + = Active Route, - = Last Active, * = Both
43
44  192.168.10.1/32      *[RSVP/7] 00:00:05, metric 10 # RSVP route only in inet.0
45                        > to 172.16.1.53 via ge-0/2/0.0, label-switched-path Livorno-to-Barcelona
46                        [OSPF/150] 00:19:06, metric 0, tag 0
47                        > to 172.16.1.53 via ge-0/2/0.0
```

Junos Tip: Combining RSVP-TE LSP prefix association with IGP shortcuts

This association of additional prefixes to RSVP-TE LSPs does not exclude the utilization of *IGP shortcuts*. In fact, they can be perfectly complemented.

The same routing result from Listing 4.9 can be obtained when combining `install` and `traffic-engineering bgp-igp`, as seen in Listing 4.10. This provides the advantage that `active` does not need to be explicitly set on each *installed prefix*, and all LSP destinations, as well as all additionally installed prefixes, are moved to inet.0.

Listing 4.10: Combining additional prefix and IGP shortcuts in Livorno

```
1  [edit protocols mpls]
2  user@Livorno# set traffic-engineering bgp-igp
3  [edit protocols mpls label-switched-path Livorno-to-Barcelona]
4  user@Livorno# set install 192.168.10.1/32
5
6  user@Livorno> show route 192.168.10.1/32 exact
7
8  inet.0: 28 destinations, 31 routes (27 active, 0 holddown, 1 hidden)
9  + = Active Route, - = Last Active, * = Both
10
11 192.168.10.1/32      *[RSVP/7] 00:00:07, metric 10 # RSVP route only in inet.0
12                       > to 172.16.1.53 via ge-0/2/0.0, label-switched-path Livorno-to-Barcelona
13                       [OSPF/150] 00:23:13, metric 0, tag 0
14                       > to 172.16.1.53 via ge-0/2/0.0
```

LDP tunneling over RSVP

LDP was originally conceived in [RFC3036] and [RFC3037] for non-traffic-engineered applications, but RSVP-TE extensions were precisely defined for that purpose at [RFC3209]. As networks grow over time, a full-mesh of RSVP-based LSPs becomes more difficult to operate and maintain due to an N^2 growth path and signaling limitations, and different scalability concepts are proposed.

One feasible implementation consists of *tunneling* LDP LSPs over RSVP-TE LSPs. LDP includes resources to support sessions between non-directly connected LSRs, and by tunneling those LSPs, the benefits of providing TE and *fast reroute* capabilities to downstream unsolicited LSPs can be achieved, combining the advantages of both worlds. The rationale is to tunnel data traffic traversing those LDP LSPs, while targeted LDP control traffic between remote neighbors remains outside this plane.

In Junos OS, this is implemented by having link-state IGPs using RSVP-TE LSPs in the SPF calculation and adding *hidden* routes to inet.3 representing each LSP. The reason for these routes to remain *hidden* is to constrain their utilization to LDP and avoid other protocols (BGP by default) being able to use those next hops for resolution. Similarly, those next hops will be disclosed if TE *shortcuts* for RSVP-TE are activated.

Also, LDP next-hop tracking is extended for these RSVP-TE LSPs: if they become active next hops for an LDP FEC, the label advertised by that remote LDP peer reachable through the RSVP-TE LSP becomes active. This means a double MPLS label *push* at the ingress LSR: the *inner* label to reach the LDP FEC and the *outer* label to reach the LDP peer via RSVP-TE LSP. This can also occur in transit between a non-extended LDP and a targeted LDP session, meaning that the programmed action can be an MPLS label *swap* and *push* for the RSVP-TE destination, instead of a double *push*.

Listing 4.11 shows the necessary configuration steps on router *Livorno* to activate LDP tunneling over the existing RSVP-TE LSP mesh from domain "Cyclone". The feature is simply activated by enabling the loopback interface for LDP protocol and adding the respective `ldp-tunneling` knob at the RSVP-TE LSPs, as shown in Lines 5, 9, and 13. Note that Junos OS presents an additional hook via `strict-targeted-hellos` to just limit targeted LDP session establishment to those LDP peers matching destinations from RSVP-TE LSPs configured with `ldp-tunneling` and do not accept other arbitrary targeted LDP sessions.

Listing 4.11: LDP tunneling activation over RSVP-TE mesh at Livorno

```
1  [edit protocols mpls]
2  user@Livorno# show
3  label-switched-path Livorno-to-Skopie {
4      to 192.168.1.2;
5      ldp-tunneling; # Activate LDP targeted session over LSP
6  }
7  label-switched-path Livorno-to-Barcelona {
8      to 192.168.1.7;
9      ldp-tunneling; # Activate LDP targeted session over LSP
10 }
11 label-switched-path Livorno-to-Honolulu {
12     to 192.168.1.21;
13     ldp-tunneling; # Activate LDP targeted session over LSP
14 }
15 interface all;
16 interface fxp0.0 {
17     disable;
18 }
19
20 [edit protocols ldp]
21 user@Livorno# show
22 track-igp-metric;
23 strict-targeted-hellos; # Limit LDP sessions to be targeted per config
24 interface lo0.0;
25
26 user@Livorno> show mpls lsp ingress
27 Ingress LSP: 3 sessions
28 To              From            State Rt P     ActivePath        LSPname
29 192.168.1.2     192.168.1.6     Up     0 *                       Livorno-to-Skopie
30 192.168.1.7     192.168.1.6     Up     1 *                       Livorno-to-Barcelona
31 192.168.1.21    192.168.1.6     Up     0 *                       Livorno-to-Honolulu
32 Total 3 displayed, Up 3, Down 0
33
34 user@Livorno> show rsvp neighbor
35 RSVP neighbor: 3 learned
36 Address            Idle Up/Dn LastChange HelloInt HelloTx/Rx MsgRcvd
37 172.16.1.33           5  1/0  2d 3:03:20        9 20316/20316 24569
38 172.16.1.90           5  3/2  1d 15:18:17        9 20313/20312 17418
39 172.16.1.53           5  2/1  1d 22:15:16        9 20237/20215 21409
```

```
40
41 user@Livorno> show ldp neighbor
42 Address            Interface         Label space ID         Hold time
43 192.168.1.2        lo0.0             192.168.1.2:0            44
44 192.168.1.7        lo0.0             192.168.1.7:0            31
45 192.168.1.21       lo0.0             192.168.1.21:0           35
46
47 user@Livorno> show ldp session
48   Address            State        Connection     Hold time
49 192.168.1.2        Operational  Open             26
50 192.168.1.7        Operational  Open             25
51 192.168.1.21       Operational  Open             24
```

When routing tables are analyzed in detail, one notices that *hidden* routes in table inet.3 are visible for RSVP-TE LSP destinations, so that these *shortcuts* are eligible for LDP route computations only. Listing 4.12 shows those routing table excerpts from the perspective of router *Livorno*. In fact, considering a label binding for the externally injected prefix 192.168.10.1/32 into domain "Cyclone" only in LDP and not as RSVP-TE LSP destination, the LDP label binding becomes active over the RSVP-TE LSP following an OSPFv2 route. Because of implicit LDP `egress-policies` at participating LSRs, router *Livorno* receives an LDP label binding for each other participating router reachable over the corresponding RSVP-TE LSP; however, because of default protocol preferences, RSVP is preferred. Line 21 shows the LDP route preferred for the externally injected prefix, installed over the RSVP-TE LSP, for the closest exit point. Lines 49, 51, 53, and 55 depict hidden routes for each destination to be considered for LDP calculations.

Listing 4.12: Routing table effect when activating LDP tunneling at Livorno

```
1  user@Livorno> show route table inet.3
2
3  inet.3: 10 destinations, 18 routes (7 active, 0 holddown, 7 hidden)
4  + = Active Route, - = Last Active, * = Both
5
6  <...>
7  192.168.1.2/32      *[RSVP/7] 2d 03:04:32, metric 64
8                       > via so-0/0/0.0, label-switched-path Livorno-to-Skopie
9                       [LDP/9] 1d 15:19:24, metric 64
10                      > via so-0/0/0.0, label-switched-path Livorno-to-Skopie
11 192.168.1.6/32      *[Direct/0] 2d 02:51:06
12                      > via lo0.0
13 192.168.1.7/32      *[RSVP/7] 1d 22:16:28, metric 10
14                      > to 172.16.1.53 via ge-0/2/0.0, label-switched-path Livorno-to-Barcelona
15                      [LDP/9] 1d 15:19:24, metric 10
16                      > to 172.16.1.53 via ge-0/2/0.0, label-switched-path Livorno-to-Barcelona
17 192.168.1.21/32     *[RSVP/7] 1d 15:19:06, metric 10
18                      > to 172.16.1.90 via ge-1/3/0.0, label-switched-path Livorno-to-Honolulu
19                      [LDP/9] 1d 15:19:06, metric 10
20                      > to 172.16.1.90 via ge-1/3/0.0, label-switched-path Livorno-to-Honolulu
21 192.168.10.1/32     *[LDP/9] 1d 15:19:24, metric 0 # LDP route over RSVP-TE LSP
22                      > to 172.16.1.53 via ge-0/2/0.0, label-switched-path Livorno-to-Barcelona
23
24 user@Livorno> show route 192.168.10.1/32 extensive table inet.3
25
26 inet.3: 10 destinations, 18 routes (7 active, 0 holddown, 7 hidden)
27 192.168.10.1/32 (2 entries, 2 announced)
28         State: <FlashAll>
29         *LDP    Preference: 9
30                 Next hop type: Router
31                 Next-hop reference count: 4
32                 Next hop: 172.16.1.53 via ge-0/2/0.0 weight 0x1, selected
33                 Label-switched-path Livorno-to-Barcelona # RSVP-TE LSP as next hop
34                 State: <Active Int>
35                 Local AS: 65001
```

```
36                 Age: 1d 15:27:07        Metric: 0
37                 Task: LDP
38                 Announcement bits (2): 2-Resolve tree 1 3-Resolve tree 2
39                 AS path: I
40
41 user@Livorno> show route table inet.3 hidden
42
43   # Hidden shortcuts for LDP computation
44
45 inet.3: 10 destinations, 18 routes (7 active, 0 holddown, 7 hidden)
46 + = Active Route, - = Last Active, * = Both
47
48 <...>
49 192.168.1.2/32      [OSPF] 2d 03:05:37, metric 64
50                     > via so-0/0/0.0, label-switched-path Livorno-to-Skopie
51 192.168.1.7/32      [OSPF] 1d 22:17:33, metric 10
52                     > to 172.16.1.53 via ge-0/2/0.0, label-switched-path Livorno-to-Barcelona
53 192.168.1.21/32     [OSPF] 1d 15:20:11, metric 10
54                     > to 172.16.1.90 via ge-1/3/0.0, label-switched-path Livorno-to-Honolulu
55 192.168.10.1/32     [OSPF] 1d 22:17:33, metric 0, tag 0
56                     > to 172.16.1.53 via ge-0/2/0.0, label-switched-path Livorno-to-Barcelona
```

The applicability of LDP tunneling for MPLS transport migration purposes resides in its simplicity to extend and merge LDP clouds: MPLS domains can be easily stitched over a transit RSVP-TE LSP core or interconnected just by activating the feature and letting LDP label bindings be populated (this also requires that IGP routes be manipulated in a similar fashion). It is a very simple mechanism to unify LDP domains and presents the advantage of interacting perfectly, and with no further resources, with non-targeted LDP sessions at the RSVP-TE ingress LSRs. Thus, LDP tunneling becomes also a practical *downstream unsolicited* label distribution implementation over an RSVP-TE LSP mesh, in which LDP label bindings can be fine-tuned via LDP egress and export policies at boundary routers.

A simplified view of these features, which determine RSVP-TE route installation and label distribution, is shown in Figure 4.3 to illustrate differences among them that lead to particular design criteria.

4.4.4 Resources for label advertisement and route resolution with Labeled BGP

Route resolution for Labeled BGP

The Labeled BGP address family (AFI 1, SAFI 4) is explicitly activated at each BGP peer with the `family inet labeled-unicast` configuration option. This knob allows each neighbor to negotiate this family type and to signal LSPs for FECs represented by IP prefixes with labels, as per [RFC3107]. Per the default action with BGP, these routes, along with their labels, are installed in Junos OS in inet.0. While this route installation can be a perfect fit for standard route resolution pointing to those next hops, Junos OS requires that MPLS VPN applications resolve over next hops present in inet.3, indicating that a label binding is present for that FEC in the dedicated table. Similarly to LDP and RSVP-TE, L-BGP routes can also be installed in inet.3 in different ways.

inet.3 as primary table for L-BGP: Junos OS allows the one-to-one relationship between BGP NLRI and table to be modified granularly. The `rib inet.3` knob for the `labeled-unicast` family associates inet.3 as the primary table into which L-BGP routes are installed along with their labels.

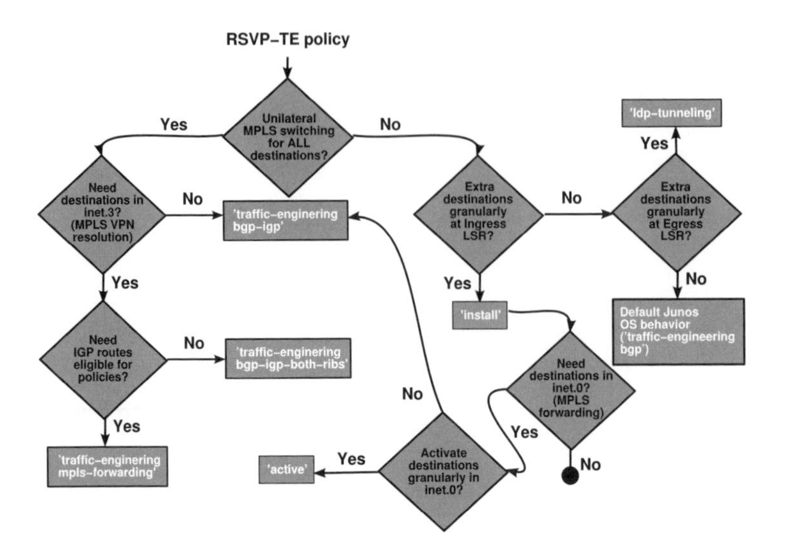

Figure 4.3: RSVP-TE design criteria and tie-breakers.

This behavior still allows standard BGP IPv4 unicast routes to use inet.0 as their corresponding table, with a unique BGP family-table liaison, and then to advertise prefixes for different purposes with different families. Junos OS also offers enough versatility with its policy language to apply different BGP TE rules on a per-family basis.

When activating this option, a behavior similar to *advertise-inactive* is internally enabled on this specific family, because it is expected to resolve those L-BGP routes (usually router loopbacks) over RSVP or LDP for proper resolution, and thus, these L-BGP routes may become inactive paths in the routing table when overlapping with RSVP-TE or LDP FECs for the same destinations.

The main advantages of using inet.3 are that other routes present in that table are directly eligible for export, and that imported routes and labels are already placed in inet.3. No further tweaks are then needed to redistribute LDP or RSVP signaled FECs into L-BGP, because they use a common container as table. On the other hand, inet.3 becomes the route resolution RIB, so any resolution apart from directly connected routes (that is, L-IBGP, confederation L-BGP, or multihop L-EBGP) requires that the corresponding next hop is present in inet.3.

It may also be desirable to place these routes in inet.0 for standard IPv4 route resolution purposes. This can be also achieved by attaching a standard `rib-group` to the family, thus importing inet.3 routes into inet.0 as well.

Copying Labeled BGP routes into inet.3: Another option offered by Junos OS is to maintain inet.0 as *primary* table for L-BGP routes, but to allow MPLS VPN resolution over them by automatically copying those routes to inet.3

The `resolve-vpn` option for the `labeled-unicast` family internally activates a `rib-group` that imports these routes both into inet.0 as primary table and inet.3 as secondary copies.

This alternative allows resolution for MPLS VPNs with the secondary copies present in inet,3 but installs and exports routes from inet.0, so they are then also eligible for standard IPv4 lookup actions.

Junos Tip: Primary routing table exclusiveness for L-BGP routes

Both L-BGP route resolution options are mutually exclusive and Junos OS enforces this decision. The `rib inet.3` option takes precedence when both are configured, with the understanding that setting a *primary* table has higher significance. The user is also notified of this behavior by means of a commit warning, as per Listing 4.13.

Listing 4.13: L-BGP route resolution exclusiveness

```
1  [edit protocols bgp]
2  user@Basel# show
3  group IBGP {
4      type internal;
5      local-address 192.168.1.10;
6      family inet {
7          unicast;
8          labeled-unicast {
9              rib {
10                 inet.3;
11             }
12             resolve-vpn;
13         }
14     }
15     export bind-local-label;
16 <...>
17 }
18 user@Basel# commit and-quit
19 [edit protocols bgp group IBGP family inet]
20   'labeled-unicast'
21     BGP: Warning: resolve-vpn keyword ignored - primary table is inet.3
22 commit complete
23 Exiting configuration mode
```

Label advertisement for Labeled BGP

As per general rule, in L-BGP the associated label is changed in Junos OS with every next-hop change for the route. This means that the standard next-hop change when transitioning from internal to external L-BGP, or vice versa, indirectly leads to a label *swap* (or *pop* if one of the labels is *implicit-null*). In other situations, such as L-IBGP route reflection, confederated L-BGP sessions with specific next-hop setting policies, and direct advertisements from an L-EBGP session to another L-EBGP peer (without a joint shared broadcast media for third-party next-hop advertisement), labels included in the NLRI are changed in Junos OS when readvertising the route, and the MPLS action (label *swap* or *pop*) is installed in the forwarding table mpls.0.

Advertising null label for connected routes: Junos OS also allows advertising an *explicit-null* label (0 for IPv4, 2 for IPv6) for directly connected routes (actually for both inet and inet6 *labeled-unicast* families).

The default behavior is to advertise the *implicit-null* label (3 for inet *labeled-unicast*) to enforce *penultimate-hop popping*. With regard to migrations, there are certain setups in which an MPLS label is always needed end to end, for instance, to enforce MPLS CoS settings between L-BGP ASBRs or, especially, to avoid an IP lookup at some point. Listing 4.14 illustrates changes in router *Livorno* before and after activating the `explicit-null` knob.

Listing 4.14: L-BGP explicit null advertisement for connected routes

```
1  user@Livorno> show route advertising-protocol bgp 172.16.1.37 192.168.1.6 extensive
2
3  inet.3: 10 destinations, 18 routes (8 active, 0 holddown, 5 hidden)
4  * 192.168.1.6/32 (1 entry, 1 announced)
5   BGP group L-EBGP type External
6       Route Label: 3 # Implicit null
7       Nexthop: Self
8       Flags: Nexthop Change
9       AS path: [64501] I
10
11 user@Livorno> configure
12 Entering configuration mode
13  [edit protocols bgp group L-EBGP]
14 user@Livorno# set family inet labeled-unicast explicit-null connected-only
15  [edit protocols bgp group L-EBGP]
16 user@Livorno# commit and-quit
17 commit complete
18 Exiting configuration mode
19 user@Livorno> show route advertising-protocol bgp 172.16.1.37 192.168.1.6 extensive
20
21 inet.3: 10 destinations, 18 routes (8 active, 0 holddown, 5 hidden)
22 * 192.168.1.6/32 (1 entry, 1 announced)
23  BGP group L-EBGP type External
24      Route Label: 0  # Explicit null
25      Nexthop: Self
26      Flags: Nexthop Change
27      AS path: [64501] I
```

Two figures provide an overview of all the design knobs in Junos OS related to L-BGP. Figure 4.4 shows a simple route injection design workflow, and Figure 4.5 shows the workflow for a label allocation scheme.

4.5 Case Study

The same company acquisition scenario from previous chapters is used in our case study here. Integrating MPLS-based applications is fairly common in a network merger scenario, and sometimes it is actually the ultimate goal of a migration.

Extending and integrating MPLS label distribution protocols over the network topology ensures that those end-to-end MPLS-based applications can be provisioned and operated among participating networks.

4.5.1 Original network

Considering the same initial setups for merging into a common service provider "Gale Internet," MPLS domains also need to be integrated. As a parallel exercise to integrating link-state IGPs and BGP, MPLS label distribution protocols will be integrated with the intention to have a final joint MPLS domain. Taking into account MPLS transport support in each of the single initial domains, as depicted in Figure 4.6, domain "Monsoon" did not previously offer any kind of MPLS service, although domain "Mistral" is using LDP as the current

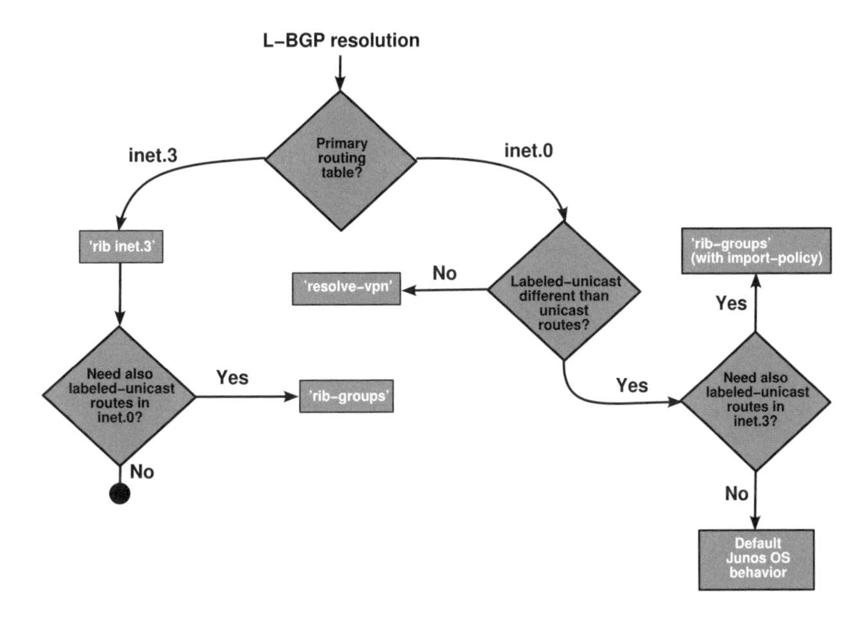

Figure 4.4: L-BGP route design options and capabilities.

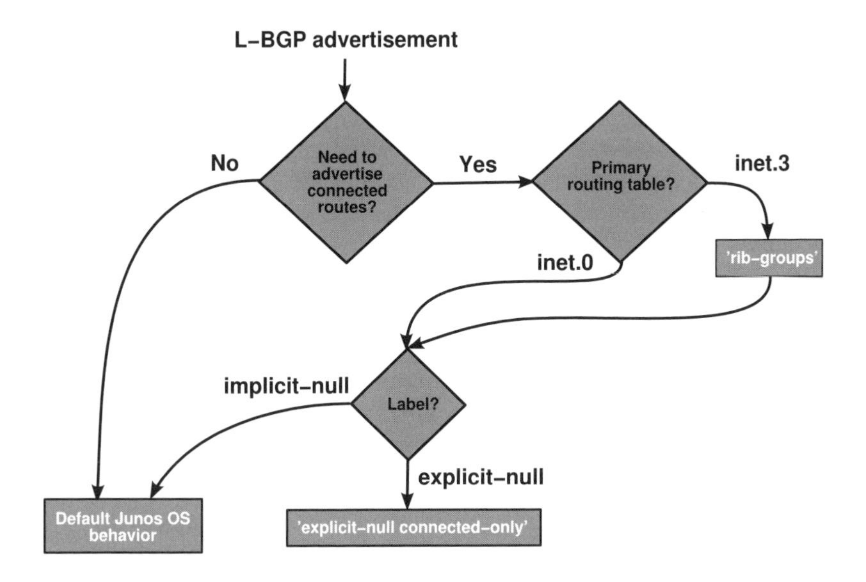

Figure 4.5: L-BGP label allocation options and capabilities.

label transport mechanism and domain "Cyclone" is based on a full mesh of MPLS RSVP-TE based LSPs with full-blown TE capabilities granted by the IGP extensions in a common OSPFv2 Area 0.

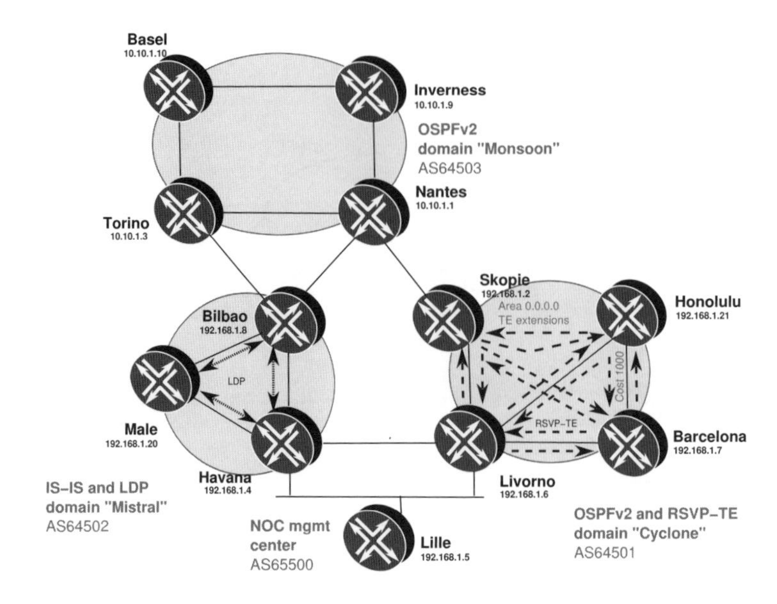

Figure 4.6: Original topology including MPLS label distribution protocols support at each domain.

In terms of BGP, the initial situation illustrated in Figure 3.9 is considered as the background scenario for eventual L-BGP deployment. Each domain is represented by a different and independent AS, which is not yet confederated or integrated, and the same mesh topology is intended to be used if needed.

Top management has also set as a parallel goal to initially integrate and connect both MPLS domains in a redundant fashion, and as final objective a joint integration into a unified MPLS domain with common rules and label distribution mechanisms.

As first milestone for the merging, it is understood that a consistent setup is created with label-binding transport across the existing domains that offer MPLS services, namely domain "Mistral" and domain "Cyclone". This setup should allow MPLS VPNs to be stretched over the recently acquired infrastructure as soon as an initial label-binding transport integration occurs. This first interconnect must also comply with the usual redundancy and resilience requirements to survive against as many single points of failure as possible.

After this first milestone, once all domains have been merged into a common link-state IGP, a gradual migration to a unified label transport protocol is scoped.

4.5.2 Target network

Considering all parallel migration exercises, the final topology includes a unique MPLS label distribution protocol in parallel to a unified link-state IGP and BGP setup. Depending on the status of other migrations, two major milestones have been defined:

- interconnect MPLS domains to be able to provide services end-to-end;
- integrate all systems under a common MPLS label distribution protocol.

The first milestone requires redundant and resilient interconnects among MPLS domains and other participating systems, as illustrated in Figure 4.7.

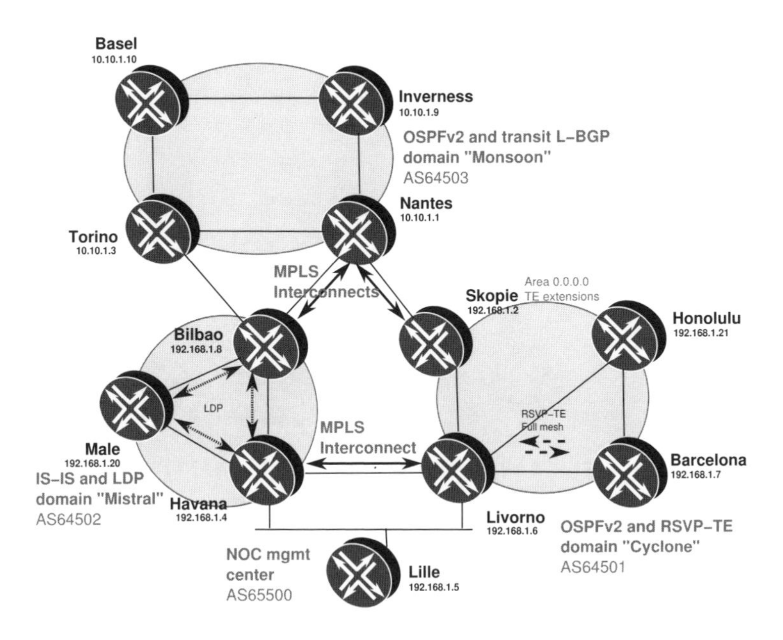

Figure 4.7: MPLS label transport interconnect scenario.

The final scenario after all migration steps requires a unified link-state IGP across the topology and can be summarized in Figure 4.8.

4.5.3 Migration strategy

The goal of this case study is to unify MPLS transport domains with two major milestones:

- interconnect the existing MPLS domains, domain "Mistral" and domain "Cyclone", in a redundant fashion for delivery of end-to-end MPLS services;

- once a unified IGP has been deployed, merge all MPLS domains into a common setup with a unified MPLS label distribution protocol and with unified mechanisms.

There are migration stages to achieve before reaching the first milestone leading to MPLS domain interconnects:

- domain "Mistral" and domain "Cyclone" will be interconnected by MPLS in a redundant fashion.

- By means of newer leased-lines to the domain "Monsoon", router *Nantes* can provide additional resilience to the interconnect, even if it is not currently offering any MPLS-based services.

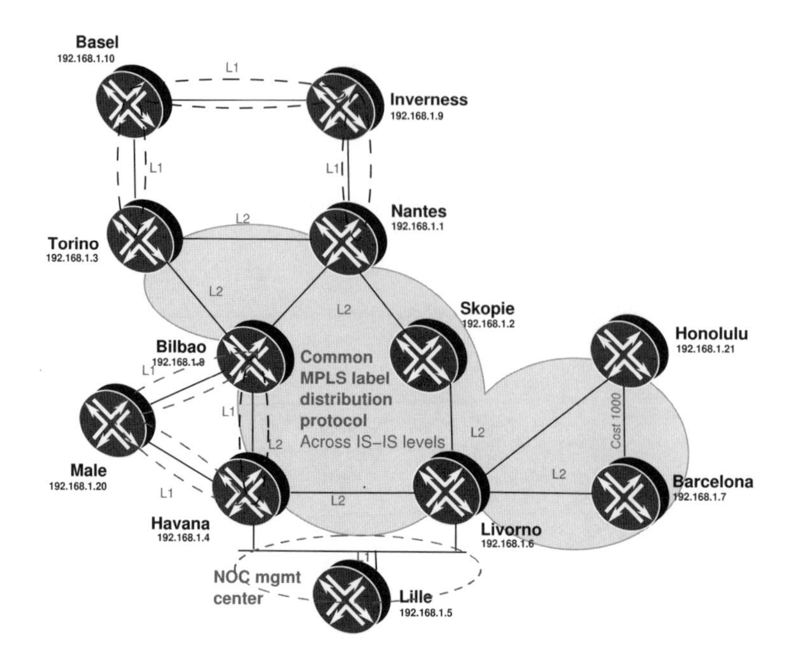

Figure 4.8: Final MPLS label distribution topology.

- Router *Lille* is the router providing transit services for the NOC, and as such, it initially offers no MPLS VPNs. However, the plan is to integrate this domain as soon as possible, because the NOC folks also need intra-VPN management. For this reason, router *Lille* will also be interconnected to be able to instantiate VPNs and thus offer managed services inside each VPN, such as CE activation and configuration.

Once all these steps have been accomplished, the final MPLS domain migration requires a unified link-state IGP on all routers and can be accomplished only when this prerequisite is met. For this reason, all changes to reach the first milestone need to remain stable and independent of any IGP changes that may occur during transient phases.

As with other migrations, redundancy and minimal impact are principal goals of the process. MPLS-based services are financially crucial for "Gale Internet," and operational turbulences with this product could have fatal consequences during the company acquisition and migration period.

Interconnecting MPLS domains

The first design decision to make is the choice for the glue to stitch both domains together in a resilient fashion.

The idea is to make use of any of the existing available MPLS label distribution protocols to distribute such label bindings across domains. Consider the existing protocols:

- *RSVP-TE*: is implemented on available information from the TED. In this case, there is no joint TED, so interdomain RSVP LSPs would not be precalculated from available information but rather, would be purely signaled across interconnects. This mechanism lacks some transparency for events in the other domain and requires additional hooks to refer to existing FECs in the IGP at each domain.

- *LDP*: usually relies on IGPs to determine the active label binding, but at this point, there is no unified link-state IGP. When supporting interarea, interdomain, or hierarchical LDP, less specific FEC route-matching is needed for selecting the label binding. This requirement imposes the need to at least redistribute some aggregates and adds opacity in a redundant interconnect, as well as requiring support for that feature.

- *L-BGP*: relies only on IGPs for internal peer reachability, which is BGP's standard behavior for internal sessions. However, BGP is decoupled from the IGP when advertising label bindings with EBGP, and Junos OS offers resources to determine whether those routes are installed in inet.0, inet.3 or both. As per standard BGP UPDATE messages propagation, the NLRI for each specific route is populated in a transparent and immediate fashion, without affecting any other label bindings from the same domain. Existing BGP attributes also enable the fine-tuning of route advertisement among domains as well as the implementation of an optimal and redundant route reachability scheme.

In this environment, *L-BGP* is elected as the vehicle of choice to transport label bindings across domains. The fundamental differences to sway the balance are its complete independence from a given IGP for granular advertisement and the previous existence of separate BGP ASs for each domain. L-BGP as label distribution protocol also ensures an easy transition by just activating the required `labeled unicast` family on existing BGP mesh topologies and adapting redistribution policies where necessary.

Furthermore, even though domain "Monsoon" does not provide any MPLS services, it is connected using the new leased lines to both domain "Mistral" and domain "Cyclone". Just by enabling external L-BGP among router *Nantes*, router *Bilbao*, and router *Skopie*, together with MPLS activation over both external links, an additional path is established without the need to redistribute routes or aggregates. This extra path enforces the redundancy requirements for the initial stage and at the same time opens the door for offering MPLS-based services even inside domain "Monsoon".

Another consideration needs to be made after discussing domain interconnection: *external label binding distribution inside each domain*. Two different approaches are feasible based on the previous choice of interconnect methodology:

- *Redistribution*: external label bindings can be redistributed at ASBRs inside each domain for further internal population. In the case of domain "Cyclone", RSVP-TE would need to be used as the substrate for another *downstream unsolicited* label distribution protocol to distribute those external bindings in a flexible manner. For domain "Mistral", because LDP follows IGP routes, these externally injected prefixes would need to be redistributed into the IGP as well.

- *Overlay*: internal L-BGP resolving over another substrate in inet.3 can be deployed over LDP and RSVP-TE, at domain "Mistral" and domain "Cyclone", respectively.

The internal L-BGP *overlay* model clearly benefits from seamless integration, because it requires no manipulation of existing internal IGPs or MPLS label distribution protocols at each domain. It also provides smooth label-binding propagation to each internal router as part of the standard route advertisement mechanisms between IBGP and EBGP, together with granular TE mechanisms inherent in BGP.

In a nutshell, L-BGP is selected as the protocol not only to interconnect domains but also to provide end-to-end label transport between devices at different sites in an *overlay* fashion over the existing, underlying label distribution protocols within each domain.

Considering the initial stage from the BGP case study in Section 3.6, `labeled unicast` is deployed as an additional family over the existing BGP virtual topologies in each domain, namely:

- full IBGP mesh in domain "Cyclone";
- route reflection with router *Male* as the reflector server for domain "Mistral".

Figure 4.9 depicts the MPLS interconnect and distribution scenario with all domains and router *Lille*, considering such BGP mesh scheme in each domain.

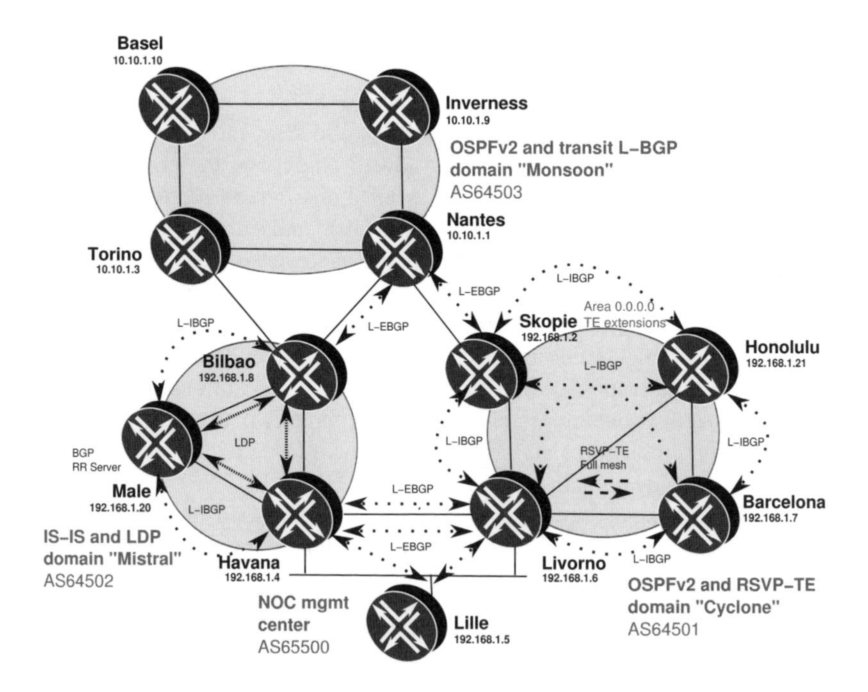

Figure 4.9: MPLS label transport interconnect scenario.

For this scenario, the activities to carry out before reaching the first milestone are:

- Deploy internal L-BGP inside domain "Mistral" over the route reflection scheme by adding the `labeled-unicast` family and external L-BGP on router *Bilbao* and router *Havana*, so they become ASBRs for external interconnects.

- Deploy internal L-BGP inside domain “Cyclone” over the full-mesh scheme by adding the `labeled-unicast` family and external L-BGP on router *Skopie* and router *Livorno*, so they become ASBRs for external interconnects.

- New leased-lines to domain “Monsoon” enable another path that can back up potential failures at either router *Havana* or router *Livorno*. By establishing external L-BGP sessions between router *Bilbao* and router *Nantes*, and between router *Nantes* and router *Skopie*, another backup transit is available between both domain “Mistral” and domain “Cyclone”.

When first interconnecting domains, note that the label-binding population inside each one is based on standard BGP route advertisement rules, both in a full-mesh or route reflection fashion. This means that inside each domain, no redistribution with any of the existing label distribution protocols is needed because they are simply used for internal L-BGP route resolution purposes.

Migration to a common LDP

Once a common and unified IGP domain has been deployed, another evaluation needs to be carried out to plan for the final unified MPLS label distribution protocol.

Discarding RSVP-TE for the previously discussed reasons, L-BGP can retain its functionality to bind domains and populate label bindings perfectly in an *overlay* model.

However, parallel changes in other protocols influence MPLS label distribution as well, namely:

- The IGP is unified in all domains, meaning that a single label distribution protocol following the IGP can be rolled out with no interconnects or redistributions.

- The diverse BGP ASs are unified into a joint global AS, meaning that previous external connections now become internal. This change is important for L-BGP in the sense that another label distribution protocol needs to be rolled out underneath all these previous external connections.

To summarize the requirements, another label distribution protocol needs to be rolled out, at least at the domain interconnects, and label bindings can follow IGP routes, thus forming a consolidated topology.

From a migration perspective, LDP can transition progressively to become the only and global MPLS label distribution protocol. While it is legitimate to retain L-BGP to transport label bindings, LDP can be deployed as IGPs are merged and can become the underlying resolution substrate for L-BGP.

All these stages for the MPLS label distribution protocol migration strategy and their interaction are shown in Figure 4.10. These stages closely depend on each other in a strict sequence or can be accomplished in parallel until reaching a milestone.

4.5.4 Stage one: Labeled BGP deployment over LDP in domain “Mistral”

As the first step, internal L-BGP needs to be rolled out in domain “Mistral”. By using the existing LDP as a feasible resolution layer for L-IBGP sessions, this deployment is as simple

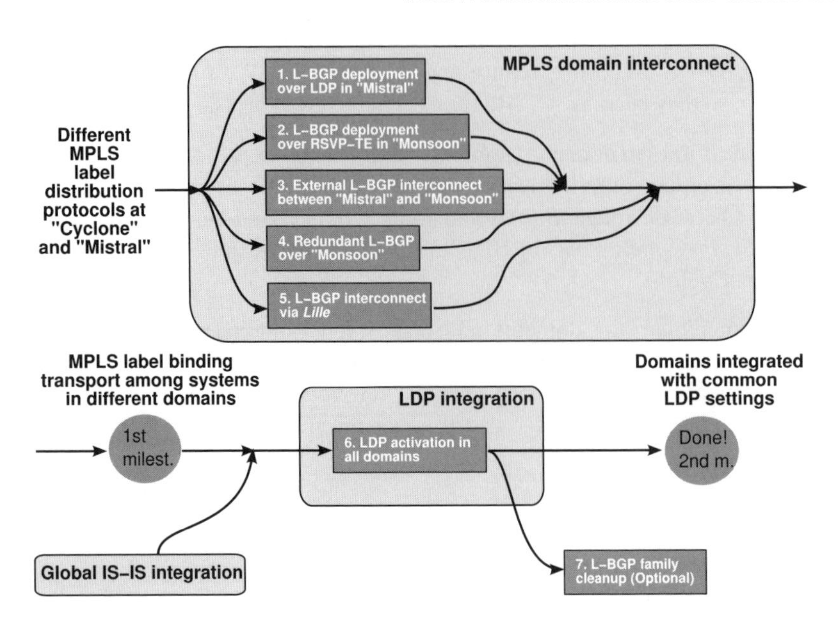

Figure 4.10: Strategy for the MPLS label distribution protocol migration.

as extending the existing BGP session infrastructure with the new `labeled-unicast` family.

Because this domain includes two ASBRs that support external L-BGP, the same family also needs to be externally activated towards domain "Cyclone" and other routers.

Listing 4.15 illustrates the full BGP configuration at router *Bilbao*. As per Lines 10 and 22, inet.3 is configured as *primary* table to install L-BGP routes without any specific leaking to inet.0. The idea is to automatically select and feed this table, which is shared with LDP and RSVP-TE in an independent fashion to inet.0.

Listing 4.15: Complete BGP configuration at Bilbao, including internal and external labeled unicast

```
1  [edit protocols bgp]
2  user@Bilbao-re0# show
3  group IBGP {
4      type internal;
5      local-address 192.168.1.8;
6      family inet {
7          unicast;
8          labeled-unicast {
9              rib {
10                 inet.3; # inet.3 as primary table
11             }
12         }
13     }
14     export [ nhs own-loopback ]; # next hop self policy for IBGP routes
15     neighbor 192.168.1.20; # Male as route Reflector Server
16 }
17 group EBGP {
18     family inet {
```

```
19          unicast;
20          labeled-unicast {
21              rib {
22                  inet.3; # inet.3 as primary table
23              }
24          }
25      }
26      export own-loopback;
27      peer-as 64503;
28      neighbor 172.16.1.6;
29  }
30
31
32  [edit policy-options policy-statement own-loopback]
33  user@Bilbao-re0# show
34  term own-loopback {
35      from {
36          protocol direct; # local loopback address
37          rib inet.3; # Enforce loopback presence in inet.3
38          route-filter 192.168.1.0/24 prefix-length-range /32-/32;
39      }
40      to rib inet.3;
41      then accept;
42  }
43
44  [edit routing-options]
45  user@Bilbao-re0# show
46  interface-routes {
47      rib-group inet int-to-inet3; # Apply rib-groups to direct routes
48  }
49  rib-groups {
50      int-to-inet3 {
51          import-rib [ inet.0 inet.3 ]; # inet.0 as primary, inet.3 as secondary
52          import-policy interface-to-inet3; # Route sharing controlled by import policy
53      }
54  }
55  router-id 192.168.1.8;
56  autonomous-system 64502;
57
58  [edit policy-options policy-statement interface-to-inet3]
59  user@Bilbao-re0# show
60  term loopback {
61      from interface lo0.0; # Just match on loopback address
62      then accept;
63  }
64  term no-more-inet3 {
65      to rib inet.3;
66      then reject; # Do not import anything else into inet.3
67  }
68  term default {
69      to rib inet.0;
70      then accept; # allow default accept for primary table inet.0
71  }
```

Note that each router needs to advertise a *label binding* for its own loopback address in L-BGP. This is achieved with two different Junos OS configuration resources:

- `rib-groups` applied to the router's own loopback address, so that the route is shared between inet.0 and inet.3, and is thus eligible for advertisement as a *labeled route* (Line 50 from Listing 4.15);

- `export` policy for all IBGP and EBGP neighbors to redistribute the local route from inet.3 (Lines 14 and 26 from Listing 4.15).

As shown in Line 52 from Listing 4.15, such route-sharing between both tables is controlled by means of a `rib-group import policy`, so that only a copy for the loopback interface address is installed and not for all direct routes.

At this stage for domain "Mistral", internal L-BGP properly resolves thanks to LDP being the underlying MPLS label distribution protocol. However, routes remain unpreferred because of default protocol preferences. Listing 4.16 illustrates the inet.3 table, considering route reflection over router *Male*, with the following observations:

- L-BGP routes remain *inactive* due to route preference against LDP (Lines 11 and 24).
- Both L-BGP loopback bindings from router *Male* and router *Havana* are *implicit-null* (Lines 13 and 26).
- Resolution for the L-BGP route for the router *Havana* loopback address, which is not directly connected by IS–IS, requires a label *push* with the LDP allocated label (Line 10). This is a clear indication of L-BGP resolving over LDP.

Listing 4.16: Internal L-BGP routes inactive resolving over LDP at Bilbao

```
1  user@Bilbao-re0> show route table inet.3 extensive |
2                   match "entr|AS:|reason|Next-hop|Label|Preference"
3  192.168.1.4/32 (2 entries, 1 announced)
4          *LDP    Preference: 9
5                  Next-hop reference count: 3
6                  Label operation: Push 299840
7                  Local AS: 64502
8           BGP    Preference: 170/-101
9                  Next-hop reference count: 1
10                 Label operation: Push 299840 # Resolving over LDP
11                 Inactive reason: Route Preference
12                 Local AS: 64502 Peer AS: 64502
13                 Route Label: 3
14 192.168.1.8/32 (1 entry, 1 announced)
15         *Direct Preference: 0
16                 Next-hop reference count: 5
17                 Local AS: 64502
18 192.168.1.20/32 (2 entries, 1 announced)
19         *LDP    Preference: 9
20                 Next-hop reference count: 3
21                 Local AS: 64502
22          BGP    Preference: 170/-101
23                 Next-hop reference count: 1
24                 Inactive reason: Route Preference
25                 Local AS: 64502 Peer AS: 64502
26                 Route Label: 3
```

Listing 4.16 shows internal L-BGP *route resolution*; the same concept needs to be evaluated for external L-BGP on the interconnects. L-BGP requires that advertised routes resolve over MPLS-enabled interfaces.

With regard to the external L-BGP activation, note that because this is single-hop L-EBGP, the external interface routes towards other domains do not need to be copied to inet.3 because no indirect resolution is needed. Listing 4.17 shows that activating the external interface for MPLS is sufficient for proper route activation.

Listing 4.17: External MPLS activation for L-EBGP at Bilbao

```
1  [edit protocols mpls]
2  user@Bilbao-re0# set interface ge-0/1/0.0 # External interface activation into MPLS
```

```
3
4 [edit interfaces]
5 user@Bilbao-re0# set ge-0/1/0.0 family mpls # Family mpls active on external interface
```

An exception worth mentioning is the case of confederated or multihop L-EBGP in which indirect next-hop resolution for the BGP peer is needed. In this case, additional configuration is compulsory for proper route installation, such as static or interface route leaking to inet.3 with the `rib-groups` knob.

Junos Tip: Symptoms of route advertisement or resolution failures with external L-BGP

A common human error consists of not properly enabling MPLS for external interfaces with L-EBGP sessions over them. Mainly because L-BGP sessions become established and routes are exchanged between peers, this failure may not be evident at first sight.

However, if the external interface is not active in the `[edit protocols mpls]` stanza, when reflecting externally received routes over the internal L-BGP, BGP label allocation failures are explicitly displayed in Junos OS when the outputs for those routes are examined carefully.

Listing 4.18 illustrates the case of routes reflected over the internal L-BGP session between router *Havana* and router *Bilbao*. Note that Line 6 explicitly refers to the reason for the label allocation failure.

Listing 4.18: L-BGP label allocation failures when external interface is not MPLS-enabled

```
1 user@Havana> show route advertising-protocol bgp 192.168.1.20 extensive
2
3 inet.3: 10 destinations, 14 routes (10 active, 0 holddown, 0 hidden)
4 * 192.168.1.2/32 (2 entries, 1 announced)
5  BGP group L-IBGP type Internal
6      BGP label allocation failure: protocols mpls not enabled on interface # No label allocation
7      Nexthop: Not advertised # Route not advertised, no MPLS action can be installed in PFE
8      Flags: Nexthop Change
9      MED: 64
10     Localpref: 100
11     AS path: [64502] 64501 I
12
13 * 192.168.1.6/32 (2 entries, 1 announced)
14  BGP group L-IBGP type Internal
15     Route Label: 305296
16     Nexthop: Self
17     Flags: Nexthop Change
18     Localpref: 100
19     AS path: [64502] 64501 I
20 <...>
21
22 user@Havana> configure
23 Entering configuration mode
24 [edit]
25 user@Havana# set protocols mpls interface so-1/0/0.0
26 [edit]
27 user@Havana# commit and-quit
28 commit complete
29 Exiting configuration mode
30 user@Havana> clear bgp neighbor 192.168.1.20 soft
31   # Soft clear outgoing advertised NLRIs
```

```
32
33 user@Havana> show route advertising-protocol bgp 192.168.1.20 extensive
34
35 inet.3: 10 destinations, 14 routes (10 active, 0 holddown, 0 hidden)
36 * 192.168.1.2/32 (2 entries, 1 announced)
37  BGP group L-IBGP type Internal
38      Route Label: 305312 # Proper label allocation (swap)
39      Nexthop: Self
40      Flags: Nexthop Change
41      MED: 64
42      Localpref: 100
43      AS path: [64502] 64501 I
44
45 * 192.168.1.6/32 (2 entries, 1 announced)
46  BGP group L-IBGP type Internal
47      Route Label: 305296  # Proper label allocation (swap)
48      Nexthop: Self
49      Flags: Nexthop Change
50      Localpref: 100
51      AS path: [64502] 64501 I
52 <...>
53
54 user@Havana> show route table mpls.0 label 305312
55
56 mpls.0: 16 destinations, 16 routes (16 active, 0 holddown, 0 hidden)
57 + = Active Route, - = Last Active, * = Both
58
59 305312              *[VPN/170] 05:36:01 # Label for L-BGP route 192.168.1.2
60                       > to 172.16.1.38 via so-1/0/0.0, Swap 305600
```

From this output, it is worth remarking that the L-BGP route for router *Livorno* was indeed readvertised internally despite the fact that the external interface was not active for MPLS. The reason for this behavior resides in the routes advertised from router *Livorno*. As per Listing 4.19, router *Livorno* advertises the L-BGP route for its loopback with an *implicit-null* label, with the intention of triggering penultimate-hop popping on router *Havana* because they are *directly connected*. Although this route can be legitimately readvertised because the associated MPLS action is a *pop* for the previous label, whatever bidirectional MPLS application established over it (such as L3VPN) would be blackholed at this point if MPLS were not enabled on the interface.

Listing 4.19: L-BGP route advertisement from Livorno and PHP MPLS pop action installation towards a non-MPLS-enabled interface

```
1  user@Havana> show route receive-protocol bgp 172.16.1.38 192.168.1.6 extensive
2
3  inet.3: 12 destinations, 22 routes (12 active, 0 holddown, 0 hidden)
4  * 192.168.1.6/32 (3 entries, 1 announced)
5       Accepted
6       Route Label: 3 # Incoming advertisement with implicit-null
7       Nexthop: 172.16.1.38
8       AS path: 64501 I
9
10 user@Havana> show route table mpls.0 label 305296
11
12 mpls.0: 16 destinations, 16 routes (16 active, 0 holddown, 0 hidden)
13 + = Active Route, - = Last Active, * = Both
14
15 305296             *[VPN/170] 05:49:24
16                      > to 172.16.1.38 via so-1/0/0.0, Pop     # Pop MPLS label towards Livorno
17 305296(S=0)        *[VPN/170] 05:44:16
18                      > to 172.16.1.38 via so-1/0/0.0, Pop     # Pop MPLS label towards Livorno
```

Omitting the `family mpls` option on the external interconnect interface is another common human error. Junos OS includes another check to enforce alignment for next-hop installation in this case. Considering the same link from the previous example, Listing 4.20 illustrates what occurs when `family mpls` is not included. Line 6 explicitly shows the reason for the label allocation failure.

Listing 4.20: L-BGP label allocation failures when external interface does not include family MPLS

```
1  user@Havana> show route advertising-protocol bgp 192.168.1.20 extensive
2
3  inet.3: 9 destinations, 13 routes (9 active, 0 holddown, 0 hidden)
4  * 192.168.1.2/32 (2 entries, 1 announced)
5   BGP group L-IBGP type Internal
6       BGP label allocation failure: family mpls not enabled on interface # No label allocation
7       Nexthop: Not advertised # Route not advertised, no MPLS action can be installed in PFE
8       Flags: Nexthop Change
9       MED: 64
10      Localpref: 100
11      AS path: [64502] 64501 I
12 <...>
13 user@Havana> configure
14 Entering configuration mode
15  [edit]
16 user@Havana# set interfaces so-1/0/0.0 family mpls
17  [edit]
18 user@Havana# commit and-quit
19 commit complete
20 Exiting configuration mode
21
22 user@Havana> clear bgp neighbor 192.168.1.8 soft
23
24 user@Havana> show route advertising-protocol bgp 192.168.1.20 extensive
25
26 inet.3: 9 destinations, 13 routes (9 active, 0 holddown, 0 hidden)
27 * 192.168.1.2/32 (2 entries, 1 announced)
28  BGP group L-IBGP type Internal
29      Route Label: 300032
30      Nexthop: Self
31      Flags: Nexthop Change
32      MED: 64
33      Localpref: 100
34      AS path: [64502] 64501 I
35 <...>
36
37 user@Havana> show route table mpls label 300032
38
39 mpls.0: 14 destinations, 14 routes (14 active, 0 holddown, 0 hidden)
40 + = Active Route, - = Last Active, * = Both
41
42 300032             *[VPN/170] 00:00:55 # Label for L-BGP route 192.168.1.2
43                     > to 172.16.1.38 via so-1/0/0.0, Swap 300272
```

Another typical error is not including a copy of the external interface route or static route leading to the next hop in inet.3. With a single-hop L-EBGP session, no indirect resolution is required, but under other circumstances where next-hop reachability is not direct, such as L-IBGP, multihop L-EBGP, or confederated-BGP (CBGP), the next hop must be present in inet.3 and routes allowing its resolution must be present as well. The principal symptom for this kind of mistake in Junos OS is a *hidden* route with *unusable path* for installation in the routing table. Listing 4.21 illustrates a case with a multihop L-EBGP session between router *Bilbao* and router *Skopie* via router *Nantes* with the *unusable path* symptom, as shown in Lines 53 and 62 for some label bindings on one side, and Lines 75 and 84 on the other BGP peer.

Note that this route resolution failure symptom is also visible in other cases, such as with L-IBGP when the corresponding next-hop route is not present in inet.3 (either via LDP, via RSVP-TE or via interface route sharing).

Listing 4.21: Multihop L-BGP route resolution failures when external interface is not leaked to inet.3

```
1  # Multihop session in Bilbao
2  [edit protocols bgp group multihop-L-EBGP]
3  user@Bilbao-re0# show
4  family inet {
5      labeled-unicast {
6          rib {
7              inet.3;
8          }
9      }
10 }
11 export own-loopback;
12 peer-as 64501;
13 neighbor 172.16.1.2 {
14     multihop {
15         ttl 3;
16     }
17 }
18 [edit routing-options static]
19 user@Bilbao-re0# show route 172.16.1.2/32
20 # Route towards next hop for multihop reachability
21 next-hop 172.16.1.6;
22
23 # Multihop session in Skopie
24 [edit protocols bgp group multihop-L-EBGP]
25 user@Skopie# show
26 import prefer-LBGP-to-LDP;
27 family inet {
28     labeled-unicast {
29         rib {
30             inet.3;
31         }
32     }
33 }
34 export from-RSVP-to-LBGP;
35 peer-as 64502;
36 neighbor 172.16.1.5 {
37     multihop {
38         ttl 3;
39     }
40 }
41 [edit routing-options static]
42 user@Skopie# show route 172.16.1.5/32
43 # Route towards next hop for multihop reachability
44 next-hop 172.16.1.1;
45
46 # Hidden inet.3 routes due to Unusable path
47 user@Bilbao-re0> show route table inet.3 hidden extensive |
48                  match "entr|hop|reason|Accepted|State"
49 192.168.1.2/32 (2 entries, 1 announced)
50                 Next hop type: Unusable
51                 Next-hop reference count: 5
52                 State: <Hidden Ext>
53                 Inactive reason: Unusable path
```

```
54                   Accepted # Valid after import policy processing
55                   Indirect next hops: 1
56                           Protocol next hop: 172.16.1.2 # Route unusable when received from Skopie
57                           Indirect next hop: 0 -
58  192.168.1.5/32 (2 entries, 1 announced)
59                   Next hop type: Unusable
60                   Next-hop reference count: 5
61                   State: <Hidden Ext>
62                   Inactive reason: Unusable path
63                   Accepted # Valid after import policy processing
64                   Indirect next hops: 1
65                           Protocol next hop: 172.16.1.2 # Route unusable when received from Skopie
66                           Indirect next hop: 0 -
67  <...>
68
69  user@Skopie> show route table inet.3 hidden extensive |
70               match "entr|hop|reason|Accepted|State"
71  192.168.1.4/32 (3 entries, 1 announced)
72                   Next hop type: Unusable
73                   Next-hop reference count: 4
74                   State: <Hidden Ext>
75                   Inactive reason: Unusable path
76                   Accepted # Valid after import policy processing
77                   Indirect next hops: 1
78                           Protocol next hop: 172.16.1.5 # Route unusable when received from Bilbao
79                           Indirect next hop: 0 -
80  192.168.1.5/32 (3 entries, 1 announced)
81                   Next hop type: Unusable
82                   Next-hop reference count: 4
83                   State: <Hidden Ext>
84                   Inactive reason: Unusable path
85                   Accepted # Valid after import policy processing
86                   Indirect next hops: 1
87                           Protocol next hop: 172.16.1.5 # Route unusable when received from Bilbao
88                           Indirect next hop: 0 -
89  <...>
90
91   # Unresolved routes in Bilbao
92  user@Bilbao-re0> show route resolution unresolved detail
93  Tree Index 1
94  192.168.1.5/32 # L-BGP route for Lille unresolved
95      Protocol Nexthop: 172.16.1.2 Push 312048
96      Indirect nexthop: 0 -
97  192.168.1.21/32 # L-BGP route for Honolulu unresolved
98      Protocol Nexthop: 172.16.1.2 Push 311808
99      Indirect nexthop: 0 -
100 192.168.1.7/32 # L-BGP route for Barcelona unresolved
101     Protocol Nexthop: 172.16.1.2 Push 311792
102     Indirect nexthop: 0 -
103 192.168.1.6/32 # L-BGP route for Livorno unresolved
104     Protocol Nexthop: 172.16.1.2 Push 311776
105     Indirect nexthop: 0 -
106 192.168.1.2/32 # L-BGP route for Skopie unresolved
107     Protocol Nexthop: 172.16.1.2
108     Indirect nexthop: 0 -
109 Tree Index 2
110 Tree Index 3
111
112  # Unresolved routes in Skopie
113 user@Skopie> show route resolution unresolved detail
114 Tree Index 1
115 192.168.1.8/32 # L-BGP route for Bilbao unresolved
116     Protocol Nexthop: 172.16.1.5
117     Indirect nexthop: 0 -
118 192.168.1.20/32 # L-BGP route for Male unresolved
119     Protocol Nexthop: 172.16.1.5 Push 301424
120     Indirect nexthop: 0 -
121 192.168.1.4/32 # L-BGP route for Havana unresolved
```

```
122      Protocol Nexthop: 172.16.1.5 Push 301408
123      Indirect nexthop: 0 -
124  192.168.1.5/32 # L-BGP route for Lille unresolved
125      Protocol Nexthop: 172.16.1.5 Push 301568
126      Indirect nexthop: 0 -
```

To resolve the failures, one needs to consider that next-hop reachability is made possible by means of a static route, which resolves over an external interface. This means that not only does this static route need to be present in inet.3, but also the direct route that enables its resolution in that table must be leaked. Listing 4.22 shows the steps needed to correct the problem.

Listing 4.22: Multihop L-BGP route resolution failures when external interface is not leaked to inet.3

```
1   # External interface route leaking with inet.3
2   [edit policy-options policy-statement interface-to-inet3]
3   user@Bilbao-re0# set term eLBGP-interface from interface ge-0/1/0.0
4   [edit policy-options policy-statement interface-to-inet3]
5   user@Bilbao-re0# set term eLBGP-interface then accept
6   [edit policy-options policy-statement interface-to-inet3]
7   user@Bilbao-re0# insert term eLBGP-interface before term no-more-inet3
8   [edit policy-options policy-statement interface-to-inet3]
9   user@Bilbao-re0# show
10  term loopback {
11      from interface lo0.0;
12      then accept;
13  }
14  term eLBGP-interface { # External interface route needs to be shared with inet.3
15      from interface ge-0/1/0.0;
16      then accept;
17  }
18  term no-more-inet3 {
19      to rib inet.3;
20      then reject;
21  }
22  term default {
23      to rib inet.0;
24      then accept;
25  }
26
27   # Additional route to BGP next hop on inet.3 for proper resolution
28   # Same could have been achieved via rib-groups with existing route in inet.0
29  [edit routing-options]
30  user@Bilbao-re0# set rib inet.3 static route 172.16.1.2/32 next-hop 172.16.1.6
31
32   # No more unusable paths, L-BGP routes properly resolved
33  user@Bilbao-re0> show route table inet.3 hidden extensive |
34                   match "entr|hop|reason|Accepted|State"
35
36  user@Bilbao-re0> show route resolution unresolved detail
37  Tree Index 1
38  Tree Index 2
39  Tree Index 3
```

Figure 4.11 illustrates the L-BGP overlay model that is placed over LDP, which is already internally present in the domain.

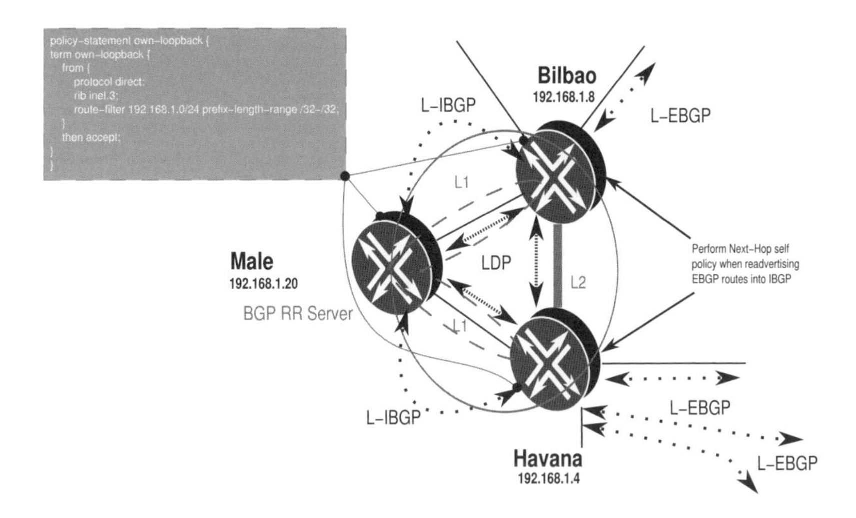

Figure 4.11: Stage one: internal L-BGP over LDP at domain "Mistral".

What If... Redistribution of LDP and L-BGP in domain "Mistral"

An alternative to implementing an overlay model of L-BGP on top of LDP is a *redistribution* approach, in which label bindings from one protocol are reinjected into the other and vice versa.

This approach can be valid when the intention is not to propagate internal L-BGP to each internal router, and a network engineer may just conceive redistribution with the internal LDP without any further manipulation than in the domain ASBRs.

Despite not requiring changes on each internal router, re-injecting routes and label bindings correctly imposes a configuration burden on the ASBRs.

Considering this alternative implementation for domain "Mistral", router *Bilbao* and router *Havana* as ASBRs for domain "Mistral" redistribute internal label bindings into external L-BGP and, in the reverse direction, to the internal LDP. In addition, an internal L-BGP session is established between both routers for added resilience.

For an LDP label binding to become active, the matching IGP route needs to be present in inet.0 as well. At a preliminary stage when IGP routes are not redistributed, an option would be to redistribute those L-BGP routes into the local IGP at both ASBRs (in fact, this is standard practice when IGPs are not merged).

Although the mutual label redistribution approach is valid, because the global migration objective includes a unified IGP, the needed substrate in inet.0 for internal label bindings to become active is possible only when a unified IGP is extended over all domains; that is, rather than injecting those external routes from L-BGP at ASBRs both into inet.0 (via external IGP routes as shown in the Junos Tip later in this section) and inet.3, those routes in inet.0 appear with the unified IS–IS domain, goal of the link-state IGP migration. At this point, the label bindings in inet.3 are deemed valid to be used beyond a single domain.

Because this domain includes two ASBRs talking external L-BGP, the potential for mutual redistribution is present. While BGP is an attribute-rich protocol with multiple NLRI-tagging mechanisms, LDP lacks any specific tagging TLV or way to administratively mark label bindings. The existence of an internal L-BGP session in this case is vital to avoid selecting internally redistributed label bindings in LDP for external backup redistribution to L-BGP at the parallel ASBR.

From the previous exercise in the BGP case study in Section 3.6, router *Male* is acting as a route reflector server within domain "Mistral" for existing IBGP sessions. In this case, the decision is to have a direct IBGP session for the `labeled-unicast` family for the following reasons:

- Router *Male* introduces an additional point of failure for communication between both ASBRs. Loop mitigation depends only on configuration and reachability between the two ASBRs.
- This is a transient stage with only two ASBRs, so this choice is easy to accommodate in a first phase and dismantle in the final scenario.
- The ability to select a different AFI/SAFI combination on a per-session basis causes no interference with the existing setup.

Figure 4.12 depicts this pitfall that occurs because LDP does not natively support any type of administrative tagging and L-BGP routes do not necessarily follow IGP selections.

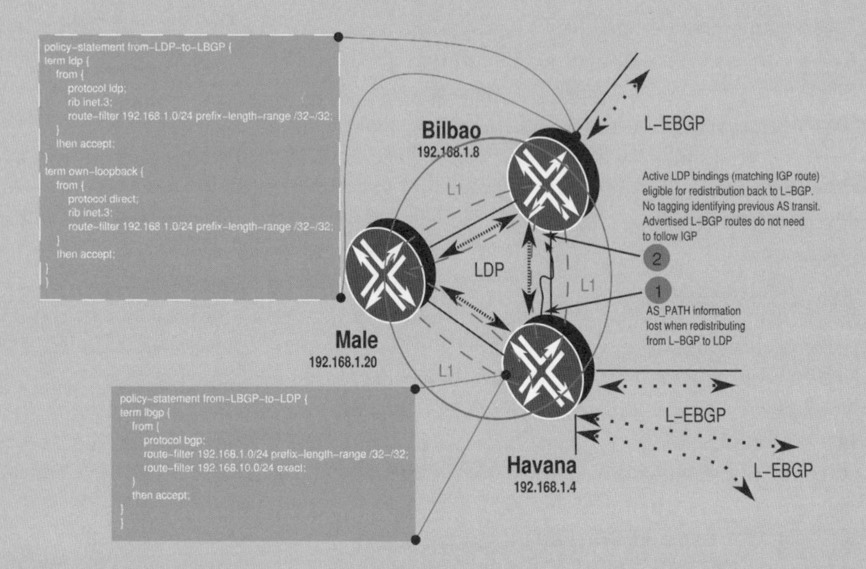

Figure 4.12: Potential for mutual redistribution without L-IBGP session at domain "Mistral".

As long as the internal L-BGP session between router *Bilbao* and router *Havana* is established between loopback addresses, it is ensured that if they remain connected, externally injected L-BGP bindings will be further populated by IBGP and must not be picked up from LDP. In the other direction, built-in BGP mechanisms ensure loop avoidance by detecting the own AS in the AS_PATH attribute (ensuring that this AS is unique in the setup). Therefore, it is basically a question of protocol preference as to whether to make L-BGP routes always preferred against LDP and, on the external side, whether to let BGP detect AS loops.

Listing 4.23 illustrates policy definition and configuration for redistribution from LDP to L-BGP at router *Bilbao*. As per Lines 10 and 22, inet.3 is configured as the *primary* table for installing L-BGP routes without any specific leaking to inet.0.

Listing 4.23: Label binding redistribution policies from LDP to L-BGP and application at Bilbao

```
1  [edit protocols bgp]
2  user@Bilbao-re0# show
3  group L-IBGP {
4      type internal;
5      local-address 192.168.1.8;
6      import prefer-LBGP-to-LDP;
7      family inet {
8          labeled-unicast {
9              rib {
10                 inet.3; # inet.3 as primary table
11             }
12         }
13     }
14     export nhs; # next hop self policy for IBGP routes
15     neighbor 192.168.1.4;
16 }
17 group L-EBGP {
18     import prefer-LBGP-to-LDP;
19     family inet {
20         labeled-unicast {
21             rib {
22                 inet.3; # inet.3 as primary table
23             }
24         }
25     }
26     export from-LDP-to-LBGP;
27     peer-as 64502;
28     neighbor 172.16.1.6;
29 }
30
31 [edit policy-options policy-statement prefer-LBGP-to-LDP]
32 user@Bilbao-re0# show
33 term prefer-LBGP {
34     then {
35         preference 7; # Lower protocol preference than LDP
36         accept;
37     }
38 }
39
40 [edit policy-options policy-statement from-LDP-to-LBGP]
41 user@Bilbao-re0# show
42 term ldp {
43     from {
44         protocol ldp;
45         rib inet.3; # Enforce presence in inet.3
46         route-filter 192.168.1.0/24 prefix-length-range /32-/32; # Standard loopback range
47     }
48     to rib inet.3;
49     then accept;
50 }
51 term own-loopback {
52     from {
```

```
53         protocol direct; # local loopback address
54         rib inet.3; # Enforce loopback presence in inet.3
55         route-filter 192.168.1.0/24 prefix-length-range /32-/32;
56     }
57     to rib inet.3;
58     then accept;
59 }
```

Figure 4.13 illustrates these changes and adaptions for proper and controlled route distribution.

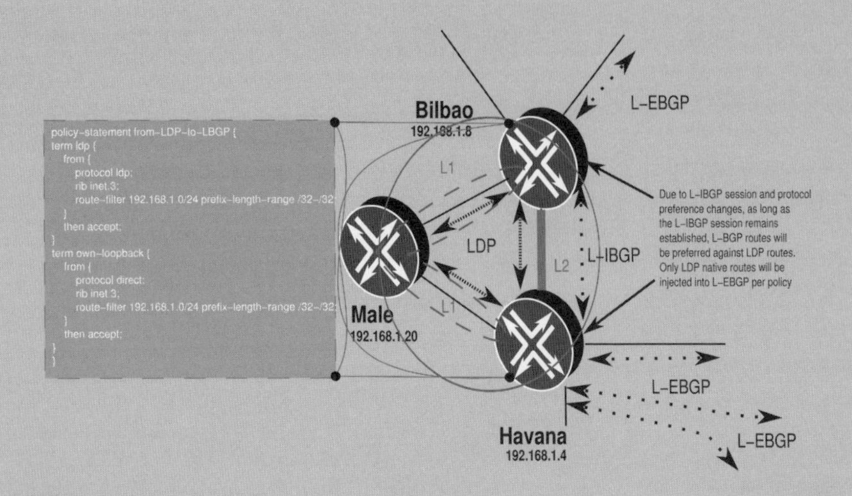

Figure 4.13: MPLS label binding redistribution from LDP to L-BGP at domain "Mistral".

As a summary, the following differences are notable when comparing *mutual redistribution* with an *overlay* model with L-BGP over LDP:

- *Mutual label-binding redistribution* requires careful tuning to avoid loop formation. LDP lacks any kind of administrative tagging and loop prevention fully relies on the IGP.
- Routes injected into the IGP may be eligible for redistribution to standard BGP at other ASBRs. Such loop formation must be artificially prevented when multiple ASBRs exist.
- Protocol preference can be tuned at ASBRs to enforce correct route selection at the internal or external side. This can be implemented by means of an internal L-BGP session between ASBRs to prevent leaked route selection for readvertisement back into L-BGP.
- An *overlay* model is independent from IGP routes, prevents loop creation with built-in BGP mechanisms, and is transparent for LDP.

Under some circumstances, a redistribution model may be considered to be a valid design approach (for instance, with a high number of internal routers where it may not be scalable to deploy L-BGP, or with internal routers that do not support that BGP family), but using an overlay implementation of L-BGP over LDP clearly benefits from both protocols and poses a much lower administrative burden.

Junos Tip: LDP egress policy out of IGP routes

When considering redistribution of label bindings from different protocols, a reader may ask: if IGPs need to be merged for internal LDP label bindings to become effective, why not redistribute routes into the IGP as well at the interconnect routers? The same ASBRs for L-BGP act as redistribution or interconnect points for IGP domains, and routes should follow the same path anyway.

The answer is obvious but needs explanation: *it cannot be done because MPLS domains are not stitched in that case.*

When selecting labeled routes for LDP `egress-policies`, even in inet.0, Junos OS understands that LDP needs to create additional and granular label bindings to *swap* labels (or *pop* if one of the existing labels is *implicit-null*). However, when selecting unlabeled routes for LDP `egress-policies` in inet.0, as would be the case for standard IGP routes, Junos OS understands that this router becomes the egress LSR for those FECs, as if there is no previously existing label binding for them.

Listing 4.24 illustrates how LDP `egress-policies` matches on active IGP routes to effectively create label bindings for them, but *implicit-null* is allocated for all of them as if router *Havana* had become the egress LSR for those LSPs and so therefore an IP lookup needs to be performed at that point.

Listing 4.24: LDP egress policy label allocation for IGP routes at Havana

```
1  [edit policy-options policy-statement from-LBGP-to-LDP]
2  user@Havana# show
3  term isis {
4      from {
5          protocol isis; # Match on active isis routes in inet.0 instead of L-BGP
6          route-filter 192.168.1.0/24 prefix-length-range /32-/32;
7          route-filter 192.168.10.1/32 exact;
8      }
9      then accept;
10 }
11
12 term own-loopback {
13     from {
14         protocol direct;
15         route-filter 192.168.1.0/24 prefix-length-range /32-/32;
16     }
17     then accept;
18 }
19 [edit policy-options policy-statement from-LBGP-to-LDP]
20 user@Havana# commit and-quit
21 commit complete
22 Exiting configuration mode
23
24 user@Havana> show ldp database
25 Input label database, 192.168.1.4:0--192.168.1.8:0
26   Label     Prefix
27  100016     192.168.1.1/32
```

```
28  100016      192.168.1.2/32
29  <...>
30
31  Output label database, 192.168.1.4:0--192.168.1.8:0
32    Label       Prefix # PHP for all FECs
33        3       192.168.1.1/32
34        3       192.168.1.2/32
35        3       192.168.1.3/32
36        3       192.168.1.4/32
37        3       192.168.1.5/32
38        3       192.168.1.6/32
39        3       192.168.1.7/32
40        3       192.168.1.8/32
41        3       192.168.1.9/32
42        3       192.168.1.10/32
43        3       192.168.1.20/32
44        3       192.168.1.21/32
45        3       192.168.10.1/32
46
47  Input label database, 192.168.1.4:0--192.168.1.20:0
48    Label       Prefix
49   299792       192.168.1.1/32
50   299792       192.168.1.2/32
51  <...>
52
53  Output label database, 192.168.1.4:0--192.168.1.20:0
54    Label       Prefix
55        3       192.168.1.1/32
56        3       192.168.1.2/32
57        3       192.168.1.3/32
58        3       192.168.1.4/32
59        3       192.168.1.5/32
60        3       192.168.1.6/32
61        3       192.168.1.7/32
62        3       192.168.1.8/32
63        3       192.168.1.9/32
64        3       192.168.1.10/32
65        3       192.168.1.20/32
66        3       192.168.1.21/32
67        3       192.168.10.1/32
```

4.5.5 Stage two: Labeled BGP deployment over RSVP-TE in domain "Cyclone"

Domain "Cyclone" is an RSVP-TE fully meshed LSP setup with full-blown TE extensions available. As per the original requirements and design decisions, it will be interconnected to domain "Mistral" via external L-BGP to advertise internal and receive external label bindings. Also, internal L-BGP is implemented in an *overlay* fashion, resolving over RSVP-TE LSPs.

Fundamentally, the main difference with the previous integration at domain "Mistral" (Section 4.5.4) is the resolution substrate for internal L-BGP: LDP at domain "Mistral", RSVP-TE at domain "Cyclone".

Because of the *downstream-on-demand* nature of RSVP-TE, a full mesh of participating devices is required when internal L-BGP is deployed. This is needed to ensure complete connectivity and resolution, and the need becomes more obvious than with LDP (which uses *downstream-unsolicited* propagation).

Therefore, internal L-BGP is rolled out in domain "Cyclone" in a parallel fashion, by simply extending the new `labeled-unicast` family over the existing BGP full mesh and locally creating a label binding for the own loopback to be advertised in L-BGP.

Because this domain also includes two ASBRs with external L-BGP, the `labeled-unicast` family also needs to be externally activated towards domain "Mistral" and other routers.

In a symmetric fashion to the activation on router *Bilbao* illustrated in Listing 4.15, Listing 4.25 describes the configuration at router *Skopie*. As with the previous case, inet.3 is configured as the *primary* table to install L-BGP routes only in inet.3 and without any copies in inet.0 (see Lines 9 and 22).

Listing 4.25: Complete BGP configuration and rib-groups at Skopie

```
1  user@Skopie# show
2  [edit protocols bgp]
3  group IBGP {
4      type internal;
5      local-address 192.168.1.2;
6      family inet {
7          labeled-unicast {
8              rib {
9                  inet.3; # inet.3 as primary table
10             }
11         }
12     }
13     export [ nhs own-loopback ]; # next hop self policy for IBGP routes
14     neighbor 192.168.1.6; # Internal full BGP mesh
15     neighbor 192.168.1.7;
16     neighbor 192.168.1.21;
17 }
18 group EBGP {
19     family inet {
20         labeled-unicast {
21             rib {
22                 inet.3; # inet.3 as primary table
23             }
24         }
25     }
26     export own-loopback;
27     neighbor 172.16.1.1 {
28         peer-as 64503;
29     }
30 }
31
32 [edit policy-options policy-statement own-loopback]
33 user@Skopie# show
34 term own-loopback {
35     from {
36         protocol direct; # local loopback address
37         rib inet.3; # Enforce loopback presence in inet.3
38         route-filter 192.168.1.0/24 prefix-length-range /32-/32;
39     }
40     to rib inet.3;
41     then accept;
42 }
43
44 [edit routing-options]
45 user@Skopie# show
46 interface-routes {
47     rib-group inet int-to-inet3; # Apply rib-groups to direct routes
48 }
49 static {
50     route 172.16.0.0/12 {
51         next-hop 172.26.26.1;
```

```
52          retain;
53          no-readvertise;
54      }
55  }
56  rib-groups {
57      int-to-inet3 {
58          import-rib [ inet.0 inet.3 ]; # inet.0 as primary, inet.3 as secondary
59          import-policy interface-to-inet3; # Route sharing controlled by import policy
60      }
61      direct {
62          import-rib [ inet.0 inet.3 ];
63      }
64  }
65  router-id 192.168.1.2;
66  autonomous-system 64501;
67
68  [edit policy-options policy-statement interface-to-inet3]
69  user@Skopie# show
70  term loopback {
71      from interface lo0.0; # Just match on loopback address
72      then accept;
73  }
74  term no-more-inet3 {
75      to rib inet.3;
76      then reject; # Do not import anything else into inet.3
77  }
78  term default {
79      to rib inet.0;
80      then accept; # allow default accept for primary table inet.0
81  }
```

Likewise, routers from domain "Cyclone" include `rib-groups` applied to their own loopback address for leaking into inet.3 (Line 57), being tailored with a `rib-group import-policy` (Line 59), and `export` policies for all BGP neighbors to advertise the the router's own label bindings (Lines 13 and 26).

Therefore, internal L-BGP resolves this time thanks to RSVP-TE acting as underlying MPLS label distribution protocol, but similarly to domain "Mistral", BGP routes remain unpreferred because of default protocol preferences. Listing 4.26 shows the inet.3 table with all internal routes from domain "Cyclone". Some observations are worth mentioning here:

- L-BGP routes remain inactive due to route preference against RSVP-TE (Lines 15, 28, and 41).
- All L-BGP loopback bindings from router *Livorno*, router *Barcelona* and router *Honolulu* are *implicit-null* (Lines 17, 30, and 43).
- All L-BGP routes resolve over the underlying RSVP-TE LSPs, requiring a label *push* with the RSVP-TE allocated label, except for router *Livorno* which is directly connected and considers *implicit-null* for RSVP-TE (Lines 14, 26, and 39).

Listing 4.26: Internal L-BGP routes inactive resolving over RSVP-TE at router *Skopie*

```
1  user@Skopie> show route table inet.3 extensive |
2               match "entr|AS:|reason|Next-hop|Label|Preference"
3  192.168.1.2/32 (1 entry, 1 announced)
4          *Direct Preference: 0
5                  Next-hop reference count: 3
6                  Local AS: 64501
7  192.168.1.6/32 (2 entries, 2 announced)
8          *RSVP   Preference: 7
```

```
 9                    Next-hop reference count: 4
10                    Label-switched-path Skopie-to-Livorno
11                    Local AS: 64501
12            BGP     Preference: 170/-101
13                    Next-hop reference count: 2
14                    Label-switched-path Skopie-to-Livorno # Resolving over RSVP-TE
15                    Inactive reason: Route Preference
16                    Local AS: 64501 Peer AS: 64501
17                    Route Label: 3
18  192.168.1.7/32 (2 entries, 2 announced)
19           *RSVP    Preference: 7
20                    Next-hop reference count: 4
21                    Label-switched-path Skopie-to-Barcelona
22                    Label operation: Push 300768
23                    Local AS: 64501
24            BGP     Preference: 170/-101
25                    Next-hop reference count: 2
26                    Label-switched-path Skopie-to-Barcelona # Resolving over RSVP-TE
27                    Label operation: Push 300768
28                    Inactive reason: Route Preference
29                    Local AS: 64501 Peer AS: 64501
30                    Route Label: 3
31  192.168.1.21/32 (2 entries, 2 announced)
32           *RSVP    Preference: 7
33                    Next-hop reference count: 4
34                    Label-switched-path Skopie-to-Honolulu
35                    Label operation: Push 301312
36                    Local AS: 64501
37            BGP     Preference: 170/-101
38                    Next-hop reference count: 2
39                    Label-switched-path Skopie-to-Honolulu # Resolving over RSVP-TE
40                    Label operation: Push 301312
41                    Inactive reason: Route Preference
42                    Local AS: 64501 Peer AS: 64501
43                    Route Label: 3
```

Such internal label bindings need to be exchanged with other routers from different domains. The same *route resolution* concept for L-BGP needs to be evaluated for external interconnects, meaning that these routes need to be exchanged over MPLS-enabled interfaces.

All external interconnects from domain "Cyclone" are single-hop L-EBGP with no requirement for indirect resolution in inet.3. Thus, in this setup, activating the external interface for MPLS is enough, together with the new family over the EBGP session. Listing 4.27 illustrates that simple activation for the case of router *Skopie*.

Listing 4.27: External MPLS activation for L-EBGP at router *Skopie*

```
1  [edit protocols mpls]
2  user@Skopie# set interface so-0/1/0.0 # External interface activation into MPLS
3
4  [edit interfaces]
5  user@Skopie# set so-0/1/0.0 family mpls # Family mpls active on external interface
```

Figure 4.14 illustrates the L-BGP overlay model over the RSVP-TE full mesh in the domain.

What If... Redistribution of RSVP-TE and L-BGP in domain "Cyclone"

At this stage, it is worth evaluating whether mutual redistribution between RSVP-TE and L-BGP label bindings could be a feasible alternative.

The main limitation is imposed by the different nature of label distribution with each protocol: *downstream on-demand* in the case of RSVP-TE and *downstream unsolicited* in the case of L-BGP.

While FECs need to be statically configured on each RSVP-TE LSP on each ingress LSR for path signaling to occur, L-BGP populates label bindings in an unsolicited fashion following the standard BGP distribution rules.

It is possible to extract RSVP-TE label bindings and inject them into L-BGP at the ASBR that is acting as ingress LSR for the RSVP-TE LSP. But how could other LSRs for those LSPs be aware of L-BGP label bindings present at the ASBRs? Each LSR would have to be configured with the FECs of existing or new LSPs, either for mapping to an existing LSP or for a downstream on-demand solicitation for new labels for each newer FEC. Mapping additional FECs over a joint RSVP-TE LSP can be achieved in Junos OS by means of the `install` knob.

However, this alternative clearly does not scale in a realistic scenario as soon as the setup requires multiple FECs.

Another option would be to deploy LDP for downstream unsolicited label distribution in an *overlay* fashion with RSVP-TE, using *LDP tunneling* techniques, and then to perform redistribution between LDP and L-BGP. This concept has a basic design flaw, because L-BGP can *already* be extended inside the domain in such an *overlay* manner without needing to introduce another protocol just for redistribution purposes.

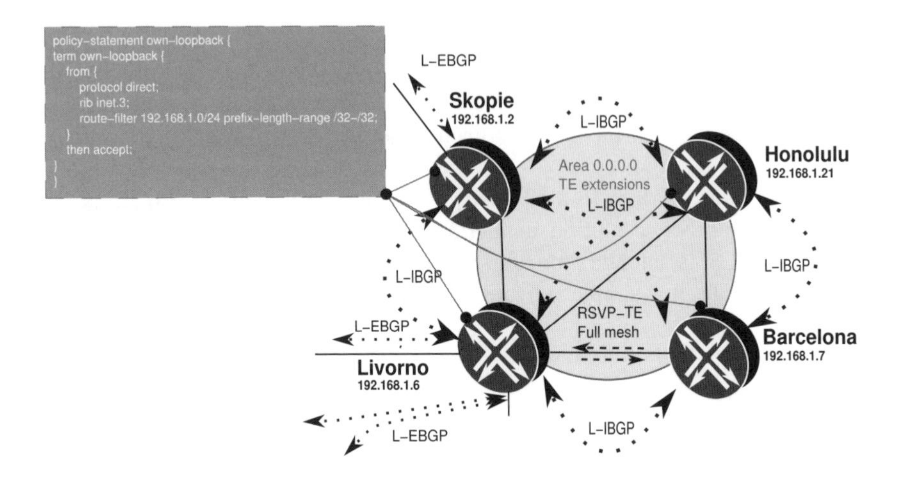

Figure 4.14: Stage two: internal L-BGP over RSVP-TE at domain “Cyclone”.

Junos Tip: Installing a secondary loopback address for RSVP-TE

Junos OS offers resources for manipulating route installation on an existing LSP for FECs represented by addresses different from the RSVP-TE destination itself.

Listing 4.28 illustrates the configuration of the `install` knob to attach additional prefixes as eligible FECs beyond a certain RSVP-TE LSP and the `no-install-to-address` feature to avoid installing the LSP end point itself, which is done for clarity purposes.

When Junos OS installs additional destinations over a common RSVP-TE LSP, metrics for the LSP egress end are inherited by all other prefixes, which aligns with the purpose of another parallel loopback identification with the same routing reachability.

Listing 4.28: LSP FEC installation migration to a secondary loopback for the LSP to Barcelona in Livorno

```
 1 [edit protocols mpls label-switched-path Livorno-to-Barcelona]
 2 user@Livorno# show
 3 no-install-to-address;
 4 to 192.168.1.7;
 5 install 192.168.1.77/32;
 6
 7 user@Livorno> show route table inet.3 192.168.1.7
 8
 9 inet.3: 11 destinations, 16 routes (10 active, 0 holddown, 1 hidden)
10 + = Active Route, - = Last Active, * = Both
11
12 192.168.1.7/32     *[LDP/9] 00:00:44, metric 10 # RSVP backup path has disappeared
13                     > to 172.16.1.53 via ge-0/2/0.0
14
15 user@Livorno> show route table inet.3 192.168.1.77
16
17 inet.3: 11 destinations, 16 routes (10 active, 0 holddown, 1 hidden)
18 + = Active Route, - = Last Active, * = Both
19
20 192.168.1.77/32    *[RSVP/10] 00:00:49, metric 10 # RSVP LSP preferred for the secondary address
21                     > to 172.16.1.53 via ge-0/2/0.0, label-switched-path Livorno-to-Barcelona
22
23 user@Livorno> show mpls lsp name Livorno-to-Barcelona extensive
24 Ingress LSP: 3 sessions
25
26 192.168.1.7
27   From: 192.168.1.6, State: Up, ActiveRoute: 0, LSPname: Livorno-to-Barcelona
28   ActivePath:  (primary)
29   LoadBalance: Random
30   Encoding type: Packet, Switching type: Packet, GPID: IPv4
31  *Primary                    State: Up, Preference: 10
32     Priorities: 7 0
33     SmartOptimizeTimer: 180
34     Computed ERO (S [L] denotes strict [loose] hops): (CSPF metric: 10)
35  172.16.1.53 S
36     Received RRO (ProtectionFlag 1=Available 2=InUse 4=B/W 8=Node 10=SoftPreempt 20=Node-ID):
37           172.16.1.53
38    16 Sep 20 19:20:55.959 CSPF: computation result accepted  172.16.1.53
39    15 Sep 20 17:24:18.289 Selected as active path
40    14 Sep 20 17:24:18.288 Record Route:  172.16.1.53
41    13 Sep 20 17:24:18.288 Up
42 <...>
```

Note that installation of the additional FEC for the secondary address and removal of the route for the primary end point has been entirely graceful, without the need to tear down the path.

At this point, MPLS services and applications could be migrated over secondary addresses. For instance, setting MP-BGP sessions among those secondary addresses automatically provides RSVP-TE inherent features for the BGP families over these sessions.

4.5.6 Stage three: Labeled BGP interconnect between domain "Cyclone" and domain "Mistral"

At this point, external L-BGP sessions are activated on router *Bilbao* and router *Havana* as ASBRs for domain "Mistral", and router *Skopie* and router *Livorno* as ASBRs for domain "Cyclone".

Router *Havana* and router *Bilbao* have two direct interconnects over different links and once these inter-AS sessions become established, label bindings from each domain are flooded inside the other.

The cornerstone of this migration methodology is visible here as a specific behavior in Junos OS: *despite being inactive paths, L-BGP routes are advertised externally and MPLS label actions are programmed in the forwarding plane.* Without requiring further manipulation, MPLS label actions derived from L-BGP are prepared and L-BGP routes are advertised over the external sessions.

Listing 4.29 shows how internal label bindings from router *Havana*, as ASBR for domain "Mistral", are advertised to router *Livorno*, and how label *swap* actions are programmed for the various labels. For instance, Lines 6, 26, and 36 show how the L-BGP label allocated for router *Bilbao* is advertised by router *Havana*, programmed in the forwarding table of router *Havana* as a *swap* action towards router *Male*, and presented as a *pop* action in router *Male* towards router *Bilbao*, respectively.

Listing 4.29: L-BGP routes advertised from Havana to Livorno

```
1  user@Havana> show route advertising-protocol bgp 172.16.1.38 extensive
2               | match "entr|Label"
3  * 192.168.1.4/32 (1 entry, 1 announced)  # Havana
4       Route Label: 3
5    192.168.1.8/32 (2 entries, 2 announced)  # Bilbao
6       Route Label: 299952
7    192.168.1.20/32 (2 entries, 2 announced)  # Male
8       Route Label: 299936
9
10 user@Havana> show route table mpls label 299936
11
12 mpls.0: 16 destinations, 16 routes (16 active, 0 holddown, 0 hidden)
13 + = Active Route, - = Last Active, * = Both
14
15 299936             *[VPN/170] 00:19:28, metric2 100, from 192.168.1.20   # Pop towards Male
16                     > to 172.16.1.85 via ge-1/1/0.0, Pop
17 299936(S=0)        *[VPN/170] 00:19:28, metric2 100, from 192.168.1.20
18                     > to 172.16.1.85 via ge-1/1/0.0, Pop
19
20
21 user@Havana> show route table mpls label 299952
22
23 mpls.0: 16 destinations, 16 routes (16 active, 0 holddown, 0 hidden)
24 + = Active Route, - = Last Active, * = Both
25
26 299952             *[VPN/170] 00:20:14, metric2 743, from 192.168.1.20 # Swap to Bilbao
27
28                     > to 172.16.1.85 via ge-1/1/0.0, Swap 299776
29
30
31 user@male-re0> show route table mpls label 299776
32
33 mpls.0: 7 destinations, 7 routes (7 active, 0 holddown, 0 hidden)
34 + = Active Route, - = Last Active, * = Both
35
36 299776             *[LDP/9] 00:16:08, metric 1 # Pop to Bilbao
37                     > via so-3/2/1.0, Pop
```

```
38 299776(S=0)         *[LDP/9] 00:16:08, metric 1
39                      > via so-3/2/1.0, Pop
```

At domain "Cyclone", a similar behavior can be observed, with the main difference being an underlying resolution substrate of single-hop RSVP-TE LSPs.

Listing 4.30 shows how internal label bindings from router *Livorno*, an ASBR for domain "Cyclone", are advertised to router *Havana*. Because these internal single-hop RSVP-TE LSPs are used for internal system reachability, all MPLS forwarding actions related to externally advertised label bindings result in a label *pop* and next-hop selection to each of these RSVP-TE LSPs when forwarding traffic internally.

Lines 17, 27, and 37 show the respective next hop installed for each of the advertised labels for router *Skopie*, router *Barcelona*, and router *Livorno* with the consequence of a label *pop* action and selection of the corresponding RSVP-TE LSP.

Listing 4.30: L-BGP routes advertised from Livorno to Havana

```
1  user@Livorno> show route advertising-protocol bgp 172.16.1.37 extensive
2                | match "entr|Label"
3    192.168.1.2/32 (2 entries, 2 announced)
4       Route Label: 301600
5  * 192.168.1.6/32 (1 entry, 1 announced)
6       Route Label: 3
7    192.168.1.7/32 (2 entries, 2 announced)
8       Route Label: 301616
9    192.168.1.21/32 (2 entries, 2 announced)
10      Route Label: 301664
11
12 user@Livorno> show route table mpls label 301600
13
14 mpls.0: 23 destinations, 23 routes (23 active, 0 holddown, 0 hidden)
15 + = Active Route, - = Last Active, * = Both
16
17 301600             *[VPN/170] 00:33:49, metric2 64, from 192.168.1.2 # Pop to Skopie
18                      > via so-0/0/0.0, label-switched-path Livorno-to-Skopie
19 301600(S=0)        *[VPN/170] 00:33:49, metric2 64, from 192.168.1.2
20                      > via so-0/0/0.0, label-switched-path Livorno-to-Skopie
21
22 user@Livorno> show route table mpls label 301616
23
24 mpls.0: 23 destinations, 23 routes (23 active, 0 holddown, 0 hidden)
25 + = Active Route, - = Last Active, * = Both
26
27 301616             *[VPN/170] 00:33:56, metric2 10, from 192.168.1.7 # Pop to Barcelona
28                      > to 172.16.1.53 via ge-0/2/0.0, label-switched-path Livorno-to-Barcelona
29 301616(S=0)        *[VPN/170] 00:33:56, metric2 10, from 192.168.1.7
30                      > to 172.16.1.53 via ge-0/2/0.0, label-switched-path Livorno-to-Barcelona
31
32 user@Livorno> show route table mpls label 301664
33
34 mpls.0: 23 destinations, 23 routes (23 active, 0 holddown, 0 hidden)
35 + = Active Route, - = Last Active, * = Both
36
37 301664             *[VPN/170] 00:34:03, metric2 10, from 192.168.1.21 # Pop to Honolulu
38                      > to 172.16.1.90 via ge-1/3/0.0, label-switched-path Livorno-to-Honolulu
39 301664(S=0)        *[VPN/170] 00:34:03, metric2 10, from 192.168.1.21
40                      > to 172.16.1.90 via ge-1/3/0.0, label-switched-path Livorno-to-Honolulu
```

At this point, it is interesting to analyze how the externally advertised label bindings from each domain are internally distributed at the other and which label actions are programmed for each of them.

Listing 4.31 illustrates how externally received label bindings are internally propagated with a label *swap* action, showing an example for the label corresponding to router *Barcelona* (Line 29).

Listing 4.31: L-BGP routes received at Havana from Livorno

```
user@Havana> show route receive-protocol bgp 172.16.1.38 extensive | match "entr|Label"
* 192.168.1.2/32 (3 entries, 1 announced)
     Route Label: 301600
* 192.168.1.6/32 (3 entries, 1 announced)
     Route Label: 3
* 192.168.1.7/32 (3 entries, 1 announced)
     Route Label: 301616
* 192.168.1.21/32 (3 entries, 1 announced)
     Route Label: 301664

user@Havana> show route advertising-protocol bgp 192.168.1.20 extensive
             | match "entr|Label"
* 192.168.1.2/32 (3 entries, 1 announced)
     Route Label: 299872
* 192.168.1.4/32 (1 entry, 1 announced)
     Route Label: 3
* 192.168.1.6/32 (3 entries, 1 announced)
     Route Label: 299888
* 192.168.1.7/32 (3 entries, 1 announced)
     Route Label: 299904
* 192.168.1.21/32 (3 entries, 1 announced)
     Route Label: 299920

user@Havana> show route table mpls label 299904

mpls.0: 16 destinations, 16 routes (16 active, 0 holddown, 0 hidden)
+ = Active Route, - = Last Active, * = Both

299904             *[VPN/170] 00:46:20
                    > to 172.16.1.38 via so-1/0/0.0, Swap 301616 # Swap external with internal label
```

Listing 4.32 shows the same behavior on router *Livorno*, considering a label *swap* action, shown here for the label corresponding to router *Bilbao* (Line 26).

Listing 4.32: L-BGP routes received at Havana from Livorno

```
user@Livorno> show route receive-protocol bgp 172.16.1.37 extensive
              | match "entr|Label"
* 192.168.1.4/32 (3 entries, 1 announced)
     Route Label: 3
* 192.168.1.8/32 (3 entries, 1 announced)
     Route Label: 299952
* 192.168.1.20/32 (3 entries, 1 announced)
     Route Label: 299936

user@Livorno> show route advertising-protocol bgp 192.168.1.7 extensive
              | match "entr|Label"
* 192.168.1.4/32 (3 entries, 1 announced)
     Route Label: 301680
* 192.168.1.6/32 (1 entry, 1 announced)
     Route Label: 3
* 192.168.1.8/32 (3 entries, 1 announced)
     Route Label: 301712
* 192.168.1.20/32 (3 entries, 1 announced)
     Route Label: 301696

user@Livorno> show route table mpls label 301712

mpls.0: 23 destinations, 23 routes (23 active, 0 holddown, 0 hidden)
+ = Active Route, - = Last Active, * = Both

```

```
26 301712          *[VPN/170] 00:49:11
27                  > to 172.16.1.37 via so-0/0/1.0, Swap 299952 # Swap external with internal label
```

Label-binding advertisements and end-to-end flows between router *Barcelona* and router *Bilbao* from previous listings in this stage are summarized in Figure 4.15 with labels corresponding to each protocol for router *Barcelona*, and Figure 4.16 for router *Bilbao*. Both figures also include MPLS label actions programmed in the forwarding plane.

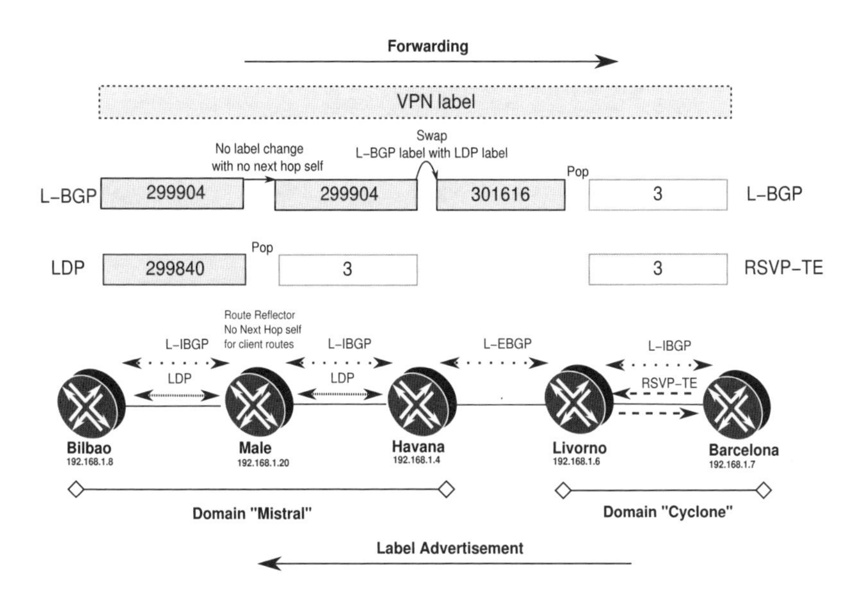

Figure 4.15: Stage three: label bindings for router *Barcelona*.

Junos Tip: Troubleshooting an L-BGP overlay model with inter-AS Option C

Once domains have been interconnected, Layer 3 VPNs can be extended over them simply by establishing multihop external BGP sessions directly among the PEs or by using route reflectors. In a nutshell, this is an implementation of the inter-AS Option C model as defined in [RFC4364] and explained later in Section 5.2.6.

As a simple example, an extended L3VPN is configured with common Route Distinguisher (RD) and Route Target (RT) on both router *Bilbao* and router *Barcelona*, as shown in Listing 4.33 for router *Bilbao*.

When establishing such a simple L3VPN for troubleshooting purposes, the following items are relevant:

- Reachability routes for EBGP peers also need to be present in inet.0 to establish the BGP session itself. This is achieved by means of `rib-groups` to copy primary inet.3 routes into inet.0 as well (Line 17).

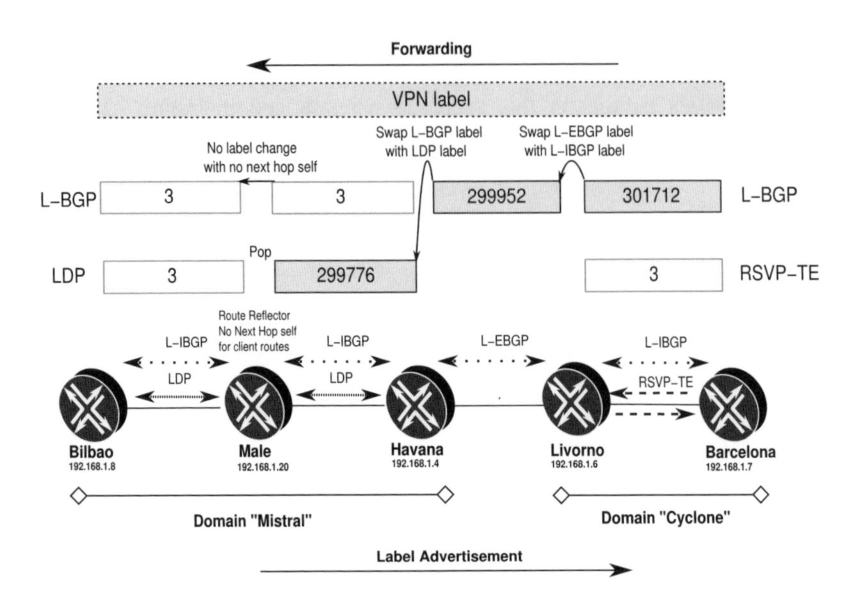

Figure 4.16: Stage three: label bindings for router *Bilbao*.

- L3VPNs are established over *multihop* EBGP sessions with extended TTL values (Line 30). Other designs may consider domain ASBRs as route reflectors for `family inet-vpn unicast`.
- A *triple* MPLS label *push* action takes place at the ingress for the *VPN* label, *L-BGP remote neighbor* reachability label, and *LDP internal ASBR* reachability label (Line 51).

Listing 4.33: Layer 3 VPN over MPLS interconnect between Barcelona and Bilbao

```
 1  [edit routing-instances NMM]
 2  user@Bilbao-re0# show
 3  instance-type vrf;
 4  interface lo0.8;
 5  route-distinguisher 65401:65401;
 6  vrf-target target:65401:65401;
 7  vrf-table-label;
 8
 9  [edit protocols bgp]
10  user@Bilbao-re0# show
11  group IBGP {
12      type internal;
13      local-address 192.168.1.8;
14      family inet {
15          unicast;
16          labeled-unicast {
17              rib-group optc; # Needed to set up extended EBGP inet-vpn session
18              rib {
19                  inet.3;
20              }
```

```
21          }
22      }
23      export [ nhs own-loopback ];
24      neighbor 192.168.1.20;
25  }
26  <...>
27  group EBGP-l3vpn {
28      type external;
29      multihop {
30          ttl 5; # extended ttl
31      }
32      local-address 192.168.1.8;
33      family inet-vpn {
34          unicast;
35      }
36      peer-as 64501;
37      neighbor 192.168.1.7;
38  }
39
40  [edit routing-options rib-groups optc]
41  user@Bilbao-re0# show
42  import-rib [ inet.3 inet.0 ]; # Copy inet.3 routes into inet.0
43
44  user@Bilbao-re0> show route table NMM
45
46  NMM.inet.0: 2 destinations, 2 routes (2 active, 0 holddown, 0 hidden)
47  + = Active Route, - = Last Active, * = Both
48
49  192.168.7.7/32      *[BGP/170] 00:15:32, localpref 100, from 192.168.1.7
50                        AS path: 64501 I
51                      > via so-1/3/1.0, Push 301824, Push 299904, Push 299840(top) # Triple push
52  192.168.8.8/32      *[Direct/0] 00:41:59
53                      > via lo0.8
54
55  user@Bilbao-re0> ping 192.168.7.7 source 192.168.8.8 routing-instance NMM
56  PING 192.168.7.7 (192.168.7.7): 56 data bytes
57  64 bytes from 192.168.7.7: icmp_seq=0 ttl=64 time=1.190 ms
58  64 bytes from 192.168.7.7: icmp_seq=1 ttl=64 time=1.111 ms
59  ^C
60  --- 192.168.7.7 ping statistics ---
61  2 packets transmitted, 2 packets received, 0% packet loss
62  round-trip min/avg/max/stddev = 1.111/1.151/1.190/0.039 ms
```

4.5.7 Stage four: Redundant label binding distribution over domain "Monsoon"

Despite not being MPLS-enabled, the leased lines over domain "Monsoon" provide an alternative and redundant interconnect path between domain "Cyclone" and domain "Mistral". In fact, router *Nantes* is strategically positioned simply to interconnect both domains locally without a specific need to extend any MPLS label distribution protocol inside the domain "Monsoon".

One virtue of an external L-BGP is that it can easily provide a redundant label distribution path that is completely independent of any IGP interconnect. LDP cannot be used to interconnect such domains at this stage, because the IGPs are not linked yet and RSVP-TE requires specific FEC installation which is not really intended in this transit label distribution scenario. Just by activating MPLS and external L-BGP from this AS, another interconnect between domain "Cyclone" and domain "Mistral" is automatically enabled.

Figure 4.17 shows a simplified form of this scenario. Listing 4.34 shows the necessary configuration steps at this dual connection, together with representative outputs for

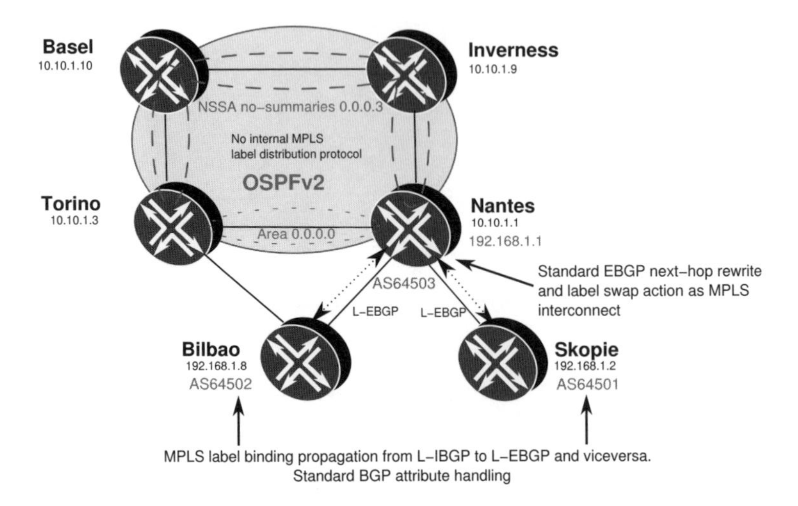

Figure 4.17: Stage four: MPLS interconnect via external L-BGP at domain "Monsoon".

exchanged routes. Note that MPLS actions are also programmed. The general action is an MPLS label *swap*, except for directly connected L-EBGP neighbors with *implicit-null* advertisements, as per Lines 132 for router *Bilbao* and 124 for router *Skopie*, which show that a label *pop* is installed in the Forwarding Engine.

Listing 4.34: MPLS interconnect between domain "Mistral" and domain "Cyclone" via Nantes

```
1  [edit protocols bgp]
2  user@Nantes# show
3  group L-EBGP {
4      type external;
5      family inet {
6          labeled-unicast {
7              rib {
8                  inet.3;
9              }
10         }
11     }
12     export own-loopback;
13     neighbor 172.16.1.5 {
14         peer-as 64502;
15     }
16     neighbor 172.16.1.2 {
17         peer-as 64501;
18     }
19 }
20
21  [edit protocols mpls]  # External interfaces active for mpls protocols
22 user@Nantes# show
23 interface so-0/1/0.0;
24 interface ge-1/3/0.0;
25
26  [edit interfaces so-0/1/0 unit 0]
27 user@Nantes# show
28 family inet {
29     address 172.16.1.1/30;
30 }
```

```
31 family iso;
32 family mpls; # active for external interconnect
33
34 [edit interfaces ge-1/3/0 unit 0]
35 user@Nantes# show
36 family inet {
37     address 172.16.1.6/30;
38 }
39 family iso;
40 family mpls; # active for external interconnect
41
42 user@Nantes> show bgp summary
43 Groups: 1 Peers: 2 Down peers: 0
44 Table          Tot Paths  Act Paths Suppressed    History Damp State    Pending
45 inet.3                14          7          0          0          0          0
46 Peer                     AS      InPkt     OutPkt    OutQ   Flaps Last Up/Dwn State|#Active/Received/
       Accepted/Damped...
47 172.16.1.2            64501       4046       4053       0       1  1d 6:22:45 Establ
48   inet.3: 4/7/7/0
49 172.16.1.5            64502       3316       3322       0       0  1d 0:47:05 Establ
50   inet.3: 3/7/7/0
51
52
53  # Received from domain ''Cyclone''
54 user@Nantes> show route receive-protocol bgp 172.16.1.2 extensive | match "entr|Label"
55 * 192.168.1.2/32 (2 entries, 1 announced)
56      Route Label: 3
57   192.168.1.4/32 (2 entries, 1 announced)
58      Route Label: 301696
59 * 192.168.1.6/32 (2 entries, 1 announced)
60      Route Label: 301120
61 * 192.168.1.7/32 (2 entries, 1 announced)
62      Route Label: 301152
63   192.168.1.8/32 (2 entries, 1 announced)
64      Route Label: 301728
65   192.168.1.20/32 (2 entries, 1 announced)
66      Route Label: 301712
67 * 192.168.1.21/32 (2 entries, 1 announced)
68      Route Label: 301504
69
70  # Received from domain ''Mistral''
71 user@Nantes> show route receive-protocol bgp 172.16.1.5 extensive | match "entr|Label"
72   192.168.1.2/32 (2 entries, 1 announced)
73      Route Label: 301888
74 * 192.168.1.4/32 (2 entries, 1 announced)
75      Route Label: 301840
76   192.168.1.6/32 (2 entries, 1 announced)
77      Route Label: 301856
78   192.168.1.7/32 (2 entries, 1 announced)
79      Route Label: 301872
80 * 192.168.1.8/32 (2 entries, 1 announced)
81      Route Label: 3
82 * 192.168.1.20/32 (2 entries, 1 announced)
83      Route Label: 301264
84   192.168.1.21/32 (2 entries, 1 announced)
85      Route Label: 301904
86
87
88  # Advertised to domain ''Cyclone''
89 user@Nantes> show route advertising-protocol bgp 172.16.1.2 extensive |
90              match "entr|Label"
91 * 192.168.1.1/32 (1 entry, 1 announced)
92      Route Label: 3
93 * 192.168.1.4/32 (2 entries, 1 announced)
94      Route Label: 310144
95 * 192.168.1.8/32 (2 entries, 1 announced)
96      Route Label: 309824
97 * 192.168.1.20/32 (2 entries, 1 announced)
98      Route Label: 309840
```

```
99
100
101  # Advertised to domain ''Mistral''
102  user@Nantes> show route advertising-protocol bgp 172.16.1.5 extensive |
103              match "entr|Label"
104  * 192.168.1.1/32 (1 entry, 1 announced)
105       Route Label: 3
106  * 192.168.1.2/32 (2 entries, 1 announced)
107       Route Label: 309696
108  * 192.168.1.6/32 (2 entries, 1 announced)
109       Route Label: 309744
110  * 192.168.1.7/32 (2 entries, 1 announced)
111       Route Label: 309760
112  * 192.168.1.21/32 (2 entries, 1 announced)
113       Route Label: 310032
114
115  user@Nantes> show route table mpls.0
116
117  mpls.0: 12 destinations, 12 routes (12 active, 0 holddown, 0 hidden)
118  + = Active Route, - = Last Active, * = Both
119
120  <...>
121  309696            *[VPN/170] 1d 00:50:29 # Skopie (directly connected)
122                     > to 172.16.1.2 via so-0/1/0.0, Pop
123  309696(S=0)       *[VPN/170] 1d 00:50:29 # Skopie (directly connected)
124                     > to 172.16.1.2 via so-0/1/0.0, Pop
125  309744            *[VPN/170] 1d 00:50:29 # Livorno
126                     > to 172.16.1.2 via so-0/1/0.0, Swap 301120
127  309760            *[VPN/170] 1d 00:50:29 # Barcelona
128                     > to 172.16.1.2 via so-0/1/0.0, Swap 301152
129  309824            *[VPN/170] 1d 00:50:29 # Bilbao (directly connected)
130                     > to 172.16.1.5 via ge-1/3/0.0, Pop
131  309824(S=0)       *[VPN/170] 1d 00:50:29 # Bilbao (directly connected)
132                     > to 172.16.1.5 via ge-1/3/0.0, Pop
133  309840            *[VPN/170] 1d 00:50:29 # Male
134                     > to 172.16.1.5 via ge-1/3/0.0, Swap 301264
135  310032            *[VPN/170] 09:50:07 # Honolulu
136                     > to 172.16.1.2 via so-0/1/0.0, Swap 301504
137  310144            *[VPN/170] 02:04:56 # Havana
138                     > to 172.16.1.5 via ge-1/3/0.0, Swap 301840
```

4.5.8 Stage five: MPLS integration and interconnect via router *Lille*

Router *Lille* is the collocated router at the NOC premises, and one initial objective is to be able to integrate it into the existing MPLS services from both domain “Cyclone” and domain “Mistral”.

L-BGP provides resources to enable this additional interconnect between the two domains and, at the same time, to activate label bindings from each IGP in a completely independent fashion. For instance, L3VPN CEs requiring specific monitoring from the NOC or SLA measurement devices based on RPM are directly reachable from the NOC even before integrating IGPs!

This exercise is parallel to the setup configured on router *Nantes* based on external L-BGP. However, there are two important remarks:

- Router *Lille* is configured with a global AS, *65500*, maintaining EBGP sessions towards both domains.
- Router *Lille* not only reflects routes between both interconnected domains (with *third-party next hops* as is standard in broadcast segments), but also advertises its own loopback so that it is present in both domains.

This second bullet is fundamental for the instantiation and integration of MPLS applications. The advertisement of the own loopback address as L-BGP route is achieved in the same fashion as with other ASBRs: by leaking the loopback address into inet.3 and enabling advertisement from there with the `rib-groups` structure.

Listing 4.35 depicts configuration changes and the outputs of the two BGP sessions towards each domain.

Listing 4.35: MPLS interconnect between domain "Mistral" and domain "Cyclone" and own integration at Lille

```
1  [edit protocols bgp]
2  user@Lille# show
3  group L-EBGP {
4      family inet {
5          labeled-unicast {
6              rib {
7                  inet.3;
8              }
9          }
10     }
11     export own-loopback;
12     neighbor 172.16.100.1 { # Havana
13         peer-as 64502;
14     }
15     neighbor 172.16.100.8 { # Livorno
16         peer-as 64501;
17     }
18 }
19
20 [edit policy-options policy-statement own-loopback]
21 user@Lille# show
22 term own-loopback {
23     from {
24         protocol direct;
25         rib inet.3;
26         route-filter 192.168.1.0/24 prefix-length-range /32-/32;
27     }
28     then accept;
29 }
30
31 [edit policy-options policy-statement interface-to-inet3]
32 user@Lille# show
33 term loopback {
34     from interface lo0.0; # Leak only loopback address to inet.3
35     then accept;
36 }
37 term no-more-inet3 {
38     to rib inet.3;
39     then reject;
40 }
41 term default {
42     to rib inet.0;
43     then accept;
44 }
45
46 [edit protocols mpls]
47 user@Lille# show
48 interface fe-0/3/0.0; # Activate external interface on MPLS
49
50 [edit interfaces fe-0/3/0]
51 user@Lille# show
52 unit 0 {
53     family inet {
54         address 172.16.100.5/24;
55     }
56     family iso;
```

```
57      family mpls;
58  }
59
60  user@Lille> show bgp summary
61  Groups: 1 Peers: 2 Down peers: 0
62  Table          Tot Paths  Act Paths Suppressed    History Damp State    Pending
63  inet.0                 0          0          0          0          0          0
64  inet.3                13          9          0          0          0          0
65  Peer               AS      InPkt     OutPkt    OutQ   Flaps Last Up/Dwn State|#Active/Received/Damped
    ...
66  172.16.100.1    64503       1286       1304       0       4     9:14:53 Establ
67    inet.3: 3/4/0
68  172.16.100.8    64501       1028       1029       0       3     7:44:59 Establ
69    inet.3: 5/8/0
70
71  user@Lille> show route advertising-protocol bgp 172.16.100.1
72
73  inet.3: 9 destinations, 16 routes (9 active, 0 holddown, 0 hidden)
74    Prefix              Nexthop           MED      Lclpref    AS path
75  * 192.168.1.2/32          Self                                     64501 I  # Skopie
76  * 192.168.1.5/32          Self                                     I   # Lille
77  * 192.168.1.6/32          Self                                     64501 I  # Livorno
78  * 192.168.1.7/32          Self                                     64501 I  # Barcelona
79  * 192.168.1.21/32         Self                                     64501 I  # Honolulu
80
81  user@Lille> show route advertising-protocol bgp 172.16.100.8
82
83  inet.3: 9 destinations, 16 routes (9 active, 0 holddown, 0 hidden)
84    Prefix              Nexthop           MED      Lclpref    AS path
85  * 192.168.1.4/32          Self                                     64502 I  # Havana
86  * 192.168.1.5/32          Self                                     I   # Lille
87  * 192.168.1.8/32          Self                                     64502 I  # Bilbao
88  * 192.168.1.20/32         Self                                     64502 I  # Male
89
90
91  user@Lille> show route receive-protocol bgp 172.16.100.1 table inet.3
92
93  inet.3: 9 destinations, 13 routes (9 active, 0 holddown, 0 hidden)
94    Prefix              Nexthop           MED      Lclpref    AS path
95  * 192.168.1.4/32          172.16.100.1                             64503 I  # Havana
96    192.168.1.6/32          172.16.100.1                             64503 64501 I  # Livorno
97  * 192.168.1.8/32          172.16.100.1      743                    64503 I  # Bilbao
98  * 192.168.1.20/32         172.16.100.1      100                    64503 I     # Male
```

Figure 4.18 depicts the interconnect setup through router *Lille*.

Junos Tip: L-BGP route selection due to "Active preferred"

Whenever router *Lille* is interconnected, a label binding for *192.168.1.5* is received from both peers and populated further, indicating that router *Lille* has been already interconnected to router *Havana* and router *Livorno*.

Both routes are received at router *Nantes* by means of different EBGP sessions from routers with different router IDs. Under these circumstances, router ID comparison is not performed by default in Junos OS if one of the external paths is active. As described in Section 3.2.8, given that both different external paths are usable and have the same preference, with same local preference, AS path lengths, origin, MED, IGP metric, and empty cluster lists, Junos OS enforces a certain *last active* path behavior in this case to avoid *persistent route oscillation* that can result from MED comparison, which can potentially lead to best-path selection loops in given scenarios.

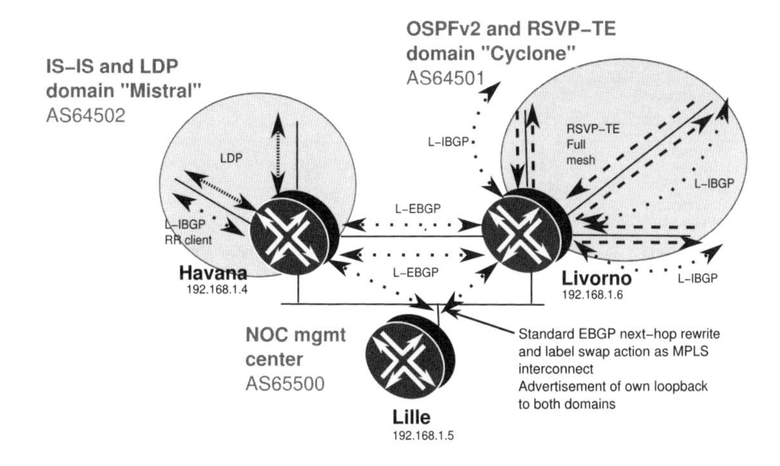

Figure 4.18: Stage five: MPLS integration and interconnect via router *Lille*.

Listing 4.36 shows how both versions of L-BGP route for *192.168.1.5* are evaluated and illustrates that the final tie-breaking item is whether the route was previously active, as shown in Line 30.

Listing 4.36: L-BGP route selection for Lille at Nantes

```
1  user@Nantes> show route 192.168.1.5/32 exact extensive table inet.3
2
3  inet.3: 10 destinations, 18 routes (10 active, 0 holddown, 0 hidden)
4  192.168.1.5/32 (2 entries, 1 announced)
5  TSI:
6  Page 0 idx 0 Type 1 val 8cfb0d8
7          *BGP    Preference: 170/-101
8                  Next hop type: Router
9                  Next-hop reference count: 2
10                 Source: 172.16.1.2
11                 Next hop: 172.16.1.2 via so-0/1/0.0, selected
12                 Label operation: Push 301744
13                 State: <Active Ext>
14                 Local AS: 64503 Peer AS: 64501
15                 Age: 2:01:47
16                 Task: BGP_64501.172.16.1.2+63387
17                 Announcement bits (1): 1-BGP RT Background
18                 AS path: 64501 65500 I
19                 Accepted
20                 Route Label: 301744
21                 Localpref: 100
22                 Router ID: 192.168.1.2
23          BGP    Preference: 170/-101
24                 Next hop type: Router
25                 Next-hop reference count: 1
26                 Source: 172.16.1.5
27                 Next hop: 172.16.1.5 via ge-1/3/0.0, selected
28                 Label operation: Push 301936
29                 State: <Ext>
30                 Inactive reason: Active preferred # Unselected
31                 Local AS: 64503 Peer AS: 64502
32                 Age: 2:01:41
33                 Task: BGP_64502.172.16.1.5+55450
```

```
34                    AS path: 64502 65500 I
35                    Accepted
36                    Route Label: 301936
37                    Localpref: 100
38                    Router ID: 192.168.1.8
```

With these MPLS interconnects and mutual label protocol redistribution, "Gale Internet" can already offer MPLS-based applications end-to-end between domain "Mistral" and domain "Cyclone" without having fully integrated their IGPs. At this point, the first milestone in the MPLS transport migration has been reached and the network is ready for a global IGP integration into a joint domain, a process that is entirely transparent to the MPLS interconnect methodology based on L-BGP and that relies only on the IGPs on each island.

4.5.9 Stage six: LDP activation in all domains

Once all MPLS domains are interconnected and there is a unified IGP across all networks, a final MPLS label distribution protocol can de deployed over all systems.

Considering the reasoning from Section 4.5.3, LDP is selected as the most suitable MPLS label distribution protocol for the final scenario.

Based on a unified link-state IGP, LDP yields at this stage consistent label binding selection with IS–IS routes. The beauty of a gradual LDP rollout is that the migration is completely smooth:

- As LDP is activated, more LDP label bindings will be preferred over L-BGP due to default protocol preferences.

- Such route overlap does not impact L-BGP-derived MPLS actions in the forwarding plane, because Junos OS programs L-BGP route-derived actions despite being inactive routes in the routing table.

Rolling out LDP not only prevents L-BGP resolution failures, but it actively replaces L-BGP as the internal label distribution protocol. Once LDP is completely deployed and because no other external routes are transported by L-BGP, default protocol preferences will simply make L-BGP unpreferred for all internal label bindings.

Domain "Cyclone" requires another consideration: L-BGP is resolving over RSVP-TE, with no LDP in the configuration. *LDP tunneling* techniques over RSVP-TE make downstream-unsolicited distribution feasible over those LSPs, emulating the existing *overlay* model, with LDP on top of RSVP-TE this time.

A preliminary requirement for the LDP rollout is a unified IGP. As described in Section 2.3, this final IGP merge consisted of several steps in each domain, namely:

- IS–IS Level 1 and Level 2 preference in domain "Cyclone" with route selection aligned, and soft dismantling of OSPFv2 neighbors with regard to RSVP-TE LSPs;

- IS–IS expansion through domain "Monsoon";

- adjustment of IS–IS authentication settings in domain "Mistral" and router *Lille*.

All these events occur in a completely transparent fashion with regard to the current MPLS transport topology. External L-BGP sessions are set on a single-hop basis and existing LDP and RSVP-TE domains are internal (the IS–IS setup keeps the same forwarding paths as OSPFv2 in domain "Cyclone").

After aligning all protocol preferences and natively activating IS–IS Level 2 in the domain interconnects, the setup is ready for LDP activation.

Junos Tip: Unadvertised effects of LDP-IGP Synchronization when stitching domains

LDP-IGP Synchronization is considered to be a best practice in networks with MPLS services and IP reachability completely aligned. For this reason, it is part of the standard configuration at many networks to activate that feature in templates or via Junos OS `groups`. However, there is an obvious but dramatic pitfall if *LDP-IGP Synchronization* is not taken into account when stitching domains.

Once the final IGP has been completely rolled out, LDP can be smoothly activated in all domain interconnects, assuming that LDP neighbor router identifiers are known through the IGP. If this LDP activation stage is needed beforehand and *LDP–IGP synchronization* is active, the feature may indirectly *maximize* the IGP metric for the interconnect link, even though the IGP adjacency is legitimate, if *router IDs* are not known between peers and the LDP session over TCP cannot be established.

This condition means that *pre-activating* LDP in those interconnect cases is harmless with regard to LDP because the session is not set up until the IGP is standardized and the router identifiers are reachable. However, if the IGP interface includes the `ldp-synchronization` knob, the collateral effect is a topology change.

Consider the interconnect case between router *Skopie* and router *Nantes*, which can establish a Level 2 IS–IS adjacency between them. Router *Nantes* is still the ASBR for the old OSPFv2 domain and is using the old non-public loopback address (which also acts, by default, as the LDP router identifier). Although IS–IS can be perfectly activated with a proper Net-ID and `family iso` activation in the interconnect, enabling LDP in this case with LDP-IGP synchronization indirectly leads to link-cost maximization. Activating LDP under these circumstances over a fully established and sane IS–IS adjacency changes the SPF tree across all domains!

Listing 4.37 summarizes the circumstances and effects on both router *Skopie* and router *Nantes*. Lines 13, 26, and 29 show standard IS–IS derived metrics in the respective old and new-style TLVs. At the moment LDP is activated and because of the TCP connection error for the LDP session, as seen in Line 64, metric values are maximized for the neighbor IS in old and new-style TLVs, as shown in Lines 92, 105, and 109 and the metric value remains maximum until the situation with LDP is corrected. Note, however, that the metric value for the related IP prefix over that interface is immune to this modification, because the feature affects only neighbor reachability, as seen in Lines 107 and 110 when compared to the previous status in Lines 27 and 30.

Listing 4.37: Indirect effects of LDP-IGP synchronization with unknown LDP router identifier

```
1 user@Skopie> show isis adjacency
2 Interface             System         L State        Hold (secs) SNPA
3 so-0/1/0.0            Nantes         2 Up                    23
```

```
4  so-1/0/0.0            Livorno        2  Up                    25
5
6  user@Skopie> show isis interface so-0/1/0.0 extensive
7  IS-IS interface database:
8  so-0/1/0.0
9    Index: 67, State: 0x6, Circuit id: 0x1, Circuit type: 2
10   LSP interval: 100 ms, CSNP interval: 15 s, Loose Hello padding
11   Adjacency advertisement: Advertise
12   Level 2
13     Adjacencies: 1, Priority: 64, Metric: 643 # Standard calculated metric
14     Hello Interval: 9.000 s, Hold Time: 27 s
15
16 user@Skopie> show isis database Nantes extensive
17              | match "IP router id|Skopie|172.16.1.0/30"
18    IS neighbor: Skopie.00                    Metric:      160
19    IP prefix: 172.16.1.0/30                  Metric:      160 Internal Up
20     IP router id: 10.10.1.1
21     IS extended neighbor: Skopie.00, Metric: default 160
22     IP extended prefix: 172.16.1.0/30 metric 160 up
23
24 user@Skopie> show isis database Skopie extensive
25              | match "IP router id|Nantes|172.16.1.0/30"
26    IS neighbor: Nantes.00                  Metric:      643  # Standard derived metric
27    IP prefix: 172.16.1.0/30                Metric:      643 Internal Up   # Standard derived
       metric
28     IP router id: 192.168.1.2
29     IS extended neighbor: Nantes.00, Metric: default 643  # Standard derived metric
30     IP extended prefix: 172.16.1.0/30 metric 643 up  # Standard derived metric
31
32 # Activating LDP over interconnect interface with LDP-IGP Synchronization
33
34 user@Skopie> configure
35 Entering configuration mode
36 [edit]
37 user@Skopie# set protocols isis interface so-0/1/0.0 ldp-synchronization
38 [edit]
39 user@Skopie# set protocols ldp interface so-0/1/0.0
40 [edit]
41 user@Skopie# commit and-quit
42 commit complete
43 Exiting configuration mode
44
45 user@Skopie> show isis adjacency
46 Interface             System         L State        Hold (secs) SNPA
47 so-0/1/0.0            Nantes         2  Up                    26
48 so-1/0/0.0            Livorno        2  Up                    25
49
50 user@Skopie> show ldp neighbor
51 Address            Interface          Label space ID         Hold time
52 192.168.1.6        lo0.0              192.168.1.6:0            32
53 192.168.1.7        lo0.0              192.168.1.7:0            31
54 192.168.1.21       lo0.0              192.168.1.21:0           43
55 172.16.1.1         so-0/1/0.0         10.10.1.1:0              10 # Neighbor detected
56
57 user@Skopie> show ldp session 10.10.1.1 extensive
58 Address: 10.10.1.1, State: Nonexistent, Connection: Closed, Hold time: 0
59   Session ID: 192.168.1.2:0--10.10.1.1:0
60   Connection retry in 0 seconds
61   Active, Maximum PDU: 4096, Hold time: 30, Neighbor count: 1
62   Neighbor types: discovered # Discovered neighbor over interface
63   Keepalive interval: 10, Connect retry interval: 1
64   Last down 00:00:00 ago; Reason: connection error # TCP connection error
65   Protection: disabled
66   Local - Restart: disabled, Helper mode: enabled
67   Remote - Restart: disabled, Helper mode: disabled
68   Local maximum neighbor reconnect time: 120000 msec
69   Local maximum neighbor recovery time: 240000 msec
70   Nonstop routing state: Not in sync
```

```
71    Next-hop addresses received:
72      so-0/1/0.0
73
74  Message type              Total                Last 5 seconds
75                     Sent  Received        Sent        Received
76  Initialization          0             0             0             0
77  Keepalive               0             0             0             0
78  Notification            0             0             0             0
79  Address                 0             0             0             0
80  Address withdraw        0             0             0             0
81  <...>
82
83  user@Skopie> show isis interface so-0/1/0.0 extensive
84  IS-IS interface database:
85  so-0/1/0.0
86    Index: 67, State: 0x6, Circuit id: 0x1, Circuit type: 2
87    LSP interval: 100 ms, CSNP interval: 15 s, Loose Hello padding
88    Adjacency advertisement: Advertise
89    LDP sync state: in hold-down, for: 00:00:06, reason: LDP down during config
90    config holdtime: infinity
91    Level 2
92      Adjacencies: 1, Priority: 64, Metric: 16777214 # Maximized cost
93      Hello Interval: 9.000 s, Hold Time: 27 s
94
95  user@Skopie> show isis database Nantes extensive |
96               match "IP router id|Skopie|172.16.1.0/30"
97     IS neighbor: Skopie.00                       Metric:       160
98     IP prefix: 172.16.1.0/30                     Metric:       160 Internal Up
99      IP router id: 10.10.1.1
100     IS extended neighbor: Skopie.00, Metric: default 160
101     IP extended prefix: 172.16.1.0/30 metric 160 up
102
103 user@Skopie> show isis database Skopie extensive |
104              match "IP router id|Nantes|172.16.1.0/30"
105    IS neighbor: Nantes.00                       Metric: 16777214  # Maximized metric for neighbor
106    IP prefix: 172.16.1.0/30                     Metric:       643 Internal Up
107      # Standard derived metric for IP prefix
108     IP router id: 192.168.1.2
109     IS extended neighbor: Nantes.00, Metric: default 16777214  # Maximized metric for neighbor
110     IP extended prefix: 172.16.1.0/30 metric 643 up   # Standard derived metric for IP prefix
```

Activating LDP over the interconnects requires adding them under the LDP protocol stanza. These interfaces are already present under `[edit protocols mpls]` and include `family mpls` in their definition, which was done earlier when external L-BGP was activated.

By enabling LDP, the implicit discovery mechanism is enabled over those links to identify neighbors for potential LDP sessions. For those systems in domain "Cyclone", there is not going to be any internal auto-discovery but rather, targeted LDP sessions on top of RSVP-TE LSPs. LDP is *tunneled* in that case over all internal RSVP-TE LSPs.

As a representative example, Listing 4.38 depicts the results of activating LDP over the interconnect and of tunneling LDP over RSVP-TE LSPs on router *Livorno*. It is worth making these observations:

- L-BGP routes do not disappear, but become unpreferred after the newer LDP paths are established.

- LDP routes point to the same peer as the previous L-BGP label bindings (Lines 52, 58, 62, 67, and 72).

- Label actions derived from L-BGP remain, and only newly allocated labels for LDP appear in the MPLS table.
- Despite being *inactive* now, L-BGP routes are exported to external EBGP neighbors with the same labels (Line 77).

This last bullet is key for a seamless migration: *the same advertised label bindings in labeled BGP persist despite the fact that the paths are locally inactive*. This is a powerful migration feature because gradual rollout of LDP activation remains transparent for L-BGP route advertisements.

Listing 4.38: LDP activation in Livorno

```
1  user@Livorno> show route table inet.3 terse
2
3  inet.3: 10 destinations, 23 routes (9 active, 0 holddown, 1 hidden)
4  + = Active Route, - = Last Active, * = Both
5
6  A Destination        P Prf   Metric 1   Metric 2  Next hop         AS path
7  * 192.168.1.1/32     B 170        100             >so-0/0/0.0      64503 I # Preferred BGP route for
        Nantes
8                       B 170        100             >172.16.1.37     64502 64503 I
9                       B 170        100             >172.16.100.1    64502 64503 I
10                      B 170        100             >172.16.100.5    65500 64502 64503 I
11 <...>
12 * 192.168.1.4/32     B 170        100             >172.16.1.37     64502 I # Preferred BGP route for
        Havana
13                      B 170        100             >172.16.100.1    64502 I
14                      B 170        100             >172.16.100.5    65500 64502 I
15 * 192.168.1.5/32     B 170        100             >172.16.100.5    65500 I # Preferred BGP route for
        Lille
16                      B 170        100             >172.16.1.37     64502 65500 I
17                      B 170        100             >172.16.100.1    64502 65500 I
18 <...>
19 * 192.168.1.8/32     B 170        100             >172.16.1.37     64502 I # Preferred BGP route for
        Bilbao
20                      B 170        100             >172.16.100.1    64502 I
21                      B 170        100             >172.16.100.5    65500 64502 I
22 * 192.168.1.20/32    B 170        100             >172.16.1.37     64502 I # Preferred BGP route for
        Male
23                      B 170        100             >172.16.100.1    64502 I
24                      B 170        100             >172.16.100.5    65500 64502 I
25 <...>
26
27   # Activating LDP over interconnect and RSVP-TE LSPs
28
29 [edit]
30 user@Livorno# show | compare
31 [edit protocols mpls label-switched-path Livorno-to-Skopie]
32 +    ldp-tunneling;
33 [edit protocols mpls label-switched-path Livorno-to-Barcelona]
34 +    ldp-tunneling;
35 [edit protocols mpls label-switched-path Livorno-to-Honolulu]
36 +    ldp-tunneling;
37 [edit protocols]
38 +   ldp {
39 +       track-igp-metric;
40 +       strict-targeted-hellos;
41 +       interface so-0/0/1.0;
42 +       interface fe-0/3/0.0;
43 +       interface lo0.0;
44 +   }
45
46 user@Livorno> show route table inet.3 terse
47
```

```
48 inet.3: 16 destinations, 38 routes (12 active, 5 holddown, 6 hidden)
49 + = Active Route, - = Last Active, * = Both
50
51 A Destination        P Prf   Metric 1   Metric 2  Next hop        AS path
52 * 192.168.1.1/32     L   9       1386              >so-0/0/1.0    # New LDP route for Nantes
53                      B 170        100              >so-0/0/0.0     64503 I
54                      B 170        100              >172.16.1.37    64502 64503 I
55                      B 170        100              >172.16.100.1   64502 64503 I
56                      B 170        100              >172.16.100.5   65500 64502 64503 I
57 <...>
58 * 192.168.1.4/32     L   9       1000              >172.16.100.1 # New LDP route for Havana
59                      B 170        100              >172.16.1.37    64502 I
60                      B 170        100              >172.16.100.1   64502 I
61                      B 170        100              >172.16.100.5   65500 64502 I
62 * 192.168.1.5/32     L   9       1000              >172.16.100.5 # New LDP route for Lille
63                      B 170        100              >172.16.100.5   65500 I
64                      B 170        100              >172.16.1.37    64502 65500 I
65                      B 170        100              >172.16.100.1   64502 65500 I
66 <...>
67 * 192.168.1.8/32     L   9       1743              >172.16.100.1 # New LDP route for Bilbao
68                      B 170        100              >172.16.1.37    64502 I
69                      B 170        100              >172.16.100.1   64502 I
70                      B 170        100              >172.16.100.5   65500 64502 I
71 <...>
72 * 192.168.1.20/32    L   9       1100              >172.16.100.1 # New LDP route for Male
73                      B 170        100              >172.16.1.37    64502 I
74                      B 170        100              >172.16.100.1   64502 I
75                      B 170        100              >172.16.100.5   65500 64502 I
76
77   # Same L-BGP routes advertised
78 user@Livorno> show route advertising-protocol bgp 172.16.1.37 extensive |
79               match "entr|Label"
80   192.168.1.1/32 (6 entries, 3 announced)
81      Route Label: 301840
82   192.168.1.2/32 (4 entries, 3 announced)
83      Route Label: 301856
84   192.168.1.5/32 (4 entries, 2 announced)
85      Route Label: 301728
86 * 192.168.1.6/32 (1 entry, 1 announced)
87      Route Label: 3
88   192.168.1.7/32 (4 entries, 3 announced)
89      Route Label: 302096
90   192.168.1.21/32 (4 entries, 3 announced)
91      Route Label: 301792
```

This transition on router *Livorno* is illustrated in Figure 4.19.

To summarize, LDP can be gracefully rolled out over all existing routers in the topology. Once all routers are fully integrated with LDP as domain-wide MPLS label distribution protocol, the `labeled-unicast` family from BGP sessions, together with supporting `rib-groups` for route sharing, can be safely dismantled.

As explained in Section 3.2.3, note that the removal of a given family from an existing BGP session triggers a renegotiation with the newly available address families.

Junos Tip: Knob *strict-targeted-hellos* and direct LDP sessions

Note that the configuration at domain "Cyclone" routers includes the `strict-targeted-hellos` knob as an additional protection mechanism against undesired targeted LDP sessions, as well as direct LDP sessions over the external interconnects in the ASBRs. Listing 4.39 reviews the LDP protocol configuration on router *Skopie*.

The `strict-targeted-hellos` knob does not restrict LDP sessions to be targeted *only*. It actually enforces that all extended targeted LDP sessions to be set up correspond to locally configured RSVP-TE destinations with the `ldp-tunneling` extension. However, this knob allows direct non-extended LDP sessions to be established over directly connected interfaces perfectly. The actions of this knob do not counteract the establishment of autodiscovered LDP sessions; the knob simply enforces constraints to LDP targeted sessions.

Listing 4.39: LDP configuration at Skopie including strict-targeted-hellos

```
1 user@Skopie> show configuration protocols ldp
2 strict-targeted-hellos;
3 egress-policy from-LBGP-to-LDP;
4 interface so-0/1/0.0;
5 interface lo0.0;
```

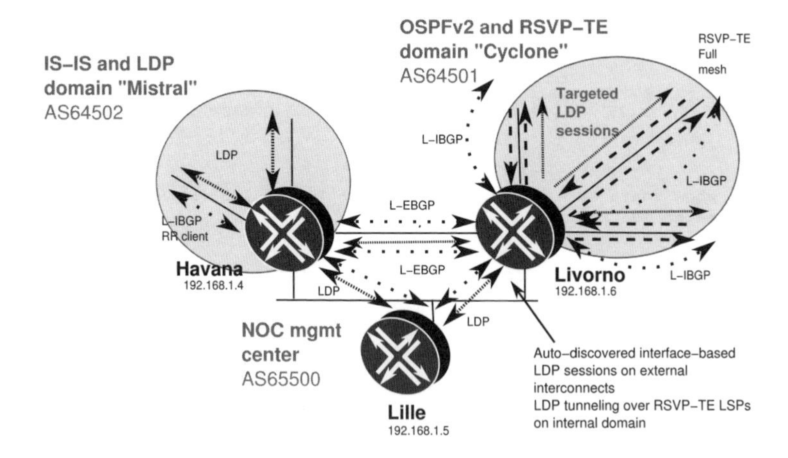

Figure 4.19: Stage six: activation of LDP in router *Livorno* as gradual label distribution transition.

4.5.10 Migration summary

Figure 4.20 summarizes the goals of the migration scenario after the integration has been completed.

The MPLS label distribution protocol migration has had several dependencies on the IGP and BGP migrations, but first by interconnecting and later transitioning the MPLS label distribution protocols, the new features have been gradually deployed in a seamless fashion.

All prerequisites have been satisfied, first to expedite the MPLS service offering across all domains and later to unify the setup with a common internal MPLS label distribution protocol.

“Gale Internet” has been able to expand their service portfolio with extended reachability while in the middle of a merger process, and customers (and obviously top management!)

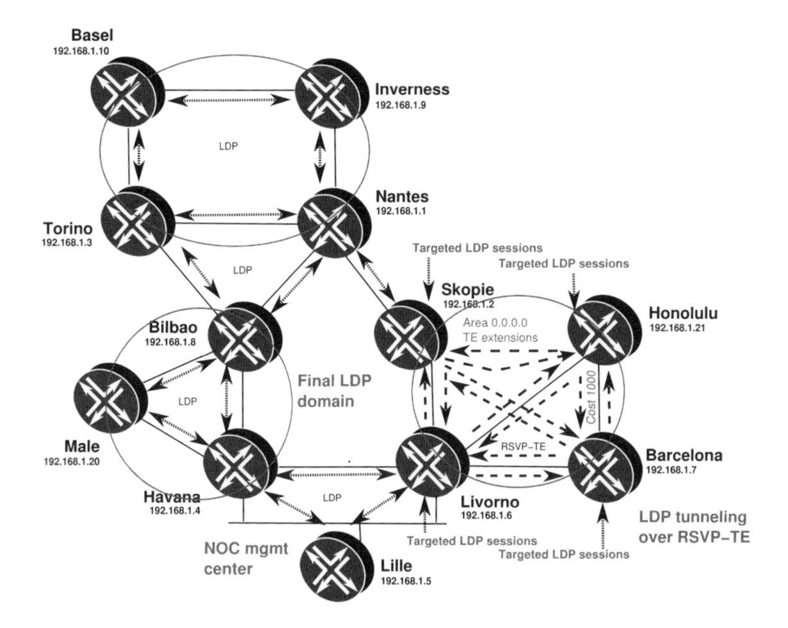

Figure 4.20: Final MPLS label distribution topology.

are impressed by this! This case study has provided several lessons and concepts learned, including the following:

- evaluation of MPLS label distribution to interconnect domains;
- analysis of several MPLS label distribution protocol strategies and dependencies with regard to IGP routes;
- evaluation of route resolution requirements for L-BGP;
- utilization of the `rib-groups` Junos OS feature for route leaking between inet.3 and inet.0;
- definition of policies to export label bindings for the routers' own loopback addresses with L-BGP;
- identification of L-BGP label allocation, incorrect configuration, and route resolution failures;
- expansion of L-IBGP inside each domain;
- LDP egress policy adaption to create label bindings for external routes in a mutual protocol redistribution model;
- analysis of potential scenarios for mutual label binding redistribution on dual-ASBR domains;
- installation of additional FECs to be bound to existing RSVP-TE LSPs;

- rollout of redundant label distribution interconnects among domains;
- identification of MPLS label action differences when announcing internally and externally L-BGP routes;
- definition of troubleshooting configuration for L3VPNs over L-BGP (Inter-AS Option C);
- advertisement of label bindings related to inactive L-BGP routes with MPLS label action installed in the forwarding plane;
- graceful activation of LDP, allowing other label distribution protocols running in the background;
- discovery of unadvertised effects of LDP-IGP synchronization when stitching domains;
- LDP-tunneling deployment over an existing RSVP-TE LSP mesh for downstream-unsolicited label distribution;
- utilization of the `strict-targeted-hellos` Junos OS feature to enforce constraints for LDP-tunneled sessions;
- as the final cleanup, gradual decommissioning of L-BGP.

Bibliography

[RFC2702] D. Awduche, J. Malcolm, J. Agogbua, M. O'Dell, and J. McManus. Requirements for Traffic Engineering Over MPLS. RFC 2702 (Informational), September 1999.

[RFC3035] B. Davie, J. Lawrence, K. McCloghrie, E. Rosen, G. Swallow, Y. Rekhter, and P. Doolan. MPLS using LDP and ATM VC Switching. RFC 3035 (Proposed Standard), January 2001.

[RFC3036] L. Andersson, P. Doolan, N. Feldman, A. Fredette, and B. Thomas. LDP Specification. RFC 3036 (Proposed Standard), January 2001. Obsoleted by RFC 5036.

[RFC3037] B. Thomas and E. Gray. LDP Applicability. RFC 3037 (Informational), January 2001.

[RFC3107] Y. Rekhter and E. Rosen. Carrying Label Information in BGP-4. RFC 3107 (Proposed Standard), May 2001.

[RFC3209] D. Awduche, L. Berger, D. Gan, T. Li, V. Srinivasan, and G. Swallow. RSVP-TE: Extensions to RSVP for LSP Tunnels. RFC 3209 (Proposed Standard), December 2001. Updated by RFCs 3936, 4420, 4874, 5151, 5420.

[RFC3212] B. Jamoussi, L. Andersson, R. Callon, R. Dantu, L. Wu, P. Doolan, T. Worster, N. Feldman, A. Fredette, M. Girish, E. Gray, J. Heinanen, T. Kilty, and A. Malis. Constraint-Based LSP Setup using LDP. RFC 3212 (Proposed Standard), January 2002. Updated by RFC 3468.

[RFC3468] L. Andersson and G. Swallow. The Multiprotocol Label Switching (MPLS) Working Group decision on MPLS signaling protocols. RFC 3468 (Informational), February 2003.

[RFC3936] K. Kompella and J. Lang. Procedures for Modifying the Resource reSerVation Protocol (RSVP). RFC 3936 (Best Current Practice), October 2004.

[RFC4090] P. Pan, G. Swallow, and A. Atlas. Fast Reroute Extensions to RSVP-TE for LSP Tunnels. RFC 4090 (Proposed Standard), May 2005.

[RFC4203] K. Kompella and Y. Rekhter. OSPF Extensions in Support of Generalized Multi-Protocol Label Switching (GMPLS). RFC 4203 (Proposed Standard), October 2005.

[RFC4205] K. Kompella and Y. Rekhter. Intermediate System to Intermediate System (IS-IS) Extensions in Support of Generalized Multi-Protocol Label Switching (GMPLS). RFC 4205 (Informational), October 2005. Obsoleted by RFC 5307.

[RFC4206] K. Kompella and Y. Rekhter. Label Switched Paths (LSP) Hierarchy with Generalized Multi-Protocol Label Switching (GMPLS) Traffic Engineering (TE). RFC 4206 (Proposed Standard), October 2005.

[RFC4364] E. Rosen and Y. Rekhter. BGP/MPLS IP Virtual Private Networks (VPNs). RFC 4364 (Proposed Standard), February 2006. Updated by RFCs 4577, 4684.

[RFC5036] L. Andersson, I. Minei, and B. Thomas. LDP Specification. RFC 5036 (Draft Standard), October 2007.

[RFC5151] A. Farrel, A. Ayyangar, and JP. Vasseur. Inter-Domain MPLS and GMPLS Traffic Engineering – Resource Reservation Protocol-Traffic Engineering (RSVP-TE) Extensions. RFC 5151 (Proposed Standard), February 2008.

[RFC5283] B. Decraene, JL. Le Roux, and I. Minei. LDP Extension for Inter-Area Label Switched Paths (LSPs). RFC 5283 (Proposed Standard), July 2008.

[RFC5543] H. Ould-Brahim, D. Fedyk, and Y. Rekhter. BGP Traffic Engineering Attribute. RFC 5543 (Proposed Standard), May 2009.

Further Reading

[1] Ina Minei and Julian Lucek. MPLS-Enabled Applications: Emerging Developments and New Technologies, June 2008. ISBN 978-0470986448.

[2] Aviva Garrett. JUNOS Cookbook, April 2006. ISBN 0-596-10014-0.

[3] Matthew C. Kolon and Jeff Doyle. Juniper Networks Routers : The Complete Reference, February 2002. ISBN 0-07-219481-2.

[4] James Sonderegger, Orin Blomberg, Kieran Milne, and Senad Palislamovic. JUNOS High Availability, August 2009. ISBN 978-0-596-52304-6.

[5] Luc De Ghein. MPLS Fundamentals, December 2006. ISBN 978-1587051975.

5

MPLS Layer 3 VPN Migrations

If MPLS has revolutionized the internetworking industry over the last ten years, BGP-based Layer 3 VPNs (L3VPNs) are probably their most popular and widespread application. In fact, the ability to construct virtual and independent Layer 3 topologies over the same network in a scalable and flexible fashion and based simply on routing paradigms has been a major factor and business driver for many service providers and enterprise networks just to start implementing MPLS.

A common exercise over the last several years at many internetworks has been a migration from a traditional flat topology to a VPN-structured MPLS environment, not only to offer additional services but also as a security mechanism or for the sake of controlled and isolated management at each L3 VPN.

MPLS BGP-based L3VPNs have gained so much popularity that different interconnect models have been precluded and evolved over time. These inter-AS interconnect types have allowed flexible disposition of Layer 3 VPNs across different backbones and providers with more granular or scalable approaches to stitch them.

[RFC2547] (now evolved to [RFC4364]) started as an appeal to the concept of defining MPLS-based L3VPNs *to support the outsourcing of IP backbone services for enterprise networks*. The idea was to benefit from MPLS features to deploy independent VPNs over the same global architecture and to have the features be completely separate without a need for specific security constructs.

BGP-based L3VPNs are based on a so-called *peer* model, in which *Customer Edge* (CE) and *Provider Edge* (PE) devices talk to each other by means of a routing protocol or construct, which can be as simple as mutual static routes or even directly connected networks. In a nutshell, a unique Route Distinguisher (RD) is added to the prefix information to grant address uniqueness, and Route Target (RT) extended communities are also added to it to determine prefix imports on remote sites. This construct is transported precisely by means of a *AFI 1, SAFI 128* BGP NLRI. The power of this concept resides in the fact that it scales the known transport capabilities of BGP, combining the protocol with the MPLS service multiplexing premise.

Network Mergers and Migrations Gonzalo Gómez Herrero and Jan Antón Bernal van der Ven

BGP-based L3VPNs have also allowed the construction of multiple any-to-any, one-to-any, or one-to-one VPN communication models simply by defining import and export rules based on the RT. Just defining the proper policies is enough to allow or block unidirectional flows between sites!

Simply by having different routing constructs for each Virtual Routing and Forwarding (VRF) table, a given degree of security is automatically enforced: unless RTs match for both import and export policies, there cannot be flows between VRF sites without common RTs (for example, part of a common VPN or *extranet*). There is no need for additional security enforcement to achieve this objective, as opposed to traditional security-based VPNs. The beauty of this concept is that any security, authentication, or encryption mechanism can be implemented if needed inside each VPN on top of the isolation provided by BGP-based MPLS L3 VPNs.

This paradigm also offers the ability to multiplex and differentiate services when associated with the L3VPNs. Because several L3VPNs can coexist in the same network, different services can naturally be mapped to different VPNs (either meshed or in arbitrary relationships) with diverse class-of-service per-hop behaviors or forwarding treatment. Junos OS even offers enough resources to map these L3VPNs to different IGP topologies (thanks to *multitopology routing*) or to naturally different LSPs, just by setting intended next hops for these BGP L3VPN NLRIs.

All this flexibility has been the trigger to migrate and integrate legacy and flat networks into L3VPNs over a common infrastructure and to interconnect them across MPLS backbones. This chapter deals with the migrations related to MPLS BGP-based L3VPNs, focusing on integrating networks as BGP-based Layer 3 VPNs and migrating through different interconnect models.

5.1 Motivations for Layer 3 VPN Migrations

As L3VPNs became more and more popular, the need to interconnect and integrate them soon arose. Ever since [RFC2547], the need to design interconnect mechanisms with protocol artifacts became clear: L3VPNs needed to span beyond a single service provider to be even more effective than a legacy network. Imagine the operational and installation costs for a multinational enterprise network being expanded across countries and continents, and compare them with a L3VPN interconnected across service providers in different regions that reuse their infrastructure to offer another VRF termination! Looking at this, how many corporate customers have major service providers gained over the last years based on this architecture?

When considering the network architectures to be integrated into L3VPNs, several aspects must be analyzed, from political reasons to service requirements. As discussed with other types of migration, some of these motivations in a real-life scenario may arise individually or in various combinations.

5.1.1 Security enforcement

The ability virtually to segregate Layer 3 topologies over a common infrastructure is tremendously powerful. Compared with plain old legacy networks, this separation is natively enforced at Layer 3 with the construction of VPNs and the granular definition of RT import and export policies at each VRF.

This segregation can leverage security policies at a given enterprise or service-providing network by implementing an additional native isolation layer between topologies, that does not impede the application of security-based authentication, protection, privacy, or confidentiality rules among endpoints.

Also, the ability to build up arbitrary communication models enforces certain security premises to allow user-to-user communication in each direction.

5.1.2 Communication flow handling

Arbitrary communication models are allowed simply by defining RT import and export policies at each VPN site.

Point-to-multipoint, point-to-point, multipoint-to-multipoint, or any kind of arbitrary mesh can be built up based on RT distribution, and this virtual topology can be administered and designed independently of routing settings on CE domains or the core backbone topology.

When integrating a legacy network as a L3VPN, more flexibility in intersite communication can be added with this inherent behavior of L3VPNs, without modifying any previous routing among internal systems.

5.1.3 Service multiplexing

A key design feature of MPLS is support for multiplexing services within the same network infrastructure, mostly on a per-label basis. This feature works not only among different VPN types and MPLS applications, but also with diverse L3 VPNs.

The concept of a single L3VPN becomes diffuse when more than one RT can be matched for a VRF import or tagged at a VRF export policy. While a L3VPN can be thought of as a subset of sites and site membership is determined by those policies, a VPN-IPv4 route with its extended communities is the ultimate, deterministic piece of information needed to establish end-to-end flows. The same VPN-IPv4 route can therefore be exported and imported into several L3VPNs, and multiple services from different L3VPNs can thus be multiplexed upon the same common piece of information.

5.1.4 Route manipulation

By combining an inter-AS connection model with additional constructs, native routes from a VPN can be populated beyond a backbone, constrained within a subset of VRFs, or aggregated at certain VPN boundaries.

This variety of interconnect options provides an added value to a L3VPN when compared with a traditional plain network. Routing information can be adapted when stitching L3VPNs back to back or when advertising prefixes out of a VRF (blocking more specific routes, for instance), independent of the communication protocol between PE and CE.

5.1.5 Routing redundancy

VPN-IPv4 uniqueness is determined by the combination of a RD and the internal IPv4 prefix. While some routing protocols running in a legacy network just select the best path for forwarding purposes, the RD data structure can provide additional identification details bound to IPv4 prefixes: VPN-IPv4 routes sharing a common IPv4 prefix can be distinguished by using alternate RDs.

This fact consumes additional resources but allows quicker convergence for alternate paths (because a backup route is maintained) and grants additional redundancy in failure cases (because the backup route does not need to be readvertised from the original PE).

5.2 Considerations for Layer 3 VPN Migrations

Not only do L3VPNs need to be evaluated for potential technical aspects in migrations, but also the possible range of protocols to be used to interconnect CE domains with PEs needs to be evaluated, considering both their advantages and drawbacks, particularly in cases in which a flat network is being newly integrated as a L3VPN, because such integration also introduces changes in the CE domain routing protocol.

Likewise, different interconnect modalities for L3VPNs must be analyzed in terms of technical features that offer advantages or pitfalls for integration scenarios.

5.2.1 Layer 3 VPNs

By default, Junos OS assigns an MPLS label to each PE–CE connection and installs that MPLS route in the respective routing table with a next hop being that logical interface. Also, the VRF-configured RD is assigned to these routes and others received from each site so that it is attached to these IPv4 prefixes when they are advertised in the form of VPN-IPv4 (or inet-vpn in Junos OS parlance) BGP NLRIs. Note that despite having different routes in the same VRF, they all share the same common RD. VPN-IPv4 Routes are exported as dictated by policies that tag them with an arbitrary number of extended communities.

[RFC4360] enforces a given structure on extended communities by defining types and subtypes, and by allowing them to specify whether they can be transitive across AS boundaries. At the time of this writing, IANA has allocated the following standard community types related to L3VPNs as per this recommendation:

- RT (2-octet AS and IPv4-address specific): route-to-VRF membership;
- Router or Site of Origin (2-octet AS and IPv4-address specific): specific site injecting routing information;
- OSPF Domain Identifier (2-octet AS and IPv4-address specific): OSPF domain for route-to-LSA translation purposes;
- BGP Data Collection (2-octet AS): standard peer relationship, or geographical or topological information;
- Source AS (2-octet AS): source AS identifier for C-multicast routes in L3VPN multicast environments;

- OSPF Router ID (IPv4-address specific): unique identifier for each instance inside the OSPF domain;
- VRF Route Import (IPv4-address specific): determination for import of C-multicast routes in a VRF in L3VPN multicast environments;
- OSPF Route Type (Opaque): identification of original OSPF Area, metric, and LSA type for translation purposes;
- Color Extended Community (Opaque): payload identification related to BGP Encapsulation SAFI;
- Encapsulation Extended Community (Opaque): encapsulation protocol identifier related to BGP Encapsulation SAFI.

RTs ultimately determine how routes are exchanged among VRFs; more specifically, how they are exported to MP-BGP and to which VRFs IPv4 prefixes are imported from MP-BGP. This process is described in detail in Section 1.2.

PEs sharing a VPN need to exchange those NLRIs, either directly or through any kind of route reflector or ASBR. When a PE announces those NLRIs, it can optionally exclude routes for VPNs not shared with peers thanks to dynamic BGP capabilities related to Outbound Route Filtering (ORF). Even if this is not done, each PE may receive VPN-IPv4 routes, but only installs the derived IPv4 prefixes in the per-VRF table if RT policies from the respective instance require their import. These routes need to be valid as well and need to pass all sanity checks, including BGP next-hop reachability and LSP existence to this endpoint by means of an MPLS label distribution protocol or a directly connected external peer, so that both MPLS labels (the outer label for next-hop reachability, and the inner label for the VPN-IPv4 route) are installed in the forwarding plane.

Note that the standard BGP route selection process takes place with VPN-IPv4 routes. Attributes that are used as tie-breakers for path determination are commonly used. This behavior introduces further route selection variables in the case of duplicate routes or multihoming, such as distributing the IGP metric among PEs or use of local preference, that were not previously present in a legacy flat network.

The VRF forwarding table is checked for incoming PE traffic through any CE interface to determine whether it must be locally switched towards another CE (in the same or another VRF that is part of the same VPN, or even the default routing instance with some configurations), locally destined to the PE control plane (such as diverse PE–CE routing protocols), or MPLS-switched to any remote PE. If the destination is remote, both inner and outer MPLS labels are pushed onto the IPv4 packet to be forwarded through the backbone.

The inner VPN label remains untouched in the forwarding path between PEs, and the MPLS backbone switches the outer label as determined by the label distribution protocol. Junos OS performs *penultimate-hop popping* (PHP) by default, so when a packet reaches such a hop, the outer label is popped before forwarding the packet to the destination PE. At this remote PE, the inner VPN label is matched against the known VRFs and the IPv4 internal payload is forwarded to the proper CE. In the case of *ultimate-hop popping*, the destination PE has to perform all these actions at the same time along with popping the outer label.

Following this review of the natural forwarding and control-plane paths, it is time to consider additional items when integrating networks by means of BGP-based L3VPNs.

Route Distinguisher (RD) selection planning for Layer 3 VPN routes

Route Distinguisher (RD) selection determines address uniqueness.

A legacy IPv4 plain topology may not have duplicate addresses. In a VPN-IPv4 setup, uniqueness is determined by RDs, that are added to the IPv4 prefix, so the risk to overlap addressing space among VPNs is determined by IPv4 address overlapping and RDs. Despite having different RTs, if VPN-IPv4 routes from different VPNs share both RD and IPv4 prefix, they overlap on a route reflector, ASBR, or PE that is importing routes from those VPNs.

A proper network-wide RD allocation scheme yields no overlaps (Junos OS even enforces a commit error for local duplicate RDs across instances), but CE sites multihomed to different PEs or VRFs may propagate their internal prefixes in the MPLS cloud with a different or a common RD.

The tradeoff with RD planning is usually *convergence* versus *memory resource occupation*. Having more than a single VPN-IP4 prefix locally representing the same IPv4 prefix with diverse RDs ensures quicker path selection in case one disappears without requiring a BGP refresh, but comes at the cost of keeping additional memory reserved for all these paths.

RDs can therefore be used to construct multiple VPN-IPv4 routes for the same IPv4 prefix, placing the IPv4 prefix in different VRFs by physically connecting the CE to the VRF or using route-sharing techniques.

In such fashion, BGP can install multiple different routes to the same end system through different VPNs, and LSPs to each destination can be implemented by establishing different packet-handling or QoS premises. This is certainly one way for differentiating treatment of diverse flows towards the same destination.

Route Target (RT) planning for Layer 3 VPN routes

A careful Route Target (RT) plan allows the implementation of any arbitrary virtual topology, leveraging the VPN concept to allow connectivity among different sites in diverse ways.

This use of RTs is considerably more powerful than plain IP networks, because they allow the crafting of multiple unidirectional flows among arbitrary VRFs without the need to implement any filtering mechanism.

For instance, the same end system can simultaneously be a CE for different VPNs. It is a question of CE routing to the desired attachment circuit to each VRF and then constructing a virtual hub-and-spoke topology with RTs to virtualize access to the same end device through different VPNs. In this case, filtering or deep packet inspection techniques can be easily enforced in one VPN, while retaining legacy and plain connections through another.

On the other hand, BGP L3VPNs naturally provide isolation between VPN topologies. Security is enforced by orthogonal VPNs not sharing any kind of RT. Just by keeping control of VRF import and export policies that match or tag RTs in each direction, differential flow treatment can be achieved on a per-VPN basis.

BGP next-hop handling for Layer 3 VPN routes

As discussed in Section 3.2.10, the BGP protocol decoupled the physical layer from the control plane by introducing a level of indirection through the *protocol next hop*. When the reachability information contains MPLS labels, this level of indirection has some implications worth exploring for cases in which an intermediate device makes a change to the next hop.

Figure 5.1 splits the two directions of traffic between the top and bottom halves. The dotted lines correspond to the routing information exchanged over a multiprotocol BGP (MP-BGP) session. Solid lines provide actual MPLS labels used by packets travelling on the forwarding plane.

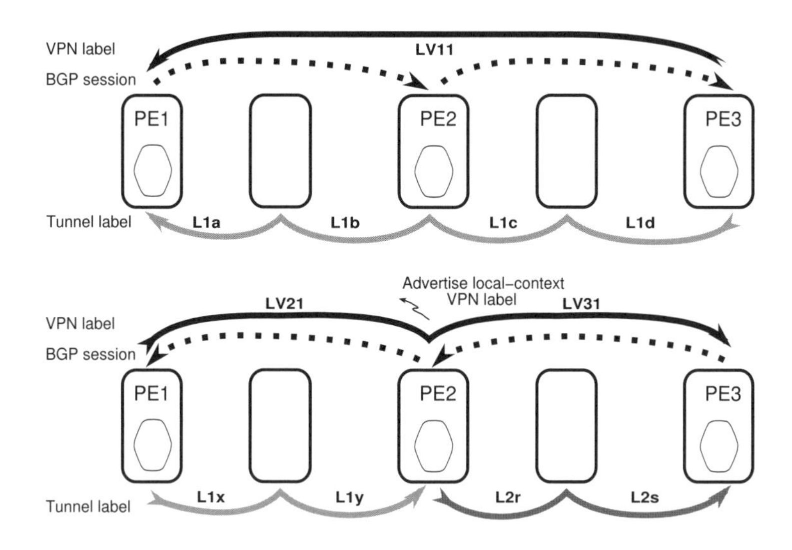

Figure 5.1: Effect of BGP next-hop change on labeled NLRIs.

The BGP session interconnects PE1, PE2, and PE3. BGP routing connectivity is accomplished from PE2 to both PE1 and PE3. Notice that the drawing does not show an AS border. PE2 might be performing the role of an IBGP route reflector at the control plane, where all devices are in the same AS. At the same time, PE2 could also be acting as a border router that has both an IBGP connection to one side and an EBGP connection (or even two EBGP connections, for each PE in different ASs) to the other side.

At the forwarding layer, transport labels $Ln\alpha$ correspond to LSP n that is built from a set of labels α. Signaling for the transport label (RSVP, LDP, or even Labeled-BGP), is omitted for clarity. The VPN labels *LVnm* correspond to the VPN labels as advertised by PE n. Only one label for each VRF is shown for illustration purposes.

The top half of Figure 5.1 illustrates a typical interconnection between two PE devices in an MPLS L3VPN network. Prefixes received on PE3 from PE1 via PE2 maintain the protocol next hop intact. The bottom half depicts the reverse situation, but with PE1 receiving prefixes from PE3 via PE2, with PE2 changing the BGP protocol next hop.

When a BGP protocol next hop is modified, the *context* of the information changes along with it. MPLS label values are meaningful only to the originator of the label (PE1 for LV11 in the top half; PE3 for LV31 and PE2 for LV21 in the bottom half). Changing the originator of a BGP update through a protocol next-hop change for an MPLS route requires the new advertisement to have an MPLS VPN label that is understood by the new originating router. If PE2 were to change the next hop to self while keeping the VPN label LV31 intact, that existing label may be misinterpreted because it might already be allocated to another service within PE2, or it might be outside of the label allocation capabilities of PE2.

Hence, the readvertising router has to allocate a *new* label and bind it to the old MPLS information in the MPLS forwarding table. At the forwarding layer in the PE1-to-PE3 direction, the transport LSP is split into two segments, requiring PE2 to perform proper MPLS label *swap* and *push* operations for incoming traffic with label LV21, namely *swap* to LV31, *push* L2r (transport to PE3).

From a pure MPLS switching perspective, the VRF present in PE2 plays no role in the discussion.

Maximum Transmission Unit (MTU) dimensioning for Layer 3 VPNs

Stretching forwarding paths inside a L3VPN requires proper Maximum Transmission Unit (MTU) dimensioning.

MPLS labels are four octets each. A common human error is lack of proper MPLS MTU adjustment inside the core to comply with internal IPv4 MTUs that may be implemented at the PE–CE interfaces or beyond the CEs. This problem is exacerbated when more MPLS labels are added to the stack, with MPLS transport overlay models or *fast reroute* implementations, for instance.

Another common failure arises when an L3VPN is extended beyond a single backbone with unified administration. When interconnecting a L3VPN with other networks, it is important to identify the internal MTU supported within each L3VPN. In the end, this factor is independent from the selection of one or the other inter-AS option for the L3VPN integration and depends rather on proper MTU dimensioning in each network.

5.2.2 RIP as PE–CE protocol

Routing Information Protocol (RIP) is a distance-vector protocol that uses the *Bellman–Ford* algorithm for route selection. It is the origin of all routing protocols and introduced the concepts of route information distribution and path computation among systems within a common domain.

Naturally, RIP has evolved to cope with modern features such as authentication and classless routing (added in RIP version 2). Although it has been clearly superseded by OSPF and IS–IS in complex networks, RIP still presents advantages in some environments because of its simplicity and low bandwidth overhead.

Many network topologies consider RIP version 2 as an adequate fit for a PE–CE protocol because of these reasons. When simple dynamic route distribution between PE and CE is required, without a strict requirement on convergence times or complicated route transport, RIP version 2 can be a valuable option.

Another typical scenario for RIP version 2 utilization includes low-end CE devices that cannot perform any kind of dynamic routing with other protocols such as BGP or OSPFv2. Because of its simplicity and easy configuration, RIP version 2 provides dynamic route advertisement, which is an advantage over simple static routes.

RIP implements different basic loop-prevention techniques, which can be considered as inherent from distance-vector protocols and less powerful than local topology computation performed with link-state routing:

- The basic *Count to Infinity* concept sets a maximum metric of 16 to represent *infinity*. Because the metric is increased as the route is populated from one neighbor to another, it is assumed that reaching such threshold means that the Update has been looped.
- The classic *Split Horizon* paradigm prevents routes learned from a neighbor from being sent back to that neighbor in own originated Updates.
- *Split Horizon with poison reverse* evolves *Split Horizon* to advertise those routes back to the neighbor they were learned from, but with their metrics set to *infinity* to immediately break potential loops.
- *Triggered Updates* in RIP attempt to speed up convergence and to correct mutual deception situations among more than a pair of routers by asynchronously sending route updates when that route changes. These updates are sent in addition to the regular RIP Updates, which also act to refresh the routing table.

Such simple mechanisms are inherent to RIP version 2, as defined in [RFC2453], and do not consider any other interactions.

However, when using RIP as the PE–CE protocol, another major consideration is needed in multihomed CE domains. As compared with other protocols that consider inherent loop-prevention mechanisms when injecting routes towards the CEs, RIP lacks any similar concept. In fact, the same problem appears in plain topologies without MPLS L3VPNs when mutual route redistribution between RIP and other protocols is performed at more than one site: RIP relies on the previous constructs for loop prevention, but cannot natively notify if a route is internally or externally injected.

Consider our well-known topology with a couple of multihomed VPN sites, as shown in Figure 5.2.

Because the default preference in Junos OS for RIP is 100, when using a standard configuration, RIP routes are preferred as active paths when compared with BGP, whose default preference in Junos OS is 170. In this scenario, the following sequence explains a first loop creation, with the inherent redistribution due to route advertisement to a CE in a L3VPN environment:

- *Livorno* advertises VPN-IPv4-redistributed routes to the CE domain as RIP Updates.
- *Havana*, as a parallel PE in the same segment, selects the RIP Updates as active paths, rather than the original VPN-IPv4 routes.
- *Livorno* is again eligible to import these VPN-IPv4 routes from *Havana* as active paths, instead of the original version. This can be fixed, though, with standard BGP mechanisms such as using the extended *Site of Origin* community and avoiding such imports.

This means that *Havana* prefers the RIP routes to the original VPN-IPv4 routes from another remote site. Note that this could happen just as easily and in a similar fashion in the other direction, depending on where VPN-IPv4 routes are received first.

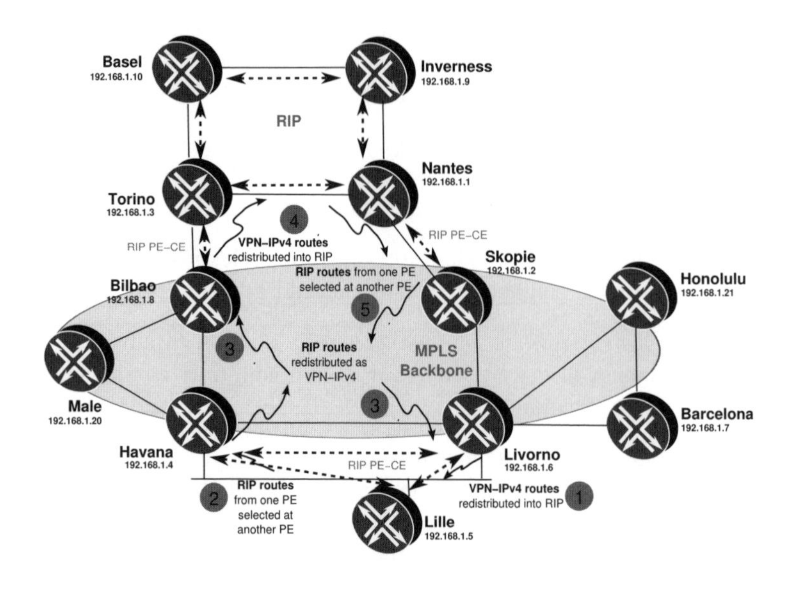

Figure 5.2: Routing *loop* formation with RIP as PE–CE.

While this path-selection behavior can easily be fixed inside the VPN site simply by blocking updates on the forwarding path (using a Junos OS firewall filter) or on the control plane (using Junos OS RIP import policies), the situation is more convoluted if these PEs that are servicing a common CE domain are not directly connected (and therefore, no source IP address for RIP updates can be matched). Following the previous sequence in the same scenario:

- *Bilbao* advertises VPN-IPv4-redistributed routes to the CE domain as RIP Updates.
- *Torino* detects such routes from the PE and populates RIP Updates inside the CE domain, increasing the metric.
- *Nantes* receives a RIP Update with those routes, increases the metric value, and propagates the route to *Skopie*.
- *Skopie*, as the parallel PE in the VPN sites, selects the RIP Updates from *Nantes* as active paths, rather than the original VPN-IPv4 routes.
- *Bilbao* is again eligible to import these VPN-IPv4 routes from *Torino* if no extended *Site of Origin* communities or similar mechanisms are used.

Note that in this case, no filters or policies can directly match on the source IPv4 address for the RIP updates, because they are propagated through the CE domain.

Inside the VPN, a possible loop-mitigation mechanism is based on using *route filtering* or *administrative tagging* to distinguish routes that have been leaked from the MPLS backbone from those that have been natively injected by the CEs.

5.2.3 OSPFv2 as PE–CE protocol

[RFC4577] was born with a migration scenario in mind, as described in the introduction and other sections in this RFC. Using OSPFv2 as a PE–CE protocol eases transitional techniques and does not require the CE domain to support or operate any protocol other than OSPFv2. As OSPFv2 is a link-state IGP based on principles that do not exactly match a BGP-based L3VPN scenario such as transparent flooding, hierarchy definition, and internetwork or intranetwork route preferences (standard route redistribution into OSPFv2 yields Type 5 and 7 LSAs), additional concepts are needed to simulate conditions of a plain OSPFv2 domain, abstracted from its mapping onto a L3VPN. The idea is to define structures that are independent of the overlay OSPFv2 domain topology.

A way to do this is to consider PEs as Area 0 ABRs by default, that is, members of the *backbone*. This area is actually a kind of *super-backbone* that can interconnect different sites from the same domain that are configured to be in Area 0. This means that in the case of CE attachment circuits, PEs are always considered to be ABRs and hence are natural exit points from that non-backbone area. In the case of CEs that are part of the backbone, they see PEs as other router members from the same area and can inject any type of summary LSAs. Other remote VRFs can be included and can be configured in the backbone. In this situation, PEs play a different role from their standard one, being able to inject Type 3 Summary LSAs for the received VPN-IPv4 routes to the locally attached CEs.

Effectively, this abstraction is a powerful migration application: the ability to partition and segregate an OSPFv2 backbone across VRFs without modifying route selection (because routes from different sites remain intra-area)! An alternative covered in [RFC2328] is the deployment of *virtual links*, but they are neither as scalable nor as flexible as an MPLS backbone with multiple L3VPNs. The *Site of Origin* extended community enforces a generic mechanism in MPLS VPNs for avoiding reinjection of routes towards a multihomed site and it may also apply here for avoiding partition repair if needed, but without its application, OSPFv2 networks can be reassembled just by modifying OSPFv2 extended communities and attributes.

From a transition perspective, those resources are tremendously useful to craft configurations and setups, and they add versatility to integration activities. Of these resources, the following concepts are important with regards to migrations.

Domain Identifiers

OSPF domain identifiers are extended communities that represent joint domains.

The glue that binds all VRF sites belonging to the same common OSPF network is the *domain identifier*. This extended community, which is attached to the routes, uniquely identifies the network to which the routes belong. Upon receiving such routes, a PE checks the attached attribute against the locally configured *domain identifier*. If there is a match with any of the local *domain identifiers*, the implicit significance is that they are intended to belong to the same OSPF network and LSA injection to the attached CEs is determined by the *route type*. If they are different, their origin is understood to be different and they are redistributed in any external LSA form.

Note that the *domain identifier* also opens the door for another migration scenario: the ability to determine local LSA types and route selection at each VRF site independently of original LSA types! Simply by attaching arbitrary *domain identifiers* when exporting VPN-IPv4 routes and planning imports matching different *domain identifiers*, route decisions at the CEs can be influenced by distinguishing different domains. This overrides the basic principles of uniform LSA types across the same area.

OSPF route type

OSPF route types are extended communities that serve as an identifier for originating area, LSA type, and external route type.

Once *domain identifiers* are matched and it is determined that the incoming route belongs to the same network, a translator PE needs to understand how this route was originally flooded in OSPFv2 by its root device. This information is necessary so as to craft the proper LSA at remote sites that are importing this route. The OSPF *route type* is another extended community that encodes the originating area identifier, LSA type, and external route type when needed.

The translation algorithm is quite simple:

- If the route is a consequence of an original Type 5 or 7 LSA (meaning that it was originally injected as an external route), the LSA type remains unchanged.

- If the route was originally part of an internal Type 1, 2, or 3 LSA, it is advertised to the local attachment circuits as a Type 3 LSA, meaning that summarization is performed for an intra-area or inter-area network as an inter-area route.

Note that Type 4 LSAs are never translated into VPN-IPv4 routes because their significance is futile: PEs always appear as ASBRs in each local VRF, and the information for the original ASBR injecting the route is natively lost in translation.

Also, the internal translation is actually independent of the original area identifier. It does not matter whether it is the same or a different area; it is always injected as a summary LSA and is presented as internal to the network.

As a rule of thumb, the global translation premise, once domains are matched, can be summarized as *keep routes that were originally external routes as external, and keep ones that were originally internal as internal summaries*. This behavior is consistent with the default OSPFv2 route preference, in which intra-area routes are preferred over inter-area routes, and these latter are preferred over external routes.

Sham links

Sham links are virtual intra-area links between peer VRFs.

It may even be necessary to inject native intra-area routing information through the backbone. This becomes especially apparent when backdoor links among CE sites exist in a topology and traffic needs to transit the backbone; that is, for intra-area routes to point to next hops in the backbone rather than through the backdoor links. Without a specific mechanism, metric adaption does not help because intra-area routes always win against inter-area, and the most preferred translation from VPN-IPv4 routes is in the form of Type 3 LSAs.

A concept called *sham links* comes to the rescue here. *Sham links* are unnumbered, point-to-point virtual constructs appearing in the PE VRF Router LSA as intra-area interconnects between VRF sites and for any eligible area (including Area 0). In a nutshell, they are virtual links between PEs inside the VPN and represent an intra-area construct. To avoid unnecessary flooding through the backbone, sham links may behave as *OSPF demand circuits* because only topological changes really need to get populated in such a scenario.

This machinery is perceived by CEs as intra-area links eligible for forwarding when compared with any other backdoor links among CE sites. The intra-area versus inter-area dichotomy can then be easily solved by adjusting metrics.

From a migration perspective, an adequate transition scenario from a legacy plain network to a L3VPN can be tuned even with a fine-grained approach of first creating a *sham link* between affected sites before decommissioning links among CEs. Rather than tearing down a backdoor link to force traffic to go through the MPLS backbone, a soft move can be performed by setting up the *sham link* and making it preferred against the internal site connection that is to be dismantled.

Another useful application in migration scenarios occurs when considering multihomed sites in which internal networks may be mutually injected to the CEs as Type 3 LSAs, but the intention is to force given flows transiting a single upstream PE. Here, a simple *sham link* between that PE and the remote site may lead to that setup, because CEs perceive that link as part of that Router LSA and hence as an intra-area route.

DN bit and VPN Route Tag

The DN bit inside the LSA Options field and the VPN Route Tag, which reutilizes the standard External Route Tag field from Type 5 and Type 7 AS and NSSA external LSAs, are identifiers for loop prevention within the backbone.

[RFC4576] grew up in parallel to [RFC4577] to provide a powerful signaling mechanism for LSAs to represent leaking from VPN-IPv4 routes into a given CE domain. It cannibalized the high-order bit from the LSA Options field (unused since its definition at [RFC2328]) to distinguish route-leaking into a CE domain at reinjected routes in the form of LSA Types 3, 5, or 7 as shown in Figure 5.3. As this remains unmodified per natural OSPFv2 LSA flooding, a redundant PE providing upstream knows that routes represented by such LSAs have been leaked from VPN-IPv4 NLRIs and do not need to be reinjected again.

Previous draft versions of [RFC4576] based the route-leaking signaling on a modification of the AS External Tag field in Type 5 and 7 LSAs. In fact, [RFC4577] defines procedures for configuring the AS External Tag and states that default values derived from the transport MPLS backbone AS should stand in for the transport through a given network. The intention of the default value definition is to help in identifying the transit MPLS backbone AS, as illustrated in Figure 5.4, so as to identify precisely how the VPN-IPv4 route has been transported and that it has indeed been transformed into an NSSA or AS External LSA towards the attached CEs and PEs in the VRF, without leaving this concept to a vendor-specific implementation. The exception to the rule occurs if the AS number is 4 bytes long. In this case, [RFC4577] requests that the value must be configured, because there is not enough space in the existing AS External Tag field for a unique AS identification.

Because the AS External Tag field is missing in Type 3 LSAs, problems were created from reinjecting already leaked intranetwork routes (imagine different *domain identifiers* at both redundant PEs, for instance). To be fully compliant and backwards compatible, [RFC4577] enforces checking and setting of both variables to indicate route-leaking and

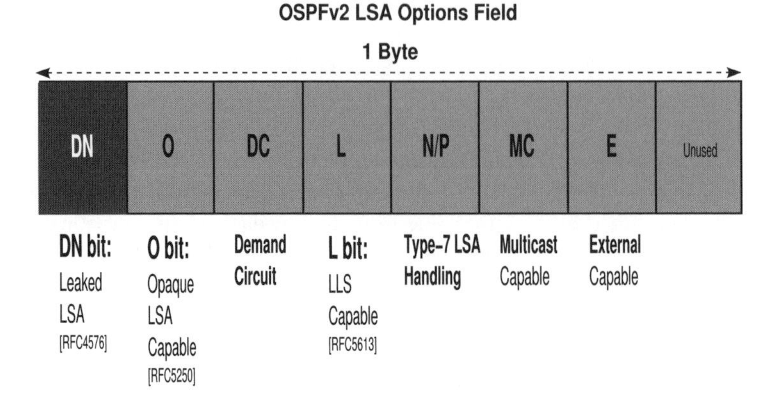

Figure 5.3: DN bit in LSA Options field.

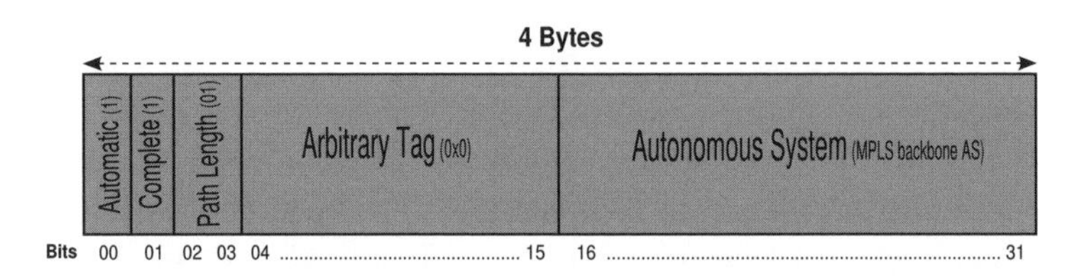

Figure 5.4: Default VPN Route Tag generation depending on MPLS backbone AS.

to avoid reinjection (Junos OS enforces such default behavior with the DN bit and the default derived values for the *VPN Route Tag*).

From a migration perspective, this loop prevention mechanism is coherent in a link-state IGP world, but goes against the principles of an MPLS L3VPN Option A interconnect: routes redistributed from a PE to the interconnect VRF are tagged with the *DN bit* (and with a *VPN Route Tag* if considered external), but this same DN bit avoids being injected by default into MP-BGP at the other PE. Therefore, when interconnecting backbones through an OSPFv2 CE domain or when the intention is to populate specific routes upstream to a certain PE, extra care and resources must be considered to avoid tagging or to allow route injection. Likewise, this route redistribution procedure is not a safeguard against interaction through any other protocol (for example, multiple redistribution) and does not cover other types of route that may be injected in OSPFv2 but that do not arrive in the VRF as VPN-IPv4 routes (such as local static routes).

OSPF metric

The OSPF metrics used in LSAs for readvertised routes are derived, by default, from MED attribute values in VPN-IPv4 routes, as indicated in [RFC4577]. If no MED is present, a predefined default is used.

From a migration perspective, it is important to ensure a smooth transition of the original metric to MED at the original advertisement site or adequate tuning of the BGP attribute. Very likely, this attribute may already be the tie-breaker in the remote VRF when selecting the active path for the VPN-IPv4 route (independent of a common or distinct *route distinguisher*) prior to translation into an OSPFv2 LSA. Accurate MED setting is another step needed to guarantee that forwarding in the MPLS backbone takes place between the same CE sites as prior to a migration, in a plain OSPFv2 topology.

Figure 5.5 illustrates a basic decision flowchart to manipulate attributes for VPN-IPv4 routes derived from the original Type 1, 2, or 3 LSAs at the root site (that is, from internal routes in the OSPF domain).

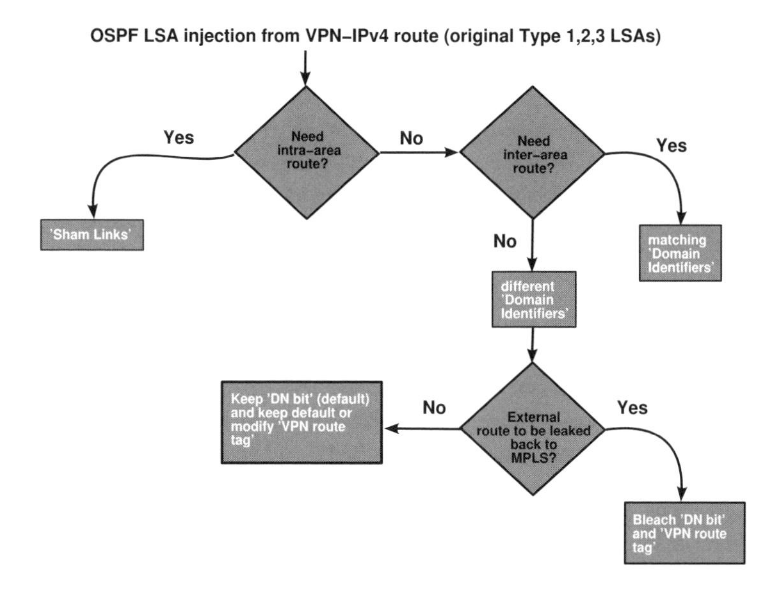

Figure 5.5: Design alternatives to inject LSAs in a L3VPN CE domain originally derived from OSPFv2 internal network routes.

5.2.4 EBGP as PE–CE protocol

Nowadays, *external BGP* is probably the most extended and commonly used interconnect option between PE and the CE domains.

EBGP between PEs and CEs is simple, because it does not require specific redistribution policies among protocols, and because it is scalable as a transport protocol, offering the same variety of TE options to steer traffic to and from each customer internal domain.

Despite not requiring any redistribution between protocols, a route transformation certainly occurs when IPv4 unicast routes from the CE domain are transformed into VPN-IPv4 routes with a RD and relevant extended communities are attached to the NLRI.

Such route transformation includes changes in BGP data structures and attributes that need to be considered from a migration perspective:

- RDs define address uniqueness across the MPLS backbone or even, in the case in which routes are advertised beyond a single transport, the AS.
- AS_PATH is expanded with the transport AS when it is advertised from one VPN site to another.
- If different VPN sites belong to the same AS (for instance, when splitting a legacy network into different VPN sites), the inherent BGP AS_PATH loop detection mechanism discards updates from other sites if no manipulation is done, because the local AS is visible in the BGP AS_PATH at border CEs.
- Depending on the specific implementation, extended communities may be propagated from PE to CE, even though the intention is to handle imports and exports out of every VRF. This is particularly relevant in the case of inter-AS Option A, in which PEs may be directly confronted with EBGP (see Section 3.2.4).
- Certain BGP attributes that may be present in IBGP no longer transition through the MPLS backbone because of the AS boundaries. This is especially relevant for LOCAL_PREFERENCE, a well-known attribute used to define TE policies *inside* an AS.
- IGP metrics towards the route next hop are a common path selection tie-breaker in most modern router implementations. Because of the implicit *next-hop rewrite* on EBGP sessions, IGP metrics for these destinations change when transiting the MPLS backbone.

When integrating a plain BGP network as part of a L3VPN, despite keeping the same protocol as the communication vehicle through the L3VPN backbone, such modifications in the protocol setup need to be analyzed. In the end, newer EBGP sessions interact with the existing setup to inject routes that originally were internal routes.

5.2.5 IBGP as PE–CE protocol

[draft-marques-ppvpn-ibgp-00] introduces an approach to transparently exchange BGP routes among CE sites.

As compared to using standard EBGP as the PE–CE protocol, in which the transport domain AS is involved (or overwritten) in the AS_PATH and other attributes such as LOCAL_PREFERENCE, ORIGINATOR, and CLUSTER_LIST get lost when transiting AS boundaries, [draft-marques-ppvpn-ibgp-00] offers an approach to blindly encapsulate and decapsulate the attributes in a new flexible attribute, *ATTRSET*, which acts as a container for the originally inserted attributes with their values (except for the NEXT_HOP, MP_REACH, and MP_UNREACH attributes). This container acts as an attribute stack, into which the original attributes are pushed by the advertising PE. They are popped at each VPN site when performing path selection in each VRF RIB, and at each site the PE's real NEXT_HOP is set before sending the IPv4 native NLRI to the end CE.

Figure 5.6 shows the handling of this attribute when routes are propagating in a sample topology.

This method provides numerous benefits when using BGP as the PE–CE protocol because, although the peer model with the PE router prevails, CEs can distribute their originally

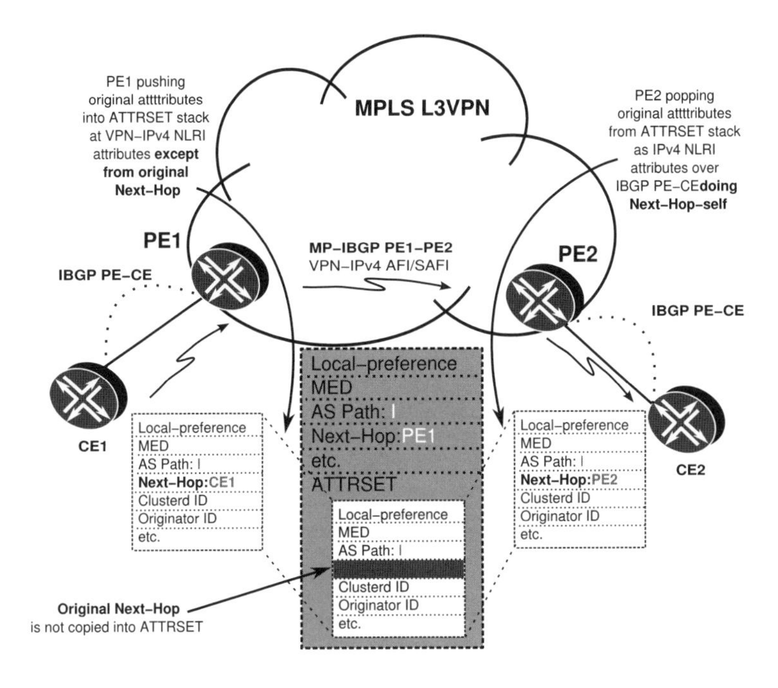

Figure 5.6: BGP attribute handling per draft-marques-ppvpn-ibgp-00.

intended TE attributes in a transparent fashion, opaque to the transport in the MPLS backbone. Considering a legacy plain network whose main route advertisement vehicle is IBGP, this feature is a powerful tool to expand regions into VRF sites easily without modifying the route distribution policies. Indirectly, PEs act as route reflection points for all routes arriving by means of VPN-IPv4 NLRIs towards each CE attachment world and therefore multiplex routing information from all VPN remote sites.

From a migration perspective, note that this attribute handling is completely orthogonal to whatever IGP or BGP hierarchies are run at each site. It merely provides transparent route redistribution, end to end, through the VPN backbone.

5.2.6 Inter-AS options

Interconnecting third-party networks using L3VPNs is addressed by simulating those third-party networks as CE devices. To minimize the impact of such a migration, the PE–CE protocol is adapted to the third-party network. However, in some cases the customer is another service provider, and the interconnection arrangement is a way to partner for additional reachability or redundancy.

Inter-AS connectivity established using any of the pre-existing PE–CE protocols follows the so-called *Option A* interconnect as defined in Section 10 of [RFC4364]. The same specification provides two additional options to improve scalability, based on the forwarding plane traffic on the interconnect: *Option B* for single-stacked MPLS traffic, and *Option C* for double-stacked MPLS traffic. These are shown in Figure 5.7. A short comparison of the various options is shown in Table 5.1.

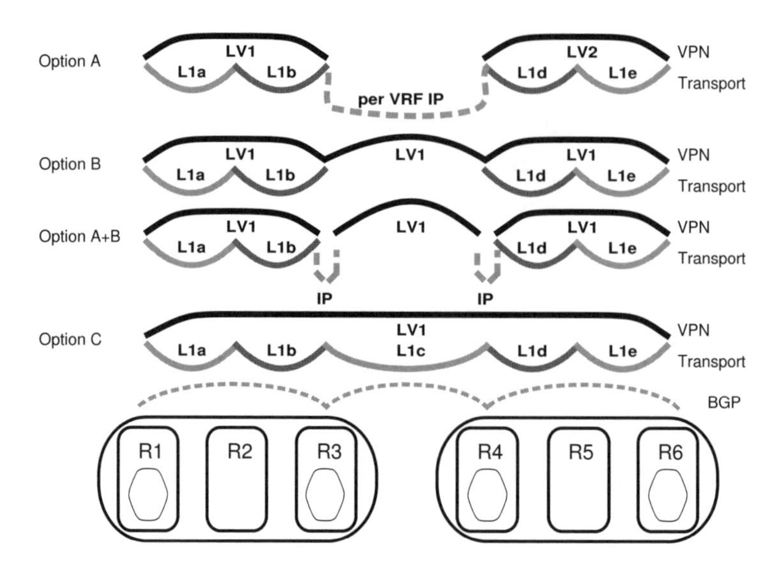

Figure 5.7: Inter-AS L3VPN interconnection alternatives.

Table 5.1 Comparison of L3VPN inter-AS interconnects

Consideration	IP (Option A)	inet-vpn (Option B)	labeled-inet (Option C)
IP features	yes	no	no
MPLS	no	1 label	2 labels
Interconnect	multiple	single	single
Scalability	per IP prefix	per VPN	per PE

Option A: IP packet as the PE–CE transport

In [RFC4364] Section 10, Option A, the interconnecting ASBR router behaves like a regular PE device. This router has complete control over redistribution and policy, and prefix aggregation within a VRF is readily available. All regular *IP features* available on a regular PE device (such as packet filtering, per-VRF statistics, and logging) are also available at this border point.

Customer-specific class-of-service requirements can be applied to each subinterface at the interconnection. Each customer can get differentiated IP precedence marking and traffic shaping. For instance, partnering with international VPNs can have traffic shaping enabled on the interconnect, limiting the amount of total off-net traffic.

Any VPNs that require interconnection must be present, which poses a scalability concern. Besides, a different PE–CE interface for each VRF can be cumbersome from a provisioning perspective, especially when the Layer 2 media is not directly connected. A typical example is the use of different logical circuits represented by combinations of VPIs and VCIs in an ATM network.

Option B: Single stacked MPLS-labeled packet as the PE–CE transport

In [RFC4364] Section 10, Option B, a first level of scalability is added by exchanging VPN-labeled traffic. Similarly to a provider with two PEs directly connected to each other, the two ASBRs exchange inet-vpn prefixes but have no transport label. Being two different providers, the BGP interconnection is of an external nature, and additional provisions for transiting of VPN traffic have to be taken into account. A similar setup can be attained by using Route Reflection for internal connections.

In an Option A, the semantics for the RT are lost because the exchange does not take place over MP-BGP. The interconnecting subinterface provides the binding between routes and target VRF. Conversely, there is a true VPN blending in Option B, requiring agreement on the RT tagging policy to be used for interconnected VPNs. Two approaches are common practice:

- Defining a new RT for shared VPNs. This requires tagging import and export policies for all PEs where this VPN has a corresponding CE attachment through a VRF, irrespective of the carrier domain. This makes sense for newly provisioned VPNs where the multidomain connectivity requirements are known. Typically, the L3VPN NLRI ends up with more than one RT community; the intradomain RT community is not changed, but rather, an additional community representing the intercarrier policy is specified.

- Provide translation at the VPN border. RT community translations are set up for selected VPNs at the ASBR as part of the inter-AS peering agreement. Most common designs centralize VPN reachability information in a couple of redundant L3VPN Route Reflectors, which become an interesting place to conduct the RT community mapping. Section 5.6.5 in the case study uses this strategy.

Option C: Dual-stacked MPLS label as the PE–CE transport

In the approach dubbed Option C, the VPN-labeled traffic is not directly connected and requires some means of end-to-end transport between the PEs. For an extensive discussion about the approach to build an MPLS transport tunnel between PEs and ASs, see the case study in Chapter 4.

The Option C alternative is used when extending all VPNs in a network or when moving PEs between networks, to minimize the impact. Each domain has labeled paths to the PEs in the other domain, so a PE homes VPNs for any of the domains in a seamless way. This is a good choice when integrating networks or moving PE routers between networks, but a good level of coordination must exist between the two groups, including the NOCs, perimeter security, management, accounting, and planning.

Two alternatives are defined to interconnect using Option C:

- Labeled-BGP. This alternative leverages the flexible policy knobs of BGP to constrain advertisement or further protect each peer from instabilities caused by the other peer.
- LDP plus an IGP of choice. Because LDP relies on a routing protocol for its label advertisement decision process, it is also required to combine FEC advertisement with an additional routing protocol.

Option A+B: Single-stacked MPLS with IP functionality

Various proposals (now expired) have been made requesting standardization for a scenario that is a combination of Options A and B, such as [draft-kulmala-l3vpn-interas-option-d]. The described proposal requires behavior changes on the ASBR device, but does not otherwise modify the interaction between systems (no new messages are defined, and existing behaviors are maintained).

In this modified approach, the end-to-end MPLS path between two PEs is broken at the ASBR border, triggering a lookup in the underlying application layer. The goal is to retain the benefits of an Option A interconnection from the application visibility perspective while leveraging the scalability enhancements provided by the Option B MPLS interconnection.

Application Note: Scaling L3VPNs with hierarchical PE

A variation of the inter-AS approach proposed as Option A+B discussed in Section 5.2.6 can be applied within the same AS, to obtain the benefits of an MPLS interconnect for scalability, while leveraging IP functionality for aggregation purposes.

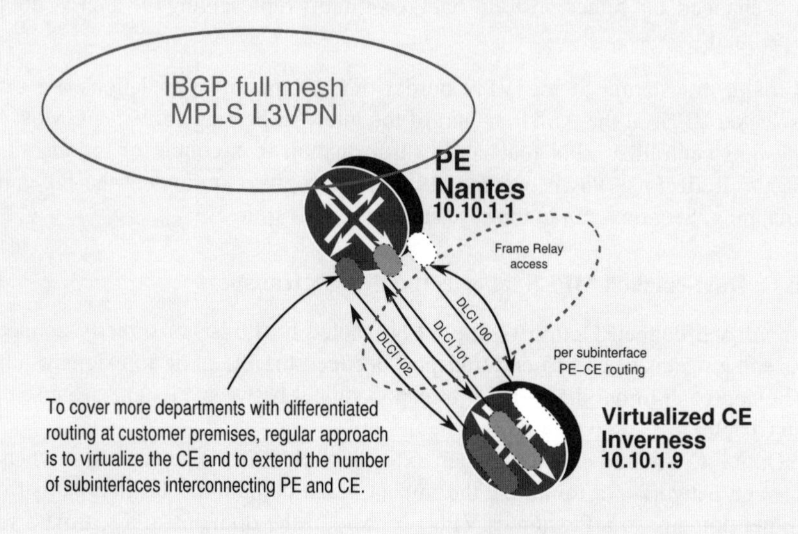

Figure 5.8: Adding low-capacity PE through aggregation.

Looking at the MPLS L3VPN network in Figure 5.8, the CE router *Inverness*, with virtualization capabilities, is a low-cost strategy because the customer requires separate connectivity for each department. The initial approach is to use *virtual-router* instances to segregate each department's routing information and to use a separate logical connection to the PE router *Nantes*. Because the CE router cannot handle all VPN information, the edge PE Nantes performs prefix aggregation, advertising only selected route blocks.

A roadblock that might prevent general deployment of this solution is the existence of a Layer 2 infrastructure (Frame Relay, ATM, or Metro Ethernet) in the interconnection. Adding a new VRF on the PE and a new virtual router on the CE also requires provisioning of a new subinterface, with the usual lead times and operational coordination.

The approach of introducing the Customer Premises device in the MPLS L3VPN (with full visibility) is very attractive from the provisioning point of view, because an MPLS pipe substitutes for the multiple set of PE–CE connections. However, this may not be desirable depending on the capabilities of the low-end CE device. As an added constraint in this example, the router *Inverness* is a small-capacity device, which can serve only a minimal set of VRFs and does not cope well maintaining an IGP of hundreds of PE devices.

The careful reader may have noticed that the first approach maps nicely to an Option A interconnect, because it is a regular PE–CE interconnection. Hiding full visibility of the whole MPLS L3VPN network while benefiting from the Option B advantages can be achieved by treating both router *Nantes* and router *Inverness* as ASBRs of an Option B interconnect. If router *Nantes* advertises itself as the next hop for all announcements, router *Inverness* does not require full topological information.

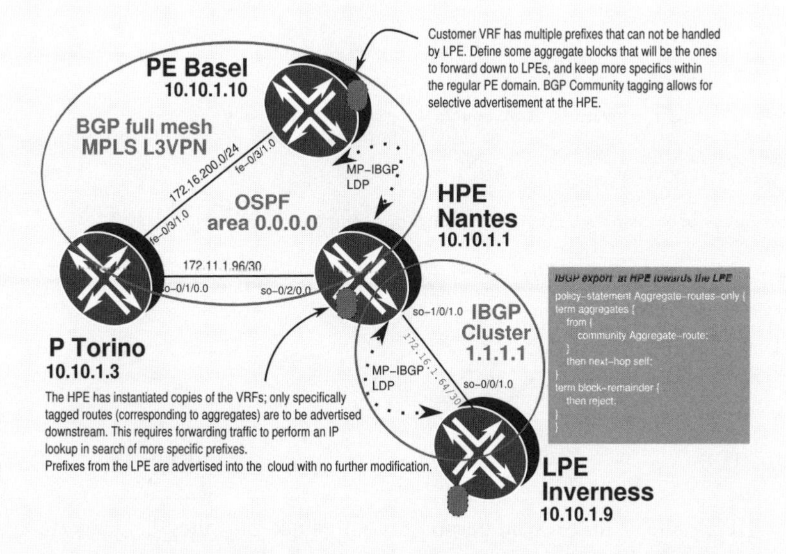

Figure 5.9: Hierarchical PE concept with route reflection.

Although the regular Option B is meant to interconnect ASs through EBGP, this Application Note provides a feasible alternative within the AS by means of route reflection, as shown in Figure 5.9. Router *Nantes* becomes a *High-end PE* (HPE) in the hierarchy, with router *Inverness* acting as the *Low-end PE* (LPE). Both HPE and LPE require inter-AS ASBR capabilities. In addition, the HPE should be able to trigger an IP lookup for traffic that follows the aggregate advertisements. In Junos OS, by default, instantiating a VRF and allocating a Table Label (using vrf-table-label or Virtual Tunnel interfaces) is enough to enable this functionality.

Note A table label VRF advertises the same table label over inet-vpn for *all* routes, whether the routes originated locally within the VRF or through readvertisement of received inet-vpn routes belonging to the VRF.

A combination of route reflection and BGP export policy controls what has to be advertised downstream to the LPE. Using BGP community tagging, the aggregates do not need to be sourced locally, giving a single point of control at the ingress PE that applies to all HPEs. If desired, use of route-filter in policy paired with proper extended RT communities can constrain advertisement of specific prefixes.

As the configuration in Listing 5.1 shows, a single inet-vpn session is established between WAN interface addresses to the HPE. Because of the inherent multihop nature of IBGP, it is necessary to leak these interface addresses into the inet.3 table properly to resolve received prefixes. Listing 4.21 in Chapter 4 on Page 338 provides more detail.

Listing 5.1: LPE Inverness setup

```
1  user@Inverness> show configuration protocols bgp
2  group IBGP {
3      type internal;
4      local-address 10.10.1.66; # Local WAN address on LPE
5      family inet-vpn {
6          unicast;
7      }
8      neighbor 10.10.1.65; # WAN address at LPE
9  }
10 user@Inverness> show configuration routing-options rib-groups
11 int-to-inet3 {
12     import-rib [ inet.0 inet.3 ];
13     import-policy interface-to-inet3; # Leak WAN interface to inet.3
14 }
15 user@Inverness> show configuration routing-options interface-routes
16 rib-group inet int-to-inet3;
17
18 user@Inverness> show configuration policy-options policy-statement
19     interface-to-inet3
20 term LPE-nh {
21     from interface so-0/0/1.0;
22     then accept;
23 }
24 term no-more-inet3 {
25     to rib inet.3;
26     then reject;
27 }
```

At the HPE router *Nantes*, configuration requires use of policy language to constrain advertisements to the LPE Inverness. A sample policy is shown in Listing 5.2.

Listing 5.2: Use of policy at HPE to constrain advertisement to LPE

```
1  user@Nantes> show configuration protocols bgp
2  group IBGP {
3      type internal;
4      local-address 10.10.1.1;
5      family inet-vpn {
```

```
 6         unicast;
 7     }
 8     export nhs;
 9     inactive: neighbor 10.10.1.3;
10     neighbor 10.10.1.10;
11 }
12 group IBGP-LPEs {
13     type internal;
14     local-address 10.10.1.65;  # WAN address of Nantes towards LPE Inverness
15     family inet-vpn {
16         unicast;
17     }
18     export [ nhs Aggregate-routes-only ];
19     cluster 10.10.1.1;  # Route reflection
20     neighbor 10.10.1.66;
21 }
22 user@Nantes> show configuration policy-options policy-statement
23      Aggregate-routes-only
24 term aggregates {
25     from community Aggregate-route;
26     then next policy;
27 }
28 term block-others { # Do not readvertise prefixes by default
29     then reject;
30 }
```

The *next-hop self* policy on Line 18 is required to ensure that all BGP reachability information on router *Inverness* can be resolved properly through the HPE router *Nantes*. Failure to add this policy would require router *Inverness* to contain a labeled path towards relevant PE loopback addresses in inet.3.

The `cluster` knob that enables reflection (Line 19) is not strictly required in the HPE to LPE advertisement direction if aggregates are configured locally on router *Nantes* instead of on the ingress PEs. For advertisements in the reverse direction, updates coming from the LPE router *Inverness* may also get only as far as the HPE, with appropriate aggregation being performed at the HPE to readvertise into the backbone. However, because scalability into the backbone should not be a concern, reflecting the routes provides direct visibility to LPE reachability information.

Notice that the policy match in Line 25 specifies routes with a community. A route filter can be applied instead.

Junos Tip: Matching prefixes for inet-vpn routes

As part of Junos OS policy-based filtering for inet-vpn routes, a `route-filter` match condition can be added to restrict further the routes to be advertised. The route-filter match ignores the RD part of the NLRI. Combining the route-filter with an extended community match is recommended to scope the match to the intended VPN. Section 5.6.5 in the case study on Page 484 uses this feature to filter more-specific routes for an aggregate route.

5.2.7 Carrier Supporting Carrier (CsC)

As a natural evolution of an IP-based L3 VPN service, [RFC4364] also allows for MPLS traffic to be contained in the VPN. The applicability of this scenario is relevant when a carrier has no local homing of the customer's VPN routes within its AS, thus optimizing resource consumption. In this scenario, the customer buys an MPLS connecting service, exchanging IP information used to build the transport tunnels across the provider VPN. A separate instance is allocated for this MPLS VPN, with properties similar to IP MPLS L3VPNs.

As with Option C described in Section 5.2.6, both L-BGP and a combination of LDP plus the IGP of choice are defined to convey the label information for these FECs.

Relying on third-party backbone infrastructure to build an MPLS L3VPN may be an attractive proposition. From a migration perspective, migrating may be as simple as carefully controlling the FEC advertisements within and outside of the domain.

In an enterprise scenario, a CsC setup may be interesting when part of the enterprise wants to build and administer VPNs independently, while leveraging the L3VPN core infrastructure. As an example, a manufacturing plant connected to the research laboratories sharing several VPNs can be hidden behind a CsC VPN supported by the IT department.

Note that the major goal of CsC is to separate the two infrastructures. Thus, this option is feasible when the inner and outer VPN layers do not require direct interconnection.

5.3 Generic Strategies for L3VPN Migrations

With all the possible expansion and interconnect models for L3VPNs, there are multiple technical alternatives for migrations. By correctly aligning and combining these alternatives, several strategies for migration can be planned and implemented.

5.3.1 Layer VPN Route Target mapping

A RT ultimately determines L3VPN membership and can be arbitrarily tagged to each VPN-IPv4 route. More than a single RT can be tagged at VRF export, and any value can be expected for VRF import.

This multiple setting and matching mechanism is a powerful vehicle for a soft transition of L3VPNs. Routes can arbitrarily include multiple RTs at a transition stage, making them eligible to be imported into more than a single VRF. Depending on how reverse flows are set up, such multiple appearance in more than one VRF ensures that a destination can be present in multiple VRF tables as long as needed.

This behavior of matching and setting multiple arbitrary RTs allows granularity on a per-route basis and is commonly deployed as a permanent design in *extranets* or other *hub-and-spoke* topologies.

5.3.2 Layer 3 VPN Route Distinguiser mapping

RDs are embedded into VPN-IPv4 prefixes. These data structures cannot be applied multiple times to the same IPv4 prefix inside the same VRF, because the combination of a RD and an IPv4 prefix create a unique VPN-IPv4 route.

Deploying common or distinct RDs across VPN sites has been a tradeoff made in several MPLS backbones. Defining granular RD (usually on a per-VRF, per-PE basis) grants another distinguishing factor to the route and therefore multiplies the paths, allowing additional resilience towards the destination in multihomed CE domains. On the other hand, this implicit route multiplication leads to additional memory consumption and has a clear scaling impact.

By using several RDs for the same root IPv4 prefix, more paths are available for selection and installation in the end device's VRF, allowing for quicker convergence when primary paths disappear.

5.3.3 Parallel L3VPNs

Another approach distinguishes different L3VPNs as *planes* between CE domains. The transition does not take place inside the MPLS backbone by keeping the same VRF, but rather, between PEs and the CE domain by including uplinks in separate VRFs.

CEs are therefore multihomed to more than a single VRF. Despite injecting the same IPv4 prefix into the PE, different RDs allow duplicate destinations within the core and allow multiple RTs in each VRF, thus virtually creating parallel topologies, which allows for a seamless migration from one VPN to another.

5.3.4 Interconnecting L3VPNs

Section 5.2.6 describes inter-AS connection options with L3VPNs. These methodologies are used to expand L3VPNs beyond a single carrier with the intention of integrating services and applications.

A migration strategy may start with a simple inter-AS Option A, and transition or combine different options to achieve the best of all worlds in common ASBRs. For instance, a simple Option A Interconnect providing IP layer operations such as filtering or accounting can be progressively expanded with an Option B for consolidated route advertisement while keeping local Option A interconnects for some VRFs that require IP header inspection.

5.4 Junos Implementation of L3VPNs

If multiple sites are connected to the same PE and they have disjoint VPNs, Junos OS implements those sites in separate routing and forwarding tables. If those local VRF sites share any VPNs, Junos OS still creates separate routing and forwarding tables for each VRF, and the routes from each site are shared by the mechanisms described in Section 1.2.3 so that traffic can be forwarded among local VRFs with a common RT or with artificially shared routes using `rib-groups`.

BGP next-hop reachability checks are implemented per default in Junos OS for VPN-IPv4 NLRIs. These checks not only require the next hops to be present and reachable in inet.0 only, but also when indirect resolution for these prefixes is required, these next hops must be identified as FECs with a valid, existing label binding in inet.3. VPN-IPv4 prefixes whose next hops are not active in inet.3 are not installed and are immediately discarded, unless `keep all` is configured, in which case they remain *hidden* in Junos OS terminology.

The VRF paradigm introduces several concepts in Junos OS that at the time of this writing are trivial but are the cornerstone of L3VPN implementations:

- The *Routing Instance* concept was leveraged: the capability exists to define routing options and protocols inside a routing instance, in a manner similar to the default, together with a proper interface association to a specific instance.

- *Dynamic RIBs* are created for each VRF instance: a separate set of routing tables is flexibly assigned to each VRF.
- Enhancement of RIB list structures to properly associate protocol-imported prefixes to RIBs for addition or deletion at each one.
- BGP mechanisms such as *Capability advertisement* and *Refresh* were first introduced with the VPN-IPv4 family to allow these routes to be able to negotiate their support with a peer, and to let the BGP peer know of the requirement to refresh and send a given set of routes, thus avoiding the need to keep all VPN-IPv4 routes (not all needed at a given time if a local matching VRF does not exist).
- BGP needed to coordinate MPLS label assignments and relinquishments for each VPN-IPv4 route to avoid overlapping and ensure proper forwarding.
- On the forwarding plane, the Packet Forwarding Engines (PFEs) needed to be ready for a minimum of a double label push or pop (excluding other mechanisms such as fast reroute variations or LDP tunneling over RSVP that may require additional labels in the stack).
- At the kernel and routing socket level, the prior interface-VRF association has to be implemented with multiple tables per address family. Ultimately, this means *multiple forwarding tables per address family* as dictated by the interfaces in each VRF to uniquely identify each route change.
- The complex forwarding structures among VRFs or between a VRF and the default routing instance described in Section 1.1 require specific next hops that can be made visible across VRFs or can directly point to another RIB and be VRF-specific when hardware lookups occur.
- The traditional *single route lookup* paradigm in Junos OS needed to be improved to allow both an MPLS VPN label lookup to the proper VRF and an IP lookup inside a VRF simultaneously on PEs for incoming traffic.

This last bullet has been discussed for many years: Junos OS-based platforms have improved mechanisms to ensure hardware-based single lookups per flow. The need to accommodate a double lookup arose later, as soon as more than a single CE interface was present on a VRF or when particular capabilities requiring IP lookups were deemed necessary.

5.4.1 MPLS label allocation for L3VPNs

By default, when advertising VRF information into the core, Junos OS allocates one VPN label to every next hop that requires forwarding within a VRF. If the prefix is not to be advertised, this allocation does not take place.

Traffic arriving over the core interface that uses this VPN label is mapped to the final next hop, and traffic is forwarded out of the VRF with no further processing.

In a nutshell, hardware-based line-rate forwarding structures in Junos OS-based platforms are based on certain key value extraction from incoming packets to perform a single lookup to determine the destination for the traffic and possibly apply other actions (such as sampling and port-mirroring). This single-lookup premise poses a challenge for incoming backbone traffic at PEs in certain scenarios:

- Outgoing firewall filtering on the PE–CE interface for policing, implementing security rules, multifield reclassification, sampling, and other actions.
- When an additional IP lookup is needed inside the VRF to reach another VRF termination over a medium shared by PE–CE for additional resolution.

Juniper Networks engineers developed a solution to allow double lookup: *table labels*. This concept is based on the allocation of a common MPLS label for *all* prefixes within each VRF.

Allocation of table labels is possible using either *Label-Switched Interfaces* (LSIs) or *Virtual Tunnel* (vt) interfaces. Once a VRF is configured for table-label functionality, all prefixes are advertised with this label by default. It is possible to change the allocation and revert to the default for each next-hop allocation.

L3 VPN table label with vrf-table-label

The virtual LSI's infrastructure enables a packet to be received directly tagged with a specific label as if it arrives on the LSIs. Indirectly, the VRF lookup has been performed and the ASICs are now ready to perform the needed IP lookup for IPv4 prefixes identified after VPN label decapsulation is performed by the LSIs. This LSI-VRF table binding is uniquely for each MPLS VPN label, which is mapped on all regular core-facing interfaces to the LSI and directly extracts IPv4 traffic inside the VRF context.

Routes inside the VRF configured with the `vrf-table-label` knob are advertised with the *same* special label allocated for that VRF. As arriving packets for the VRF are passed through the LSI structure and MPLS labels are detached, an IP lookup is performed as if they were arriving on the VRF through the LSI, permitting the configuration of arbitrary PE–CE egress filtering to implement the same policies and rules that could be achieved for any local traffic between two PE–CE interfaces inside a VRF. Listing 5.3 shows related resources visible through the Junos OS CLI for LSI interfaces associated with each VRF (Line 16) and how a VPN table label is allocated in the mpls.0 routing table with a fake route (Line 25): it issues a proper lookup for traffic in the table referred as *Next-Table*.

Listing 5.3: Simple VRF-table-label implementation at router *Livorno* for route lookup

```
1  user@Livorno> show configuration routing-instances
2  nmm {
3      instance-type vrf;
4      interface fe-0/3/0.0;
5      route-distinguisher 65000:65000;
6      vrf-target target:65000:65000;
7      vrf-table-label; # Table label allocation for this VRF
8  }
9
10 user@Livorno> show interfaces routing-instance nmm terse
11 Interface               Admin Link Proto    Local                 Remote
12 fe-0/3/0.0              up    up   inet     172.16.100.8/24
13                                             172.16.100.108/24
14                                    iso
15                                    inet6    fe80::214:f6ff:fe85:405d/64
16 lsi.0                   up    up   inet     # LSI interface associated with VRF nmm table-label
17                                    iso
```

```
18                                  inet6
19
20  user@Livorno> show route table mpls extensive
21
22  mpls.0: 1 destinations, 1 routes (1 active, 0 holddown, 0 hidden)
23  16 (1 entry, 0 announced)
24          *VPN    Preference: 0
25                  Next table: nmm.inet.0 # Action for VRF nmm table-label received lookup
26                  Label operation: Pop
27                  Next-hop reference count: 1
28                  State: <Active NotInstall Int Ext>
29                  Age: 38:48
30                  Task: RT
31                  AS path: I
```

This implementation concept leads to two different effects:

- Because a VRF lookup is indirectly performed with the LSI association, full IP lookup capabilities are available inside the VRF.
- Assigning the same VRF table label for all IPv4 prefixes from a VRF reduces consumption of MPLS labels.

Junos Tip: VRF EXP classification with vrf-table-label

When the `vrf-table-label` feature is activated, the default MPLS EXP classifier is applied to the routing instance. Indirectly, the complete VPN table label is represented by the LSI, so in the ultimate hop to the PE, the inner label is directly matched with the LSI, and EXP bits are checked for an initial classification on the LSI.

If the Juniper Networks router PIC containing the PE–P interface is installed on any type of Enhanced FPC, this default indirect multifield classification on the LSI can be overridden with an arbitrary customer EXP classifier, the one intended for VPN label EXP inspection (usually the same Multifield classification from the core). Listing 5.4 depicts the changes needed on top of Listing 5.3 for a VRF-specific EXP classification and how to inspect the results. Line 14 indicates that the EXP-based Multifield classification is actually applied on the allocated LSI to that VRF.

Listing 5.4: EXP classifier with VRF-table-label implementation at router *Livorno*

```
1  user@Livorno> show configuration class-of-service classifiers exp NMM
2  import default;
3  forwarding-class assured-forwarding {
4      loss-priority low code-points [ 000 001 010 011 100 101 ];
5  }
6
7  user@Livorno> show configuration class-of-service routing-instances nmm
8  classifiers {
9      exp NMM;
10 }
11 user@Livorno> show class-of-service routing-instance
12 Routing instance: nmm
13
14   Logical interface: lsi.0, Index: 69 # LSI allocated to VRF nmm with EXP classification
15     Object          Name          Type          Index
16     Classifier      NMM           exp           58329
```

This EXP-based classification can also be enabled for other VPN types, such as VPLS LSIs, on MX-series platforms. Note that this feature is currently not supported whenever VRFs are defined inside logical systems.

Assigning a common VRF table label should be considered a best practice from an implementation perspective: IPv4 prefixes should already be unique or duplicated inside a VRF, and the IPv4 prefix part of a VPN-IPv4 route should already disambiguate destinations once the VRF is reached. In other words, having a common VRF table label does not affect forwarding to destinations inside the VRF and saves numerous resources in terms of MPLS labels and next-hop allocations. This can be helpful from a migration perspective, but also the opposite intention could be an ultimate goal when implementing L3VPNs: avoid an IP lookup in a PE–CE termination and blindly forward traffic to the CE. Scenarios such as a hub-and-spoke extranet or an inter-AS Option A interconnect in which traffic needs to be sent to a CE inside a particular VRF site without looking into the real IP destination address inside can be easily implemented by avoiding the `vrf-table-label` concept.

L3VPN table label with virtual tunnel interfaces

Another internal VRF lookup alternative is available in devices that have a Tunnel PIC: the definition of a virtual tunnel interface associated with each VRF. The fundamentals of this structure are based on traffic first transiting through the virtual loopback tunnel to detach the MPLS VPN label and then issuing a proper IP lookup after the loop.

These virtual loopback tunnel logical interfaces are defined following the Tunnel PIC location, need to include `family inet` and `family mpls`, and need to be included in the VRF for each logical unit for proper lookup. Listing 5.5 shows the configuration snippet needed and the effects in terms of VRF-related interfaces. Line 4 highlights the need to include the full logical vt- extension inside the VRF. Note that this feature is mutually exclusive with `vrf-table-label`, and an LSI is no longer allocated for the VRF.

Listing 5.5: Simple virtual loopback tunnel implementation at router *Livorno* for route lookup

```
1  user@Livorno> show configuration routing-instances nmm
2  instance-type vrf;
3  interface fe-0/3/0.0;
4  interface vt-1/2/0.0; # VT logical interface matching Tunnel PIC location
5  route-distinguisher 65000:65000;
6  vrf-target target:65000:65000;
7
8  user@Livorno> show configuration interfaces vt-1/2/0.0
9  family inet;
10 family mpls;
11
12 user@Livorno> show interfaces routing-instance nmm terse
13 Interface               Admin Link Proto    Local                 Remote
14 fe-0/3/0.0              up    up   inet     172.16.100.8/24
15                                             172.16.100.108/24
16                                  iso
17                                  inet6    fe80::214:f6ff:fe85:405d/64
18 vt-1/2/0.0              up    up   inet     # VT interface, no longer LSI
19
20 user@Livorno> show interfaces vt-1/2/0.0
21   Logical interface vt-1/2/0.0 (Index 86) (SNMP ifIndex 245)
22     Flags: Point-To-Point 0x4000 Encapsulation: Virtual-loopback-tunnel # Reserved vt tunnel
```

```
23      Input packets : 0
24      Output packets: 0
25      Protocol inet, MTU: Unlimited
26        Flags: None
```

Compared with `vrf-table-label`, virtual loopback tunnels require no extra resources from the PIC or the FPC and they do not enforce a particular MPLS label allocation scheme. This means that they can be implemented independently of the PE–P interface and PIC type. However, this comes at the cost of including a Tunnel PIC for this purpose.

Modifying the label allocation behavior

The configuration knobs discussed previously modify the allocation behavior on a per-VRF basis to provide a specific functionality. For some application scenarios this granularity is considered too coarse. As Listing 5.6 shows, the granularity can be adjusted through a policy language action. This extension allows, on a per-prefix basis, the specification of the VPN label *mode* to use when advertising a prefix from a VRF.

Listing 5.6: Label allocation strategies

```
1 user@Havana# set policy-options policy-statement allocate-label then label-allocation ?
2 Possible completions:
3   per-nexthop          Set per-nexthop label allocation mode
4   per-table            Set per-table label allocation mode
```

Constraint checks are in effect to allow the per-table allocation mode only for VRFs that have a table-label capability (that is, using either a virtual tunnel interface or a table label). The policy can be applied as part of the VRF export policy or, to keep a per-VRF generic approach, the *mode* can also be set through policy by applying it at the `routing-options label` level under the relevant routing-instance, as shown in Listing 5.7.

Listing 5.7: Per VRF reference to label allocation strategy

```
1 user@Havana# set routing-instances NMM-Big routing-options label ?
2 <...>
3 + allocation             Label allocation policy
```

Junos Tip: Using per-prefix MPLS labels for balancing

Junos OS includes hidden (and thus unsupported) knobs to enable per-prefix label allocation. The traditional demand for this type of allocation was to facilitate balancing of MPLS traffic within a VPN. A generalized solution following a different approach has been proposed using *entropy* labels, as described in [MPLS-ENTROPY], which allows for more label information to be added specifically for this purpose.

Newer hardware can inspect beyond the multiple MPLS headers in the packet. Detection of the Bottom of Stack (BoS) bit in the MPLS header flags the start of the payload information. If the first byte following the MPLS header matches IPv4 (0x45) or the first four bits match IPv6 (0x6), further IP header information is extracted and fed into the hashing engine for balancing purposes.

Application Note: Selective label allocation for hub-and-spoke VPN

The per-next-hop default label allocation scheme in Junos OS is perfectly suited for hub-and-spoke VPNs with a single VRF, in which spoke traffic is directed to the hub, thus bypassing a best-match IP prefix lookup. When the hub PE has also locally attached CEs that are on a broadcast media, as shown in Figure 5.10, it is necessary to activate IP functionality to allow ARP resolution to occur.

The conflict of needing MPLS forwarding for the hub site and at the same time leveraging IP functionality for the CEs on the same VRF can be resolved either by moving the hub site onto a separate VRF with table-label functionality or by leveraging the per-next-hop allocation capability feature.

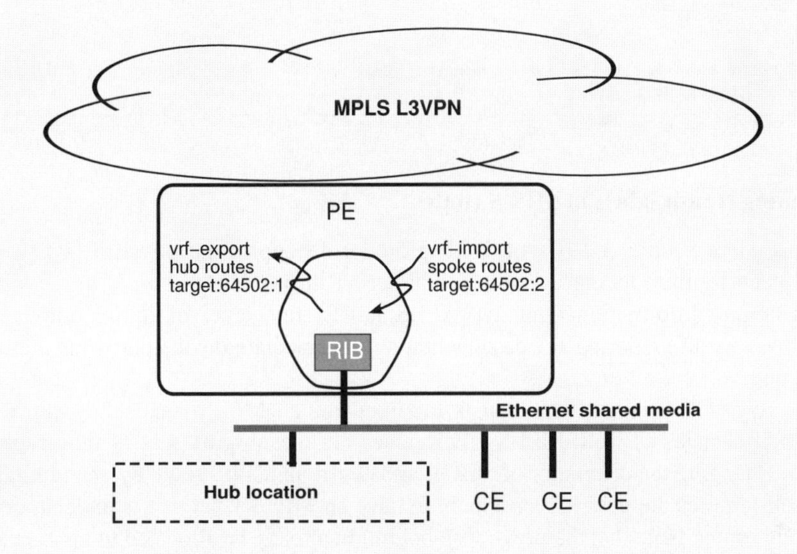

Figure 5.10: Hub-and-spoke requiring ARP for the local CE.

Listing 5.8 shows a sample configuration for a Hub VRF with table-label functionality enabled (Line 6) that is overridden globally to per-next-hop mode (Line 9). Besides the usual vrf-export policy that tags the hub prefixes, an additional vrf-export policy reactivates the table-label capability for direct routes (Line 5), allowing for ARP resolution while maintaining per-next-hop functionality for the remaining prefixes towards the hub site.

Listing 5.8: Hub-and-spoke with per-table policy for direct routes

```
1 user@Havana# show routing-instances NMM-Hub
2 instance-type vrf;
3 interface so-1/0/0.100;
4 vrf-target import target:64502:2; # Import Spoke community
5 vrf-export [ table-for-direct tag-hub ]
6 vrf-table-label; # Use of table label by default
7 routing-options {
8     label {
9         allocation default-nexthop; # Default to next-hop allocation
```

```
10|     }
11| }
12| user@Havana# show policy-options policy-statement table-for-direct
13| term direct {
14|     from protocol direct;
15|     then label-allocation per-table;
16| }
17|
18| user@Havana# show policy-options policy-statement default-nexthop
19| term default {
20|     then {
21|         label-allocation per-nexthop;
22|         accept;
23|     }
24| }
25| user@Havana# show policy-options policy-statement tag-hub
26| then {
27|     community add Hub;
28|     accept;
29| }
30| user@Havana# show policy-options community Hub
31| members target:64502:1;
```

Substituting transit labels in MPLS routes

As discussed in Section 3.2.10, the BGP protocol decouples the physical layer from the control plane by introducing a level of indirection through the *protocol next hop*. When the reachability information contains MPLS labels, this level of indirection has some implications worth exploring for cases when an intermediate device performs a change in the next hop.

When the BGP protocol next hop is modified for an MPLS route, the *context* of the information changes as well. MPLS label values are meaningful only to the originator of the label. Changing the originator of a BGP update for an MPLS route by changing the next hop to self requires the new advertisement to have an MPLS label that is understood by the originating router. Note that the previous label might already be allocated to another service, or it might be outside the label pool capabilities of this router. Hence, the readvertising router has to allocate a *new* label and bind this label to the old MPLS information in the MPLS forwarding table.

In Junos OS, this new VPN label advertised is normally taken from the global pool and is bound to the incoming label with an MPLS label *swap* operation. Traffic coming in with this label is MPLS-switched and the advertised label is changed with the original label.

If the label to be advertised is the table label allocated to one of the VRFs, traffic is received on the IP VRF and an IP lookup takes place. This effectively breaks the end-to-end MPLS path and enables IP transit functionality.

In Junos OS, it is possible to control the allocation of VPN labels for transit MPLS routes, as shown in Listing 5.9. The associated policy should *accept* or *reject* that the advertised label is a table label instead of a regular label for a particular prefix.

Listing 5.9: Per-VRF reference to transit label substitution strategy

```
1| user@Havana# set routing-instances NMM-Big routing-options label ?
2| <...>
3| + substitution          Label substitution policy
```

Junos Tip: Choosing a table label with extranets

In overlapping VPN scenarios, a single VPN prefix has more than one RT community attached. If the prefix is installed in multiple VRFs and more than one VRF has a table label, which of the table labels should be used in the advertisement?

Junos OS enforces deterministic behavior by sorting communities and VRF instance names lexicographically (that is, alphabetically) and choosing the table label of the first VRF with a table label that matches the first community in the list.

Once the table label has been selected, the substitution policy is inspected.

Figure 5.11 summarizes the decision process from a design perspective when an additional lookup inside a VRF is intended.

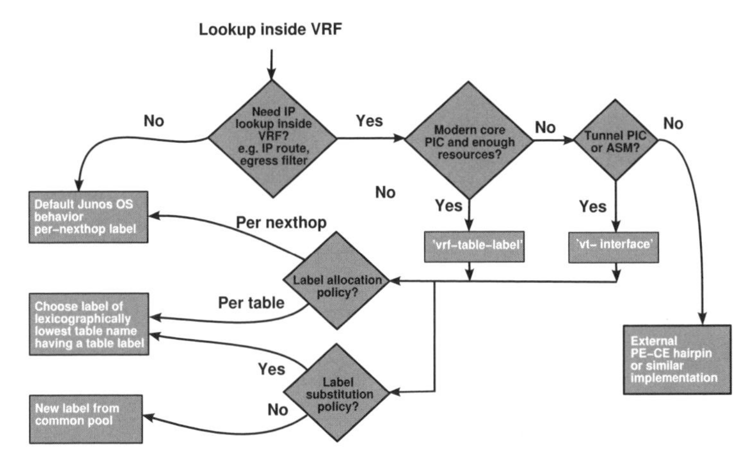

Figure 5.11: Design alternatives to perform lookup inside a VRF.

5.5 Resources for L3VPN Migrations

After general strategies and routing implementations with regards to L3VPNs have been analyzed, the local and proprietary features offered by most operating systems with this respect must be considered.

Junos OS offers numerous configuration options related to standard protocol structures that provide added value with regards to migration activities. This section analyzes the most relevant Junos OS resources for L3VPN migrations, examining the features for each PE–CE protocol.

5.5.1 RIP PE–CE resources

Despite not presenting relevant differences or newer features as compared to RIP in a flat network or not considering any specific standard for RIP as a PE–CE protocol, certain resources from the standard RIP version 2, as defined in [RFC2453], must be taken into account differently when integrating this protocol to interact with CEs in a L3 VPN.

Among others, *administrative tagging* can become an interesting loop-prevention mechanism in CE domains.

Administrative tagging

[RFC2453] enforces that the Route Tag field be preserved and readvertised with a route. The original intention was to use this field as a distinguisher between *internal* and *external* RIP routes.

While this field can be used arbitrarily, it is certainly helpful in redistribution scenarios, particularly to address potential routing loop formation, as shown in Figure 5.2 on Page 384.

As with other protocols, the Junos OS Route `Tag` attribute is automatically populated into the Route Tag field when exporting RIP Updates that include those prefixes.

Application Note: Mitigate loop formation with a multihomed RIP network into a L3VPN with Route Tag fields

The advantage of the Route Tag field in RIP Updates is that it is maintained while propagating the route update and while the metric increases. Therefore, a practical application consists of using this identifier as unique distinguisher for leaking information.

In a nutshell, PEs may set a specific value on a RIP Update towards the CE domain and can use that value to identify routes that originated outside the local VPN site but that were leaked from the MPLS backbone.

If administrative control on such action is maintained, this tool can be more scalable than policies based on route-filtering or firewall filters matching on peer PEs on a common LAN.

Consider a legacy RIP network such as the one in Figure 5.12, in which each participating device injects a external prefix into RIP. This network is intended to be integrated as a L3VPN, first by segregating *Lille* into a separate VRF.

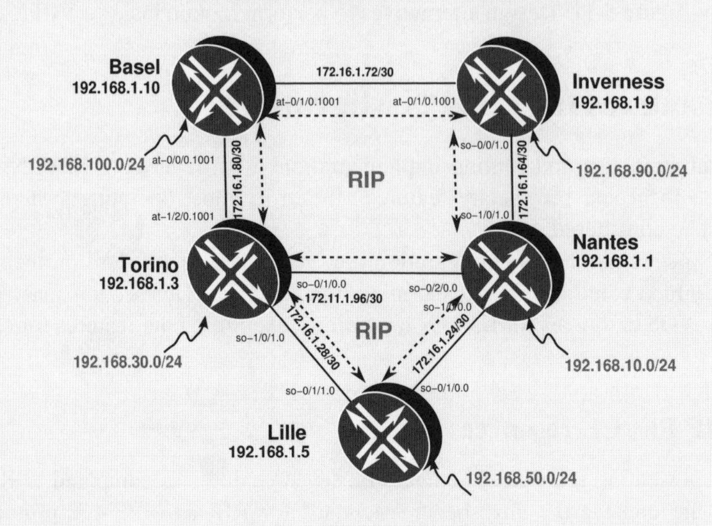

Figure 5.12: Original RIP network to be integrated in a L3VPN.

If such segregation is performed while dual-homing both *Lille* and the rest of the network to represent similar parallel paths to the original setup, as shown on Figure 5.13, less than optimal routing can be easily observed when inspecting original prefixes.

Focusing on the address segment injected by router *Nantes* as a legitimate PE configured with default protocol preferences, Listing 5.10 shows active paths for this destination in all participating PEs in the network. The following situation can be observed:

1. Router *Nantes* advertises the native segment in RIP.

2. Router *Bilbao* and router *Skopie* select the local RIP route as the active path because of default protocol preference and redistribute this prefix as VPN-IPv4 route into MP-BGP.

3. This route is eligible to be imported at other VRFs in the network sharing the same RT.

4. Router *Livorno* selects the BGP route from router *Skopie* because of its lower MED, if it arrives first via MP-BGP than internally through the CE domain.

5. Under these circumstances, router *Havana* selects the RIP route advertised by router *Livorno* internally in the CE domain connecting to *Lille*. This is because of the default protocol preference values and can be considered less-than-optimal routing because the direct path to the source PE remains as backup.

6. Router *Havana* readvertises this *fake* local route back into MP-BGP considering it local. Router *Bilbao* keeps selecting the local RIP route directly from router *Nantes* as the active path because of the default protocol preference.

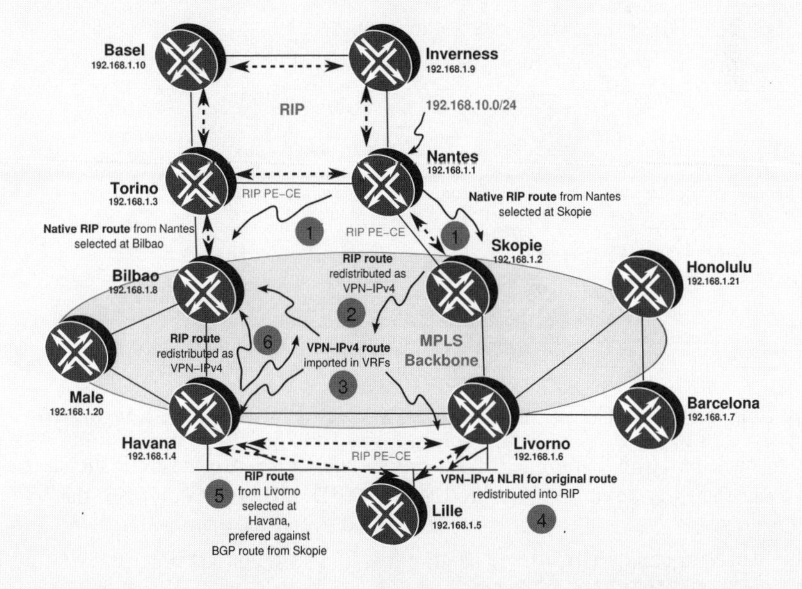

Figure 5.13: Less-than-optimal routing with default RIP as PE–CE protocol.

To summarize, RIP paths visible through a backdoor link inside a VRF remain preferred if no manipulation is done. This case does not properly illustrate a loop for two reasons:

- Router *Bilbao* receives that route natively in RIP. Other routes from different sites would be accepted here through BGP paths that would be readvertised back using RIP.
- Junos OS translates the RIP metric into the BGP MED attribute when translating routes as VPN-IPv4 by default. No policies manipulate MEDs in this example. This keeps the BGP route path from router *Skopie* as preferred on router *Livorno*, as compared with the route from router *Havana*, which has a higher MED (Line 16).

Listing 5.10: Active paths for Nantes routes at all PEs

```
1  user@Skopie> show route 192.168.10.0/24 table NMM terse
2
3  NMM.inet.0: 8 destinations, 13 routes (8 active, 0 holddown, 0 hidden)
4  + = Active Route, - = Last Active, * = Both
5
6  A Destination        P Prf   Metric 1   Metric 2  Next hop        AS path
7  * 192.168.10.0/24    R 100          2             >172.16.1.1
8                       B 170        100          3 >so-1/0/0.0     I
9
10 user@Livorno> show route 192.168.10.0/24 table NMM terse
11
12 NMM.inet.0: 8 destinations, 13 routes (8 active, 0 holddown, 0 hidden)
13 + = Active Route, - = Last Active, * = Both
14  # BGP MED is tie-breaker here
15 A Destination        P Prf   Metric 1   Metric 2  Next hop        AS path
16 * 192.168.10.0/24    B 170        100          2 >so-0/0/0.0     I
17                      B 170        100          3 >so-0/0/1.0     I
18
19 user@Havana> show route 192.168.10.0/24 table NMM terse
20
21 NMM.inet.0: 8 destinations, 12 routes (8 active, 0 holddown, 0 hidden)
22 + = Active Route, - = Last Active, * = Both
23
24 A Destination        P Prf   Metric 1   Metric 2  Next hop        AS path
25 * 192.168.10.0/24    R 100          2             >172.16.100.8
26                      B 170        100          3 >172.16.1.85    I
27
28 user@Bilbao-re0> show route 192.168.10.0/24 table NMM terse
29
30 NMM.inet.0: 8 destinations, 18 routes (8 active, 0 holddown, 0 hidden)
31 + = Active Route, - = Last Active, * = Both
32
33 A Destination        P Prf   Metric 1   Metric 2  Next hop        AS path
34 * 192.168.10.0/24    R 100          3             >172.16.1.101
35                      B 170        100          2 >at-1/2/1.0     I
36                      B 170        100          2 >so-1/3/1.0     I
```

Junos Tip: Default MED population for VPN-IPv4 NLRIs with RIP metric

As seen on Line 16 in Listing 5.10, Junos OS injects the original RIP metric from the IPv4 route that was being redistributed as the MED attribute value from the VPN-IPv4 NLRI.

This behavior can be considered as another safeguard against potential routing loops, because transiting the MPLS backbone does not set a lower metric value by default if such a route is readvertised in RIP.

As example in the previous topology, imagine a transient situation in which both PEs (router *Skopie* and router *Bilbao*) are much closer internally through the CE domain in terms of RIP metrics than the original source in the same CE domain, and such a route is not protected with a *Site of Origin* attribute. If the BGP NLRI is received with much lower MED values, this could eventually lead to the other PE preferring the backdoor path from the other PE instead of the original advertisement.

Defining a simple policy to block or deprefer routes from the peer PE through the CE domain is an adequate tactic to improve routing and to avoid any potential loops. Using a common `tag` value when exporting the routes in RIP and matching that field for import from the CE domain ensures that a PE knows whenever a route has been leaked from the backbone.

Listing 5.11 illustrates changes needed in *Livorno* as arbitrary PE to accomplish such policies.

Listing 5.11: Configuration changes for route optimization on router *Livorno*

```
1  [edit routing-instances NMM protocols rip]
2  user@Livorno# show
3  group CE {
4      export NMM-bgp-to-rip;
5      import NMM-rip-to-bgp;
6      neighbor fe-0/3/0.0;
7  }
8
9  [edit policy-options policy-statement NMM-rip-to-bgp]
10 user@Livorno# show
11 term rip {
12     from {
13         protocol rip;
14         tag 64502;
15     }
16     then {
17         preference 200; # Higher preference than default BGP
18     }
19 }
20
21 [edit policy-options policy-statement NMM-bgp-to-rip]
22 user@Livorno# show
23 term bgp {
24     from protocol bgp;
25     then {
26         tag 64502;
27         accept; # Tag all redistributed routes from BGP
28     }
29 }
30
31 [edit policy-options policy-statement NMM-vrf-export]
32 user@Livorno# show
33 term block-PE {
34     from {
35         protocol rip;
36         tag 64502;
37     }
38     then reject; # Do not export routes into BGP learned from other PE
39 }
40 term rip {
41     from protocol rip;
42     then {
43         community add target:64501:64501;
44         accept;
```

```
45        }
46  }
```

Instead of directly rejecting routes from the parallel PE through the backdoor path, the design option is to unprefer such backup paths but to keep them eligible by means of setting a preference higher than BGP. The VRF export policy is tuned in this case so that these routes are not readvertised into BGP, where they could possibly create a loop.

This design ensures that the root BGP path is preferred as long as it remains available and that backup paths through the CE domain are eligible for forwarding installation but are not redistributed into MP-BGP.

Focusing again on the address segment injected by router *Nantes* after these configuration changes in all PEs, Listing 5.12 illustrates the changes. Lines 18 and 31 show how paths for the same destinations learned from RIP from the parallel PE are now kept as backups with higher preference.

Listing 5.12: Active paths for Nantes routes at all PEs after tagging routes

```
1  user@Skopie> show route 192.168.10.0/24 table NMM terse
2
3  NMM.inet.0: 8 destinations, 12 routes (8 active, 0 holddown, 0 hidden)
4  + = Active Route, - = Last Active, * = Both
5
6  A Destination        P Prf   Metric 1   Metric 2  Next hop        AS path
7  * 192.168.10.0/24    R 100          2             >172.16.1.1
8                       B 170        100          3  >so-1/0/0.0      I
9
10 user@Livorno> show route 192.168.10.0/24 table NMM terse
11
12 NMM.inet.0: 8 destinations, 16 routes (8 active, 0 holddown, 0 hidden)
13 + = Active Route, - = Last Active, * = Both
14
15 A Destination        P Prf   Metric 1   Metric 2  Next hop        AS path
16 * 192.168.10.0/24    B 170        100          2  >so-0/0/0.0      I
17                      B 170        100          3  >so-0/0/1.0      I
18                      R 200          2             >172.16.100.1 # RIP route from Havana
19
20 user@Livorno> show route 192.168.10.0/24 protocol rip extensive | match Tag
21 RIP route tag 64502; no poison reverse
22                 Age: 7:42    Metric: 2    Tag: 64502 # Tag present in RIP route
23
24 user@Havana> show route 192.168.10.0/24 table NMM terse
25
26 NMM.inet.0: 8 destinations, 12 routes (8 active, 0 holddown, 0 hidden)
27 + = Active Route, - = Last Active, * = Both
28
29 A Destination        P Prf   Metric 1   Metric 2  Next hop        AS path
30 * 192.168.10.0/24    B 170        100          3  >172.16.1.85     I
31                      R 200          2             >172.16.100.8  # RIP route from Livorno
32
33 user@Havana> show route 192.168.10.0/24 protocol rip extensive | match Tag
34 RIP route tag 64502; no poison reverse
35                 Age: 1:07:29    Metric: 2    Tag: 64502 # Tag present in RIP route
36
37 user@Bilbao-re0> show route 192.168.10.0/24 table NMM terse
38
39 NMM.inet.0: 8 destinations, 13 routes (8 active, 0 holddown, 0 hidden)
40 + = Active Route, - = Last Active, * = Both
41
42 A Destination        P Prf   Metric 1   Metric 2  Next hop        AS path
43 * 192.168.10.0/24    R 100          3             >172.16.1.101 # Original path from Nantes
44                      B 170        100          2  >at-1/2/1.0      I
```

These changes are illustrated on Figure 5.14.

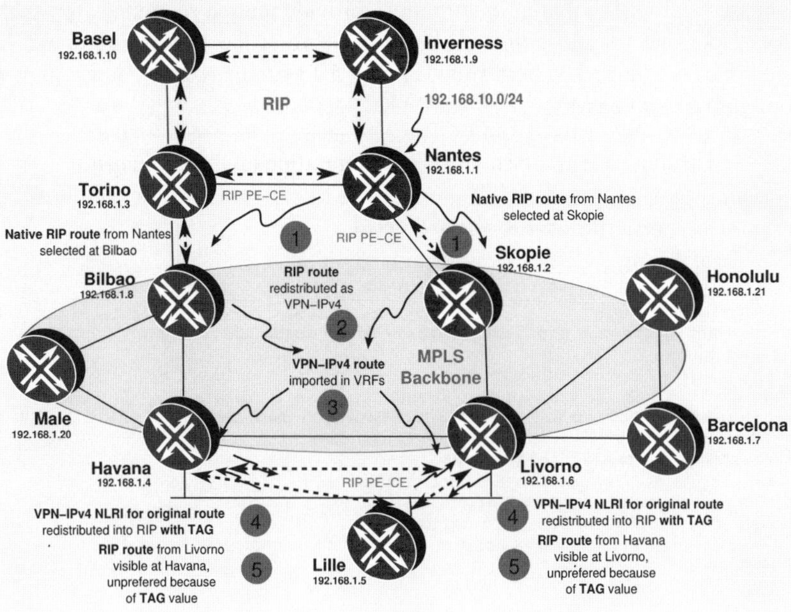

Figure 5.14: Optimal routing with administrative tagging for RIP as PE–CE.

5.5.2 OSPFv2 PE–CE resources

Junos OS offers several configuration features related to OSPFv2 PE–CE mechanisms defined in [RFC4576] and [RFC4577]. Such features are not only limited to comply with standards, but also offer interesting tools for migrations when using OSPFv2 as the PE–CE protocol.

Sham links present specific structures to emulate backdoor connections over a L3VPN. The OSPF Domain Identifier extended community attribute ultimately determines how a route is going to be redistributed towards attached CEs, mainly as Type 3 or as Type 5 or 7 LSA. Both the DN bit and VPN Route Tag are also tools to steer route leaking beyond a VRF, affecting redistribution of Type 3, 5, and 7 LSAs from a CE domain to the MPLS backbone.

From a migration perspective, these elements are tools to control the scope and redistribution of VPN-IPv4 routes into and possibly beyond a VRF. Junos OS offers a wide configuration range, even allowing some of these behaviors to be disabled under controlled circumstances.

Sham links

Sham links are an adequate migration tool to replace backdoor links among CEs with backbone connections. [RFC4577] enforces that whenever sham links are present in a topology, they should be equally eligible as intra-area links.

In Junos OS, when configuring a `sham-link`, a local database of `sham-link` entries is created in each instance. These entries are indexed by remote `sham-link` end point addresses and OSPFv2 area identifiers to provide unique references for each virtual connection. Note that for *sham links* to become active at the forwarding plane, both end points need to be configured on both sides. Listing 5.13 summarizes the basic steps in Junos OS to configure `sham-links`.

Listing 5.13: Basic sham-link configuration on router *Livorno*

```
1  [edit routing-instances NMM]
2  user@Livorno# set protocols ospf sham-link local ?
3  Possible completions:
4    <local>               Local sham link endpoint address # Local identifier for all sham links
5
6  [edit routing-instances NMM]
7  user@Livorno# set protocols ospf area 0 sham-link-remote 192.168.2.2 ?
8  Possible completions:     # Configuration options for each Sham-Link destination
9    <[Enter]>             Execute this command
10 + apply-groups          Groups from which to inherit configuration data
11 + apply-groups-except   Don't inherit configuration data from these groups
12   demand-circuit        Interface functions as a demand circuit
13   flood-reduction       Enable flood reduction
14   ipsec-sa              IPSec security association name
15   metric                Sham link metric (1..65535)
16 > topology              Topology specific attributes
17   |                     Pipe through a command
```

Each *sham link* is virtually represented as a logical interface to the routing subsystem. This virtual logical interface is created for each *sham link* database entry whenever the end point address is reachable through a destination over the MPLS L3VPN. The decision to bring up or tear down these logical interfaces is made as a result of monitoring the destination for the end point and inspecting its reachability. However, logical interfaces are not removed from the system as long as the configuration remains, even if the destination has long been unreachable, because the intent is to bring this virtual circuit up as soon as the end point is visible beyond a certain PE VRF. In fact, a dummy physical interface is created to group all the logical interfaces representing *sham links* with a specific encapsulation for identification purposes with OSPFv2. In Junos OS, all these interface indexes remain known only to the routing subsystem.

Thus, OSPFv2 is notified whenever a `sham-link` virtual interface becomes active and triggers the specific OSPFv2 data structure construction. A monitoring process (daemon) is programmed with the remote end-point address destination in specific `sham-link` constructs to identify remotely received packets over such a *sham link*. Once such a virtual adjacency is established, it is advertised as an unnumbered point-to-point link type in each Type 1 Router LSA. This link-type entry contains the SNMP MIB index for the *sham link* from a reserved range, a mechanism similar to what is done for other unnumbered point-to-point links, which are uniquely referred to by their *ifIndex*.

In the OSPFv2 database description packet exchange process, the `sham-link` MTU is set to zero, assuming unknown values for the transit path between PEs. This mechanism is similar to parallel OSPFv2 constructs, such as *virtual links* or standard tunnel interfaces.

If the virtual *sham–link* interface becomes the natural next hop for active paths, when performing SPF calculations, the router does not add these routes to the RIB, because it is expected that there is a VPN-IPv4 covering prefix for this destination, which therefore ensures that the MPLS backbone path inside the L3VPN is preferred, as is intended with its definition. In the case of Equal-Cost Multiple Paths (ECMP) between the *sham link* and the backdoor paths, they are also not added because the intent is to prefer *sham links* over equal metric values.

Application Note: Using sham links to integrate a legacy OSPFv2 network into a L3VPN

A very common exercise at Service Providers nowadays is dismantling legacy OSPFv2 networks with the intention of integrating regions as VRF sites across a common multiservice MPLS backbone in which connectivity is maintained by means of L3VPN resources. The ability to speak OSPFv2 between the PE and each of the legacy CE sites eases integration tasks tremendously without the need to use another protocol as vehicle for connectivity and redistribution, and while, at the same time, retaining OSPFv2 database and route control as a result of protocol peering with each of the PEs.

Considering a scenario like the one represented in Figure 5.15, the idea is to implement useful Junos OS and standard configuration resources to progressively segregate routers as different VRF sites to be integrated under a common VPN. As the most representative situation, router *Lille* is first going to be segregated and interconnected over more than one MPLS backbone, as shown in Figure 5.16. Needless to say, such a transition is intended to be carried out as smoothly as possible.

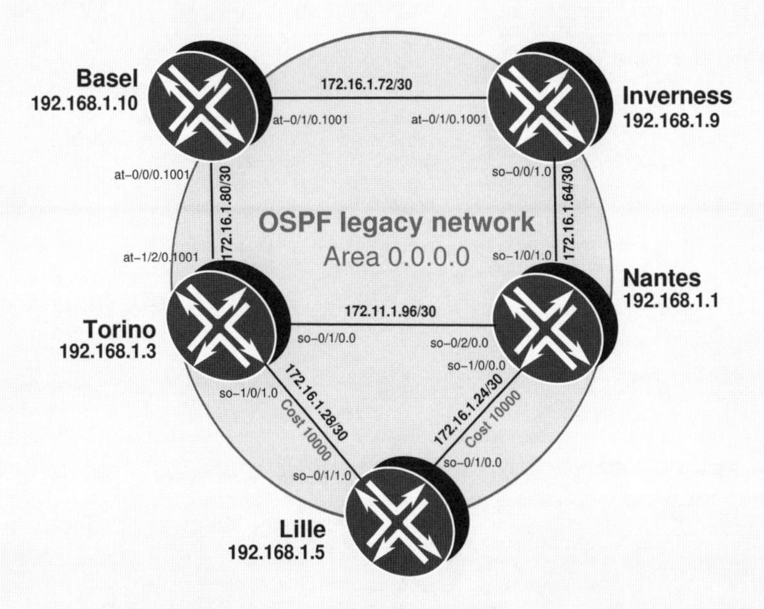

Figure 5.15: Representative legacy OSPFv2 network to be integrated in an L3VPN.

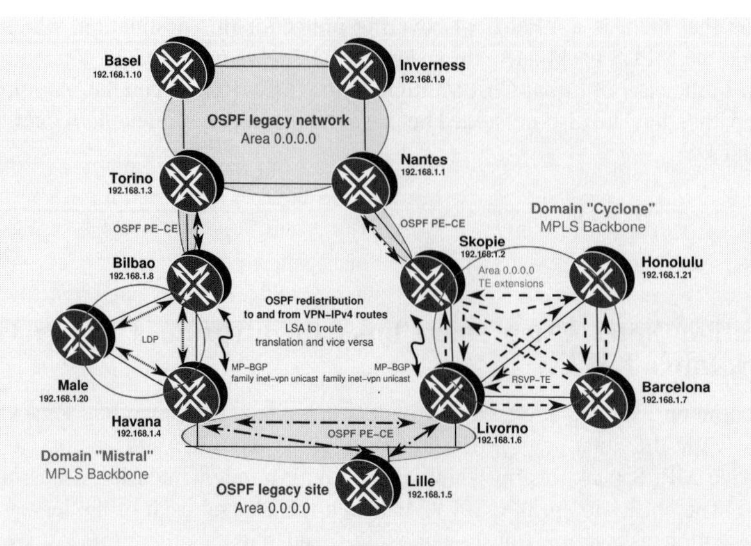

Figure 5.16: Migrated legacy OSPFv2 network over an L3VPN.

At this initial stage, the OSPFv2 database contains only LSAs representative of the legacy networks routers and links, as shown in Listing 5.14 from router *Lille*'s perspective, including link costs to the rest of the network (assuming that bidirectional metrics are the same). Note that Opaque LSAs with TE extensions are considered to be a substantial part of the legacy network.

Listing 5.14: Initial OSPFv2 database and link costs at Lille prior to L3VPN integration

```
1  user@Lille> show ospf database
2
3      OSPF database, Area 0.0.0.0
4   Type       ID               Adv Rtr           Seq      Age  Opt  Cksum  Len
5  Router   192.168.1.1      192.168.1.1      0x8000002f    23  0x22 0x1d26 132  # Nantes
6  Router   192.168.1.3      192.168.1.3      0x80000023    24  0x22 0xfdd8 120  # Torino
7  Router  *192.168.1.5      192.168.1.5      0x80000029    23  0x22 0xd541  96  # Lille
8  Router   192.168.1.9      192.168.1.9      0x8000001c    24  0x22 0xa9ae  96  # Inverness
9  Router   192.168.1.10     192.168.1.10     0x8000001a    25  0x22 0xb775  84  # Basel
10 OpaqArea 1.0.0.1          192.168.1.1      0x80000018    23  0x22 0x82be  28
11 OpaqArea 1.0.0.1          192.168.1.3      0x80000017    24  0x22 0x8cb1  28
12 OpaqArea*1.0.0.1          192.168.1.5      0x80000018    23  0x22 0x92a6  28
13 OpaqArea 1.0.0.1          192.168.1.9      0x80000018    24  0x22 0xa28e  28
14 OpaqArea 1.0.0.1          192.168.1.10     0x80000018    25  0x22 0xa688  28
15
16 user@Lille> show ospf interface extensive | match "PtToPt|Cost"
17 so-0/1/0.0           PtToPt  0.0.0.0          0.0.0.0          0.0.0.0              1
18   Type: P2P, Address: 0.0.0.0, Mask: 0.0.0.0, MTU: 4470, Cost: 10000
19   Topology default (ID 0) -> Cost: 10000
20 so-0/1/0.0           PtToPt  0.0.0.0          0.0.0.0          0.0.0.0              0
21   Type: P2P, Address: 172.16.1.26, Mask: 255.255.255.252, MTU: 4470, Cost: 10000
22   Topology default (ID 0) -> Passive, Cost: 10000
23 so-0/1/1.0           PtToPt  0.0.0.0          0.0.0.0          0.0.0.0              1
24   Type: P2P, Address: 0.0.0.0, Mask: 0.0.0.0, MTU: 4470, Cost: 10000
25   Topology default (ID 0) -> Cost: 10000
26 so-0/1/1.0           PtToPt  0.0.0.0          0.0.0.0          0.0.0.0              0
27   Type: P2P, Address: 172.16.1.30, Mask: 255.255.255.252, MTU: 4470, Cost: 10000
28   Topology default (ID 0) -> Passive, Cost: 10000
```

The intent is progressively to replace the natural uplinks for router *Lille* with connections to a PE in an L3VPN setup. Because they are intra-area links and the intent is to provide parallel upstreams, a feasible approach is to use *sham links* as a natural replacement for intra-area connections during the first migration. Basically, a *sham link* that is parallel to a natural intra-area link should be equally eligible from an SPF computation perspective, and in that case the *sham link* cost becomes a tie-breaker, while natural LSA translation mechanisms based on different *OSPF Route Types* yield Type 3, 5, or 7 LSAs and therefore are less preferred than the internal link in the backbone area.

This first migration step is summarized in Figure 5.17. Note that *sham links* are set up between PEs inside the VRF and should appear at router *Lille* as virtual intra-area links.

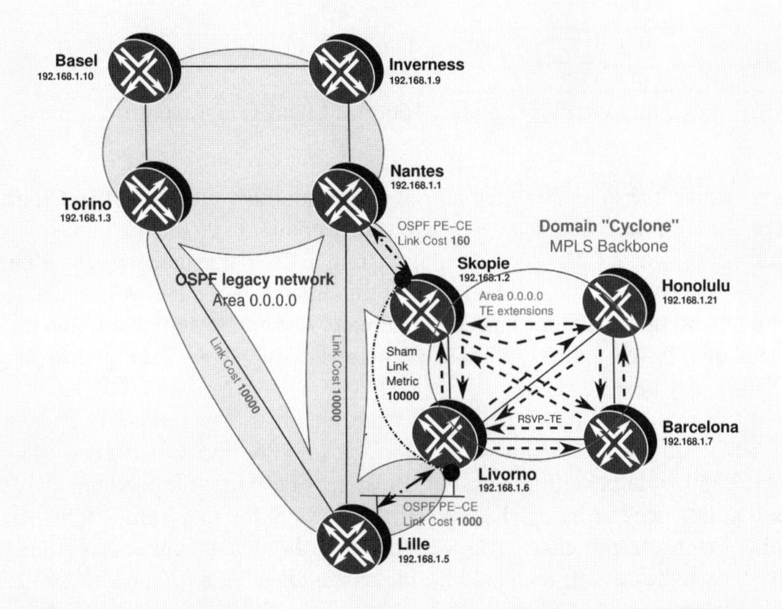

Figure 5.17: Stage one: prepare parallel sham link over L3VPN network.

For such a setup, the basic configuration (even considering a NULL *Domain Identifier*) is illustrated in Listing 5.15. As a precaution during initial activation, a higher calculated metric should be derived in paths over the *sham links* so as to check that they are established before actually forwarding any traffic over them. (If not configured, a value of 1 is determined by default and therefore all flows may automatically travel over the *sham link*.)

Considering the costs for both links between router *Lille* and the rest of a common Area 0, the direct circuit *Lille-Nantes* injects a metric value of 10000, while a default reference results on 1000 for the FastEthernet link *Lille-Livorno* and 160 for the OC-12 link *Skopie-Nantes*. The metric for the *sham link* may not be arbitrary at this stage, if it is intended to first deploy the *sham link* as a backup path. The link costs can be summarized as follows:

$$\overbrace{\textit{Lille–Nantes}\text{ OSPFv2 path cost}} < \overbrace{\textit{Lille–Nantes}\text{ MPLS path cost}} \tag{5.1}$$

$$\overbrace{\textit{Lille–Nantes}\text{ internal OSPFv2 path}} = \overline{\text{Lille–Nantes}} = 10000 \tag{5.2}$$

$$\overbrace{\textit{Lille–Nantes}\text{ MPLS path}} = \overline{\text{Sham-Link Livorno–Skopie}} + \overbrace{\text{Uplinks to }\textit{Livorno}\text{ and }\textit{Skopie}} \tag{5.3}$$

$$\overbrace{\textit{Lille–Nantes}\text{ MPLS path}} = \overline{\text{Sham-Link Livorno–Skopie}} + \overline{\text{Lille–Livorno}} + \overline{\text{Skopie–Nantes}} \tag{5.4}$$

$$\overbrace{\textit{Lille–Nantes}\text{ MPLS path}} = \overline{\text{Sham-Link Livorno–Skopie}} + 1160 \tag{5.5}$$

$$\overbrace{\textit{Lille–Nantes}\text{ internal OSPFv2 path}} = 10000 < \overline{\text{Sham-Link Livorno–Skopie}} + 1160 \tag{5.6}$$

In other words, the *sham-link* metric needs to be carefully selected so that CE backdoors remain preferred. It has been decided to arbitrarily configure the metric value of 10000 on both ends, (as shown on Lines 15 and 36) in Listing 5.15, so that the internal path inside the legacy OSPFv2 network remains preferred at this first stage. Note that in other scenarios, costs through internal backdoor links could already result in lower values than just adding metrics for the PE–CE uplinks, without even considering *sham links*, and further metric manipulation may be needed.

It is worth remarking that local and remote identifiers for *sham-link* end points are locally configured loopback addresses inside the L3VPN that do not need to be visible at CEs. The rationale behind this design decision is based on the fundamentals of *sham links*: end points need to be reachable over the L3VPN for *sham links* to be eligible for forwarding. With the presence of a backdoor and the default preference values at this stage, if these addresses are injected into the native OSPFv2 backbone, those destination addresses become preferred through the backbone links (even if they are redistributed with an AS external LSA) and *sham links* are not set up. This leads to a design decision in which the addresses used to establish *sham links* reside virtually outside the OSPFv2 network and are used only for signalling purposes.

Listing 5.15: Basic sham-link configuration on router *Skopie* and router *Livorno* as transit PEs

```
1  user@Skopie> show configuration routing-instances
2  NMM {
3      instance-type vrf;
4      interface so-0/1/0.0;
5      interface lo0.2;
6      route-distinguisher 64501:64501;
7      vrf-export NMM-vrf-export;
8      vrf-target target:64501:64501;
9      vrf-table-label;
10     protocols {
11         ospf {
12             reference-bandwidth 100g;
13             sham-link local 192.168.2.2; # Global identifiers for sham-link local addresses
```

```
14              area 0.0.0.0 {
15                  sham-link-remote 192.168.6.6 metric 10000; # Higher-metric end point
16                  interface so-0/1/0.0;
17              }
18          }
19      }
20  }
21
22  user@Livorno> show configuration routing-instances
23  NMM {
24      instance-type vrf;
25      interface fe-0/3/0.0;
26      interface lo0.6;
27      route-distinguisher 64501:64501;
28      vrf-export NMM-vrf-export;
29      vrf-target target:64501:64501;
30      vrf-table-label;
31      protocols {
32          ospf {
33              reference-bandwidth 100g;
34              sham-link local 192.168.6.6; # Global identifiers for sham-link local addresses
35              area 0.0.0.0 {
36                  sham-link-remote 192.168.2.2 metric 10000; # Higher-metric end point
37                  interface fe-0/3/0.0;
38              }
39          }
40      }
41  }
42
43  user@Livorno> show configuration policy-options policy-statement NMM-vrf-export
44  term ospf {
45      from protocol ospf;
46      then {
47          community add target:64501:64501;
48          accept;
49      }
50  }
51  term direct {
52      from protocol direct;
53      then {
54          community add target:64501:64501;
55          accept;
56      }
57  }
```

From these configuration snippets, it is also worth noting that there are *no export policies* in OSPFv2 inside the VRF for remotely received VPN-IPv4 routes to be flooded to the CEs. The intent is to flood these routes over *sham link*s in this simple example, which has only two VRF sites, because the *sham link* actually *bridges* both locations in the OSPFv2 topology. Reinjecting routes in OSPFv2 that are already native in the topology as a result of the backdoor links adds additional churn and increases the chance of a loop to be created. If another VRF site without a *sham-link* termination were to be included in the same L3VPN, additional policies (for example, using different RTs, standard communities, or administrative tagging) would need to be created to allow AS External LSAs to be generated at each PE acting as ASBR only for those specific routes.

Listing 5.16 depicts how *sham links* are represented in the topology: an unnumbered point-to-point-type link appears in each Type 1 Router LSA and a virtual neighbor is constructed to signify this virtual connection through the backbone. Lines 3 and 18 show how such virtual neighbors are present in Junos OS at each side once reachability is checked. In newly received Type 1 Router LSAs for the PEs in the OSPFv2 database,

Lines 45 and 64 show how *sham links* are presented as unnumbered point-to-point link types in each Type 1 Router LSA.

Listing 5.16: Sham-link representation at router *Skopie* and router *Livorno* as transit PEs

```
1  user@Skopie> show ospf neighbor instance NMM extensive
2  Address          Interface              State     ID               Pri  Dead
3  192.168.6.6      shamlink.0             Full      192.168.6.6        0    34
4    Area 0.0.0.0, opt 0x42, DR 0.0.0.0, BDR 0.0.0.0
5    Up 00:02:00, adjacent 00:02:00      # Virtual representation of sham link
6    Topology default (ID 0) -> Bidirectional
7  172.16.1.1       so-0/1/0.0             Full      192.168.1.1      128    35
8    Area 0.0.0.0, opt 0x42, DR 0.0.0.0, BDR 0.0.0.0
9    Up 00:02:11, adjacent 00:02:11
10   Topology default (ID 0) -> Bidirectional
11
12 user@Livorno> show ospf neighbor instance NMM extensive
13 Address          Interface              State     ID               Pri  Dead
14 172.16.100.5     fe-0/3/0.0             Full      192.168.1.5      128    33
15   Area 0.0.0.0, opt 0x42, DR 172.16.100.5, BDR 172.16.100.8
16   Up 00:05:44, adjacent 00:05:40
17   Topology default (ID 0) -> Bidirectional
18 192.168.2.2      shamlink.0             Full      192.168.2.2        0    37
19   Area 0.0.0.0, opt 0x42, DR 0.0.0.0, BDR 0.0.0.0
20   Up 00:05:46, adjacent 00:05:46 # Virtual representation of sham link
21   Topology default (ID 0) -> Bidirectional
22
23 user@Livorno> show ospf database instance NMM
24
25     OSPF database, Area 0.0.0.0
26  Type       ID               Adv Rtr           Seq      Age  Opt  Cksum  Len
27 Router   192.168.1.1      192.168.1.1      0x8000003d  2314  0x22 0x852  144
28 Router   192.168.1.3      192.168.1.3      0x8000002e  2801  0x22 0xe7e3 120
29 Router   192.168.1.5      192.168.1.5      0x8000003a   440  0x22 0x6279  96
30 Router   192.168.1.9      192.168.1.9      0x80000027  2802  0x22 0x93b9  96
31 Router   192.168.1.10     192.168.1.10     0x80000025  2802  0x22 0xa180  84
32  # Router-LSA for Skopie inside L3VPN
33 Router   192.168.2.2      192.168.2.2      0x80000011  2037  0x22 0x5570  60
34  # Router-LSA for Livorno inside L3VPN
35 Router  *192.168.6.6      192.168.6.6      0x80000010  2006  0x22 0x5f06  48
36 Network  172.16.100.5     192.168.1.5      0x8000000f  2002  0x22 0xe2ce  32
37 <...>
38
39 user@Livorno> show ospf database instance NMM lsa-id 192.168.2.2 extensive
40
41     OSPF database, Area 0.0.0.0
42  Type       ID               Adv Rtr           Seq      Age  Opt  Cksum  Len
43 Router   192.168.2.2      192.168.2.2      0x80000011  2064  0x22 0x5570  60
44   bits 0x1, link count 3
45   id 192.168.6.6, data 128.1.0.0, Type PointToPoint (1)     # ifIndex for sham link
46     Topology count: 0, Default metric: 10000      # Metric for sham link
47   id 192.168.1.1, data 172.16.1.2, Type PointToPoint (1)
48     Topology count: 0, Default metric: 160   # Metric derived from reference-bd 100g
49   id 172.16.1.0, data 255.255.255.252, Type Stub (3)
50     Topology count: 0, Default metric: 160   # Metric derived from reference-bd 100g
51  <...>
52   Aging timer 00:25:35
53   Installed 00:34:21 ago, expires in 00:25:36, sent 00:34:19 ago
54   Last changed 09:44:21 ago, Change count: 4
55
56 user@Livorno> show ospf database instance NMM lsa-id 192.168.6.6 extensive
57
58     OSPF database, Area 0.0.0.0
59  Type       ID               Adv Rtr           Seq      Age  Opt  Cksum  Len
60 Router  *192.168.6.6      192.168.6.6      0x80000010  2040  0x22 0x5f06  48
61   bits 0x1, link count 2
62   id 172.16.100.5, data 172.16.100.8, Type Transit (2)
```

```
63      Topology count: 0, Default metric: 1000  # Metric derived from reference-bd 100g
64    id 192.168.2.2, data 128.1.0.0, Type PointToPoint (1)      # ifIndex for sham link
65      Topology count: 0, Default metric: 10000   # Metric for sham link
66    <...>
67    Aging timer 00:25:59
68    Installed 00:34:00 ago, expires in 00:26:00, sent 00:33:58 ago
69    Last changed 09:44:00 ago, Change count: 4, Ours
```

Once the *sham link* is properly checked and established, the setup is ready for a soft transition to prefer one of router *Lille*'s uplinks through the L3VPN. Whenever a local routing decision is to be made and a *sham link* is present, the latter is selected based on an equal-path metric. However, the existence of the *sham link* remains unknown to the CEs: they just see a point-to-point connection between routers, as signalled by their LSAs. This means that in order to softly migrate traffic over the L3VPN without affecting any other flows going over the uplink to router *Nantes*, the path-calculated cost over the L3VPN must be *lower* than the native internal uplink. At this moment, the metric for the *sham link* must be adapted to revert the equation:

$$\overbrace{\textit{Lille–Nantes}\text{ OSPFv2 path cost}} > \overbrace{\textit{Lille–Nantes}\text{ MPLS path cost}} \tag{5.7}$$

$$\overbrace{\textit{Lille–Nantes}\text{ internal OSPFv2 path}} = \overline{\text{Lille–Nantes}} = 10000 \tag{5.8}$$

$$\overbrace{\textit{Lille–Nantes}\text{ MPLS path}} = \overline{\text{Sham-link Livorno–Skopie}} + 1160 \tag{5.9}$$

$$\overbrace{\textit{Lille–Nantes}\text{ internal OSPFv2 path}} = 10000 > \overline{\text{Sham-link Livorno–Skopie}} + 1160 \tag{5.10}$$

The most suitable approach for migrating traffic in this setup is to decrease the link metric for the *sham link* from 10000 to any value lower than 8840 in both directions. In order to avoid any influence on traffic flows over the parallel uplink between router *Lille* and router *Torino*, a value of 8839 ensures traffic shift over the L3VPN for the *Lille–Nantes* connection with minimal changes. This second migration step is based simply on metric adaption on such *sham link* and is illustrated in Figure 5.18.

When modifying such values in the configuration, path calculations promptly converge over the L3VPN, as shown in Listing 5.17.

Listing 5.17: Bidirectional metric adaption in sham link at Skopie and Livorno

```
1   user@Nantes> show route 192.168.1.5
2
3   inet.0: 34 destinations, 39 routes (34 active, 0 holddown, 0 hidden)
4   + = Active Route, - = Last Active, * = Both
5
6   192.168.1.5/32      *[OSPF/10] 00:09:47, metric 10000
7                        > via so-1/0/0.0 # Lille preferred over native uplink
8
9   user@Lille> show route 192.168.1.1
10
11  inet.0: 27 destinations, 29 routes (27 active, 0 holddown, 0 hidden)
12  + = Active Route, - = Last Active, * = Both
13
14  192.168.1.1/32      *[OSPF/10] 00:09:41, metric 10000
15                       > via so-0/1/0.0 # Nantes preferred over native uplink
16
17  [edit routing-instances NMM protocols ospf]
```

```
18 gonzalo@Skopie# set area 0.0.0.0 sham-link-remote 192.168.6.6 metric 8839
19 [edit routing-instances NMM protocols ospf]
20 gonzalo@Livorno# set area 0.0.0.0 sham-link-remote 192.168.2.2 metric 8839
21
22 user@Lille> show route 192.168.1.1
23
24 inet.0: 27 destinations, 29 routes (27 active, 0 holddown, 0 hidden)
25 + = Active Route, - = Last Active, * = Both
26
27 192.168.1.1/32      *[OSPF/10] 00:00:06, metric 9999
28                      > to 172.16.100.8 via fe-0/3/0.0 # Path to Nantes now pointing to sham link
29
30 user@Nantes> show route 192.168.1.5
31
32 inet.0: 34 destinations, 39 routes (34 active, 0 holddown, 0 hidden)
33 + = Active Route, - = Last Active, * = Both
34
35 192.168.1.5/32      *[OSPF/10] 00:00:03, metric 9999
36                      > via so-0/1/0.0 # Path to Lille now pointing to sham link
```

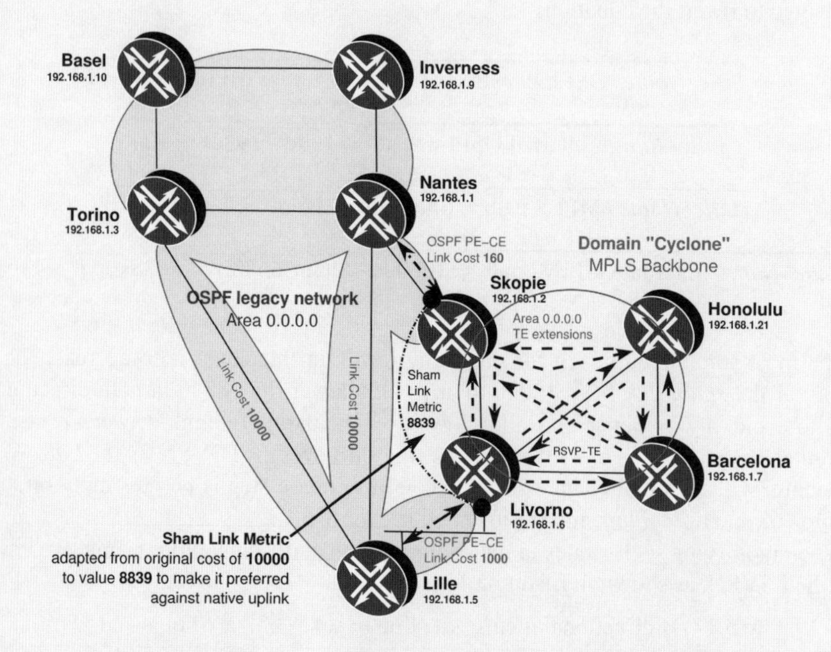

Figure 5.18: Stage two: adapt metric for Skopie–Livorno sham link.

After the metric adoption, the second native uplink internal to the legacy network between router *Nantes* and router *Lille* can be safely disconnected. This is checked by confirming active path next-hop selection for all OSPFv2 routes on router *Nantes* and router *Lille* and confirming that the relevant routes now point towards the L3VPN uplinks and that no routes elect the former connection, as shown in Listing 5.18.

Listing 5.18: Route selection towards sham link on router *Lille* and router *Nantes*

```
1 user@Nantes> show ospf neighbor
2 Address          Interface              State     ID               Pri  Dead
3 172.16.1.2       so-0/1/0.0             Full      192.168.2.2      128    34
```

```
 4 172.16.1.97      so-0/2/0.0          Full     192.168.1.3      128    38
 5 172.16.1.26      so-1/0/0.0          Full     192.168.1.5      128    37 # Lille
 6 172.16.1.66      so-1/0/1.0          Full     192.168.1.9      128    33
 7
 8 user@Nantes> show route protocol ospf table inet.0 terse active-path
 9
10 inet.0: 34 destinations, 39 routes (34 active, 0 holddown, 0 hidden)
11 + = Active Route, - = Last Active, * = Both
12
13 A Destination        P Prf   Metric 1   Metric 2  Next hop        AS path
14 <...>
15 * 192.168.1.3/32     O  10        160             >so-0/2/0.0
16 * 192.168.1.5/32     O  10       9999             >so-0/1/0.0   # Lille preferred beyond Skopie
17 * 192.168.1.9/32     O  10        643             >so-1/0/1.0
18 * 192.168.1.10/32    O  10        320             >so-0/2/0.0
19 <...>
20
21 user@Lille> show ospf neighbor
22 Address          Interface           State    ID               Pri  Dead
23 172.16.100.8     fe-0/3/0.0          Full     192.168.6.6      128    32
24 172.16.1.25      so-0/1/0.0          Full     192.168.1.1      128    37    # Nantes
25 172.16.1.29      so-0/1/1.0          Full     192.168.1.3      128    39
26
27 user@Lille> show route protocol ospf terse table inet.0 active-path
28
29 inet.0: 27 destinations, 29 routes (27 active, 0 holddown, 0 hidden)
30 + = Active Route, - = Last Active, * = Both
31
32 A Destination        P Prf   Metric 1   Metric 2  Next hop        AS path
33 <...>
34 * 192.168.1.1/32     O  10       9999            >172.16.100.8  # Nantes preferred beyond Livorno
35 * 192.168.1.3/32     O  10      10000            >so-0/1/1.0
36 * 192.168.1.9/32     O  10      10642            >172.16.100.8
37 * 192.168.1.10/32    O  10      10160            >so-0/1/1.0
38 <...>
```

Minimal metric differences in the path cost calculation over the *sham link* ensure that no previous flows over the second uplink between router *Torino* and router *Lille* are diverted over the newly established *sham link*.

Tearing down the OSPF neighbor relationship between both routers over the legacy link is seamless at this point. Figure 5.19 shows the scenario after decommissioning the link from the old topology.

A second redundant uplink between router *Torino* and router *Lille* still needs to be removed. Using the same *sham link* resources, the migration strategy is the same: a progressive deployment of a parallel *sham link*, with metric tuning for minimal cost differences. For the sake of illustration, another well-known domain is going to be used as second MPLS backbone service provider: domain “Mistral” from our known topology. Actually, utilization of such *sham link* mechanisms is independent of using one or many L3VPN providers. The scenario would have been similar with domain “Cyclone” as a single provider, but the intent is to enforce the main rationale behind *sham links*: virtual point-to-point OSPF L3VPN-specific structures to exclusively bind end points. Such a construct makes standard MP-BGP to PE–CE protocol redistribution useless because the tunnel is constructed inside the L3VPN and routing information prefers the tunnel as the eligible next hop because it is considered to be an intra-area connection.

Figure 5.20 now includes domain “Mistral” as a redundant L3VPN service provider with OSPF PE–CE connections in the same domain.

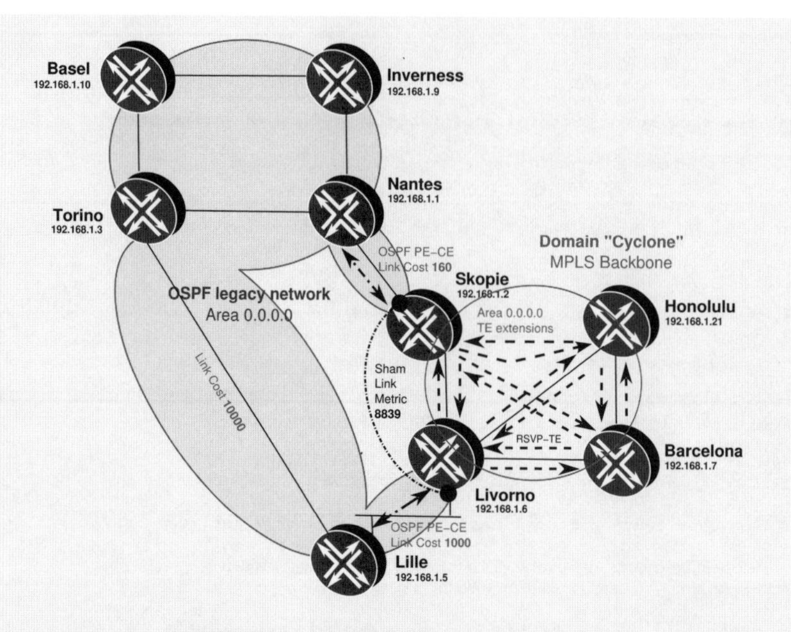

Figure 5.19: Stage three: decommissioning Lille–Nantes uplink.

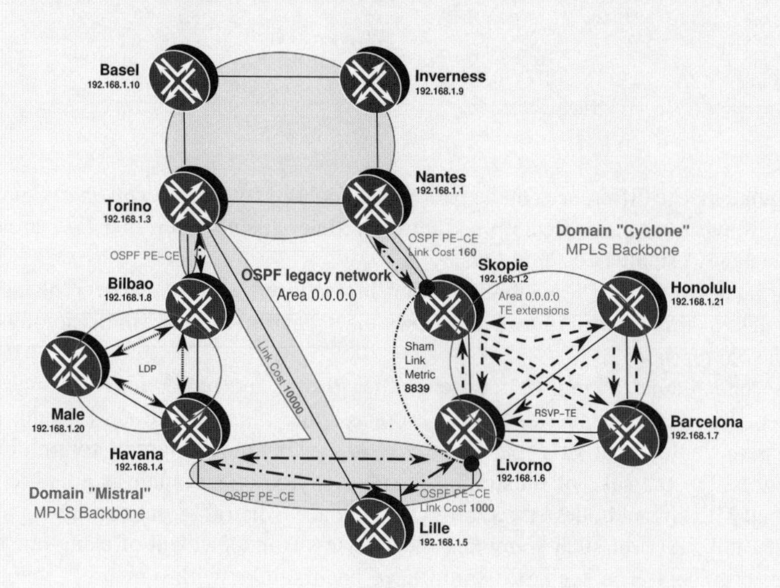

Figure 5.20: Stage four: connecting domain "Mistral" as second L3VPN supplier.

The previous debate applies to this situation and considering that the topology is almost symmetric, a parallel *sham link* is established between private loopback addresses inside the L3VPN, specifically between router *Bilbao* and router *Havana*. Metrics are adjusted in a completely parallel fashion as in the domain "Cyclone"'s *sham link*. To keep complete symmetry, only the ATM OC-3 link between router *Torino* and *Bilbao* includes a cost of 160 to completely replicate the OC-12 *Nantes-Skopie* uplink metric. Listing 5.19 includes snippets for the relevant changes on *Havana* and *Bilbao* as the PEs providing transit services for the OSPF domain with a parallel *sham link*, applying same design criteria as with router *Livorno* and router *Skopie*.

Listing 5.19: VRF and sham-link configuration on Havana and Bilbao

```
1  user@Havana> show configuration routing-instances
2  NMM {
3      instance-type vrf;
4      interface fe-0/3/0.0;
5      interface lo0.4;
6      route-distinguisher 64503:64503;
7      vrf-export NMM-vrf-export;
8      vrf-target target:64503:64503;
9      vrf-table-label;
10     routing-options {
11         router-id 192.168.4.4;
12     }
13     protocols {
14         ospf {
15             reference-bandwidth 100g;
16             sham-link local 192.168.4.4;
17             area 0.0.0.0 {
18                 sham-link-remote 192.168.8.8 metric 8839; # Same metric for parallel sham link
19                 interface lo0.4 {
20                     passive;
21                 }
22                 interface fe-0/3/0.0;
23             }
24         }
25     }
26 }
27
28 user@Bilbao-re0> show configuration routing-instances
29 NMM {
30     instance-type vrf;
31     interface at-1/2/0.0;
32     interface lo0.8;
33     route-distinguisher 64503:64503;
34     vrf-export NMM-vrf-export;
35     vrf-target target:64503:64503;
36     vrf-table-label;
37     routing-options {
38         router-id 192.168.8.8;
39     }
40     protocols {
41         ospf {
42             reference-bandwidth 100g;
43             sham-link local 192.168.8.8;
44             area 0.0.0.0 {
45                 sham-link-remote 192.168.4.4 metric 8839; # Same metric for parallel sham link
46                 interface at-1/2/0.0 {
47                     metric 160;
48                 }
49                 interface lo0.8 {
50                     passive;
51                 }
52             }
53         }
54     }
55 }
```

The setup is depicted on Figure 5.21, which includes parallel *sham-link* settlement and metric adaption.

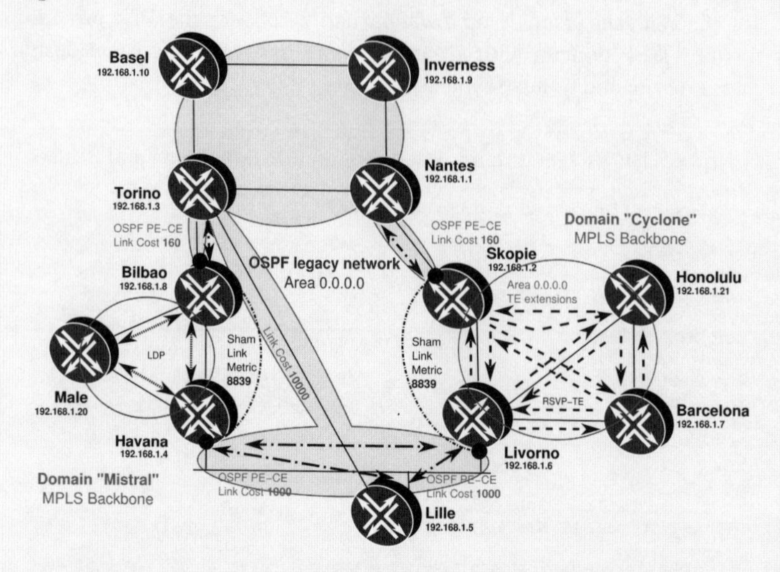

Figure 5.21: Stage five: sham link and metric adaption over domain "Mistral".

With the calculated metric settings, no traffic should flow over the remaining internal uplink between router *Torino* and router *Lille*, similar to what was achieved previously with domain "Cyclone." Inspecting the routing tables on both directly attached CEs, as shown in Listing 5.20, confirms the successful smooth migration over the L3VPN.

Listing 5.20: Routing tables on Nantes, Torino, and Lille after the final sham-link metric adaption

```
1  user@Nantes> show route protocol ospf active-path table inet.0 terse
2
3  inet.0: 34 destinations, 38 routes (34 active, 0 holddown, 0 hidden)
4  + = Active Route, - = Last Active, * = Both
5
6  A Destination        P Prf   Metric 1   Metric 2  Next hop        AS path
7  <...>
8  * 192.168.1.3/32     O  10        160             >so-0/2/0.0     # Torino over direct link
9  * 192.168.1.5/32     O  10       9999             >so-0/1/0.0     # Lille over L3VPN to Skopie
10 * 192.168.1.9/32     O  10        643             >so-1/0/1.0     # Inverness over direct link
11 * 192.168.1.10/32    O  10        320             >so-0/2/0.0     # Basel over Torino
12 * 224.0.0.5/32       O  10          1              MultiRecv
13
14 user@Torino> show route protocol ospf active-path table inet.0 terse
15
16 inet.0: 30 destinations, 35 routes (30 active, 0 holddown, 0 hidden)
17 + = Active Route, - = Last Active, * = Both
18
19 A Destination        P Prf   Metric 1   Metric 2  Next hop        AS path
20 <...>
21 * 192.168.1.1/32     O  10        160             >so-0/1/0.0     # Nantes over direct link
22 * 192.168.1.5/32     O  10       9999             >at-0/0/1.0     # Lille over L3VPN to Bilbao
23 * 192.168.1.9/32     O  10        803             >so-0/1/0.0     # Inverness balanced over Nantes
```

```
24                                                     at-1/2/0.1001 #   and Inverness
25 * 192.168.1.10/32    O 10        160                >at-1/2/0.1001 # Inverness over direct link
26 * 224.0.0.5/32       O 10          1                MultiRecv
27
28 user@Lille> show route protocol ospf active-path table inet.0 terse
29
30 inet.0: 27 destinations, 28 routes (27 active, 0 holddown, 0 hidden)
31 + = Active Route, - = Last Active, * = Both
32
33 A Destination        P Prf   Metric 1   Metric 2  Next hop        AS path
34 <...>
35 * 192.168.1.1/32     O 10       9999              >172.16.100.8 # Nantes over L3VPN to Livorno
36 * 192.168.1.3/32     O 10       9999              >172.16.100.1 # Torino over L3VPN to Havana
37 * 192.168.1.9/32     O 10      10642              >172.16.100.8 # Inverness over uplink to Livorno
38 * 192.168.1.10/32    O 10      10159              >172.16.100.1 # Basel over L3VPN to Havana
39 * 224.0.0.5/32       O 10          1               MultiRecv
```

At this point, the remaining internal link between router *Torino* and router *Lille* can be safely decommissioned because no active paths are established over it. This last CE backdoor is dismantled, as shown in Figure 5.22, and the OSPFv2 network is virtually redundantly connected to router *Lille*, but is now physically split.

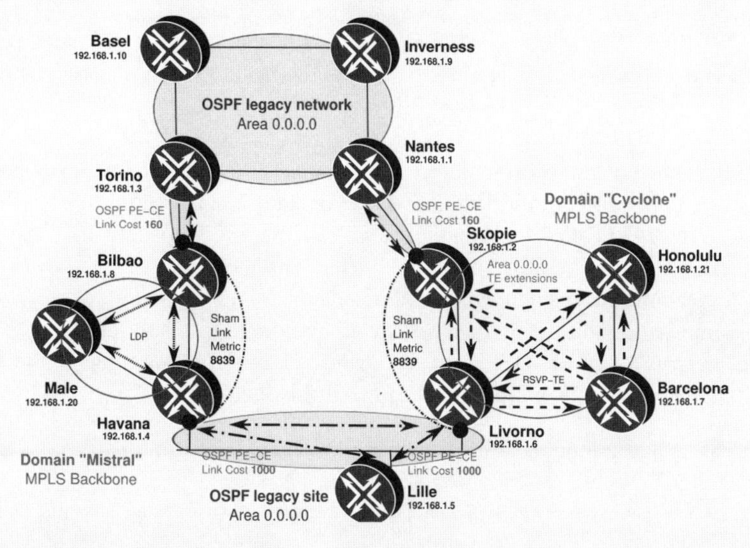

Figure 5.22: Stage six: decommissioning the last link Torino–Lille and segmenting the OSPFv2 network.

Sham links were designed in [RFC4576] as virtual interconnects to be most preferred as next hops compared to existing CE backdoors. In the current stage, all backdoors from router *Lille* to the rest of the OSPFv2 network have been gracefully dismantled. The rationale for using *sham links* to interconnect OSPF islands is no longer necessary, because all backdoors have been disconnected. Standard route-to-LSA translation, as per [RFC4576], can be tuned to achieve a similar objective without needing to maintain point-to-point structures and *sham-link* end points.

The Application Note on Page 431 continues this example, showing how to dismantle these now unnecessary *sham links* and how to achieve a proper route-to-LSA translation with `domain-ID` planning.

Junos Tip: Sham-link end point address advertisement and selection

Sham links require that next hops be visible through the L3VPN to be properly established. Listing 5.21 depicts resolution at each PE inside the VRF for the *sham-link* end points.

Listing 5.21: Sham-link end-point resolution on Skopie and Livorno as transit PEs

```
1  user@Skopie> show route 192.168.6.6 table NMM
2
3  NMM.inet.0: 23 destinations, 41 routes (23 active, 0 holddown, 0 hidden)
4  + = Active Route, - = Last Active, * = Both
5  192.168.6.6/32      *[BGP/170] 17:55:24, localpref 100, from 192.168.1.6
6                         AS path: I
7                       > via so-1/0/0.0, label-switched-path Skopie-to-Livorno
8
9  user@Livorno> show route 192.168.2.2 table NMM
10 NMM.inet.0: 23 destinations, 41 routes (23 active, 0 holddown, 1 hidden)
11 + = Active Route, - = Last Active, * = Both
12
13 192.168.2.2/32      *[BGP/170] 17:55:16, localpref 100, from 192.168.1.2
14                        AS path: I
15                      > via so-0/0/0.0, label-switched-path Livorno-to-Skopie
```

This requirement is intended to detect active paths for the destinations beyond a remote PE over the L3VPN and not for destinations beyond any CE backdoor. Junos OS enforces Section 4.2.7.1 of [RFC4577] by explicitly not advertising stub networks that represent *sham-link* end point addresses when configuring `sham-link local`, even when the covering logical interface is active in OSPF inside the VRF.

In the same setup, before any link cost manipulation, even when setting such loopback interfaces as `passive` in OSPF, Type 1 Router LSAs do not include the addresses representing the *sham-link* endpoints, as shown in Listing 5.22.

Listing 5.22: Futile loopback interface passivation in OSPFv2 at Skopie and Livorno

```
1  [edit]
2  user@Skopie# set routing-instances NMM protocols ospf area 0 interface lo0.2 passive
3  [edit]
4  user@Livorno# set routing-instances NMM protocols ospf area 0 interface lo0.6
5      passive
6
7  user@Skopie> show ospf database instance NMM lsa-id 192.168.2.2 extensive
8
9      OSPF database, Area 0.0.0.0
10  Type        ID              Adv Rtr          Seq        Age  Opt  Cksum  Len
11 Router  *192.168.2.2      192.168.2.2      0x80000064  1695  0x22 0xaec3  60
12   bits 0x1, link count 3
13   id 192.168.6.6, data 128.1.0.0, Type PointToPoint (1)  # Sham link to Livorno
14     Topology count: 0, Default metric: 10000
15   id 192.168.1.1, data 172.16.1.2, Type PointToPoint (1)  # P2P link to CE Nantes
16     Topology count: 0, Default metric: 160
17   id 172.16.1.0, data 255.255.255.252, Type Stub (3)  # Stub network for P2P prefix
18     Topology count: 0, Default metric: 160
19   Gen timer 00:21:45
20   Aging timer 00:31:45
```

```
21    Installed 00:28:15 ago, expires in 00:31:45, sent 00:28:13 ago
22    Last changed 02:08:15 ago, Change count: 17, Ours
23
24 user@Livorno> show ospf database instance NMM lsa-id 192.168.6.6 extensive
25
26     OSPF database, Area 0.0.0.0
27  Type       ID               Adv Rtr          Seq      Age  Opt  Cksum  Len
28 Router  *192.168.6.6      192.168.6.6      0x80000059  2278  0x22 0xb0b   48
29   bits 0x3, link count 2
30   id 172.16.100.8, data 172.16.100.8, Type Transit (2)   # Transit network to CE Lille
31     Topology count: 0, Default metric: 1000
32   id 192.168.2.2, data 128.1.0.1, Type PointToPoint (1)  # Sham Link to Skopie
33     Topology count: 0, Default metric: 10000
34   Gen timer 00:01:45
35   Aging timer 00:22:01
36   Installed 00:37:58 ago, expires in 00:22:02, sent 00:37:56 ago
37   Last changed 02:00:03 ago, Change count: 21, Ours
```

On the other hand, setting *sham-link* end point addresses to match those from the PE–CE interface is futile when the CE backdoor is active, because such end points are internally known through the legacy OSPFv2 domain and default preferences result in OSPF internal routes being preferred over VPN-IPv4 BGP paths. Listing 5.23 shows that the failed *sham-link* end point address resolution over internal OSPF routes causes the *sham link* not to become active.

Listing 5.23: Sham-link end-point wrong resolution through internal OSPF paths on Skopie and Livorno

```
1  [edit routing-instances NMM]
2  user@Skopie# show | compare
3  [edit routing-instances NMM protocols ospf]
4  -    sham-link local 192.168.2.2;
5  +    sham-link local 172.16.1.2;
6  [edit routing-instances NMM protocols ospf area 0.0.0.0]
7  +      sham-link-remote 172.16.100.8 metric 10000;
8  -      sham-link-remote 192.168.6.6 metric 10000;
9
10 [edit routing-instances NMM]
11 user@Livorno# show | compare
12 [edit routing-instances NMM protocols ospf]
13 -    sham-link local 192.168.6.6;
14 +    sham-link local 172.16.100.8;
15 [edit routing-instances NMM protocols ospf area 0.0.0.0]
16 +      sham-link-remote 172.16.1.2 metric 10000;
17 -      sham-link-remote 192.168.2.2 metric 10000;
18
19 user@Skopie> show route 172.16.100.8 table NMM
20
21 NMM.inet.0: 20 destinations, 39 routes (20 active, 0 holddown, 0 hidden)
22 + = Active Route, - = Last Active, * = Both
23
24 172.16.100.0/24    *[OSPF/10] 00:01:50, metric 11160  # Sham link destination internally preferred
25                     > via so-0/1/0.0
26                     [BGP/170] 00:01:50, localpref 100, from 192.168.1.6
27                       AS path: I
28                     > via so-1/0/0.0, label-switched-path Skopie-to-Livorno
29
30 user@Skopie> show ospf neighbor instance NMM
31   # No more Sham-Links
32 Address          Interface              State     ID               Pri  Dead
33 172.16.1.1       so-0/1/0.0             Full      192.168.1.1      128    37
34
35 user@Livorno> show route 172.16.1.2 table NMM
36
37 NMM.inet.0: 20 destinations, 38 routes (20 active, 0 holddown, 0 hidden)
```

```
38 + = Active Route, - = Last Active, * = Both
39
40 172.16.1.0/30      *[OSPF/10] 00:01:11, metric 11160 # Sham-Link destination internally preferred
41                     > to 172.16.100.5 via fe-0/3/0.0
42                     [BGP/170] 00:01:07, localpref 100, from 192.168.1.2
43                       AS path: I
44                     > via so-0/0/0.0, label-switched-path Livorno-to-Skopie
45
46 user@Livorno> show ospf neighbor instance NMM
47   # No more Sham-Links
48 Address          Interface              State     ID               Pri  Dead
49 172.16.100.5     fe-0/3/0.0             Full      192.168.1.5      128    39
```

To summarize, because Junos OS enforces this deliberate route suppression towards attached CEs for *sham links* to be established, a best practice is to define specific loopback addresses as end points for these *sham links* or to use existing loopback addresses inside each VRF, ensuring that the disappearance of the route in the OSPF database does not lead to any side effects at any CE with other services.

OSPF route handling

Junos OS complies with the default behaviors defined in [RFC4576] and [RFC4577] by enforcing a NULL *Domain Identifier* if no specific value is configured.

The `domain-id` knob allows local setting of an OSPF Domain Identifier on a per-VRF basis to check incoming VPN-IPv4 routes and determine how they are redistributed locally. Adding the OSPF Domain Identifier extended community to advertised VPN-IPv4 routes is controlled with policy language (usually as part of the `vrf-export` chain), in which such identifiers must be explicitly added to the routes before being added for advertisement on the backbone.

The absence of any `domain-id` configuration is interpreted in Junos OS as a NULL domain, but a good practice is to configure explicit OSPF Domain Identifiers among intended sites to avoid accidental redistribution effects from other domains.

Junos OS supports both AS- and IPv4-address-like extended communities for the OSPF Domain Identifier, also implementing autocompletion of the Administrative Field to a value of 0 in case it is not configured. Listing 5.24 shows an example of how to configure both exports by adding the AS-like OSPF Domain Identifier extended community and by performing import checks against the local AS-like OSPF Domain Identifier. Using IPv4-address-like OSPF Domain Identifiers is entirely similar.

Listing 5.24: Configuring an AS-like Domain Identifier on a routing instance on Livorno

```
1 [edit routing-instances NMM] # Incoming checks against local domain-id
2 user@Livorno# set protocols ospf domain-id 64502:1
3
4 [edit] # Definition of the extended community
5 user@Livorno# set policy-options community domain-id:64502:1 members domain-id:64502:1
6
7 [edit policy-options policy-statement NMM-vrf-export term ospf]  # Add extended community on export
8 user@Livorno# set then community add domain-id:64502:1
```

Junos Tip: Configuring Domain Identifiers

A common operator mistake with the activation of Domain Identifiers in Junos OS is assuming that just configuring the `domain-id` in the ospf protocol stanza, that is, just with the command from Line 2 in Listing 5.24, is sufficient.

That snippet just activates a particular `domain-id` to compare against *Domain Identifier* extended communities from imported routes and to determine the LSA translation type derived from the route according to [RFC4577]. However, it does not perform tagging of extended communities with that `domain-id` for advertised routes. It controls inbound checking only, but does not perform outbound tagging.

Adding OSPF Domain Identifier extended communities is done the same way as with other extended communities in Junos OS: by defining a community and adding it in a policy term, as shown in the rest of Listing 5.24. Note that the rich policy language in Junos OS allows an administrator to even add more than a single *Domain Identifier* to advertised routes.

An interesting corner case for a migration case study appears when setting `domain-id disable`. The following effects appear when crafting this setup with that value:

- The backbone becomes detached from the VRF for local ABR translations between areas in the same VRF, which is relevant for cases with default summary injection in stub areas.
- Unilateral translation of MP-BGP routes into LSAs 5 occurs, independent of the Domain Identifier values present in received routes.
- Type 3 LSAs from a local non-backbone area are selected for translation into MP-BGP.

In a nutshell, configuring Domain Identifiers is an indirect way to enforce that all routes are redistributed towards the CE as External LSAs, independently of incoming VPN-IPv4 Domain Identifier values.

Application Note: Dismantling sham links and selecting adequate Domain Identifiers for route translation

Following the example and setup illustrated in the Application Note 5.5.2 for the last stage, shown in Figure 5.22, the idea is to maintain a joint domain and keep the same area identification understandable across L3VPN islands by relating it to a common domain. Potential routing information accuracy loss that may occur when dismantling *sham links* and injecting Type 3 LSAs for all internal routes from each L3VPN island to each VRF is negligible because there are now no backdoors and it is not possible to detect the same routes internally in the area through Type 1 or Type 2 LSAs.

Imagine a multiservices MPLS VPN backbone with more than one OSPF domain that is expanded across VPNs. Adequate Domain Identifier planning and settings are essential for proper area identification and route translation. Following a considered best practice, a common AS-like Domain Identifier “64502:1” is configured in the VRF on all PEs, as shown in Listing 5.24 for router *Livorno*.

However, an additional step is needed for proper route redistribution: an export policy on each PE that redistributes backbone routes into OSPF towards the CEs. This is shown for router *Livorno* in Listing 5.25 for illustrative purposes, but it is done similarly on all other PEs.

Listing 5.25: Activating OSPF export policies in Livorno

```
1  [edit policy-options policy-statement NMM-bgp-to-ospf]
2  user@Livorno# show
3  term bgp {
4      from {
5          protocol bgp;
6      }
7      then accept;
8  }
9
10 [edit routing-instances NMM protocols ospf]
11 user@Livorno# set export NMM-bgp-to-ospf
```

The setup depicted in Figure 5.22 is extended over two different MPLS VPN backbones, but the following examples are equally representative of the case in which one or more MPLS backbones provides connectivity to the OSPF domain.

After deploying *OSPF Domain Identifiers* and export policies while keeping *sham links*, another OSPF database inspection is needed. Comparing the new situation in Listing 5.26 with the previous OSPF database in Listing 5.16, no relevant changes are visible.

Listing 5.26: OSPF database from the domain after activating Domain Identifiers and before dismantling sham links

```
1  user@Livorno> show ospf database instance NMM
2
3      OSPF database, Area 0.0.0.0
4   Type      ID              Adv Rtr          Seq         Age  Opt  Cksum  Len
5  Router    192.168.1.1     192.168.1.1     0x80000028  1337  0x22 0x796b 108
6  Router    192.168.1.3     192.168.1.3     0x8000003f  1174  0x22 0x87aa 120
7  Router    192.168.1.5     192.168.1.5     0x8000003c   148  0x22 0x2348  48
8  Router    192.168.1.9     192.168.1.9     0x80000023  1338  0x22 0x9bb5  96
9  Router    192.168.1.10    192.168.1.10    0x80000022  1339  0x22 0xa77d  84
10 Router    192.168.2.2     192.168.2.2     0x80000042  1336  0x22 0x7c5c  60
11 Router    192.168.4.4     192.168.4.4     0x80000059  1076  0x22 0xf4cb  48
12 Router   *192.168.6.6     192.168.6.6     0x80000060  1335  0x22 0x2b4   48
13 Router    192.168.8.8     192.168.8.8     0x80000009  1080  0x22 0x1f14  60
14 Network   172.16.100.5    192.168.1.5     0x80000027  1336  0x22 0x1d07  36
15 OpaqArea 1.0.0.1          192.168.1.1     0x8000001d  1337  0x22 0x78c3  28
16 OpaqArea 1.0.0.1          192.168.1.3     0x8000001d  1338  0x22 0x80b7  28
17 OpaqArea 1.0.0.1          192.168.1.5     0x8000002b  1336  0x22 0x6cb9  28
18 OpaqArea 1.0.0.1          192.168.1.9     0x8000001d  1338  0x22 0x9893  28
19 OpaqArea 1.0.0.1          192.168.1.10    0x8000001d  1339  0x22 0x9c8d  28
```

Smart readers may wonder why no new LSAs appear in the database now that OSPF export policies are active in each PE's VRF. Latest Junos OS releases implement additional loop-prevention safeguard checks when exporting routes from MP-BGP to OSPF when active *sham links* are present in the setup: BGP routes are not exported to OSPF if a sham link is available; however, the routes remain locally active on the PE for forwarding purposes. The rationale behind this implementation is to avoid the creation of additional forwarding loops that could be formed when simultaneously populating routes from the remote end through the *sham link* as a consequence of MP-BGP-to-OSPF redistribution in the PE.

This safeguard check is visible in the form of OSPF *hidden* routes in each table on each PE representing the MP-BGP routes that should be exported to OSPF, but whose redistribution is hindered by this prevention mechanism, as seen in Listing 5.27 for routes learned over the *sham link* on router *Livorno* and router *Skopie* pointing to the remote end. The end result is to keep MP-BGP routes active on the PE for forwarding purposes over the backbone while impeding their redistribution to OSPF because of the overlapping OSPF hidden route in the RIB that points to the *sham link*. Similarly, those hidden OSPF routes that point to the *sham link* are not eligible for MP-BGP redistribution either to avoid creating the loop in the other direction. Note that this also benefits from Junos OS default protocol preferences being applied (OSPF is preferred over BGP).

Listing 5.27: Hidden OSPF routes learned over the sham link on Livorno and Skopie

```
1  user@Skopie> show route protocol ospf hidden table NMM
2
3  NMM.inet.0: 18 destinations, 21 routes (18 active, 0 holddown, 2 hidden)
4  + = Active Route, - = Last Active, * = Both
5
6  172.16.100.0/24     [OSPF] 01:01:39, metric 9839
7                      > via shamlink.0
8  192.168.1.5/32      [OSPF] 01:01:39, metric 9839 # Lille
9                      > via shamlink.0
10
11 user@Livorno> show route protocol ospf hidden table NMM
12
13 NMM.inet.0: 18 destinations, 30 routes (18 active, 0 holddown, 12 hidden)
14 + = Active Route, - = Last Active, * = Both
15
16 <...>
17 172.16.1.0/30       [OSPF] 01:02:12, metric 8999
18                     > via shamlink.0
19 172.16.1.64/30      [OSPF] 01:02:12, metric 9642
20                     > via shamlink.0
21 <...>
22 172.16.1.100/30     [OSPF] 00:38:50, metric 9319
23                     > via shamlink.0
24 192.168.1.1/32      [OSPF] 01:02:12, metric 8999  # Nantes
25                     > via shamlink.0
26 192.168.1.3/32      [OSPF] 01:02:12, metric 9159  # Torino
27                     > via shamlink.0
28 192.168.1.9/32      [OSPF] 01:02:12, metric 9642  # Inverness
29                     > via shamlink.0
30 192.168.1.10/32     [OSPF] 01:02:12, metric 9319  # Basel
31                     > via shamlink.0
```

Therefore, the modification is guaranteed to be completely hitless for the setup, because the OSPFv2 database from the domain remains unmodified, as shown in Listing 5.26.

The advantage of performing the previous activation is that it allows to check that communities have been properly added on each PE, without modifying routing tables, just by inspecting VPN-IPv4 advertised routes, as shown in Listing 5.28 for the redistributed OSPF routes to the backbone, without effectively impacting the setup. Lines 7, 11, 15, and 19 show properly set extended communities on advertised routes from router *Skopie* representing Router LSAs for router *Nantes*, router *Torino*, *Inverness*, and *Basel*. Line 26 shows the same for router *Livorno* representing router *Lille*. On the other backbone, the parallel effect is visible on Lines 34, 38, 42, and 46 for *Bilbao* and on Line 53 for *Havana* representing the same LSAs. Note that if the setup were a joint or interconnected backbone, the effect would be the same; only the AS paths would change accordingly.

Listing 5.28: Extended community inspection after activating Domain Identifiers

```
1  user@Skopie> show route advertising-protocol bgp 192.168.1.6 192.168.1.0/24
2              extensive | match "entry|Commun|path"
3  <...>
4  * 192.168.1.1/32 (1 entry, 1 announced)  # Nantes
5       AS path: [64501] I
6       # Common Domain-ID, Area 0 Router-LSA
7       Communities: target:64501:64501 domain-id:64502:1 rte-type:0.0.0.0:1:0
8  * 192.168.1.3/32 (1 entry, 1 announced)  # Torino
9       AS path: [64501] I
10      # Common Domain-ID, Area 0 Router-LSA
11      Communities: target:64501:64501 domain-id:64502:1 rte-type:0.0.0.0:1:0
12 * 192.168.1.9/32 (1 entry, 1 announced)  # Inverness
13      AS path: [64501] I
14      # Common Domain-ID, Area 0 Router-LSA
15      Communities: target:64501:64501 domain-id:64502:1 rte-type:0.0.0.0:1:0
16 * 192.168.1.10/32 (1 entry, 1 announced)  # Basel
17      AS path: [64501] I
18      # Common Domain-ID, Area 0 Router-LSA
19      Communities: target:64501:64501 domain-id:64502:1 rte-type:0.0.0.0:1:0
20
21 user@Livorno> show route advertising-protocol bgp 192.168.1.2 192.168.1.5/32
22      extensive | match "entry|Commun|path"
23 * 192.168.1.5/32 (1 entry, 1 announced)  # Lille
24      AS path: [64501] I
25      # Common Domain-ID, Area 0 Router-LSA
26      Communities: target:64501:64501 domain-id:64502:1 rte-type:0.0.0.0:1:0
27
28 user@Bilbao-re0> show route advertising-protocol bgp 192.168.1.4
29     extensive | match "entry|Commun|path"
30 <...>
31 * 192.168.1.1/32 (1 entry, 1 announced)  # Nantes
32      AS path: [64503] I
33      # Common Domain-ID, Area 0 Router-LSA
34      Communities: target:64503:64503 domain-id:64502:1 rte-type:0.0.0.0:1:0
35 * 192.168.1.3/32 (1 entry, 1 announced)  # Torino
36      AS path: [64503] I
37      # Common Domain-ID, Area 0 Router-LSA
38      Communities: target:64503:64503 domain-id:64502:1 rte-type:0.0.0.0:1:0
39 * 192.168.1.9/32 (1 entry, 1 announced)  # Inverness
40      AS path: [64503] I
41      # Common Domain-ID, Area 0 Router-LSA
42      Communities: target:64503:64503 domain-id:64502:1 rte-type:0.0.0.0:1:0
43 * 192.168.1.10/32 (1 entry, 1 announced)  # Basel
44      AS path: [64503] I
45      # Common Domain-ID, Area 0 Router-LSA
46      Communities: target:64503:64503 domain-id:64502:1 rte-type:0.0.0.0:1:0
47
48 user@Havana> show route advertising-protocol bgp 192.168.1.8 192.168.1.5/32
49             extensive | match "entry|Commun|path"
50 * 192.168.1.5/32 (1 entry, 1 announced)  # Lille
51      AS path: [64503] I
52      # Common Domain-ID, Area 0 Router-LSA
53      Communities: target:64503:64503 domain-id:64502:1 rte-type:0.0.0.0:1:0
```

The current situation, including *Domain Identifier* checking and advertisement, is depicted in Figure 5.23.

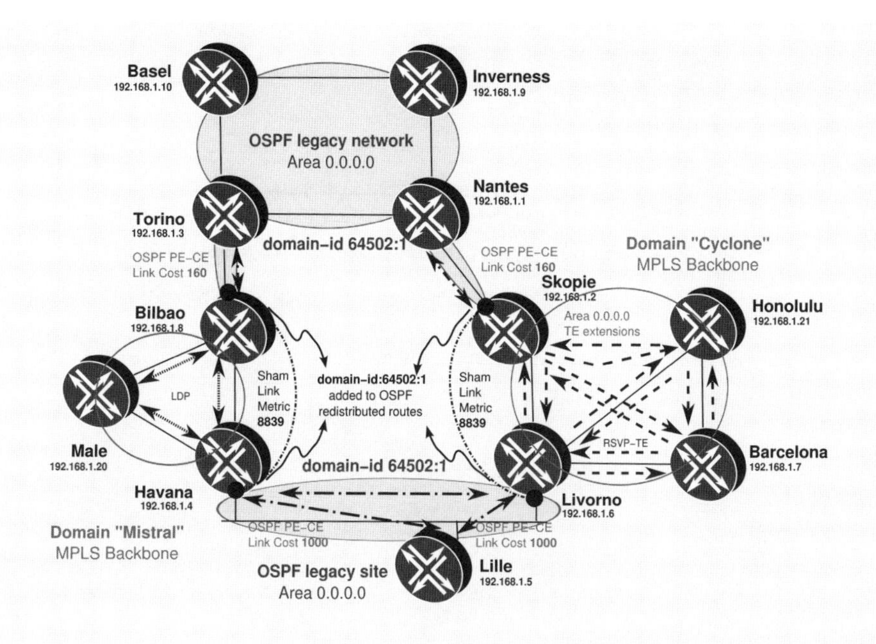

Figure 5.23: Stage seven: Domain Identifier activation.

Once OSPF Domain Identifier tagging has been ensured, *sham links* can be dismantled. Even though both of them can be deleted simultaneously, a good network engineer must consider a transient stage. Thus, this exercise first deletes the *sham link* over domain "Mistral" (forwarding and traffic balancing considerations apart) to evaluate potential issues. Listing 5.29 shows the configuration changes and effects in the OSPF database after this operation. In inspecting the OSPF database, the only new AS external LSAs that appear on Lines 49 and 51 refer to the parallel *sham-link* end points for router *Skopie* and router *Livorno*, which were used previously only for signaling purposes and were not previously redistributed to OSPF because they were visible beyond the parallel *sham link*. No other major changes are visible. As expected, Lines 58 and 78 reveal that point-to-point link types representing the dismantled *sham link* disappear from the Router LSAs.

Listing 5.29: Dismantling sham links on Havana and Bilbao

```
1  [edit routing-instances NMM protocols ospf]
2  user@Havana# delete sham-link local
3  [edit routing-instances NMM protocols ospf]
4  user@Havana# delete area 0.0.0.0 sham-link-remote 192.168.8.8
5
6  Sep  3 18:05:11  Havana rpd[1221]: RPD_OSPF_NBRDOWN: OSPF neighbor 192.168.8.8
7       (realm ospf-v2 shamlink.0 area 0.0.0.0) state changed from Full to Down because of KillNbr
8       (event reason: interface went down)
9  Sep  3 18:05:11  Havana rpd[1221]: EVENT Delete UpDown shamlink.0 index 2147549184
10      <PointToPoint Localup>
11 Sep  3 18:05:11  Havana rpd[1221]: EVENT <Delete UpDown> shamlink.0 index 2147549184
```

```
12      <PointToPoint>
13
14 [edit routing-instances NMM protocols ospf]
15 user@Bilbao-re0# delete sham-link local
16 [edit routing-instances NMM protocols ospf]
17 user@Bilbao-re0# delete area 0.0.0.0 sham-link-remote 192.168.4.4
18
19 Sep  3 18:05:15  Bilbao-re1 rpd[1127]: RPD_OSPF_NBRDOWN: OSPF neighbor 192.168.4.4
20      (realm ospf-v2 shamlink.0 area 0.0.0.0) state changed from Full to Down because of KillNbr
21      (event reason: interface went down)
22 Sep  3 18:05:15  Bilbao-re1 rpd[1127]: EVENT Delete UpDown shamlink.0 index 2147549184
23      <PointToPoint Localup>
24 Sep  3 18:05:15  Bilbao-re1 rpd[1127]: EVENT <Delete UpDown> shamlink.0 index 2147549184
25      <PointToPoint>
26
27 user@Bilbao-re0> show ospf database instance NMM
28
29     OSPF database, Area 0.0.0.0
30  Type       ID               Adv Rtr          Seq      Age  Opt  Cksum  Len
31 Router   192.168.1.1      192.168.1.1      0x8000002f  1759  0x22 0x6b72 108
32 Router   192.168.1.3      192.168.1.3      0x80000046  1595  0x22 0x79b1 120
33 Router   192.168.1.5      192.168.1.5      0x8000004d  1006  0x22 0x159   48
34 Router   192.168.1.9      192.168.1.9      0x8000002a  1760  0x22 0x8dbc  96
35 Router   192.168.1.10     192.168.1.10     0x80000029  1759  0x22 0x9984  84
36 Router   192.168.2.2      192.168.2.2      0x8000004d  1159  0x22 0x6667  60
37 Router   192.168.4.4      192.168.4.4      0x80000004    24  0x22 0x6737  48
38 Router   192.168.6.6      192.168.6.6      0x80000003   998  0x22 0xbc57  48
39 Router  *192.168.8.8      192.168.8.8      0x80000014    16  0x22 0x841c  60
40 Network  172.16.100.5     192.168.1.5      0x80000002  1006  0x22 0x77d1  36
41 OpaqArea 1.0.0.1          192.168.1.1      0x80000024  1759  0x22 0x6aca  28
42 OpaqArea 1.0.0.1          192.168.1.3      0x80000024  1758  0x22 0x72be  28
43 OpaqArea 1.0.0.1          192.168.1.5      0x80000032  2289  0x22 0x5ec0  28
44 OpaqArea 1.0.0.1          192.168.1.9      0x80000024  1760  0x22 0x8a9a  28
45 OpaqArea 1.0.0.1          192.168.1.10     0x80000024  1759  0x22 0x8e94  28
46     OSPF AS SCOPE link state database
47  Type       ID               Adv Rtr          Seq      Age  Opt  Cksum  Len
48 Extern   192.168.2.2      192.168.6.6      0x80000001  1161  0xa2 0x8606  36
49  # Redistribution of Sham-Link endpoint
50 Extern   192.168.6.6      192.168.2.2      0x80000008  2019  0xa2 0x582d  36
51  # Redistribution of Sham-Link endpoint
52
53 user@Bilbao-re0> show ospf database instance NMM lsa-id 192.168.8.8 extensive
54
55     OSPF database, Area 0.0.0.0
56  Type       ID               Adv Rtr          Seq      Age  Opt  Cksum  Len
57 Router  *192.168.8.8      192.168.8.8      0x80000014   555  0x22 0x841c  60
58   bits 0x3, link count 3  # Missing Point-to-Point link for sham link
59   id 192.168.1.3, data 172.16.1.102, Type PointToPoint (1)
60     Topology count: 0, Default metric: 160
61   id 172.16.1.100, data 255.255.255.252, Type Stub (3)
62     Topology count: 0, Default metric: 160
63   id 192.168.8.8, data 255.255.255.255, Type Stub (3)
64     Topology count: 0, Default metric: 0
65   Topology default (ID 0)
66     Type: PointToPoint, Node ID: 192.168.1.3
67       Metric: 160, Bidirectional
68   Gen timer 00:40:45
69   Aging timer 00:50:45
70   Installed 00:09:15 ago, expires in 00:50:45, sent 00:09:15 ago
71   Last changed 00:09:15 ago, Change count: 7, Ours
72
73 user@Havana> show ospf database instance NMM lsa-id 192.168.4.4 extensive
74
75     OSPF database, Area 0.0.0.0
76  Type       ID               Adv Rtr          Seq      Age  Opt  Cksum  Len
77 Router  *192.168.4.4      192.168.4.4      0x80000004   606  0x22 0x6737  48
78   bits 0x3, link count 2 # Missing Point-to-Point link for sham link
79   id 172.16.100.5, data 172.16.100.1, Type Transit (2)
```

```
80        Topology count: 0, Default metric: 1000
81      id 192.168.4.4, data 255.255.255.255, Type Stub (3)
82        Topology count: 0, Default metric: 0
83      Topology default (ID 0)
84        Type: Transit, Node ID: 172.16.100.5
85          Metric: 481, Bidirectional
86      Gen timer 00:39:53
87      Aging timer 00:49:53
88      Installed 00:10:06 ago, expires in 00:49:54, sent 00:10:06 ago
89      Last changed 00:10:06 ago, Change count: 3, Ours
```

The reason for this smooth dismantling with no unexpected side effects is because of default OSPF route selection policies and Junos OS default preferences, together with the built-in loop-prevention mechanisms in the latest Junos OS releases when *sham links* coexist with export policies for common destinations. Listing 5.30 shows route selection at router *Bilbao* and router *Havana* for internal routers in the remote island, namely router *Nantes* and router *Lille*. The output shows that the internal route preference remains for OSPF and also points to the remaining *sham link* on Lines 8 and 20; that is to say, the existing *sham link* keeps islands stitched in the same area, and Router LSAs remain visible there. When *Bilbao* and *Havana* evaluate the route for translation, the internal OSPF path is preferred because of Junos OS default preferences.

Listing 5.30: Route selection for remote CEs after dismantling sham links in Havana and Bilbao

```
1  user@Bilbao-re0> show route 192.168.1.5 exact table NMM
2    # Route for Lille
3
4  NMM.inet.0: 18 destinations, 35 routes (18 active, 0 holddown, 0 hidden)
5  + = Active Route, - = Last Active, * = Both
6
7  192.168.1.5/32     *[OSPF/10] 00:12:03, metric 10319
8                      > via at-1/2/0.0   # Internal next hop towards sham link
9                      [BGP/170] 00:12:00, MED 1000, localpref 100, from 192.168.1.4
10                       AS path: I
11                     > via so-1/3/1.0, Push 16, Push 299824(top)
12
13 user@Havana> show route 192.168.1.1 exact table NMM
14   # Route for Nantes
15
16 NMM.inet.0: 18 destinations, 34 routes (18 active, 0 holddown, 0 hidden)
17 + = Active Route, - = Last Active, * = Both
18
19 192.168.1.1/32     *[OSPF/10] 00:12:55, metric 9999
20                     > to 172.16.100.8 via fe-0/3/0.0   # Internal next hop towards sham link
21                     [BGP/170] 00:12:55, MED 320, localpref 100, from 192.168.1.8
22                       AS path: I
23                     > to 172.16.1.85 via ge-1/1/0.0, Push 16, Push 299904(top)
```

The chances that a transient routing loop might occur are minimized as a result of this analysis. This transient scenario is illustrated in Figure 5.24.

Nevertheless, all traffic between both sites is flowing over the *sham link* as compared with the previous load-balancing scenario, so the remaining *sham link* between router *Skopie* and router *Livorno* must be torn down as well for an equivalent effect. Listing 5.31 shows the modifications for this last removal.

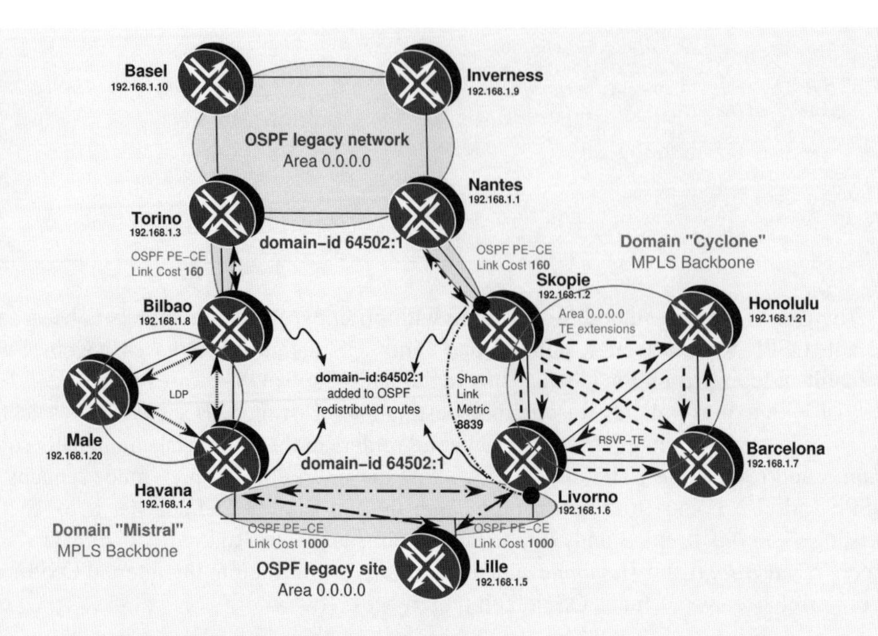

Figure 5.24: Stage eight: sham-link decommissioning over domain "Mistral".

Listing 5.31: Dismantling sham links on Skopie and Livorno

```
1  [edit routing-instances NMM protocols ospf]
2  user@Livorno# delete sham-link local
3  [edit routing-instances NMM protocols ospf]
4  user@Livorno# delete area 0 sham-link-remote 192.168.2.2
5
6  Sep  4 07:17:02.862  Livorno rpd[1321]: RPD_OSPF_NBRDOWN: OSPF neighbor 192.168.2.2
7       (realm ospf-v2 shamlink.0 area 0.0.0.0) state changed from Full to Down because of KillNbr
8       (event reason: interface went down)
9  Sep  4 07:17:02.864  Livorno rpd[1321]: EVENT Delete UpDown shamlink.0 index 2147549184
10      <PointToPoint Localup>
11 Sep  4 07:17:02.864  Livorno rpd[1321]: EVENT <Delete UpDown> shamlink.0 index 2147549184
12      <PointToPoint>
13
14 [edit routing-instances NMM protocols ospf]
15 user@Skopie# delete sham-link local
16 [edit routing-instances NMM protocols ospf]
17 user@Skopie# delete area 0 sham-link-remote 192.168.6.6
18
19 Sep  4 07:16:57  Skopie rpd[26607]: RPD_OSPF_NBRDOWN: OSPF neighbor 192.168.6.6
20      (realm ospf-v2 shamlink.0 area 0.0.0.0) state changed from Full to Down because of KillNbr
21      (event reason: interface went down)
22 Sep  4 07:16:57  Skopie rpd[26607]: EVENT Delete UpDown shamlink.0 index 2147549184
23      <PointToPoint Localup>
24 Sep  4 07:16:57  Skopie rpd[26607]: EVENT <Delete UpDown> shamlink.0 index 2147549184
25      <PointToPoint>
```

This latest step has actually been the real transition to uniform route redistribution and translation. Because there is no longer an artificial backdoor created by any *sham link*, two different but related OSPFv2 databases need to be inspected and analyzed. Listing 5.32 and

Listing 5.33 show an overview of each OSPFv2 database (after issuing a `clear ospf database purge instance NMM` in each segment for illustration purposes).

Listing 5.32: OSPFv2 database at Livorno after decommissioning the sham link

```
1  user@Livorno> show ospf database instance NMM
2
3      OSPF database, Area 0.0.0.0
4   Type       ID               Adv Rtr           Seq      Age  Opt  Cksum  Len
5  Router   192.168.1.5      192.168.1.5      0x8000004f  212  0x22 0xfc5b  48  # Lille
6  Router   192.168.4.4      192.168.4.4      0x80000005  212  0x22 0x6538  48  # Havana
7  Router  *192.168.6.6      192.168.6.6      0x80000005  211  0x22 0xd9b0  48  # Livorno
8  Network  172.16.100.5     192.168.1.5      0x80000003  212  0x22 0x75d2  36
9  <...>
10 Summary  172.16.1.80      192.168.4.4      0x80000002  212  0xa2 0x8578  28
11 Summary *172.16.1.80      192.168.6.6      0x80000002  211  0xa2 0xb1a7  28
12 Summary  172.16.1.96      192.168.4.4      0x80000002  212  0xa2 0xe409  28
13 Summary *172.16.1.96      192.168.6.6      0x80000002  211  0xa2 0xca1f  28
14 Summary *172.16.1.100     192.168.6.6      0x80000002  211  0xa2 0xe85c  28
15 Summary  192.168.1.1      192.168.4.4      0x80000002  212  0xa2 0x8419  28  # Nantes
16 Summary *192.168.1.1      192.168.6.6      0x80000002  211  0xa2 0x2515  28
17 Summary  192.168.1.3      192.168.4.4      0x80000002  212  0xa2 0x2b11  28  # Torino
18 Summary *192.168.1.3      192.168.6.6      0x80000002  211  0xa2 0x5641  28
19 Summary  192.168.1.9      192.168.4.4      0x80000002  212  0xa2 0x69a6  28  # Inverness
20 Summary *192.168.1.9      192.168.6.6      0x80000002  211  0xa2 0x9a3   28
21 Summary  192.168.1.10     192.168.4.4      0x80000002  212  0xa2 0x2a6a  28  # Basel
22 Summary *192.168.1.10     192.168.6.6      0x80000002  211  0xa2 0x5699  28
23 Summary  192.168.2.2      192.168.4.4      0x80000002  212  0xa2 0xb545  28  # Skopie
24 Summary *192.168.8.8      192.168.6.6      0x80000002  211  0xa2 0x1dcd  28  # Bilbao
25 OpaqArea 1.0.0.1          192.168.1.5      0x80000034  212  0x22 0x5ac2  28
26     # TE extensions from Lille
27     OSPF AS SCOPE link state database
28  Type       ID               Adv Rtr           Seq      Age  Opt  Cksum  Len
29 Extern  *172.16.1.0       192.168.6.6      0x80000002  211  0xa2 0xbd80  36
30     # Direct route from Skopie for PE-CE segment
31 Extern   172.16.1.100     192.168.4.4      0x80000002  212  0xa2 0x10cb  36
32     # Direct route from Bilbao for PE-CE segment
33 Extern  *192.168.2.2      192.168.6.6      0x80000002  211  0xa2 0x8407  36
34     # Endpoint from Skopie
35 Extern   192.168.8.8      192.168.4.4      0x80000002  212  0xa2 0x443d  36
36     # Endpoint from Bilbao
```

Listing 5.33: OSPFv2 database at Skopie after decommissioning the sham link

```
1  user@Skopie> show ospf database instance NMM
2
3      OSPF database, Area 0.0.0.0
4   Type       ID               Adv Rtr           Seq      Age  Opt  Cksum  Len
5  Router   192.168.1.1      192.168.1.1      0x80000031   25  0x22 0x6774 108  # Nantes
6  Router   192.168.1.3      192.168.1.3      0x80000048   26  0x22 0x75b3 120  # Torino
7  Router   192.168.1.9      192.168.1.9      0x8000002c   26  0x22 0x89be  96  # Inverness
8  Router   192.168.1.10     192.168.1.10     0x8000002b   28  0x22 0x9586  84  # Basel
9  Router  *192.168.2.2      192.168.2.2      0x8000004f   24  0x22 0x90c3  60  # Skopie
10 Router   192.168.8.8      192.168.8.8      0x80000016   28  0x22 0x801e  60  # Bilbao
11 Summary *192.168.1.5      192.168.2.2      0x80000002   24  0xa2 0xc635  28  # Lille
12 Summary  192.168.1.5      192.168.8.8      0x80000002   28  0xa2 0x7877  28
13 Summary *192.168.4.4      192.168.2.2      0x80000002   24  0xa2 0xaf4a  28  # Havana
14 Summary  192.168.6.6      192.168.8.8      0x80000002   28  0xa2 0x37b2  28  # Livorno
15 OpaqArea 1.0.0.1          192.168.1.1      0x80000026   25  0x22 0x66cc  28
16 OpaqArea 1.0.0.1          192.168.1.3      0x80000026   26  0x22 0x6ec0  28
17 OpaqArea 1.0.0.1          192.168.1.9      0x80000026   26  0x22 0x869c  28
18 OpaqArea 1.0.0.1          192.168.1.10     0x80000026   28  0x22 0x8a96  28
19     OSPF AS SCOPE link state database
20  Type       ID               Adv Rtr           Seq      Age  Opt  Cksum  Len
21 Extern  *172.16.100.0     192.168.2.2      0x80000002   24  0xa2 0xbe21  36
22     # LAN segment for Lille
23 Extern   172.16.100.0     192.168.8.8      0x80000002   28  0xa2 0x943d  36
```

```
24 Extern  192.168.4.4     192.168.8.8     0x80000002   28  0xa2 0x641d  36 # Havana
25 Extern *192.168.6.6     192.168.2.2     0x8000000a   24  0xa2 0x542f  36 # Livorno
```

The following conclusions can be made after inspecting both OSPFv2 databases:

- Only Type 1 and 2 Router and Network LSAs from each OSPFv2 island are visible in each database.
- All CE loopback and transit addresses visible by OSPF at each PE are properly redistributed back to the other island as Type 3 Summary LSAs, with proper identification of the OSPF Domain Identifier and OSPF Route Type (Lines 10 to 24 and Lines 11 to 14, respectively).
- Opaque LSA information remains local in each OSPF island without any advertisement and translation mechanism. It is now time for each MPLS backbone to perform TE in the core!
- Present AS External LSAs correspond to direct routes that are not being tagged with any OSPF Domain Identifiers (Lines 30 to 36 and Lines 22 to 25, respectively). Note that this is considered to be a pure transport exercise with no PE communication to CE at the other island. If communication were necessary, it would be convenient to tag the PE loopback addresses with the common extended communities for optimal routing purposes, because these loopback addresses are now advertised by the parallel PE to BGP and are preferred over that path as Type 3 Summary LSAs.

Finally, the original parallel routing needs to be inspected from each segregated router, namely, router *Lille*, router *Torino*, and router *Nantes*, to confirm that traffic forwarding keeps selecting the same paths, as shown in Listing 5.34. Note that router *Lille* detects both router *Torino* and router *Nantes* as being equally distant over the best-path uplink (Lines 6 and 8), which was the case in the original topology. The exact same metric is calculated in the opposite direction towards router *Lille* from each one of the routers (Lines 44 and 31). The main change for OSPF routes on the CEs, therefore, is that the LSA translation has resulted in inter-area routes instead of the original intra-area routes. However, keeping symmetry throughout the whole exercise has minimized the disruption.

Listing 5.34: Routing table inspection on Lille, Torino, and Nantes

```
1  user@Lille> show route 192.168.1.0/24
2
3  inet.0: 37 destinations, 38 routes (37 active, 0 holddown, 0 hidden)
4  + = Active Route, - = Last Active, * = Both
5
6  192.168.1.1/32     *[OSPF/10] 00:44:17, metric 1160 # Nantes
7                      > to 172.16.100.8 via fe-0/3/0.0
8  192.168.1.3/32     *[OSPF/10] 00:44:17, metric 1160 # Torino
9                      > to 172.16.100.1 via fe-0/3/0.0
10 192.168.1.5/32     *[Direct/0] 3w4d 09:45:29  # Lille
11                     > via lo0.0
12 192.168.1.9/32     *[OSPF/10] 00:44:17, metric 1803  # Inverness
13                     > to 172.16.100.8 via fe-0/3/0.0
14 192.168.1.10/32    *[OSPF/10] 00:44:17, metric 1320  # Basel
15                     > to 172.16.100.1 via fe-0/3/0.0
16
17 user@Lille> show ospf route | match "Prefix|Type|192.168.1"
18 Prefix              Path   Route        NH   Metric NextHop        Nexthop
```

```
19                    Type  Type       Type         Interface      addr/label
20  192.168.1.1/32    Inter Network    IP     1160  fe-0/3/0.0     172.16.100.8 # Nantes
21  192.168.1.3/32    Inter Network    IP     1160  fe-0/3/0.0     172.16.100.1 # Torino
22  192.168.1.5/32    Intra Network    IP        0  lo0.0
23  192.168.1.9/32    Inter Network    IP     1803  fe-0/3/0.0     172.16.100.8 # Inverness
24  192.168.1.10/32   Inter Network    IP     1320  fe-0/3/0.0     172.16.100.1 # Basel
25
26  user@Torino> show route 192.168.1.5
27
28  inet.0: 33 destinations, 38 routes (33 active, 0 holddown, 0 hidden)
29  + = Active Route, - = Last Active, * = Both
30
31  192.168.1.5/32     *[OSPF/10] 00:46:46, metric 1160  # Lille
32                      > via at-0/0/1.0
33
34  user@Torino> show ospf route | match "Prefix|Type|192.168.1.5"
35  Prefix            Path  Route      NH    Metric NextHop        Nexthop
36                    Type  Type       Type         Interface      addr/label
37  192.168.1.5/32    Inter Network    IP     1160  at-0/0/1.0  # Lille
38
39  user@Nantes> show route 192.168.1.5
40
41  inet.0: 36 destinations, 40 routes (36 active, 0 holddown, 0 hidden)
42  + = Active Route, - = Last Active, * = Both
43
44  192.168.1.5/32     *[OSPF/10] 00:48:06, metric 1160 # Lille
45                      > via so-0/1/0.0
46
47  user@Nantes> show ospf route | match "Prefix|Type|192.168.1.5"
48  Prefix            Path  Route      NH    Metric NextHop        Nexthop
49                    Type  Type       Type         Interface      addr/label
50  192.168.1.5/32    Inter Network    IP     1160  so-0/1/0.0 # Lille
```

This final setup is represented in Figure 5.25, which shows that the topology has finally emulated the original uplinks for the legacy network over MPLS backbones keeping symmetric and parallel costs, after carrying out smooth transitions first using *sham links* first and then using proper LSA translation.

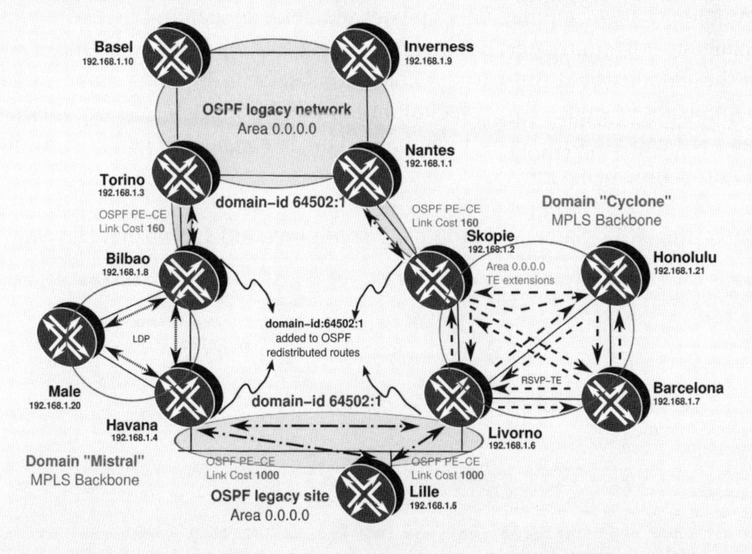

Figure 5.25: Stage nine: sham-link decommissioning over domain "Cyclone" and final setup.

domain-vpn-tag

Junos OS complies with standard mechanisms defined in [RFC4576] and [RFC4577] with regard to the DN bit and VPN Route Tag by setting automatically derived values to be fully compatible with earlier leak-prevention implementations that made use only of the VPN Route Tag.

The `domain-vpn-tag` configuration knob allows the setting of specific External Route Tag values in Type 5 and 7 LSAs injected towards the CEs. This knob allows arbitrary values to be configured following certain administrative policies, for instance. Note that the value set with the knob cannot be overridden with an additional OSPF export policy inside the VRF: `domain-vpn-tag` takes precedence and sets the definitive value for the External Route Tag field in Junos OS.

A standard scenario that makes use of this implementation is a PE multihomed topology in which route leaking towards the backbone is controlled by setting common or disjoint OSPF VPN Route Tags. For legacy PEs that may not understand the DN bit but do comply with previous versions of [RFC4577], or when multihoming the setup to more than one MPLS transport backbone (with different AS numbers resulting in different default OSPF VPN Route Tag values), this feature is tremendously helpful.

Another possible application is crafting an External Route Tag value with the `domain-vpn-tag` knob at the PEs. Legacy CEs that comply with [RFC1403], Section 4.4 for old-fashioned BGP-OSPF interaction can generate AS paths derived from the External Route Tag value further down the path as advertisements within EBGP sessions.

Despite the fact that the PE can mark the DN bit for all advertisements towards the CE domain, considering the topology from Figure 5.25, Listing 5.35 illustrates the changes on router *Livorno* and router *Havana* in a given redistributed AS External LSA after implementing a common VPN Route Tag. Note that the LSAs originally include default AS-derived values in the Tag field, as shown in Figure 5.4 on Lines 7 and 19. Also note that the configuration change automatically updates the field, as seen in Lines 31 and 39.

This change is not strictly needed to control route leaking because both routers mark the DN bit in the Options field prior to the change, as seen in Figure 5.3 and in Listing 5.35 with the Options byte value "0xa2" including the DN bit marking, before the modification (Lines 4 and 16) and after it (Lines 30 and 38). This is considered here just to illustrate the backward-compatible behavior.

Listing 5.35: Same VPN route tag configured in Livorno and Havana

```
1  user@Havana> show ospf database instance NMM lsa-id 172.16.1.100 extensive external
2      OSPF AS SCOPE link state database
3   Type       ID               Adv Rtr           Seq      Age  Opt  Cksum  Len
4  Extern  *172.16.1.100    192.168.4.4      0x80000001  1860  0xa2 0x12ca  36  # DN bit set
5    mask 255.255.255.252
6    Topology default (ID 0)
7      Type: 2, Metric: 0, Fwd addr: 0.0.0.0, Tag: 208.0.251.247   # AS-derived VPN Route Tag
8    Gen timer 00:05:04
9    Aging timer 00:29:00
10   Installed 00:31:00 ago, expires in 00:29:00, sent 00:31:00 ago
11   Last changed 00:31:00 ago, Change count: 1, Ours
12
13 user@Livorno> show ospf database instance NMM lsa-id 172.16.1.0 external extensive
14     OSPF AS SCOPE link state database
15  Type       ID               Adv Rtr           Seq      Age  Opt  Cksum  Len
16 Extern  *172.16.1.0      192.168.6.6      0x80000003  2647  0xa2 0xbb81  36  # DN bit set
17   mask 255.255.255.252
```

```
18   Topology default (ID 0)
19     Type: 2, Metric: 0, Fwd addr: 0.0.0.0, Tag: 208.0.251.245    # AS-derived VPN Route Tag
20   Gen timer 00:02:45
21   Aging timer 00:15:53
22   Installed 00:44:07 ago, expires in 00:15:53, sent 00:44:05 ago
23   Last changed 01:33:31 ago, Change count: 1, Ours
24
25  [edit routing-instances NMM protocols ospf]
26 user@Havana# set domain-vpn-tag 64502
27
28 user@Havana> show ospf database instance NMM lsa-id 172.16.1.100 extensive external |
29              match "Extern|Tag:"
30 Extern  *172.16.1.100     192.168.4.4      0x80000002    63  0xa2 0xc3ea  36  # DN bit set
31     Type: 2, Metric: 0, Fwd addr: 0.0.0.0, Tag: 0.0.251.246 # VPN Route Tag 64502
32
33  [edit routing-instances NMM protocols ospf]
34 user@Livorno# set domain-vpn-tag 64502
35
36 user@Livorno> show ospf database instance NMM lsa-id 172.16.1.0 external extensive |
37               match "Extern|Tag"
38 Extern  *172.16.1.0       192.168.6.6      0x80000004    46  0xa2 0x8f7c  36  # DN bit set
39     Type: 2, Metric: 0, Fwd addr: 0.0.0.0, Tag: 0.0.251.246 # VPN route tag 64502
```

Another interesting corner case for controlling route leaking takes place when setting `domain-vpn-tag 0`. This configuration has the following consequences:

- "Bleaching" (zeroing) of External Route Tag field on Type 5 and 7 LSAs;
- Untagging of DN bit in Options field on Type 5 and 7 LSAs.

To sum up, `domain-vpn-tag 0` removes all distinguishers for PE–CE leaking from any External LSA type and overrides the standard loop protection behavior on OSPFv2 CE domains to allow redistribution through other PEs towards their backbone in the same setup.

Note that Type 3 Summary LSAs representing inter-area networks after translation are not affected by this specific setting in Junos OS and keep the DN bit set after redistribution (there is no such Route Tag field in Summary LSAs). It is understood that if a VPN-IPv4 route is translated into a Type 3 Summary LSA, it belongs to the same domain (sharing the same *Domain Identifier* between VPN sites).

It goes without saying that a smart network engineer must think more than twice before implementing this knob, because it disables an implicit loop-prevention mechanism and opens a door for redistribution and forwarding issues!

Junos Tip: Inter-AS Option A based on OSPF Domain Identifier mismatch, DN bit and VPN Route Tag bleaching

[RFC4577] describes resources for using OSPF as a PE–CE communication protocol inside a VRF, but doing this certainly impedes PE–PE communication within a VRF (that is to say, with an inter-AS Option A interconnect) with the already discussed built-in loop-blocking mechanisms. This case certainly requires careful design and should also consider the domain-flooding scope of OSPFv2 as a link-state routing protocol.

However, there can be some scenarios in which a network designer may need to interconnect two MPLS backbones through an OSPF island inside a VPN site or to allow route advertisement beyond a PE for the purpose of installing a backup path. In such setups, LSAs advertised towards the CE domain must cope with the following requirements:

- DN bit must not be set;
- VPN Route Tag must not be the same across PEs.

The `domain-vpn-tag 0` configuration allows bleaching of both the DN bit and Tag field, but acts on Type 5 or Type 7 LSAs. If VPN sites belong to a common OSPF domain, Type 3 LSAs (such as inter-area routes) are not affected by this change. This means that an artificial *OSPF Domain Identifier* mismatch must be issued to force all routes to be injected towards the attached CEs and PEs in the VPN site as *external* LSAs. While different *OSPF Domain Identifier* values can be configured at participating CEs, the advantage of using `domain-id disable` is that it always and unilaterally detaches the PE from the backbone for that purpose, leading to the generation of Type 5 and Type 7 LSAs, independent of any *OSPF Domain Identifier*, either null or any specific value, of incoming routes.

Note that the bidirectional configuration imposed in Junos OS for setting (using export policies) and matching (using the `domain-id` knob) *OSPF Domain Identifiers* also offers additional possibilities for further tweaking of route advertisements, because a given PE can tag a specific OSPF Domain Identifier on exported routes while matching another value (or completely disabling *Domain Identifier* matching), thereby creating unidirectional rules.

Consider a derivation from the well-known topology in Figure 5.25, in which it is now desired to have the CE segment shared among router *Lille*, router *Livorno*, and router *Havana* also act as a backup path for domain "Cyclone" to reach VPN routes from the OSPFv2 legacy network.

By removing the *Domain Identifier* checks and bleaching the DN bit and VPN Route Tag fields on *Havana*, all received VPN-IPv4 routes are translated as AS External LSAs inside the VPN site, with the following actions:

- Because `domain-id disable` is set for incoming route checks, *OSPF Domain Identifiers* from received routes are ignored and these routes are translated as either Type 5 or Type 7 LSAs.
- Because the only Area configured in the VRF site is 0.0.0.0, the translation results into Type 5 LSAs.
- The DN bit is not set and the VPN route tag is bleached as a result of the `domain-vpn-tag 0` configuration.
- Router *Livorno* imports those Type 5 LSAs for internal readvertisement to MP-BGP because none of the [RFC4577] loop-control mechanisms are set.

The configuration changes and effects on Havana are described in Listing 5.36. When comparing this listing with Listing 5.32, the expected changes introduced by the configuration changes become clearer: routes that were previously injected as Type 3 Summary LSAs with matching *OSPF Domain Identifiers* and proper route translation (Lines 10 to 24 from Listing 5.32) are now visible as Type 5 AS External LSAs (Lines 18 to 29 from Listing 5.36).

Listing 5.36: Disabling Domain Identifier checks and DN bit and VPN route tagging on Havana

```
[edit routing-instances NMM protocols ospf]
user@Havana# set domain-vpn-tag 0
[edit routing-instances NMM protocols ospf]
user@Havana# set domain-id disable

user@Havana> show ospf database instance NMM

    OSPF database, Area 0.0.0.0
 Type       ID               Adv Rtr           Seq      Age  Opt  Cksum  Len
Router   192.168.1.5      192.168.1.5      0x8000005c    12  0x22 0xe268  48
Router  *192.168.4.4      192.168.4.4      0x80000014    11  0x22 0x444b  48
Router   192.168.6.6      192.168.6.6      0x80000011    12  0x22 0xc1bc  48
Network  172.16.100.5     192.168.1.5      0x80000010    12  0x22 0x5bdf  36
OpaqArea 1.0.0.1          192.168.1.5      0x80000036  1097  0x22 0x56c4  28
    OSPF AS SCOPE link state database
 Type       ID               Adv Rtr           Seq      Age  Opt  Cksum  Len
<...>
Extern  *172.16.1.72      192.168.4.4      0x80000001    11  0x22 0xc6b   36
Extern  *172.16.1.80      192.168.4.4      0x80000001    11  0x22 0x866e  36
Extern  *172.16.1.96      192.168.4.4      0x80000001    11  0x22 0xe5fe  36
Extern  *172.16.1.100     192.168.4.4      0x80000004    11  0x22 0x2cf2  36
Extern  *192.168.1.1      192.168.4.4      0x80000001    11  0x22 0x850f  36  # Nantes
Extern  *192.168.1.3      192.168.4.4      0x80000001    11  0x22 0x2c07  36  # Torino
Extern  *192.168.1.9      192.168.4.4      0x80000001    11  0x22 0x6a9c  36  # Inverness
Extern  *192.168.1.10     192.168.4.4      0x80000001    11  0x22 0x2b60  36  # Basel
 # Mgmt Loopback for Skopie
Extern  *192.168.2.2      192.168.4.4      0x80000001    11  0x22 0xb63b  36
 # Mgmt Loopback for Bilbao
Extern  *192.168.8.8      192.168.4.4      0x80000004    11  0x22 0x6064  36
```

Listing 5.37 reveals further details with regard to the data on the parallel PE in the same network segment:

- Options and External Route Tag field are untagged (Options field "0x22" instead of "0xa2") and bleached (Tag: 0.0.0.0) (for example, see Lines 12 and 13 for the AS External LSA representing router *Nantes* loopback);
- As a consequence, those routes from external LSAs are advertised internally in the MPLS backbone domain "Cyclone" towards other neighbors. As a curiosity, note that the original OSPF metric cost is kept from the original translation and is passed to the next transformation into the VPN-IPv4 BGP route MED attribute value.

Listing 5.37: AS External LSA inspection and route advertisement on Livorno after changes on Havana

```
user@Livorno> show ospf database instance NMM external extensive
    | match "Extern|Tag"
<...>
Extern   172.16.1.72      192.168.4.4      0x80000001   161  0x22 0xc6b   36
    Type: 2, Metric: 963, Fwd addr: 0.0.0.0, Tag: 0.0.0.0
Extern   172.16.1.80      192.168.4.4      0x80000001   161  0x22 0x866e  36
    Type: 2, Metric: 320, Fwd addr: 0.0.0.0, Tag: 0.0.0.0
Extern   172.16.1.96      192.168.4.4      0x80000001   161  0x22 0xe5fe  36
    Type: 2, Metric: 320, Fwd addr: 0.0.0.0, Tag: 0.0.0.0
Extern   172.16.1.100     192.168.4.4      0x80000004   161  0x22 0x2cf2  36
    Type: 2, Metric: 0, Fwd addr: 0.0.0.0, Tag: 0.0.0.0
Extern   192.168.1.1      192.168.4.4      0x80000001   161  0x22 0x850f  36  # Nantes
    Type: 2, Metric: 320, Fwd addr: 0.0.0.0, Tag: 0.0.0.0
Extern   192.168.1.3      192.168.4.4      0x80000001   161  0x22 0x2c07  36  # Torino
    Type: 2, Metric: 160, Fwd addr: 0.0.0.0, Tag: 0.0.0.0
```

```
16 Extern   192.168.1.9      192.168.4.4      0x80000001  161  0x22 0x6a9c  36  # Inverness
17     Type: 2, Metric: 963, Fwd addr: 0.0.0.0, Tag: 0.0.0.0
18 Extern   192.168.1.10     192.168.4.4      0x80000001  161  0x22 0x2b60  36 # Basel
19     Type: 2, Metric: 320, Fwd addr: 0.0.0.0, Tag: 0.0.0.0
20  # Mgmt Loopback for Skopie
21 Extern   192.168.2.2      192.168.4.4      0x80000001  161  0x22 0xb63b  36
22     Type: 2, Metric: 480, Fwd addr: 0.0.0.0, Tag: 0.0.0.0
23  # Mgmt Loopback for Bilbao
24 Extern   192.168.8.8      192.168.4.4      0x80000004  161  0x22 0x6064  36
25     Type: 2, Metric: 0, Fwd addr: 0.0.0.0, Tag: 0.0.0.0
26
27 user@Livorno> show route advertising-protocol bgp 192.168.1.2 table NMM
28
29 NMM.inet.0: 20 destinations, 34 routes (20 active, 0 holddown, 0 hidden)
30   Prefix            Nexthop            MED     Lclpref    AS path
31 <...>
32 * 172.16.1.72/30          Self                 963      100        I
33 * 172.16.1.80/30          Self                 320      100        I
34 * 172.16.1.96/30          Self                 320      100        I
35 * 172.16.1.100/30         Self                 0        100        I
36 * 172.16.100.0/24         Self                          100        I
37 * 192.168.1.1/32          Self                 320      100        I
38 * 192.168.1.3/32          Self                 160      100        I
39 * 192.168.1.5/32          Self                 481      100        I
40 * 192.168.1.9/32          Self                 963      100        I
41 * 192.168.1.10/32         Self                 320      100        I
42 * 192.168.2.2/32          Self                 480      100        I
43 * 192.168.4.4/32          Self                 481      100        I
44 * 192.168.6.6/32          Self                          100        I
45 * 192.168.8.8/32          Self                 0        100        I
```

This situation is illustrated in Figure 5.26.

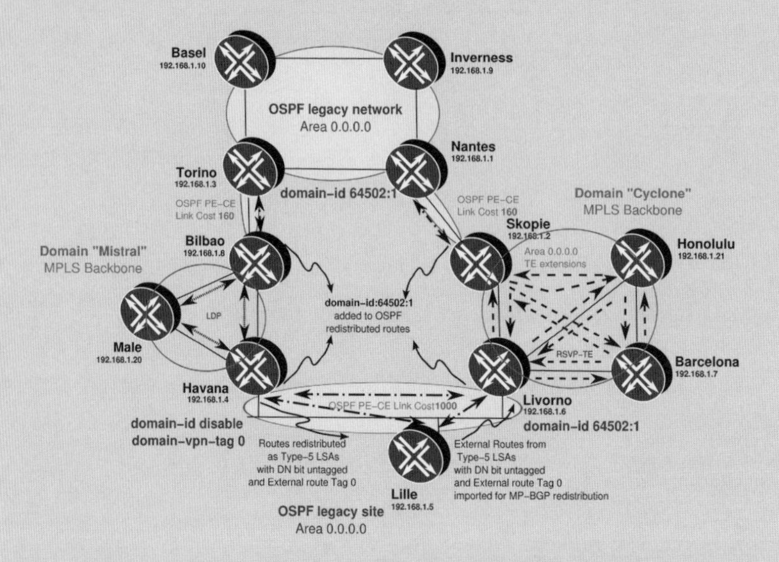

Figure 5.26: OSPF DN bit and VPN Route Tag bleaching, and Domain Identifier check removal at Havana.

In a nutshell, all other PEs participating in the same VPN at domain “Cyclone” can see two different active paths for the same original OSPF routes. It is a question of performing some BGP TE as desired inside the domain to ensure that paths advertised from router

Skopie are always preferred when present (this can be as simple as defining different `vrf-export` policies to manipulate the BGP attributes).

On router *Skopie*, the default protocol preferences from Junos OS ensure that no routing loop is formed, because OSPF paths are always preferred over BGP paths from router *Livorno* whenever they are present, as seen in Listing 5.38. Because all selected routes for leaking on router *Havana* are originally advertised as Type 1 or Type 2 LSAs, this also confirms that no routing loops should be formed on router *Skopie* when transient routing churn occurs, because intra-area routes are always preferred in OSPF over external routes as soon as they become present.

Listing 5.38: AS External LSA inspection and route advertisement on Livorno after changes on Skopie

```
 1 user@Skopie> show route table NMM | no-more
 2
 3 NMM.inet.0: 20 destinations, 35 routes (20 active, 0 holddown, 0 hidden)
 4 + = Active Route, - = Last Active, * = Both
 5
 6 <...>
 7 192.168.1.1/32     *[OSPF/10] 10:14:04, metric 160  # Nantes
 8                     > via so-0/1/0.0
 9                     [BGP/170] 00:03:24, MED 320, localpref 100, from 192.168.1.6
10                       AS path: I
11                     > via so-1/0/0.0, label-switched-path Skopie-to-Livorno
12 192.168.1.3/32     *[OSPF/10] 10:14:04, metric 320 # Torino
13                     > via so-0/1/0.0
14                     [BGP/170] 00:03:24, MED 160, localpref 100, from 192.168.1.6
15                       AS path: I
16                     > via so-1/0/0.0, label-switched-path Skopie-to-Livorno
17 192.168.1.5/32     *[BGP/170] 00:07:54, MED 481, localpref 100, from 192.168.1.6  # Lille
18                       AS path: I
19                     > via so-1/0/0.0, label-switched-path Skopie-to-Livorno
20 <...>
21 192.168.8.8/32     *[OSPF/10] 02:05:39, metric 480  # Bilbao mgmt loopback
22                     > via so-0/1/0.0
23                     [BGP/170] 00:06:19, MED 0, localpref 100, from 192.168.1.6
24                       AS path: I
25                     > via so-1/0/0.0, label-switched-path Skopie-to-Livorno
26 224.0.0.5/32       *[OSPF/10] 10:15:32, metric 1
27                       MultiRecv
```

To be clear, this exercise has focused on route leaking in one direction for simplicity. Most applications require bidirectional flows, and thus a similar exercise should be performed in the other direction if required by the service (for example, similar `domain-id disable` and `domain-vpn-tag 0` in router *Livorno* for reverse backup paths).

Junos Tip: Controlling route propagation from a CE with the VPN Route Tag

Another interesting application related to VPN Route Tags appears when considering the OSPFv2 protocol resources that [RFC4577] and [RFC4576] are reutilizating for PE–CE communications:

- DN bit reuses the high-order bit of the LSA Options field, which is unused as per [RFC2328], Section A.2.

- VPN Route Tag reuses the 32-bit External Route tag that [RFC2328], Section A.4.5 described to be used to propagate information across a domain.

Thus, the *External Route Tag* field is available to any router participating in OSPFv2, because the tag was designed for administrative tagging. On the other hand, [RFC4577] requests that any participating PE check and overwrite this field for backward-compatibility purposes if it detects a loop.

To briefly summarize, any participating CE can influence MP-BGP route redistribution for locally originated Type 5 and Type 7 LSAs by setting external tags to match those from selected PEs! This ability provides another indirect route advertisement mechanism that remains external to the PEs and allows further possibilities for VPN customers to influence routing in an MPLS backbone.

Consider another variation of the scenario in Figure 5.25, in which router *Inverness* as a CE is externally injecting 192.168.90.0/24 by means of a Type 5 LSA. The prerequisite in this case is that traffic towards that destination uniquely flows through domain "Mistral". Although this can be controlled on PEs with OSPFv2 `import` or `vrf-export` policies, it is requested that such TE be remotely controlled by the CE and be kept transparent to the PE configuration.

By making use of the mechanism described in this Junos Tip, router *Inverness* could craft the *External Route Tag* field in the Type 5 LSA for 192.168.90.0/24 so that it matches the VPN Route Tag from domain "Cyclone" so that this LSA is rejected in that domain, but is kept for legitimate redistribution to MP-BGP at domain "Mistral".

Assuming default VPN Route Tag values in this scenario, derived from the AS number as shown in Figure 5.4, router *Inverness* needs to set a decimal value of 3489725429 in the External Route Tag field to cause the *VPN Route Tag* to have the value corresponding to AS 64501.This can be controlled simply with the OSPFv2 `export` policy on router *Inverness*, as shown in Listing 5.39, because the tag-setting action of the policy not only modifies the internal route attribute in the routing table, but also sets the External Route Tag field with that value in case it is exported into any type of AS external LSAs, as shown in Line 14.

Listing 5.39: External route tag setting on Inverness matching domain "Cyclone" VPN route tag

```
1  [edit policy-options policy-statement NMM-static-to-ospf]
2  user@Inverness# show
3  term static {
4      from protocol static;
5      then {
6          tag 3489725429; # External Route Tag value matching ''Cyclone'' domain's VPN Route Tag
7          accept;
8      }
9  }
10 user@Inverness> show ospf database instance NMM external lsa-id 192.168.90.0
11                 extensive | match "Extern|Tag"
12 Extern   192.168.90.0     192.168.1.9      0x80000001   130  0x22 0x6751  36
13     Type: 2, Metric: 0, Fwd addr: 0.0.0.0, Tag: 208.0.251.245
14     # ''Cyclone'' domain's VPN Route Tag
```

Once the crafted Type 5 LSA is originated on router *Inverness*, which is an ASBR, routing table inspection on both PEs, router *Bilbao* and router *Skopie*, shows that despite detecting the Type 5 LSA, the corresponding route is installed only on router *Bilbao*, which selects it for redistribution as a VPN-IPv4 route. The route is not even installed on router *Skopie*, which considers that the matching *VPN Route Tag* indicates that the route is an already leaked route into OSPFv2 and thus assumes that an MP-BGP path for it should also be available. This is illustrated in Listing 5.40 for both router *Bilbao* and router *Skopie*, which shows that no route-matching occurs on router *Skopie* and also shows how the VPN-IPv4-redistributed route is advertised by router *Bilbao* with the proper OSPF Domain Identifier and OSPF Route Type (Line 26).

Listing 5.40: OSPPFv2 database and routing table inspection on Bilbao and Skopie for the crafted external route from Inverness

```
1  user@Bilbao-re0> show ospf database instance NMM external lsa-id 192.168.90.0
2      extensive
3     OSPF AS SCOPE link state database
4   Type       ID              Adv Rtr          Seq         Age  Opt  Cksum  Len
5  Extern   192.168.90.0    192.168.1.9     0x80000001    24  0x22 0x6751  36
6    mask 255.255.255.0
7    Topology default (ID 0)
8      Type: 2, Metric: 0, Fwd addr: 0.0.0.0, Tag: 208.0.251.245
9    Aging timer 00:59:36
10   Installed 00:00:21 ago, expires in 00:59:36
11   Last changed 00:00:21 ago, Change count: 1
12
13 user@Bilbao-re0> show route advertising-protocol bgp 192.168.1.4 192.168.90.0
14     extensive
15   # Route selected for MP-BGP redistribution
16 NMM.inet.0: 21 destinations, 22 routes (21 active, 0 holddown, 0 hidden)
17 * 192.168.90.0/24 (1 entry, 1 announced)  # External route from Inverness
18  BGP group IBGP type Internal
19      Route Distinguisher: 64503:64503
20      VPN Label: 16
21      Nexthop: Self
22      Flags: Nexthop Change
23      MED: 0
24      Localpref: 100
25      AS path: [64503] I
26      Communities: target:64503:64503 domain-id:64502:1 rte-type:0.0.0.0:5:1
27      # Extended Communities referring to Type 5 LSA
28
29 user@Skopie> show ospf database instance NMM external lsa-id 192.168.90.0 extensive
30     OSPF AS SCOPE link state database
31  Type       ID              Adv Rtr          Seq         Age  Opt  Cksum  Len
32 Extern   192.168.90.0    192.168.1.9     0x80000001   130  0x22 0x6751  36
33   mask 255.255.255.0
34   Topology default (ID 0)
35     Type: 2, Metric: 0, Fwd addr: 0.0.0.0, Tag: 208.0.251.245
36   Aging timer 00:57:50
37   Installed 00:02:08 ago, expires in 00:57:50
38   Last changed 00:02:08 ago, Change count: 1
39
40 user@Skopie> show route 192.168.90.0/24 exact table NMM
41   # No matches on the external route
42 user@Skopie>
```

This example has considered a scenario with two different MPLS backbones. However, in the case of a joint core providing transit services, the same objective can be easily achieved by setting different *VPN Route Tags* on redundantly connected PEs.

`Route-Type-community`

[RFC4577] recommends using the 2-byte type field 0x306 value to represent the OSPF Route Type attribute, and at the time of this writing, [IANA-extended-communities] assigns the same value to OSPF Route Type extended communities.

Previous implementations of the OSPF Route Type attribute were based on a different value for the 2-byte type field: 0x8000. [RFC4577] states that this value should also be accepted for backwards compatibility. From a network integration perspective, the challenge is to get legacy devices that implement earlier versions of [RFC4577] to properly understand the OSPF Route Type from other VPN locations. In this case, use of this previous value may be required until heterogeneous support is confirmed across all PEs that are members of the VPN.

Junos OS supports the arbitrary setting of one or the other extended community type values, as shown in Listing 5.41.

Listing 5.41: Configuration options for the extended route type community for compatibility purposes

```
1 [edit routing-instances NMM protocols ospf]
2 user@Livorno# set route-type-community ?
3 Possible completions:
4   iana                 BGP extended community value used is 0x0306
5   vendor               Vendor BGP extended community value used is 0x8000
```

5.5.3 BGP PE–CE resources

Considering BGP as the most scalable, and probably most used, protocol for PE–CE communication, Junos OS is specifically committed to extend all plain BGP features that can be configured in the main routing instance so that they can also be configured within a VRF where applicable.

Section 3.5 describes the most relevant BGP features for migrations. It is worth considering some of them again from the perspective of using them for PE–CE communication.

`local-as`

Section 3.5.2 describes the wide range of available commands in Junos OS to manipulate AS insertion in the AS_PATH. The `local-as` knob in Junos OS allows an external representation of a different AS for peers when negotiating and establishing BGP sessions, with several related suffixes to steer AS_PATH creation. Those resources are also available inside a VRF in Junos OS and therefore can be used for scenarios with EBGP as the PE–CE protocol.

This flexible handling of ASs allows for administrative manipulation of a given AS_PATH, in which ASs can be artificially inserted, expanded, or suppressed. From all available options included in Section 3.5.2, two are particularly relevant for an environment with EBGP as the PE–CE protocol and transport over an MPLS backbone:

- `local-as private` keeps the `local-as` for BGP session handling and route advertisement towards peers, but this AS is suppressed from AS_PATHs towards the backbone. This allows the AS identified with the `local-as` command in the VRF for AS_PATHs to be stripped from routes advertised towards the backbone;

- `local-as no-prepend-global-as` does not include the global AS number when advertising routes out of the network. This allows the AS corresponding to the MPLS transport backbone for AS_PATHs to be stripped from routes advertised towards the CE. Note that the `local-as` value configured in the VRF BGP session remains present in the AS_PATH; only the MPLS transport AS is suppressed.

By selecting these `local-as` options, arbitrary AS numbers can be suppressed from the AS_PATH and transit through the MPLS core AS can remain hidden from the CE, allowing utilization of arbitrary AS_PATH policies just based on CE domain ASs for traffic engineering.

Junos Tip: Integrate route distribution into a L3VPN without prepending the transport AS

A combination of `local-as private` and `local-as no-prepend-global-as` allows the suppression of ASs of the transport backbones and VRF `local-as` values for AS_PATHs transmitted upstream to the backbone.

The `local-as` feature provides a clear benefit when the intent is to keep route distribution between ASs transparent for transport through an MPLS service provider. A migration scenario in which a legacy network that bases customer and service route redistribution on a given interaction of EBGP sessions can also make use of this feature as part of being integrated into a L3VPN. It is also feasible to inject other prefixes into one or another VRF site, thus avoiding the insertion of the MPLS backbone AS, and with or without a minimal need to manipulate BGP policies on any of the legacy routers, now considered CEs in this scenario. This L3VPN can act both as a replacement for the old direct EBGP interconnect and also as an uplink for routes from other sites.

With the network topology set up as defined in Figure 5.27, considering a full IBGP mesh on some routers and simultaneous EBGP sessions to router *Lille* on router *Nantes* and router *Torino*, one of the backdoors is going to be migrated to become part of a L3VPN.

The exchanged BGP routes are considered the main focus for the migration, and it is considered essential that they are kept independent of the transit through the service provider. In fact, router *Nantes* implements some TE policies based on `as-path-prepend` twice with its own AS for routes further from router *Torino* and router *Basel*, and router *Torino* does the same for router *Nantes* and router *Inverness*.

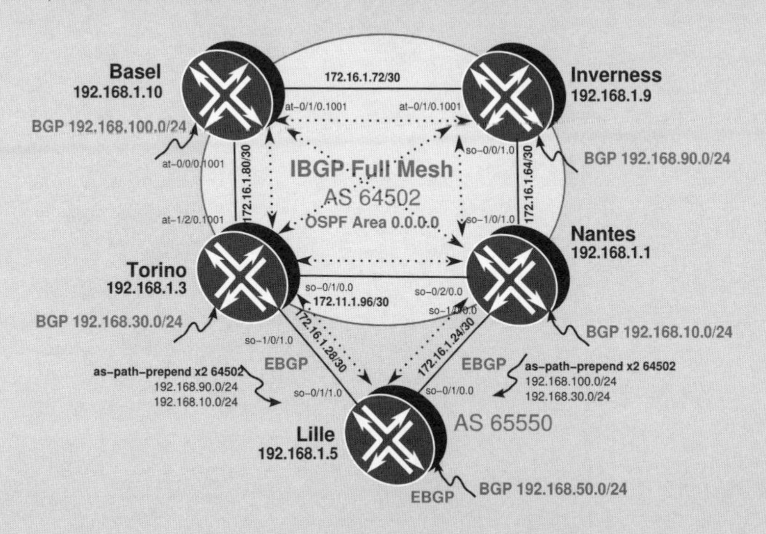

Figure 5.27: Representative legacy BGP network to be integrated in a L3VPN.

Listing 5.42 depicts the routing table including AS_PATH for each of the received routes. With this setup, route transit over the L3VPN leads to a modification in the TE strategies due to additional ASs that could appear in the AS_PATH, including, among others, the one for the MPLS backbone.

Listing 5.42: Initial BGP routing table on Lille prior to L3VPN integration

```
1  user@Lille> show route protocol bgp terse table inet.0
2
3  inet.0: 24 destinations, 29 routes (24 active, 0 holddown, 0 hidden)
4  + = Active Route, - = Last Active, * = Both
5
6  A Destination        P Prf   Metric 1   Metric 2  Next hop        AS path
7  * 192.168.10.0/24    B 170        100             >172.16.1.25    64502 I
8                       B 170        100             >172.16.1.29    64502 64502 64502 I
9  * 192.168.30.0/24    B 170        100             >172.16.1.29    64502 I
10                      B 170        100             >172.16.1.25    64502 64502 64502 I
11 * 192.168.90.0/24    B 170        100             >172.16.1.25    64502 I
12                      B 170        100             >172.16.1.29    64502 64502 64502 I
13 * 192.168.100.0/24   B 170        100             >172.16.1.29    64502 I
14                      B 170        100             >172.16.1.25    64502 64502 64502 I
15
```

By making use of the `local-as private` and the `no-prepend-global-as` configuration options when integrating one of the uplinks as part of the L3VPN, only the final `local-as` for the egress VRF is added to the AS_PATH when advertising routes to the CE.

A migration scenario using both commands is conceivable when interconnecting PEs with router *Nantes* and router *Lille*, as shown in Figure 5.28.

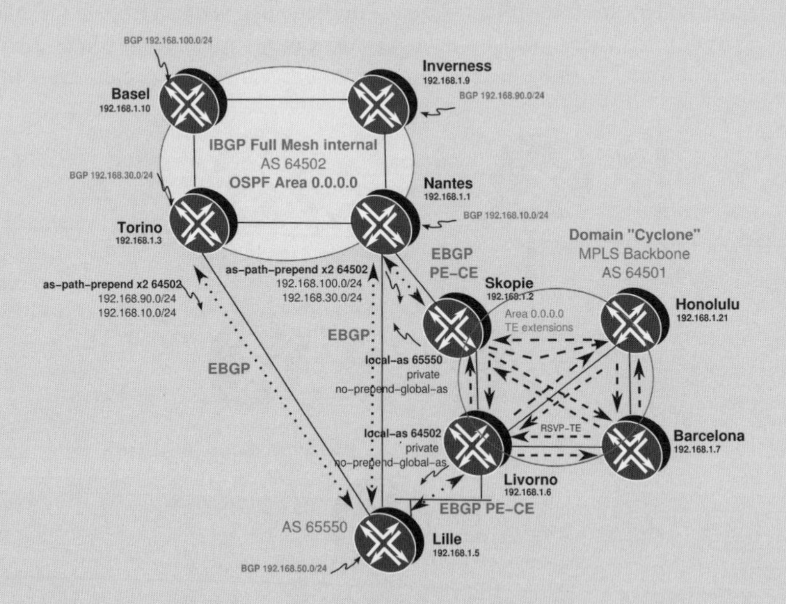

Figure 5.28: Interconnect domain “Cyclone” PEs with router *Lille* and router *Nantes*.

Listing 5.43 shows the configuration changes needed on *Livorno*, one of the PEs, to activate the `local-as private` and the `no-prepend-global-as` together with EBGP peering towards router *Lille*. It is worth noting the following items:

- The `no-prepend-global-as` option is active on the *local* AS used with the CE (64502), which is still used for BGP session establishment and control purposes (Line 10).

- Advertised routes towards the CE do not include the MPLS backbone provider AS (64501) because of the `no-prepend-global-as` knob (Line 15).

- Router *Lille* selects backup paths for destinations over the L3VPN instead of from router *Torino* (Line 22).

Listing 5.43: EBGP routes and no-prepend-global-as configuration in router *Livorno* and router *Lille*

```
1  [edit routing-instances NMM protocols bgp]
2  user@Livorno# show
3  group EBGP-PE-CE {
4      type external;
5      local-address 172.16.100.8;
6      family inet {
7          unicast;
8      }
9      peer-as 65550;
10     local-as 64502 private no-prepend-global-as; # Private and no-prepend-global-as
11     neighbor 172.16.100.5;
12 }
13
14 user@Livorno> show route advertising-protocol bgp 172.16.100.5 table NMM
15  # AS 64501 not included in AS_PATH
16 NMM.inet.0: 7 destinations, 12 routes (7 active, 0 holddown, 0 hidden)
17   Prefix                  Nexthop              MED     Lclpref    AS path
18 * 192.168.10.0/24         Self                                    64502 I
19 * 192.168.90.0/24         Self                                    64502 I
20
21 user@Lille> show route protocol bgp terse table inet.0
22  # Backup paths for preferred routes over L3VPN instead of Torino
23 inet.0: 24 destinations, 33 routes (24 active, 0 holddown, 0 hidden)
24 + = Active Route, - = Last Active, * = Both
25
26 A Destination        P Prf   Metric 1   Metric 2  Next hop        AS path
27 * 192.168.10.0/24    B 170        100             >172.16.1.25    64502 I
28                      B 170        100             >172.16.100.8   64502 64502 I # Over VPN
29                      B 170        100             >172.16.1.29    64502 64502 64502 I
30 * 192.168.30.0/24    B 170        100             >172.16.1.29    64502 I
31                      B 170        100             >172.16.1.25    64502 64502 64502 I
32                      B 170        100             >172.16.100.8   64502 64502 64502 I # Over VPN
33 * 192.168.90.0/24    B 170        100             >172.16.1.25    64502 I
34                      B 170        100             >172.16.100.8   64502 64502 I # Over VPN
35                      B 170        100             >172.16.1.29    64502 64502 64502 I
36 * 192.168.100.0/24   B 170        100             >172.16.1.29    64502 I
37                      B 170        100             >172.16.1.25    64502 64502 64502 I
38                      B 170        100             >172.16.100.8   64502 64502 64502 I # Over VPN
```

Paths through the L3VPN are now eligible to become active after the direct backdoor between router *Lille* and router *Nantes* is disconnected.

When using both `local-as private` and `no-prepend-global-as` commands, not all ASs can be suppressed from AS_PATHs. The `local-as` corresponding to the egress VRF is added when advertising the route to the CE, so there is a minimum of one additional hop. However, the MPLS backbone's own AS is not included in any direction.

independent-domain

[draft-marques-ppvpn-ibgp-00] has been strongly supported in Junos OS since its conception in the form of *independent domains*. This draft allows a PE to participate in a given VPN AS domain while acting as a transport for routes that keep the original BGP attributes and without making them appear in the Service Provider network.

Junos OS implements this feature simply by enabling the `independent-domain` knob and IBGP peering from the PE, as shown in Listing 5.44. Note that the `independent-domain` knob binds the customer AS only to a different domain (Line 10) and that IBGP sessions between the PE and CE can be simply based on interface addressing. In terms of VRF export policies, BGP routes need to be matched for export in this scenario.

Listing 5.44: Basic PE configuration for independent-domain on Livorno

```
1  user@Livorno> show configuration routing-instances NMM
2  instance-type vrf;
3  interface fe-0/3/0.0;
4  route-distinguisher 64501:64501;
5  vrf-export NMM-vrf-export;
6  vrf-target target:64501:64501;
7  vrf-table-label;
8  routing-options {
9      router-id 192.168.6.6;
10     autonomous-system 64502 independent-domain; # Independent-domain knob
11 }
12 protocols {
13     bgp {
14         group IBGP-PE-CE {
15             type internal; # IBGP session bound to interface addresses
16             local-address 172.16.100.8;
17             family inet {
18                 unicast;
19             }
20             neighbor 172.16.100.5;
21         }
22     }
23 }
```

When `independent-domain` is activated in Junos OS, two capabilities are added:

- support for multiple AS configuration domains;
- encoding of the original BGP attributes in the *ATTRSET* BGP path attribute for transit through the Service Provider, as shown in Figure 5.6.

The first item means that from a routing loop control perspective, a different and independent instance is created and bound to this customer AS. Each of these customer ASs is therefore considered to be local. Loop detection purpose is bounded within different and independent instances (note that the Junos OS `local-as loops` knob acts on a per-independent-domain basis), orthogonal to the master instance as well. The master independent-domain instance includes all other local ASs including its own and those ASs defined in instances without the `independent-domain` knob.

Indirectly, this means that any `local-as` information that may be included in a given stanza is domain-specific and relevant only for AS_PATH loop-detection purposes inside that *independent domain*. Thus, if a route migrates across these AS-path domains using *rib-groups* or *auto-export*, it is necessary to perform AS prepending on the AS_PATH as if it were a virtual EBGP session between the domains to ensure that the AS_PATH is correct and consistent for loop detection.

From a migration perspective, `independent-domain` provides a twofold benefit:

- Service Providers can transport customer routes without their AS being present in the customer routes' AS_PATH.
- Original BGP attributes are kept without need for specific policies on PEs.

Both these items are key distinguishers for this feature because they make customer BGP policies and TE almost completely transparent and independent from the transport through the Service Providers. Interim scenarios in which a given AS shares transit through traditional Internet peering and an MPLS L3VPN, or transition topologies in which the intent is not to modify the original BGP attributes can benefit greatly from this concept because they allow end customers to steer traffic directly without any further interaction with the transit provider.

Application Note: Integrating route distribution and attributes into an L3VPN with independent domains

Because the `independent-domain` knob allows transparent forwarding of the original BGP attributes from a given CE, a migration scenario in which a legacy network that bases customer and service route redistribution on any arbitrary BGP mesh can benefit from this feature to migrate backdoor links into a L3VPN and therefore disseminate routing information over different networks, even across more than one MPLS backbone.

While this feature can act as a transparent replacement for a backdoor link, the main advantage of an implementation based on an L3VPN (as opposed to a parallel implementation with any L2VPN or VPLS flavor) is that other routes from remote VPN sites can be granularly controlled for VRF import and therefore multiplexed with the existing replacement. In a nutshell, a traditional L3VPN can also act as a transparent substitute for a backdoor link from another network.

If application flows are based on BGP routes, it is not strictly necessary to interact with any IGP from the PE, but only directly with BGP. The `independent-domain` feature allows the original TE attributes to be kept from the CEs while encapsulating traffic inside an MPLS L3VPN.

Consider our well-known topology constructed as shown in Figure 5.29 with an original full IBGP mesh among participating routers based on a common OSPFv2 backbone in which each device externally peers with another customer AS with plain unicast IPv4 routes, importing them with their respective BGP attributes. The migration goal is to dismantle both backdoor links between router *Lille* and router *Nantes*, and router *Lille* and router *Torino*, while integrating the transit together with other routes into a given L3VPN in a smooth and non-intrusive fashion.

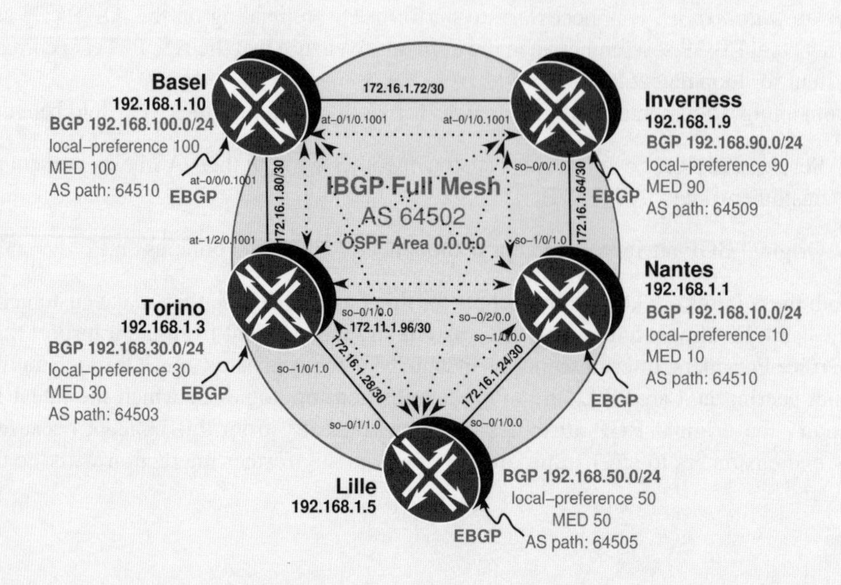

Figure 5.29: Representative legacy IBGP full-mesh network to be integrated in a L3VPN.

These routes represent customer address space and sensitive applications and are the main focus of the integration. Therefore, it is considered vital to maintain all BGP attributes from the original end customer ASs so as not to change the original TE policies on the routers, and it is assumed that physical transits are derived over the new PEs. There is also a certain risk that traffic-engineering policies from the transport Service Provider might overlap those from the original domain.

Because each participating router in the original full mesh injects the external routes with a different LOCAL_PREFERENCE, MED, and AS_PATH, the `independent-domain` feature becomes the perfect fit for transporting them in a scalable fashion without requiring crafted policies on the transport domain PEs. Independent domains directly skip any potential conflicts because the transport TE policies are therefore based on BGP attributes attached to the internal VPN-v4 NLRIs rather than on the encapsulated attributes in the *ATTRSET*.

Figure 5.30 illustrates the objective scenario in which router *Lille* is physically isolated from the original backbone, but remains connected and integrated into a joint L3VPN that is used as a transparent transport.

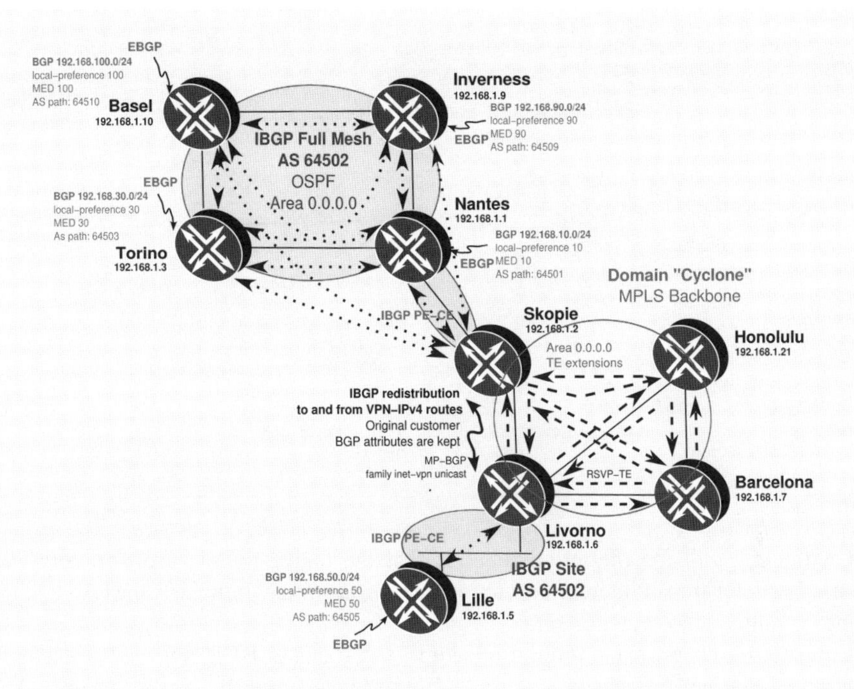

Figure 5.30: Migrated legacy IBGP full-mesh network over a L3VPN.

A global configuration and routing table overview is provided in Listing 5.45 from router *Lille*'s perspective, including IGP metrics to each of the other peers in the network, and the MED and local-preference values as designed.

Listing 5.45: Initial BGP configuration and routing table on Lille prior to L3VPN integration

```
 1 user@Lille> show configuration protocols bgp group IBGP
 2 type internal;
 3 local-address 192.168.1.5;
 4 family inet {
 5     unicast;
 6 }
 7 export redist-static;
 8 neighbor 192.168.1.1;  # Nantes
 9 neighbor 192.168.1.3;  # Torino
10 neighbor 192.168.1.10; # Basel
11 neighbor 192.168.1.9;  # Inverness
12
13 user@Lille> show route protocol bgp table inet.0 extensive |
14           match "entr|Protocol next hop|Localpref|metric|AS path"
15 192.168.10.0/24 (1 entry, 1 announced) # Address space from Nantes
16                 Protocol next hop: 192.168.1.1
17                 Age: 2:56   Metric: 10  Metric2: 643 # Local preference and MED from Nantes
18                 AS path: 64501 I
19                 Localpref: 10
20                         Protocol next hop: 192.168.1.1 Metric: 643
21               Metric: 643             Node path count: 1
22 192.168.30.0/24 (1 entry, 1 announced) # Address space from Torino
```

```
23                  Protocol next hop: 192.168.1.3
24                  Age: 4:06   Metric: 30  Metric2: 643 # Local preference and MED from Torino
25                  AS path: 64503 I
26                  Localpref: 30
27                          Protocol next hop: 192.168.1.3 Metric: 643
28                Metric: 643             Node path count: 1
29  <...>
30  192.168.90.0/24 (1 entry, 1 announced) # Address space from Inverness
31                  Protocol next hop: 192.168.1.9
32                  Age: 6:37   Metric: 90  Metric2: 1286 # Local preference and MED from Inverness
33                  AS path: 64509 I
34                  Localpref: 90
35                          Protocol next hop: 192.168.1.9 Metric: 1286
36                Metric: 1286            Node path count: 1
37  192.168.100.0/24 (1 entry, 1 announced) # Address space from Basel
38                  Protocol next hop: 192.168.1.10
39                  Age: 2:27   Metric: 100     Metric2: 803 # Local preference and MED from Basel
40                  AS path: 64510 I
41                  Localpref: 100
42                          Protocol next hop: 192.168.1.10 Metric: 803
43                Metric: 803             Node path count: 1
44
```

Continuing with the migration requires that the native uplinks from router *Lille* be replaced with the transit interconnect to router *Livorno*, and within the same L3VPN, router *Nantes* is migrated to router *Skopie*.

Compared with previous migration setups, the focus is BGP route advertisement. It is assumed that newer transits lead to a different forwarding behavior, but inter-AS TE policies must remain the same. In this case, the BGP route advertisement must retain the same attributes and criteria.

A first legitimate consideration is the BGP virtual topology into which the PEs need to be inserted. Figure 5.29 represents a full IBGP mesh among routers with no specific route reflection. Indirectly, inserting a PE with `independent-domain` means that routes received from a CE are readvertised to the backbone, and vice versa, so reflection is performed *using IBGP and with no specific BGP attributes related to BGP route reflection*. While the intent is actually to include the reflection in this migration, PEs need to be able to gather all IBGP routes from each VPN site in which they participate.

Although router *Lille* as a standalone segregated site can receive all routes from router *Livorno* without further propagation down the path to any IBGP peer, only router *Nantes* is physically connected to router *Skopie* at the remaining major site.

The traditional IBGP reflection paradigm for how to mesh routers or reflect routes is presented here. In a simplistic fashion, two alternatives are available for this major VPN site:

- *Route Reflection:* directly attached CE (router *Nantes*) performs route reflection towards the PE (router *Skopie*).

- *Full mesh:* PE (router *Skopie*) is fully meshed with all CEs (router *Nantes*, router *Torino*, router *Inverness* and router *Basel*) in the final VPN site.

Next-hop reachability is also a legitimate concern. [draft-marques-ppvpn-ibgp-00] imposes a *next-hop* substitution when advertising routes from the PE to the CE and extracting the original attributes from the transport *ATTRSET*.

Depending on the selected virtual BGP topology into which router *Skopie* gets meshed, the original next hop may or may not need to be advertised:

- *Route reflection:* router *Nantes* performs the next-hop rewrite together with route reflection towards router *Skopie* and towards the rest of the domain.

- *Full mesh:* either router *Skopie* is engaged in the internal IGP or bidirectional reachability is ensured not only with router *Nantes* but also with other routers at the major site.

The first option is based on [RFC4456] *route reflection* and presents the advantage of inserting a clear demarcation point at the PE boundary to the directly attached CEs, without influencing any other internal routers in the major VPN island. Router *Skopie* does not need to participate in the corresponding IGP or make its BGP next hop reachable beyond router *Nantes*, which even allows for address overlapping with that transit interface. A main disadvantage is that router *Nantes* becomes a route reflector server with router *Skopie* being its client, requiring a next-hop change in each direction and inserting additional attributes inherent to route reflection, as per [RFC4456], that are not present in the original path: ORIGINATOR_ID and CLUSTER_LIST. Their influence is not only limited to loop detection, but CLUSTER_LIST length is also considered in Junos OS as a route selection tie-breaker at some point.

The second option does not attach any further BGP attributes to the original routes, keeping the root *Attribute-Set*, but requires further protocol manipulation on router *Skopie* and eventually, router *Nantes*. In fact, router *Skopie* gets fully inserted in the major VPN island IBGP mesh, which poses a challenge of proper *next-hop* reachability. This requires that either router *Skopie* participates in the IGP as well, or that equivalent bidirectional BGP neighbor reachability is ensured by setting the transit interface as IGP passive on router *Nantes* and by configuring a static route on router *Skopie* for the reverse direction, or equivalent constructs.

Both options are illustrated in Figure 5.31.

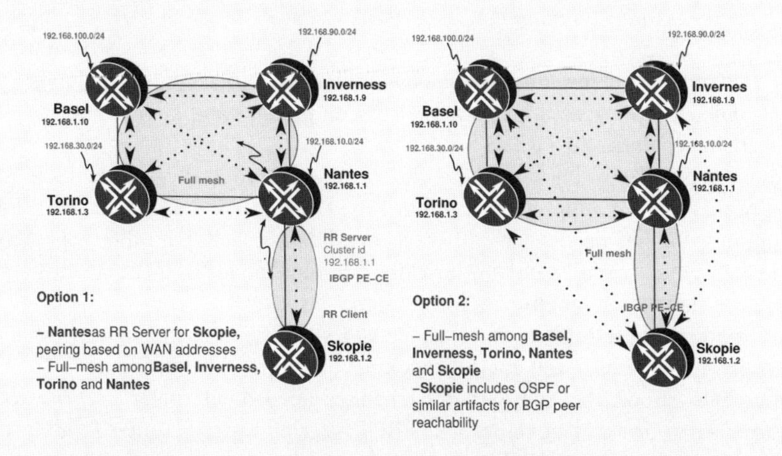

Figure 5.31: IBGP mesh alternatives for integration into a L3VPN.

As usual, our criteria are that the topology under study be future-proof and simple. From this perspective, the second option based on an IBGP mesh is selected for the following reasons:

- The migration setup presents only four internal routers that require minimal changes for the new BGP neighbor router *Skopie*.
- Because the upstream connection is single-homed, router *Nantes* can easily set the PE–CE interface as OSPF passive to ensure the best and optimal routing.
- Because of the addressing scheme used, router *Skopie* does not need to be active in OSPF, and a simple reverse static route for the loopback range suffices for BGP-peer and next-hop reachability.

Note that all these considerations are not really needed in the case of simple VPN sites, as seen with router *Lille*, in which PEs can directly peer to CEs without further propagation to other IBGP peers.

Once the IBGP virtual topology has been evaluated and chosen, planning the uplink transition requires further thinking. A traditional approach in this book has been first to perform migrations only for control-plane checks before activating the newer scenario for forwarding: in this setup, first activate the IBGP uplinks to the PEs but keep routes unpreferred for installation into the forwarding table, and then just get them installed as active paths after successful inspection of the BGP advertisement.

The inherent *next-hop* substitution dictated by [draft-marques-ppvpn-ibgp-00] has a clear influence here, because all original BGP attributes are disclosed again in the advertisement towards the CE, except for the *next hop*. The consequence is therefore that IGP metrics become dependent on the distance to the next hop calculated by each CE. This fact is the cornerstone of the ultimate BGP route selection transition in this scenario: while keeping the same topology and legacy BGP attributes, *IGP metrics to the PE are the decision makers on each CE to select either the path advertised by the PE or the path through the legacy backdoors to the destination.*

In a nutshell, create the adequate policies in order not to select PE's advertised routes in CEs first. Planning this initial transient stage must take into account that Junos OS allows BGP attributes to be tweaked inside the VRF with policies and knobs. These override the original attributes in the ATTRSET.

Additional policies or knobs can therefore be used on each PE to make their advertised routes towards the CEs unpreferred. Arbitrarily, the Junos OS `metric-out` knob with a maximum value has been selected here as a valid approach, but other approaches, such as using the `local-preference` knob for outbound advertisement or setting a specific export policy, are equally legitimate for the same purpose.

Thus, the first migration step interconnects PEs to CEs in each island with unpreferred BGP advertisements for initial control checking, as summarized in Figure 5.32.

Listing 5.46 shows the configuration changes needed on router *Livorno* to activate `independent-domain` together with IBGP peering towards router *Lille*. In this case, it suffices to have a single IBGP session towards router *Lille* as the solitary CE. This snippet includes MED magnification for outbound advertisements as a preventive measure for initial comparisons (Line 5).

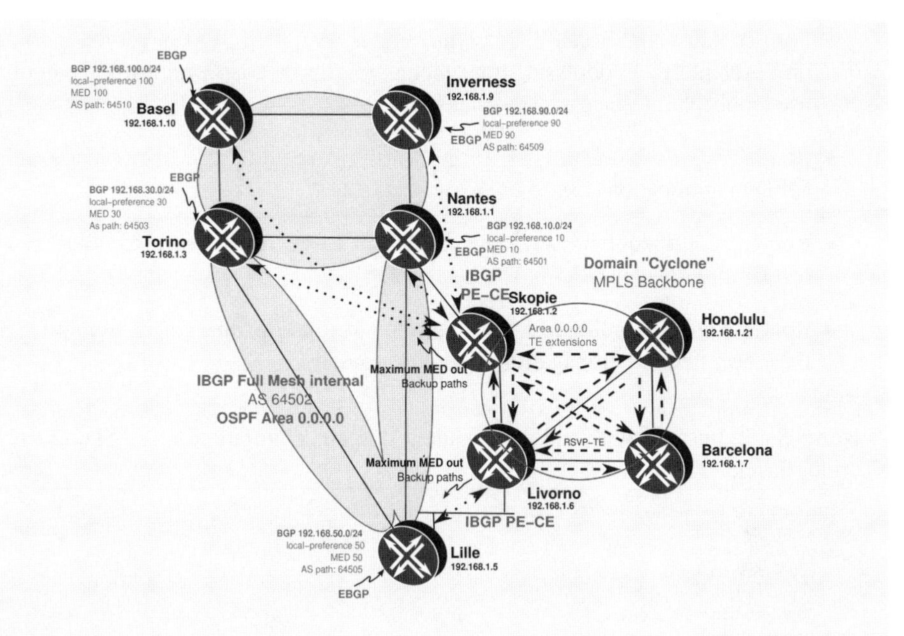

Figure 5.32: Stage one: interconnect domain "Cyclone" PEs and make PE-advertised routes unpreferred.

Listing 5.46: Initial IBGP and independent-domain configuration on router *Lille* and router *Livorno*

```
 1  [edit routing-instances NMM protocols bgp]
 2 user@Livorno# show
 3 group IBGP-PE-CE {
 4     type internal;
 5     metric-out 4294967295; # MED magnification for backup paths
 6     local-address 172.16.100.8;
 7     family inet {
 8         unicast;
 9     }
10     neighbor 172.16.100.5;
11 }
12
13  [edit routing-instances NMM routing-options]
14 user@Livorno# show
15 autonomous-system 64502 independent-domain; # Independent-domain creation
16
17  [edit protocols bgp group IBGP-PE-CE]
18 user@Lille# show
19 type internal;
20 local-address 172.16.100.5;
21 family inet {
22     unicast;
23 }
24 neighbor 172.16.100.8;
```

However, the remaining major VPN site requires a more verbose configuration. Listing 5.47 illustrates the new configuration on router *Skopie* and all other internal routers, which, following the previous discussion, can be summarized with these points:

- For simplicity and because the connection is unique, BGP peer addressing is based on WAN interface addressing for router *Skopie* and router *Nantes*, while other routers use loopback interface addresses as local.
- Router *Skopie* activates IBGP peering with `independent-domain` and outbound MED magnification (Line 5).
- Router *Skopie* configures a static route not to be readvertised for BGP neighbor reachability inside the VPN site (Line 19).
- Router *Nantes* sets the upstream interface to router *Skopie* as passive in OSPF for BGP peer address propagation inside the domain.

Listing 5.47: Initial IBGP and independent-domain configuration on Skopie, Nantes, Torino, Basel, and Inverness

```
1  [edit routing-instances NMM protocols bgp]
2  user@Skopie# show
3  group IBGP-PE-CE {
4      type internal;
5      metric-out 4294967295;  # MED magnification for backup paths
6      local-address 172.16.1.2;
7      family inet {
8          unicast;
9      }
10     neighbor 172.16.1.1;  # Nantes
11     neighbor 192.168.1.3;  # Torino
12     neighbor 192.168.1.9;  # Inverness
13     neighbor 192.168.1.10;  # Basel
14 }
15
16  [edit routing-instances NMM routing-options]
17 user@Skopie# show
18 static {
19     route 192.168.1.0/24 {  # Route for reverse reachability towards all internal routers
20         next-hop so-0/1/0.0;
21         no-readvertise;
22     }
23 }
24 autonomous-system 64502 independent-domain;  # Independent-domain creation
25
26  [edit protocols ospf]
27 user@Nantes# set area 0 interface so-0/1/0.0 passive  # WAN interface to Skopie
28  [edit protocols bgp group IBGP-PE-CE]
29 user@Nantes# show
30 type internal;
31 local-address 172.16.1.1;
32 family inet {
33     unicast;
34 }
35 neighbor 172.16.1.2; # Skopie
36
37  [edit protocols bgp]
38 user@Torino# set group IBGP neighbor 172.16.1.2  # Skopie
39  [edit protocols bgp]
40 user@Inverness# set group IBGP neighbor 172.16.1.2  # Skopie
41  [edit protocols bgp]
42 user@Basel# set group IBGP neighbor 172.16.1.2  # Skopie
```

After BGP sessions are established, advertised routes need to be inspected for proper comparison. Listing 5.48 illustrates the initial inspection on router *Lille*, with the following remarks:

- Outbound MED magnification is properly performed while keeping other original attributes such as LOCAL_PREF and AS_PATH.
- Advertised routes from router *Livorno* remain as backup paths as a result of MED comparison.
- IGP metric (*Metric2*) is 0 for router *Livorno* as a directly connected PE.

Listing 5.48: IBGP-received routes in Lille from Livorno

```
1  user@Lille> show route receive-protocol bgp 172.16.100.8 table inet.0
2
3  inet.0: 29 destinations, 35 routes (29 active, 0 holddown, 0 hidden)
4    Prefix          Nexthop          MED      Lclpref    AS path
5    192.168.10.0/24       172.16.100.8        4294967295 10       64501 I  # Maximum MED Out
6    192.168.30.0/24       172.16.100.8        4294967295 30       64503 I  # Maximum MED Out
7    192.168.90.0/24       172.16.100.8        4294967295 90       64509 I  # Maximum MED Out
8    192.168.100.0/24      172.16.100.8        4294967295 100      64510 I  # Maximum MED Out
9
10 user@Lille> show route protocol bgp detail |
11             match "entr|Protocol next hop|Localpref|metric|AS path"
12 192.168.10.0/24 (2 entries, 1 announced)
13                 Protocol next hop: 192.168.1.1
14                 Age: 2:23:39    Metric: 10  Metric2: 643
15                 AS path: 64501 I
16                 Localpref: 10
17                 Protocol next hop: 172.16.100.8
18                 Inactive reason: Not Best in its group - Route Metric or MED comparison
19                 Age: 1:51   Metric: 4294967295  Metric2: 0  # Minimal IGP Metric
20                 AS path: 64501 I
21                 Localpref: 10 # Same original Localpref
22 192.168.30.0/24 (2 entries, 1 announced)
23                 Protocol next hop: 192.168.1.3
24                 Age: 2:27:19    Metric: 30  Metric2: 643
25                 AS path: 64503 I
26                 Localpref: 30
27                 Protocol next hop: 172.16.100.8
28                 Inactive reason: Not Best in its group - Route Metric or MED comparison
29                 Age: 35:36  Metric: 4294967295  Metric2: 0  # Minimal IGP Metric
30                 AS path: 64503 I
31                 Localpref: 30 # Same original Localpref
32 <...>
33 192.168.90.0/24 (2 entries, 1 announced)
34                 Protocol next hop: 192.168.1.9
35                 Age: 2:26:43    Metric: 90  Metric2: 1286
36                 AS path: 64509 I
37                 Localpref: 90
38                 Protocol next hop: 172.16.100.8
39                 Inactive reason: Not Best in its group - Route Metric or MED comparison
40                 Age: 35:36  Metric: 4294967295  Metric2: 0  # Minimal IGP Metric
41                 AS path: 64509 I
42                 Localpref: 90 # Same original Localpref
43 192.168.100.0/24 (2 entries, 1 announced)
44                 Protocol next hop: 192.168.1.10
45                 Age: 2:26:17    Metric: 100     Metric2: 803
46                 AS path: 64510 I
47                 Localpref: 100
48                 Protocol next hop: 172.16.100.8
49                 Inactive reason: Not Best in its group - Route Metric or MED comparison
50                 Age: 35:36  Metric: 4294967295  Metric2: 0 # Minimal IGP Metric
51                 AS path: 64510 I
52                 Localpref: 100  # Same original Localpref
```

Inspecting a sample route on router *Livorno* coming over the backbone also reveals how the *ATTRSET* is built and detected, as seen in Listing 5.49.

Listing 5.49: IBGP-received routes on Lille from Livorno

```
1  user@Livorno> show route receive-protocol bgp 192.168.1.2 detail |
2               table NMM 192.168.10.0/24
3
4  NMM.inet.0: 7 destinations, 7 routes (7 active, 0 holddown, 0 hidden)
5  * 192.168.10.0/24 (1 entry, 1 announced)
6       Import Accepted
7       Route Distinguisher: 64501:64501
8       VPN Label: 16
9       Nexthop: 192.168.1.2
10      Localpref: 100
11      AS path: I
12      Communities: target:64501:64501
13      AttrSet AS: 64502
14          MED: 10
15          Localpref: 10
16          AS path: 64501 I
17
18 user@Livorno> show route advertising-protocol bgp 172.16.100.5 detail table NMM
       192.168.10.0/24
19
20 NMM.inet.0: 7 destinations, 7 routes (7 active, 0 holddown, 0 hidden)
21 * 192.168.10.0/24 (1 entry, 1 announced)
22  BGP group IBGP-PE-CE type Internal
23      Nexthop: Self
24      Flags: Nexthop Change
25      MED: 4294967295
26      Localpref: 10
27      AS path: [64502] 64501 I
```

Junos Tip: AS path loop avoidance by means of independent-domains

Another application of the `independent-domain` feature is to create different *AS-path domains* for AS loop checking without using any `local-as loops` knobs.

Reviewing Listing 5.49 again, on Lines 16 and 27 *64501* appears as the final customer AS from which this route is originated.

However, *64501* is also the internal AS used in the Service Provider transport backbone (Line 5), which is configured without any particular allowance for *AS loops*, as seen in Listing 5.50.

Listing 5.50: Internal AS configuration on Livorno

```
1 user@Livorno> show bgp neighbor 192.168.1.2 | match "^Peer"
2 Peer: 192.168.1.2+55701 AS 64501 Local: 192.168.1.6+179 AS 64501
3
4 user@Livorno> show configuration routing-options autonomous-system
5 64501;
6
7 user@Livorno> show configuration routing-instances NMM routing-options
8 autonomous-system 64502 independent-domain;
```

Effectively, the transport Service Provider is not performing any kind of AS path loop checking in their main instance on this customer's AS_PATHs, as a consequence of the additional domain creation triggered by the `independent-domain` feature. Those checks are therefore carried out on a per-`independent-domain` basis, in this case independently in the VRF and in the main routing instance.

Junos Tip: AS path loop detection per family

The `local-as loops` knob specifies the maximum number of times that a local AS can appear in AS_PATH attributes from routes received from a BGP peer.

Junos OS considers each AS configured under `[edit routing-options]` or related to the `local-as loops` knob to be *local* for loop detection purposes. All the ASs across the routing instances are considered to be part of a common *AS-path domain* or *Independent domain*, and Junos OS enforces that the *loops* value is therefore common to all instances, because AS paths are shared across all of them and loop checking is performed on a per *AS-path domain* basis. The main routing table is also considered to be part of the same *AS-path domain*, and the same *loops* value needs to be set if this feature is needed.

However, Junos OS also offers the option of configuring a *loops* count for a particular address family for a specific BGP neighbor. In this case, arbitrary values are tolerated and can be distinct from the *loops* count in the *AS-path domain*.

From a migration perspective, this functionality requires that *loops* are granularly set for each neighbor and AFI/SAFI combination, but relaxes the need to enforce the same value across all instances part of the same *AS-path domain*.

Listing 5.51 shows both configuration commands to set `local-as loops` on a per-instance basis, requiring the same *loops* value in all other instances inside the *AS-path domain* (Line 1) and to set the `loops` count for each family on each BGP neighbor (Line 6).

Listing 5.51: AS loops configuration on Livorno

```
1  [edit routing-instances NMM protocols bgp] # Local-as loops
2  user@Livorno# set group EBGP-PE-CE local-as 64502 loops ?
3  Possible completions:
4    <loops>                Maximum number of times this AS can be in an AS path (1..10)
5
6  [edit routing-instances NMM protocols bgp] # Per-family loops
7  user@Livorno# set group EBGP-PE-CE neighbor 172.16.100.5 family inet unicast loops ?
8  Possible completions:
9    <loops>                AS-Path loop count (1..10)
```

A parallel route inspection is performed on the other side of the VPN, namely, by router *Nantes*, as shown in Listing 5.52, with these results:

- Outbound MED magnification is properly performed while keeping other original attributes such as LOCAL_PREF and AS_PATH.
- The advertised route from router *Skopie* remains as a backup path as a result of MED comparison. Because the MED is consistent throughout the setup, the same criterion applies to other internal routers.
- IGP metric (*Metric2*) is 0 for router *Skopie* as a directly connected PE.

Listing 5.52: IBGP-received routes on router *Lille* from router *Livorno*

```
1  user@Nantes> show route receive-protocol bgp 172.16.1.2 table inet.0
2
3  inet.0: 36 destinations, 41 routes (36 active, 0 holddown, 0 hidden)
```

```
4    Prefix           Nexthop             MED       Lclpref     AS path
5    192.168.50.0/24          172.16.1.2             4294967295 50        64505 I
6
7  user@Nantes> show route 192.168.50.0/24 detail
8                | match "entr|Protocol next hop|Localpref|metric|AS path"
9  192.168.50.0/24 (2 entries, 1 announced)
10                  Protocol next hop: 192.168.1.5
11                  Age: 2:55:00    Metric: 50  Metric2: 643
12                  AS path: 64505 I
13                  Localpref: 50
14                  Protocol next hop: 172.16.1.2
15                  Inactive reason: Not Best in its group - Route Metric or MED comparison
16                  Age: 46:11  Metric: 4294967295  Metric2: 0
17                  AS path: 64505 I
18                  Localpref: 50
```

All backup paths have now been evaluated and the setup is considered ready for an active transition over the L3VPN. This migration step can be applied in each direction simply by removing the `metric-out` knob with the maximum metric and letting the IGP metric decide active paths for routes received from the PE.

These configuration changes and a similar route inspection are visible in Listing 5.53 and Figure 5.33.

Listing 5.53: Route selection changes after removing maximum MED from Livorno and Skopie

```
1   [edit routing-instances NMM protocols bgp]
2  user@Livorno# delete group IBGP-PE-CE metric-out
3   [edit routing-instances NMM protocols bgp]
4  user@Skopie# delete group IBGP-PE-CE metric-out
5
6  user@Lille> show route protocol bgp detail
7              | match "entr|Protocol next hop|Localpref|metric|AS path"
8  192.168.10.0/24 (2 entries, 1 announced)
9                  Protocol next hop: 172.16.100.8
10                 Age: 1:52   Metric: 10  Metric2: 0
11                 AS path: 64501 I
12                 Localpref: 10
13                 Protocol next hop: 192.168.1.1
14                 Inactive reason: Not Best in its group - IGP metric
15                 Age: 3:04:14    Metric: 10  Metric2: 643
16                 AS path: 64501 I
17                 Localpref: 10
18 192.168.30.0/24 (2 entries, 1 announced)
19                 Protocol next hop: 172.16.100.8
20                 Age: 1:52   Metric: 30  Metric2: 0
21                 AS path: 64503 I
22                 Localpref: 30
23                 Protocol next hop: 192.168.1.3
24                 Inactive reason: Not Best in its group - IGP metric
25                 Age: 3:07:54    Metric: 30  Metric2: 643
26                 AS path: 64503 I
27                 Localpref: 30
28 192.168.90.0/24 (2 entries, 1 announced)
29                 Protocol next hop: 172.16.100.8
30                 Age: 1:52   Metric: 90  Metric2: 0
31                 AS path: 64509 I
32                 Localpref: 90
33                 Protocol next hop: 192.168.1.9
34                 Inactive reason: Not Best in its group - IGP metric
35                 Age: 3:07:18    Metric: 90  Metric2: 1286
36                 AS path: 64509 I
37                 Localpref: 90
38 192.168.100.0/24 (2 entries, 1 announced)
```

```
            Protocol next hop: 172.16.100.8
            Age: 1:52   Metric: 100    Metric2: 0
            AS path: 64510 I
            Localpref: 100
            Protocol next hop: 192.168.1.10
            Inactive reason: Not Best in its group - IGP metric
            Age: 3:06:52   Metric: 100    Metric2: 803
            AS path: 64510 I
            Localpref: 100

user@Nantes> show route 192.168.50.0/24 detail |
        match "entr|Protocol next hop|Localpref|metric|AS path"
192.168.50.0/24 (2 entries, 1 announced)
            Protocol next hop: 172.16.1.2
            Age: 1:49   Metric: 50  Metric2: 0
            AS path: 64505 I
            Localpref: 50
            Protocol next hop: 192.168.1.5
            Inactive reason: Not Best in its group - IGP metric
            Age: 3:08:45   Metric: 50  Metric2: 643
            AS path: 64505 I
            Localpref: 50
```

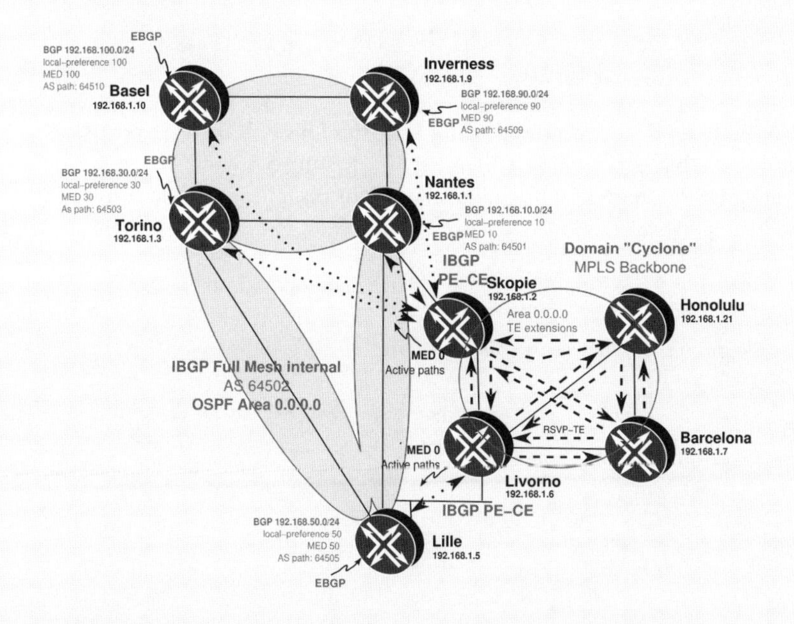

Figure 5.33: Stage two: make PE-advertised routes preferred.

All active paths in the setup have selected router *Skopie* and router *Livorno* as next hops. This can also be seen by checking on router *Lille*, router *Inverness*, and router *Skopie* that no remaining routes select any backdoor link as next hop, as shown in Listing 5.54.

Listing 5.54: No remaining routes over backdoor links on Lille, Inverness, and Skopie

```
user@Lille> show ospf neighbor
Address          Interface              State     ID               Pri  Dead
172.16.1.25      so-0/1/0.0             Full      192.168.1.1      128    36
172.16.1.29      so-0/1/1.0             Full      192.168.1.3      128    37
```

```
 5
 6 user@Lille> show route next-hop 172.16.1.29 table inet.0
 7
 8 inet.0: 29 destinations, 35 routes (29 active, 0 holddown, 0 hidden)
 9
10 user@Lille> show route next-hop 172.16.1.25 table inet.0
11
12 inet.0: 29 destinations, 35 routes (29 active, 0 holddown, 0 hidden)
13
14 user@Nantes> show ospf neighbor
15 Address          Interface              State     ID               Pri  Dead
16 172.16.1.97      so-0/2/0.0             Full      192.168.1.3      128    36
17 172.16.1.26      so-1/0/0.0             Full      192.168.1.5      128    32
18 172.16.1.66      so-1/0/1.0             Full      192.168.1.9      128    32
19
20 user@Nantes> show route next-hop 172.16.1.26 table inet.0
21
22 inet.0: 36 destinations, 41 routes (36 active, 0 holddown, 0 hidden)
23
24 user@Torino> show ospf neighbor
25 Address          Interface              State     ID               Pri  Dead
26 172.16.1.82      at-1/2/0.1001          Full      192.168.1.10     128    36
27 172.16.1.98      so-0/1/0.0             Full      192.168.1.1      128    37
28 172.16.1.30      so-1/0/1.0             Full      192.168.1.5      128    38
29
30 user@Torino> show route next-hop 172.16.1.30 table inet.0
31
32 inet.0: 34 destinations, 39 routes (34 active, 0 holddown, 0 hidden)
```

Because all services are now flowing over the L3VPN, there is no risk when finally dismantling the backdoor links. This is seen in Listing 5.55 and Figure 5.34 after simply deactivating OSPF on router *Lille*: all backup paths disappear, but the active routes towards the backbone persist.

Listing 5.55: Final OSPF deactivation and decommissioning on Lille

```
 1 [edit]
 2 user@Lille# deactivate protocols ospf
 3
 4 user@Lille> show route protocol bgp detail |
 5            match "entr|Protocol next hop|Localpref|metric|AS path"
 6 192.168.10.0/24 (2 entries, 1 announced)
 7                 Protocol next hop: 172.16.100.8
 8                 Age: 21:08  Metric: 10  Metric2: 0
 9                 AS path: 64501 I
10                 Localpref: 10
11 192.168.30.0/24 (2 entries, 1 announced)
12                 Protocol next hop: 172.16.100.8
13                 Age: 21:08  Metric: 30  Metric2: 0
14                 AS path: 64503 I
15                 Localpref: 30
16 192.168.90.0/24 (2 entries, 1 announced)
17                 Protocol next hop: 172.16.100.8
18                 Age: 21:08  Metric: 90  Metric2: 0
19                 AS path: 64509 I
20                 Localpref: 90
21 192.168.100.0/24 (2 entries, 1 announced)
22                 Protocol next hop: 172.16.100.8
23                 Age: 21:08  Metric: 100    Metric2: 0
24                 AS path: 64510 I
25                 Localpref: 100
26
27 user@Nantes> show route 192.168.50.0/24 detail |
28            match "entr|Protocol next hop|Localpref|metric|AS path"
29 192.168.50.0/24 (1 entry, 1 announced)
30                 Protocol next hop: 172.16.1.2
```

```
31          Age: 21:38  Metric: 50  Metric2: 0
32          AS path: 64505 I
33          Localpref: 50
```

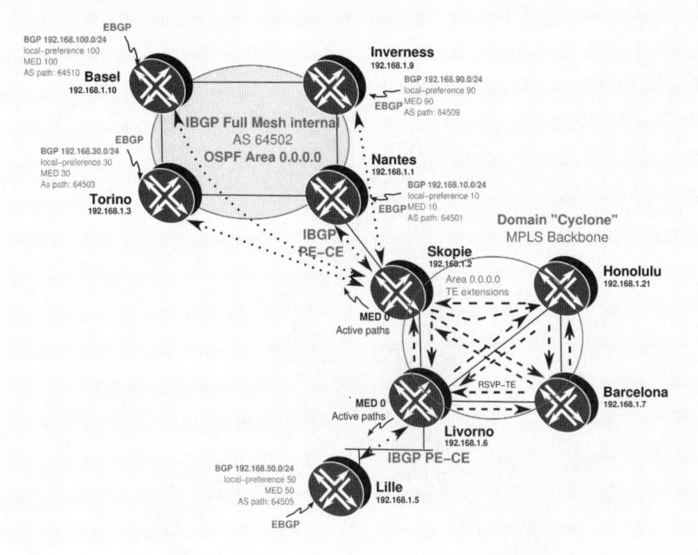

Figure 5.34: Stage three: dismantle backdoors.

Junos Tip: Integrating standard VPN routes with independent-domains

Consider the final setup from Figure 5.30 with *Barcelona* injecting in the same VRF, another internal route, namely 192.168.70.0/24.

This route is considered internal to the same L3VPN and is advertised to CEs, equally eligible as sites from the former domain, as seen in Listing 5.56.

Listing 5.56: Advertisement of internal routes in the same L3VPN on Skopie and Livorno

```
 1 user@Skopie> show route advertising-protocol bgp 172.16.1.1
 2
 3 NMM.inet.0: 9 destinations, 9 routes (9 active, 0 holddown, 0 hidden)
 4   Prefix              Nexthop          MED      Lclpref    AS path
 5 * 192.168.50.0/24         Self                  50         50          64505 I
 6 * 192.168.70.0/24         Self                             100         64501 I  # Internal route
 7
 8 user@Livorno> show route advertising-protocol bgp 172.16.100.5
 9
10 NMM.inet.0: 8 destinations, 8 routes (8 active, 0 holddown, 0 hidden)
11   Prefix              Nexthop          MED      Lclpref    AS path
12 * 192.168.10.0/24         Self                  10         10          64501 I
13 * 192.168.30.0/24         Self                  30         30          64503 I
14 * 192.168.70.0/24         Self                             100         64501 I # Internal route
15 * 192.168.90.0/24         Self                  90         90          64509 I
16 * 192.168.100.0/24        Self                  100        100         64510 I
```

The beauty of using the `independent-domain` feature in this scenario is that it easily allows integration and expansion of such scenarios while keeping control over route distribution and attribute manipulation.

5.6 Case Study

To put into practice the fundamental concepts related to L3VPN migrations, this case study focuses on the different L3VPN interconnection options available when merging domains. Both the inter-AS options discussed in Section 5.2.6 and Carrier Supporting Carrier (CsC) in Section 5.2.7 are implemented in this case study.

The topology being used through this book has been slightly enhanced, as shown in Figure 5.35. Two physical interconnections between router *Livorno* and router *Havana* assist in simulating multiple peering options. Connectivity to the CsC provider is available on the four ASBRs in domain "Monsoon."

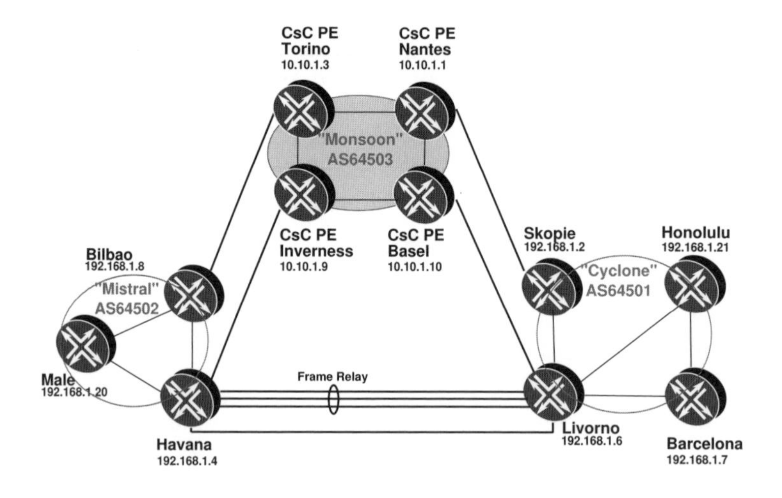

Figure 5.35: Modified topology for Inter-AS case study.

The first section describes the original network. This is followed by a discussion of the target network to provide a view of the final scenario. A migration strategy section discusses considerations and approaches. Finally, details are given for each migration stage. Only one VPN, *NMM*, is considered throughout the case study, with one PE in each domain, but this can obviously be generalized.

5.6.1 Original network

The existing scenario is a successful L3VPN service being provided by domain "Cyclone". The Marketing department of the common service provider "Gale Internet" sees the integration of domain "Mistral" as a good opportunity to capture more business by extending VPNs to remote locations in which domain "Mistral" has a presence.

Additional information related to the L3VPN service for VPN *NMM* has to be considered:

- Domain "Cyclone" provides L3VPN services using many PE routers, of which router *Barcelona* is a representative device. At the same time, domain "Mistral" also provides a VPN service, with router *Male* being a representative PE.

- VPN *NMM* is present on router *Barcelona*. CE attachments for this VPN lie behind the PE routers and are not shown. This full-mesh VPN uses `route-target target:64501:64501` on domain "Cyclone" to exchange its prefixes.
- As a representative case of the desire to extend this VPN, PE router *Male* in domain "Mistral" also built this customer VPN. Because of separate provisioning systems and operations processes, the short-term agreement between domain "Mistral" and domain "Cyclone" is to set up independent VPNs at each domain, with domain "Mistral" using `route-target target:64502:64502` for this VPN, which is also called *NMM*.
- The volume of routing information from PE routers in domain "Mistral" for this VPN is to be controlled by aggregating route blocks on router *Havana* towards router *Livorno*.
- Inbound traffic accounting and policing for those aggregates is desired on router *Havana*.

5.6.2 Target network

The end goal of the migration is to leverage backbone MPLS transport connectivity from a third party that offers CsC services, as shown in Figure 5.36. PEs in domain "Cyclone" interconnect to PEs in domain "Mistral," with a combined Option B for domain "Mistral" and Option C for domain "Cyclone," over the CsC setup in domain "Monsoon."

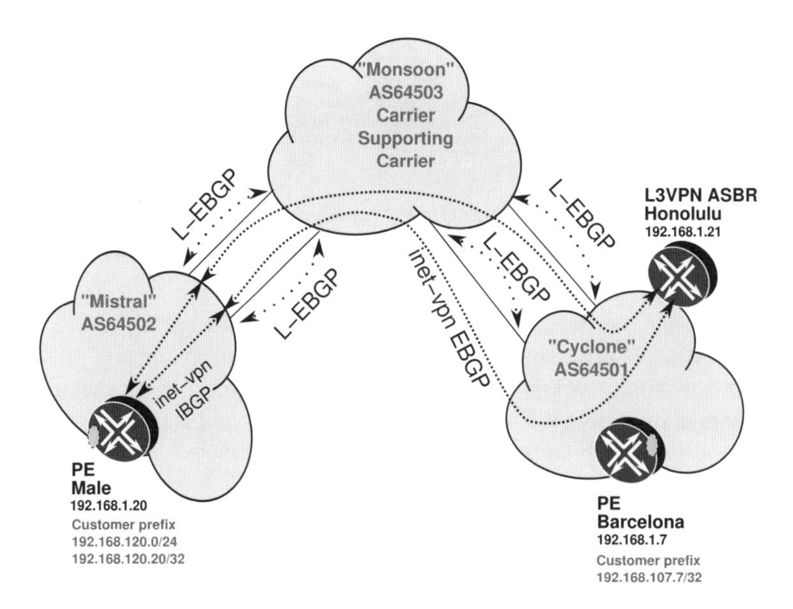

Figure 5.36: Target network migrated with Options B and C over CsC.

5.6.3 Migration strategy

Migration planning

For illustration purposes, the unification of the L3VPN service occurs in several stages, covering each of the interconnecting options. There is little time dependency among the stages.

The proposed approach for the migration timeline is shown in Figure 5.37.

- In a first stage, a rapid extension of this service over domain "Mistral" is achieved using an inter-AS Option A arrangement for all VPNs that have to be shared. To keep the example focused, only the VPN *NMM* is considered.

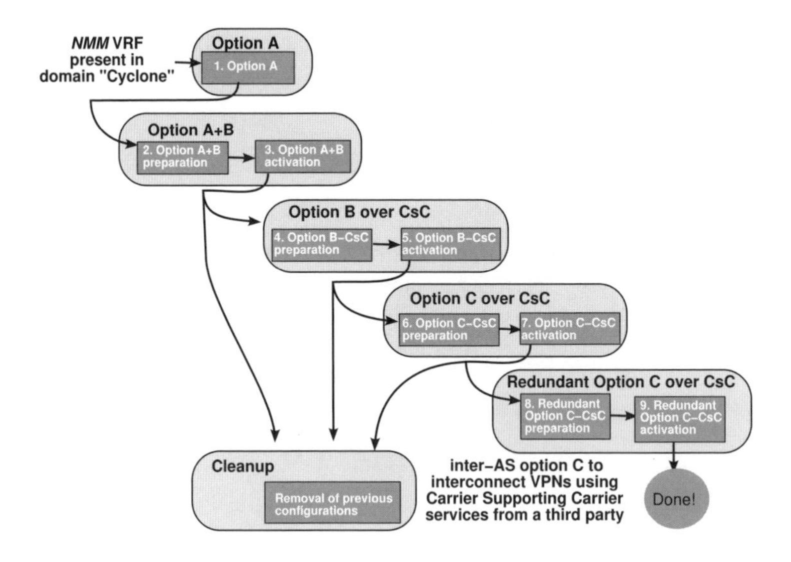

Figure 5.37: Approach to interconnect *NMM* L3VPN between ASs.

- In a second stage, scalable service evolution prompts a change in peering arrangement to Option B while maintaining service independence in each domain. Existing service requirements force an Option A+B configuration at the border router Havana for selected prefixes within the L3VPN *NMM*. This stage completes the preparatory work required to keep both Option A and Option B interconnects active, with preference given to the existing Option A interconnect.

- Stage three activates the Option B interconnect and removes the VRF instance on router *Livorno*. For router *Havana*, however, the *NMM* VRF has to stay to provide an anchoring point for the IP functions of aggregation and accounting, as is discussed later.

- Transmission cost optimization is introduced in stage four, by retaining the services of a third-party domain as a CsC. Preparatory work to bring up the session with a lower preference ensures that traffic stays on the existing direct interconnect between router *Livorno* and router *Havana*.

- Stage five switches preferences on the Option B interconnect over to the CsC service, redirecting the traffic on the direct Option B interconnect between router *Livorno* and router *Havana* to free up the direct interconnect.

- In a sixth stage, generalization of the connectivity from domain "Mistral" to multiple PEs in domain "Cyclone" is achieved with preparatory work to migrate the Option B peering at router *Livorno* to an Option C interconnect. Reachability to all PEs within domain "Cyclone" is required for redistribution of MPLS transport labels. To avoid a mesh of BGP sessions for L3VPN from all the PE routers in domain "Cyclone" to router *Havana*, the router *Honolulu* is established as the ASBR, with a multihop EBGP L3VPN peering session between router *Havana* and router *Honolulu*. Business requirements force router *Havana* to remain as the ASBR for domain "Mistral" so that this router can continue to hide the internal VPN topology from domain "Mistral" by being a relay point for VPN information. Router *Havana* does so by setting next-hop self on the advertisements for VPN *NMM*.

- The activation of the Option C interconnect is analyzed in stage seven.

- To showcase possible redundancy issues, the eighth stage establishes an additional Option C interconnect between router *Bilbao* and router *Honolulu* that crosses router *Skopie* using the CsC infrastructure.

- As the final stage, stage nine enables this additional session in parallel with the existing Option C interconnects, providing a final setup with Option C connectivity to domain "Cyclone" with the L3VPN peering from router *Honolulu* to both router *Havana* and router *Bilbao*, which act as ASBRs for domain "Mistral".

By choosing the right BGP attributes to control route preferences, all transitions in this case study can be controlled from a single point. This point is chosen to be router *Havana*. Table 5.2 provides a summary of the various stages and the preferences given to advertisements on each of the sessions. The first value represents the LOCAL_PREF to set for received prefixes; the second value is the MED BGP attribute for advertised prefixes. All the modifications are done on router *Havana* except for the last column (Opt-C-CsC-Redundant), which is configured on router *Bilbao*.

5.6.4 Stage one: Inter-AS Option A deployment

As the most straightforward interconnection mechanism, the VRF for VPN *NMM* is instantiated on router *Havana* and router *Livorno* and a dedicated subinterface connection is reserved. Figure 5.38 depicts this interconnection stage, which provides a common VPN service.

The selected location to place the inter-AS Option A interconnection is between router *Livorno* and router *Havana*. Listing 5.57 provides the baseline configuration. VPN *NMM* leverages one of the multiple subinterfaces that join the various VPNs across the inter-AS binding. Besides the PE–CE subinterface connection, a specific per-VRF loopback interface

Table 5.2 Preference settings for controlling the migration stages

Stage	Description	A	B-direct	B–CsC	C–CsC	C–CsC–Dual
1	Initial A	100				
		4000				
2	Build B Direct		50			
			4500			
3	Switch to B Direct		200			
			3000			
4	Build B over CsC			150		
				3500		
5	Switch to B over CsC			300		
				2000		
6	Build C over CsC				250	
					2500	
7	Switch to C over CsC				450	
					500	
8	Build redundant C					400
						1000
9	Balance with redundant C				400	
					1000	

LOCAL_PREF and MED for router *Havana* at every stage. Stage 8 involves router *Bilbao*.

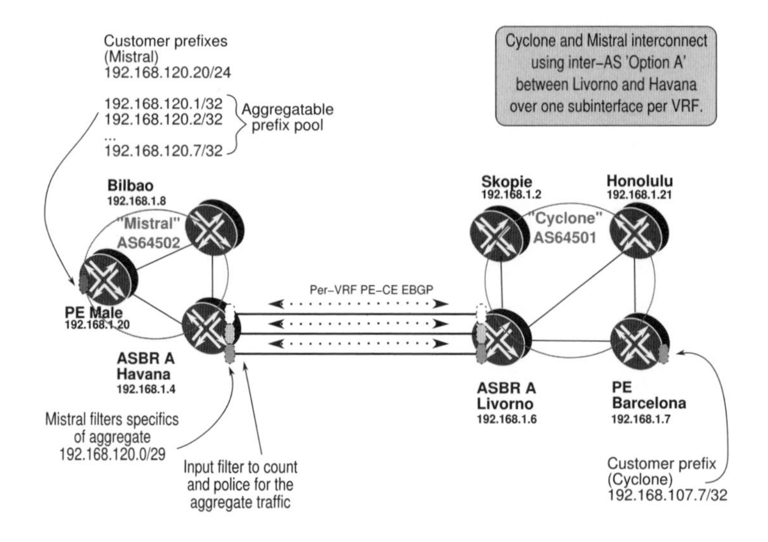

Figure 5.38: Stage one: Inter-AS Option A topology.

is added for troubleshooting purposes on Line 4. This loopback address is visible only within the domain. Notice that no specific BGP policy is applied to advertise the loopback address in the interconnection in the BGP configuration block that starts on Line 7. Following default BGP policy rules, only BGP routes are advertised to the peer.

Listing 5.57: Baseline configuration for Inter-AS Option A connectivity

```
1  user@Havana> show configuration routing-instances NMM
2  instance-type vrf;
3  interface so-1/0/0.100;              # PE-CE subinterface for the interconnect
4  inactive: interface lo0.104;         # VRF loopback used for troubleshooting
5  vrf-target target:64502:64502;       # Full-mesh RT
6  protocols {
7      bgp {
8          group Cyclone {
9              peer-as 64501;
10             neighbor 172.16.250.38;  # PE-CE subinterface peering
11         }
12     }
13 }
14 user@Livorno> show configuration routing-instances NMM
15 instance-type vrf;
16 interface so-0/0/1.100;              # PE-CE subinterface on interconnect
17 inactive: interface lo0.106;         # VRF loopback used for troubleshooting
18 vrf-target target:64501:64501;       # Full-mesh RT
19 protocols {
20     bgp {
21         group Mistral {
22             peer-as 64502;
23             neighbor 172.16.250.37;  # On PE-CE subinterface peering
24         }
25     }
26 }
```

Listing 5.58 shows successful connectivity within the VPN between prefixes on router *Male* and router *Barcelona*. Notice that the connectivity at Lines 5 and 14 have no MPLS label reflecting the plain IP interconnection. The existence of a single MPLS label in the trace on Line 4 is a consequence of the reduced setup that has no Provider (P) routers and the implicit-null behavior when signaling transport labels.

Listing 5.58: Inter-AS Option A connectivity verification

```
1  user@Barcelona-re0> traceroute routing-instance NMM source barcelona-NMM male-NMM
2  traceroute to male-NMM (192.168.120.20) from barcelona-NMM, 30 hops max, 40 byte packets
3   1  livorno-ge020 (172.16.1.54)  1.112 ms  0.917 ms  0.872 ms
4       MPLS Label=305568 CoS=0 TTL=1 S=1
5   2  havana-so100100 (172.16.250.37)  0.792 ms  0.725 ms  0.702 ms
6   3  male-ge4240 (172.16.1.85)  0.801 ms  0.809 ms  0.763 ms
7       MPLS Label=299808 CoS=0 TTL=1 S=1
8   4  male-NMM (192.168.120.20)  0.781 ms  0.828 ms  0.727 ms
9
10 user@male-re0> traceroute routing-instance NMM barcelona-NMM source male-NMM
11 traceroute to barcelona-NMM (192.168.107.7) from male-NMM, 30 hops max, 40 byte packets
12  1  havana-ge1100 (172.16.1.86)  0.744 ms  0.644 ms  0.569 ms
13      MPLS Label=301408 CoS=0 TTL=1 S=1
14  2  livorno-so001100 (172.16.250.38)  0.492 ms  10.015 ms  0.444 ms
15  3  barcelona-ge120 (172.16.1.53)  0.830 ms  0.668 ms  0.639 ms
16      MPLS Label=302304 CoS=0 TTL=1 S=1
17  4  barcelona-NMM (192.168.107.7)  0.631 ms  0.598 ms  0.591 ms
```

Additional considerations in this setup are worth mentioning:

- Concerns about the amount of routing information sourced from sites in domain "Mistral" prompt aggregation of a very specific route block at Havana.

- A trust relationship exists between domain "Cyclone" and domain "Mistral" that does not require any specific traffic IP processing in domain "Cyclone" at the border router *Livorno*. However, inbound traffic from domain "Cyclone" to domain "Mistral" to those aggregated destinations is considered low priority, and a specific bandwidth policing function is to be established inbound at router *Havana*. It is also desirable to count traffic to these destinations for accounting purposes.

- The impact of default community handling in BGP PE–CE must be considered. In Junos OS, RT communities attached to VPN prefixes are not filtered by default.

Prefix aggregation at the border router

PE router *Male* receives a large set of prefixes that can be aggregated in simpler blocks. Domain "Cyclone" prefers to keep its control plane scalability under control, and agrees with the administrators of domain "Mistral" that only aggregates are to be advertised over the interconnection. Although the aggregation could happen at the PE router *Male* in this oversimplified setup, the assumption is that different prefix blocks forming the aggregatable space are to be sourced by multiple PE routers within domain "Mistral". Listing 5.59 shows how the PE–CE interconnect has an outbound policy (Line 14) to advertise a locally generated aggregate (Line 7). Therefore, the ASBR border router *Havana*, acting as PE, holds aggregate blocks and filters more specific routes (Line 27) using routing policy in its advertisements to router *Livorno*, as shown on Line 49.

Listing 5.59: Filtering more-specific routes in aggregate advertisements

```
1  user@Havana> show configuration routing-instances NMM
2  instance-type vrf;
3  interface so-1/0/0.100;          # PE-CE subinterface on interconnect
4  inactive: interface lo0.104;     # VRF loopback used for troubleshooting
5  vrf-target target:64502:64502;   # Full-mesh RT
6  routing-options {
7      aggregate {
8          route 192.168.120.0/29 discard;
9      }
10 }
11 protocols {
12     bgp {
13         group Cyclone {
14             export Aggregate;
15             peer-as 64501;
16             neighbor 172.16.250.38; # On PE-CE subinterface peering
17         }
18     }
19 }
20
21 user@Havana> show configuration policy-options policy-statement Aggregate
22 term aggregate {
23     from protocol aggregate; # Advertise aggregate
24     then accept;
25 }
26 term no-specifics {
27     from aggregate-contributor; # Filter out more-specific routes
28     then reject;
29 }
30 user@Havana> show route table NMM 192.168.120/29 terse
31
32 NMM.inet.0: 15 destinations, 16 routes (15 active, 0 holddown, 0 hidden)
33 + = Active Route, - = Last Active, * = Both
34
```

```
35 A Destination        P Prf   Metric 1   Metric 2  Next hop        AS path
36 * 192.168.120.0/29   A 130                         Discard
37 * 192.168.120.1/32   B 170        100             >172.16.1.85     I
38 * 192.168.120.2/32   B 170        100             >172.16.1.85     I
39 * 192.168.120.3/32   B 170        100             >172.16.1.85     I
40 * 192.168.120.4/32   B 170        100             >172.16.1.85     I
41 * 192.168.120.5/32   B 170        100             >172.16.1.85     I
42 * 192.168.120.6/32   B 170        100             >172.16.1.85     I
43 * 192.168.120.7/32   B 170        100             >172.16.1.85     I
44
45 user@Havana> show route advertising-protocol bgp 172.16.250.38
46
47 NMM.inet.0: 15 destinations, 16 routes (15 active, 0 holddown, 0 hidden)
48   Prefix                 Nexthop              MED     Lclpref    AS path
49 * 192.168.120.0/29       Self                                    I # Aggregate
50 * 192.168.120.20/32      Self                                    I # Male prefix outside aggregate
```

IP accounting functions

Listing 5.60 provides the relevant configuration that illustrates how traffic is being accounted and limited for the aggregate. A sample traffic pattern test starting at Line 32 and the resulting statistics on Line 40 validate this setup. Notice the requirement for input counting and policing does not impose any special IP lookup function on egress, so a table-label is not required.

Listing 5.60: Counting and filtering traffic towards an aggregate route

```
1  user@Havana> show configuration firewall
2  family inet {
3      filter count-and-police {
4          term aggregate {
5              from {
6                  destination-address {
7                      192.168.120.0/29;
8                  }
9              }
10             then {
11                 policer 1m;
12                 count aggregate-in;
13             }
14         }
15         term rest {
16             then accept;
17         }
18     }
19 }
20 policer 1m {
21     if-exceeding {
22         bandwidth-limit 1m;
23         burst-size-limit 10k;
24     }
25     then discard;
26 }
27
28 user@Havana> show interfaces filters so-1/0/0.100
29 Interface       Admin Link Proto Input Filter         Output Filter
30 so-1/0/0.100    up    up   inet  count-and-police
31
32 user@Barcelona-re0> ping routing-instance NMM source barcelona-NMM male-aggr
33                     rapid count 1000 size 1400
34 PING male-aggr (192.168.120.1): 56 data bytes
35 !!!.!!!.!!!! <...>
36 --- male-aggr ping statistics ---
37 1000 packets transmitted, 902 packets received, 9% packet loss
```

```
38 round-trip min/avg/max/stddev = 1.641/1.988/43.545/2.090 ms
39
40 user@Havana> show firewall filter count-and-police
41 Filter: count-and-police
42 Counters:
43 Name                                                Bytes              Packets
44 aggregate-in                                      1288056                  902
45 Policers:
46 Name                                              Packets
47 1m-aggregate                                           98
```

Extended communities in the BGP PE–CE

For Option A interconnects, and as a general security measure, it is important to make sure that RT communities are not attached to prefixes coming from CEs. By default, Junos OS gives no special treatment to route policy related to communities, and as such passes them along unchanged.

Listing 5.61, Line 15, shows the resulting advertisement by router *Livorno* when no specific policy is set up outbound, with an extended community attached, as per Line 22. The result is that routers in domain "Mistral" end up with an undesired RT community and act accordingly (Line 43).

Listing 5.61: Advertising IP prefixes with extended communities

```
1  user@Livorno> show configuration routing-instances
2  NMM {
3      instance-type vrf;
4      interface so-0/0/1.100;
5      interface lo0.106;
6      vrf-target target:64501:64501;
7      protocols {
8          bgp {
9              group Mistral {
10                 export accept-direct; # No special community handling
11                 peer-as 64502;
12                 neighbor 172.16.250.37;
13 }   }   }     }
14 user@Livorno> show route advertising-protocol bgp 172.16.250.37 extensive
15     192.168.107.7
16
17 NMM.inet.0: 7 destinations, 7 routes (7 active, 0 holddown, 0 hidden)
18 * 192.168.107.7/32 (1 entry, 1 announced)
19  BGP group Mistral type External
20      Nexthop: Self
21      AS path: [64501] I
22      Communities: target:64501:64501 # Prefix Tagged with RT community
23
24 user@male-re0> show route table bgp detail | find 192.168.107.7
25 192.168.1.4:6:192.168.107.7/32 (1 entry, 1 announced)
26         *BGP    Preference: 170/-101
27                 Route Distinguisher: 192.168.1.4:6
28                 Next hop type: Indirect
29                 Next-hop reference count: 9
30                 Source: 192.168.1.4
31                 Next hop type: Router, Next hop index: 812
32                 Next hop: 172.16.1.86 via ge-4/2/4.0, selected
33                 Label operation: Push 301456
34                 Protocol next hop: 192.168.1.4
35                 Push 301456
36                 Indirect next hop: 8c2c2a0 1048580
37                 State: <Active Int Ext>
38                 Local AS: 64502 Peer AS: 64502
39                 Age: 3:19        Metric2: 10
```

```
40                 Task: BGP_64502.192.168.1.4+179
41                 Announcement bits (1): 0-BGP RT Background
42                 AS path: 64501 I
43                 Communities: target:64501:64501 target:64502:64502 # Additional RT
44                 Import Accepted
45                 VPN Label: 301456
46                 Localpref: 100
47                 Router ID: 192.168.1.4
48                 Secondary Tables: NMM.inet.0
```

A best practice to protect the carrier domain is to filter extended communities using inbound routing policy, as shown in Listing 5.62. It is also good practice to facilitate this protection for the peer by cleaning communities towards the peer. Notice on Line 19 that the update contains the RT community, but filtering is effective and the route installed in the VRF table does not contain it, as shown on Line 21.

Listing 5.62: Protecting Option A interconnect from RT communities received from CE

```
1  user@Havana> show configuration policy-options
2  ...
3  policy-statement clean-RT {
4      then {
5          community delete extended;
6      }
7  }
8  community extended members [ target:*:* origin:*:* ];
9
10 user@Havana# set routing-instances NMM protocols bgp group Cyclone import clean-RT
11
12 user@Havana> show route receive-protocol bgp 172.16.250.38 detail 192.168.107.7/32
13
14 NMM.inet.0: 15 destinations, 16 routes (15 active, 0 holddown, 0 hidden)
15 * 192.168.107.7/32 (1 entry, 1 announced)
16      Accepted
17      Nexthop: 172.16.250.38
18      AS path: 64501 I
19      Communities: target:64501:64501 # RT community present in regular BGP update
20
21 user@Havana> show route table NMM 192.168.107.7/32 detail
22
23 NMM.inet.0: 15 destinations, 16 routes (15 active, 0 holddown, 0 hidden)
24 192.168.107.7/32 (1 entry, 1 announced)
25         *BGP    Preference: 170/-101
26                 Next hop type: Router, Next hop index: 527
27                 Next-hop reference count: 6
28                 Source: 172.16.250.38
29                 Next hop: 172.16.250.38 via so-1/0/0.100, selected
30                 State: <Active Ext>
31                 Peer AS: 64501
32                 Age: 15:32
33                 Task: BGP_64501.172.16.250.38+64636
34                 Announcement bits (2): 0-KRT 2-BGP RT Background
35                 AS path: 64501 I
36                 Accepted
37                 Localpref: 100
38                 Router ID: 192.168.106.6
```

5.6.5 Stage two: Inter-AS Option B deployment

With successful rollout of the Option A interconnect, additional VPNs can be extended over the interconnect. However, for every new VPN desiring interconnection, a new subinterface has to be allocated, and instantiation of the VRF on both sides of the interconnection needs to occur.

Provisioning quickly becomes a burden, and a decision is made to adopt a more scalable approach, focusing only on the local interconnect first. The alternative chosen is Option B, as shown in Figure 5.39, which sets up an MPLS interconnect on a connection that has no subinterfaces and removes the requirement of instantiating VRFs at the borders.

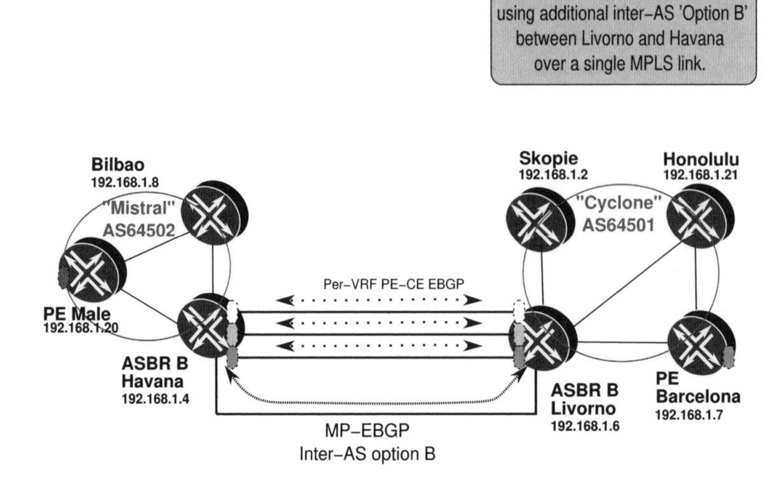

Figure 5.39: Stage two: Inter-AS Option B topology.

Migration to this type of interconnection must be designed with some considerations in mind:

- A smooth transition calls for parallel Option A and Option B sessions to validate connectivity at the control plane first, before switching service to the new interconnect. To ease migration control, decisions are centralized at router *Havana* by setting of the LOCAL_PREF BGP attribute for inbound updates and the MED BGP attribute to influence outbound updates, as per Table 5.2 on Page 474.

- Shared VPNs have to agree on the RT community to use. Of the approaches described in Section 5.2.6, mapping of RT is implemented. Using policy language in the outbound direction, an additional RT is added at the ASBR to match the AS of the peer. To avoid adding the same community multiple times, this addition is performed only if the community is not already present.

- The aggregation point on router *Havana* for L3VPN *NMM* filters more specific routes as contributing members of the aggregate. In a regular Option B scenario, each route is a separate NLRI formed from the RD and a prefix and that populates bgp.l3vpn.0. In this interconnect option, the concept of a contributing route does not exist; therefore, more specific routes bear no relation to the aggregate.

- The aggregation point on router *Havana* for L3VPN *NMM* advertises a VPN label for a route that represents a different next hop. Each more specific route probably has a different VPN label, so using the VPN label of the aggregate for forwarding might send traffic to the wrong more specific prefix. It is necessary to enable an Option A+B behavior, at least for the aggregate prefix.

- L3VPN *NMM* has a requirement for inbound accounting on router *Havana*. This can be implemented by leveraging the same Option A+B configuration. The VPN label for the aggregate route maps to an IP lookup at the VRF, allowing for a forwarding table filter to be applied.

Bringup of depreferred Option B

To ease incremental transition, the current interface interconnecting router *Havana* and router *Livorno* that has one subinterface for each VRF is maintained, and an additional link is set up to establish pure MPLS connectivity. The extended configuration illustrated in Listing 5.63 assumes that a new interface has been provisioned with MPLS capabilities and establishes a new BGP peering group for family inet-vpn named *EBGP-Cyclone-vpn-Livorno-WAN*. Provisioning configuration includes forcing the BGP attributes to deprefer any updates in both directions over this session as shown on Lines 3 and 8.

Listing 5.63: New Option B inet-vpn session with depreference

```
 1 user@Havana> show configuration protocols bgp group EBGP-Cyclone-vpn-Livorno-WAN
 2 type external;
 3 metric-out 4500;                # Depreference outbound updates
 4 local-address 172.16.100.1;
 5 family inet-vpn {
 6     unicast;
 7 }
 8 import LP50;                    # Depreference inbound updates
 9 peer-as 64501;
10 neighbor 172.16.100.8;
11
12 user@Livorno> show configuration protocols bgp group EBGP-Mistral-vpn-Havana-WAN
13 type external;
14 local-address 172.16.100.8;
15 family inet-vpn {
16     unicast;
17 }
18 peer-as 64502;
19 neighbor 172.16.100.1;
```

Because this is the first EBGP group for inet-vpn prefixes, a switch in path selection mode triggers a restart on all inet-vpn sessions. Section 1.2.2 provides more information about this behavior.

Listing 5.64 shows the resulting information being exchanged over the inet-vpn session as received on each end. Notice that prefixes advertised by router *Livorno* have a diminished local preference value of 50 and that router *Livorno* receives advertisements with a higher metric value of 4500 to ensure that prefixes are not preferred.

Comparing this output with the Option A interconnect in Listing 5.59, Line 45, the more specific prefixes are now advertised along with the aggregate (Line 25 to Line 37).

Listing 5.64: NLRI exchange after bringup of Option B

```
1  user@Havana> show route table bgp source-gateway 172.16.100.8
2
3  bgp.l3vpn.0: 15 destinations, 15 routes (15 active, 0 holddown, 0 hidden)
4  + = Active Route, - = Last Active, * = Both
5
6  192.168.1.6:4:172.16.250.36/30
7                     *[BGP/170] 00:04:16, localpref 50
8                       AS path: 64501 I
9                     > to 172.16.100.8 via fe-0/3/0.0, Push 305056
10 192.168.1.7:14:192.168.107.7/32
11                    *[BGP/170] 00:04:16, localpref 50
12                      AS path: 64501 I
13                    > to 172.16.100.8 via fe-0/3/0.0, Push 305600
14
15 user@Livorno> show route receive-protocol bgp 172.16.100.1 table bgp
16
17 bgp.l3vpn.0: 16 destinations, 16 routes (16 active, 0 holddown, 0 hidden)
18   Prefix                 Nexthop          MED     Lclpref    AS path
19   192.168.1.4:6:172.16.250.36/30
20 *                        172.16.100.1     4500               64502 I
21   192.168.1.4:6:192.168.104.4/32
22 *                        172.16.100.1     4500               64502 I
23   192.168.1.4:6:192.168.120.0/29
24 *                        172.16.100.1     4500               64502 I
25   192.168.1.20:10:192.168.120.1/32
26 *                        172.16.100.1     4500               64502 I
27   192.168.1.20:10:192.168.120.2/32
28 *                        172.16.100.1     4500               64502 I
29   192.168.1.20:10:192.168.120.3/32
30 *                        172.16.100.1     4500               64502 I
31   192.168.1.20:10:192.168.120.4/32
32 *                        172.16.100.1     4500               64502 I
33   192.168.1.20:10:192.168.120.5/32
34 *                        172.16.100.1     4500               64502 I
35   192.168.1.20:10:192.168.120.6/32
36 *                        172.16.100.1     4500               64502 I
37   192.168.1.20:10:192.168.120.7/32
38 *                        172.16.100.1     4500               64502 I
39   192.168.1.20:10:192.168.120.20/32
40 *                        172.16.100.1     4500               64502 I
```

RT community mapping

No information from the other domain is populating the local VRF despite the information being exchanged properly over the Option B interconnect because of a mismatch in the RT communities. Listing 5.65 shows the mapping policy that is implemented (Line 2 and Line 15) and later applied (Line 28 and Line 33) to the configuration, as well as a sample prefix with both communities on Line 46 and Line 56.

Listing 5.65: RT mapping at the borders

```
1  [edit]
2  user@Havana# show policy-options policy-statement map-RT-to-Cyclone
3  term NMM-checkBoth {
4      from community CycloneAndMistral-NMM;
5      then next policy;
6  }
7  term NMM-addCyclone {
8      from community Mistral-NMM;
9      then {
10         community add Cyclone-NMM;
11     }
```

```
12 }
13
14  [edit]
15 user@Livorno# show policy-options policy-statement map-RT-to-Mistral
16 term NMM-checkBoth {
17     from community CycloneAndMistral-NMM;
18     then next policy;
19 }
20 term NMM-addMistral {
21     from community Cyclone-NMM;
22     then {
23         community add Mistral-NMM;
24     }
25 }
26
27  [edit protocols bgp]
28 user@Havana# show group EBGP-Cyclone-vpn-Livorno-WAN | compare rollback 1
29  [edit protocols bgp group EBGP-Cyclone-vpn-Livorno-WAN]
30 +    export map-RT-to-Cyclone;
31
32  [edit protocols bgp]
33 user@Livorno# show group EBGP-Mistral-vpn-Havana-WAN | compare rollback 1
34  [edit protocols bgp group EBGP-Mistral-vpn-Havana-WAN]
35 +    export map-RT-to-Mistral;
36
37 user@Havana> show route advertising-protocol bgp 172.16.100.8 detail | find 120
38 * 192.168.1.4:6:192.168.120.0/29 (1 entry, 1 announced)
39  BGP group EBGP-Cyclone-vpn-Livorno-WAN type Externa
40      Route Distinguisher: 192.168.1.4:6
41      VPN Label: 301616
42      Nexthop: Self
43      Flags: Nexthop Change
44      MED: 4500
45      AS path: [64502] I (LocalAgg)
46      Communities: target:64501:64501 target:64502:64502
47 <...>
48
49 user@Havana> show route receive-protocol bgp 172.16.100.8 detail table bgp | find 107
50 * 192.168.1.7:14:192.168.107.7/32 (1 entry, 1 announced)
51      Import Accepted
52      Route Distinguisher: 192.168.1.7:14
53      VPN Label: 305600
54      Nexthop: 172.16.100.8
55      AS path: 64501 I
56      Communities: target:64501:64501 target:64502:64502
```

The result of activating this configuration is shown in Listing 5.66 for a prefix from router *Barcelona* present on router *Havana* with two feasible paths, one for each interconnection type. Notice on Line 9 that the inactive path has a lower local preference setting.

Listing 5.66: Sample population of NMM VRF with depreferred Option B information

```
1  user@Havana> show route table NMM 192.168.107.7
2
3  NMM.inet.0: 13 destinations, 15 routes (13 active, 0 holddown, 0 hidden)
4  + = Active Route, - = Last Active, * = Both
5
6  192.168.107.7/32   *[BGP/170] 01:46:20, localpref 100
7                        AS path: 64501 I
8                      > to 172.16.250.38 via so-1/0/0.100 # Over PE-CE option A
9                      [BGP/170] 00:00:30, localpref 50
10                       AS path: 64501 I
11                     > to 172.16.100.8 via fe-0/3/0.0, Push 305600 # Over MPLS interconnect
```

Filtering more-specific routes for an aggregate

The resulting advertised information shown in Listing 5.65, Line 15 on Page 482 for router *Livorno* still includes the aggregate routes. Filtering for these is done in the Option A interconnect at the routing-instance level (Listing 5.59, Line 14 on Page 476), and a similar policy has to be applied to the advertisement over the VPN connection. However, in the Option B, the aggregate route becomes another route, decoupled from the contributing ones. It is not feasible to use the aggregate-contributor condition (Listing 5.59 Line 27).

The configuration shown in Listing 5.67 allows matching on the route-filter to filter more specific prefixes than the aggregate at the ASBR Option B interconnect. As the RD part is ignored in the match, an additional RT community match constrains this filtering to the *NMM* VRF. The result of this configuration change is shown on Line 18, which reports that filtering works correctly advertising only the aggregate and a route outside of the aggregate block (Line 26).

Listing 5.67: Using route-filter to filter VPN aggregate contributors

```
1  user@Havana> show configuration policy-options policy-statement block-specifics
2  term members {
3      from {
4          community Mistral-NMM;
5          route-filter 192.168.120.0/29 longer;
6      }
7      then reject;
8  }
9
10 user@Havana> show configuration policy-options community Mistral-NMM
11 members target:64502:64502;
12
13 user@Havana> show configuration | compare rollback 1
14 [edit protocols bgp group EBGP-Cyclone-vpn-Livorno-WAN]
15 -     export map-RT-to-Cyclone;
16 +     export [ map-RT-to-Cyclone block-specifics ];
17
18 user@Havana> show route advertising-protocol bgp 172.16.100.8
19
20 bgp.l3vpn.0: 14 destinations, 14 routes (14 active, 0 holddown, 0 hidden)
21   Prefix                  Nexthop              MED     Lclpref    AS path
22   192.168.1.4:6:172.16.250.36/30
23 *                         Self                 4500               I
24   192.168.1.4:6:192.168.120.0/29
25 *                         Self                 4500               I
26   192.168.1.20:10:192.168.120.20/32
27 *                         Self                 4500               I
```

Forwarding for aggregated prefixes in transit

An aggregate route within a VPN represents multiple final destinations, namely each of the contributing routes. The contributing routes may be sourced from different VRFs in different PEs with distinct VPN labels. If forwarding is performed by MPLS, information about the more specific destination is hidden behind the aggregate prefix in which a single VPN label is associated.

The IP lookup function supports checking for the more-specific route, thus providing optimal cost forwarding. Therefore, VRF *NMM* enables local table-label allocation for the aggregate. Only the aggregate route requires this IP lookup, so a selective policy as shown in Listing 5.68 is configured, with the following components:

- `vrf-table-label` is enabled for the entire VRF. This automatically enables advertising of the table-label in all outgoing announcements with this RT, both local and transit.
- `label allocation` policy on Line 12 as defined on Line 27 and Line 36 reverses the default for all prefixes except the aggregate.
- `label substitution` policy on Line 13 does not perform any substitution for transit labels, as per the policy definition on Line 44. This ensures that the combined functionality for Option A+B is enabled only for the aggregate route. Other prefixes are processed in a standard Option B fashion.

Listing 5.68: Advertising a table-label for the aggregate route

```
1  user@Havana> show configuration routing-instances NMM
2  instance-type vrf;
3  interface so-1/0/0.100;
4  interface lo0.104;
5  vrf-target target:64502:64502;
6  vrf-table-label;        # Allocate a table-label
7  routing-options {
8      aggregate {
9          route 192.168.120.0/29 discard;
10     }
11     label {
12         allocation [ aggregate-allocation default-nexthop ]; # Only the aggregate
13         substitution none;                                   # Keep transit routes on pure MPLS
14     }
15 }
16 protocols {
17     bgp {
18         group Cyclone {
19             import clean-RT;
20             export Aggregate; # Advertisement on Option A; does not affect Option B session
21             peer-as 64501;
22             neighbor 172.16.250.38;
23         }
24     }
25 }
26
27 user@Havana> show configuration policy-options policy-statement aggregate-allocation
28 term aggregate {
29     from protocol aggregate;
30     then {
31         label-allocation per-table;
32         accept;
33     }
34 }
35
36 user@Havana> show configuration policy-options policy-statement default-nexthop
37 term default {
38     then {
39         label-allocation per-nexthop;
40         accept;
41     }
42 }
43
44 user@Havana> show configuration policy-options policy-statement none
45 then reject;
```

The introduction of the `vrf-table-label` command changes VPN label advertisement for all prefixes so that an IP lookup function is performed. The above changes yield an

IP function only for a restricted set of prefixes, namely the locally originated aggregate. This action can be verified by looking at the advertising information in Listing 5.69 for both the IBGP (Line 1) and EBGP sessions (Line 15).

Notice that the prefix of router *Barcelona* is advertised twice to router *Male*. The RD part constructed with the loopback address of the router shows that one of the copies is locally originated on router *Havana* as part of the Option A interconnect (Line 6), while the other prefix is learned over the Option B session (Line 12).

Listing 5.69: Directing the aggregate only to transit IP function

```
1  user@Havana> show route advertising-protocol bgp 192.168.1.20 detail | match "\*|VPN"
2
3  bgp.l3vpn.0: 14 destinations, 14 routes (14 active, 0 holddown, 0 hidden)
4  * 192.168.1.4:6:172.16.250.36/30 (1 entry, 1 announced)
5       VPN Label: 301472
6  * 192.168.1.4:6:192.168.107.7/32 (1 entry, 1 announced)
7       VPN Label: 301504
8  * 192.168.1.4:6:192.168.120.0/29 (1 entry, 1 announced)
9       VPN Label: 16                                    # Table-Label. IP lookup in force
10 * 192.168.1.6:4:172.16.250.36/30 (1 entry, 1 announced)
11      VPN Label: 301520
12 * 192.168.1.7:14:192.168.107.7/32 (1 entry, 1 announced)
13      VPN Label: 301552
14
15 user@Havana> show route advertising-protocol bgp 172.16.100.8 detail
16
17 bgp.l3vpn.0: 14 destinations, 14 routes (14 active, 0 holddown, 0 hidden)
18 * 192.168.1.4:6:172.16.250.36/30 (1 entry, 1 announced)
19  BGP group EBGP-Cyclone-vpn-Livorno-WAN type Externa
20      Route Distinguisher: 192.168.1.4:6
21      VPN Label: 301472
22      Nexthop: Self
23      Flags: Nexthop Change
24      MED: 4500
25      AS path: [64502] I
26      Communities: target:64501:64501 target:64502:64502
27
28 * 192.168.1.4:6:192.168.120.0/29 (1 entry, 1 announced)
29  BGP group EBGP-Cyclone-vpn-Livorno-WAN type Externa
30      Route Distinguisher: 192.168.1.4:6
31      VPN Label: 16                                    # Table-Label. IP lookup in force
32      Nexthop: Self
33      Flags: Nexthop Change
34      MED: 4500
35      AS path: [64502] I (LocalAgg)
36      Communities: target:64501:64501 target:64502:64502
37
38 * 192.168.1.20:10:192.168.120.20/32 (1 entry, 1 announced)
39  BGP group EBGP-Cyclone-vpn-Livorno-WAN type Externa
40      Route Distinguisher: 192.168.1.20:10
41      VPN Label: 301584
42      Nexthop: Self
43      Flags: Nexthop Change
44      MED: 4500
45      AS path: [64502] I
46      Communities: target:64501:64501 target:64502:64502
47
48 user@Havana> show route table mpls label 16
49
50 mpls.0: 11 destinations, 11 routes (11 active, 0 holddown, 0 hidden)
51 + = Active Route, - = Last Active, * = Both
52
53 16                 *[VPN/0] 17:33:13
54                        to table NMM.inet.0, Pop
```

Allowing for IP accounting functions in Option B

The accounting functions already present on the Option A PE–CE subinterface have to be migrated to the new interconnect. IP lookup functions are available on a per-VRF basis, so the pre-existing input filter has to be mapped to the VRF as an input forwarding table filter. With the previous label substitution procedures, only the aggregate requires an IP lookup function, and the policing can be done for the VRF as a whole. Listing 5.70 shows the additional configuration for this simple scenario. The original interface filter (Line 3) is duplicated as a table filter (Line 4) and is applied to the *NMM* VRF forwarding table on Line 24.

Listing 5.70: Moving filtering from interface to VRF

```
1  user@Havana> show configuration | compare rollback 1
2  [edit firewall family inet]
3        filter count-and-police { ... }  # Interface filter on the PE-CE interface
4  +     filter count-and-police-table {  # Table-filter definition for the VRF
5  +         term aggregate {
6  +             from {
7  +                 destination-address {
8  +                     192.168.120.0/29;
9  +                 }
10 +             }
11 +             then {
12 +                 policer 1m;
13 +                 count aggregate-in;
14 +             }
15 +         }
16 +         term rest {
17 +             then accept;
18 +         }
19 +     }
20 [edit routing-instances NMM]
21 +    forwarding-options {
22 +        family inet {
23 +            filter {
24 +                input count-and-police-table; # Apply table-filter
25 +            }
26 +        }
27 +    }
```

A forwarding test with the newly implemented filter is shown in Listing 5.71, which shows that all the traffic currently being carried by the Option A interconnect is passing through the subinterface filter. The end result is a two level counting/policing stage, triggering counting at two places, namely the PE–CE interface at the Option A interconnect (Line 12) and the input filter before lookup in the *NMM* VRF table (Line 20). Cleaning up the Option A configuration removes the subinterface filter.

Listing 5.71: Forwarding test with both filters in place

```
1  user@Barcelona-re0> ping routing-instance NMM source barcelona-NMM male-aggr
2                      rapid count 1000 size 1400
3  PING male-aggr (192.168.120.1): 1400 data bytes
4  !!!!!!!!!.!!!!!!!!
5  <...>
6  --- male-aggr ping statistics ---
7  1000 packets transmitted, 902 packets received, 9% packet loss
8  round-trip min/avg/max/stddev = 1.637/1.962/42.015/1.973 ms
9
10 user@Havana> show firewall
11
12 Filter: count-and-police            # PE-CE input interface filter
13 Counters:
```

```
14 Name                                             Bytes              Packets
15 aggregate-in                                   1288056                  902
16 Policers:
17 Name                                           Packets
18 1m-aggregate                                        98
19
20 Filter: count-and-police-table     # NMM VRF input table filter
21 Counters:
22 Name                                             Bytes              Packets
23 aggregate-in                                   1288056                  902
24 Policers:
25 Name                                           Packets
26 1m-aggregate                                         0
```

Junos Tip: Traffic affected by a forwarding table filter

A forwarding table filter is processed for every packet that has to be looked up, regardless of the incoming interface. In a general case, the simple filter used in this case study would need to be tailored to ensure that the directionality of traffic is taken into account, by adding an input interface match condition.

In this particular scenario, given that only the aggregate prefix is being advertised with a table-label, no other traffic is expected to hit the filter after the transition from Option A has concluded, so the input interface match condition is not used.

More detail about the possible placements of a firewall filter is given in Section 1.3.2 on Page 19.

5.6.6 Stage three: Inter-AS Option B activation

The relevant configuration is in place at both ASBR router *Livorno* and router *Havana*, albeit with depreferred routes. It is time to confirm at the control plane that everything is learned as expected. Only when routing information learned over the Option A interconnect can be made less interesting is it time to switch over the forwarding to the Option B interconnect.

Verification of initial state

Three prefixes are relevant at this point: the aggregate block 192.168.120.0/29 in domain "Mistral" and the two prefixes representing the end customers at each side, 192.168.120.20/32 for router *Male* and 192.168.107.7/32 for router *Barcelona*. The next-hop interface for each of the prefixes is outlined in Listing 5.72 from both sides of the inter-AS connection on router *Havana* and router *Livorno*.

Besides the contributing routes, router *Havana* shows the additional route from the Option B prefix belonging to router *Barcelona* as less preferred (local preference 50) on Line 10, with a lower metric value (MED 4500) for prefixes from domain "Mistral" as learned by router *Livorno* on Line 38 and Line 44. The aggregate prefix 192.168.120.0/29 received on router *Livorno* correctly shows a table-label on Line 40.

Listing 5.72: Prefix state prior to Option B activation

```
1
2 user@Havana> show route table NMM
3
4 NMM.inet.0: 12 destinations, 14 routes (12 active, 0 holddown, 0 hidden)
5 + = Active Route, - = Last Active, * = Both
6
```

```
 7 192.168.107.7/32    *[BGP/170] 03:57:54, localpref 100
 8                        AS path: 64501 I
 9                      > to 172.16.250.38 via so-1/0/0.100 # Option A preferred
10                      [BGP/170] 01:34:56, localpref 50
11                        AS path: 64501 I
12                      > to 172.16.100.8 via fe-0/3/0.0, Push 305600
13 192.168.120.0/29    *[Aggregate/130] 1d 04:03:19
14                        Discard
15 192.168.120.1/32    *[BGP/170] 00:00:12, localpref 100, from 192.168.1.20
16                        AS path: I
17                      > to 172.16.1.85 via ge-1/1/0.0, Push 299808
18 <...>
19 192.168.120.7/32    *[BGP/170] 00:00:12, localpref 100, from 192.168.1.20
20                        AS path: I
21                      > to 172.16.1.85 via ge-1/1/0.0, Push 299824
22 192.168.120.20/32   *[BGP/170] 00:00:12, localpref 100, from 192.168.1.20
23                        AS path: I
24                      > to 172.16.1.85 via ge-1/1/0.0, Push 299808
25
26 user@Livorno> show route table NMM
27
28 NMM.inet.0: 5 destinations, 8 routes (5 active, 0 holddown, 0 hidden)
29 + = Active Route, - = Last Active, * = Both
30
31 <...>
32 192.168.107.7/32    *[BGP/170] 03:28:05, localpref 100, from 192.168.1.7
33                        AS path: I
34                      > to 172.16.1.53 via ge-0/2/0.0, label-switched-path Livorno-to-Barcelona
35 192.168.120.0/29    *[BGP/170] 02:26:51, MED 4000, localpref 100 # Better MED for option A
36                        AS path: 64502 I
37                      > to 172.16.250.37 via so-0/0/1.100
38                      [BGP/170] 01:36:14, MED 4500, localpref 100 # Worse MED for option B
39                        AS path: 64502 I
40                      > to 172.16.100.1 via fe-0/3/0.0, Push 16
41 192.168.120.20/32   *[BGP/170] 02:26:51, MED 4000, localpref 100 # Better MED for option A
42                        AS path: 64502 I
43                      > to 172.16.250.37 via so-0/0/1.100
44                      [BGP/170] 02:13:22, MED 4500, localpref 100 # Worse MED for option B
45                        AS path: 64502 I
46                      > to 172.16.100.1 via fe-0/3/0.0, Push 301584
```

Activation of Option B

Listing 5.73 modifies preferences as instructed in Table 5.3. The better LOCAL_PREF and MED BGP settings on the direct Option B session are changed as shown on Line 1, with the new table outputs showing a reverse situation from that in the premigration stage, with best-path selection now preferring the inter-AS Option B routes (Line 16, Line 57, and Line 63).

Table 5.3 Change in preference settings to switch to option B

Stage	Description	A	B–direct	B–CsC	C–CsC	C–CsC–Dual
1	Initial A	100 4000				
2	Build B Direct		50 4500			
3	Switch to B Direct		200 3000			

Listing 5.73: Prefix state after Option B activation

```
user@Havana> show configuration | compare rollback 1
[edit protocols bgp group EBGP-Cyclone-vpn-Livorno-WAN]  # Inter-AS option B interconnection
-      metric-out 4500;
+      metric-out 3000;                                  # Better MED
-      import LP50;
+      import LP200;                                     # Better LOCAL_PREF

user@Havana> show route table NMM

NMM.inet.0: 12 destinations, 14 routes (12 active, 0 holddown, 0 hidden)
+ = Active Route, - = Last Active, * = Both

<...>
192.168.107.7/32   *[BGP/170] 00:02:06, localpref 200
                      AS path: 64501 I
                    > to 172.16.100.8 via fe-0/3/0.0, Push 305600 # Option B preferred
                    [BGP/170] 04:25:54, localpref 100
                      AS path: 64501 I
                    > to 172.16.250.38 via so-1/0/0.100
192.168.120.0/29   *[Aggregate/130] 1d 04:49:36
                      Discard
192.168.120.1/32   *[BGP/170] 00:02:06, localpref 100, from 192.168.1.20
                      AS path: I
                    > to 172.16.1.85 via ge-1/1/0.0, Push 299808
192.168.120.2/32   *[BGP/170] 00:02:06, localpref 100, from 192.168.1.20
                      AS path: I
                    > to 172.16.1.85 via ge-1/1/0.0, Push 299824
192.168.120.3/32   *[BGP/170] 00:02:06, localpref 100, from 192.168.1.20
                      AS path: I
                    > to 172.16.1.85 via ge-1/1/0.0, Push 299824
192.168.120.4/32   *[BGP/170] 00:02:06, localpref 100, from 192.168.1.20
                      AS path: I
                    > to 172.16.1.85 via ge-1/1/0.0, Push 299824
192.168.120.5/32   *[BGP/170] 00:02:06, localpref 100, from 192.168.1.20
                      AS path: I
                    > to 172.16.1.85 via ge-1/1/0.0, Push 299824
192.168.120.6/32   *[BGP/170] 00:02:06, localpref 100, from 192.168.1.20
                      AS path: I
                    > to 172.16.1.85 via ge-1/1/0.0, Push 299824
192.168.120.7/32   *[BGP/170] 00:02:06, localpref 100, from 192.168.1.20
                      AS path: I
                    > to 172.16.1.85 via ge-1/1/0.0, Push 299824
192.168.120.20/32  *[BGP/170] 00:02:06, localpref 100, from 192.168.1.20
                      AS path: I
                    > to 172.16.1.85 via ge-1/1/0.0, Push 299808

user@Livorno> show route table NMM

NMM.inet.0: 5 destinations, 8 routes (5 active, 0 holddown, 0 hidden)
+ = Active Route, - = Last Active, * = Both

192.168.107.7/32   *[BGP/170] 02:41:17, localpref 100, from 192.168.1.7
                      AS path: I
                    > to 172.16.1.53 via ge-0/2/0.0, label-switched-path Livorno-to-Barcelona
192.168.120.0/29   *[BGP/170] 00:03:19, MED 3000, localpref 100
                      AS path: 64502 I
                    > to 172.16.100.1 via fe-0/3/0.0, Push 16      # Table-Label IP lookup
                    [BGP/170] 02:54:46, MED 4000, localpref 100
                      AS path: 64502 I
                    > to 172.16.250.37 via so-0/0/1.100
192.168.120.20/32  *[BGP/170] 00:03:19, MED 3000, localpref 100
                      AS path: 64502 I
                    > to 172.16.100.1 via fe-0/3/0.0, Push 301584 # Option B preferred
                    [BGP/170] 02:54:46, MED 4000, localpref 100
                      AS path: 64502 I
                    > to 172.16.250.37 via so-0/0/1.100
```

Both traceroute and forwarding through the firewall filter confirm that traffic is taking the Option B interconnect, as illustrated in Listing 5.74. No traffic is counted on the subinterface filter on Line 18 and an end-to-end path with MPLS labels is present for every hop (on Line 31) as well as the reverse path (on Line 41). Notice that only one label is present at every hop because Option B is stitching labels at the VPN level.

Listing 5.74: Connectivity verification after transition to Option B

```
1
2  user@Barcelona-re0> ping routing-instance NMM source barcelona-NMM male-aggr
3                      rapid count 1000 size 1400
4  PING male-aggr (192.168.120.1): 1400 data bytes
5  !!!!!!!!!.!!!!!!!!!.!
6  <...>
7  --- male-aggr ping statistics ---
8  1000 packets transmitted, 903 packets received, 9% packet loss
9  round-trip min/avg/max/stddev = 2.015/2.433/35.796/2.328 ms
10
11 user@Havana> show firewall
12
13 Filter: __default_bpdu_filter__
14
15 Filter: count-and-police
16 Counters:
17 Name                                              Bytes              Packets
18 aggregate-in                                          0                    0
19 Policers:
20 Name                                            Packets
21 1m-aggregate                                          0
22
23 Filter: count-and-police-table
24 Counters:
25 Name                                              Bytes              Packets
26 aggregate-in                                    1288056                  902
27 Policers:
28 Name                                            Packets
29 1m-aggregate                                         98
30
31 user@Barcelona-re0> traceroute routing-instance NMM source barcelona-NMM male-NMM
32 traceroute to male-NMM (192.168.120.20) from barcelona-NMM, 30 hops max, 40 byte packets
33  1  livorno-ge020 (172.16.1.54)  1.509 ms  1.261 ms  1.218 ms
34      MPLS Label=305696 CoS=0 TTL=1 S=1
35  2  havana-fe0300 (172.16.100.1)  1.353 ms  1.224 ms  1.165 ms
36      MPLS Label=301584 CoS=0 TTL=1 S=1
37  3  male-ge4240 (172.16.1.85)  0.854 ms  0.849 ms  0.816 ms
38      MPLS Label=299808 CoS=0 TTL=1 S=1
39  4  male-NMM (192.168.120.20)  0.799 ms  0.777 ms  0.763 ms
40
41 user@male-re0> traceroute routing-instance NMM source male-NMM barcelona-NMM
42 traceroute to barcelona-NMM (192.168.107.7) from male-NMM, 30 hops max, 40 byte packets
43  1  havana-ge1100 (172.16.1.86)  1.438 ms  1.129 ms  1.095 ms
44      MPLS Label=301552 CoS=0 TTL=1 S=1
45  2  livorno-fe030 (172.16.100.8)  1.228 ms  1.080 ms  1.088 ms
46      MPLS Label=305600 CoS=0 TTL=1 S=1
47  3  barcelona-ge120 (172.16.1.53)  0.726 ms  0.690 ms  0.687 ms
48      MPLS Label=302304 CoS=0 TTL=1 S=1
49  4  barcelona-NMM (192.168.107.7)  0.670 ms  0.643 ms  0.659 ms
50
```

Decommissioning of Option A setup

After the preference change, the Option A interconnect does not carry any traffic. As soon as a similar strategy is in place for the remaining VRFs on the interconnect, the link can be decommissioned. The VRF *NMM* can be completely removed from router *Livorno* because all traffic is now being switched using MPLS label swapping. On router *Havana*, however, the *NMM* VRF needs to remain, because it is the anchoring point for the IP functions of deaggregation and traffic accounting. Listing 5.75 shows the VRF configuration after removing the subinterface and BGP PE–CE protocol.

Listing 5.75: Final instance configuration on router *Havana* with Option B

```
1  user@Havana> show configuration routing-instances
2  NMM {
3      # No need to define any interface on this VRF
4      instance-type vrf;
5      vrf-target target:64502:64502;
6      vrf-table-label;                   # Table-label needs to stay enabled
7      routing-options {
8          aggregate {                    # Maintain anchoring point for the aggregate route
9              route 192.168.120.0/29 discard;
10         }
11         label {                        # Apply table-label only to the aggregate
12             allocation [ aggregate-allocation default-nexthop ];
13             substitution none;
14         }
15     }
16     forwarding-options {
17         family inet {
18             filter {
19                 input count-and-police-table; # VRF is anchoring point for table filter
20             }
21         }
22     }
23 }
```

Although removal of the PE–CE interface can be performed safely after migration to the Option B interconnect, route preferences have been carefully selected to allow for the route to remain without becoming active.

For illustration purposes, the NMM instance and its companion PE–CE session are kept active but not preferred throughout the remainder of this case study.

5.6.7 Stage four: Bringup of redundant Option B over CsC

In an attempt to achieve cost savings in transmission capacity, management wants to leverage the availability of CsC services in the marketplace. The Option B interconnection between router *Havana* and router *Livorno* is to be extended through the CsC service, and the existing direct interconnect between router *Havana* and router *Livorno* is to be decommissioned. Figure 5.40 illustrates the bringup of the new Option B connection using the CsC carrier in domain "Monsoon."

Additional factors to consider in this migration follow:

- Existence of a CsC carrier providing Labeled BGP (L-BGP) interconnection service means that this carrier has to instantiate a VRF for the VPN (*NMM-CsC*) that holds labeled paths between the ASBRs in domain "Cyclone" and domain "Mistral".

- Similar to the work done in the previous stage between router *Havana* and router *Livorno*, the same two routers need to exchange VPN prefixes over the indirect CsC connection. An MPLS transport connection through a CsC interconnect is required. L-BGP is leveraged for this connection.

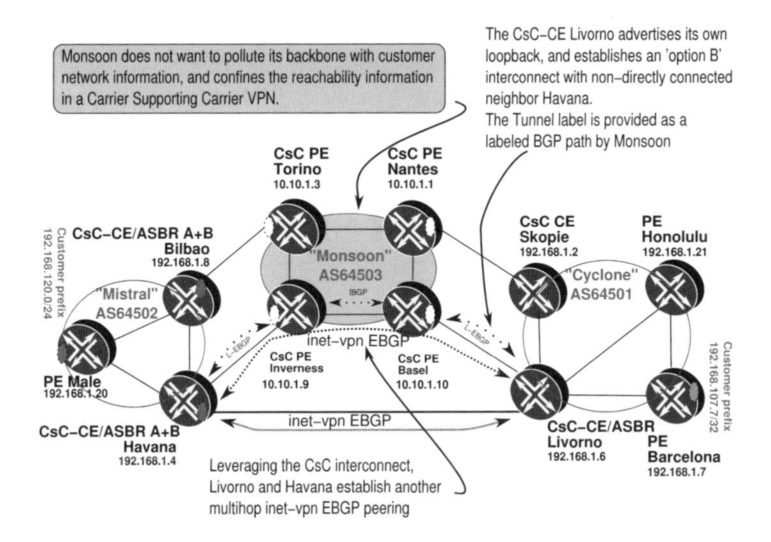

Figure 5.40: Stage four: Inter-AS Option B over CsC topology.

- A multihop L3VPN peering session is established through this interconnect between router *Havana* and router *Livorno* over the CsC service.
- As in previous stages, per-prefix preference control provides granular transition capabilities, as described in Table 5.2.

CsC setup

The setup for this scenario, as shown in Listing 5.76, includes Labeled BGP as PE–CE protocol as per Line 11. The MPLS protocol has to be enabled on the PE–CE interface to allow label binding to occur. The same configuration is replicated across the four PEs in domain "Monsoon" that are providing the CsC service.

Listing 5.76: VRF on router *Inverness* providing CSC service to router *Havana*

```
1  user@Inverness> show configuration routing-instances NMM-CsC
2  instance-type vrf;
3  interface so-0/0/3.0;
4  interface lo0.109;
5  vrf-target target:64503:64503;
6  protocols {
7      bgp {
8          group EBGP-Mistral-labeledBGP {
9              type external;
10             family inet {
11                 labeled-unicast;
12             }
13             neighbor 172.16.1.69 {
14                 peer-as 64502;
15             }
16         }
17     }
18 }
```

Bringup of labeled sessions between ASBR routers

The configuration to establish the L-BGP sessions to the CsC carrier is shown in Listing 5.77 for router *Havana*. Details on the mechanics of the L-BGP arrangements are discussed at length in the case study in Chapter 4, starting on Page 324.

Listing 5.77: Havana L-BGP configuration to CsC–PE Inverness

```
1  user@Havana> show configuration protocols bgp group EBGP-Monsoon-lBGP-Inverness
2  type external;
3  local-address 172.16.1.69;
4  family inet {
5      labeled-unicast {
6          resolve-vpn;
7      }
8  }
9  export loopback;
10 peer-as 64503;
11 neighbor 172.16.1.70;
```

There is no need for router *Havana* to propagate the received loopbacks inside the domain, so it suffices to leak the routes learned over the L-BGP session into inet.0 while enabling L3VPN resolution (Line 6). Having reachability from inet.0 is required to allow BGP to establish the TCP session to router *Livorno*.

A similar configuration is set-up on router *Livorno*, as shown in Listing 5.78. In preparation for the change to an Option C interconnect, the loopbacks are kept in inet.3 for further distribution inside domain "Cyclone,"as shown on Line 8.

Listing 5.78: Router *Livorno* L-BGP configuration to CsC–PE router *Basel*

```
1  user@Livorno> show configuration protocols bgp group EBGP-Monsoon-lbgp-basel
2  type external;
3  local-address 172.16.200.8;
4  family inet {
5      labeled-unicast {
6          rib-group leak-to-inet0; # Need to establish BGP L3VPN connections
7          rib {
8              inet.3;              # Useful for later move to option C
9          }
10     }
11 }
12 export Cyclone-PE-Loopback;
13 peer-as 64503;
14 neighbor 172.16.200.10;
15
16 user@Livorno> show configuration policy-options policy-statement Cyclone-PE-Loopback
17 term 1 {
18     from {
19         protocol direct;
20         route-filter 192.168.1.0/24 orlonger;
21     }
22     then accept;
23 }
24 term 2 {
25     then reject;
26 }
```

Listing 5.79 shows the loopbacks of the ASBR in domain "Mistral" as learned over the L-BGP connections by the ASBR in domain "Cyclone," and vice versa. Notice that the CsC setup has an *any-to-any* connectivity policy, so both ASBRs in domain "Cyclone" receive the loopback of both ASBRs in domain "Mistral".

Listing 5.79: Livorno and Havana receiving loopbacks over L-BGP

```
1  user@Livorno> show route receive-protocol bgp 172.16.200.10 192.168.1/24
2  
3  inet.0: 43 destinations, 46 routes (42 active, 0 holddown, 1 hidden)
4    Prefix                  Nexthop              MED     Lclpref    AS path
5  * 192.168.1.4/32          172.16.200.10                           64503 64502 I
6  
7  inet.3: 16 destinations, 26 routes (12 active, 0 holddown, 7 hidden)
8    Prefix                  Nexthop              MED     Lclpref    AS path
9  * 192.168.1.4/32          172.16.200.10                           64503 64502 I
10 
11 user@Havana> show route receive-protocol bgp 172.16.1.70 192.168.1/24
12 
13 inet.0: 45 destinations, 48 routes (45 active, 0 holddown, 0 hidden)
14   Prefix                  Nexthop              MED     Lclpref    AS path
15 * 192.168.1.6/32          172.16.1.70                             64503 64501 I
16 
17 inet.3: 10 destinations, 10 routes (10 active, 0 holddown, 0 hidden)
18   Prefix                  Nexthop              MED     Lclpref    AS path
19 * 192.168.1.6/32          172.16.1.70                             64503 64501 I
```

Bringup of L3VPN session over CsC

Listing 5.80 provides the additional configuration required on router *Livorno* and router *Havana* to establish a multihop E-BGP session that uses the labeled paths advertised over the CsC connection. Notice that LOCAL_PREF (Line 8) and MED (Line 3) follow Table 5.2 to ensure that this new session does not attract any traffic at this stage.

Listing 5.80: Setup of L3VPN session over CsC on router *Livorno* and router *Havana*

```
1  user@Havana> show configuration protocols bgp group EBGP-Cyclone-vpn-Livorno-loopback
2  type external;
3  metric-out 3500; # Worse MED than direct option B
4  multihop {       # Need to instantiate multihop session
5      ttl 10;
6  }
7  local-address 192.168.1.4;
8  import LP150;    # Worse local preference than direct Option B
9  family inet-vpn {
10     unicast;
11 }
12 export [ map-RT-to-Cyclone block-specifics ]; # Same policies as direct option B
13 vpn-apply-export;
14 peer-as 64501;
15 neighbor 192.168.1.6;    # Use loopback learned over CsC
16 
17 user@Livorno> show configuration protocols bgp group EBGP-Mistral-vpn-Havana-loopback
18 type external;
19 multihop {      # Need to instantiate multihop session
20     ttl 10;
21 }
22 local-address 192.168.1.6;
23 family inet-vpn {
24     unicast;
25 }
26 export map-RT-to-Mistral; # Same policies as direct option B session
27 peer-as 64502;
28 neighbor 192.168.1.4;     # Use loopback learned over CsC
```

Junos Tip: L3VPN EBGP sessions require Multihop

When the BGP peer is not directly connected, it is said to be multiple hops away. For IBGP, it is assumed that this is the common case and no special configuration is required. For an EBGP session, such as the case between router *Livorno* and router *Havana*, specific multihop configuration is required. Failure to configure this statement for a non-directly connected EBGP peer leaves the session *Idle*. Listing 5.81 shows the state on router *Livorno* before and after enabling multihop.

Listing 5.81: Livorno to Havana L3VPN session over CsC requires multihop

```
1  user@Livorno# run show bgp summary | match 192.168.1.4
2  192.168.1.4          64502          0          0       0       0      49:30 Idle
3
4  user@Livorno# show protocols bgp group EBGP-Mistral-vpn-Havana-loopback
5  type external;
6  local-address 192.168.1.6;
7  family inet-vpn {
8      unicast;
9  }
10 export map-RT-to-Mistral;
11 peer-as 64502;
12 neighbor 192.168.1.4;
13
14 [edit]
15 user@Livorno# set protocols bgp group EBGP-Mistral-vpn-Havana-loopback multihop
16     ttl 10
17
18 [edit]
19 user@Livorno# show | compare
20 [edit protocols bgp group EBGP-Mistral-vpn-Havana-loopback]
21 +     multihop {
22 +         ttl 10;
23 +     }
24
25 [edit]
26 user@Livorno# run show bgp summary | match 192.168.1.4
27 192.168.1.4          64502          4          5       0       0          8 Establ
```

5.6.8 Stage five: Activation of Option B over CsC

At this stage, the configuration is ready with two L3VPN sessions between router *Havana* and router *Livorno*: one direct session and one multihop session that crosses the CsC domain.

Verification of initial state

Listing 5.82 shows the BGP routes from ASBR router *Havana*. The prefix 192.168.107.7/32 belonging to router *Barcelona* lists three possible alternatives: over the direct Option B session (Line 6), over the CsC to router *Livorno* (Line 9), and as a result of the lower-preference Option A interconnect (Line 12). The priority set is for the existing Option B session because of the LOCAL_PREF attribute. Similar output can be observed for router *Livorno*, with path selection based on the MED BGP attribute instead (Line 21, Line 24, and Line 27).

An interesting observation on Line 11 is that the double *push* operation for the route over the CsC interconnect has the same VPN label as the direct Option B session on Line 8 because the ASBR router *Livorno* is the same for both.

Listing 5.82: Havana and Livorno L3VPN paths for Barcelona and Male route before switchover

```
1  user@Havana> show route 192.168.107.7
2
3  NMM.inet.0: 12 destinations, 15 routes (12 active, 0 holddown, 0 hidden)
4  + = Active Route, - = Last Active, * = Both
5
6  192.168.107.7/32    *[BGP/170] 00:06:26, localpref 200 # Direct option B session
7                        AS path: 64501 I
8                      > to 172.16.100.8 via fe-0/3/0.0, Push 305600
9                       [BGP/170] 00:06:15, localpref 150, from 192.168.1.6
10                       AS path: 64501 I
11                     > to 172.16.1.70 via so-1/0/3.0, Push 305600, Push 320464(top)
12                      [BGP/170] 05:18:04, localpref 100 # Inter-AS option A PE-CE
13                       AS path: 64501 I
14                     > to 172.16.250.38 via so-1/0/0.100
15
16 user@Livorno> show route 192.168.120.20
17
18 NMM.inet.0: 5 destinations, 11 routes (5 active, 0 holddown, 0 hidden)
19 + = Active Route, - = Last Active, * = Both
20
21 192.168.120.20/32  *[BGP/170] 00:09:02, MED 3000, localpref 100
22                       AS path: 64502 I
23                     > to 172.16.100.1 via fe-0/3/0.0, Push 301584
24                      [BGP/170] 00:08:50, MED 3500, localpref 100, from 192.168.1.4
25                       AS path: 64502 I
26                     > to 172.16.200.10 via fe-0/3/1.0, Push 301584, Push 319360(top)
27                      [BGP/170] 03:48:18, MED 4000, localpref 100
28                       AS path: 64502 I
29                     > to 172.16.250.37 via so-0/0/1.100
```

Activation of Option B over CsC

Activation consists of tweaking the BGP attributes on router *Havana* to ensure that the metrics for the direct Option B session are worse, as shown in Table 5.4. Listing 5.83 shows the configuration changes and the resulting switch in BGP best-path selection for router *Livorno* on Line 9 and for router *Havana* on Line 27 for the prefixes from router *Male* and router *Barcelona*, respectively.

Listing 5.83: Enable active forwarding over CsC

```
1  user@Havana> show configuration | compare rollback 1
2  [edit protocols bgp group EBGP-Cyclone-vpn-Livorno-loopback]
3  -     metric-out 3500;
4  +     metric-out 2000;
5  -     import LP150;
6  +     import LP300;
7
8   # VPN Prefix from Male
9  user@Livorno> show route 192.168.120.20
10
11 user@Livorno> show route 192.168.120.20
12
13 NMM.inet.0: 5 destinations, 11 routes (5 active, 0 holddown, 0 hidden)
14 + = Active Route, - = Last Active, * = Both
15
```

```
16 192.168.120.20/32  *[BGP/170] 00:01:04, MED 2000, localpref 100, from 192.168.1.4 # Havana over CsC
17                       AS path: 64502 I
18                     > to 172.16.200.10 via fe-0/3/1.0, Push 301584, Push 319360(top)
19                     [BGP/170] 00:07:21, MED 3000, localpref 100 # Havana over direct option B
20                       AS path: 64502 I
21                     > to 172.16.100.1 via fe-0/3/0.0, Push 301584
22                     [BGP/170] 03:56:41, MED 4000, localpref 100 # Havana over direct option A
23                       AS path: 64502 I
24                     > to 172.16.250.37 via so-0/0/1.100
25
26  # VPN Prefix from Barcelona
27 user@Havana> show route 192.168.107.7
28
29 NMM.inet.0: 15 destinations, 24 routes (15 active, 0 holddown, 0 hidden)
30 + = Active Route, - = Last Active, * = Both
31
32 192.168.107.7/32   *[BGP/170] 00:07:40, localpref 100, from 192.168.1.6 # Livorno
33                       AS path: 64501 I
34                     > to 172.16.1.70 via so-1/0/3.0, Push 303504, Push 319712(top)
35                     [BGP/170] 00:07:40, localpref 100, from 192.168.1.20 # Male (from Bilbao)
36                       AS path: 64501 I
37                     > to 172.16.1.85 via ge-1/1/0.0, Push 303424, Push 299776(top)
38                     [BGP/170] 00:07:40, localpref 10 # Livorno over direct Option B
39                       AS path: 64501 I
40                     > to 172.16.100.8 via fe-0/3/0.0, Push 303504
41                     [BGP/170] 10:21:01, localpref 1  # Livorno over direct Option A
42                       AS path: 64501 I
43                     > to 172.16.250.38 via so-1/0/0.100
```

Table 5.4 Preferring option B over CsC

Stage	Description	A	B Direct	B–CsC	C–CsC	C–CsC–Dual
3	Switch to B Direct		200			
			3000			
4	Build B over CsC			150		
				3500		
5	Switch to B over CsC			300		
				2000		

Validation of the new route is shown in the traceroute output in Listing 5.84 from router *Male* to router *Barcelona* (Line 1), and in the opposite direction on Line 13.

Listing 5.84: Forwarding verification using CsC after switchover

```
1  user@male-re1> traceroute routing-instance NMM source male-NMM barcelona-NMM
2  traceroute to barcelona-NMM (192.168.107.7) from male-NMM, 30 hops max, 40 byte packets
3   1  havana-ge1100 (172.16.1.86)  1.545 ms  1.254 ms  1.207 ms
4       MPLS Label=301648 CoS=0 TTL=1 S=1
5   2  inverness-so0030 (172.16.1.70)  1.611 ms  1.927 ms  1.490 ms
6       MPLS Label=320464 CoS=0 TTL=1 S=0
7       MPLS Label=305600 CoS=0 TTL=1 S=1
8   3  basel-at0101001 (10.10.1.74)  1.988 ms  1.958 ms  1.992 ms
9       MPLS Label=319152 CoS=0 TTL=1 S=0
10      MPLS Label=305600 CoS=0 TTL=2 S=1 # TTL of VPN label to be exposed
11  4  barcelona-NMM (192.168.107.7)  1.475 ms  1.402 ms  9.995 ms
12
13 user@Barcelona-re0> traceroute routing-instance NMM source barcelona-NMM male-NMM
14 traceroute to male-NMM (192.168.120.20) from barcelona-NMM, 30 hops max, 40 byte packets
15  1  livorno-ge020 (172.16.1.54)  1.537 ms  1.370 ms  1.357 ms
```

```
16        MPLS Label=305824 CoS=0 TTL=1 S=1
17  2  basel-fe0310 (172.16.200.10)  2.627 ms  2.229 ms  1.499 ms
18        MPLS Label=319360 CoS=0 TTL=1 S=0
19        MPLS Label=301584 CoS=0 TTL=1 S=1
20  3  inverness-at0101001 (10.10.1.73)  2.111 ms  2.281 ms  1.971 ms
21        MPLS Label=320496 CoS=0 TTL=1 S=0
22        MPLS Label=301584 CoS=0 TTL=2 S=1 # TTL of VPN label to be exposed
23  4  male-NMM (192.168.120.20)  2.053 ms  2.059 ms  2.026 ms
24
```

By comparing both traceroute outputs in Listing 5.84, the careful reader may have noticed that the penultimate hop is missing. Focusing on the traceroute starting on Line 1, Line 10 shows that the response given by router *Basel* includes a VPN label with a time-to-live (TTL) of 2.

Router *Basel* is the penultimate hop of the labeled BGP transport path between router *Havana* and router *Livorno*. Because of the penultimate-hop behavior, the VPN label is exposed with a TTL that does not expire at the next hop router *Livorno*.

Decommissioning of direct Option B setup

As usual, with no traffic on the previous interconnections, decommissioning the BGP session can be done at any time, by deleting the relevant BGP group. For the purposes of this case study, the direct Option B session is kept because it is better than the direct Option A session, but worse than the Option B session over CsC.

5.6.9 Stage Six: Inter-AS Option C deployment in domain "Cyclone"

In preparation for scaling the solution beyond domain "Cyclone," management decides to provide domain "Mistral" with all the PE loopback addressing information from domain "Cyclone" by means of labeled paths. The requirement is for domain "Mistral" to peer with the infrastructure ASBR router *Honolulu* to obtain VPN information. The new solution is depicted in Figure 5.41.

This stage provides an example of the evolution of using an Option B interconnection as the peering between router *Havana* and router *Livorno* to a more general Option C deployment on router *Livorno* that uses L-BGP to bind the transport labels.

Using an L-BGP session towards router *Havana*, router *Livorno* leaks internal L-BGP labeled path information from all known PEs, becoming a pure transit router for labeled paths and providing no additional VPN information. Router *Havana* also peers over a multihop session with the infrastructure border router *Honolulu* to obtain L3VPN information. Router *Honolulu* propagates PE reachability information from L3VPN belonging to domain "Cyclone," but does *not* change the next hop of these advertisements, allowing for a shortest-path connectivity.

The following considerations have to be taken into account:

- As a concentrator for all L3VPN information, the ASBR router *Havana* has to learn labeled paths to all PEs in domain "Cyclone." This next-hop information is required to validate L3VPN prefixes learned from within the domain. The existing MPLS transport setup within domain "Cyclone" provides these labeled paths. A full-mesh L-BGP over RSVP is made available for this. Border router *Livorno* advertises these labeled paths

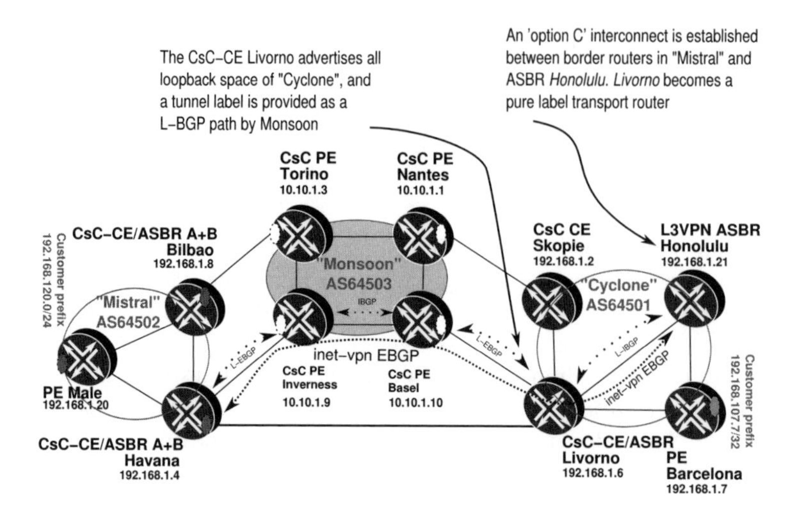

Figure 5.41: Stage six: Inter-AS Option C topology.

by means of external L-BGP to the CsC service, which propagates the information to domain "Mistral".

- The change from Option B to Option C requires changing the BGP peering in domain "Cyclone" from target router *Livorno*, moving to router *Honolulu* for L3VPN information. Notice that router *Havana* reuses the infrastructure L-BGP session from router *Livorno* to learn reachability to router *Honolulu* and piggybacks the L3VPN session. The L3VPN information over this session is depreferred in this initial deployment phase.

- VPN updates from router *Honolulu* should keep the original next hop to allow shortest-path forwarding. As a L3VPN ASBR peering with an external AS, there is a next-hop change by default along with a new VPN label for every prefix relayed. Note that router *Honolulu* is an infrastructure router placed deep into domain "Cyclone." A change in next hop for advertised prefixes would forward traffic over to router *Honolulu*, which is undesired.

- As in the previous stages, once all control channels are established, router *Havana* controls the transition by changing route preferences inbound and outbound as per Table 5.2.

Leaking labeled paths

The existing L-BGP peering on router *Havana* can be reused directly to receive reachability information for router *Honolulu* and the PE routers within domain "Cyclone." A modification on the border router *Livorno* is required to leak not only its own loopback, but also all the loopbacks in the domain. Listing 5.85 shows the policy change and the resulting advertisement of all the loopbacks.

Advertising the loopback space in domain "Cyclone" to router *Havana* is performed with the policy on Line 18. Instead of advertising only the local direct route representing the loopback interface, the whole loopback range from domain "Cyclone" is leaked.

Listing 5.85: Configuration on Livorno to set up a labeled EBGP session to Havana

```
1  [edit]
2  user@Livorno# run show route advertising-protocol bgp 172.16.200.10
3
4  inet.3: 16 destinations, 26 routes (12 active, 0 holddown, 7 hidden)
5    Prefix                  Nexthop              MED     Lclpref    AS path
6  * 192.168.1.6/32          Self                                    I
7
8  [edit policy-options policy-statement Cyclone-PE-Loopback]
9  user@Livorno# deactivate term 1 from protocol
10
11 [edit]
12 user@Livorno# show | compare
13 [edit policy-options policy-statement Cyclone-PE-Loopback term 1 from]
14 -     protocol direct;
15 +     inactive: protocol direct;
16
17 [edit]
18 user@Livorno# show policy-options policy-statement Cyclone-PE-Loopback
19 term 1 {
20     from {
21         inactive: protocol direct;
22         route-filter 192.168.1.0/24 orlonger; # Entire loopback range in domain ''Cyclone''
23     }
24     then accept;
25 }
26 term 2 {
27     then reject;
28 }
29
30 [edit]
31 user@Livorno> show route advertising-protocol bgp 172.16.200.10
32
33 inet.3: 16 destinations, 26 routes (12 active, 0 holddown, 7 hidden)
34   Prefix                  Nexthop              MED     Lclpref    AS path
35 * 192.168.1.2/32          Self                 64                 I
36 * 192.168.1.6/32          Self                                    I
37 * 192.168.1.7/32          Self                 10                 I
38 * 192.168.1.21/32         Self                 10                 I
```

The configuration on router *Livorno* has to be modified further to incorporate the received loopback from domain "Mistral" into L-BGP, to allow PEs within domain "Cyclone" to have direct visibility of the ASBR. Interestingly enough, because all MPLS transport labels are kept at the L-BGP layer, automatic redistribution from external L-BGP to internal L-BGP occurs. Details are outlined in Listing 5.86. Notice in Line 16 that an automatic *next-hop self* operation takes place for labeled NLRIs that are readvertised into the domain.

Listing 5.86: Injecting Havana loopback labeled path into domain "Cyclone"

```
1  user@Livorno> show route receive-protocol bgp 172.16.200.10 192.168.1/24
2
3  inet.0: 43 destinations, 46 routes (42 active, 0 holddown, 1 hidden)
4    Prefix                  Nexthop              MED     Lclpref    AS path
5  * 192.168.1.4/32          172.16.200.10                           64503 64502 I
6
7  inet.3: 16 destinations, 26 routes (12 active, 0 holddown, 7 hidden)
8    Prefix                  Nexthop              MED     Lclpref    AS path
9  * 192.168.1.4/32          172.16.200.10                           64503 64502 I
10
```

```
11 # advertising prefixes from EBGP to IBGP (Honolulu)
12 user@Livorno> show route advertising-protocol bgp 192.168.1.21 192.168.1/24
13
14 inet.3: 16 destinations, 26 routes (12 active, 0 holddown, 7 hidden)
15   Prefix                  Nexthop              MED     Lclpref    AS path
16 * 192.168.1.4/32          Self                         100        64503 64502 I
```

Multihop peering to L3VPN ASBR Honolulu

After the labeled path is available in both directions, it is feasible to establish a L3VPN peering session with router *Honolulu*. This is illustrated in Listing 5.87 for router *Havana*. Line 4 indicates that the external peering is not directly connected, requiring resolution of the protocol next hop following the labeled path built in Section 5.6.9. Notice in Line 12 that the same services of RT mapping and more-specific filtering described in Section 5.6.5 take place.

Listing 5.87: Multihop L3VPN session to Honolulu

```
1  user@Havana> show configuration protocols bgp group EBGP-Cyclone-vpn-ASBR-Honolulu
2  type external;
3  metric-out 2500; # Worse than existing option B over CsC
4  multihop {        # Multihop session
5      ttl 10;
6  }
7  local-address 192.168.1.4;
8  import LP250;     # Worse than existing option B over CsC
9  family inet-vpn {
10     unicast;
11 }
12 export [ block-specifics map-RT-to-Cyclone ]; # Same service as regular inet-vpn
13 vpn-apply-export;
14 peer-as 64501;
15 neighbor 192.168.1.21;
```

Retaining original protocol next hop on ASBR Honolulu

The most interesting addition to this subcase is the desire to retain best-path forwarding while concentrating L3VPN interaction outside domain "Cyclone" on a single ASBR (the same solution can be applied to a pair of ASBRs for redundancy). Listing 5.88 shows the relevant BGP configuration for router *Honolulu*, with Line 5 indicating that updates should be propagated without following the default behavior of next-hop change. Again, Line 19 shows the mapping of RT from the point of view of domain "Cyclone."

Listing 5.88: Multihop L3VPN session on router *Honolulu* with no next-hop change

```
1  user@honolulu-re0> show configuration protocols bgp group EBGP-Mistral-vpn
2  type external;
3  multihop {
4      ttl 10;
5      no-nexthop-change;
6  }
7  local-address 192.168.1.21;
8  family inet {
9      labeled-unicast {
10         rib-group leak-to-inet0;
11         rib {
12             inet.3;
13         }
```

```
14        }
15    }
16    family inet-vpn {
17        unicast;
18    }
19    export map-RT-to-Mistral; # Same policies as on router Livorno
20    peer-as 64502;
21    neighbor 192.168.1.4;
22    inactive: neighbor 192.168.1.8; # Item for stage 8
```

5.6.10 Stage Seven: Inter-AS Option C activation

Verification can take place when all routes have come up. Interestingly in this scenario, and unlike the case in Section 5.6.6, the anchoring point for the routes before and after the change is not the same. While router *Havana* can be used as the single point of control, the L3VPN routes land either at router *Livorno* or router *Honolulu*.

Preferring the routes on router *Livorno* results in the routes received from router *Honolulu* being inactive; conversely, a better preference for router *Honolulu* routes has an impact on router *Livorno*.

Verification of initial state

Listing 5.89 shows the status before the transition. With all things being equal, for router *Havana* the labeled routes to router *Honolulu* are farther away than the directly connected router *Livorno*. This explains the worse preference on Line 9 for reachability to router *Barcelona*. The specific MED BGP attribute tweaking on router *Honolulu* depreferences the L3VPN routes advertised to domain "Cyclone," thus controlling best-path selection for traffic towards domain "Mistral".

Listing 5.89: Routes on Option B before transition to Option C

```
1   user@Havana> show route 192.168.107.7
2
3   NMM.inet.0: 12 destinations, 17 routes (12 active, 0 holddown, 0 hidden)
4   + = Active Route, - = Last Active, * = Both
5
6   192.168.107.7/32    *[BGP/170] 00:04:50, localpref 300, from 192.168.1.6 # Option B over CsC route
7                         AS path: 64501 I
8                       > to 172.16.1.70 via so-1/0/3.0, Push 305600, Push 320464(top)
9                        [BGP/170] 00:04:50, localpref 250, from 192.168.1.21 # Option C route
10                        AS path: 64501 I
11                      > to 172.16.1.70 via so-1/0/3.0, Push 302304, Push 320528(top)
12                       [BGP/170] 00:04:50, localpref 200              # Lingering option B route
13                        AS path: 64501 I
14                      > to 172.16.100.8 via fe-0/3/0.0, Push 305600
15                       [BGP/170] 05:56:14, localpref 100              # Lingering option A route
16                        AS path: 64501 I
17                      > to 172.16.250.38 via so-1/0/0.100
18
19  user@honolulu-re0> show route table bgp | find 192.168.120.20
20  192.168.1.20:10:192.168.120.20/32
21                      *[BGP/170] 00:30:07, MED 2000, localpref 100, from 192.168.1.6
22                      # Option B route via Livorno
23                        AS path: 64502 I
24                      > to 172.16.1.89 via ge-5/2/0.0, label-switched-path Honolulu-to-Livorno
25                       [BGP/170] 00:08:09, MED 2500, localpref 100, from 192.168.1.4
26                       # Option C worse because of MED
27                        AS path: 64502 I
28                      > to 172.16.1.89 via ge-5/2/0.0, label-switched-path Honolulu-to-Livorno
```

Activation of inter-AS Option C

Transitioning consists of preferring the L3VPN routes from the multihop L3VPN session. This is achieved by changing preferences, as summarized in Table 5.5, with the configuration shown in Listing 5.90. Line 13 shows that the Option C path is preferred, and the missing output in Line 28 proves that router *Livorno* prefers this path over its local counterpart.

Table 5.5 Switching preferences to activate Option C

Stage	Description	A	B Direct	B–CsC	C–CsC	C–CsC–Dual
5	Switch to B over CsC			300		
				2000		
6	Build C over CsC				250	
					2500	
7	Switch to C over CsC				450	
					500	

Listing 5.90: Changing policy on router *Havana* to transition to Option C

```
user@Havana> show configuration | compare rollback 1
[edit protocols bgp group EBGP-Cyclone-vpn-ASBR-Honolulu]
-     metric-out 2500;
+     metric-out 500;
-     import LP250;
+     import LP450;

user@Havana> show route 192.168.107.7

NMM.inet.0: 12 destinations, 17 routes (12 active, 0 holddown, 0 hidden)
+ = Active Route, - = Last Active, * = Both

192.168.107.7/32    *[BGP/170] 00:00:41, localpref 450, from 192.168.1.21 # Option C
                      AS path: 64501 I
                    > to 172.16.1.70 via so-1/0/3.0, Push 302304, Push 320528(top)
                     [BGP/170] 00:00:41, localpref 300, from 192.168.1.6  # Option B over CsC
                      AS path: 64501 I
                    > to 172.16.1.70 via so-1/0/3.0, Push 305600, Push 320464(top)
                     [BGP/170] 00:00:41, localpref 200                    # Option B direct
                      AS path: 64501 I
                    > to 172.16.100.8 via fe-0/3/0.0, Push 305600
                     [BGP/170] 06:12:58, localpref 100                    # Option A
                      AS path: 64501 I
                    > to 172.16.250.38 via so-1/0/0.100

user@honolulu-re0>  show route table bgp | find 192.168.120.20
192.168.1.20:10:192.168.120.20/32
                    *[BGP/170] 00:02:22, MED 500, localpref 100, from 192.168.1.4
                      AS path: 64502 I
                    > to 172.16.1.89 via ge-5/2/0.0, label-switched-path Honolulu-to-Livorno
```

Double MPLS label *push* operations from router *Barcelona* to router *Male* on the first hop on Line 4 and in the reverse direction on Line 14 in Listing 5.91 confirm that forwarding is working as expected.

Listing 5.91: Forwarding with Option C enabled

```
user@Barcelona-re0> traceroute routing-instance NMM source barcelona-NMM male-NMM
traceroute to male-NMM (192.168.120.20) from barcelona-NMM, 30 hops max, 40 byte packets
 1  livorno-ge020 (172.16.1.54)  2.203 ms  1.930 ms  1.977 ms
     MPLS Label=305792 CoS=0 TTL=1 S=0
     MPLS Label=301584 CoS=0 TTL=1 S=1
 2  basel-fe0310 (172.16.200.10)  2.096 ms  1.728 ms  2.007 ms
     MPLS Label=319360 CoS=0 TTL=1 S=0
     MPLS Label=301584 CoS=0 TTL=2 S=1
 3  inverness-at0101001 (10.10.1.73)  2.143 ms  1.794 ms  1.921 ms
     MPLS Label=320496 CoS=0 TTL=1 S=0
     MPLS Label=301584 CoS=0 TTL=3 S=1
 4  male-NMM (192.168.120.20)  10.450 ms  1.203 ms  1.510 ms

 user@male-re1> traceroute routing-instance NMM source male-NMM barcelona-NMM

traceroute to barcelona-NMM (192.168.107.7) from male-NMM, 30 hops max, 40 byte packets
 1  havana-ge1100 (172.16.1.86)  1.536 ms  1.225 ms  1.247 ms
     MPLS Label=301680 CoS=0 TTL=1 S=1
 2  inverness-so0030 (172.16.1.70)  1.937 ms  1.924 ms  1.989 ms
     MPLS Label=320528 CoS=0 TTL=1 S=0
     MPLS Label=302304 CoS=0 TTL=1 S=1
 3  basel-at0101001 (10.10.1.74)  1.986 ms  1.461 ms  1.490 ms
     MPLS Label=319392 CoS=0 TTL=1 S=0
     MPLS Label=302304 CoS=0 TTL=2 S=1
 4  livorno-fe031 (172.16.200.8)  1.986 ms  1.410 ms  1.491 ms
     MPLS Label=305872 CoS=0 TTL=1 S=0
     MPLS Label=302304 CoS=0 TTL=3 S=1
 5  barcelona-ge120 (172.16.1.53)  1.095 ms  0.843 ms  0.990 ms
     MPLS Label=302304 CoS=0 TTL=1 S=1
 6  barcelona-NMM (192.168.107.7)  0.931 ms  0.900 ms  1.493 ms
```

Decommissioning the old setup

With the preferences properly in place, bringing down the Option B setup requires removing the L3VPN BGP peering connection on both router *Livorno* and router *Havana*. As discussed in Section 5.6.9, the new Option C session has been established on a different peering address; therefore, there is no impact to the service in this cleanup stage once traffic is using the new session.

5.6.11 Stage Eight: Build redundant Option C

Recent incidents within domain "Mistral" make router *Havana* a dangerous single point of failure. Management decides to provide domain "Cyclone" with two interconnect points for the L3VPN, leveraging the CsC service to connect to router *Bilbao*. The new solution is depicted in Figure 5.42.

This stage offers an example of introducing redundancy in the interconnect. Care has to be taken to consider possible loop conditions for the redistribution of labeled paths over the third-party carrier. Because the carrier is offering L-BGP services, the embedded AS_PATH attribute takes care of routing loops. The L-BGP setup within domain "Cyclone" suffices to keep the looping information coherent.

In this stage, router *Bilbao* becomes an ASBR and mimics the function provided by router *Havana*. A L3VPN peering to router *Honolulu* is set up over the CsC carrier through router *Skopie*.

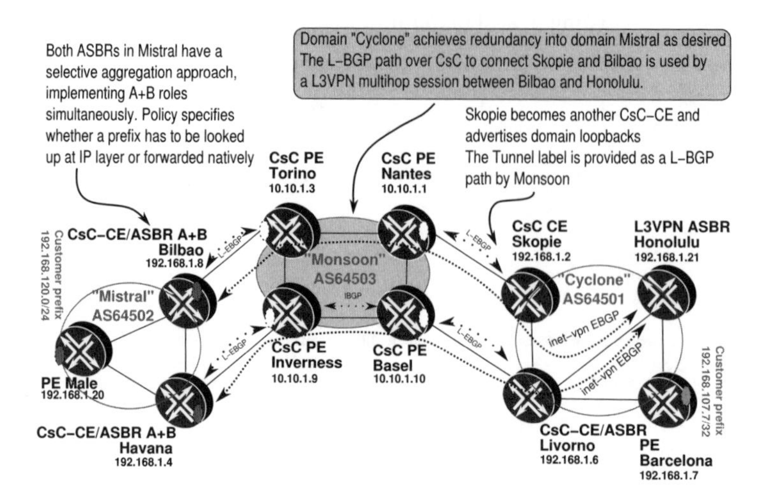

Figure 5.42: Stage eight: Redundant Inter-AS Option C topology.

Additional factors to consider in this migration follow:

- The configuration on router *Havana* is replicated on router *Bilbao*, including prefix Aggregation and IP functions. A parallel session is setup over the other end of the domain "Mistral" to provide redundancy. In addition, RT Community mappings need to be ported to the new Option C sessions crossing the CsC.

- Similar to the work done in the previous stage in Section 5.6.9 between router *Havana* and router *Livorno*, router *Skopie* and router *Bilbao* need to setup a CsC connection. L-BGP is leveraged for this connection.

- A multihop L3VPN peering session is established through this interconnect between router *Bilbao* and router *Honolulu* over the CsC service. The availability of the CsC service prompts a similar setup between router *Havana* and router *Honolulu*.

- As in previous stages, per-prefix preference control provides granular transition capabilities. To retain control on router *Havana*, the initial preferences in the configuration on router *Bilbao* are purposely different from those in previous stages. Still, the overall process follows Table 5.2.

Bringup of Bilbao as ASBR

The routing instance configuration on router *Havana* with no PE–CE interface is mirrored on router *Bilbao*, because the same aggregation functionality is desired. The *NMM* VRF has to be set up, with the label allocation and substitution policies, and advertising of more-specific routes needs to be restricted. Listing 5.92 provides the final instance configuration on router *Bilbao*.

Listing 5.92: Bilbao NMM instance to become ASBR with IP functions

```
1  user@Bilbao-re0> show configuration routing-instances NMM
2  instance-type vrf;
3  inactive: interface lo0.108;
4  vrf-target target:64502:64502;
5  vrf-table-label;
6  routing-options {
7      aggregate {
8          route 192.168.120.0/29 discard;
9      }
10     label {
11         allocation [ aggregate-allocation default-nexthop ];
12         substitution none;
13     }
14 }
15 forwarding-options {
16     family inet {
17         filter {
18             input count-and-police-table;
19         }
20     }
21 }
```

Establishing L-BGP sessions

A labeled session is configured between router *Bilbao* and router *Torino*, along with a corresponding session between router *Basel* and router *Skopie*. Listing 5.93 illustrates the configuration required to bring up the labeled session and also shows the loopbacks received by router *Bilbao* from the CsC mesh.

Listing 5.93: Bilbao configuration mirroring Havana as ASBR to a CsC

```
1  user@Bilbao-re0> show configuration protocols bgp group group EBGP-Monsoon-labeledBGP
2  group EBGP-Monsoon-labeledBGP {
3      type external;
4      local-address 172.16.1.102;
5      family inet {
6          labeled-unicast {
7              inactive: rib {
8                  inet.3;
9              }
10             resolve-vpn;
11         }
12     }
13     export loopback;
14     peer-as 64503;
15     neighbor 172.16.1.101;
16 }
17
18 user@Bilbao-re0> show route receive-protocol bgp 172.16.1.101 192.168.1/24
19
20 inet.0: 31 destinations, 31 routes (31 active, 0 holddown, 0 hidden)
21   Prefix                  Nexthop              MED     Lclpref    AS path
22 * 192.168.1.2/32          172.16.1.101                            64503 64501 I
23 * 192.168.1.6/32          172.16.1.101                            64503 64501 I
24 * 192.168.1.7/32          172.16.1.101                            64503 64501 I
25 * 192.168.1.21/32         172.16.1.101                            64503 64501 I
26
27 inet.3: 13 destinations, 13 routes (13 active, 0 holddown, 0 hidden)
28   Prefix                  Nexthop              MED     Lclpref    AS path
29 * 192.168.1.2/32          172.16.1.101                            64503 64501 I
30 * 192.168.1.6/32          172.16.1.101                            64503 64501 I
31 * 192.168.1.7/32          172.16.1.101                            64503 64501 I
32 * 192.168.1.21/32         172.16.1.101                            64503 64501 I
```

Multihop L3VPN

The MED setting on Bilbao is set to be temporarily worse to keep L3VPN on the existing CsC Option C connection. Listing 5.94 summarizes router *Bilbao* BGP settings, showing the slightly worse MED (Line 5) and LOCAL_PREF (Line 10) attributes than the currently active Option C peering on router *Havana*.

Because of the LOCAL_PREF settings, router *Bilbao* prefers router *Havana* for its connectivity requirements to domain "Cyclone."

Listing 5.94: Bilbao configuration mirroring Havana as ASBR to a CsC

```
1  user@Bilbao-re0> show configuration protocols bgp group group
2      EBGP-Cyclone-vpn-ASBR-Honolulu
3  group EBGP-Cyclone-vpn-ASBR-Honolulu {
4      type external;
5      metric-out 1000;     # Initial MED for L3VPN prefixes
6      multihop {
7          ttl 10;
8      }
9      local-address 192.168.1.8;
10     import LP400;        # Preference for received L3VPN prefixes
11     family inet-vpn {
12         unicast;
13     }
14     export map-RT-to-Cyclone;
15     peer-as 64501;
16     neighbor 192.168.1.21;
17 }
18
19 user@Bilbao-re0> show route 192.168.107.7
20
21 NMM.inet.0: 11 destinations, 15 routes (11 active, 0 holddown, 0 hidden)
22 + = Active Route, - = Last Active, * = Both
23
24 192.168.107.7/32   *[BGP/170] 00:09:58, localpref 450, from 192.168.1.20 # From Havana via Male
25                       AS path: 64501 I
26                     > via so-1/3/1.0, Push 301680, Push 299792(top)
27                     [BGP/170] 00:09:58, localpref 400, from 192.168.1.21 # Worse from Honolulu
28                       AS path: 64501 I
29                     > to 172.16.1.101 via at-1/2/0.0, Push 302304, Push 310832(top)
```

Router *Honolulu* prefers the route from router *Havana* because of the MED BGP attribute setting, as shown in Listing 5.95. Notice that no VRF is instantiated on router *Honolulu*, so the bgp.l3vpn table has to be consulted instead.

Listing 5.95: Routes for router *Barcelona* and router *Male* before switchover to Option C

```
1  user@honolulu-re0> show route table bgp | find 192.168.120.20
2  192.168.1.20:10:192.168.120.20/32
3                     *[BGP/170] 01:03:17, MED 500, localpref 100, from 192.168.1.4  # Havana
4                       AS path: 64502 I
5                     > to 172.16.1.89 via ge-5/2/0.0, label-switched-path Honolulu-to-Livorno
6                     [BGP/170] 00:22:42, MED 1000, localpref 100, from 192.168.1.8 # Bilbao
7                       AS path: 64502 I
8                     > to 172.16.1.89 via ge-5/2/0.0, label-switched-path Honolulu-to-Livorno
9
10 user@Bilbao-re0> show route table NMM 192.168.107.7
11
12 NMM.inet.0: 11 destinations, 15 routes (11 active, 0 holddown, 0 hidden)
13 + = Active Route, - = Last Active, * = Both
14
15 192.168.107.7/32   *[BGP/170] 00:09:58, localpref 450, from 192.168.1.20 # Havana via Male
16                       AS path: 64501 I
```

```
                    > via so-1/3/1.0, Push 301680, Push 299792(top)
                    [BGP/170] 00:09:58, localpref 400, from 192.168.1.21 # Honolulu
                      AS path: 64501 I
                    > to 172.16.1.101 via at-1/2/0.0, Push 302304, Push 310832(top)
```

Junos Tip: Inter-AS Option C and load balancing

Incorporating additional L3VPN sessions provides redundancy at the transport layer, but does not change the VPN prefix exchange. An exception worth making is the aggregate route that is present on both ASBRs in domain "Mistral". Listing 5.96 illustrates how the aggregate prefix is originated at both ASBRs, and as such, two different prefixes are propagated to domain "Cyclone."

Listing 5.96: Aggregate route originated in multiple ASBR

```
user@honolulu-re0> show route table bgp | match 192.168.120.0
192.168.1.4:6:192.168.120.0/29      # Aggregate from Havana
192.168.1.8:3:192.168.120.0/29      # Aggregate from Bilbao
```

Note that while a regular PE can have multiple inter-AS Option C labeled paths, the inet-vpn route is unique.

5.6.12 Stage Nine: Activation of redundant Option C

At this final stage, the two Option C peerings are established. Preferences are modified to make both alternatives feasible, as shown in Table 5.6, to showcase a possible redundant scenario.

Table 5.6 Activating redundant Option C connections

Stage	Description	A	B Direct	B–CsC	C–CsC	C–CsC–Dual
7	Switch to C over CsC				450 500	
8	Build redundant C					400 1000
9	Balance with redundant C				400 1000	

Notice that buildout of parallel session occurs on router *Bilbao* but is controlled from router *Havana*.

Activation of a parallel Option C

To confirm redundancy for prefixes over both L3VPN sessions, the same preference settings set on router *Bilbao* are configured on router *Havana*, as shown in Listing 5.97.

Listing 5.97: Configuring same preferences as Bilbao to achieve balancing

```
 1 user@Havana> show configuration | compare rollback 1
 2 [edit protocols bgp group EBGP-Cyclone-vpn-ASBR-Honolulu]
 3 -     metric-out 500;
 4 +     metric-out 1000;
 5 -     import LP450;
 6 +     import LP400;
 7
 8 user@Bilbao-re0> show route 192.168.107.7
 9
10 NMM.inet.0: 11 destinations, 15 routes (11 active, 0 holddown, 0 hidden)
11 + = Active Route, - = Last Active, * = Both
12
13 192.168.107.7/32   *[BGP/170] 00:28:02, localpref 400, from 192.168.1.21 # Same preference
14                       AS path: 64501 I
15                     > to 172.16.1.101 via at-1/2/0.0, Push 302304, Push 310832(top)
16                     [BGP/170] 00:01:13, localpref 400, from 192.168.1.20 # Same preference
17                       AS path: 64501 I
18                     > via so-1/3/1.0, Push 301680, Push 299792(top)
19
20 user@Bilbao-re0> show route 192.168.107.7 detail | match "Pref|proto|Reason"
21         *BGP    Preference: 170/-401
22                 Protocol next hop: 192.168.1.7 # Barcelona; Honolulu does not change next hop
23                 Localpref: 400
24          BGP    Preference: 170/-401
25                 Protocol next hop: 192.168.1.4 # Havana sets next-hop self
26                 Inactive reason: Not Best in its group - Interior > Exterior > Exterior via Interior
27                 Localpref: 400
28
29 user@honolulu-re0> show route table bgp detail | find 192.168.120.20 |
30                    match "metric|protocol|reason"
31                 Protocol next hop: 192.168.1.4
32                 Age: 6:37       Metric: 1000    Metric2: 10
33                 Protocol next hop: 192.168.1.8
34                 Inactive reason: Not Best in its group - Active preferred
35                 Age: 40:08      Metric: 1000    Metric2: 10
```

As a validation test, the L3VPN session on router *Havana* is torn down. Traffic follows the alternate path, as shown on Listing 5.98.

Listing 5.98: Verifying redundancy when bringing down the L3VPN session on Havana

```
 1 [edit]
 2 user@Havana# deactivate protocols bgp group EBGP-Cyclone-vpn-ASBR-Honolulu
 3
 4 user@Barcelona-re0> traceroute routing-instance NMM source barcelona-NMM male-NMM
 5 traceroute to male-NMM (192.168.120.20) from barcelona-NMM, 30 hops max, 40 byte packets
 6  1  livorno-ge020 (172.16.1.54)  1.980 ms  2.362 ms  1.485 ms
 7      MPLS Label=305904 CoS=0 TTL=1 S=0
 8      MPLS Label=303840 CoS=0 TTL=1 S=1
 9  2  basel-fe0310 (172.16.200.10)  2.111 ms  1.722 ms  1.472 ms
10      MPLS Label=319424 CoS=0 TTL=1 S=0
11      MPLS Label=303840 CoS=0 TTL=2 S=1
12  3  torino-at1201001 (10.10.1.81)  1.658 ms  1.473 ms  1.517 ms
13      MPLS Label=310848 CoS=0 TTL=1 S=0
14      MPLS Label=303840 CoS=0 TTL=3 S=1
15  4  bilbao-at1200 (172.16.1.102)  1.656 ms  1.602 ms  1.502 ms
16      MPLS Label=303840 CoS=0 TTL=1 S=1
17  5  male-so3210 (172.16.1.77)  1.115 ms  1.052 ms  1.409 ms
18      MPLS Label=299808 CoS=0 TTL=1 S=1
19  6  male-NMM (192.168.120.20)  1.863 ms  1.565 ms  1.446 ms
20
21 user@male-re1> traceroute routing-instance NMM source male-NMM barcelona-NMM
22 traceroute to barcelona-NMM (192.168.107.7) from male-NMM, 30 hops max, 40 byte packets
23  1  bilbao-so1310 (172.16.1.78)  1.526 ms  1.294 ms  1.268 ms
24      MPLS Label=303920 CoS=0 TTL=1 S=1
25  2  torino-at0010 (172.16.1.101)  1.470 ms  1.324 ms  1.485 ms
```

```
26        MPLS Label=310832 CoS=0 TTL=1 S=0
27        MPLS Label=302304 CoS=0 TTL=1 S=1
28   3  basel-at0001001 (10.10.1.82)  2.047 ms  1.767 ms  1.991 ms
29        MPLS Label=319392 CoS=0 TTL=1 S=0
30        MPLS Label=302304 CoS=0 TTL=2 S=1
31   4  livorno-fe031 (172.16.200.8)  2.063 ms  1.824 ms  1.985 ms
32        MPLS Label=305872 CoS=0 TTL=1 S=0
33        MPLS Label=302304 CoS=0 TTL=3 S=1
34   5  barcelona-ge120 (172.16.1.53)  1.635 ms  1.372 ms  1.494 ms
35        MPLS Label=302304 CoS=0 TTL=1 S=1
36   6  barcelona-NMM (192.168.107.7)  1.921 ms  1.399 ms  1.494 ms
```

5.6.13 Migration summary

The case study has moved through the possible inter-AS options in an incremental fashion. Some final remarks are worth mentioning:

- The use case covered was for a single VPN named *NMM*. Granularity of BGP advertisements is a per-prefix function. The different interconnecting options can be mixed and matched, with proper preference selection giving the final decision regarding which forwarding plane to use. Listing 5.99 represents all BGP sessions simultaneously established on router *Havana*. Through careful policy control, it is possible to define which prefixes use which type of interconnection.

- BGP next-hop change for L3VPN prefixes defines the type of interconnect. If no next-hop change occurs and an EBGP multihop L3VPN session is established, an Option C is being considered. Notice that the change in next hop is, again, a per-prefix function. It would be perfectly feasible to build labeled paths for all PE in domain "Mistral" and establish an Option B interconnect with next-hop self on router *Havana* purely for the aggregate route, while the remaining prefixes could use the Option C interconnection.

- BGP routing advertisement is unidirectional. The case study covered a symmetric setup for an interconnect option, but it may be feasible to set different arrangements for each direction. For instance, in Section 5.6.9, router *Honolulu* could set BGP next-hop self on all advertisements within the domain, forcing all traffic leaving the domain to pass through router *Honolulu*.

Listing 5.99: All interconnect options available on router *Havana*

```
1  user@Havana> show bgp summary
2  Groups: 6 Peers: 6 Down peers: 0
3  Table          Tot Paths  Act Paths Suppressed    History Damp State    Pending
4  inet.0                10          9          0          0          0          0
5  inet.3                10         10          0          0          0          0
6  bgp.l3vpn.0           15         11          0          0          0          0
7  Peer                     AS      InPkt     OutPkt    OutQ   Flaps Last Up/Dwn State|#Active/Received/
     Accepted/Damped...
8   # To CsC-PE Inverness (CsC)
9  172.16.1.70          64503        387        381       0       0     2:51:47 Establ
10   inet.0: 9/10/10/0
11  # To ASBR Livorno (B-direct)
12 172.16.100.8         64501        837        854       0       0     6:19:21 Establ
13   bgp.l3vpn.0: 0/2/2/0
14   NMM.inet.0: 0/1/1/0
15  # To PE-CE peer Livorno (A)
16 172.16.250.38        64501       1043       1053       0       2     7:53:03 Establ
```

```
17    NMM.inet.0: 0/2/2/0
18  # To labeled peer Livorno (B-CSC)
19  192.168.1.6          64501        354        357        0        0     2:39:31 Establ
20    bgp.l3vpn.0: 0/2/2/0
21    NMM.inet.0: 0/1/1/0
22  # To IBGP peer Male (regular VPN)
23  192.168.1.20         64502        848        860        0        3     6:18:59 Establ
24    bgp.l3vpn.0: 9/9/9/0
25    NMM.inet.0: 8/9/9/0
26  # To ASBR Honolulu (C-CsC)
27  192.168.1.21         64501          3          6        0        0          14 Establ
28    bgp.l3vpn.0: 2/2/2/0
29    NMM.inet.0: 1/2/2/0
```

Route attributes control the preferred path to be chosen. This can be generalized to any VPN. Note that this flexibility is inherent to the L3VPN technology, and allows for a rich service offering to accommodate different end customer requirements.

Bibliography

[IANA-extended-communities] Internet Assigned Numbers Authority – IANA. http://www.iana.org/assignments/bgp-extended-communities, Border Gateway Protocol (BGP) data collection standard communities, August 2009.

[RFC1403] K. Varadhan. BGP OSPF Interaction. RFC 1403 (Historic), January 1993.

[RFC2328] J. Moy. OSPF Version 2. RFC 2328 (Standard), April 1998.

[RFC2453] G. Malkin. RIP Version 2. RFC 2453 (Standard), November 1998. Updated by RFC 4822.

[RFC2547] E. Rosen and Y. Rekhter. BGP/MPLS VPNs. RFC 2547 (Informational), March 1999. Obsoleted by RFC 4364.

[RFC4360] S. Sangli, D. Tappan, and Y. Rekhter. BGP Extended Communities Attribute. RFC 4360 (Proposed Standard), February 2006.

[RFC4364] E. Rosen and Y. Rekhter. BGP/MPLS IP Virtual Private Networks (VPNs). RFC 4364 (Proposed Standard), February 2006. Updated by RFCs 4577, 4684.

[RFC4456] T. Bates, E. Chen, and R. Chandra. BGP Route Reflection: An Alternative to Full Mesh Internal BGP (IBGP). RFC 4456 (Draft Standard), April 2006.

[RFC4576] E. Rosen, P. Psenak, and P. Pillay-Esnault. Using a Link State Advertisement (LSA) Options Bit to Prevent Looping in BGP/MPLS IP Virtual Private Networks (VPNs). RFC 4576 (Proposed Standard), June 2006.

[RFC4577] E. Rosen, P. Psenak, and P. Pillay-Esnault. OSPF as the Provider/Customer Edge Protocol for BGP/MPLS IP Virtual Private Networks (VPNs). RFC 4577 (Proposed Standard), June 2006.

[draft-kulmala-l3vpn-interas-option-d] M. Kulmala, J. Soini, J. Guichard, R. Hanzl, and M. Halstead. Internet draft, ASBR VRF context for BGP/MPLS IP VPN, February 2006.

[draft-marques-ppvpn-ibgp-00] P. Marques, R. Raszuk, D. Tappan, and L. Martini. Internet draft, RFC2547bis networks using internal BGP as PE-CE protocol, April 2003.

Further Reading

[1] Aviva Garrett. JUNOS Cookbook, April 2006. ISBN 0-596-10014-0.

[2] Ivan Pepelnjak and Jim Guichard. MPLS and VPN Architectures, November 2000. ISBN 978-1587050022.

[3] Ivan Pepelnjak, Jim Guichard, and Jeff Apcar. MPLS and VPN Architectures, Volume II, June 2003. ISBN 978-1587051128.

Index

Network Mergers and Migrations Gonzalo Gómez Herrero and Jan Antón Bernal van der Ven

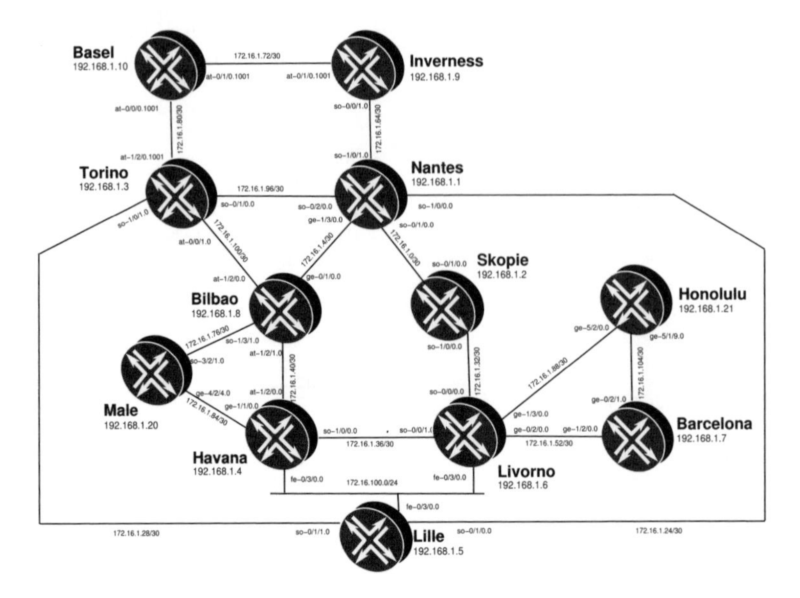

Quick Reference: Basic network topology